Thomas M. Klapötke
Chemistry of High-Energy Materials

Also of interest

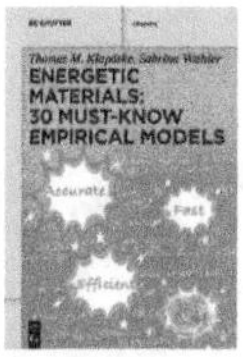

Energetic Materials: 30 Must-Know Empirical Models
Klapötke. Wahler, 2025
ISBN 978-3-11-109602-5, e-ISBN (PDF) 978-3-11-109702-2,
e-ISBN (EPUB) 978-3-11-109782-4

Combustible Organic Materials.
Determination and Prediction of Combustion Properties
Mohammad Hossein Keshavarz, 2022
ISBN 978-3-11-078204-2, e-ISBN (PDF) 978-3-11-078213-4,
e-ISBN (EPUB) 978-3-11-078225-7

Toxicity: 77 Must-Know Predictions of Organic Compounds.
Including Ionic Liquids
Mohammad Hossein Keshavarz, 2023
ISBN 978-3-11-118912-3, e-ISBN (PDF) 978-3-11-118967-3,
e-ISBN (EPUB) 978-3-11-119092-1

The Properties of Energetic Materials.
Sensitivity, Physical and Thermodynamic Properties
Mohammad Hossein Keshavarz, Thomas M. Klapötke, 2021
ISBN 978-3-11-074012-7, e-ISBN (PDF) 978-3-11-074015-8,
e-ISBN (EPUB) 978-3-11-074024-0

Set Energetic Materials Encyclopedia: Volume 1–3
Thomas M. Klapötke, 2021
ISBN 978-3-11-067465-1

Thomas M. Klapötke

Chemistry of High-Energy Materials

Explosives, Propellants, Pyrotechnics

7th, Revised and Extended Edition

DE GRUYTER

Author
Prof. Dr. Dr. h.c. Thomas M. Klapötke
Ludwig-Maximilians-Universität
Department of Chemistry
Haus D
Butenandtstr. 5-13
81377 Munich
Germany
Thomas.M.Klapoetke@cup.uni-muenchen.de

ISBN 978-3-11-144698-1
e-ISBN (PDF) 978-3-11-144708-7
e-ISBN (EPUB) 978-3-11-144713-1

Library of Congress Control Number: 2024952504

Bibliographic information published by the Deutsche Nationalbibliothek
The Deutsche Nationalbibliothek lists this publication in the Deutsche Nationalbibliografie; detailed bibliographic data are available on the Internet at http://dnb.dnb.de.

Cover image: The cover picture shows the detonation of a new thermobaric formulation developed at the Military Technical Institute, Belgrade in collaboration with LMU. © By courtesy of Prof. Dr. Danica Bajic (nee Simic).
Typesetting: Integra Software Services Pvt. Ltd.

www.degruyter.com
Questions about General Product Safety Regulation:
productsafety@degruyterbrill.com

“We will not waver; we will not tire; we will not falter;
And we will not fail. Peace and freedom will prevail.”

G. W. Bush,
Presidential Address to the Nation,
October 7th 2001

Preface to this 7th English edition

Everything which has been said in the preface to the first German, as well as in the first to sixth English editions still holds true, and essentially does not need any addition or correction. In this revised and extended seventh edition in English, the manuscript has been updated and various recent aspects of energetic materials have been added:

(i) Some errors which were unfortunately present in the sixth edition have been corrected and the references have also been updated where appropriate.
(ii) The following chapters have been updated and significantly extended:
Historical Overview (Classification) (1.1), 1.2.4 (Oxidizers), Propellant Charges (2.3), Critical Diameter (3), Thermodynamics (4.2.1), Electric Detonators (5.4), Sensitivities (6.1), Trauzl Test (6.6), Plate Dent Test (7.4), Future Energetics (new explosives and melt-casts, 9.6), Safe Handling of Energetic Materials in the Laboratory (11), Thermobaric Weapons (13.1), Toxicity Measurements (13.7)
(iii) Seven new chapters on Flash Compositions (2.5.6), EXPLO5: Capabilities and Background (4.2.7), Gun Barrel Design (4.2.8) and Continuous Flow Chemistry (13.12) have been added. Also, three new short chapters on test methods have been included: Sand Test (6.7), Ballistic Mortar Test (6.8) and Cylinder Test (6.9).
(iv) There are over 50 new figures and many new tables in this 7th edition.
(v) Finally, the number of study questions (and answers on the De Gruyter web site) at the end of the book has also been expanded.

In addition to the people thanked in the German and first six English editions, the author would like to thank Prof. Dr. Jürgen Evers (LMU) and Joe Backofen (BRIGS) for many inspired discussions and help with this seventh edition. Thanks also to Lukas Bauer, Marcus Lommel and Jennifer Heidrich for their help and support with the manuscript.

The authors is also indebted to and thanks the following colleagues for a year-long great collaboration and for many inspired discussions also with respect to topics of this book: Prof. Dr. Danica Bajic (nee Simic) (Military Technical Institute, Belgrade), Prof. Dr. Traian Rotario (MTA, Bucharest), Prof. Dr. Stanislaw Cudzilo (WAT, Warsaw), Prof. Dr. Muhamed Suceska (Univ. of Zagreb), Prof. Dr. Mario Dobrilovic (Univ. of Zagreb), and Dr. Lothar Reich (WTD91, Meppen).

The author would also like to thank Dr. Karin Sora and Dr. Ria Sengbusch (De Gruyter) for the excellent collaboration, and Mr. Matthias Wand for his great assistance with the production of this book.

Munich, January 2025 — Thomas M. Klapötke

https://doi.org/10.1515/9783111447087-202

Preface to this 6th English edition

Everything which has been said in the preface to the first German, as well as in the first to fifth English editions still holds true, and essentially does not need any addition or correction. In this revised and extended sixth edition in English, the manuscript has been updated and various recent aspects of energetic materials have been added:

(vi) Some errors which were unfortunately present in the fifth edition have been corrected and the references have also been updated where appropriate.

(vii) The following chapters have been updated and significantly extended: RDX/HMX (1.1), new oxidizers (1.2.4), detonations, craters and shock-waves (1.3), tertiary explosives (2.1), rocket propellants and double-base propellants (2.4), empirical methods (3), combustion parameters (4.2.3), TXX-50 as rocket propellant (4.2.4), aging and critical temperature (6.7), energetic materials of the future (12), Gurney model (7.3), thermobaric weapons (13.1), and airbags (13.6).

(viii) Five new chapters on the detonation velocity of mixtures of secondary explosives with non-explosive liquids (7.5), energetic MOFs (13.8), civil explosives and boosters (13.9), structured reactive materials (13.10) and 3D printing of explosives (13.11) have been added.

In addition to the people thanked in the German and first five English editions, the author would like to thank Ms. Jasmin Lechner (LMU), Mr. Philip Kneisl (Explosive Examiners LLC), Mr. Joe Backofen (BRIGS Co.), Dr. Ernst-Christian Koch (Lutradyn), Prof. Dr. Jean'ne Shreeve (University of Idaho), Prof. Dr. Stanislaw Cudziło (Military University of Technology, Warsaw), as well as Prof. Maria D. M. C. Ribeiro da Silva and Dr. Ana L. R. Silva (both University of Porto) for many inspired discussions and help with this sixth edition.

The author would also like to thank Dr. Bettina Noto (deGruyter) for the excellent collaboration.

Munich, January 2022 Thomas M. Klapötke

https://doi.org/10.1515/9783111447087-203

Preface to the 5th English edition

Everything which has been said in the preface to the first German, as well as in the first to fourth English editions still holds true, and essentially does not need any addition or correction. In this revised fifth edition in English, the manuscript has been updated and various recent aspects of energetic materials have been added:

(i) Some errors which were unfortunately present in the fourth edition have been corrected and the references have also been updated where appropriate.
(ii) Chapters on the critical diameter, colored smokes, delay compositions, sensitivities, gun erosion, TKX-50, liners, thermobaric weapons and nitrocellulose have been updated.
(iii) Four new chapters on Aging (6.7), Air Bags (13.6), Spent Acid (10.4) and Toxicity Measurements (13.7) have been added.

In addition to the people thanked in the German and first four English editions, the author would like to thank Dr. Manfred Bohn (ICT), Dr. Johann Glück (ZF), Dr. Paul Wanninger (PWX), Mr. Moritz Mühlemann (BIAZZI), Mr. Marcel Holler (Bayern Chemie), Ms. Cornelia Unger (LMU) and Ms. Teresa Küblböck (LMU) for many inspired discussions.

Munich, July 2019 Thomas M. Klapötke

https://doi.org/10.1515/9783111447087-204

Preface to the 4th English edition

Everything which has been said in the preface to the first German, as well as in the first to third English editions, still holds and essentially does not need any addition or correction. In this revised fourth edition in English, the manuscript has been updated and various recent aspects of energetic materials have been added:

(i) Some errors, which were unfortunately present in the third edition, have been corrected, and the references have also been updated where appropriate.

(ii) The chapters on shaped charges, visible pyrotechnics, smokes (addition of P_3N_5), primary explosives (MTX-1), composite and double-base propellants, and Gurney energy have been updated.

(iii) Four new chapters on Electric Detonators (5.4), Detonation velocity from Laser Induced Air Shock (9.7), Thermally Stable Explosives (9.8), and Explosive Welding (13.5) have been added.

In addition to the people thanked in the German and first three English editions, the author would like to thank Professor Dr. Mohammad H. Keshavarz, Dr. Tomasz Witkowski, Col. Dr. Ahmed Elbeih, Cpt. Mohamed Abd-Elghany, M.Sc. and Lt. Andreea Voicu, M.Sc. for many inspired discussions.

Munich, July 2017 Thomas M. Klapötke

https://doi.org/10.1515/9783111447087-205

Preface to the 3rd English edition

Everything which has been said in the preface to the first German and first and second English editions still holds and essentially does not need any addition or correction. In this revised third edition in English the manuscript has been up-dated and various recent aspects of energetic materials have been added:

(i) some errors which unfortunately occurred in the first and second editions have been corrected and the references have also been updated where appropriate.

(ii) The chapters on critical diameters, delay compositions, visible light (blue) pyrotechnics, polymer-bonded explosives (PBX), HNS, thermodynamic calculations, DNAN, smoke (yellow) formulations and high-nitrogen compounds have been updated.

(iii) Five new short chapters on Ignition and Initiation (Chapters 5.2 and 5.3), the Plate Dent Test (chapter 7.4), Underwater Explosions (chapter 7.5) and the Trauzl Test (chapter 6.6) have been added.

In addition to the people thanked in the German and first and second English editions, the author would like to thank Dr. Vladimir Golubev and Tomasz Witkowski (both LMU) for many inspired discussions concerning hydrocode calculations. The author is also indebted to and thanks Dr. Manuel Joas (DynITEC, Troisdorf, Germany) for his help with the preparation of Chapters 5.2 and 5.3.

Munich, October 2015 Thomas M. Klapötke

https://doi.org/10.1515/9783111447087-206

Preface to the 2nd English edition

Everything said in the preface to the first German and first English editions still holds and essentially does not need any addition or correction. In this revised second edition in English we have up-dated the manuscript and added some recent aspects of energetic materials:

(i) We have tried to correct some mistakes which can not be avoided in a first edition and also updated the references where appropriate.
(ii) The chapters on Ionic Liquids, Primary Explosives, NIR formulations, Smoke Compositions and High-Nitrogen Compounds were updated.
(iii) Two new short chapters on Co-Crystallization (9.5) and Future Energetic Materials (9.6) have been added.

In addition to the people thanked in the German and first English edition, the author would like to thank Dr. Jesse Sabatini and Dr. Karl Oyler (ARDEC, Picatinny Arsenal, NJ) for many inspired discussions concerning pyrotechnics.

Munich, May 2012 Thomas M. Klapötke

https://doi.org/10.1515/9783111447087-207

Preface to the first English edition

Everything said in the preface to the first German edition remains valid and essentially does not need any addition or correction. There are several reasons for translating this book into English:

- The corresponding lecture series at LMU is now given in English in the postgraduate M.Sc. classes, to account for the growing number of foreign students and also to familiarize German students with the English technical terms.
- To make the book available to a larger readership world-wide.
- To provide a basis for the author's lecture series at the University of Maryland, College Park.

We have tried to correct some omissions and errors which can not be avoided in a first edition and have also updated the references where appropriate. In addition, five new chapters on Combustion (Ch. 1.4), NIR formulations (Ch. 2.5.5), the Gurney Model (Ch. 7.3), dinitroguanidine chemistry (Ch. 9.4) and nanothermites (Ch. 13.3) have been included in the English edition. The chapter on calculated combustion parameters (Ch. 4.2.3) has been extended.

In addition to the people thanked in the German edition, the author would like to thank Dr. Ernst-Christian Koch (NATO, MSIAC, Brussels) for pointing out various mistakes and inconsistencies in the first German edition. For inspired discussions concerning the Gurney model special thanks goes to Joe Backofen (BRIGS Co., Oak Hill). Dr. Anthony Bellamy, Dr. Michael Cartwright (Cranfield University), Neha Mehta, Dr. Reddy Damavarapu and Gary Chen (ARDEC) and Dr. Jörg Stierstorfer (LMU) are thanked for ongoing discussions concerning secondary and primary explosives.

The author also thanks Mr. Davin Piercey, B.Sc. for corrections and for writing the new chapter on nanothermites, Dr. Christiane Rotter for her help preparing the English figures and Dr. Xaver Steemann for his help with the chapter on detonation theory and the new combustion chapter. The author thanks the staff of de Gruyter for the good collaboration preparing the final manuscript.

Munich, January 2011 Thomas M. Klapötke

https://doi.org/10.1515/9783111447087-208

Preface to the first English edition

Preface to the first German edition

This book is based on a lecture course which has been given by the author for more than 10 years at the Ludwig-Maximilian University Munich (LMU) in the post-graduate Master lecture series, to introduce the reader to the chemistry of highly energetic materials. This book also reflects the research interests of the author. It was decided to entitle the book "Chemistry of High-Energy Materials" and not simply "Chemistry of Explosives" because we also wanted to include pyrotechnics, propellant charges and rocket propellants into the discussion. On purpose we do not give a comprehensive historical overview and we also refrained from extensive mathematical deductions. Instead we want to focus on the basics of chemical explosives and we want to provide an overview of recent developments in the research of energetic materials.

This book is concerned with both the civil applications of high-energy materials (e.g. propellants for carrier or satellite launch rockets and satellite propulsion systems) as well as the many military aspects. In the latter area there have been many challenges for energetic materials scientists in recent days some of which are listed below:

- In contrast to classical targets, in the on-going global war on terrorTab. 4.10
- 4water (GWT), new targets such as tunnels, caves and remote desert or mountain areas have become important.
- The efficient and immediate response to time critical targets (targets that move) has become increasingly important for an effective defense strategy.
- Particularly important is the increased precision ("we want to hit and not to miss the target", Adam Cumming, DSTL, Sevenoaks, U.K.), in order to avoid collateral damage as much as possible. In this context, an effective coupling with the target is essential. This is particularly important since some evil regimes often purposely co-localize military targets with civilian centers (e.g. military bases near hospitals or settlements).
- The interest in insensitive munitions (IM) is still one of the biggest and most important challenges in the research of new highly energetic materials.
- The large area of increasing the survivability (for example by introducing smokeless propellants and propellant charges, reduced signatures of rocket motors and last but not least, by increasing the energy density) is another vast area of huge challenge for modern synthetic chemistry.
- Last but not least, ecological aspects have become more and more important. For example, on-going research is trying to find suitable lead-free primary explosives in order to replace lead azide and lead styphnate in primary compositions. Moreover, RDX shows significant eco- and human-toxicity and research is underway to find suitable alternatives for this widely used high explosive. Finally, in the area of rocket propulsion and pyrotechnical compositions, replacements for toxic ammonium perchlorate (replaces iodide in the thyroid gland) which is currently

https://doi.org/10.1515/9783111447087-209

used as an oxidizer are urgently needed. Despite all this, the performance and sensitivity of a high-energy material are almost always the key-factors that determine the application of such materials – and exactly this makes research in this area a great challenge for synthetically oriented chemists.

The most important aspect of this book and the corresponding lecture series at LMU Munich, is to prevent and stop the already on-going loss of experience, knowledge and know-how in the area of the synthesis and safe handling of highly energetic compounds. There is an on-going demand in society for safe and reliable propellants, propellant charges, pyrotechnics and explosives in both the military and civilian sector. And there is no one better suited to provide this expertise than well trained and educated preparative chemists.

Last but not least, the author wants to thank those who have helped to make this book project a success. For many inspired discussions and suggestions the authors wants to thank the following colleagues and friends: Dr. Betsy M. Rice, Dr. Brad Forch and Dr. Ed Byrd (US Army Research Laboratory, Aberdeen, MD), Prof. Dr. Manfred Held (EADS, TDW, Schrobenhausen), Dr. Ernst-Christian Koch (NATO MSIAC, Brussels), Dr. Miloslav Krupka (OZM, Czech Republic), Dr. Muhamed Sucesca (Brodarski Institute, Zagreb, Croatia), Prof. Dr. Konstantin Karaghiosoff (LMU Munich), Prof. Dr. Jürgen Evers (LMU Munich), as well as many of the past and present co-workers of the authors research group in Munich without their help this project could not have been completed.

The author is also indebted to and thanks Dipl.-Chem. Norbert Mayr (LMU Munich) for his support with many hard- and soft-ware problems, Ms. Carmen Nowak and Ms. Irene S. Scheckenbach (LMU Munich) for generating many figures and for reading a difficult manuscript. The author particularly wants to thank Dr. Stephanie Dawson (de Gruyter) for the excellent and efficient collaboration.

Munich, July 2009 Thomas M. Klapötke

Contents

1 Introduction

1.1 Historical overview

In this chapter we do not want to be exhaustive in scope, but rather to focus on some of the most important milestones in the chemistry of explosives (Tab. 1.1). The development of energetic materials began with the accidental discovery of **blackpowder** in China (~ 220 BC). In Europe this important discovery remained dormant until the thirteenth and fourteenth centuries, when the English monk Roger Bacon (1249) and the German monk Berthold Schwarz (1320) started to research the properties of blackpowder. At the end of the thirteenth century, blackpowder was finally introduced into the military world. However, it was not until 1425 that Corning greatly improved the production methods and blackpowder (or gunpowder) was then introduced as a propellant charge for smaller and later also for large caliber guns. Even today, ca. 100,000 pounds of blackpowder are still used in the US military per year.

One application in which BP is still in use is in “time blasting fuses”. They were invented as early as 1831, and contain approx. 4.7 g/m BP in the core (see Fig. 1.1a). The burn rate is about 135 m/s, and since they are RF safe (radio frequency), they are very popular with the military.

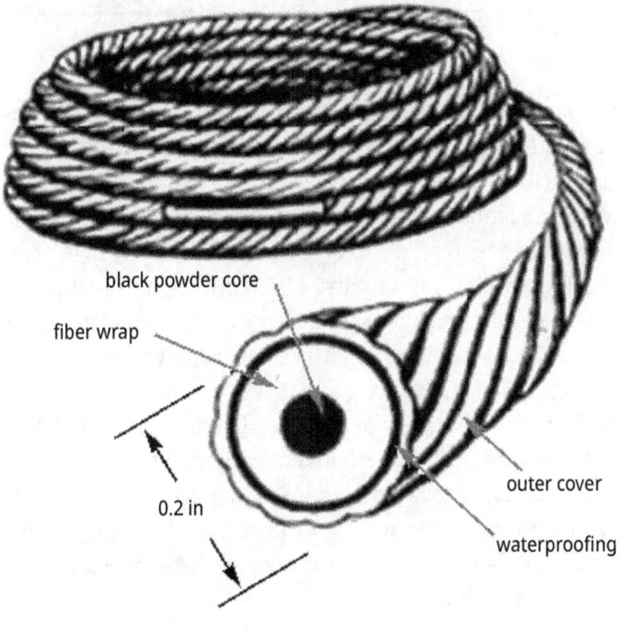

Fig. 1.1a: Time Blasting Fuse.

https://doi.org/10.1515/9783111447087-001

Tab. 1.1: Historical overview of some important secondary explosives.

Substance	Acronym	Development	Application	Density/g cm^{-3}	Explosive power[a]
Blackpowder	BP	1250–1320	1425–1900	ca. 1.0	
Nitroglycerine	NG	1863	In propellant charges	1.60	170
Dynamite	Dy	1867	Civil/commercial only	varies	varies
Picric acid	PA	1885–1888	WW I	1.77	100
Nitroguanidine	NQ	1877	most in TLPs	1.71	99
Trinitrotoluene	TNT	1880	WW I	1.64	116
Nitropenta	PETN	1894	WW II	1.77	167
Hexogen	RDX	1920–1940	WW II	1.81	169
Octogen	HMX	1943	WW II	1.91 (β polymorph)	169
Hexanitrostilbene	HNS	1913	1966	1.74	
Triaminotrinitro benzene	TATB	1888	1978	1.93	
HNIW	CL-20	1987	Under evaluation	2.1 (ε polymorph)	

[a] rel. to PA.

The next milestone was the first small-scale synthesis of **nitroglycerine (NG)** by the Italian chemist Ascanio Sobrero (1846). Later, in 1863 Imanuel Nobel and his son Alfred commercialized NG production in a small factory near Stockholm (Tab. 1.1). NG is produced by running highly concentrated, almost anhydrous, and nearly chemically pure glycerin into a highly concentrated mixture of nitric and sulfuric acids (HNO_3 / H_2SO_4), while cooling and stirring the mixture efficiently. At the end of the reaction, the nitroglycerine and acid mixture is transferred into a separator, where the NG is separated by gravity. Afterwards, washing processes using water and alkaline soda solution remove any residual acid.

Nitroglycerin (NG) showed itself to be extremely dangerous for blasting. Alfred Nobel scrambled to find a solution, and in doing so, invented dynamite. His solution was to solidify the NG by adding diatomaceous earth (mineral powder). It was not understood why this worked. It is now known that liquid HEs are sensitized by *cavitation bubble collapse*. Liquid HEs are dangerous, because they are often insensitive in pure form, but highly sensitized by aeration.

Initially NG was very difficult to handle because of its high impact sensitivity and unreliable initation by blackpowder. Among many other accidents, one explosion in 1864 destroyed the Nobel factory completely, killing Alfred's brother Emil. In the same year,

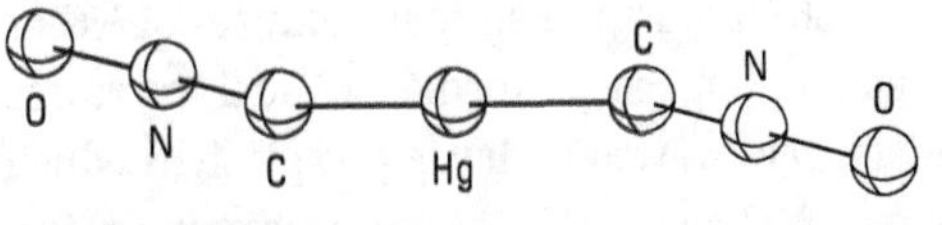

Fig. 1.1b: Molecular structure of mercury fulminate, $Hg(CNO)_2$.

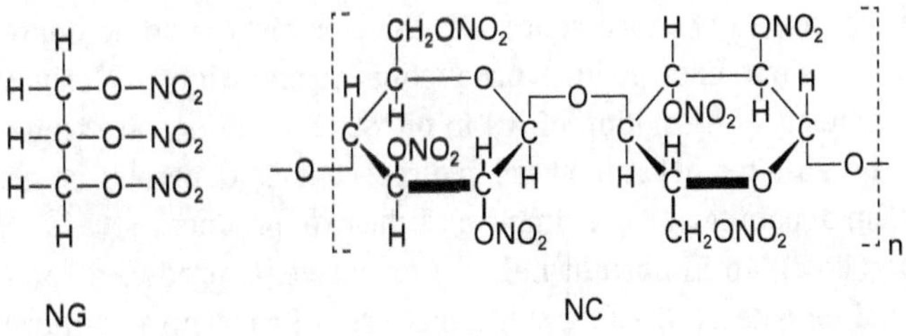

Fig. 1.2: Molecular structures of nitroglycerin (NG) and nitrocellulose (NC).

Alfred Nobel invented the metal blasting cap detonator, and replaced blackpowder with **mercury fulminate (MF)**, $Hg(CNO)_2$. Although the Swedish-German Scientist Johann Kunkel von Löwenstern had described $Hg(CNO)_2$ as far back as in the seventeenth century, it did not have any practical application prior to Alfred Nobel's blasting caps. It is interesting to mention that it was not until the year 2007 that the molecular structure of $Hg(CNO)_2$ was elucidated by the LMU research team (Fig. 1.1b) [1, 2]. Literature also reports the thermal transformation of MF, which, according to the below equation, forms a new mercury containing explosive product which is reported to be stable up to 120 °C.

$$3\,Hg(CNO)_2 \rightarrow Hg_3(C_2N_2O_2)_3$$

After another devastating explosion in 1866 which completely destroyed the NG factory, Alfred Nobel focused on the safe handling of NG explosives. In order to reduce the sensitivity, Nobel mixed NG (75%) with an absorbent clay called "Kieselguhr" (25%). This mixture called "Guhr Dynamite" was patented in 1867. Despite the great success of dynamite in the civil sector, this formulation has never found significant application or use in the military sector.

One of the great advantages of NG (Fig. 1.2) in comparison to blackpowder (75% KNO_3, 10% S_8, 15% charcoal) is that it contains both the fuel and oxidizer in the same molecule which guarantees optimal contact between both components, whereas in blackpowder, the oxidizer (KNO_3) and the fuel (S_8, charcoal) have to be physically mixed.

At the same time as NG was being researched and formulated several other research groups (Schönbein, Basel and Böttger, Frankfurt-am-Main) worked on the nitration of cellulose to produce **nitrocellulose (NC)**. In 1875 Alfred Nobel discovered that when NC is formulated with NG, they form a gel. This gel was further refined to produce blasting gelatine, gelatine dynamite and later in 1888 ballistite (49% NC, 49% NG, 2% benzene and camphor), which was the first smokeless powder (Cordite which was developed in 1889 in Britain, had a very similar composition). In 1867 it was proven that mixtures of NG or dynamite and ammonium nitrate (AN) showed enhanced performance. Such mixtures were used in the civil sector. In 1950, manufacturers started to develop explosives which were waterproof and solely contained the less hazardous AN. The most prominent formulation was ANFO (Ammonium Nitrate Fuel Oil) which found extensive use in commercial areas (mining, quarries etc.). Since the 1970s, aluminum and monomethylamine were added to such formulations to produce gelled explosives which could detonate more easily. More recent developments include production of emulsion explosives which contain suspended droplets of a solution of AN in oil. Such emulsions are water proof, yet readily detonate because the AN and oil are in direct contact. Generally, emulsion explosives are safer than dynamite and are simple and cheap to produce.

2,4,6-Trinitrotoluene (TNT), an important military explosive, is produced commercially by the nitration of toluene. In the first stage, mixtures of concentrated nitric and sulfuric acids (mixed acid) are used to produce an isomeric mixture of dinitrotoluenes. Mixtures of nitric acid and oleum (sulfuric acid containing up to 44% free sulfur trioxide) are subsequently used to convert the dinitrotoluene mixture to TNT. However, one major problem in this overall process is that unwanted asymmetrical TNT isomers are produced in addition to symmetrical 2,4,6-TNT (Fig. 1.2a). These isomers are generated by nitration occurring in the 3 or meta- position of the toluene ring in quantities of up to 3–5%. If the final 2,4,6-TNT product is contaminated with these isomers, then it has too low a melting point for military use. Therefore, any asymmetrical isomers must be removed in the TNT production process by treatment with bisulfite ($NaHSO_3$).

Fortunately, asymmetrical TNT derivatives readily react with $NaHSO_3$ solution ("sellite") to form the corresponding sulfonic acids via nitro group displacement. Derivatives with nitro groups in *ortho* positions with respect to each other are more prone to nucleophilic attack of sodium sulfite. The sulfonic acid derivatives can then be easily removed by washing with water, resulting in so-called "red water" (Fig. 1.2b). Yellow water results from washing of crude, acidic TNT obtained after nitration with HNO_3/H_2SO_4, whereas red water results after treatment with sodium sulfite, depending on the exact type and amounts of the by-products (Fig. 1.2c). This "red water" by-product waste-stream is difficult and costly to treat (Fig. 1.2c). Compounds present in the red water waste are known to be toxic, carcinogenic and mutagenic. In addition to the yellow and red water waste streams, a further problem with current TNT manufacturing processes is the generation of the undesirable by-products tetrani-

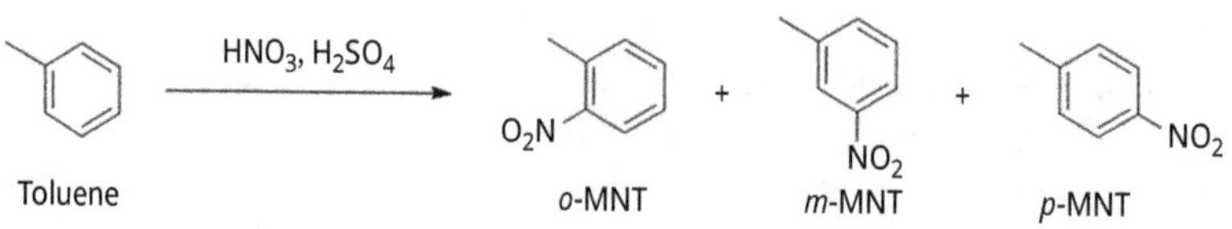

Fig. 1.2a: The first step in the nitration of toluene yields up to 3–5% of the unwanted *m*-nitrotoluene which eventually yields the asymmetrical TNT.

Fig. 1.2b: Redwater pollution at a TNT production facility.

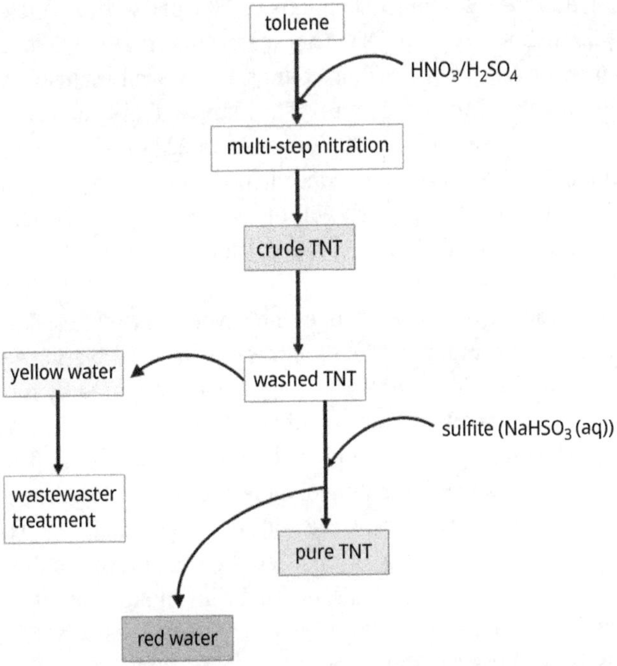

Fig. 1.2c: The Sellite process generates red water due to the sulfite reaction with asymmetrical TNT.

tromethane ($C(NO_2)_4$), and nitrogen oxide (NO_x) gases, which require waste treatment which adds additional extra costs.

Therefore, for future TNT production, there are two main strategies which could be employed to improve the environmental impact of TNT production:

(i) Find an alternative TNT synthesis which does not produce side products of unwanted nitro-toluene compounds. This would eliminate the necessity of the subsequent sulfite reaction. If complete elimination of unwanted nitro compounds is not possible, the concentration of these compounds in the first nitration step should be less than 1%.

(ii) Find an alternative work-up process that destroys all unwanted products in an efficient and ecologically acceptable process.

As a consequence of the geopolitical situation in 2024, there is a severe shortage of the explosives TNT (melt cast), nitroguanidine (NQ), nitrocellulose (NC) and RDX. It would therefore be advantageous to increase the production capacity of TNT in Western Europe which is presently only produced by Nitro-Chem (Poland) and also in Romania in the future.

As stated above, at present, the factory located near the city of Bydgoszcz in Poland, is the last surviving TNT plant in Europe or North America. It's run by a state-owned company, Nitro-Chem, and makes about 10,000 tons of TNT per year. A single 155 mm round typically requires about 10 kg of TNT. That means that the 10,000 tons of TNT would be enough to provide for about 1 million rounds, if it was all exclusively used for 155 mm shells. In the US alone, the production of 155 mm artillery shells has increased dramatically over the last years (Fig. 1.2d) However, in 2025 the US Army awarded a contract to Repkon USA for the design, construction and operation of a TNT production facility in Graham, Kentucky, which will be the first to produce TNT domestically since 1986. The contract is worth $ 435 million [National Defense, January 2025, p. 10].

Picric acid (PA) was first reported in 1742 by Glauber, however it was not used as an explosive until the late nineteenth century (1885–1888), when it replaced blackpowder in nearly all military operations world-wide (Fig. 1.3). PA is best prepared by dissolving phenol in sulfuric acid and subsequent nitration of the resulting of phenol-2,4-disulfonic acid with nitric acid. The direct nitration of phenol with nitric acid is not possible because the oxidizing HNO_3 decomposes the phenol molecule. Since the sulfonation is reversible, the —SO_3H groups can then be replaced with —NO_2 groups by refluxing the disulfonic acid in concentrated nitric acid. In this step the third nitro group is introduced as well. Although pure PA can be handled safely, a disadvantage of PA is its tendency to form impact sensitive metal salts (picrates, primary explosives) when in direct contact with shell walls. PA was used as a grenade and as mine filling.

Tetryl was developed at the end of the nineteenth century (Fig. 1.3) and represents the first explosive of the nitroamino (short: nitramino) type. Tetryl is best obtained by dissolving monomethylaniline in sulfuric acid and then pouring the solution into nitric acid, while cooling the process.

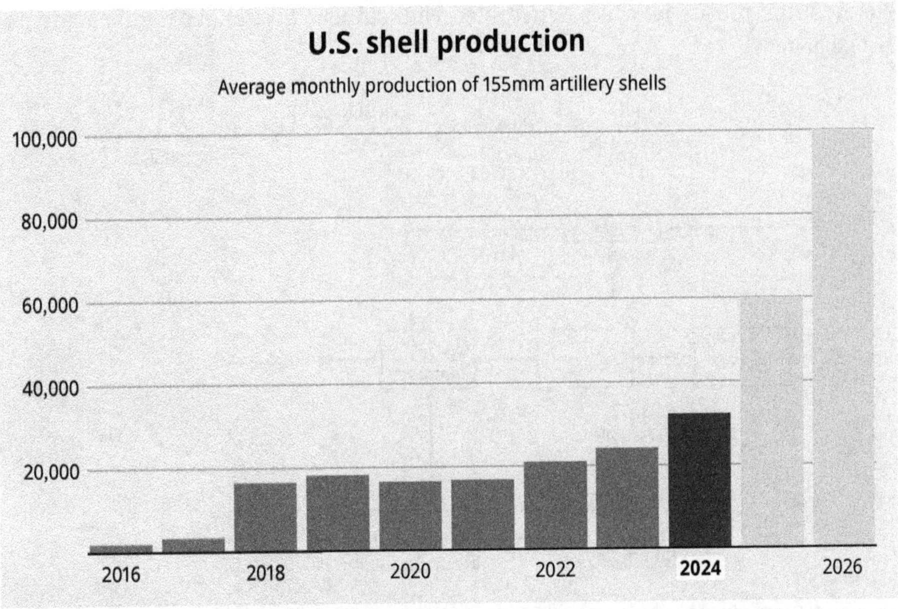

Fig. 1.2d: US shell production (source: https://www.reuters.com/investigates/special-report/ukraine-crisis-artillery/).

The above mentioned disadvantages of PA are overcome by the introduction of **trinitrotoluene (TNT)**. Pure 2,4,6-TNT was first prepared by Hepp (Fig. 1.3) and its structure was determined by Claus and Becker in 1883. In the early twentieth century TNT almost completely replaced PA and became the standard explosive during WW I. TNT is produced by the nitration of toluene with mixed nitric and sulfuric acid. For military purposes, TNT must be free of any isomer other than the 2,4,6-isomer. This is achieved by recrystallization from organic solvents or from 62% nitric acid. TNT is still one of the most important explosives for blasting charges today. Charges are produced through casting and pressing. However, cast charges of TNT often show sensitivity issues and do not comply with the modern insensitive munition requirements (IM). For this reason alternatives to TNT have been suggested. One of these replacements for TNT is NTO (filler) combined with 2,4-dinitroanisole (DNAN, binder).

Nitroguanidine (NQ) was first prepared by Jousselin in 1887 (Fig. 1.3). However, during WW I and WW II it only found limited use, for example in formulations with AN in grenade for mortars. In more recent days NQ has been used as a component in triple-base propellants together with NC and NG. One advantage of the triple-base propellants is that unlike double-base propellants the muzzle flash is reduced. The introduction of about 50% of NQ to a propellant composition also results in a reduction of the combustion temperature and consequently reduced erosion and increased lifetime of the gun. NQ can be prepared from dicyandiamide and ammonium nitrate via guanidinium nitrate which is dehydrated with sulfuric acid under the formation of NQ (Scheme 1):

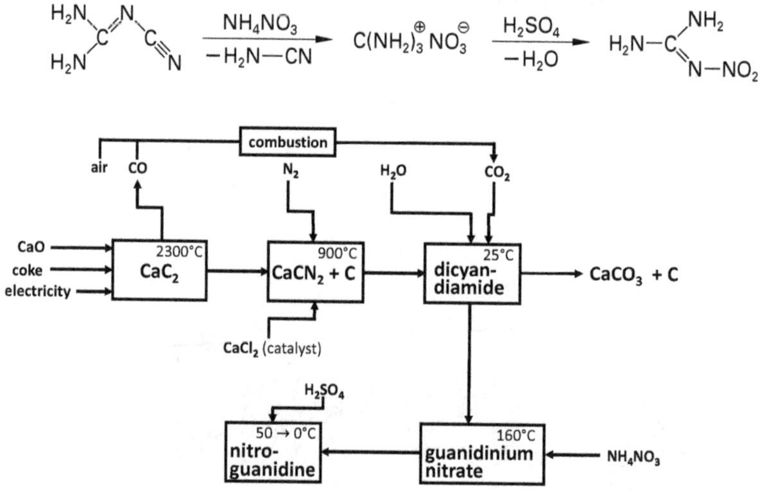

Scheme 1: Production of nitroguanidine (NQ) (modified from: E. C. Koch, *Prop. Expl. Pyrotech.*, 2019, 44, 267–292).

Nitroguanidine is a versatile ingredient for insensitive, gun, rocket, base bleed, and gas generator propellants. Gun propellant formulations containing NQ show relatively low combustion/explosion temperatures, with only minimum muzzle flash, as well as reduced erosion. Their performance spans from very lean formulations giving some f = 750 J/g, up to f = 1,320 J/g for formulations containing extra RDX and an energetic binder of the GAP-type. In rocket propellants in combination with oxidizers (AP or ADN), specific impulses range from 185–267 s. NQ-based propellants are mechanically rigid due to the fibrous nature of NQ, are only mildly sensitive to impact, and yield favorable IM signatures. The weak basicity and strong hydrogen bonds in NQ are the origin for its unmatched thermal stabilization and desensitization of formulations which contain other energetic materials. An excellent review on this topic has recently been published by E.-C. Koch (*Propellants Explos. Pyrotech.* **2021**, *46*, 1–34).

The most widely used explosives in WW II other than TNT were hexogen (RDX) and pentaerythritol tetranitrate (nitropenta, PETN) (Fig. 1.3). Since PETN is more sensitive and chemically less stable than RDX, RDX was (and is) the most commonly used high explosive. PETN is a powerful high explosive and has a great shattering effect (brisance). It is used in grenades, blasting caps, detonation cords and boosters. PETN is not used in its pure form because it is too sensitive. A formulation of 50% TNT and 50% PETN is known as "pentolite". In combination with plasticized nitrocellulose PETN is used to form polymer bonded explosives (PBX). The military application of PETN has largely been replaced by RDX. PETN is prepared by introducing pentaery-

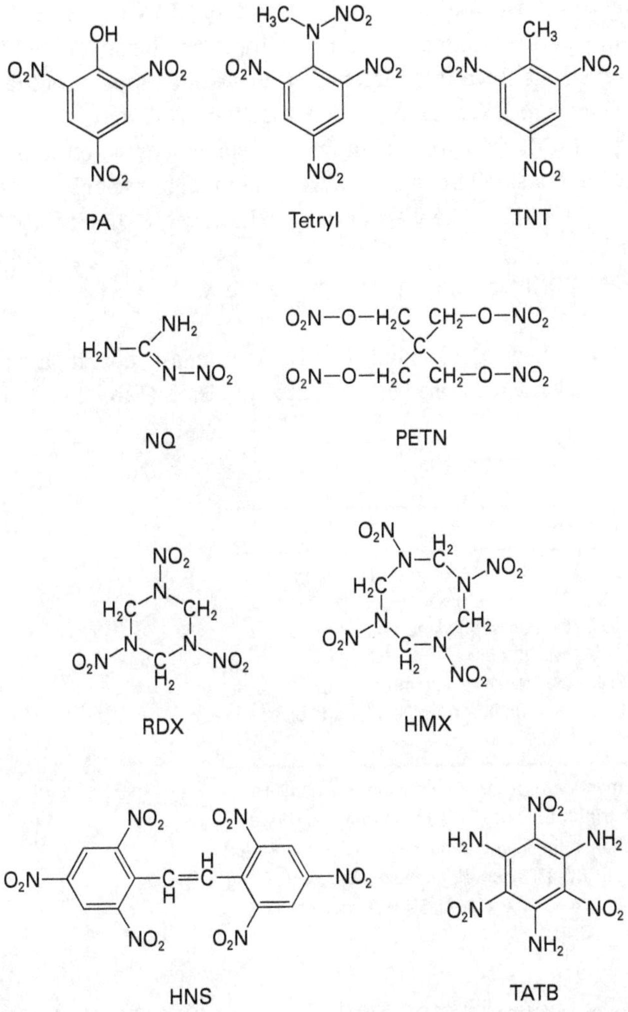

Fig. 1.3: Molecular structures of picric acid (PA), tetryl, trinitrotoluene (TNT), nitroguanidine (NQ), pentaerythritol tetranitrate (PETN), hexogen (RDX), octogen (HMX), hexanitrostilbene (HNS) and triaminotrinitrobenzene (TATB).

thritol into concentrated nitric acid while cooling and stirring the mixture efficiently. The then formed bulk of PETN crystallizes out of the acid solution. The solution is then diluted to about 70% HNO_3 in order to precipitate the remaining product. The washed crude product is purified by recrystallization from acetone.

Hexogen (RDX) was first prepared in 1899 by Henning for medicinal use. (*N.B.* NG and PETN are also used in medicine to treat angina pectoris. The principal action of these nitrate esters is vasodilation (i.e. widening of the blood vessels). This effect arises because in the body the nitrate esters are converted to nitric oxide (NO) by mitochondrial aldehyde dehydrogenase, and nitric oxide is a natural vasodilator.) In 1920, Herz prepared RDX for the first time by the direct nitration of hexamethylene tetramine. Shortly afterwards Hale (Picatinny Arsenal, NJ) developed a process that formed RDX in 68% yield. The two processes most widely used in WW II were:

1. Bachmann process (type B RDX) (KA process) in which hexamethylene tetramine dinitrate reacts with AN and a small amount of nitric acid in an acetic anhydride medium to form RDX (type B RDX). The yields are high, however, 8–12% of HMX form as a side product.
2. Woolwich process (type A RDX) essentially produces pure RDX.

After WW II **octogen (HMX)** started to become available. Until today, most high explosive compositions for military use are based on TNT, RDX and HMX (Tab. 1.2).

Tab. 1.2: Composition of some high explosive formulations.

Name	Composition
Composition A	88.3% RDX, 11.7% non-energetic plasticizers
Composition B	60% RDX, 39% TNT, 1% binder (wax)
Composition C4	90% RDX, 10% polyisobutylene
Octol	75% HMX, 25% RDX
Torpex[a]	42% RDX, 40% TNT, 18% aluminum
PBXN-109	64% RDX, 20% aluminum, 16% binder and plasticizer
OKFOL	96.5% HMX, 3.5% wax

[a]An Australian improved development of torpex is known under the name H6 and also contains hexogen (RDX), trinitrotoluene (TNT) and aluminum. H6 was used as a high explosive formulation in the MOAB bomb (Massive Ordnance Air Blast bomb). MOAB (also known as GBU-43/B) is with a load of approx. 9,500 kg high explosive formulation (30% TNT, 45% RDX, 20% Al, 5% wax) one of the largest conventional bombs ever used.

As stated above, the industrial synthesis of RDX usually employs either the Woolwich (U.K., Type A RDX) or Bachmann process (USA, Type B RDX). The Woolwich process is a direct nitrolysis, in which hexamine is directly nitrated with nitric acid at 25–30 °C to produce RDX in 70–75% yield as shown in the following equation:

$$(CH_2)_6N_4 + 6\,HNO_3(98\,\%) \rightarrow RDX + 3\,CO_2 + 2\,N_2 + 6\,H_2O$$

Bachmann however, developed a process in which hexamine reacts with a mixture of ammonium nitrate and nitric acid in the presence of acetic anhydride at 75 °C to produce RDX in ca. 70% yields:

$$(CH_2)_6N_4 + 4\,HNO_3(98\,\%) + 2\,NH_4NO_3 + 6\,(CH_3CO)_2O \rightarrow 2\,RDX + 12\,CH_3COOH$$

The Bachmann process uses less nitric acid (expensive), but produces RDX with impurities of 8–12% HMX. The Woolwich process usually yields RDX with HMX impurities of < 5% but uses more HNO_3 (more expensive). Based on the content of HMX impurity, the RDX is classified either as Type I RDX (up to 5% HMX) or Type II RDX (contains 4–17% HMX). Table 1.2a lists different categories of RDX.

Tab. 1.2a: RDX categories.

Manufacturer	Grade	Type	Expected % HMX
Ordnance Systems Inc.	–	II	4–17
Eurenco, France	I-RDX [a]		< 0.5
Dyno Nobel	Type II	II	4–17
Australian Defence Industries (ADI)	Grade A	I	< 0.5
Royal Ordnance Defence	Grade A	I	< 0.5

[a] I-RDX stands for insensitive RDX

The synthesis of HMX according to the Bachmann process can be summarized as follows:

$$2\,(CH_2)_6N_4 + 8\,HNO_3(98\,\%) + 4\,NH_4NO_3 + 6\,(CH_3CO)_2O \rightarrow 3\,HMX + 12\,CH_3COOH + 6\,H_2O$$

Ammonium nitrate is an essential reagent for directing the course of the nitrolysis to form RDX or HMX.

The hexamine used in the Bachmann process can undergo two different types of cleavage:

1. Compounds containing three amino nitrogen atoms are formed: RDX, linear trinitramine
2. Compounds containing four amino nitrogen atoms are formed: HMX, DPT, linear tetranitramine

The first path (→ RDX) is favored by high acidity and/or a high acidity of the nitrating agent, whereas the second path (→ HMX) is favored by lower acidity and/or lower acidity of the nitrating agent. Furthermore, the concentration of ammonium nitrate has an influence on the yield of HMX, whereby the optimum ratio is 2–3 moles of ammonium nitrate to 1 mole hexamine.

In addition, the temperature of the reaction is important, since performing the nitration reaction at 44 °C favors the formation of HMX, whereas at 65 °C RDX is predominantly formed.

Typical reaction ratios are shown in Tab. 1.2b.

Since 1966 **hexanitrostilbene (HNS)** and since 1978 **triaminotrinitrobenzene (TATB)** are produced commercially (Fig. 1.3). Both secondary explosives show excellent thermal stabilities and are therefore of great interest for the NAVY (fuel fires) and for hot

Tab. 1.2b: Typical ratios of reactants used for the synthesis of RDX and HMX.

reactants	lbs. for RDX production	lbs. for HMX production
ammonium nitrate	17.2	{11.0}
98% nitric acid	13.6	
hexamine	9.2	17.0
acetic acid	15.0	18.0
acetic anhydride	45.0	54.0

deep oil drilling applications (Fig. 1.3). Especially HNS is known as a heat- and radiation-resistant explosive which is used in heat-resistant explosives in the oil industry. The brisance of HNS is lower than that of RDX, but the melting point of approx. 320 °C is much higher. HNS can directly be prepared from trinitrotoluene through oxidation with sodium hypochlorite in a methanol/THF solution:

$$2\,C_6H_2(NO_2)_3CH_3 + 2\,NaOCl \rightarrow C_6H_2(NO_2)_3{-}CH{=}CH{-}C_6H_2(NO_2)_3 + 2\,H_2O + 2\,NaCl$$

Since oil deposits which are located closer to the surface are becoming rare, deeper oil reserves now have to be explored where (unfortunately) higher temperatures are involved. Therefore, there is an ongoing search for explosives which are even more thermally stable (decomposition temperatures > 320 °C) than HNS, but at the same time show better performance (Tab. 1.2c). Higher thermal stabilities usually result in compounds with lower sensitivities which are therefore safer to handle.

According to J. P. Agrawal, new energetic materials with high thermal stabilities can be achieved by incorporating the following points in the compounds:
- Salt formation
- Introduction of amino groups
- Introduction of conjugation
- Condensation with a triazole ring

Two possible replacements for HNS which are presently under investigation are PYX and PATO.

Various picryl and picrylamino substituted 1,2,4–triazoles which were formed by condensing 1,2,4-triazole or amino-1,2,4-triazole with picryl chloride (1-chloro -2,4,6-trinitrobenzene) were studied in detail by Coburn & Jackson. One of these molecules is PATO (3-picrylamino-1,2,4-triazole), a well known, thermally-stable explosive, which is obtained by the condensation of picryl chloride with 3-amino-1,2,4-triazole (Fig. 1.3a). Another promising candidate for a high-temperature explosive is PYX (Fig. 1.3a). The synthesis for PYX is shown in Fig. 1.3b.

Tab. 1.2c: Desired properties of potential HNS replacements.

Thermal stability	No changes after 100 h at 260 °C
Detonation velocity	> 7,500 m/s
Specific energy*	> 975 kJ/kg
Impact sensitivity	> 7.4 J
Friction sensitivity	> 235 N
Total costs	< 500 Euro/kg
Critical diameter	≥ HNS

*specific energy. $F = p_e \cdot V = n \cdot R \cdot T$

Agrawal et al. reported the synthesis of BTDAONAB (Fig. 1.3c) which does not melt below 550 °C and is considered to be a better and thermally more stable explosive than TATB. According to the authors, this material has a very-low impact (21 J), no friction sensitivity (> 360 N) and is thermally stable up to 550 °C. These reported properties makes BTDAONAB superior to all of the nitro-aromatic compounds which have been discussed. BTDAONAB has a VoD of 8,300 m/s while TATB is about 8,000 m/s [Agrawal et al., *Ind. J. Eng. & Mater Sci.*, **2004**, *11*, 516–520; Agrawal et al., Central *Europ. J. Energ. Mat.* **2012**, *9(3)*, 273–290.]

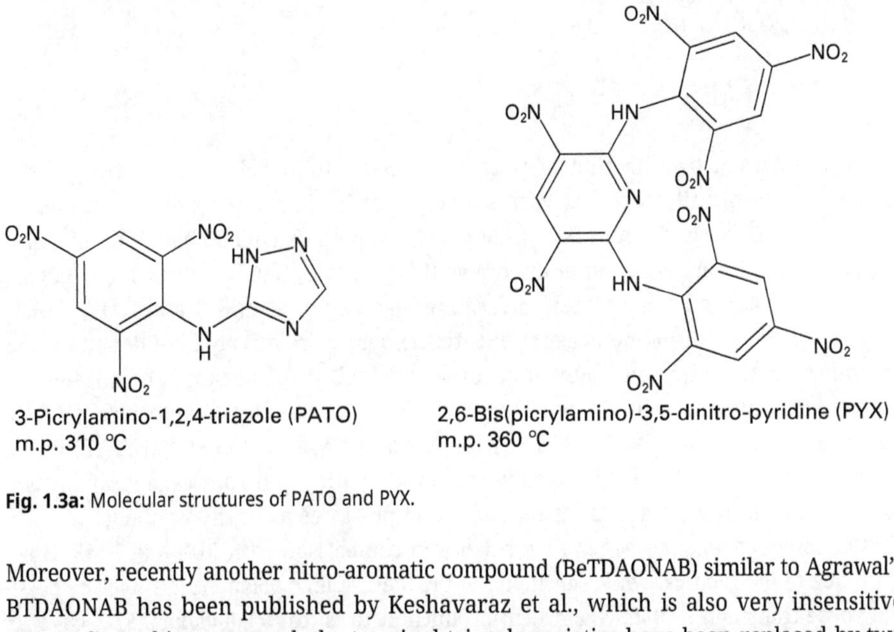

3-Picrylamino-1,2,4-triazole (PATO)
m.p. 310 °C

2,6-Bis(picrylamino)-3,5-dinitro-pyridine (PYX)
m.p. 360 °C

Fig. 1.3a: Molecular structures of PATO and PYX.

Moreover, recently another nitro-aromatic compound (BeTDAONAB) similar to Agrawal's BTDAONAB has been published by Keshavaraz et al., which is also very insensitive (Fig. 1.3d). In this compound, the terminal triazole moieties have been replaced by two more energetic (more endothermic) tetrazole units [Keshavaraz et al., Central *Europ. J. Energ. Mat.* **2013**, *10(4)*, 455; Keshavaraz et al., *Propellants, Explos. Pyrotech.*, **2015**, *40*, 886–891. DOI: 10.1002/prep.201500017]. Table 1.2d shows a comparison of the thermal and explosive properties of TATB, HNS, BTDAONAB and BeTDAONAB.

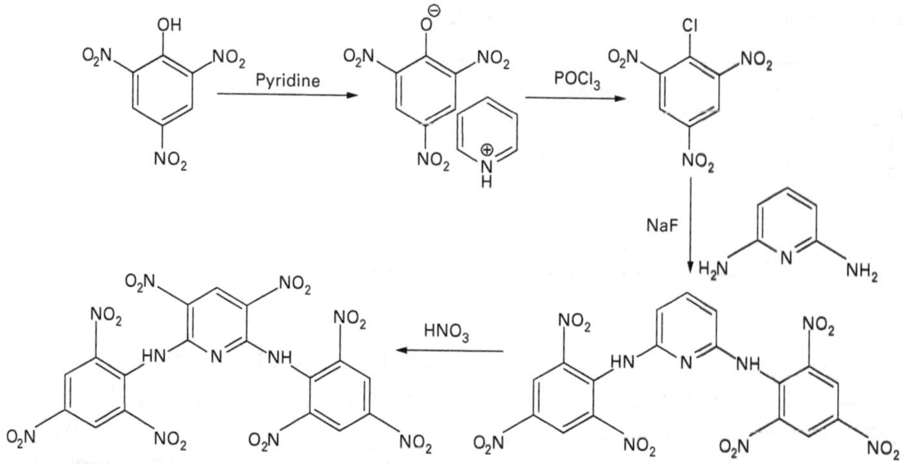

Fig. 1.3b: Synthetic route for PYX.

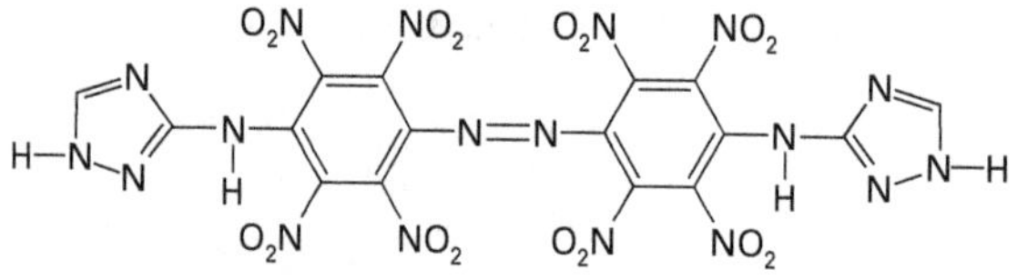

Fig. 1.3c: Molecular structure of BTDAONAB.

TATB is obtained by nitration of trichlorobenzene followed by a reaction of the formed trichlorotrinitro benzene with ammonia gas in benzene or xylene solution.

As shown above, the number of chemical compounds which have been used for high explosive formulations until after WW II is relatively small (Tabs. 1.1 and 1.2). As seen in Tabs. 1.1 and 1.2, the best performing high explosives (RDX and HMX; TNT is only used because of its melt-cast applications) possess relatively-high densities and contain oxidizer (nitro and nitrato groups) and fuel (C—H backbone) combined in one and the same molecule. One of the most powerful new high explosives is **CL-20** which was first synthesized in 1987 by the Naval Air Warfare Center (NAWF) China Lake (Fig. 1.7, Tab. 1.1). CL-20 is a cage compound with significant cage strain which also contains nitramine groups as oxidizers and possesses a density of about 2 g cm^{-3}. This already explains the better performance in comparison with RDX and HMX. However, due to the relatively high sensitivity of the (desirable) ε polymorph as well as possible phase transition problems and high production costs (ca. \$ 1,000/lb; cf. RDX ca. \$ 20.—/kg; GAP ca. \$ 500/kg) so far CL-20's wide and general application has not been established.

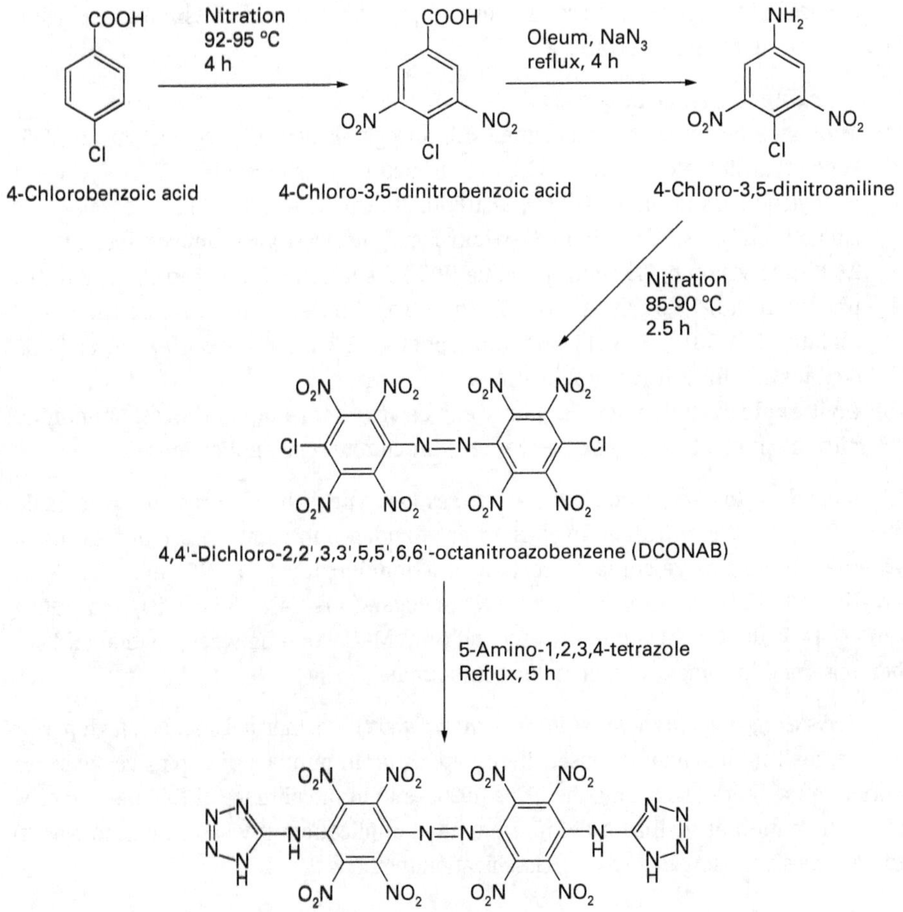

Fig. 1.3d: Synthetic route for the synthesis of BeTDAONAB.

Tab. 1.2d: Comparative data of the thermal and explosive properties of TATB, HNS, BTDAONAB and BeTDAONAB.

Property	TATB	HNS	BTDAONAB	BeTDAONAB
Density / g/cc	1.94	1.74	1.97	1.98
Sensitivity to temperature / °C	360	318	350	260
DTA (exo) / °C	360	353	550	275
DSC (exo) / °C	371	350		268
Ω_{CO} / %	−18.6	−17.8	−6.8	−5.9
IS / J	50	5	21	21
FS / N	> 353	240	353	362
VoD / m s^{-1}	7,900	7,600	8,600	8,700
$p_{C\text{-}J}$ / kbar	273	244	341	354

According to Agrawal, there are several possible ways of classifying explosives and important among these are:

(i) ***According to their end-use*** e. g.

(a) **Military explosives** [Nitrocellulose (NC), Nitroglycerine (NG), Trinitrotoluene (TNT), Pentaerythritol tetranitrate (PETN), Cyclotrimethylene trinitramine (RDX), Cyclotetramethylene tetranitramine (HMX), 2,4,6,8,10,12-Hexanitro-2,4,6,8,10,12-hexaazaisowurtzitane (CL-20), 1,3,5-Triamino-2,4,6-trinitrobenzene (TATB), Hexanitrostilbene (HNS), 2,6-Bis(picrylamino)-3,5-dinitropyridine (PYX), 2,4,6-Tris(3′,5′-diamino-2′,4′,6′-trinitrophenylamino)-1,3,5-triazine (PL-1), Dihydroxylammonium-5,5′-bistetrazole-1,1′-diolate (TKX-50), 5,5′-Bis(2,4,6-trinitrophenyl)-2,2′-bi(1,3,4-oxadiazole) (TKX-55) etc.] mainly for military applications

(b) **Civil explosives** (dynamites, slurry explosives, emulsion explosives, ammonium nitrate – fuel oil (ANFO) explosives etc.) for commercial applications

(ii) ***According to their chemical structures*** i. e. type of bonds present in the explosive. Using this classification, explosives are divided into eight classes depending on whether the explosive contains certain functional groups: (i) $-NO_2$ and $-ONO_2$ (ii) –N=N– and –N=N=N– (iii) $-NX_2$ where X = halogen (iv) –C=N– (v) $-OClO_2$ and $-OClO_3$ (vi) –O–O– and –O–O–O– (vii) –C≡C– and (viii) M–C, bond between a metal and carbon observed in some organometallic compounds.

(iii) ***According to a single most important property*** i. e. thermal stability, high performance, melting behavior or insensitivity etc. A large number of explosives were reported in the literature during the 1970s–1990s, and in the absence of full and extensive characterization as well as a definite intended application, it was difficult to classify these explosives using the existing classification methods.

Agrawal also proposed a new method of classification in which he classified HEMs as the following:

A. ***Thermally stable or heat-resistant explosives:*** 1,3-Diamino-2,4,6-trinitrobenzene (DATB), 1,3,5-Triamino-2,4,6-trinitrobenzene (TATB), Hexanitrostilbene (HNS), 3-Picrylamino-1,2,4-triazole (PATO), 3,3′-Diamino-2,2′,4,4′,6,6′-hexanitrodiphenyl (DIPAM), *N*2,*N*4,*N*6-Tripicryl melamine (TPM), 2,6-Bis(picrylamino)-3,5-dinitropyridine (PYX), 2,2′,2″,4,4′,-4″,6,6′,6″-Nonanitroterphenyl (NONA), 1,3-Bis(1,2,4-triazol-3-amino)-2,4,6-trinitrobenzene (BTATNB), *N*,*N*′-Bis(1,2,4-triazol-3-yl)-4,4′-diamino-2,2′,3,3′,5,5′,6,6′-octanitroazobenzene (BTDAONAB), 2,4,6-Tris(3′,5′-diamino-2′,4′,6′-trinitrophenylamino)-1,3,5-triazine (PL-1), 5,5′-Bis(2,4,6-trinitrophenyl)-2,2′-bi(1,3,4-oxadiazole) (TKX-55), *N*5,*N*5′-(1,2,4,5-tetrazine-3,6-diyl)-bis(1*H*-1,2,4-triazole-3,5-diamine), 1,4-Bis(2′,4′,6′-trinitrophenyl)-3,6-dinitro[4,3-*c*] pyrazole (DTNBDNP) etc.

B. ***High performance*** (high density & high detonation velocity) ***explosives:*** Cyclotrimethylene trinitramine (RDX), Cyclotetramethylene tetranitramine (HMX), 1,3,4,6-Tetranitroglycouril (TNGU), 1-(3,5-Dinitro-1*H*-pyrazol-4-yl)-3-nitro-1*H*-1,2,4-triazol-5-amine

(HCPT), Tetranitropropanediurea (TNPDU), 2,4,6,8,10,12-Hexanitro-2,4,6,8,10,12-hexaazaisowurtzitane (HNIW/CL-20), 2,5,7,9-Tetranitro-2,5,7,9-tetraazabicyclo[4.3.0]nonane-8-one (TNABN), 1,3,5,5-Tetranitrohexahydropyrimidine (DNNC), Dihydroxylammonium-5,5′-bistetrazole-1,1′-diolate (TKX-50) etc.

C. ***Melt-castable explosives***: Trinitrotoluene (TNT), 2,4,6-Tris(2-nitroxyethylnitramino)-1,3,5-triazine (Tris-X), 1,3,3-Trinitroazetidine (TNAZ), 4,4′-Dinitro-3,3′-bifurazan (DNBF), 2,4-Dinitroanisole (DNAN), 1-Methyl-2,4,5-trinitroimidazole (MTNI), Bis(nitroxymethylisoxazolyl) furoxan (SMCE-2) and Bis(1,2,4-oxadiazolyl) furoxan (MCEE-5) etc.

D. ***Insensitive high explosives (IHEs):*** TATB, 7-Amino-4,6-dinitro-benzofuroxan (ADNBF), 3-Nitro-1,2,4-triazol-5-one (NTO), 1,4-Dinitroglycouril (DINGU), 5-amino-3-nitro -1*H*-1,2,4-triazole (ANTA), *trans*-1,4,5,8-Tetranitro-1,4,5,8-tetraazadecalin (TNAD), 1,1-Diamino-2,2-dinitroethylene (FOX-7), *N*-Guanylurea salt of dinitramide (FOX-12) etc.

E. ***Energetic binders***: Polynitropolyphenylenes (PNPs), Glycidylazide polymer (GAP), Nitrated hydroxy-terminated polybutadiene (NHTPB), Poly(3-nitratomethyl-3-methyloxetane) [Poly(NiMMO)]; Poly(glycidyl nitrate) [(Poly(GlyN)] etc. and plasticizers: Nitroglycerin (NG), Ethyleneglycol dinitrate (EGDN), Diethylene glycol dinitrate (DEGDN), Triethylene glycol dinitrate (TEGDN), 1,2,4-Butanetrioltrinitrate (BTTN), Bis(2,2-dinitropropyl) acetal/formal (BDNPA/F), *N*-Butyl-*N*-(2-nitroxyethyl) nitramine (Bu-NENA), Bis(2-azidoethyl) adipate (BAEA) etc.

F. ***Novel energetic materials***: materials synthesized with the use of dinitrogen pentoxide (N_2O_5): NHTPB, Poly(NiMMO), Poly(GlyN), Ammonium dinitramide (ADN), TNGU, etc.

Explosives that show improved high-temperature properties are termed thermally stable or heat-resistant explosives. Thermally stable explosives have received special attention recently, because of their ability to withstand the high temperatures and low pressures encountered in space environments. Thermally stable explosives are used for specialized applications such as in:

- Oil & gas industry: deep drilling and perforation of oil & gas wells etc.
- Space applications: stage separation in space rockets, deorbitation, seismic experiments on the moon's surface, etc.
- Nuclear applications: explosive lenses, implosion bombs, etc.
- High speed guided missiles & hypersonic[#)] rockets: more recently, explosives with $T_{dec} > 375$ °C have also gained special interest for this purpose.

Based on an analysis of the structure and of possible synthetic routes, Agrawal proposed that there are currently four general approaches for the synthesis of thermally stable explosives:

(i) Salt formation
(ii) Introduction of amino group/s

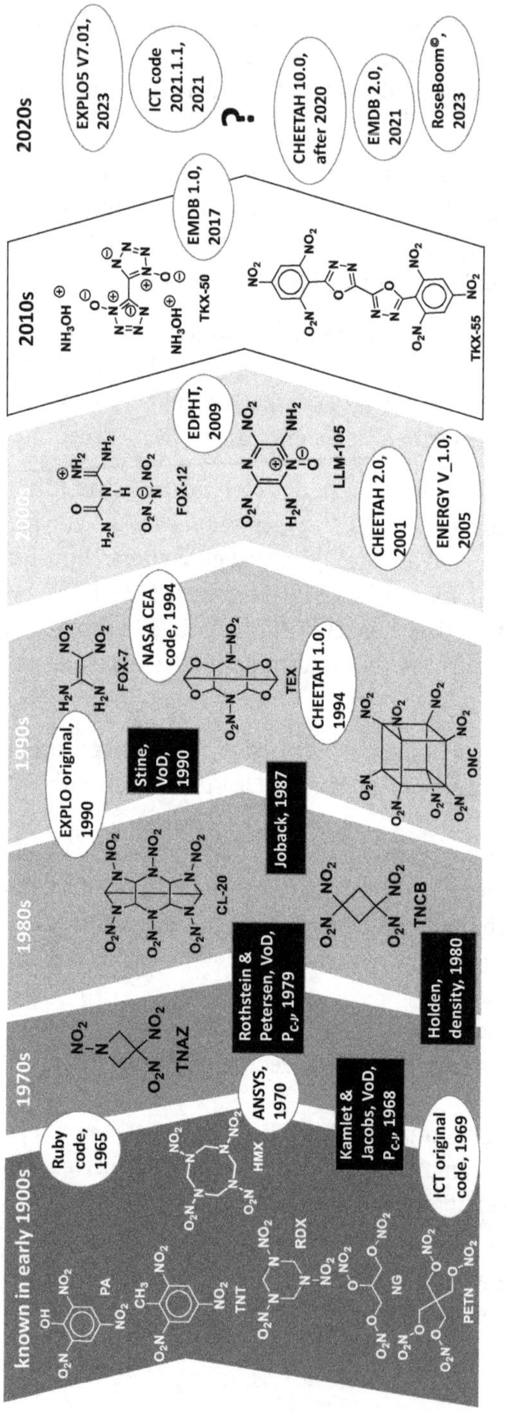

Fig. 1.3e: An extremely basic overview of developments in energetic materials research showing some selected landmark compounds as well as computational advances.

(iii) Incorporation of conjugation
(iv) Condensation with a triazole ring

#)
A hypersonic missile usually reaches speeds more than Mach 5. Hypersonic boost-glide missiles consist of a rocket motor, which accelerates the missile to a high speed, and a glide body containing the warhead. The rocket booster initially accelerates a hypersonic cruise missile to speeds approaching Mach 4, then boosts and maintains a higher speed throughout its flight using a jet engine called a supersonic combustion ramjet, or scramjet, that can take it toward Mach 5.

Figure 1.3e shows a very brief overview of advances in energetic materials research over the past decades as a schematic diagram. Only single compounds have been included and not formulations or TBXs.

Literature

J. P. Agrawal, V. S. Dodke, Validation of Approaches (Salt Formation & Introduction of Amino Group/s) for Imparting/Improving Thermal Stability of Explosives, Part 1, *Propellants Explos. Pyrotech.*, **2021**, *46*, 1–12.

J. P. Agrawal, V. S. Dodke, Some Novel High Energy Materials for Improved Performance, *Z. Anorg. Allg. Chem.*, **2021**, *647*, 1856–1882.

1.2 New developments

1.2.1 Polymer-bonded explosives

Since about 1950 polymer-bonded (or plastic-bonded) explosives (PBX) have been developed in order to reduce sensitivity and to facilitate safe and easy handling. PBX also shows improved processibility and mechanical properties. In such materials the crystalline explosive is embedded in a rubber-like polymeric matrix. One of the most prominent examples of a PBX is **Semtex**. Semtex was invented in 1966 by Stanislav Brebera, a chemist who worked for VCHZ Synthesia in Semtin (hence the name Semtex), a suburb of Pardubice in the Czech Republic. Semtex consists of varying ratios of PETN and RDX. Usually polyisobutylene is used for the polymeric matrix, and phthalic acid *n*-octylester is the plasticizer. Other polymer matrices which have been introduced are polyurethane, polyvinyl alcohol, PTFE (Teflon), Viton, Kel-F, PBAN, HTPB and various polyesters. PBAN (polybutadiene acrylonitrile) was used in the space shuttle composite propellant.

Often, however, problems can arise when combining the polar explosive (RDX) with the non-polar polymeric binder (*e.g.* polybutadiene or polypropylene). In order to overcome such problems, additives are used to facilitate mixing and intermolecular interactions. One of such polar additives is dantacol (DHE) (Fig. 1.4).

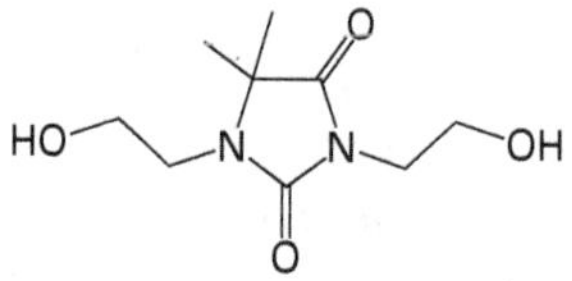

Fig. 1.4: Structure of dantacol (DHE).

One disadvantage of the polymer-bonded explosives of the first generation, is that the non-energetic binder (polymer) and plasticizer lessened the performance. To overcome this problem energetic binders and plasticizers have been developed. The most prominent examples for **energetic binders** are (Fig. 1.5a):
- poly-GLYN, poly(glycidyl)nitrate
- poly-NIMMO, poly(3-nitratomethyl-3-methyl-oxetane)
- GAP, glycidylazide polymer
- poly-AMMO, poly(3-azidomethyl-3-methyl-oxetane)
- poly-BAMO, poly(3,3-bis-azidomethyl-oxetane).

Examples for **energetic plasticizers** are (Fig. 1.5b):
- NENA derivatives, alkylnitratoethylnitramine
- EGDN, ethyleneglycoldinitrate
- MTN, metrioltrinitrate
- BTTN, butane-1,2,4-trioltrinitrate.

Typical energetic oxetane monomers are shown in Fig. 1.5a and their physical properties are summarized in Tab. 1.2e.

Figure 1.5b shows a general synthesis route for various energetic polymers using the ring-opening polymerization (ROP) strategy.

For binders in particular – but also for plasticizers – it is important to know the glass transition temperature. The value of the glass transition temperature should be as low as possible but at least −50 °C. If the temperature of a polymer drops below T_g, it behaves in an increasingly brittle manner. As the temperature rises above T_g, the polymer becomes more rubber-like. Therefore, knowledge of T_g is essential in the selection of materials for various applications. In general, values of T_g well below room temperature correspond to elastomers and values above room temperature to rigid, structural polymers.

In a more quantitative approach for the characterization of the liquid-glass transition phenomenon and T_g, it should be noted that in cooling an amorphous material from the liquid state, there is no abrupt change in volume such as that which occurs on cooling a crystalline material below its freezing point, T_f. Instead, at the glass transition temperature, T_g, there is a change in the slope of the curve of specific volume vs. tem-

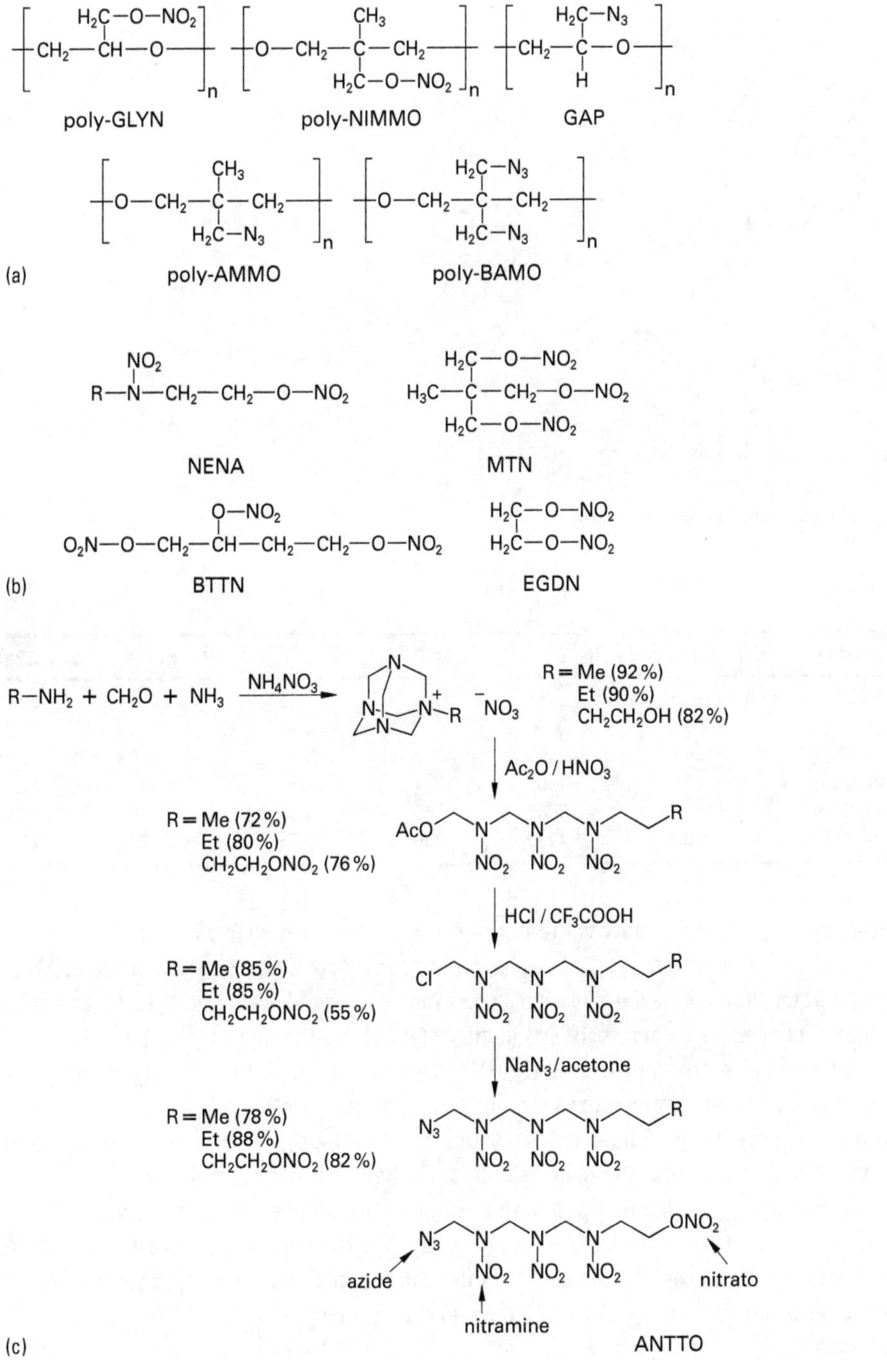

Fig. 1.5: Energetic binders (a) and energetic plasticizers (b) Synthesis of the NENA compound, ANTTO (c) (ANTTO = azido nitrato trinito triaza octane).

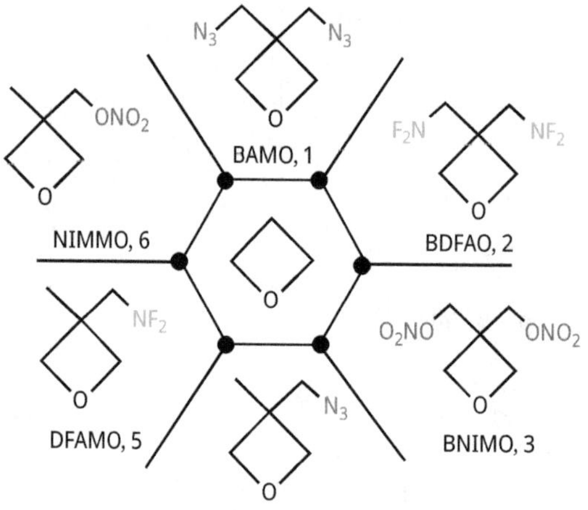

Fig. 1.5a: Energetic oxetane monomers.

Tab. 1.2e: Physical properties of energetic oxetane monomers.

monomer	$\Omega(CO_2)$ / %	HOF / J g^{-1}	T_{dec} /°C	ρ / g cm^{-3}	VoD / m s^{-1}	p_{CJ} / GPa
AMMO (4)	−170.1	784	139	1.06	5660	11.1
BAMO (1)	−123.8	2660	160	1.22	6770	17.3
NIMMO (6)	−165.5	−1468	203	1.7	6280	14.7
BNIMO (3)	−109.4	−1305	–	–	7850	26.9
DFAMO (5)	−157.7	−2380	215	1.08	4780	8.6
BDFAO (2)	−110.6	−1790	230	1.65	7090	21.6

perature, moving from a low value in the glassy state to a higher value in the rubbery state over a range of temperatures. This comparison between a crystalline material (1) and an amorphous material (2) is illustrated in the figure below. Note that the intersection of the two straight line segments of curve (2) defines the quantity T_g (Fig. 1.5c).

Differential scanning calorimetry (DSC) can be used to determine experimentally the glass transition temperature. The glass transition process is illustrated in Fig. 1.5d for a glassy polymer which does not crystallize and is being slowly heated from a temperature below T_g. Here, the drop which is marked T_g at its midpoint, represents the increase in energy which is supplied to the sample to maintain it at the same temperature as the reference material. This is necessary due to the relatively rapid increase in the heat capacity of the sample as its temperature is increases pass T_g. The addition of heat energy corresponds to the endothermal direction.

The typical formulation of a cast polymer-bonded explosive is shown below (reproduced with kind permission of Dr. Paul Wanninger):

Percentage	Ingredient	Comment
85–90%	AP, RDX, HMX, Al	
10–15%	HTPB, plasticizer	HTPB is the binder
0.2%	Lecithin	Lecithin is used to reduce the rheological yield point
0.2%	Antioxidants	An antioxidant is added for long-life stability
10^{-3}–10^{-4}%	Catalyst	The amount of catalyst determines the curing time
0.05–0.1%	Bonding agent	Different bonding agents for RDX, HMX and AP or Al are recommended to improve the mechanical properties

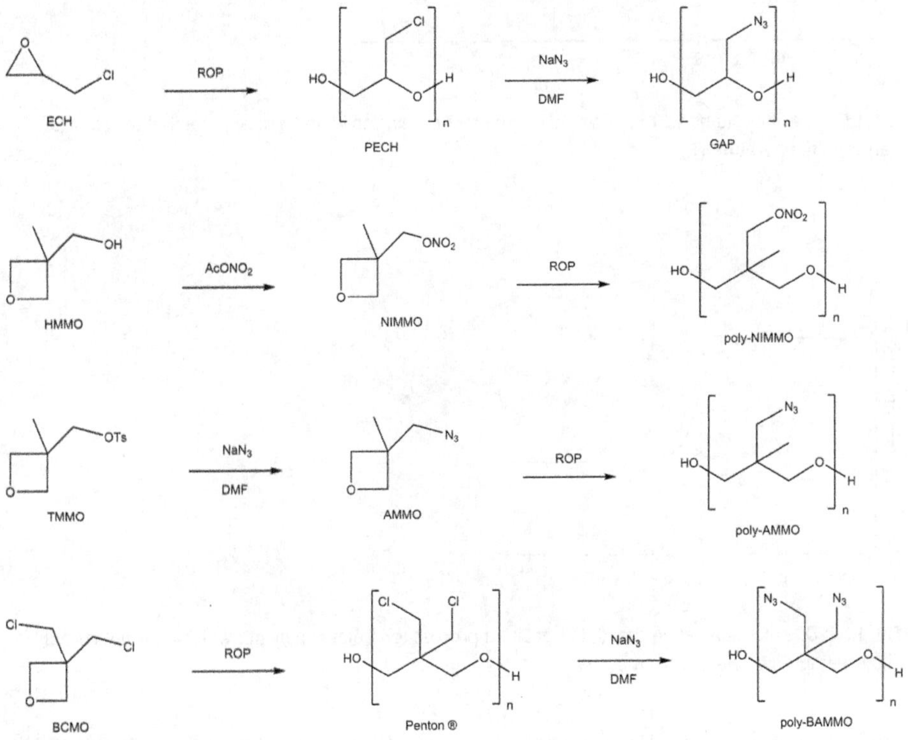

Fig. 1.5b: Ring-opening reactions (ROP) for the synthesis of energetic polymers.

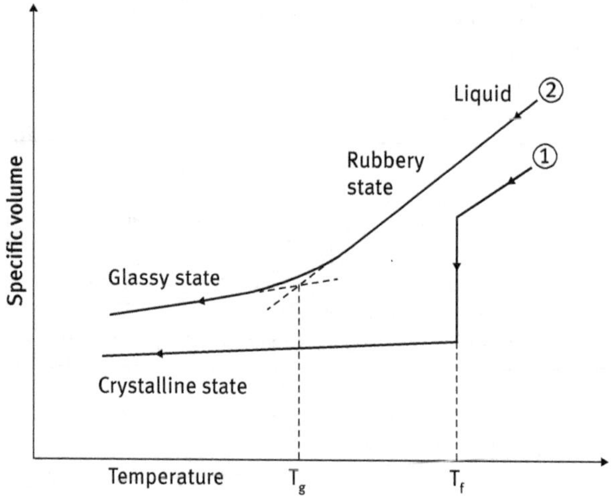

Fig. 1.5c: Specific volume vs. temperature plot for a crystalline solid and a glassy material with a glass transition temperature (T_g).

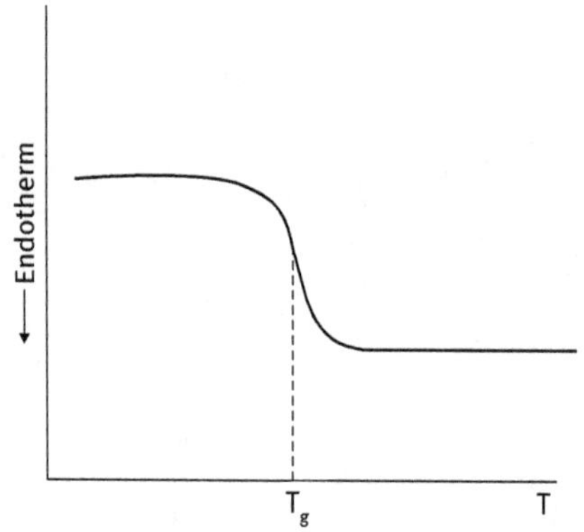

Fig. 1.5d: DSC plot illustrating the glass transition process for a glassy polymer which does not crystallize and is being slowly heated from below T_g.

Typical properties of melt cast, cast, and pressed PBX formulations are shown below (reproduced with kind permission of Dr. Paul Wanninger):

	Melt cast PBX	**Cast PBX**	**Pressed PBX**
Ingredients	TNT, RDX, HMX, NTO	RDX, HMX + polymer	RDX, HMX + thermoplast
Technical process	Melting and casting	Mixing under vacuum and casting under vacuum, polymerization	HE + binder + solvent, pressing at up to 1,500 bar

Density (in % of TMD)	98	99.5	97–98
Density gradient	High	Very low	Varies
Vulnerability	High	Very low	Medium
Homogenicity	Low	High	Medium
Melting point (°C)	Melting 70–80	Decomposition ca. 180	Decomposition ca. 180

1.2.2 New high (Secondary) explosives

New secondary explosives which are currently under research, development or testing include 5-nitro-1,2,4-triazol-3-one (NTO), 1,3,3-trinitroazetidine (TNAZ), hexanitrohexaazaisowurtzitane (HNIW, CL-20) and octanitrocubane (ONC) (Fig. 1.7). **NTO** has already found application as a very insensitive compound in gas generators for automobile inflatable air bags and in some polymer-bonded explosive formulations. (*N.B.* Initially NaN_3 was used in air bag systems, however, nowadays guanidinium nitrate is often used in combination with oxidizers such as AN in some non-azide automotive inflators.) It is used to enhance burning at low flame temperatures. Low flame temperatures are desired in order to reduce the formation of NO_x gasses in inflators. NTO is usually produced in a two-step process from semicarbazide hydrochloride with formic acid via the intermediate formation of 1,2,4-triazol-5-one (TO) and subsequent nitration with 70% nitric acid:

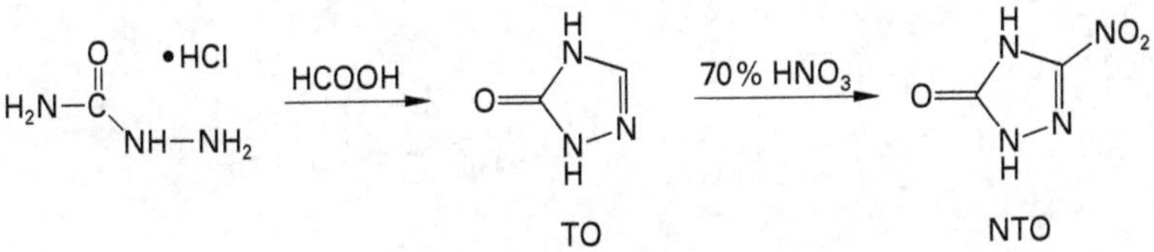

Another interesting new and neutral high explosive is BiNTO, which can be synthesized as shown in the below equation from commercially available NTO in a one-step reaction.

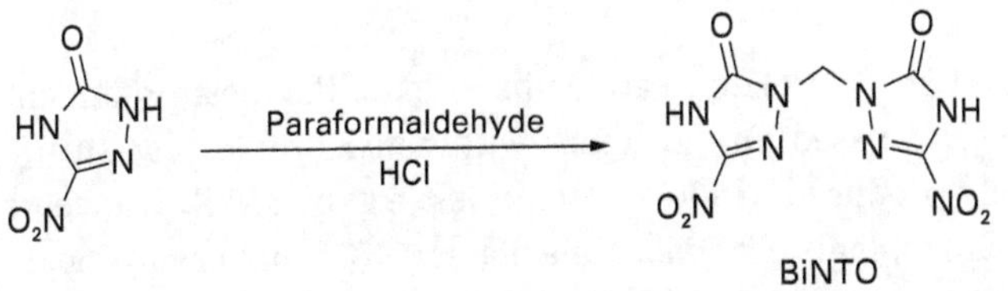

TNAZ was first synthesized in 1983 and has a strained four-membered ring backbone with both C-nitro and nitramine (N—NO_2) functionalities. There are various routes that yield TNAZ all of which consist of several reaction steps. One possible synthesis of TNAZ is shown in Fig. 1.6. It starts from epichlorohydrine and ^{t}Bu-amine. As far as the author of this book is aware, there has been no wide-spread use for TNAZ so far.

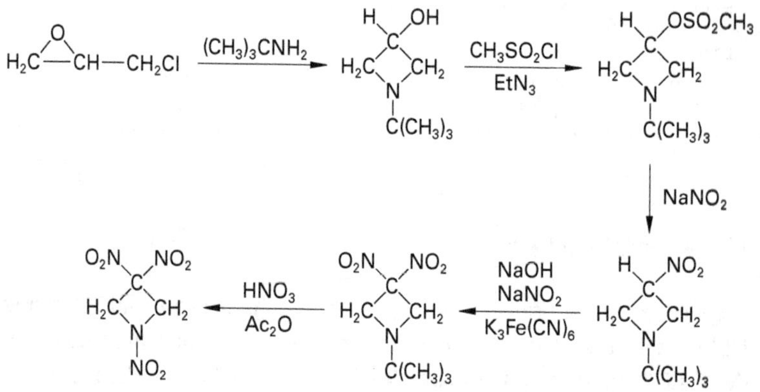

Fig. 1.6: Synthesis of 1,3,3-trinitroazetidine (TNAZ).

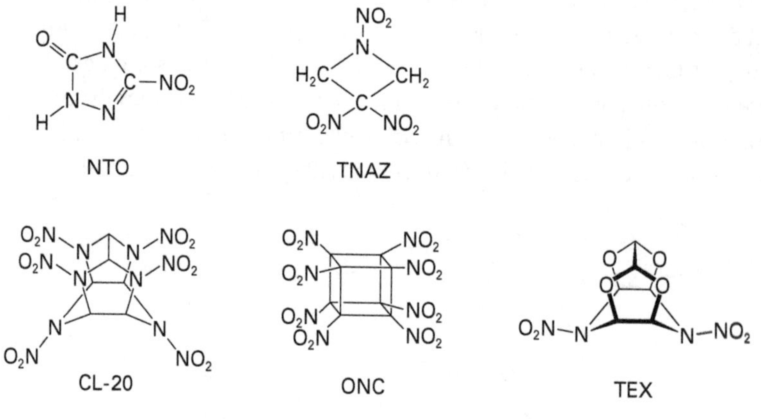

Fig. 1.7: Molecular structures of 5-nitro-1,2,4-triazol-3-one (NTO), 1,3,3-trinitroazetidine (TNAZ), hexanitrohexaazaisowurtzitane (CL-20), octanitrocubane (ONC) and 4,10-dinitro-2,6,8,12-tetraoxa-4,10-diazaisowurtzitane (TEX).

CL-20 (1987, A. Nielsen) and ONC (1997, Eaton) are without doubt the most prominent recent examples of explosives based on molecules with considerable cage-strain. While CL-20 is now already produced in 100 kg quantities (e.g. by SNPE, France or Thiokol, USA, ca. \$ 1000.—/lb) on industrial pilot scale plants, ONC is only available on a mg to g scale because of its very difficult synthesis. Despite the great enthusiasm for CL-20 since its discovery over 20 years ago, it has to be mentioned that even today most of the high explosive formulations are based on RDX (see Tab. 1.2). There are

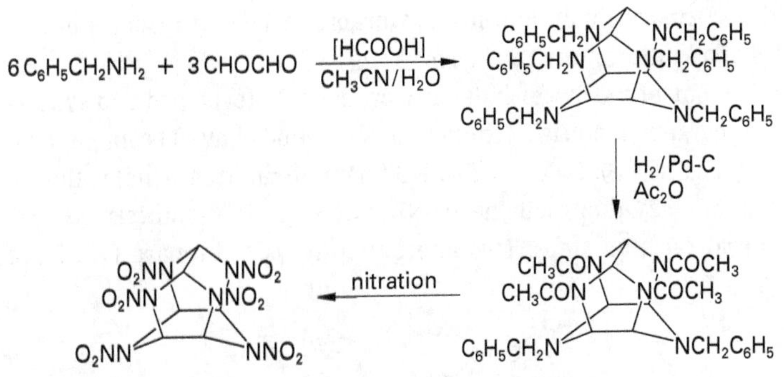

Fig. 1.8: Synthesis of hexanitrohexaazaisowurtzitane (CL-20).

several reasons why CL-20 despite its great performance has not yet been introduced successfully:
- CL-20 is much more expensive than the relatively cheap RDX
- CL-20 has some sensitivity issues (see insentitive munitions)
- CL-20 exists in several polymorphic forms and the desired ε polymorph (because of its high density and detonation velocity) is thermodynamically not the most stable one.

Interconversion of the ε form into a more stable but perhaps also more sensitive other polymorph would result in a loss of performance and an increase in sentitivity.

CL-20 is obtained by the condension of glyoxal with benzylamine in an acid catalyzed reaction to yield hexabenzylhexaazaisowurtzitane (Fig. 1.8). Afterwards the benzyl groups are replaced under reducing conditions (Pd-C catalyst) by easily removable acetyl substituents. Nitration to form CL-20 takes place in the final reaction step.

Tab. 1.3: Characteristic performance and sensitivity data of FOX-7 and FOX-12 in comparison with RDX.

	FOX-7	FOX-12	RDX
Detonation pressure, p_{C-J} / kbar	340	260	347
Detonation velocity, D / m s^{-1}	8,870	7,900	8,750
Impact sensitivity / J	25	> 90	7.5
Friction sensitivity / N	> 350	> 352	120
ESD / J	ca. 4.5	> 3	0.2

In the past, EURENCO produced CL-20 on a semi-industrial scale (pilot scale). However, this production was stopped, predominantly due to the cost and the sensitivity of CL-20. The process EURENCO used was the method of Nielson and co-workers, in which a catalyst was used to produce TADBIW, which was subsequently nitrated to produce CL-20.

Another very insensitive high explosive which is structurally closely related to CL-20 is 4,10-dinitro-2,6,8,12-tetraoxa-4,10-diazaisowurtzitane (**TEX**, see Fig. 1.7), which was first described by Ramakrishnan and his co-workers in 1990. It displays one of the highest densities of all nitramines (2.008 g cm^{-3}) [1c].

The chemist N. Latypov of the Swedish defense agency FOI developed and synthesized two other new energetic materials. These two compounds have become known as **FOX-7** and **FOX-12** (Fig. 1.9 (a)). FOX-7 or DADNE (diamino dinitroethene) is the covalent molecule 1,1-diamino-2,2-dinitroethene: $(O_2N)_2C{=}C(NH_2)_2$. The synthesis of FOX-7 always includes several reaction steps. Two alternative ways to prepare FOX-7 are

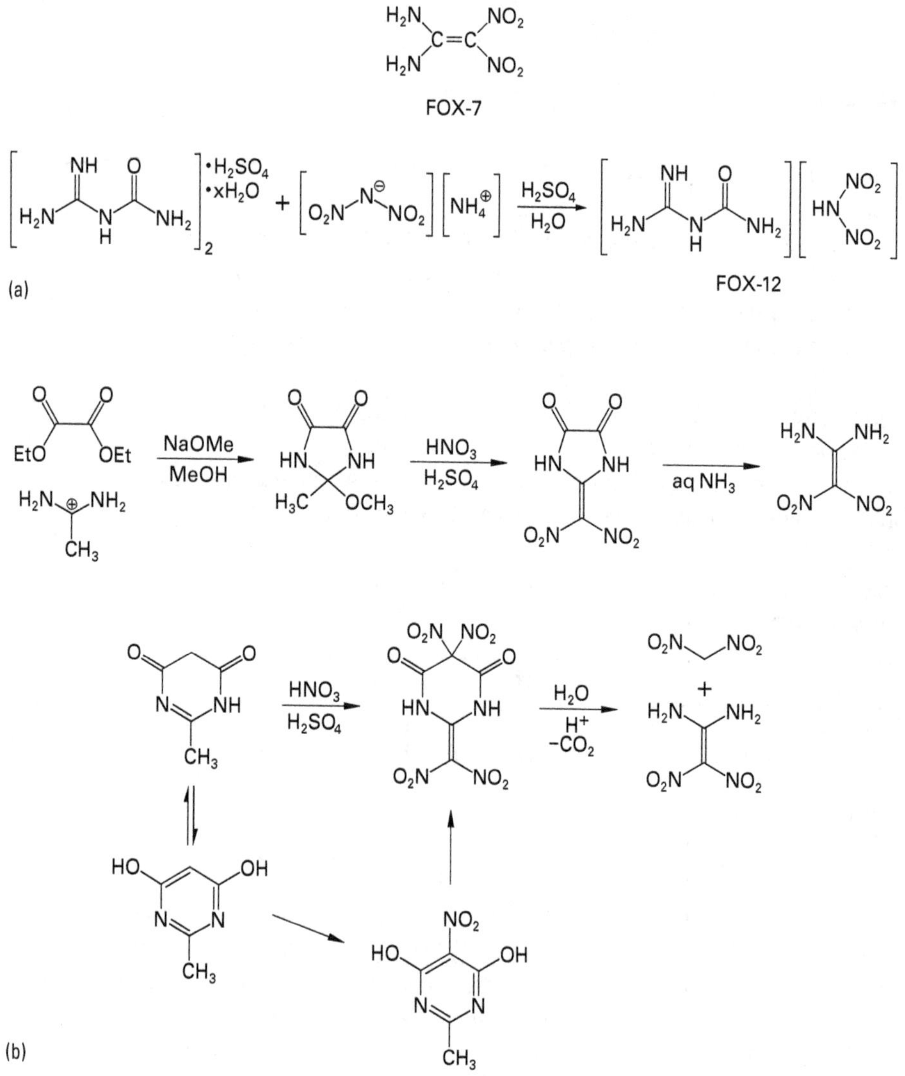

Fig. 1.9: Molecular structures of FOX-7 and FOX-12 (a). Two alternative synthetic routes for the synthesis of FOX-7 (b).

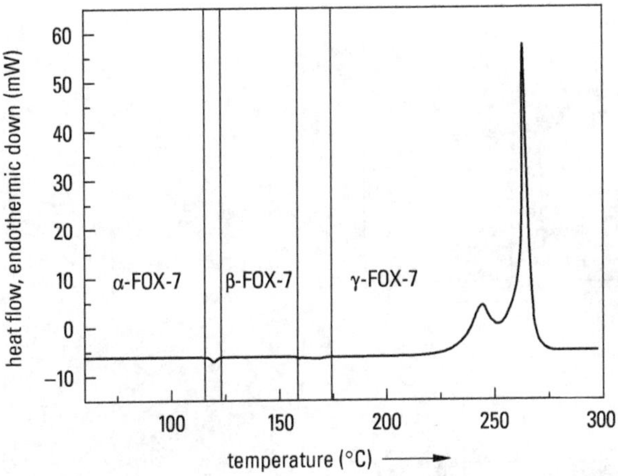

Fig. 1.10: DSC-plot of FOX-7.

shown in Fig. 1.9 (b). FOX-12 or GUDN (guanylurea dinitramide) is the dinitramide of guanylurea: $[H_2N—C(=NH_2)—NH—C(O)—NH_2]^+[N(NO_2)_2]^-$.

It is interesting that FOX-7 has the same C/H/N/O ratio as RDX or HMX. Although neither FOX-7 nor (and in particular not) FOX-12 meet RDX in terms of performance (detonation velocity and detonation pressure). Both compounds are much less sensitive than RDX and might be of interest due to their insensitive munition (IM) properties. Table 1.3 shows the most characteristic performance and sensitivity data of FOX-7 and FOX-12 in comparison with RDX.

FOX-7 exists in at least three different polymorphic forms (α, β and γ). The α modification converts reversibly into the β form at 389 K (Fig. 1.10) [2]. At 435 K the β polymorph converts into the γ phase and this interconversion is not reversible. The γ form can be quenched at 200 K. When heated, the γ form decomposes at 504 K. Structurally, the three polymorphs are closely related and quite similar, with the planarity of the individual FOX-7 layers increasing from α via β to γ (i.e. γ posesses the most planar layers) (Fig. 1.11).

Another member of the family of nitramine explosives is the compound dinitroglycoluril (DINGU) which was first reported as early as 1888. The reaction between glyoxal (O=CH—CH=O) and urea yields glycoluril which can be nitrated with 100% nitric acid to produce DINGU. Further nitration with a mixture of HNO_3/N_2O_5 yields the corresponding tetramine SORGUYL. The latter compound is of interest because of its high density (2.01 g cm^{-3}) and its high detonation velocity (9,150 m s^{-1}) (Fig. 1.12). SORGUYL belongs to the class of cyclic dinitroureas. These compounds generally show a higher hydrolytic activity and may therefore be of interest as "self-remediating" energetic materials.

A new neutral nitrimino-functionalized high explosive which was first mentioned in 1951 (*J. Am. Chem. Soc.* **1951**, *73*, 4443, see also Damavarapu et al., OPR & D, **2016**, 683–686) and which was recently suggested as a RDX replacement in C4 and Comp.B by Damavarapu (ARDEC) is bisnitraminotriazinone (DNAM). This compound has a melting point of

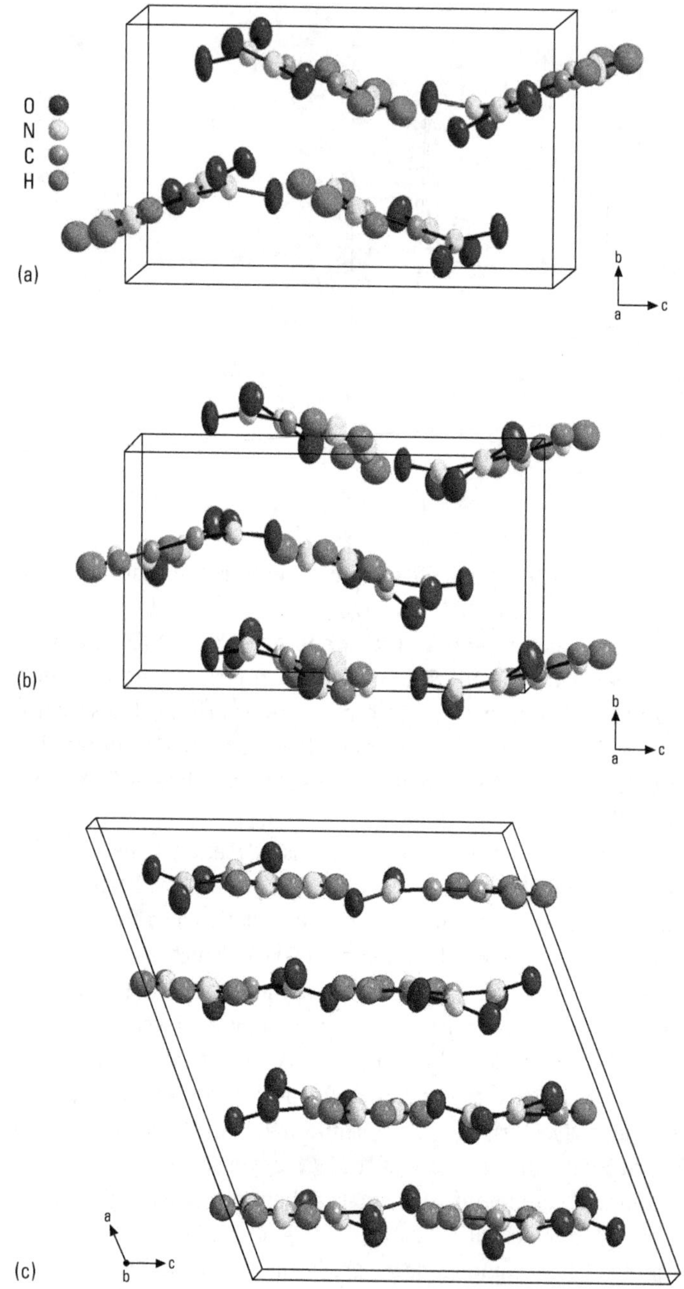

Fig. 1.11: Crystalline packing of α-FOX-7 (a), β-FOX-7 (b) and γ-FOX-7 (c).

228 °C and a remarkably high density of 1.998 g/cc. Due to the high density and the not too negative enthalpy of formation ($\Delta H^\circ_f = -111$ kJ mol^{-1}) DNAM has a detonation velocity of 9,200 m s^{-1} but still desirably low sensitivities (IS = 82.5 cm, FS = 216 N, ESD = 0.25 J). The synthesis of DNAM can be achieved in 50–60% yield by nitration of melamine using in-situ generated $AcONO_2$ as the effective nitrating agent (Fig. 1.13a), or by direct nitration of melamine. One possible concern about DNAM is that the compound hydrolyzes rapidly at 80 °C with liberation of nitrous oxide. At room temperature, the hydrolysis requires one to two days and is acid catalyzed.

The reaction of DNAM with $NaHCO_3$, CsOH and $Sr(OH)_2 \cdot 8\,H_2O$ yields the corresponding mono-deprotonated salts NaDNAM, CsDNAM and $Sr(DNAM)_2$, respectively.

Pyrazine derivatives are six-membered heterocyclic compounds containing two nitrogen atoms in the ring system. As high-nitrogen heterocylic compounds, they have an ideal structure for energetic materials (EMs). Some of them have a high enthalpy of formation, appropriate thermal stability and good safety characteristics. The basic structure of energetic pyrazine compounds is that of 3,5-dinitro-2,6-diaminopyrazine (**I** in Fig. 1.13). One of the most prominent members in this family is the 1-oxide, 3,5-dinitro-2,6-pyrazinediamine-1-oxide (also known as **LLM-105**, Fig. 1.13). LLM-105 has a high density of 1.92 g cm^{-3} and it shows a detonation velocity of 8,730 m s^{-1} and a deto-

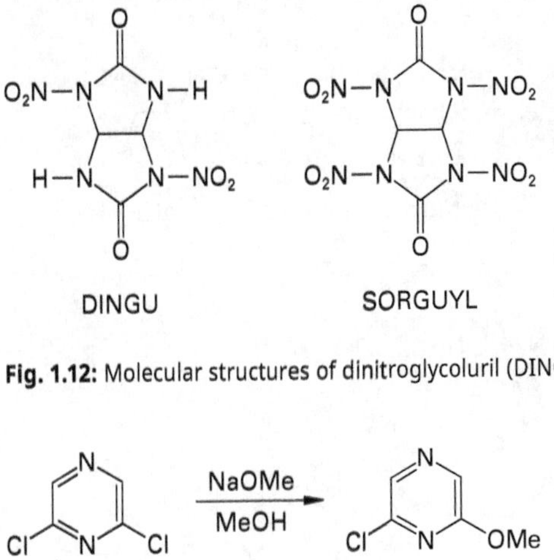

Fig. 1.12: Molecular structures of dinitroglycoluril (DINGU) and the corresponding tetramine (SORGUYL).

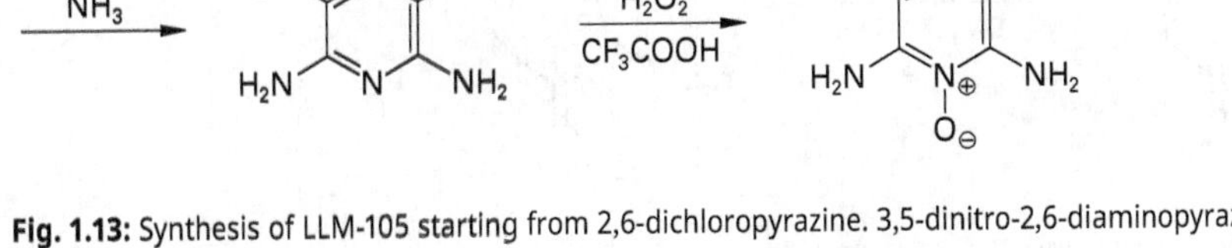

Fig. 1.13: Synthesis of LLM-105 starting from 2,6-dichloropyrazine. 3,5-dinitro-2,6-diaminopyrazine is oxidized to 3,5-dinitro-2,6-pyrazinediamine-1-oxide (LLM-105) in the final step.

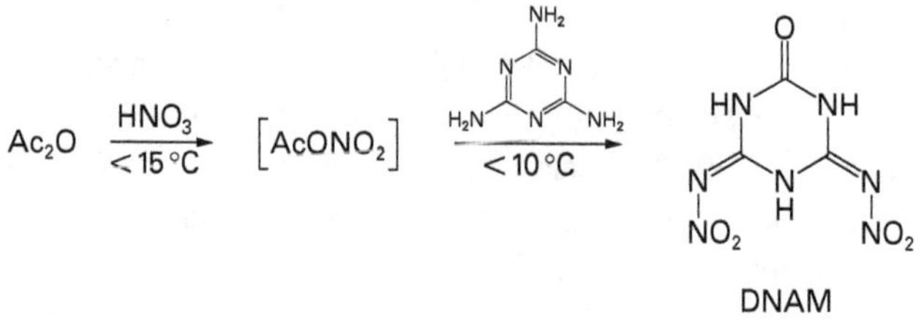

Fig. 1.13a: Synthesis of DNAM.

nation pressure of 359 kbar which are comparable to those of RDX (density = 1.80 g cm^{-3}, exptl. values: VoD = 8,750 m s^{-1}, $P_{C\text{-}J}$ = 347 kbar). LLM-105 is a lot less impact sensitive than RDX and is not sensitive towards electrostatics and friction [1d].

The synthesis of LLM-116 is shown in Fig. 1.13c.

Another *N*-oxide which has recently been suggested by Chavez et al. (LANL) as an insensitive high explosive is 3,3′-diaminoazoxy furazan (DAAF). Though the detonation velocity and detonation pressure of DAAF are rather low (7,930 m s^{-1}, 306 kbar @ 1.685 g/cc), the low sensitivity (IS > 320 cm, FS > 360 N) and a critical diameter of < 3 mm make this compound promising. The synthesis of DAAF is shown in Fig 1.13b [see also E.-C. Koch, Propellants, Explosives, Pyrotech, **2016**, *41*, 529–538].

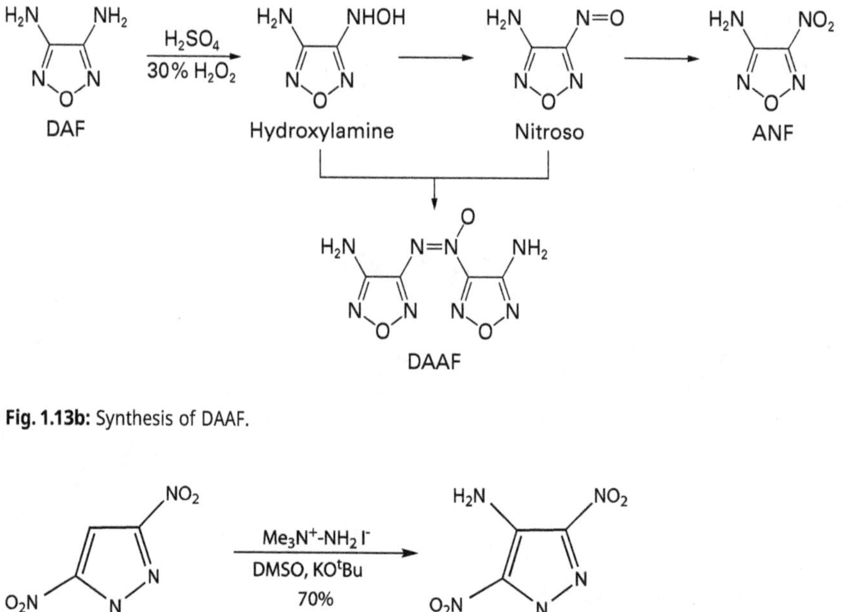

Fig. 1.13b: Synthesis of DAAF.

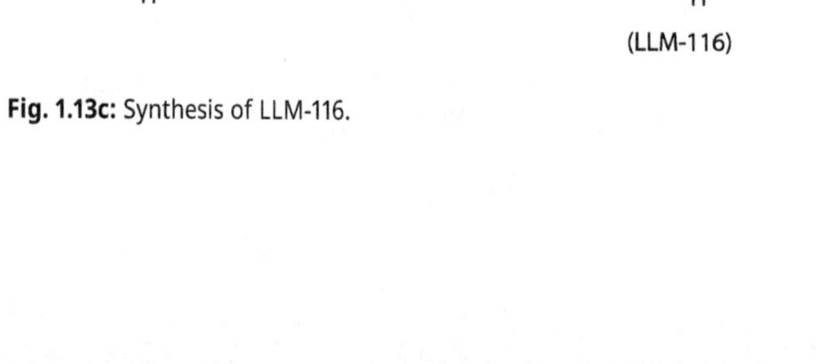

Fig. 1.13c: Synthesis of LLM-116.

There are various methods to prepare LLM-105. Most methods start from commercially available 2,6-dichloropyrazine (Fig. 1.13) and oxidize dinitropyrazinediamine in the final step to the 1-oxide (LLM-105).

Organic peroxides are another class of explosives which has been researched recently. This class of explosives (organic, covalent peroxides) includes the following compounds:

- H_2O_2
- peracids, R—C(O)—OOH
- peresters, R—C(O)—OO—R′
- perethers, R—O—O-R′
- peracetals, $R'_2C(OOR)_2$.
- peroxyanhydride, R—C(O)—O—O—C(O)—R′

Triacetone triperoxide (TATP, Fig. 1.14) is formed from acetone in sulfuric acid solution when acted upon by 45% (or lower concentration) hydrogen peroxide (the acid acts as a catalyst). Like most other organic peroxides TATP has a very high impact (0.3 J), friction (0.1 N) and thermal sensitivity. TATP has the characteristics of a primary explosive. For this reason and because of its tendency to sublime (high volatility) it is not used in practice (apart from terrorist and suicide bomber activities).

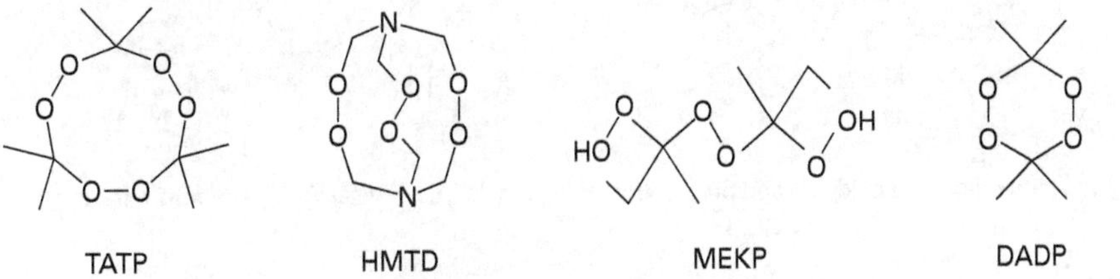

Fig. 1.14: Molecular structures of triacetone triperoxide (TATP), hexamethylene triperoxide diamine (HMTD), methyl ethyl ketone peroxide (MEKP) and diacetone diperoxide (DADP).

Because of the use of TATP by terrorists, a reliable and fast detection of this material is desirable. In addition to conventional analytical methods such as mass spectrometry and UV (ultraviolet) spectroscopy, specially trained explosive detection dogs (EDD) play an important role in the detection of organic peroxides. However, fully trained EDDs are expensive (up to $ 60 k) and can only work for 4 h per day. Although the high vapor pressure helps the dogs to detect the material, it is also a disadvantage because of the limited time-span in which the dog is able to find it (traces may sublime and disappear forever). Matrices in which the compounds can be imbedded are sought after for the safe training of explosive detection dogs. These matrices should not have any volatility, or any characteristic smell for the dogs. In this respect zeolites may be of interest [1e, f]. The ongoing problem with zeolites is that they need to be loaded with solutions, and the solvents (e.g. acetone) may not completely vaporize before the peroxide.

Typical organic peroxides, which have been or may be used by terrorists are so-called homemade explosives (HMEs): triacetone triperoxide (TATP), hexamethylene triperoxide diamine (HMTD), methyl ethyl ketone peroxide (MEKP) and diacetone diperoxide (DADP) (Fig. 1.14).

The following class of *N*-oxide compounds is considerably more stable than the above mentioned peroxides. For example, the oxidation of 3,3′-azobis(6-amino-1,2,4,5-tetrazine) in H_2O_2 / CH_2Cl_2 in the presence of trifluoroacetic acid anhydride yields the corresponding *N*-oxide (Fig. 1.15). This compound has a desirable high density and only modest impact and friction sensitivity.

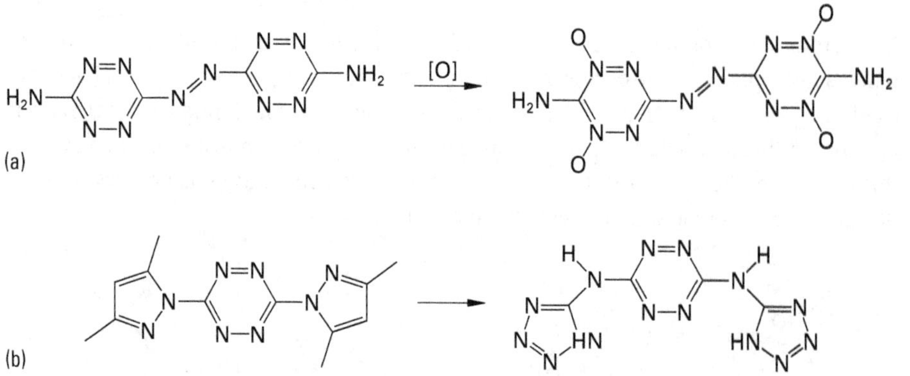

Fig. 1.15: Synthesis of an *N*-oxide (a) and preparation of 3,6-bis(1*H*-1,2,3,4-tetrazole-5-ylamino)-*s*-tetrazine (b).

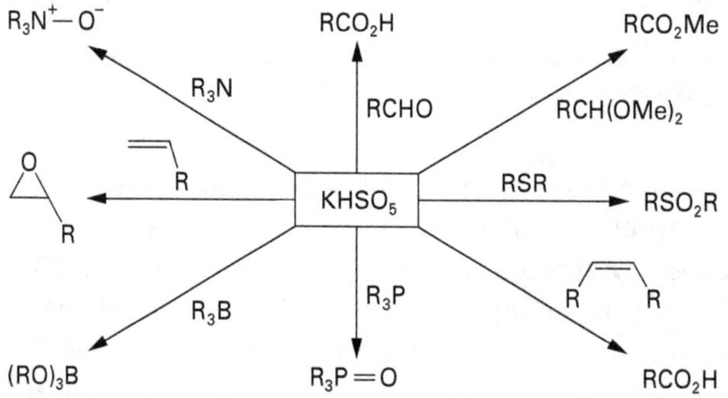

Fig. 1.16: Oxidation reactions with Oxone® as the oxidant (see for example: B. R. Travis, M. Sivakumar, G. O. Hollist, B. Borhan, Org. Lett. **2003**, 5, 1031–1034).

Another oxidizing reagent that has proven useful at introducing *N*-oxides is commercially available Oxone® (2 $KHSO_5 \cdot KHSO_4 \cdot K_2SO_4$). The active ingredient in this oxidizing agent is potassium peroxomonosulfate, $KHSO_5$, which is a salt of Caro's acid, H_2SO_5. Examples of oxidation reactions involving Oxone® are shown in Figure 1.16, including the interconversion of an amine (R_3N) into an *N*-oxide. (*N.B.* Sometimes, mCPBA [meta-chloro perbenzoic acid] or CF_3COOH/H_2O_2 are also used as an oxidizing agent for forming *N*-oxides.). See: *Crystals* **2022**, *12*(10), 1354; https://doi.org/10.3390/cryst12101354.

Another tetrazine derivative, 3,6-bis(1*H*-1,2,3,4-tetrazole-5-ylamino)-*s*-tetrazine, has recently been prepared from *bis*(pyrazolyl)tetrazine (Fig. 1.15). It is interesting to note that the tetrazine derivatives potentially form strong intermolecular interactions via *π*-stacking. This can influence many of the physical properties in a positive way, for example by reducing the electrostatic sensitivity.

1.2.3 New primary explosives

In early days Alfred Nobel already replaced mercury fulminate (MF, see above), which he had introduced into blasting caps, with the safer to handle primary explosives lead azide (LA) and lead styphnate (LS) (Fig. 1.17). However, the long-term use of LA and LS has caused considerable lead contamination in military training grounds which has stimulated world-wide activities in the search for replacements that are heavy-metal free. In 2016, the US military still used ca. 2000–3,000 kg LA per year. In 2006 Huynh und Hiskey published a paper proposing iron and copper complexes of the type $[cat]_2^+[M^{II}(NT)_4(H_2O)_2]$ ($[cat]^+ = NH_4^+, Na^+; M = Fe, Cu$; NT = 5-nitrotetrazolate) as environmentally friendly, "green" primary explosives (Fig. 1.17) [3].

In 2007 the LMU Munich research group reported on the compound copper *bis*(1-methyl-5-nitriminotetrazolate) with similarly promising properties (Fig. 1.17) [4]. Because they have only been discovered recently, none of the above mentioned complexes has found application yet, but they appear to have substantial potential as lead-free primary explosives.

Another environmentally compatible primary explosive is copper(I) 5-nitrotetrazolate (Fig. 1.17). This compound has been developed under the name of **DBX-1** by Pacific Scientific EMC and is a suitable replacement for lead azide. DBX-1 is thermally stable up to 325 °C (DSC). The impact sensitivity of DBX-1 is 0.04 J (ball-drop instrument) compared with 0.05 J for LA. The compound is stable at 180 °C for 24 h in air and for 2 months at 70 °C. DBX-1 can be obtained from NaNT and Cu(I)Cl in HCl/H_2O solution at a higher temperature. However, the best preparation for DBX-1 in a yield of 80–90% is shown in the following equation where sodium ascorbate, $NaC_6H_7O_6$, is used as the reducing agent:

$$CuCl_2 + NaNT \xrightarrow{\text{reducing agent, } H_2O\text{, 15 min, } \Delta T} DBX-1$$

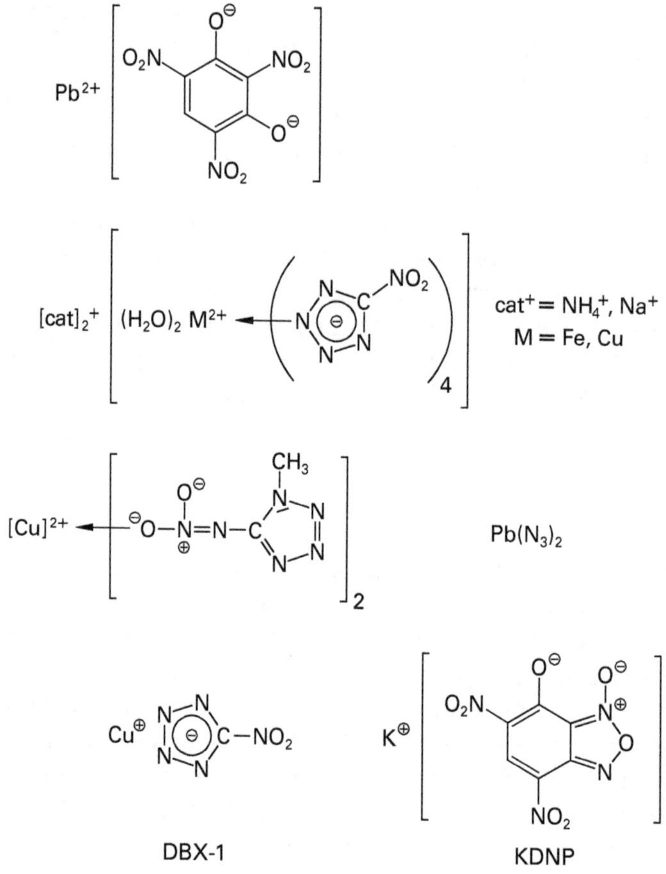

Fig. 1.17: Molecular structures of lead styphnate (LS), lead azide (LA), an iron and copper nitrotetrazolate complexes as well as copper(I) 5-nitrotetrazolate (DBX-1) and potassium 7-hydroxy-6-dinitrobenzofuroxane (KDNP).

The mercury salt of NT, Hg(NT)$_2$, is also known under the name DXN−1 or DXW−1.

A possible replacement for lead styphnate is potassium-7-hydroxy-6-dinitrobenzofuroxane (**KDNP**) (Fig. 1.17). KDNP is a furoxane ring containing explosive and can best be prepared from commercially available bromoanisol according to the following equation. The KN_3 substitutes the Br atom in the final reaction step and also removes the methyl group:

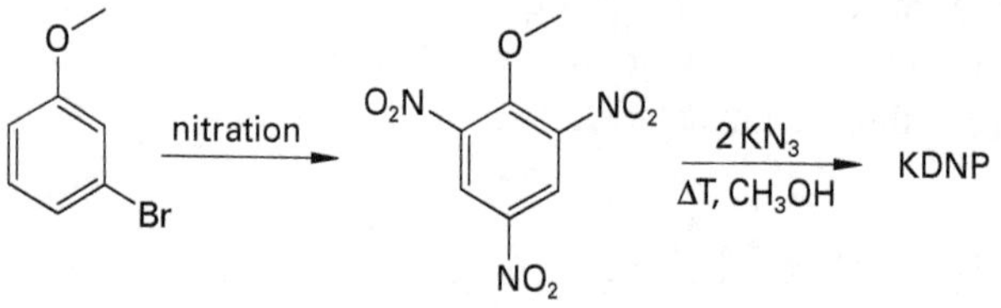

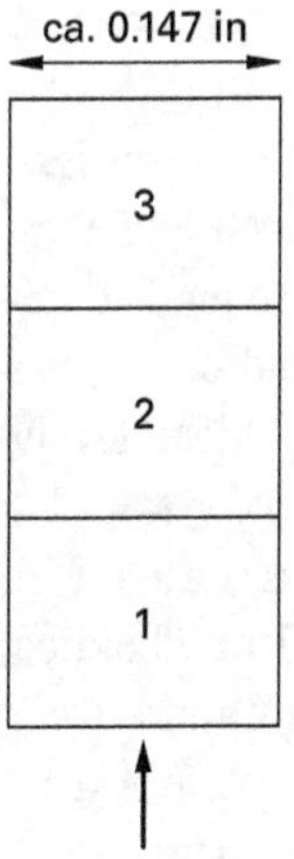

Fig. 1.18: Typical design of a stab detonator; 1: initiating charge, stab mix, e.g. NOL –130 (LA, LS, tetrazene, Sb_2S_3, $Ba(NO_3)_2$); 2: transfer charge (LA); 3: output charge (RDX).

A typical stab detonator (Fig. 1.18) consists of three main components:
1. initiating mixture or initiating charge (initiated by a bridgewire),
2. transfer charge: primary explosive (usually LA),
3. output charge: secondary explosive (usually RDX).

A typical composition for the initiating charge is:
20% LA
40% LS (basic)
5% Tetrazene
20% Barium nitrate
15% Antimony sulfide, Sb_2S_3

It is therefore desirable to find suitable heavy metal-free replacements for both lead azide and lead styphnate. Current research is addressing this problem. The following replacements in stab detonators are presently being researched:

1. initiating charge	LA	→	DBX-1
	LS	→	KDNP
2. transfer charge:	LA	→	triazine triazide (TTA) or APX
3. output charge:	RDX	→	PETN or BTAT

Primary explosives are substances which show a very rapid transition from deflagration to detonation and generate a shock-wave which makes transfer of the detonation to a (less sensitive) secondary explosive possible. Lead azide and lead styphnate are the most commonly used primary explosives today. However, the long-term use of these compounds (which contain the toxic heavy metal lead) has caused considerable lead contamination in military training grounds. Costly clean-up operations require a lot of money that could better be spent improving the defense capability of the country's forces. A recent article published on December 4th 2012 in the Washington Post (https://www.washingtonpost.com/blogs/federal-eye/wp/2012/12/03/new-report-warns-of-high-lead-risk-for-military-firing-range-workers/) entitled "Defense Dept. Standards On Lead Exposure Faulted" stated: ". . . it has found overwhelming evidence that 30 year-old federal standards governing lead exposure at Department of Defense firing ranges and other sites are inadequate to protect workers from ailments associated with high blood lead levels, including problems with the nervous system, kidney, heart and reproductive system."

Devices using lead primary explosives – from primers for bullets to detonators for mining – are manufactured in the tens of millions every year in the United States. In the US alone, over 750 lbs. of lead azide are consumed every year for military use.

Researchers from LMU Munich have now synthesized in collaboration with ARDEC at Picatinny Arsenal, N.J. a compound named K_2DNABT (Fig. 1.18a), a new heavy metal-free primary explosive which has essentially the same sensitivity (impact, friction and electrostatic sensitivity) as that of lead azide, but does not contain toxic lead. Instead of lead, it contains the ecologically and toxicologically benign element potassium instead. Preliminary experimental detonation tests (dent-plate tests) and high-level computations have shown that the performance of K_2DNABT even exceeds that of lead azide. Therefore, there is great hope that toxic lead azide and / or lead styphnate can be replaced in munitions and detonators with this physiologically and ecologically benign compound.

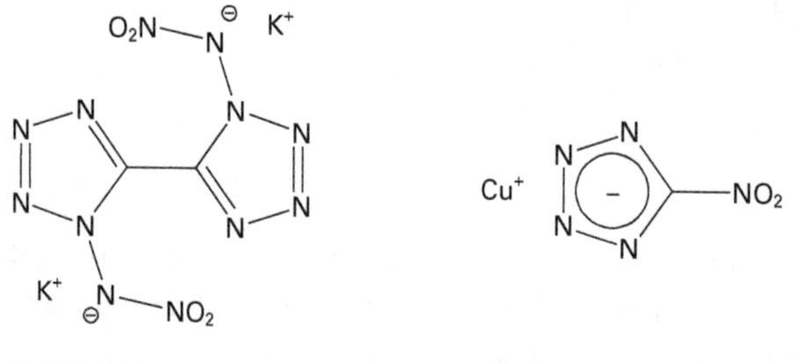

Fig. 1.18a: Chemical structures of K_2DNABT and DBX-1.

In theory, unprotected 1,1′-diamino-5,5′-bistetrazole can be nitrated. However, the amination of 5,5′-bistetrazole is a procedure which results in only low yields and also requires considerable effort, therefore an alternative route was developed. The bisnitrilimine would appear to be a suitable precursor, however, unfortunately unprotected bisnitrilimine is not known and only the corresponding diphenyl derivative is known. Therefore another derivative was prepared which contains a more easily removable protecting group than the phenyl group. The synthetic process for the synthesis of K_2DNABT starts from the easily preparable dimethylcarbonate. This is reacted with hydrazine hydrate to form the carbazate **1**. The subsequent condensation reaction with half an equivalent of glyoxal forms compound **2**, which is subsequently oxidized with NCS (*N*-chlorosuccinimide) to the corresponding chloride. Substitution with sodium azide offers the diazide (in only 38% yields) which then is cyclized with hydrochloric acid in ethereal suspension. The carboxymethyl protected 1,1′-diamino-5,5′-bistetrazole is then gently nitrated with N_2O_5 (Fig. 1.18b).

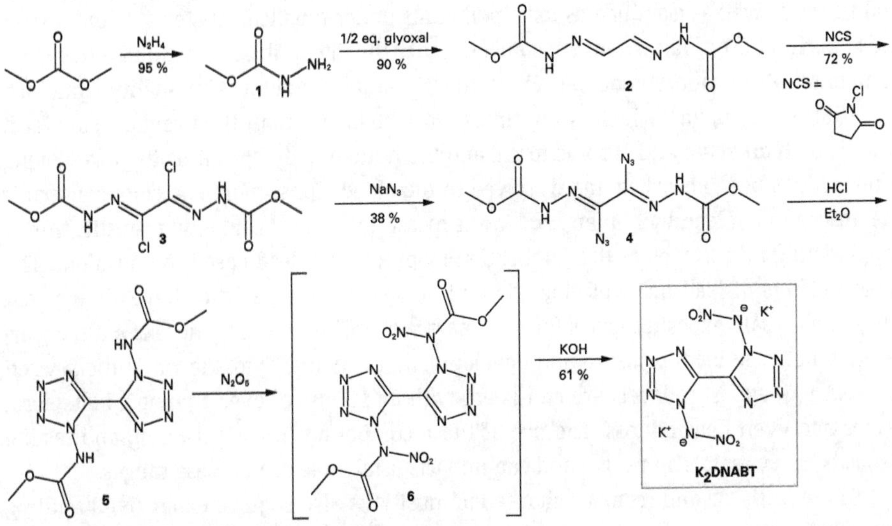

Fig. 1.18b: Synthetic pathway for the formation of K_2DNABT.

An alkaline aquatic work-up with KOH precipitates dipotassium 1,1′-dinitramino-5,5′-bistetrazolate. The products of the individual stages can be purified by recrystallization, or used as obtained. No column chromatography must be used. Fortunately K_2DNABT shows low water solubility, which (i) facilitates its isolation and purification and (ii) avoids future toxicity problems due to potential ground water pollution.

A primary explosive is an explosive that is extremely sensitive to stimuli such as impact, friction, heat or electrostatic discharge. Only very small amounts of energy are required to initiate such a material. Generally, primary explosives are considered to be materials which are more sensitive than PETN. Primary explosives are described as being the initiating materials that initiate less sensitive energetic materials such as secondary explosives (e.g. RDX/HMX) or propellants. A small quantity – usually only milligrams – is required to initiate a larger charge of explosive which is safer to handle. Primary explosives are widely used in primers, detonators and blasting caps. The most commonly used primary explosives are lead azide and lead styphnate. Lead azide is the more powerful of the two and is used as a pure substance typically in detonators as a transfer charge, or in formulations for initiation mixes (e.g. NOL-130). Lead styphnate is mostly used in formulations for initiation and primer mixes and is rarely found as a neat material in applications.

Although lead azide has been used extensively for decades, it is a very poisonous material that reacts with copper, zinc or alloys containing these metals, forming other azides that can be highly sensitive and dangerous to handle. Furthermore, lead-based materials have been clearly found to cause environmental and health related problems. Lead-based materials are included on the EPA Toxic Chemical List (EPA List of 17 Toxic Chemicals). They are additionally regulated under the Clean Air Act as Title II Hazardous Air Pollutants, as well as classified as toxic pollutants under the Clean Water Act, and are on the Superfund list of hazardous substances. Under the Clean Air Act, the USEPA (US Environmental Protection Agency) revised the National Ambient Air Quality Standard (NAAQS) to 0.15 $\mu g/m^3$, which is ten times more stringent than the previous standard. Lead is both an acute and chronic toxin, and the human body has difficulty in removing it once it has been absorbed and dissolved in the blood. Consequently, a chief concern is the absorption of lead by humans who are exposed to the lead-containing constituents of the initiating mix, as well as the combustion by-products of lead-based compositions. The health effects of lead are well documented however, recent studies have shown that there are no safe exposure levels for lead, particularly for children. There is a direct correlation between lead exposure and development, including IQ loss (even at the revised lead NAAQS, exposure levels are consistent with an IQ loss of over 2 points), behavioral issues and even hearing loss. The use of these compounds during training and testing deposits heavy metals on ranges and can impact sustainable use of these ranges.

These initiator and transfer charge compositions also require expensive handling procedures during production and disposal. The manufacturing of any lead-based primary explosives, such as lead azide or lead styphnate, results in the production of significant quantities of highly toxic, hazardous waste. In addition, the handling and storage of these compounds is also a concern.

Due to its environmental and health impact, there is a need to develop green primary explosives, to replace lead-based compounds. Current research seeks to replace lead azide and lead-based formulations in commonly used detonators, and in the M55 stab detonators and percussion primers, such as M115, M39, M42, etc. in particular. One important strategy for developing lead-free alternatives focuses on the use of high-nitrogen compounds. High-nitrogen compounds are widely considered to be viable, environmental-friendly energetic, since the predominant detonation product is non-toxic nitrogen (N_2) gas. At the same time, high-nitrogen compounds possess high enthalpies of formation (ΔH_f), which lead to high energy outputs. In developing/testing viable lead replacements, the following criteria must be considered for new compounds. The material:

- must be safe to handle and possess a rapid deflagration to detonation transition
- be thermally stable to temperatures above 150 °C and must have a melting point greater than 90 °C
- should possess high detonation performance and sensitivity
- should have long term chemical stability
- should not contain toxic heavy metals or other known toxins
- be easy to synthesize and affordable.

For primer performance testing, primers are sealed in an air-tight test apparatus and initiated by dropping an 8 or 16 oz steel ball onto the primary explosive. Pressure transducers measure the output (Fig. 1.18c).

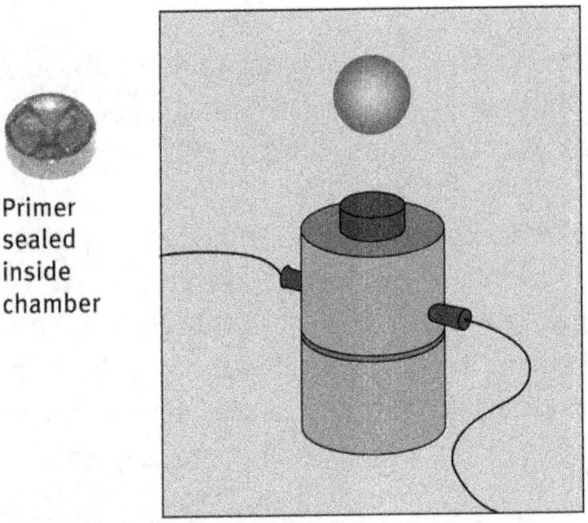

Fig. 1.18c: Primer performance test set-up.

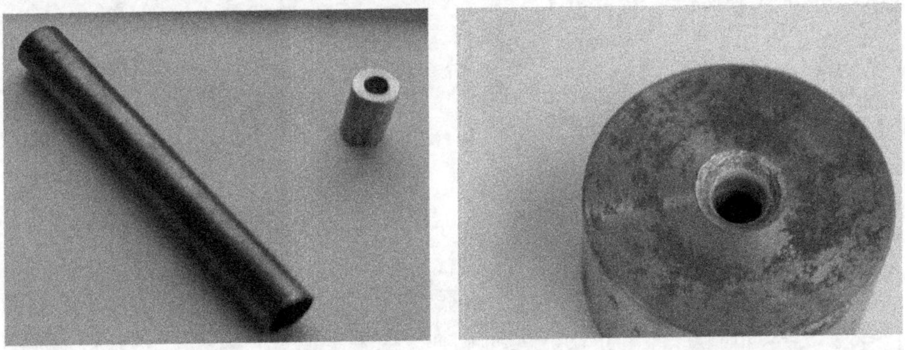

Fig. 1.18d: Performance of K_2DNABT/RDX after initiation by an electrical igniter.

The sensitivity of K_2DNABT was tested and was found to be very sensitive to impact, friction and ESD, as all primary explosives are. K_2DNABT is more sensitive compared to lead azide, lead styphnate and DBX-1 (as shown in Tab. 1.3a).

Tab. 1.3a: Comparison of the sensitivities of K_2DNABT with other primary explosives.

Sample	Impact / in	Friction / N	ESD / mJ	Density / g cm^{-3}	VoD / m s^{-1}
LA	7 – 11	0.1	4.7	4.8	5,300
LS	5	0.1	0.2	3.00	4,900
DBX-1	4	0.1	3.1	2.58	ca. 7,000
K_2DNABT	2	0.1	0.1	2.2	ca. 8,330

To compare the performance of K_2DNABT with that of lead azide, 40 mg of K_2DNBT was loaded in a small aluminum holder and 1 g of RDX was pressed in a standard detonator copper shell. This was initiated by an electrical igniter. The performance of K_2DNABT/RDX was shown to pass the requirements expected of a primary explosive mixture as can be seen in Fig. 1.18d.

A modified Small-Scale Shock Reactivity Test (SSRT) was performed with K_2DNABT as well as with lead azide for comparison. In these tests, 500 mg of each compound was ignited by an igniter and it was found that K_2DNABT showed more indentation than lead azide on the witness plate (Fig 1.18e).

When performed electrically, K_2DNABT is comparable to lead styphnate in performance. Primer mixes have been also been formulated and require testing.

Future work requires optimization of the system in order for K_2DNABT to be considered as a replacement for lead azide in detonators. From the initial results, K_2DNABT appears to be a good possible replacement for lead styphnate when initiated via hot-wire or electric bridgewire. Formulations will also need to be optimized in the future, with respect to finding the correct particle size.

Fig. 1.18e: Witness plate of SSRT of lead azide (left) and K_2DNABT (right).

An image of the Fischer test is shown in Fig. 1.18f.

Another promising and thermally stable (Tab. 1.3b) lead-free primary explosive is copper(II) 5-chlorotetrazolate (CuClT, PSI & LMU). The synthesis is achieved in a one-step reaction (Fig. 1.18a) starting from commercially available aminotetrazole. CuClT can then be further converted into the synthetically useful compounds sodium chlorotatrazolate and chlorotetrazole (Fig. 1.18h).

In the area of metal-free primary explosives, covalently bound azides are often advantageous. Although these compound soften do not show the high thermal stability of metal complexes, some may have an application as LA replacements in transfer charges (see Fig. 1.18). Two of the most promising candidates are triazido triazine (triazine tria-

Fig. 1.18f: Fischer test of K_2DNABT.

zide, TTA, ARDEC, see above) and diazidoglyoxime (DAGL, LMU, Tab. 1.3c). The latter one can be prepared according to Fig. 1.18h.

Tab. 1.3b: Energetic properties of copper(II) chlorotetrazolate (CuClT).

	CuClT
Formula	$C_2Cl_2CuN_8$
Molecular mass [g mol^{-1}]	270.53
Impact sensitivity [J]	1
Friction sensitivity [N]	< 5
ESD [J]	0.025
N [%]	41.42
$T_{dec.}$ [°C]	289

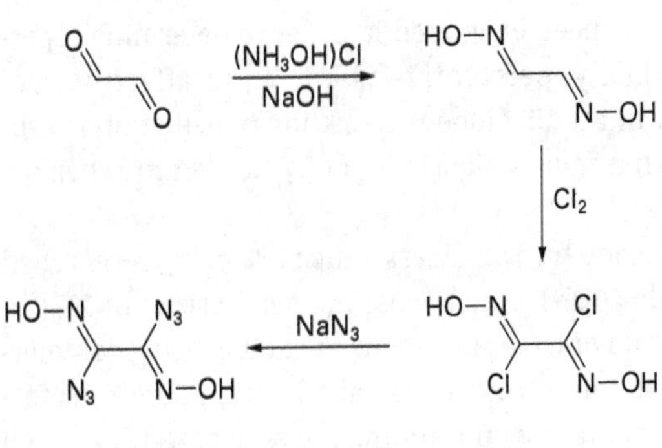

Fig. 1.18g: Synthesis of diazidoglyoxime (DAGL).

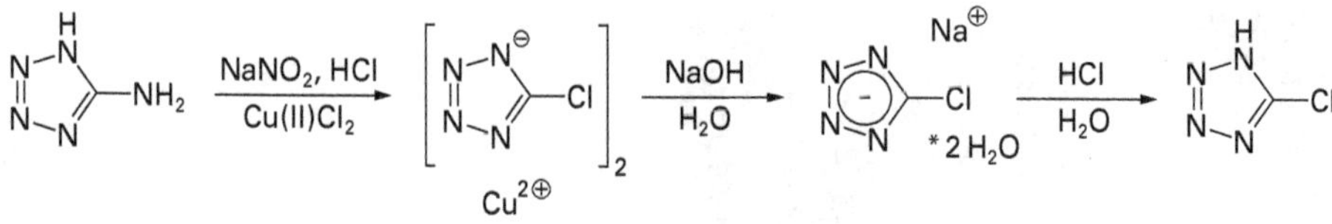

Fig. 1.18h: Synthesis of copper(II) chlorotetrazolate (CuClT).

Tab. 1.3c: Energetic properties of TTA and DAGL in comparison with LA.

	TTA	DAGL	LA
Formula	C_3N_{12}	$C_2H_2N_8O_2$	N_6Pb
Molecular mass [g mol^{-1}]	204.1	170.1	291.3
Impact sensitivity [J]	1.3	1.5	2.5–4
Friction sensitivity [N]	< 0.5	< 5	0.1–1
ESD [J]	< 0.36 (?)	0.007	0.005
$T_{dec.}$ [°C]	187 (m.p. 94)	170	315

1.2.4 New oxidizers for solid rocket motors

Solid propellants of essentially all solid rocket boosters are based on a mixture of aluminum (Al, fuel) and ammonium perchlorate (AP, oxidizer).

Ammonium perchlorate (AP) has applications in munitions, primarily as an oxidizer for solid rocket and missile propellants. It is also used as an air bag inflator in the automotive industry, in fireworks, and appears as a contaminant of agricultural fertilizers. As a result of these uses and ammonium perchlorate's high solubility, chemical stability, and persistence, it has become widely distributed in surface and ground water systems. There is little information about the effects of perchlorate on the aquatic life, but it is known that perchlorate is an endocrine disrupting chemical that interferes with normal thyroid function which impacts both growth and development in vertebrates. Because perchlorate competes for iodine binding sites in thyroids, adding iodine to culture water has been examined in order to determine if perchlorate effects can be mitigated. Finally, perchlorate is known to affect normal pigmentation of amphibian embryos. In the US alone the cost for remediation is estimated to be several billion dollars. That money that is urgently needed in other defense areas [5–7].

The currently most promising chlorine-free oxidizers which are being researched at present are ammonium dinitramide (ADN), which was first synthesized in 1971 in Russia (Oleg Lukyanov, Zelinsky Institute of Organic Chemistry) and is being commercialized today by EURENCO, as well as the nitroformate salts hydrazinium nitroformate (HNF, APP, Netherlands) and triaminoguanidinium nitroformate (Germany) (Fig. 1.19) [8]. The salt hydroxylammonium nitrate, $HO—NH_3^+ NO_3^-$ (HAN) is also of interest. However, all four compounds possess relatively-low decomposition temperatures and TAGNF only has a positive oxygen balance with respect to CO (not to CO_2).

While ADN has the best oxygen balance (Ω_{CO2} = 25.8%, *cf.* AP 34.0%) of all presently discussed AP replacements it still has some stability issues with respect to binder compatibility and thermal stability ($T_{dec.}$ = 127 °C). Thermal decomposition of ADN is

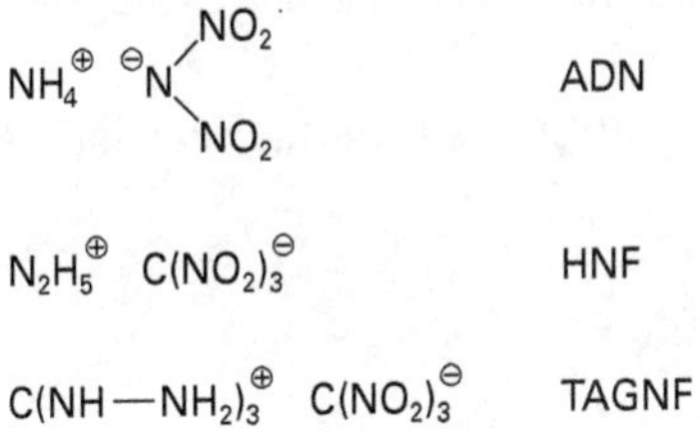

Fig. 1.19: Molecular structures of ammonium dinitramide (ADN), hydrazinium nitroformate (HNF) and triaminoguanidinium nitroformate (TAGNF).

observed at 127 °C after complete melting at 91.5 °C. The main decomposition pathway is based on the formation of NH_4NO_3 and N_2O followed by the thermal decomposition of NH_4NO_3 to N_2O and H_2O at higher temperatures. Side-reactions forming NO_2, NO, NH_3, N_2 and O_2 are described and a mechanism for the acid-catalyzed decomposition of hydrogen dinitramide dissociation product of ADN, has been proposed.

As far as it is publically known, ADN is already in use in certain Russian systems, *e.g.* Topol-M (second and third stages), SS-24 (first and third stages) and SS-20N.

Alternatively, ammonium nitrate (AN, Ω_{CO_2} = 20.0%, begins decomposition at m.p. = 169.9 °C, complete decomposition at 210 °C) has been discussed, however this compound has severe burn rate issues. Furthermore, AN is hygroscopic and shows phase transitions from one polymorph to another at 125.2 °C, 84.2 °C, 32.3 °C and –16.9 °C. Phase stabilized ammonium nitrate (PSAN) and spray crystallized AN (SCAN) are special qualities provided by ICT.

Another recently suggested organic oxidizer is TNC-NO_2 which has an oxygen balance of Ω_{CO_2} = 14.9% and a thermal stability of up to 153 °C. TNC-NO_2 can be synthesized by direct nitration of TNC (2,2,2-trinitroethylcarbamate) using mixed acid:

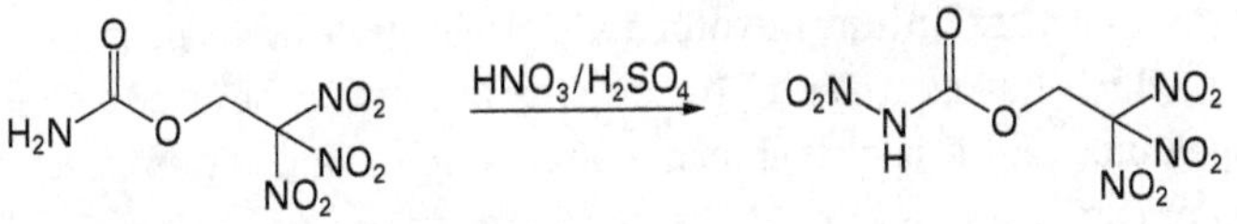

The further new nitroethyl compounds based on boron esters are tris-(2-nitroethyl) borate and tris-(2,2,2-trinitroethyl)borate. Especially the trinitroethyl derivative is a suitable candidate for high energy density oxidizers and for smoke-free, green coloring agents in pyrotechnic compositions. Tris-(2-nitroethyl)borate and tris-(2,2,2-trinitroethyl) borate can be obtained from boron oxide with 2-nitroethanol and 2,2,2-trinitroethanol, respectively:

$$B_2O_3 \xrightarrow[-3\,H_2O]{25\text{–}60\,°C/12\,h} \begin{cases} \xrightarrow{HOCH_2CH_2NO_2} B[OCH_2CH_2NO_2]_3 \quad \mathbf{2} \\ \xrightarrow{HOCH_2C(NO_2)_3} B[OCH_2C(NO_2)_3]_3 \quad \mathbf{3} \end{cases}$$

The oxygen balance of **2** is –59.70% and of **3** is + 13.07%. The density for **3** was determined to be 1.982 g cm^{-3}, which is a quite high value. DSC measurements revealed an exothermic decomposition at 216 °C for **2** and 161 °C for **3**.

The starting material 2,2,2-trinitroethanol (**1**) was prepared from the reaction of trinitromethane with formaldehyde (see Fig. 9.18).
Sensitivity data for **3**.

Grain size	< 100 µm
Impact sensitivity	15 J
Friction sensitivity	144 N
Electrostatic sensitivity	0.5 J

In general, elemental boron and boron compounds have been the subject of many studies involving propulsion systems due to the very high heats of combustion. However, due to some combustion problems, the theoretical benefits of boron and its compounds have rarely been delivered. The main limitations for boron combustion are (i) ignition delay due to the presence of an oxide layer on the surface of the boron particle (only for elemental boron) and (ii) the energy release during the combustion process in hydrogen containing gases is significantly lowered due to the formation of HBO_2 or "HOBO" as it is commonly referred to. The slow oxidation of "HOBO" to the preferred product of combustion B_2O_3 often results in the particles leaving the propulsion-system prior to releasing all of their energy thus resulting in less than expected performance.

Nitroform (trinitromethane, NF) is a valuable starting material for the production of propellant and explosive components due to its high oxygen content. It was first obtained as the ammonium salt by Shiskov in 1857. The acidic hydrogen atom facilitates the formation of derivatives, since deprotonation can readily occur with bases to form, for example, hydrazinium nitroformate (HNF) [69]. HNF possesses many useful properties which makes it useful for application: it can be used as an oxidizer in rocket propellants, and it is also chlorine-free, which is a desirable property in modern propellant compositions [70, 71].

NF is also a highly useful starting material in the synthesis of many other chlorine-free oxidizers, such as TNC-NO_2 and tris(trinitroethyl) borate. Whereas HNF is directly synthesized from NF, the oxidizers TNC-NO_2 and tris(trinitroethyl) borate are prepared from trinitroethanol (TNE). TNE can easily be synthesized in two steps from an aqueous solution of nitroform with formaldehyde:

$$H{-}C(NO_2)_3 + CH_2O \rightarrow (O_2N)_3C{-}CH_2{-}OH + 8.56\ kcal\ mol^{-1} \quad (1)$$

NF participates in many reactions such as addition reactions [72], condensation reactions [73], and substitution reactions [74]. NF reacts with unsaturated compounds to form multi-nitroalkane derivatives, and therefore, it can be used for the synthesis of an explosive composition containing the nitroform group such as 1-(2,2,2-trinitroethylamino)-2-nitroguanidines [75] or 1,1,1,3-tetranitro-3-azabutane [76].

Since it has proven itself to be a crucially important raw material in the synthesis of many energetic compounds, synthetic routes for the formation of NF have attracted

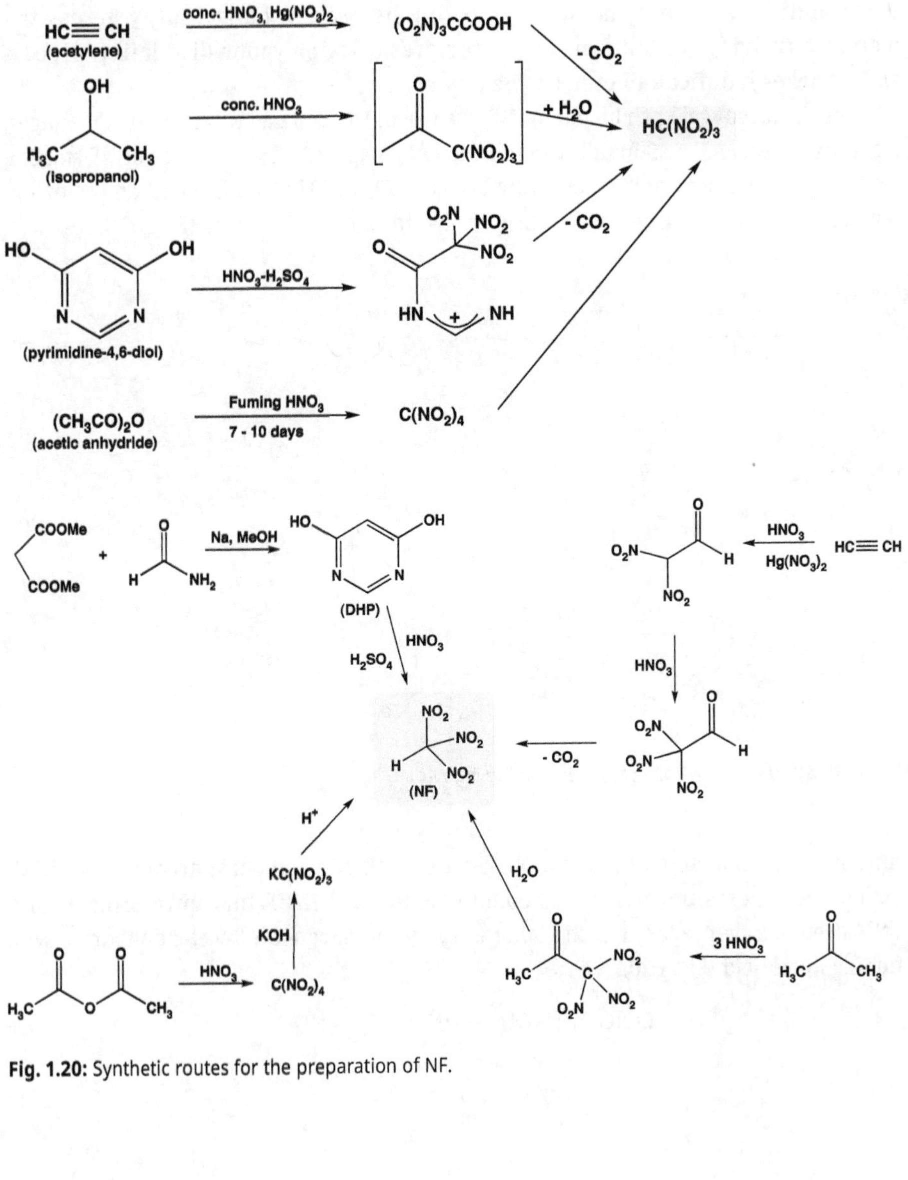

Fig. 1.20: Synthetic routes for the preparation of NF.

considerable attention and have been explored, for example, by the nitration of various substrates including acetylene, acetic anhydride, pyrimidine-4,6-diol, and isopropanol (Fig. 1.20).

The most investigated preparative method of nitroform in the literature is the nitration of acetylene with nitric acid. However, this method requires the use of expensive and toxic mercury nitrate as a catalyst, which is clearly detrimental from the economic and environmental point of view. Other preparative methods for nitroform which have been reported in the literature include the nitration of other suitable compounds, such as acetone or isopropanol, the hydrolysis of tetranitromethane, or the nitration-hydrolysis of pyrimidine-4,6-dione [77, 78]. Due to the volatility and flammability of acetone, methods involving its use are dangerous, whereas the method involving tetranitromethane requires a water vapor distillation process which makes it difficult to prepare the raw materials.

For the large-scale production of NF, the use of isopropanol has certain advantages, e.g. safety, cost, and sustainability (environmental aspects). The synthesis of NF starting from isopropanol was first reported by Frankel [79] in 1978, and is based on the oxidation, nitration, and hydrolysis of isopropanol with fuming nitric acid (Fig. 1.21).

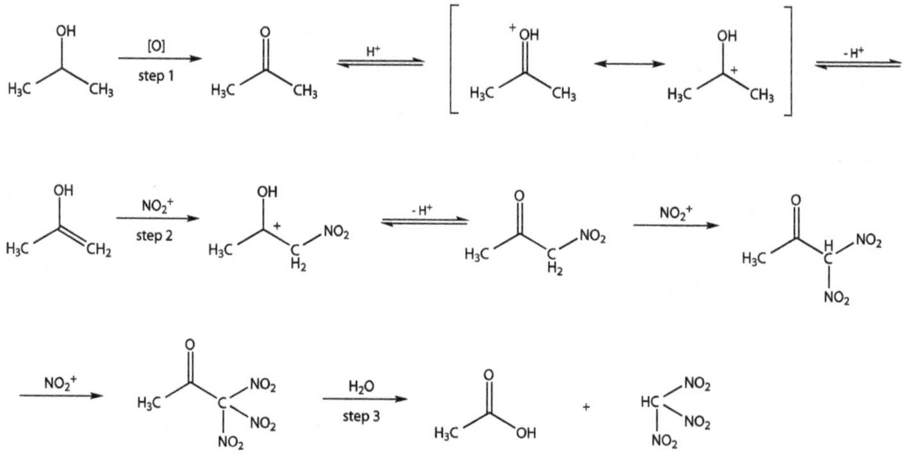

Fig. 1.21: Synthesis of NF from isopropanol in fuming nitric acid.

Alternatively, on a laboratory scale, NF can be obtained from tetranitromethane (TNM) which decomposes under alkaline conditions to yield HNO_3 and nitroformate salts. TNM itself can be prepared in the laboratory by the nitration of acetic anhydride with fuming nitric acid (Fig. 1.20).

$$C(NO_2)_4 + KOH \rightarrow HNO_3 + K^+C(NO_2)_3^- \qquad (2)$$

Another convenient synthesis which can be used to obtain potassium nitroformate (which can be subsequently converted into TNE) on a laboratory scale is the nitration of barbituric acid as shown in Fig. 1.22.

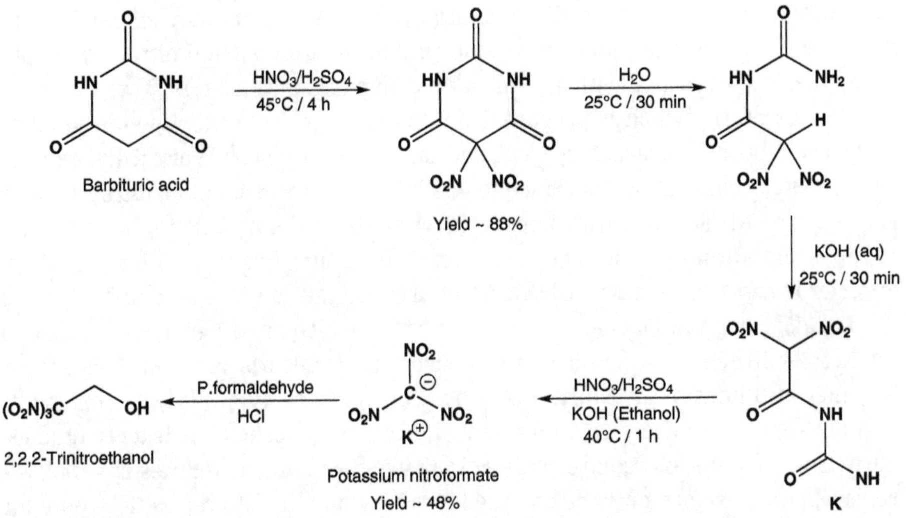

Fig. 1.22: Synthesis of TNE via potassium nitroformate.

All of the problems mentioned above with the known synthetic routes for the preparation of NF demanded new, safer and more efficient methods to be developed for the synthesis of NF. In the last few years, several new methods have been reported. In 2003, A. Langlet et al. published a patent [80] concerned with the preparation of nitroform (NF) by nitrating a starting material which is dissolved or suspended in sulfuric acid. A nitrating agent consisting of nitric acid, nitrate salts or nitrogen pentoxide is added to the sulfuric acid solution/suspension at temperatures in the range of –10 °C to + 80 °C. The starting material can be selected from one of the following classes of compounds:

a) a gem-dinitroacetyl urea;

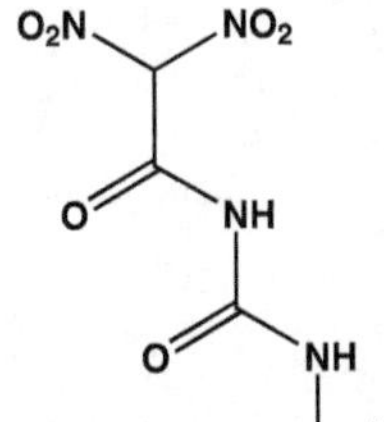

X, in which X is H or a group 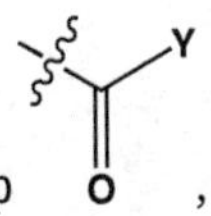 , wherein Y is an alkoxy or amino group, and salts thereof;

b) gem-dinitroacetyl guanidine

c) 4,6-hydroxypyrimidine

The sulfuric acid can have a concentration of 70–100%, preferably 95%, and the nitric acid is added as a concentrated acid (85–100%). The molar ratio between the nitrating acid and the substrate may be in the range 2.0–6.0 : 1. Good yields of NF have been obtained with a molar ratio as low as 3 : 1, which (in combination with the reuse of sulfuric acid) makes the method economically attractive. After completion of the nitration, the product (the gem-trinitro compound) is hydrolyzed and nitroform is split off by mixing the reaction with an aqueous medium (e.g. crushed ice).

The nitroform which is prepared can be extracted from the reaction mixture using a suitable polar extracting agent, such as methylene chloride or diethyl ether. A neutralizing agent can then be added to precipitate the corresponding nitroform salt. Extraction of the gem-trinitro compound directly from the nitration mixture is also possible, and nitroform can then be subsequently obtained using a hydrolysis procedure, for instance, by adding a base. The sulfuric acid may then be used several times.

In 2015, Hong-Yan Lu et al. published an article which presented a new method to synthesize nitroform by the nitrolysis of cucurbituril. Cucurbituril (CB[6]) is a cyclic oligomer of 6 units of glycoluril linked by 12 methylene bridges [81]. In this study, CB[n] (n = 5–8) was nitrolyzed with fuming nitric acid in acetic anhydride to produce nitroform. This process requires only mild reaction conditions, reduces risk, is inexpensive, and therefore must be regarded as a new and useful preparative route for the formation of nitroform.

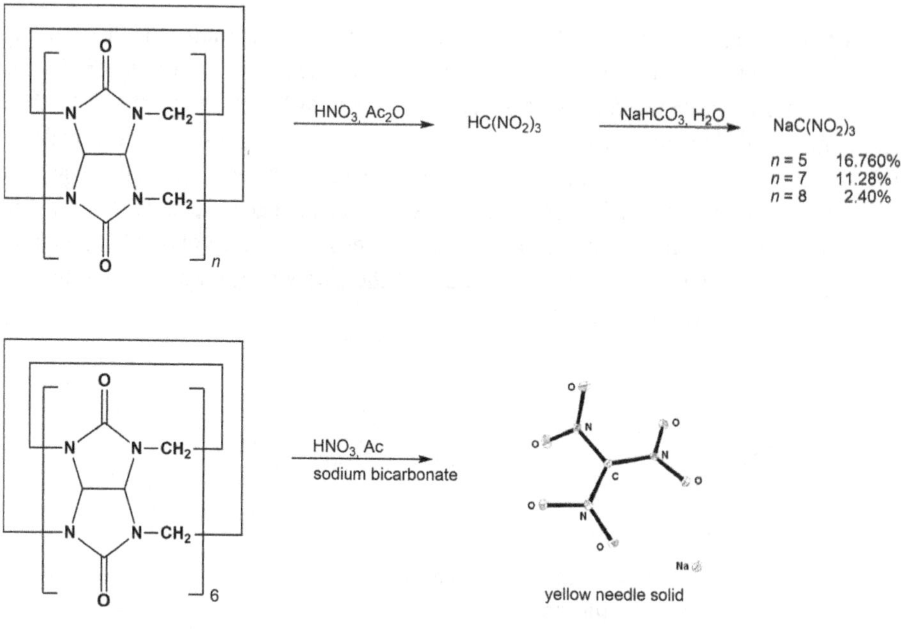

Fig. 1.23: Nitrolysis of CB.

In short, there is a reason why AP has been used as the most important oxidizer for composite propellants for decades. It has the advantage of being completely convertible to gaseous reaction products and has an oxygen balance of $\Omega_{CO2} = 34\%$. It can easily be prepared by neutralizing ammonia with perchloric acid and is purified by crystallization. AP is stable at room temperature but decomposes at measurable rates at temperatures greater than about 150 °C. At decomposition temperatures below approximately 300 °C, AP undergoes an autocatalytic reaction which ceases after about 30% decomposition. This is usually called the low-temperature reaction. The residue is quite stable at these temperatures unless rejuvenated by sublimation, recrystallization, or mechanical disturbance. At temperatures above 350 °C, the high-temperature decomposition occurs; this reaction is not autocatalytic and decomposition is complete. Concurrently with these decomposition reactions, AP also undergoes dissociative sublimation.

A new polynitro derivative of an azo-bridged triazole (**3**) was synthesized by J. Shreeve (*J. Mater. Chem. A*, **2021**, *9*, 24903-24908) in two steps from *N*-alkyl-functionalized 3-amino-5-nitro-1,2,4-triazole (**1**) in a high yield of 74.5%. Compound **3** is stable for long periods of time at room temperature, and has an excellent density and thermal stability ($T_{dec.} = 175$ °C) in comparison with most polynitro compounds. This is attributed to the azo group introduced into the conjugated system. An analysis of the noncovalent interactions shows that the dimeric stabilization of **3** is dominated by π-π stacking interactions, which are helpful in enhancing the density and stability. Additionally, **3** exhibits excellent explosive performance and has a specific impulse of 241 s, making it an interesting candidate for use as a high-energy-density oxidizer material ($\Omega_{CO} = 29\%$) to replace AP in solid rocket propellants and missiles.

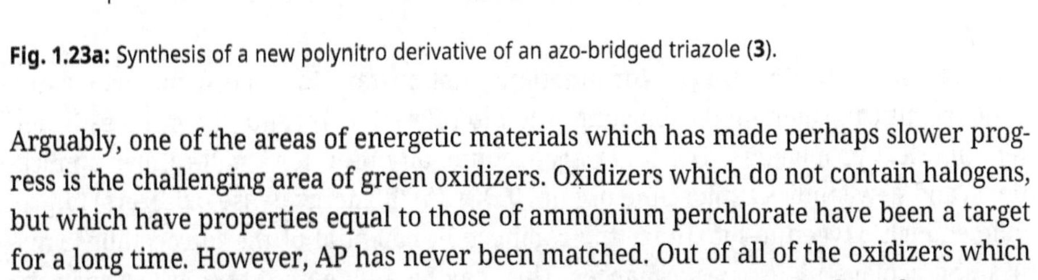

Fig. 1.23a: Synthesis of a new polynitro derivative of an azo-bridged triazole (**3**).

Arguably, one of the areas of energetic materials which has made perhaps slower progress is the challenging area of green oxidizers. Oxidizers which do not contain halogens, but which have properties equal to those of ammonium perchlorate have been a target for a long time. However, AP has never been matched. Out of all of the oxidizers which have been investigated by the energetic materials group at LMU, the best three examples are probably bis(2,2,2-trinitroethyl)oxalate [BTNEO] (**1**), 2,2,2-trinitroethyl-nitrocarbamate [TNENC] (**2**) and 2,2,2-trinitroethyl formate [TNEF] (**3**) (Fig. 1.23b) [1–3].

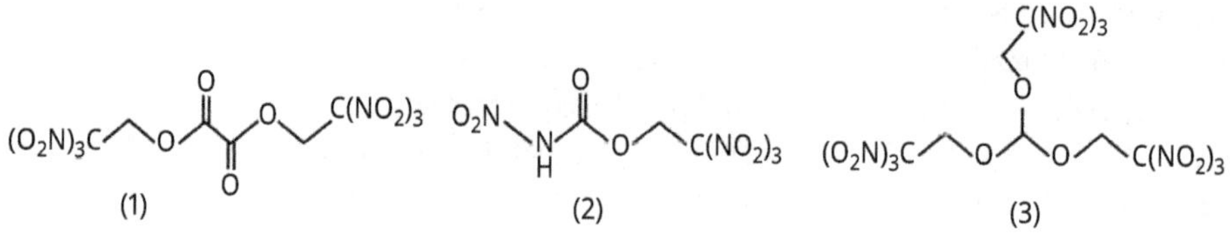

Fig. 1.23b: New CHNO oxidizers.

Common to all three of these compounds is the absence of any halogen, presence of oxygen-rich -$C(NO_2)_3$ groups, as well as incorporation of O atoms into the carbon skeleton. All three compounds possess a positive oxygen balance with respect to CO and CO_2 and have acceptable sensitivity parameters (Tab. 1.3d). Unfortunately, the decomposition temperature of the nitrocarbamate **(2)** is only 153 °C which is too low for application. Therefore, the oxalate **(1)** and the formate **(3)** are the more promising candidates for replacing ammonium perchlorate and potassium perchlorate in thermobaric formulations. TNEF (3) has a highly positive oxygen balance of 30.4% (with respect to CO), a high density of 1.81 g cm^{-3} and a decomposition temperature of 192 °C. Therefore, TNEF was chosen to be investigated further, and has already been scaled up in the lab to 100 g scale.[4]

Tab. 1.3d: Relevant data for the three oxidizers BTNEO (1), TNENC (2) and TNEF (3) [5].

	BTNEO (1)	**TNENC (2)**	**TNEF (3)**
Formula	$C_6H_4N_6O_{16}$	$C_3H_3N_5O_{10}$	$C_7H_7N_9O_{21}$
M / g mol^{-1}	416	269	553
IS / J	10	10	5
FS / N	> 360	96	9
ESD / J	0.7	0.1	0.2
Ω(CO) / %	15.4	32.7	30.4
Ω(CO_2) / %	7.7	14.9	10.1
m.p. / °C	115	109	128
T_{dec} / °C	186	153	192
ρ / g cm^{-3}	1.84	1.73	1.81
$\Delta H°_f$ / kJ mol^{-1}	−688	−366	−519

The feasibility of high-energy formulations that sustain high performances while using significant amounts of burning rate modifiers has recently been investigated for spherical propellants. The latter are traditionally used for small-caliber applications, and new studies explore the use of HEDOs (high-energy dense oxidizers) to provide enough oxygen to maximize the combustion potential of the nitrocellulose matrix and achieve higher performances. This may be a potential route to broaden the applicability of spherical propellants to medium-caliber applications. Tris(2,2,2-trinitroethoxy)methane, also known as TNEF or trinitroethyl formate[4], and SMX (1,4-dinitrato-2,3-dinitro-2,3-bis(nitratomethylene) butane[6]) were synthesized and characterized on a hundred-gram scale. Spherical and extruded propellants were prepared containing increasing amounts of TNEF or SMX and using traditional and green stabilizers [7,8]. The energetic materials were investigated in terms of stability, compatibility and burning rate.

[1] *Recent advances in new oxidizers for solid rocket propulsion*, D. Trache, T. M. Klapötke, L. Maiz, M. Abd-Elghany, L. T. DeLuca, *Green Chem.*, **2017**, *19*, 4711–4736.
[2] *Polynitro Carbamates and N-Nitrocarbamates*, M. B. Frankel, *J. Chem. Eng. Data*, **1962**, *7*, 410.
[3] a) *Process for acetal preparation*, M. E. Hill, K. G. Shipp, US Patent US3526667, **1970**; b) *Synthesis of 2 R-2,2-dinitroethanol orthoesters in ionic liquids*, A. B. Sheremetev, I. L. Yudin, *Mendeleev Commun.*, **2005**, *15*, 204–205; c) *Acetal preparation in sulfuric acid*, K. G. Shipp, M. E. Hill, *J. Org. Chem.*, **1966**, *31*, 853–856.
[4] *An Optimized & Scaled-up Synthetic Procedure for Trinitroethyl Formate TNEF*, D. E. Dosch, K. Andrade, T. M. Klapötke, B. Krumm, *Prop. Explos. Pyrotech.*, **2021**, *46*, 895–898.
[5] *Energetic Materials Encyclopedia*, Volumes 1–3, 2nd edn., T. M. Klapötke, De Gruyter, Berlin/Boston, **2021**.
[6] D. E. Chavez, M. A. Hiskey, D. L. Naud, D. Parrish. Synthesis of an energetic nitrate ester. Angew Chem Int Ed Engl **2008**, 47, 8307–8309. 10.1002/anie.200803648
[7] A. Dejeaifve, R. Dobson. Tocopherol stabilisers for nitrocellulose-based propellants. PCT/EP2016/053948 **2016**, European Patent Office, The Hague.
[8] A. Dejeaifve, A. Sarbach, B. Roduit, P. Folly, R. Dobson. Making Progress Towards »Green« Propellants Part II. Propellants, Explosives, Pyrotechnics **2020**, *45*, 1185-1193. 10.1002/prep.202000059.

1.2.5 Initial characterization of new energetic materials

Once a new energetic material (high explosive, primary explosive, oxidizer) has been synthesized in the lab, various characterization and evaluation methods need to be applied before one can consider up-scaling of the material and the preparation of formulations. The most important characterization and evaluation methods are summarized in Tab. 1.3e.

Tab. 1.3e: Synthesis, characterization & evaluation of new EMs (before formulation).

	Synthesis & characterization:
	– IS, FS, ESD (grain size) – CHN analysis – X-ray, density or pycnometry – ^{1}H, $^{13}C\{^{1}H\}$, $^{14}N/^{15}N$ NMR; IR/Raman – DSC (5 K/min), TGA, m.p., b.p., T_{dec} – calcd: VoD, $p_{C\text{-}J}$, I_{sp}, Q_{ex}, . . . – Phase transitions / polymorphs (powder X-ray, DSC) – aquatic toxicity – better, cheaper ($/kg), higher yield, safer prep?

Tab. 1.3e (continued)

	Synthesis & characterization:	
	– hydrolysis, degradation, detonation/combustion products – shape and habit of crystals (electron microscopy) – vapor pressure measurement	
Secondary explosives:	**Primary explosives:**	**Oxidizers:**
1. VST 2. Compatibilities (DSC & VST)# 3. VoD (exptl.) 4. Seel shell (Koenen) 5. Trauzl 6. Under-water ex. 7. SSRT 8. FCO 9. SCO 10. GAP 11. Isothermal long term 12. Bomb calorimetry 13. ANSYS: fragment velocity, E_{kin}	1. Laser initiation 2. Initiation cap (RDX, PETN, HNS) 3. Electromag. ball-drop primer test 4. M55 stab 5. Compatibilities*	1. VST 2. Compatibilities (DSC & VST)+ 3. Bomb calorimetry 4. FCO 5. SCO 6. Isothermal long term

#CuO, SS, Al, binder & plasticizer, DNAN, TNT, NTO, RDX, HMX
*RDX, PETN, Al, SS, Cu, CuO, LA, LS, $Ba(NO_3)_2$, Tetracene, Sb_2S_3
+HTPB, binder, plasticizer, Al, SS, NC, NQ, GAP

1.3 Definitions

In accordance to the ASTM international definition (American Society for Testing and Materials: http://www.astm.org/), an **energetic material** is defined as a compound or mixture of substances which contains both the fuel and the oxidizer and reacts readily with the release of energy and gas. Examples of energetic materials are:
- primary explosives
- secondary explosives
- gun propellant charges
- rocket fuels
- pyrotechnics, e.g. signal torches und illuminants, smoke and fog generators, decoys, incendiary devices, gas generators (air bags) and delay compositions.

Energetic materials can be initiated using thermal, mechanical or electrostatic ignition sources and do not need atmospheric oxygen to maintain the exothermic reaction.

A material which could be considered to be a potential explosive, is a chemical compound or a mixture of substances in a metastable state, which is capable of a quick chemical reaction, without requiring additional reaction partners (e.g. atmospheric oxygen). In order to estimate how explosive a substance is, the Berthelot-Rot value B_R can be used (equation 1). Here ρ_0 (in kg m^{-3}) is the density of the potential explosive, V_0 (in m^3 kg^{-1}) is the volume of smoke and Q_V (in kJ kg^{-1}) is the heat of explosion.

$$B_R\left[\mathrm{kJm^{-3}}\right] = p_0^2 V_0\, Q_V \qquad (1)$$

Table 1.4 shows the Berthelot-Rot values for some well-known explosives. In general, compounds with a B_R value higher or equal to that of the "Oppauer salt" (55% NH_4NO_3, 45% $(NH_4)_2SO_4$) – the detonation of which in 1921 caused a massive and catastrophic explosion with over 1,000 fatalities – are considered potentially explosive.

Sporadic accidents resulting from the improper storage of civil explosives in recent years have resulted in great losses to the personal safety and property of many civilians, and have threatened social stability where these accidents have occurred. On August 4th, 2020, a massive accident involving stored ammonium nitrate at the Port of Beirut killed at least 190 people, in addition to injuring a further 6,500 people. On August 12th, 2015, an accident involving ammonium nitrate in the Binhai New Area of Tianjin which was the result of a fire resulted in the deaths of 165 people, with an additional eight people reported as missing, and 798 injured. The direct economic loss due to this accident was estimated to be in the region of CNY 6.866 billion.

However, it is not only the unsuitable storage of ammonium nitrate which can result in dangerous accidents. Another issue that can occur when storing ammonium nitrate is the formation of the compound tetraamine copper nitrate (TACN), which is explosive. TACN and copper nitrate can form, when copper or copper alloys are in contact with ammonium nitrate. Usually, the product of corrosion due to the reaction of copper with ammonium nitrate is blue and is likely to be copper nitrate which is not dangerous. However, the explosive tetraamine copper nitrate (TACN) is formed when air, moisture, ammonia, copper and electrical currents combine. It is deep purple in color, in contrast to the blue and green colors of copper nitrates. If these blue/green copper nitrates are exposed to air, purple TACN will begin to form underneath. Unfortunately, TACN is an impact-sensitive explosive that shows increasing sensitivity to impact as it becomes drier, with the impact required to initiate TACN being reported to be equivalent to dropping a 2 kg weight from a height of approximately 20 cm which corresponds to an IS of only 4 J.

Tab. 1.4: Berthelot-Rot values for potentially explosive substances.

Potentially explosive substance	ρ_0 / 10^3 kg m^{-3}	V_0 / m^3 kg^{-1}	Q_v / kJ kg^{-1}	B_R/ kJ m^{-3}
HMX	1.96	0.927	5253	18707
NG	1.60	0.782	6218	12448
TNT	1.65	0.975	3612	9588
Gun powder	1.87	0.274	3040	2913
Hydrazine	1.00	1.993	1785	3558
Oppauer salt	1.10	0.920	1072	1193

In contrast to the potentially explosive substances discussed above, explosion hazards are defined as follows. Hazardous explosive substances are compounds or mixtures of substances which give a positive result to at least one of the following tests:
- steel-sleeve test (Koenen test) with a nozzle plate with a 2 mm diameter hole
- impact sensitivity (using the BAM drop-hammer); more sensitive than 39 J
- friction sensitivity; more sensitive than 353 N.

The enthalpy of decomposition is typically related to the mass of the chemical substance and is expressed in J/g. Some typical values are: −1740 J/g for nitrobenzene, −1882 J/g for 4-nitroaniline, −800 J/g for azobenzene, −830 J/g for 3,5-dimethoxyaniline, −920 J/g for maleic anhydride, −406 J/g for glucose, and −835 J/g for sodium azide.

Internationally, the classification of an energetic substance based on its enthalpy of decomposition is regulated by the "Recommendations on the Transport of Dangerous Goods – Handbook of Tests and Criteria" by the UN. There, much lower thresholds are set to simplify hazardous materials testing:
- Up to 300 J/g for self-reactive substances of Class 51
- Up to 500 J/g for explosive substances of Class 1
- Up to 800 J/g and a (negative) detonation propagation test.

If all these tests are passed, the substance remains in Class 4.1 for flammable solids. See also: Angew Chem 1995, 107, 1284–1301.

Energetic materials which contain the fuel as well as the oxidizer ignite on raising the temperature past the ignitation temperature (due to the supply of external heat), forming a flame which generates heat as a result of the exothermic chemical reaction which is greater than the heat loss of the surroundings. In contrast to simple combustion (e.g. a candle = fuel) which occurs without an increase in pressure and uses oxygen from the atmosphere, in a deflagration (oxidizer and fuel combined in the energetic material) pressure is formed (Fig. 1.24). The term **Deflagration** means that when using a mixture of fuel and oxidizer, a flame is propagated which is less than the speed of sound, but which still can be heard. The (linear) burn rate (r in m s^{-1}) is therefore the speed with which the flame (zone of reaction is on the surface) moves through the unreacted material. Since the temperature increases with increasing pressure (p), the burn rate is also pressure dependent (equation 2). It is strongly influenced by the composition of the energetic material, as well as the confinement. In this context, β is a coefficient ($\beta = f(T)$) and α is the index of the burn rate which describes the pressure dependency. The index α is < 1 for energetic materials which deflagrate (usually 0.2 . . . 0.6 for propellants), and > 1 for detonating explosives.

$$r = \beta p^{\alpha} \quad (2)$$

A flame in the gas above an energetic solid conducts and radiates heat to the solid surface, which evaporates fuel. The fuel in turn feeds the flame – a cooperative system. There is an equilibrium standoff distance which represents a balance between these two effects. It is a function of pressure.

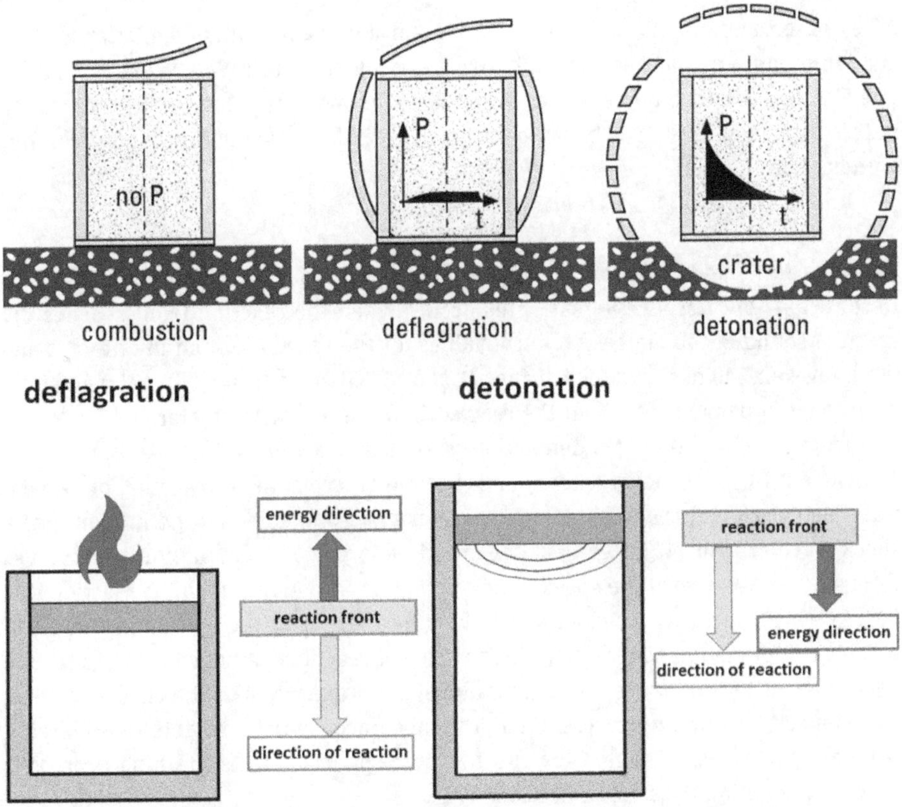

Fig. 1.24: Schematic representation of the pressure-over-time diagram of a combustion, a deflagration and a detonation. (This diagram is reproduced with slight modification from the original of Prof. Dr. Manfred Held, who is herewith thanked for his permission to reproduce this.).

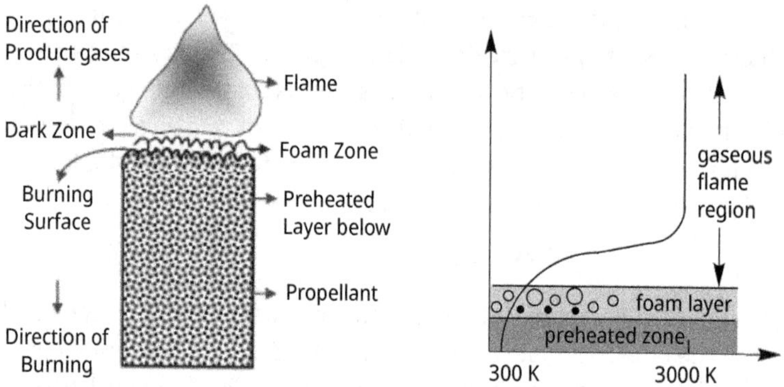

Fig. 1.25a: Self-sustained combustion in a stagnant environment (not to scale); x-axis: T (K).

A key observation is that a flame will not enter a crack (δ) unless the latter is large enough to support the necessary physical reaction structure (Figs. 1.25a and 1.25b). The standoff distance decreases as *P* increases, allowing the flame to enter smaller spaces as *P* increases. This behavior is given by Balyaev's equation (empirical, but with basis in theory):

$$\text{If } \dot{m}_a = aP^{\beta}, \text{ then } \delta^2 p_{cr}^{1+2\beta} = \text{k}$$

Under certain conditions (e.g. strong confinement), a deflagration can change to a **Detonation** (but not vice versa!). This occurs when the reaction front reaches the speed of sound in the material (typical values for the speed of sound in different materials are: air 340 m s^{-1}, water 1,484 m s^{-1}, glass 5,300 m s^{-1}, iron 5,170 m s^{-1}) and then propagates supersonically from the reacted into the unreacted material. The transition from a deflagration to a detonation is known as a **Deflagration-to-Detonation Transition** (DDT). The term detonation is used to describe the propagation of a chemical reaction zone through an energetic material accompanied by the influence of a shock-wave at a speed faster than the speed of sound. The detonation zone moves through the explosive at the speed of detonation *D*, perpendicular to the reaction surface and with a constant velocity. All of the properties of the system are uniform within the detonation zone. When these chemical reactions occur with the release of heat at a constant pressure and temperature, the propagation of the shock-wave becomes a self-sustaining process. Chemical substances which are able to undergo a DDT, are described as explosives and the corresponding self-sustaining process is called a detonation (Figs. 1.24 and 1.26).

The theory of the shock-wave is mathematically well founded. In order to obtain a clear idea about the origin of a shock-wave, it is useful to think about the following simple experiment: A pipe that is very long to the right and closed by a piston on the left is filled with air. If the piston is now moved with very low speed (δω), a weak compression wave in the gas is now generated, which progresses to the right with the speed of

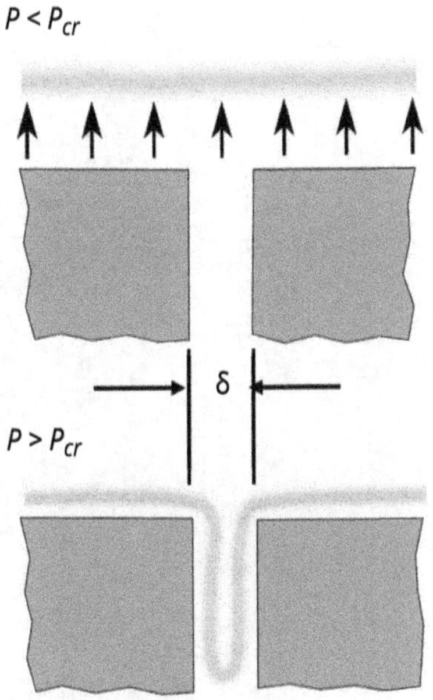

Fig. 1.25b: Flame intrusion into cracks.

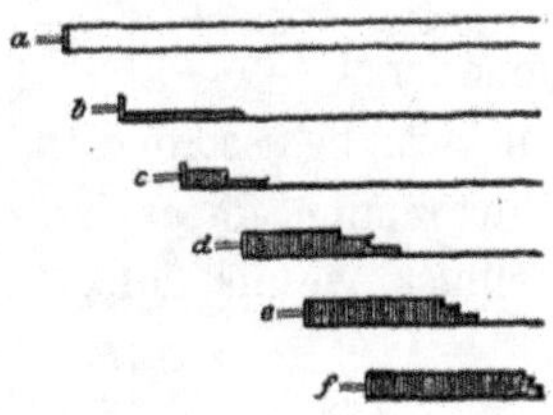

Fig. 1.25c: Generation of a shock-wave.

sound $c = (\gamma RT)^{0.5}$ (Fig. 1.25c (a)). At a certain moment (b), the air to the right of the wave-front is unchanged and at rest, while it is adiabatically compressed by an amount $d\rho$ between the wave-front and the piston and has the speed $d\omega$. If the speed of the piston is now increased again by the amount $d\omega$, a second compression wave is generated in the air mass mentioned above as a result, which runs after the first one (c). By repeating this procedure frequently, the piston will be brought to the finite speed ω. In this way, a step-shaped wave-front in the gas mass has been created, at the top of which, the gas particles also have the velocity ω. Now, the further fate of the wave-front is considered. First of all, the upper steps of the stairs have a greater speed relative to the pipe than the lower ones. This is because on one hand, the temperature (and therefore the speed of sound) is higher there, and on the other hand, the gas itself has higher flow speed at higher level. The result is that the individual stages will push themselves together in further course, so that the wave front will become steeper and steeper (e and f). It therefore is impossible to overlook what will happen when the steepness of the ascent becomes infinite after a certain time.

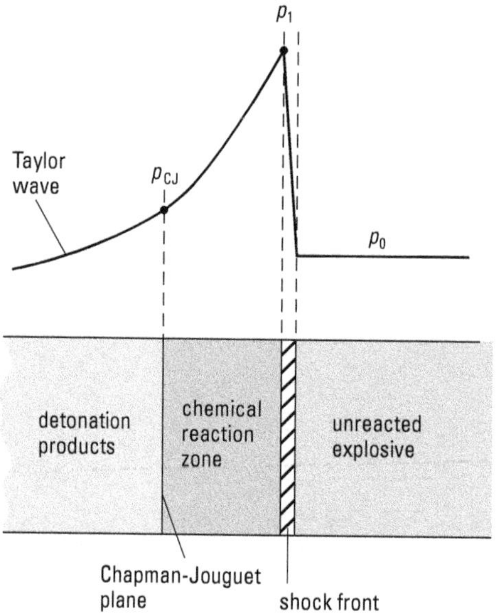

Fig. 1.26: Schematic representation of the detonation process and the structure of the detonation wave.

As is shown in Fig. 1.24, a crater is usually formed after a detonation.

There are relatively simple formulas in the literature which can be used to calculate the size (diameter and depth) of a crater formed by a detonation, for example (I. D. Bjelovuk, S. Jaramaz, P. Elek, D. Micković, L. Kričak, Tehnički vjesnik, **2015**, *22*, 227–232):

$$M_{ex} = 10.65\ V_K^{0.5} \tag{a}$$

where M_{ex} is the mass of converted explosive in kg (in TNT equivalents), and V_K is the crater volume in m^3.

The ammonium nitrate (AN) detonation that occurred in August 2020 in a storage facility at Beirut harbor will now be considered. If it is assumed that 1 kg of AN corresponds to about 0.750 kg of TNT (AN has only 75% of the power of TNT), the detonation of 2,750 t of AN in Beirut would correspond to a mass of $M_{ex} = 2063.5$ t of TNT. According to the formula in eq. (a), this would result in a crater volume of $V_K = 3.75 \cdot 10^4\ m^3$. If it is further assumed that the crater corresponds to a hemisphere (spherical volume = $4/3\ \pi\ r^3$), this would result in a radius of 26 m. This means that the depth of the crater should be about 26 m and its diameter about 52 m.

Another approach correlates the crater diameter D (in m) with the mass W (in kg) of the converted explosive (TNT equivalents) (equations b and c) (D. Ambrosini, B. Luccioni, R. Danesi, Mecanica Computacional, **2003**, *XXII*, 678–692).

The following diagrams show that the explosive can either lie on the ground (top) or can be dug into the ground like a hemisphere (bottom). The difference between these two scenarios is taken into account by the empirically fitted factors 0.42 and 0.61 respectively:

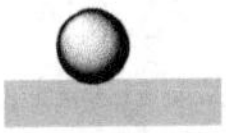

$$D\,[\mathrm{m}] = 0.42\ W\,[\mathrm{kg}]^{1/3} \tag{b}$$

$$D\,[\mathrm{m}] = 0.61\ W\,[\mathrm{kg}]^{1/3} \tag{c}$$

Using this method, for the explosion in the port of Beirut discussed above, a crater diameter of 95 m and therefore an estimated depth of 48 m is obtained.

Therefore, it can be expected, that the crater diameter should be 52–95 m and the depth 26–48 m. The values actually observed corresponded to a diameter of 75 m and a depth of 40 m, which are exactly the mean values of the two methods. The explosion in Beirut therefore had an output in the low kt range, which corresponds to the value for small nuclear explosions. For comparison, it is interesting to note that the first North Korean nuclear weapon test in 2006 had an output of approx. 0.6 kt TNT, and had therefore probably a lower output than that of the explosion in Beirut.

One side effect of a bomb blast may be the creation of a crater. Cratering is dependent on the bomb size, warhead type, fuze type and configuration, soil type and composition, building size, and other variables. Table 1.4a provides an overview of the expected cratering effects of aerial bombs.

Tab. 1.4a: Approximate crater size depending on the bomb mass.

Bomb mass class/kg	**Surface attack (width × depth)/m**		**Subsurface (width × depth)/m**	
	Heavy soil	**Light soil**	**Heavy soil**	**Light soil**
23	1.65 × 0.55	2.75 × 0.9	3.6 × 1.1	6 × 1.8
45	2 × 0.75	3.25 × 1.2	4.5 × 1.45	7.5 × 2.4
113	2.7 × 0.9	4.5 × 1.5	6 × 1.8	10 × 3
227	3.3 × 1.1	5.5 × 1.8	8 × 2.2	13.7 × 3.7
454	4.8 × 1.5	8 × 2.4	10 × 3	17 × 4.9
907	5.5 × 1.6	9 × 2.7	11.7 × 3.3	19.5 × 5.5

Source: ARESM Explosive Weapons in Populated Areas, 2016:
https://www.icrc.org/en/document/explosive-weapons-populated-areas-use-effects.

Under the influence of the dynamic properties of the shock-wave, a thin layer of still unreacted explosive from the original specific volume V_0 ($V_0 = 1/\rho_0$) along with the shock adiabat of the corresponding explosive (or Hugoniot-Adiabat) is compressed to volume V_1 (Fig. 1.27). The pressure increases from p_0 to p_1 as a consequence of the dynamic compression, which subsequently results in an increase of the temperature in the thin compressed layer of the explosive (Figs. 1.26 and 1.27), which causes initiation of the chemical reaction. At the end of the chemical reaction, the specific volume and pressure have the values V_2 and p_2. At the end of the reaction zone the reaction products are in chemical equilibrium ($\Delta G = 0$) and the gaseous reaction products move at the local speed of sound, c. Mass, impulse and energy conservation apply. The requirement of a chemical equilibrium between the reaction products is known as the Chapman-Jouguet condition. This state corresponds to the point of the detonation products on the shock adiabat (Figs. 1.27 and 1.27a). At this point it is important to emphasize again, that in a deflagration the propagation of the reaction occurs as a result of thermal processes, whereas in the considerably faster detonation, a shock-wave mechanism is found.

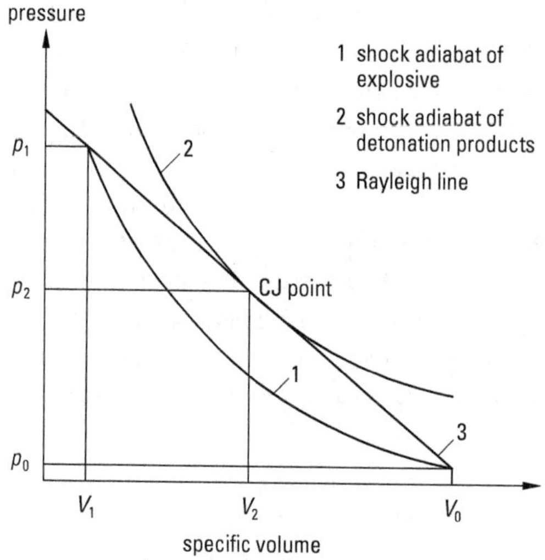

Fig. 1.27: Shock adiabat for an explosive and the detonation products (detonation in stationary state).

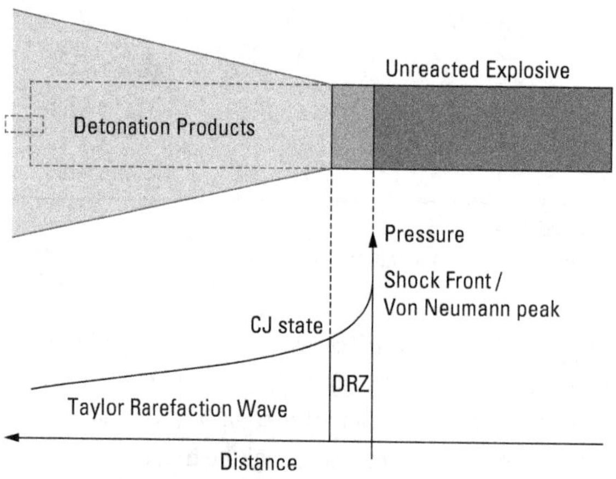

Fig. 1.27a: Structure of a stationary detonation wave (ideal detonation) in the pressure-distance plane according to the Zeldovich-Neumann-Doring (ZND) detonation model (DRZ = detonation reaction zone).

Following the detonation model for the stationary state, the points (V_0, p_0), (V_1, p_1) and (V_2, p_2) lie on one line (Fig. 1.27), which is known as the **Rayleigh line**. The slope of the Rayleigh line is determined by the detonation velocity of the explosive. In accordance with Chapman and Jouguet's theory, the Rayleigh line is tangent to the shock adiabat of the detonation products and at this point corresponds to the end of the chemical reaction (V_2, p_2). This point is therefore also known as the **Chapman-Jouguet Point** (C-J point). At the C-J point an equilibrium is reached and the speed of the reaction products corresponds to that of the detonation velocity *D*. Now the gaseous products can expand and a dilution or Taylor wave occurs (Fig. 1.26). Concrete values for typical speeds of reactions and mass flow rates for combustion, deflagration and detonation are summarized in Tab. 1.5.

Tab. 1.5: Reaction types of an energetic material with Q_{ex} = 1,000 kcal kg^{-1}.

Reaction type	Reaction speed / m s^{-1}	Mass flow / m^3 s^{-1}	Gaseous products / m^3 s^{-1}	Reaction time / s m^{-3}
Combustion	10^{-3}–10^{-2}	10^{-3}–10^{-2}	1–10	10^2–10^3
Deflagration	10^2	10^2	10^5	10^{-2}
Detonation	10^4	10^4	10^7	10^{-4}

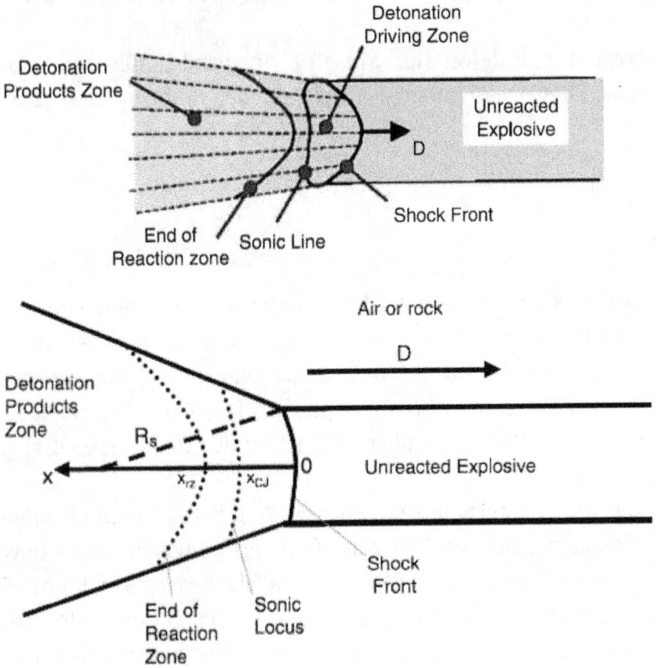

Fig. 1.28: Representation of a non-ideal detonation (see for example: S. Esen, Rock Mech.-Rock Engng **2008**, *41*, 46–497).

It is important to stress that the above description is a simplified one, and is only valid for ideal explosives. In an ideal explosive, there is a plane shock front (Fig. 1.26), whereas in a non-ideal detonation, the shock front is somewhat curved (Fig. 1.28), and the flow of the reaction products significantly diverge. A consequence of this is that reactions in a non-ideal detonation are never complete in the detonation driving zone (DDZ) – i.e. between the shock front and the sonic line. The detonation (a detonation is a reactive shock – shock followed by chemical reactions) driving reaction terminates at the "sonic line" and contributes to supporting the detonation process.

For an ideal explosive at the C-J point (at the end of the chemical reaction zone), conversion of the explosive into the products corresponding to the thermochemical equilibrium is complete (1.0), whereas for a non-ideal explosive the conversion is < 1. Thus, for a non-ideal explosive, the C-J point is also called the "sonic point" (end of DDZ) (Fig. 1.28), and where the Rayleigh line has the same slope as the Hugoniot of the detonation products formed at the sonic point.

For both ideal and non-ideal detonations, at the C-J point the "flow" (detonation velocity, D – particle velocity, U_p) corresponds to sound velocity, C_o:

$$D - U_p = C_o$$

1.4 Combustion, deflagration, detonation – a short introduction

Fire, combustion, deflagration, and detonation are all terms used to describe exothermic oxidation reactions. They are distinguishable from each other by their reaction velocities.

1.4.1 Fire and combustion

Generally speaking, combustion is a heat-releasing chemical oxidation reaction, which is often accompanied by the occurrence of a flame. Flame temperatures differ widely from approx. 2,000 K for an open flame to approx. 3,000 K for the flame of an acetylene cutting torch. Remarkably, the heats of combustion of explosives is generally lower than those of common fuels (pine wood with 12.9% water: 4.422 kcal/g, acetylene: 11.923 kcal/g, dynamite 75%: 1.290 kcal/g).

Fire is a simple example of a combustion reaction, despite it being less predictable and slower than pure combustion reactions. There are four factors which decide how big a fire can get: fuel, oxygen, heat, and uninhibited chemical chain reactions. If one of these factors is restricted or even absent, the fire will become extinct. For example, lean fuel-air mixtures cannot be ignited because the ratio of fuel is too low and rich mixtures lack the sufficient oxygen ratio to be ignitable. Water as a fire extinguisher has a cooling effect and it therefore stops burning through heat restriction and heat transfer. During burning, a reducing agent (fuel, e.g. gasoline) reacts with oxygen. When heating an ignitable liquid, prior to reaching the actual boiling point of the liquid, a considerable amount of fumes can be observed on the surface of the liquid. As these gases mix with air, ignitable mixtures can develop. The temperature at which an ignitable fume layer is found directly on the liquid's surface is called the flash point of the substance. Liquids at temperatures exceeding the flash point are easily set on fire by a flame or spark. The flash point is usually found at temperatures lower than the boiling point. As the boiling liquid is heated further, the temperature of the heated fume-air mixture may become high enough to evoke a spontaneous oxidation. At this temperature, the fume-air mix will ignite itself spontaneously. The corresponding temperature is called the autoignition temperature (usually above 300 °C, for some compounds, e.g. diethyl ether, significantly lower). A comparable effect can be observed when water is added to overheated oil. As the oil is spilled in form of a finely dispersed cloud, an oil-air mixture is obtained at a temperature enabling rapid oxidation. The resulting hot oil cloud will ignite spontaneously. Thus when heating an ignitable liquid, the temperatures first reach the flash point, then the boiling point, and finally the autoignition point. In order to burn, liquid fuels must be evaporated to maximize contact with oxygen (air) and solid fuels must be sublimed first (e.g. camphor) or pyrolyzed into ignitable gases or fumes.

After a sufficient heat source starts the fire, it propagates through the fuel by heat carrying the uninhibited chemical chain reaction. Heat is transferred by convection, conduction and radiation. Convection is based on the lower densities of heated gases resulting in a gas flow. This is the base for the so-called Zone Model to predict heat flow and fire growth in rooms and spaces. In the Zone Model (e.g. CFAST program; Fire Growth and Smoke Transport Modeling with CFAST (version 6.1.1), National Institute of Standards and Technology (NIST), http://cfast.nist.gov), a room is divided into a cold gas layer (zone) at the floor and a hot zone at the ceiling. Heat transfer by conductivity is understood either as thin, when heat transfer within the object is faster than the rate at which the surface of the object is changed by heat transfer, or as thick, when there is a temperature gradient within the object and the unexposed surface has no significant influence on the heat transfer into the exposed surface. In thin materials (e.g. sheet metal), the temperature can be described by the one-dimensional heat transfer equation:

$$(\delta \rho C_\mathrm{p}) \frac{\partial T}{\partial x} = \dot{q}''_\mathrm{rad} + \dot{q}''_\mathrm{con} + \dot{q}''_\mathrm{e}$$

with the heat transfer rates $\dot{q}''_i$ (i = rad for radiation, con for convection and e for background loss) and the measurable material thickness δ, density ρ and heat capacity C_p. For thick materials, heat transfer within the material can be described one-dimensionally by:

$$\rho C_\mathrm{p} \frac{\partial T}{\partial t} = k \frac{\partial^2 T}{\partial x^2}$$

Where k is the thermal conductivity of the material. Heat transfer by radiation can be described by the model of black body radiation and Boltzmann's equation for the thermal energy emission of an ideal radiator

$$E_b = \sigma T^4$$

with the Boltzmann constant of σ. Unfortunately, in reality, most burning objects do not behave like ideal black body radiators and do not completely absorb thermal energy.

When a fire is started in the open the burning is fuel-limited. Since oxygen is abundant in the air, the rate at which fuel is supplied to the fire limits its growth. Even for fires in confined areas (rooms), the early stage of fire is fuel limited. The heat release of a fire is dependent on time. In the early stage of burning, the heat release will grow with the square of time until reaching a steady state. When the fire is extinguished, e.g. when all fuel or oxygen available is consumed, the heat release will decrease to zero. The heat release of a fire can be estimated from the height of its flames. In confined spaces, the initial stage of fuel limited burn will grow into a fully developed fire, wherein all burnable fuels in the room are burning. The fully developed fire will terminally be oxygen-limited as the amount of oxygen available to the chemical chain reactions will constrain the fire. The transition to a fuel-rich oxygen-limited fire will be indicated by a flashover, the transition to uniform burning of all burnable substances in the room, preceded by rollover, flaming-out and off-gassing. In the final oxygen-limited stage, more and more under-oxidized combustion products like CO, HCN and other potentially hazardous compounds are detected in the combustion product mixture.

Combustion reactions are excessively used in propellant systems. The impulse of the gaseous combustion products is used to propel a payload. For obvious technical reasons, the burning temperature of a rocket engine is of interest. The adiabatic flame temperature of combustion (T_{ad}) is the temperature at which reactants and products do not differ in enthalpy. The enthalpies of the components of the system have to be calculated from their standard enthalpy by adding of the enthalpy caused by heating to T_{ad}. This is where the specific heat capacity C_v comes into play. Unfortunately, the component ratios of the system are functions of the temperature, necessitating the use of iterative calculations of T_{ad}.

1.4.2 Deflagration and detonation

Combustion reactions can be self-accelerating. The acceleration of chemical reactions is achieved by two routes: large increases in active particles and branching of exothermic chain reactions. Active particles, like radicals, are formed by the dissociation of molecules, a process expected to increasingly take place at elevated temperatures.

The more active particles are generated, the more collisions between active particles can be observed which causes an acceleration of the overall reaction rate. Branching of chain reactions may occur whenever a single reaction produces more active particles than there were previously in this reaction. However, active particles are continuously de-activated by inevitable collisions. Chain branching then is in concurrence with chain termination. If the rate of chain branching is greater than the rate of chain termination, the reaction will be accelerated and explosion conditions can be obtained. If the energy release of the reaction exceeds the thermal heat loss, thermal explosion can be achieved. Since the loss of active particles does not require activation energy, the temperature's influence on the rate of chain branching is much bigger than it would be on chain termination. Taking the rates of active center creation, chain branching and chain termination into account, the critical temperature, below which explosion is impossible, is to be found.

When starting the reaction, there is an induction period, in which active particles are built up in the system. After the induction time, the number of active particles continues to rise without limit. From this time on, deflagration is observed which may develop into detonation.

Deflagration is the exothermic chemical reaction proceeding at subsonic speed, a detonation proceeds in a supersonic detonation wave. While deflagration can be initiated by mild energy release, e.g. a spark, detonation is triggered by localized explosions caused by the impact of a shock wave. Rapid release of a large amount of energy is typically needed to initiate detonations. However, there often is a transition from deflagration to detonation (a transition from detonation to deflagration is not apparent) (Fig. 1.29). This deflagration-to-detonation transition (DDT) is relevant because of its destructive potential. Experimental observations indicate that DDT rarely takes place in unconfined propagating flames. The presence of walls or other spatial confinements of the propagating flame greatly enhance the probability of DDT. In an unconfined burning, the pressure of the gaseous reaction products, which might form small shock waves, are able to expand freely without pressurizing the unburned composition. As the burning system becomes more and more confined, e.g. in a tube, the shock waves caused by the expansion of product gases can reflect from the walls, then form turbulences and fold to convulsions. These reflections, turbulences and convulsions may increase the flame velocity. The flame, which is thus accelerated by itself, can act with the flow of the product gases as a piston in the refined system generating pressure and shock waves in the unreacted material. This may again contribute to an even faster and exponential self-acceleration of the flame until a miniature explosion prompts the onset of detonation. The shock wave must be strong enough to raise the temperature in the system up to roughly 1,500 K in order to initiate the reaction. For Chapman-Jouguet type detonations, Mach numbers of ~ 6/7 are found for the leading shock wave; DDT takes place when the velocity of the reaction wave approaches the local speed of sound. Experimentally, a tulip-shaped or spiked flame front appears shortly before the transition to detonation is achieved. As soon as the

DDT has taken place, the reaction proceeds through the unreacted material in a self-sustained detonation.

Strictly speaking, a *detonation* is based on the almost instantaneous release of energy, whereas an *explosion* is the result of a bursting container, in which large amounts of gas are released, so that the bursting results in a transient gas release – not throughput as with a rocket engine, but its change over time as a result of a rapidly increasing outlet opening – which is responsible for the bang.

While a *deflagration* can be related to the *release of energy* per time (power = energy/time [J/s]), a *detonation* is related to the instantaneous release of energy (energy [J]).

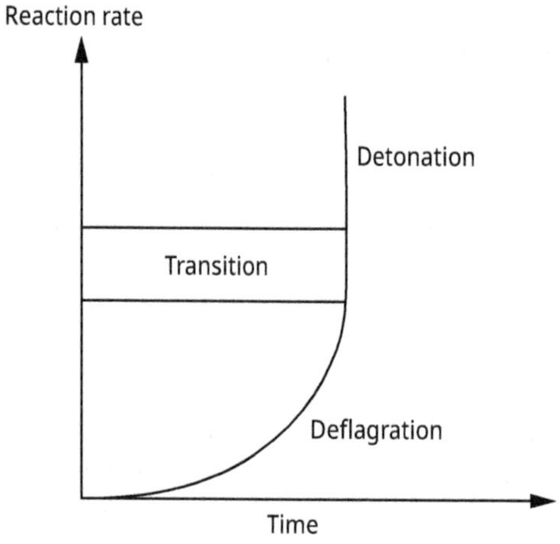

Fig. 1.29: Schematic presentation of DDT (reaction rate over time).

Check your knowledge: see chapter 14 for study questions.

2 Classification of energetic materials

Energetic materials which derive their energy from a chemical reaction (in contrast to a nuclear reaction) can be classified according to their use as shown in Fig. 2.1.

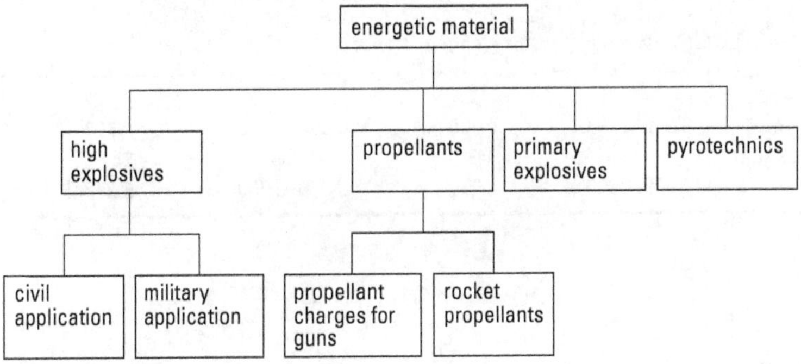

Fig. 2.1: Classification of energetic materials.

2.1 Primary explosives

Primary explosives are substances which unlike secondary explosives show a very rapid transition from combustion (or deflagration) to detonation and are considerably more sensitive towards heat, impact or friction than secondary explosives. Primary explosives generate either a large amount of heat or a shockwave which makes the transfer of the detonation to a less sensitive secondary explosive possible. They are therefore used as initiators for secondary booster charges (e.g. in detonators), main charges or propellants. Although primary explosives (e.g. $Pb(N_3)_2$) are considerably more sensitive than secondary explosives (e.g. RDX), their detonation velocities, detonation pressures and heat of explosions are as a rule, generally lower than those of secondary explosives (Tab. 2.1).

Typical primary explosives are lead azide and lead styphnate (see Fig. 1.17). The latter one is less powerful than LA but easier to initiate. Tetrazene (Fig. 2.2) is often added to the latter in order to enhance the response (sensitizer) (*N.B.* mercury fulminate used to be used as a sensitizer). Tetrazene is an effective primer which decomposes without leaving any residue behind. It has been introduced as an additive to erosion-free primers based on lead trinitroresorcinate. Unfortunately, tetrazene is hydrolytically not entirely stable and in long-term studies decomposes at temperatures above 90 °C. Diazodinitrophenol (Fig. 2.2) is also a primary explosive and is primarily used in the USA. However, the compound quickly darkens in sunlight. SAcN (silver acetylide nitrate, $Ag_2C_2 \cdot 6\ AgNO_3$), mainly in combination with PETN, has also shown to be useful for use in detonators [9].

https://doi.org/10.1515/9783111447087-002

Tab. 2.1: Typical sensitivity and performance data of primary and secondary explosives.

	Typical primary explosives	$Pb(N_3)_2$	Typical secondary explosives	RDX
Sensitivity data				
Impact sensitivity / J	≤ 4	2.5–4	≥ 4	7.4
Friction sensitivity / N	≤ 10	< 1	≥ 50	120
ESD / J	0.002–0.020	0.005	≥ 0.1	0.2
Performance data				
Detonation velocity / m s^{-1}	3,500–5,500	4,600–5,100	6,500–9,000	8,750
Detonation pressure / kbar		343	210–390	347
Heat of explosion / kJ kg^{-1}	1,000–2000	1,639	5,000–6,000	5,277 (H_2O(g))

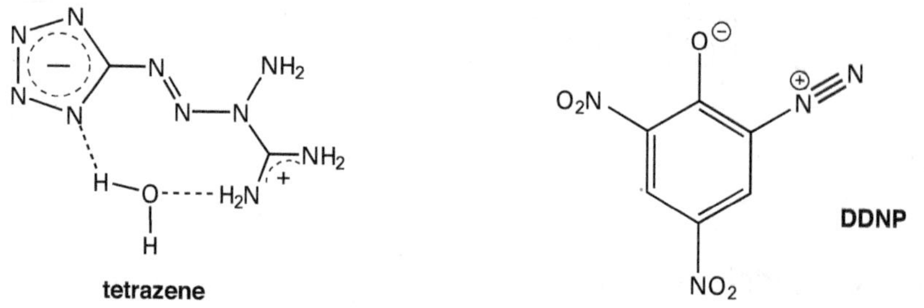

Fig. 2.2: Molecular structures of tetrazene and diazodinitrophenol (DDNP).

Tetrazene (see above) is widely used in ordnance systems as a sensitizer of primer mixes and is used in both percussion and stab applications. It has low thermal stability (decomposition at 90 °C within six days, forming two equivalents of aminotetrazole, 5-AT) and hydrolytic stability compared with other components of primer mixes. Therefore, there currently exists the need to find a replacement for tetrazene, which shows enhanced stability characteristics. MTX-1 (1-[(2 E)-3-(1*H*-tetrazol-5-yl)triaz-2-en-1-ylidene]methanediamine) meets these criteria and shows great promise as a tetrazene replacement [J. W. Fronabarger, M. D. Williams, A. G. Stern, D. A. Parrish, *Central European Journal of Energetic Materials*, **2016**, *13*(1), 33–52]. Preliminary testing of this material has confirmed that MTX-1 has safety and performance properties which are similar to those of tetrazene, but its chemical properties (including thermal and hydrolytic stability) are better than those of tetrazene (Fig. 2.2a). MTX-1 has been successfully evaluated against tetrazene in a variety of chemical/output tests, including comparative testing in the PVU-12 primer.

HNS is a temperature stable secondary explosive, which is particularly useful for blasting in very hot oil deposits, because it is stable to approx. 320 °C. Problems in this area however relate to the initiator, since HNS is relatively difficult to initiate. The most useful initiator is cadmium azide, $Cd(N_3)_2$, ($T_{dec.}$ ca. 295 °C). However, since cad-

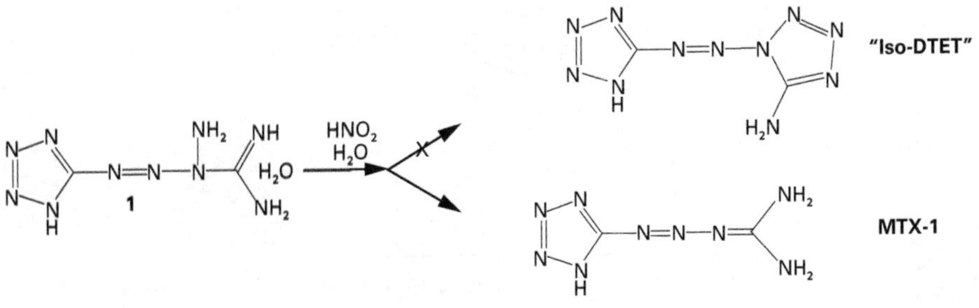

Fig. 2.2a: Synthesis of MTX-1.

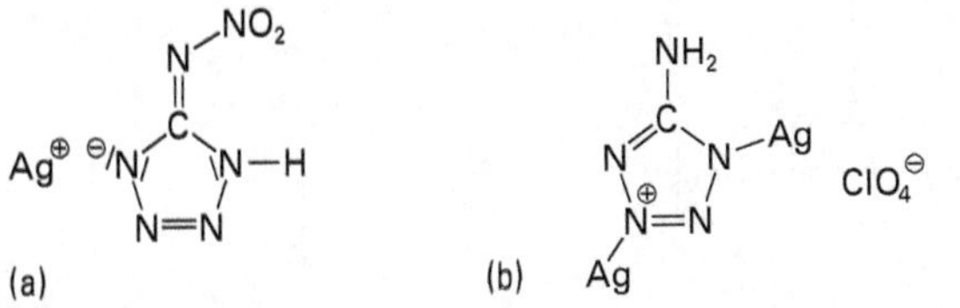

Fig. 2.3: Structures of silver nitraminotetrazolate and disilver(aminotetrazole) perchlorate.

mium is toxic, alternatives are currently being sought. The two most promising compounds to date to replace Cd$(N_3)_2$ are silver nitriminotetrazolate ($T_{dec.}$ = 366 °C) and di (silveraminotetrazole) perchlorate ($T_{dec.}$ = 319 °C) (Fig. 2.3) [10].

Calcium nitriminotetrazolate has shown huge potential due to its large thermal stability (T_{dec} = 360 °C) and relatively low sensitivity (impact sensitivity: 50 J, friction sensitivity 112 N, ESD = 0.15 J), but at the same time good initiation behavior with commercial pyroelectric initiators (Fig. 2.4). The corresponding cadmium salt is also a good primary explosive.

While the environmental impact of cadmium azide in deep oil deposits is relatively low, the long-term use of Pb$(N_3)_2$ and lead styphnate in military training grounds has resulted in considerable lead contamination (see Ch. 1.2.3, see Fig. 1.17). "On demand lead azide" (ODLA) is available from the reaction of lead acetate and sodium azide (Fig. 2.3a, Tab. 2.1a). The recently introduced iron and copper complexes of the type $[Cat]_2\,[M^{II}(NT)_4(H_2O)_2]$ ($[Cat]^+$ = NH_4^+, Na^+; M = Fe, Cu; NT = 5-nitrotetrazolate) as "green" primary explosives [3] are relatively easily obtained and show similar initiator properties as those of lead azide (Tab. 2.2).

2.2 High (Secondary) explosives

Secondary explosives (also known as "high explosives" = HEs) unlike primary explosives cannot be initiated simply through heat or shock. In order to initiate primary explosives have to be used, whereby the shockwave of the primary explosive initiates

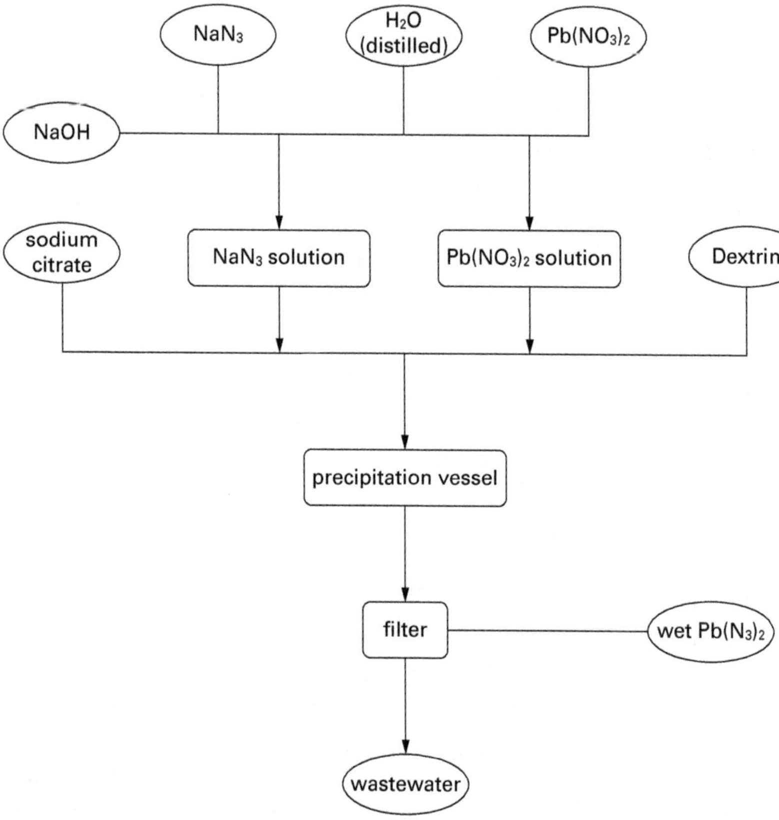

Fig. 2.3a: Manufacturing process for lead azide.

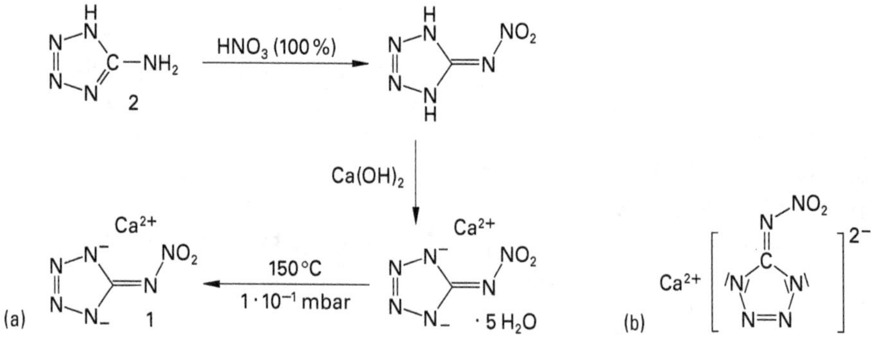

Fig. 2.4: Synthesis (a) and structure (b) of calcium (nitriminotetrazolate).

Tab. 2.1a: Key types of lead(II) azide suitable for industrial applications.

Type of lead azide	Product characteristics
Service (SLA)	Service lead azide (SLA) was developed by the British and does not involve a coating agent; instead, it makes use of the addition of acetic acid and sodium carbonate, which provides a nucleus to precipitate lead azide in a less-hazardous (compared to needle-shaped crystals) spherical morphology. SLA has a higher explosive performance than dextrinated lead azide (DLA), or RD1333/special purpose lead azide (SPLA), but is somewhat more sensitive.
Colloidal	Colloidal lead azide is the purest form of lead azide used in the United States, and is made by mixing dilute solutions of lead and azide salts. It has a very fine particle size of ~ 5 μm, making it ideal for its main application, which is as a coating on electric bridgewires for commercial electric detonators. It is known to be extremely sensitive to electrostatic discharge (ESD).
Dextrinated (DLA)	Dextrinated lead azide (DLA) is considered to be the safest-to-handle form and the most common type now used commercially. It was developed in the United States in 1931 as a solution to the numerous accidental explosions associated with attempts to manufacture pure lead azide. The key feature is the incorporation of dextrin (a short-chained, starch-based polysaccharide), which helps to desensitize the explosive by preventing the formation of large fragile crystals, albeit at the cost of performance and increased hygroscopicity.
RD1333 (SPLA)	RD1333 and special purpose lead azide (SPLA). Although involving somewhat different processing pathways, both (British developed) RD1333 and (the later U. S. developed) SPLA are similar in that they have nearly identical performance requirements/ specifications and make use of the sodium salt of carboxymethylcellulose as the desensitizing agent. As such, in the United States, the two types are often used interchangeably. The materials were developed to meet the need for lead azide, which performed better than DLA (especially in smaller detonators), but which maintained some of its safe handling characteristics. SPLA makes up a large portion of the current U. S. military stockpile of lead azide, which has existed since the 1960s.
On-demand (ODLA)	On-demand lead azide (ODLA) is a much more recent development compared to the others described in this table. It is prepared by a military-qualified process, developed by the U. S. Army Armament Research, Development and Engineering Center, Picatinny Arsenal, NJ, which produces lead azide that meets RD1333 specifications and which is considered to be equivalent to RD1333 and SPLA. The main advantage is that it is produced in an on-demand, continuous fashion, therefore avoiding the hazards associated with handling large scale batches of the material. The small footprint and low cost of the processing equipment also means that it can be placed close to item production lines, which further reduces the need for the expensive and dangerous transport of lead azide on public roadways. ODLA was qualified by the U. S. Army in 2012 and is currently being evaluated in larger-scale loading operations.

Tab. 2.2: Properties of lead azide (LA) and lead styphnate (LS) in comparison to new "green" primary explosives.

Primary explosive	T_{dec} / °C	Impact sensitivity / J	ESD / J	Density / g cm^{-3}	Detonation velocity / m s^{-1}
NH_4FeNT	255	3	> 0.36	2.2	7,700
NaFeNT	250	3	> 0.36	2.2	not determined
NH_4CuNT	265	3	> 0.36	2.0	7,400
NaCuNT	259	3	> 0.36	2.1	not determined
LA	315	2.4	0.005	4.8	5,500
LS	282	3.4	0.0002	3.0	5,200

the secondary. However, the performance of the secondary explosive is usually higher than that of the primary (see Tab. 2.1). Typical currently used secondary explosives are TNT, RDX, HMX, NQ and TATB (see also Tab. 1.2) and, for civil applications, HNS and NG e.g. in the form of dynamite for commercial use.

The current trends in the research of secondary explosives can be arranged into three branches (Fig. 2.5):
- Higher performance
- Lower sensitivity (insensitive munition, IM)
- Lower toxicity of the explosives, their biological degradation products and the detonation products.

A higher performance for secondary explosives is of fundamental importance and always desired. The main performance criteria are:
1. heat of explosion Q (in kJ kg^{-1})
2. detonation velocity D (m s^{-1})
3. detonation pressure P (in kbar)

and less importantly,
4. explosion temperature T (K) and
5. volume of gas released V per kg explosive (in l kg^{-1}).

Depending on the location of the mission, one parameter can be more important than the other. For example, the performance (or brisance) of an explosive cannot be described by one parameter alone. The term **brisance** describes the destructive fragmentation effect of a charge on its immediate vicinity. The higher the charge density (corresponding to the energy per volume) and the detonation velocity (corresponding to the reaction rate), the more brisant the explosive is. Furthermore, the detonation velocity as well as the detonation pressure increase with increasing density. According to Kast, the brisance value (B) is defined as the product of the loading density (ρ), the specific energy (F for "force of an explosive") and the detonation velocity D:

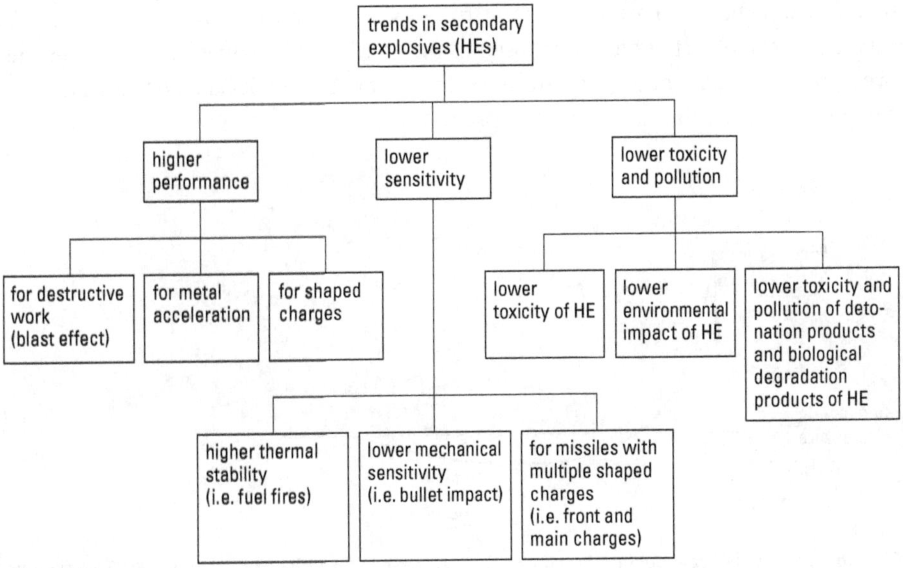

Fig. 2.5: Trends for new secondary explosives.

$$\text{Brisance: } B = \rho \times F \times D$$

The specific energy ("force") of an explosive (F) on the other hand can be calculated according to the general equation of state for gases:

$$\text{Specific energy: } F = p_e V = n\ R\ T$$

Here p_e is the maximum pressure through the explosion (not to be confused with the much higher values for the detonation pressure $p_{C\text{-}J}$ at the C-J point), V is the volume of detonation gases (in l kg^{-1} or m^3 kg^{-1}), n is the number of moles of gas formed by the explosion per kilogram of explosive, R is the gas constant and T is the temperature of the explosion. The specific energy has the units J kg^{-1}. Consequently, the brisance can be given using the unit kg s^{-3}.

Generally, the specific energy of secondary explosives is higher than that for propellant charges. This is due to the fact that in propellant charges the temperature of combustion is kept as low as possible in order to protect the barrel by preventing the formation of iron carbide (from the CO in the combustion). For shaped charges, a high brisance and therefore a high specific energy and also high loading densities are particularly important. For typical general purpose bombs, a maximum heat of explosion (Q) and gaseous products (V) should be achieved.

As we have already seen above (see Fig. 1.24), the pressure vs. time diagram shows that the explosion pressure (not to be confused with the by magnitudes higher detonation pressure $p_{C\text{-}J}$) generated on detonation increases instantly and decreases exponentially with time. Figure 2.6 shows the typical time dependence of a blast wave. When the detonation occurs at time $t = 0$, the shock front reaches the "observer" i.e. the object, which is exposed to the shock front at time t_a. Afterwards, the pressure falls exponentially and for a certain period of time it falls beneath the pressure of its surroundings (atmospheric pressure).

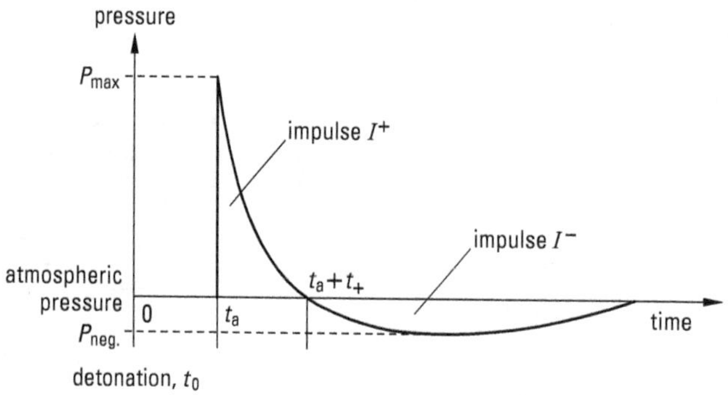

Fig. 2.6: Typical time dependent progression of a shock wave in the vicinity of a detonation. (This diagram is reproduced with slight modification from the original of Prof. Dr. Manfred Held, who is herewith thanked for his permission to reproduce this.).

As we can see in Fig. 2.7, after a detonation, the shock wave naturally requires a certain amount of time t ($t_1 < t_2 < t_3 < t_4$), to reach a certain point. However, the maximum pressure p of the shockwave also decreases as the distance from the center of the detonation increases.

The shock front is formed by the fusion of the incident and reflected shock fronts from an explosion. The term is generally used with reference to a blast wave which is propagated in the air and reflected at the surface of the earth. In an ideal case, the Mach Stem is perpendicular to the reflecting surface and slightly convex (forward). The Mach Stem is also sometimes referred to as the Mach front.

If the explosion occurs above the ground (e.g. MOAB), when the expanding blast wave strikes the surface of the earth, it is reflected off the ground to form a second shock wave which travels behind the first (Fig. 2.7a). This reflected wave travels faster than the first (or incident) shock wave, since it is traveling through air already, moving at high speed due to the passage of the incident wave. The reflected blast wave merges with the incident shock wave to form a single wave, known as the Mach Stem (Fig. 2.7a). The overpressure at the front of the Mach wave is generally about twice as large as that at the direct blast wave front.

Generally, it can be said that the damaging effect of a shockwave produced by a detonation is proportional to its impulse (impulse = mass × velocity of the gaseous explosion products) and its maximum pressure, with the impulse being the most influential factor at smaller distances and the pressure being most important at larger distances. As a "rule of thumb", the distance D, which offers a chance of survival, is

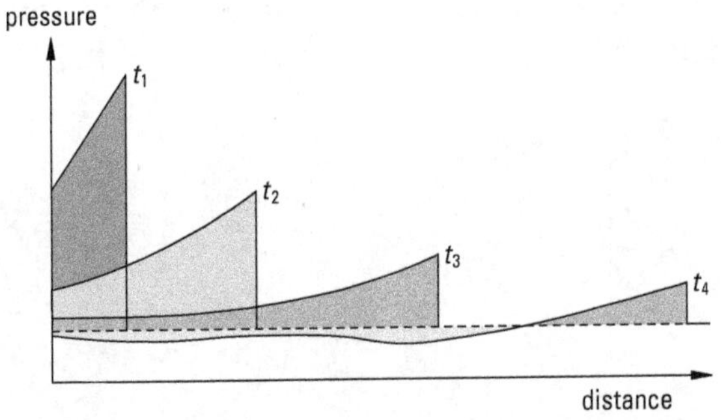

Fig. 2.7: Pressure-distance-dependence of a progressing shock wave at different points in time t ($t_1 < t_2 < t_3 < t_4$) after a detonation. (This diagram is reproduced with slight modification from the original of Prof. Dr. Manfred Held, who is herewith thanked for his permission to reproduce this.).

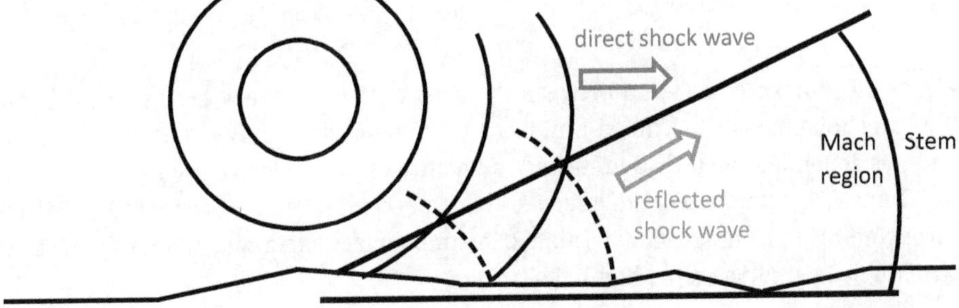

Fig. 2.7a: Formation of a Mach Stem.

proportional to the cube root of the mass w of an explosive, while for typical secondary explosives at larger distances, the proportionality constant is approximately 2:

$$D = c\,w^{0.33} \approx 2\,w^{0.33}$$

It is important to note, that this approximation is only based on the pressure. The impulse of the shockwave and the fragment impact (e.g. from confined charges or fume-cupboard shields in lab) are *not* taken into consideration.

In order to work safely with highly energetic materials in the chemical laboratory the following rules must be obeyed:

- amounts used are to be kept as small as possible
- the distance from the experiment is to be kept as large as possible (filled vessels are not to be transported by hand, but with well fitted tongs or clamps instead)
- mechanical manipulators are to be used when possible, particularly for larger quantities
- vessels are never to be enclosed by hand (confinement)

- protective clothing (gloves, leather or Kevlar vest, ear protectors, face shield, anti-electrostatic shoes) is to be worn at all times.

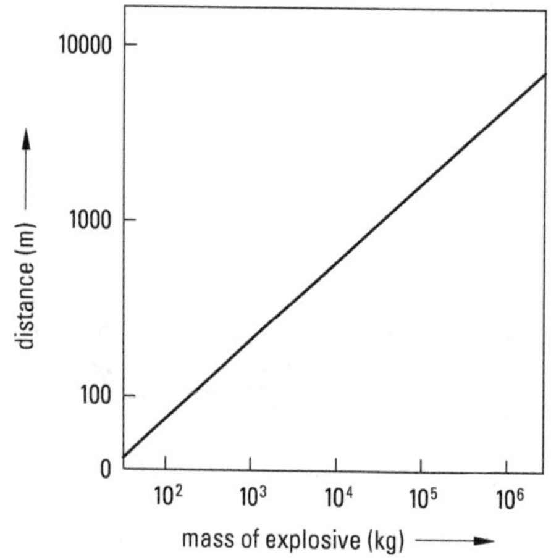

Fig. 2.8: "Safe" distance for occupied buildings from the detonation of a secondary explosive.

Figure 2.8 shows a correlation between the mass of the secondary explosive used and the "safe" distance for occupied buildings. Of course this is only a very rough guide and it is dependent on the building and the nature of the explosive used.

Table 2.2a summarizes the effects of blast overpressure and blast wind on structures and on the human body. Table 2.2b summarizes the approximate damage to structures depending on the bomb mass.

Tab. 2.2a: Effects of blast overpressure and blast wind on structures and on the human body.

Peak overpressure/bar	Max. wind speed/km h^{-1}	Effects on structures	Effects on the human body
0.07	61	Window glass shatters	Light injuries from fragments
0.14	113	Moderate damage to houses (windows and doors blown out and damage to roofs)	People injured by flying glass and debris
0.21	164	Residential structures collapse	Serious injuries are common, fatalities may occur
0.34	262	Most buildings collapse	injuries are universal, fatalities are widespread
0.69	473	Reinforced concrete buildings are severely damaged or demolished	Most prople are killed

Tab. 2.2a (continued)

Peak overpressure/bar	Max. wind speed/km h^{-1}	Effects on structures	Effects on the human body
1.38	808	Heavy built concrete buildings are severely damaged or demolished	Fatalities approach 100%

Source: ARESM Explosive Weapons in Populated Areas, 2016:. https://www.icrc.org/en/document/explosive-weapons-populated-areas-use-effects.

Tab. 2.2b: Approximate damage to structures based on bomb mass.

Bomb mass class/kg	Damage to brick buildings (radius in m, from point of impact)			Fragmentation range/m
	Destroyed	Damaged beyond repair	Damaged/inhabitable	
23	6	15	60	890
113	20	40	180	1100
227	30	60	290	1250
454	40	90	430	1400
907	90	180	880	1550

Source: ARESM Explosive Weapons in Populated Areas, 2016:https://www.icrc.org/en/document/explosive-weapons-populated-areas-use-effects.

A collateral damage estimate (CDE) is part of the targeting process for military forces. It aims to predict the incidental (or collateral) civilian casualties and damage to civilian objects, such as buildings and infrastructure. The CDE methodology must answer the following questions:

1. Can the target be positively identified?
2. Are there protected objects, non-combatants, or environmental concerns within the effects range of the weapon?
3. Can collateral concerns be mitigated by employing a different weapon or method of engagement?
4. If not, how many non-combatants are estimated to be killed or injured in the attack?
5. Are the collateral effects excessive in relation to the expected military gain?

In addition to formulations of secondary explosives (see Tab. 1.2), metallized mixtures are sometimes used as well. Metals such as beryllium, magnesium or aluminum which are air resistant, but at the same time easily oxidized in very exothermic reactions are suitable. In practice, aluminum is used almost exclusively. Since most formulations possess a negative oxygen balance, the aluminum does not contribute to raising the heat of detonation in atmospheric explosions a lot, but it combusts afterwards (post-detonation combustion, often a large fireball) using the surrounding air which increases the visibility (psychological effect) considerably. This is different for underwater explosions, e.g. torpedoes (see Tab. 1.2, TORPEX), because here the aluminum obtains the oxygen from the (liquid) water, which contains oxygen in a high density. Metallized formulations are also suitable for special missions (e.g. caves, tunnels e.g. in GWT), since in enclosed spaces the oxygen in the air needed to breathe for survival vanishes (see Ch. 13). Figure 2.8a shows for comparison the flame propagation in the first 100 ms after a TNT and an AE (aluminized explosive = HMX/Al) explosion.

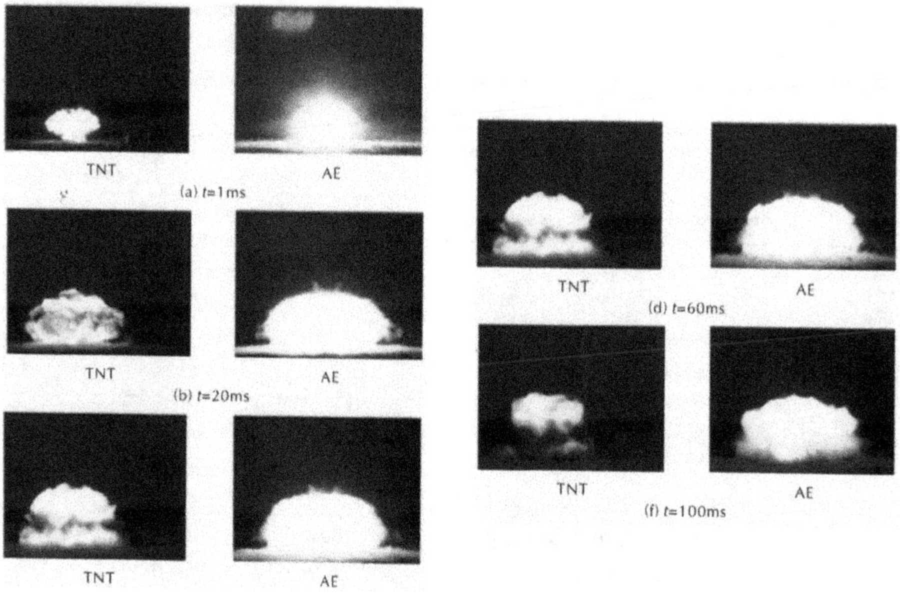

Fig. 2.8a: Flame propagation after explosion of TNT and AE (HMX/Al).

In attempts to strive for better performance, safety aspects (lower sensitivities) cannot be ignored. For example, by using the formulations Composition B and octol (see Tab. 1.2) instead of TNT, the performance can be increased significantly (Fig. 2.1). However, the sensitivity increases as well which causes safety to decrease accordingly. The goal of current research is to develop considerably less sensitive secondary explosives, which offer maximum performance and a high safety standard (Fig. 2.9).

Particularly important in the area of explosives for torpedoes, missiles, war heads and bombs which are transported by plane or submarine is, that if a fuel fire should occur, the charge is not thermally initiated. Insensitive, but high performing explosives also play a large role for shaped charges with several charges (front- and main charges) e.g. for operations against tanks, which are protected by Explosive Reactive Armour (ERA). Last but not least, insensitive also means that no accident can occur as a result of friendly fire or through enemy attack.

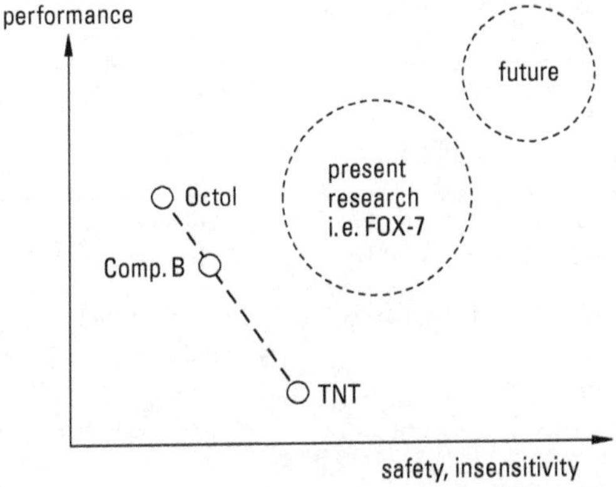

Fig. 2.9: Safety (insensitivity) and performance.

There still is a need for intensive research (Fig. 2.1) in the area of the latter aspect of new secondary explosives (and not only for those!) in terms of low toxicity and environmental impact. Conventional energetic materials (TNT, RDX), which are currently used by NATO armies for training purposes on their training grounds, have an extremely negative ecological impact if they end up unwanted and uncontrolled in the environment as unexploded munition (UXOs = unexploded ordnance), or through low-order as well as high-order detonations. This necessitates expensive and very time-consuming remediation and detoxification operations. The development of alternative energetic materials, which are environmentally compatible, but still fulfil the performance and insensitivity requirements of the NATO armies (see above), i.e. safe handling, is an important step in the direction of developing ecologically compatible, non-toxic and sustainable highly energetic materials. For example, RDX is toxic when ingested or inhaled in large quantities, causing seizures, nausea and vomiting. The EPA (Environmental Protection Agency) has suggested a limit of 2 $\mu g\ l^{-1}$ for RDX in drinking water. However, this value is marginally overstepped in some areas in the vicinity of troop training grounds. Limits for RDX have also been determined for the workplace by NIOSH (National Institute for Occupational Safety and Health). These are in general 3 $\mu g\ m^{-3}$ air and 1.5 $\mu g\ m^{-3}$ air for a 40 h working week.

Table 2.3 shows a summary of examples of some ecological and toxicological problem areas of highly energetic materials.

Particularly in the USA, but also in Germany, great efforts are being made by different institutions to reduce the environmental impact caused by the use of highly energetic materials. Leaders in this area are SERDP (Strategic Environmental Research and Development Program: http://www.serdp.org/) as well as ESTCP (Environmental Security Technology Certification Program: http://www.estcp.org/).

Some possible replacements for existing formulations based on DNAN are summarized in Tab. 2.3a.

Tab. 2.3: Ecological and toxicological problem areas of highly energetic materials.

Class of energetic material	Example	Problem		Possible solution
		Environment	Human	
Primary explosive	$Pb(N_3)_2$	Pb contamination of training grounds	Heavy metal toxicity of Pb	Pb-free primary explosives, e.g. Fe
Secondary explosive	RDX	RDX and degradation products are toxic for plants, microorganisms and microbes (earthworms)	Kidney toxin	New high-N, high-O compounds
	TNT	TNT and degradation products are ecologically toxic		New melt-castable high explosives, e.g. IMX-101: DNAN = dinitroanisole (binder) + NTO (filler) + NQ
	Comp. B	Contains RDX and TNT		IMX-104: DNAN (binder) + NTO and RDX (filler)
Pyrotechnics	Ba (green)	Heavy metal	Heavy metal[a]	Insoluble Ba compounds; Ba-free colorants, e.g. Cu
	ClO_4^- as Chlorine source		Decrease of thyroxin synthesis due to inhibition of iodine storage[b]	
Pyrotechnics	Pb resorcinate burn rate catalyst in minimum smoke missiles	Heavy metal	Heavy metal[a]	Cu, Bi based burn rate modifiers; reactive metal nano-powders as burn rate catalysts/ modifiers
Solid rocket propellant	AP	HCl in atmosphere, ozone layer	Decrease of thyroxin synthesis due to inhibition of iodine storage	$CaSi_2$ additives as HCl scrubbers; new HEDOs, e.g. ADN
Monopropellant	N_2H_4		Carcinogenic	new propellants, e.g. DMAZ

Tab. 2.3 (continued)

Class of energetic material	Example	Problem		Possible solution
		Environment	Human	
Bipropellant	MMH		Carcinogenic	New bipropellants, e.g. DMAZ
	HNO_3	NO_x	Toxic	e.g. 100% H_2O_2

[a]causes hypertension, muscular weakness, cardiotoxic effects, gastrointestinal effects
[b]EPA has recommended a max. level of 24.5 ppb ClO_4^- in drinking water

Tab. 2.3a: Possible replacements for existing formulations based on DNAN.

Formulation	Ingredients	Replaces	Applications
IMX-101	DNAN 43% NTO 20% NQ 37%	TNT	Artillery and other large caliber munitions
IMX-104	DNAN 32% NTO 53% RDX 15%	Comp B	Mortar applications
PAX-21	RDX 36% DNAN 34% AP 30%	Comp B	Mortar applications
PAX-25	RDX 20% DNAN 60% AP 20%	Comp B	Mortar applications
PAX-26	DNAN, Al, AP, MNA	Tritonal	General purpose bomb
PAX-28 DNAN 40% Al 20%	RDX 20% DNAN 40% Al 20% AP 20%	Comp B	Warheads

Tab. 2.3b: Comparison of the detonation velocities and critical diameters of possible replacements for TNT and Comp B.

Formulation	VoD / km s^{-1}	Critical diameter / mm
Composition B	7.98	4.3
HMX-104	7.4	22.2
TNT (cast clear)		3.5
TATB	7.79	6.4

Tab. 2.3b (continued)

Formulation	VoD / km s^{-1}	Critical diameter / mm
IMX-101	6.9	66.0
MCX-6100	7.2	19.8
Baratol	4.96	43.2
Comp-B3		3.81
$Pb(N_3)_2$ (LA)		0.5
RDX/wax (95/5)		4.5

In this context it should also be mentioned that 3,4-dinitropyrazole (DNP) is an interesting melt-cast explosive (m.p. 87 °C, $T_{dec.}$ = 276 °C) which can easily be prepared from nitropyrazole (Fig. 2.9a). Due to the relatively high density of 1.79 g cm^{-3}, DNP has good detonation parameters (VoD = 8,115 m s^{-1}, $p_{C\text{-}J}$ = 294 kbar) and may therefore be a potential replacement for Comp. B (RDX/TNT, VoD = 7,969 m s^{-1}, $p_{C\text{-}J}$ = 292 kbar). Therefore, DNP has great potential as a new melt-pour base. DNP (3,4-dinitropyrazole) is now technically mature, and has been consistently synthesized at BAE SYSTEMS on the 100 g scale with 65% crude yield and 99.9% HPLC purity but with ~ 5% remaining acids.

Another very promising melt-pour explosive is propyl nitroguanidine (PrNQ). The compound melts at 99 °C and has TNT-like performance. The synthesis is easy and straight-forward from commercially available NQ (Fig. 2.9b).

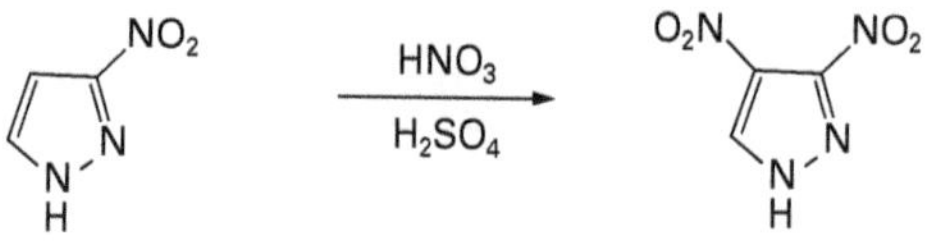

Fig. 2.9a: Synthesis of 3,4-Dinitropyrazole (DNP).

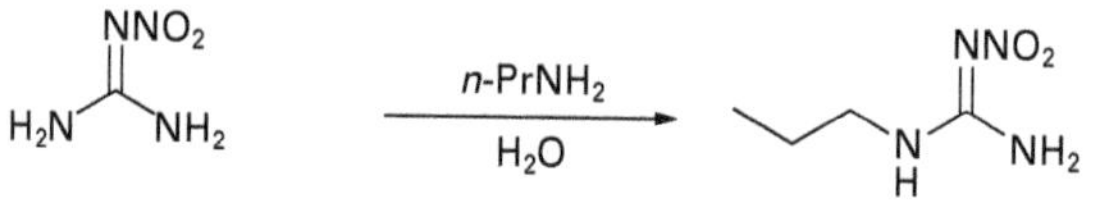

Fig. 2.9b: Synthesis of propyl nitroguanidine (PrNQ).

While traditionally general purpose bombs (GP bombs) are filled with sensitive tritonal (80% TNT, 20% Al powder) or H6 (44% RDX & NC, 29.5% TNT, 21% Al powder, 5% paraffin wax, 0.5% $CaCl_2$) and do not pass all IM requirements, PrNQ may emerge as a novel ingredient (melt-cast) for IM formulations for use in GP bombs.

Two of the most promising replacements for TNT and Comp B are IMX-101 (TNT replacement) and IMX-104 (Comp B replacement) (Tab. 2.3b). IMX stands for "Insensitive Munitions eXplosives". **IMX-101** is a high-performance insensitive high explosive composite mixture developed by BAE systems (Holston) and the US Army to replace TNT in artillery shells. A typical 155 mm artillery ammunition is shown in Fig. 2.9c. **IMX-104** is an insensitive melt-pour explosive to replace Composition B for mortar applications and was also developed by the Holston Army Ammunition Plant (BAE). The main ingredients are:

IMX-101: 43.5% 2,4-dinitroanisole (DNAN), 19.7% 3-nitro-1,2,4-triazol-5-one (NTO) and 36.8% nitroguanidine (NQ)
IMX-104: 32% DNAN, 15% RDX, 53% NTO.

Other interesting new formulations include MCX-6100 (NTO 51%, DNAN 32%, RDX 17%; melt-cast for IM, Chemring Nobel) and PBXN-9 (HMX 92%, HyTemp 2%, DOA 6%; shaped charges, Chemring Nobel). For shaped charge applications, formulations with HMX (82%) and GAP or PolyNimmo (18%) can also be used.

2.2.1 Tertiary explosives

Tertiary explosives (also called blasting agents), are so insensitive to shock that they cannot be reliably detonated by practical quantities of primary explosives, and instead require an intermediate explosive booster of a secondary explosive. These are often used to increase safety and also because of the typically lower material costs and easier handling. The largest consumers of tertiary explosives are large-scale mining and construction operations. Most tertiaries include a fuel and an oxidizer. ANFO can be a tertiary explosive if its reaction rate is slow.

2.3 Propellant charges

The oldest known propellant charge is blackpowder (or gunpowder) which is a mixture of 75% KNO_3, 10% sulfur and 15% charcoal dust. It is an easy to ignite powder with a burn rate of 600–800 m s^{-1}, which is still used today as a charge in military and civil pyrotechnic munitions. NC powder was discovered in 1846 by Schönbein and burns almost without any residue. Its burn rate at atmospheric pressure is 0.06–0.1 m s^{-1}, which is considerably slower than blackpowder, as long as no pressure is applied to the propellant charge on burning. In addition to the linear burn rate r (in m s^{-1}), the mass-weighted burning rate m (in g s^{-1}) is of interest as well (A: surface, in m^2; ρ: density, in g m^{-3}):

$$m = r\,A\,\rho$$

Single-base nitrocellulose powder is the oldest of all types of NC powders and is often referred to as “smokeless” powder. It is prepared by reacting cellulose with nitric acid.

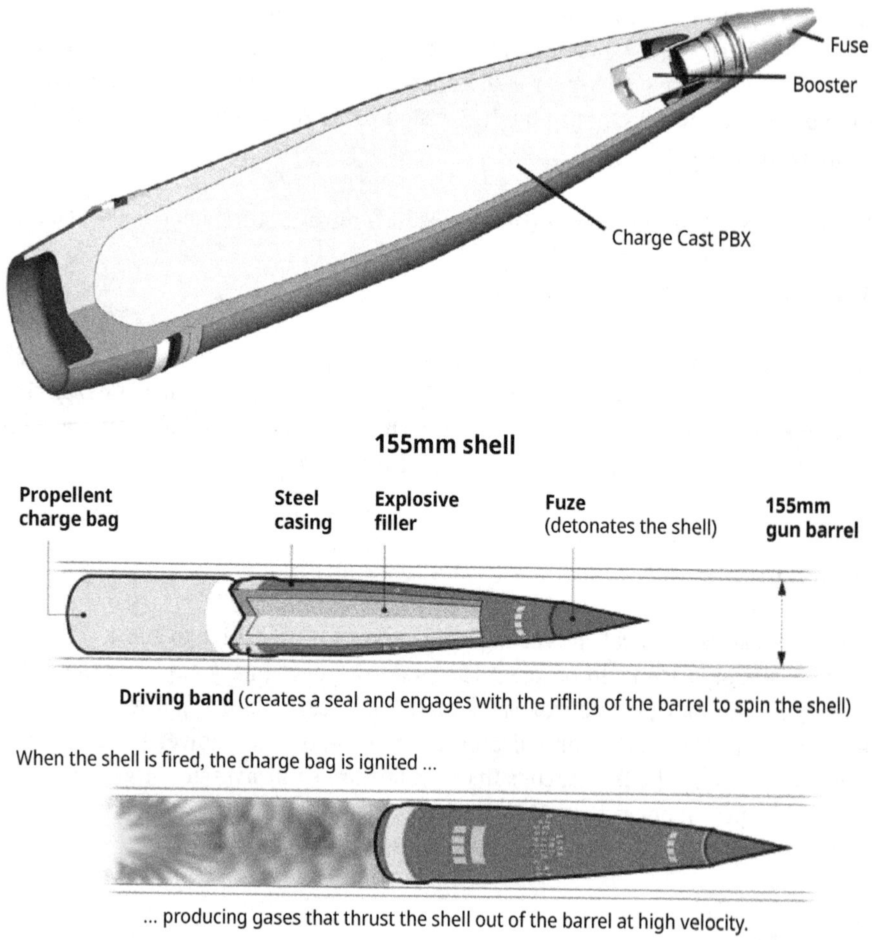

Fig. 2.9c: Illustration of important aspects of the 155 mm shell.
(source: https://www.reuters.com/investigates/special-report/ukraine-crisis-artillery/).

Depending on the acid concentration used, NC is formed with different degrees of nitration (e.g. 11.5–14.0%) (Fig. 2.10). In addition to the nitrocellulose base, double- and triple-base propellants contain further substances such as nitroglycerine and nitroguanidine.

The dependence of the enthalpy of formation of nitrocellulose on the percentage nitrogen content of NC (degree of nitration) is shown in Fig. 2.10 [1 g].

It is also interesting to mention that the conversion of cellulose into nitrocellulose can nicely be seen in the corresponding IR spectra due to the occurrence of the asymmetric NO_2 stretching mode (ca. 1550 cm^{-1}) and the symmetric stretch at about 1300 cm^{-1} (Fig. 2.10a)

While single-base propellant charges (NC) are used in weapons from pistols to artillery weapons, the higher performance double-base propellant charges (NC + NG) are

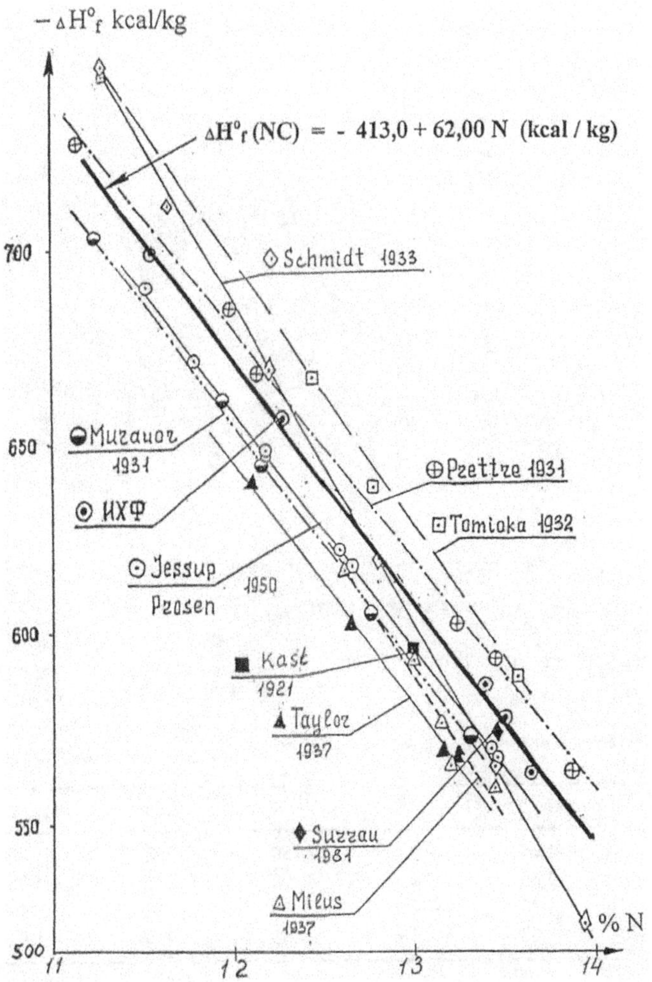

Fig. 2.10: The dependence of the enthalpy of formation of NC on the percentage nitrogen content of the NC.

used predominantly in pistols and mortars. The disadvantage of the double-base powder is the strong erosion of the gun barrel (see below), resulting from the significantly higher combustion temperatures as well as the appearance of a muzzle flash because of the explosion of some of the combustion gases upon contact with air. To prevent erosion and the muzzle flash, a triple-base powder (NC + NG + NQ) with an NQ content of up to 50% is used, particularly in large caliber tank and NAVY weapons. However, the performance of triple-base powders is lower than that of double-base powders (Tab. 2.4). In a triple-base powder, particularly in large tank and NAVY cannons, NQ is replaced by RDX in order to increase the performance. However, the barrels suffer increased erosion problems again due to the significantly higher combustion temperatures.

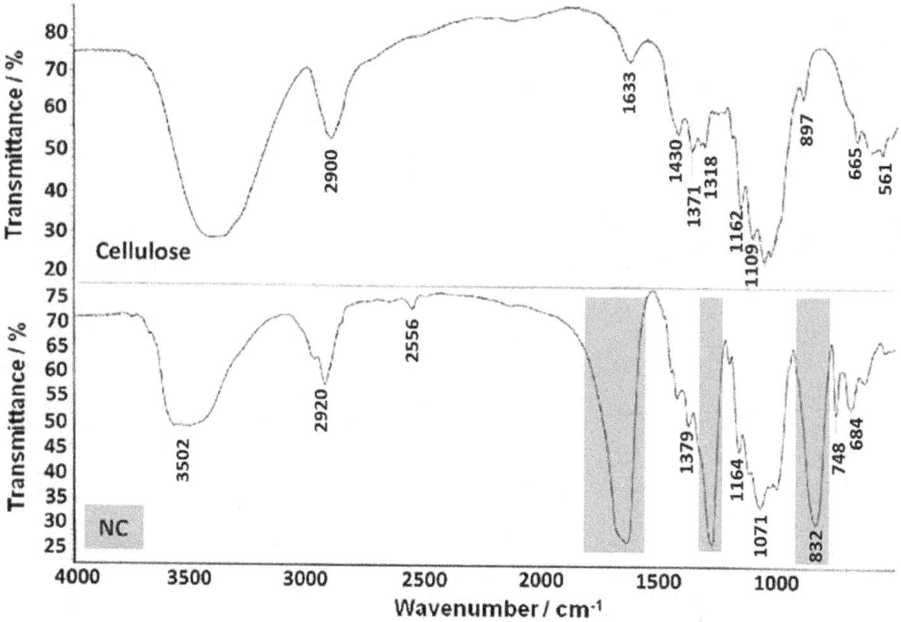

Fig. 2.10a: IR spectra of cellulose (top) and nitrocellulose (bottom).

Tab. 2.4: Parameters for single, double and triple-base propellant charges.

Propellant charge	ρ/g cm^{-3}	Ω / %	T_c / K	I_{sp} / s	N_2 / CO
NC[a]	1.66	−30.2	2,750	232	0.3
NC[a] / NG (50 : 50)	1.63	−13.3	3,308	249	0.7
NC[a] / NG / NQ (25 : 25 : 50)	1.70	−22.0	2,683	236	1.4
Hy-At / ADN (50 : 50)	1.68	−24.7	2,653	254	6.3

[a]NC-13.3 (N content 13.3%)

Double-base propellants are typically composed of a mixture of nitrocellulose (NC) and nitroglycerin (NG). To achieve the desired thermodynamic, ballistic, mechanical, and aging properties, additives such as plasticizers (e.g. adipates, phthalates), burn rate modifiers (e.g. copper/lead salts, polyacetylenes), stabilizers (e.g. nitrodiphenylamine, centralite, acardite), and processing aids (e.g. wax, graphite) are incorporated. These are summarized in Table 2.4a for a homogeneous propellant in which both the oxidizer and fuel are present in one single compound such as in nitroglycerin.

Tab. 2.4a: Characteristic composition of a double-base propellant. Homogeneous propellant: oxidizer and fuel are present in a single uniform compound.

Basic substance (actual propellant)	Additives
Nitrocellulose (NC) (solid)	Stabilizer (NO_2 binding)
Nitroglycerin (NG) (liquid)	Plasticizer
	Burn rate modifiers
	Processing aids

Depending on the processing and manufacturing method, double-base propellants can be classified as being either pressed or cast, as illustrated in the schematic Fig. 2.10b. The manufacturing process for pressed propellant charges is shown in Fig. 2.10b. After kneading and mixing the raw materials (a water or alcohol moistened mixture of nitrocellulose and nitroglycerin) with the additives, the product undergoes extensive drying and gelatinization through friction and simultaneous rolling processes. This is followed by the formation of pressed rolls, which are reshaped into propellant strands using a press. After mechanical finishing, insulation is applied.

For cast propellants of this class, a fine-grained nitrocellulose powder to which the other additives have already been added (base grain, casting powder) is used. This powder is placed into a mold, and the voids are filled with the casting liquid, which is a mixture of nitroglycerin and plasticizer liquid. During the subsequent gelatinization process, the individual grains bond together, resulting in a macroscopically homogeneous body.

Table 2.4b summarizes the standard formulations for typical pressed double-base propellants.

Tab. 2.4b: Composition and performance values of typical pressed double-base propellants.

Propellant A (composition, %)	Propellant B (composition, %)
NC (51.2)	NC (49.8)
NG (43.0)	NG (34.7)
Diethylphthalate (3.2)	Diethylphthalate (10.6)
Potassium sulfate (1.3)	Nitrodiphenylamine (2.0)
Ethylcentralite (1.0)	Lead silicate (1.5)
Soot (0.2)	Lead ethylhexoate (1.4)
Wax (0.1)	

As the nitroglycerin content increases, it can be observed that the performance of the propellant improves. However, at very high nitroglycerin levels, the mechanical properties change, and this means that the practical nitroglycerin content is limited to about 50%. To further enhance performance, additional substances, such as aluminum powder or explosives, may sometimes be added.

To improve the mechanical properties of double-base propellants, a binder can be added, resulting in modified double-base propellants (known in English as Elastomer Modified Double-Base propellants (EMDBs) or Cross-Linked Double-Base propellants (CLDBs)). Unfortunately, the terminology used in this area is not entirely consistent.

A typical plant for the production of double-base propellants is shown in Fig. 2.10b.

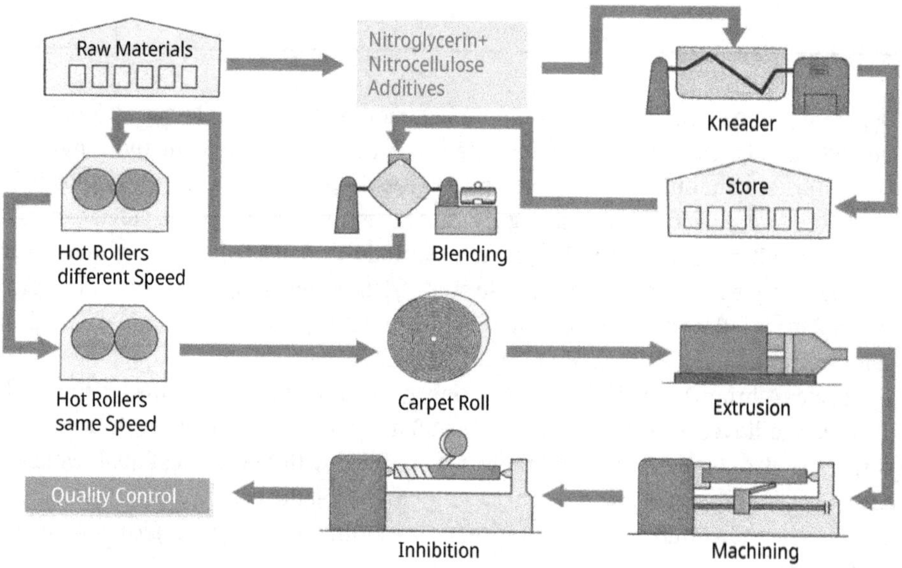

Fig. 2.10b: Double-base propellant plant (with friendly permission of Dr. Paul Wanninger).

The velocity v of a projectile on ejection from the gun barrel can roughly be estimated using the following equation:

$$v = \left(\frac{2mQ\eta}{M}\right)^{0.5}$$

$$Q = \Delta U_f^o \text{ (prop. charge)} - \Delta U_f^o \text{(combustion products)}$$

Where m is the mass of the propellant charge (in g), M is the mass of the projectile (in g), Q is the heat of combustion (in J g^{-1}) and η is a constant, which is specific for the system. Here, the energy released in Q is once again of crucial importance.

Figure 2.10c shows a schematic representation of gun propulsion.

The erosion problems in guns are generally caused by the formation of iron carbide (Fe from the gun barrel, C from CO) at high temperatures. Modern research on propellant charges is therefore focused on the development of powders which burn at lowest possible temperatures, but show good performance (see Ch. 4.2.3). Moreover, the N_2/CO ratio, which lies at approximately 0.5 for conventional propellant charges,

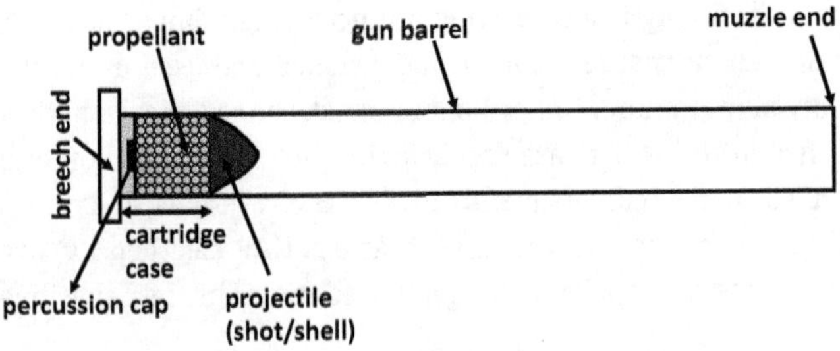

Fig. 2.10c: Gun propulsion.

should be increased as much as possible. The formation of N_2 instead of CO should also strongly reduce the amount of erosion, since iron nitride has a higher melting point than iron carbide (Fe_3C) and furthermore, it forms a solid protective layer on the inside of the gun barrel. New studies have shown that the introduction of propellant charges which are based on nitrogen-rich compounds such as e.g. triaminoguanidinium azotetrazolate (TAGzT) result in a considerably better N_2/CO ratio and the life-span of high caliber NAVY cannons can be increased by up-to a factor of 4 [11]. The synthesis of azotetrazolate salts (e.g. Na_2ZT) occurs in a facile route from aminotetrazole by oxidation using $KMnO_4$ in alkaline conditions (NaOH). *Note:* the reduction of Na_2ZT in acidic conditions (HOAc) using Mg forms the neutral compound bistetrazolylhydrazine (BTH).

Also hydrazinium 5-aminotetrazolate (Hy-At) appears to show good potential in this context (Tab. 2.4) [12] (Fig. 2.10d).

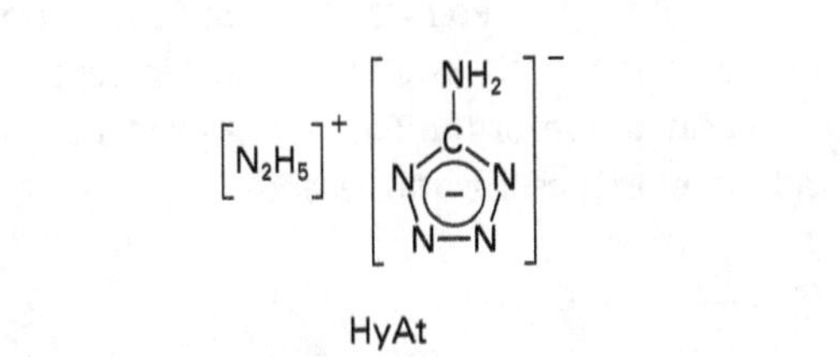

Fig. 2.10d: Molecular structures of triaminoguanidinium azotetrazolate (TAGzT) and hydrazinium 5-aminotetrazolate (Hy-At).

For propellant charges the insensitivity is also playing an increasingly important role. Since approximately 1970 propellant charges have been developed and used under the title **LOVA** (low-vulnerability ammunition). Under bullet impact, shaped charge impact or fire they respond with fire in the worst case scenario, but not in deflagration and definitely not in detonation. As energetic fillers mainly RDX or HMX were used. As energetic binders, energy and gas generating polymers such as e.g. GAP (see Fig. 1.4) are particularly suitable. Such composite propellant charges are considerably less sensitive than powders based on NC.

2.4 Rocket propellants

Rocket propellants are similar to the propellant charge powders discussed above, because they combust in a controlled manner and do not detonate. However, propellant charge powders burn considerably quicker than rocket propellants, which results in a significantly higher pressure for gun propellants in comparison with rocket propellants. Typical pressures in the combustion chambers of rockets are 70 bar, which can be compared with up to 4,000 bar for large artillery and NAVY cannons.

As is the case for propellant charges, the specific impulse for rocket propellants is also one of the most important performance parameters.

The **specific impulse** (I_{sp}) is the change in the impulse (impulse = mass × velocity or force × time) per mass unit of the propellant. It is an important performance parameter of rocket motors and shows the effective velocity of the combustion gases when leaving the nozzle, and is therefore a measure for the effectiveness of a propellant composition.

$$I_{sp} = \frac{\bar{F} \cdot t_b}{m} = \frac{1}{m} \cdot \int_0^{t_b} F(t) dt$$

The force F is the time dependent **thrust** F (t), or the average thrust $\bar{F}$; t_b is the combustion time (in s) and m is the mass of propellant (in kg). Therefore, the specific impulse has the units N s kg^{-1} or m s^{-1}. Sometimes, predominantly in English speaking areas, the specific impulse is given based on the gravitation of the Earth g (g = 9.81 m s^{-2}) which is on the mass of the propellant, and therefore has the unit seconds (s):

$$I^{*}_{sp} = \frac{I_{sp}}{g}$$

The specific impulse can be formulated exactly as follows:

$$I_{sp} = \sqrt{\frac{2\gamma R T_c}{(\gamma - 1)M}}$$

Whereby γ is the ratio of the specific heat capacities of the gas mixture, R is the gas constant, T_c is the temperature (K) in the combustion chamber and M is the average molecular weight (kg mol^{-1}) of the formed combustion gases:

$$\gamma = \frac{C_p}{C_V}$$

$$I^*_{sp} = \frac{1}{g}\sqrt{\frac{2\gamma R T_c}{(\gamma - 1)M}}$$

The average thrust of a rocket $\bar{F}$ can in accordance with the equation above be given simply as:

$$\bar{F} = I_{sp}\frac{\Delta m}{\Delta t}$$

Where I_{sp} is the specific impulse in (m s^{-1}), Δm is the mass of used propellant (in kg) and Δt is the duration of burning of the engine (in s). Therefore, the thrust has the unit kg ms^{-2} or N and corresponds to a force.

In the following discussion we want to predominantly use the English term I^*_{sp}, which is smaller than the I_{sp} by a factor of approximately 10 and has the units s. Typical values for the specific impulse of solid boosters are 250 s, whereas for bipropellants they are found at approx. 450 s. For chemistry it is important that the specific impulse is proportional to the square root of the ratio of the combustion chamber's temperature T_c and the average molecular mass of the combustion product M:

$$I_{sp} \propto \sqrt{\frac{T_c}{M}}$$

It is important in the discussion of the specific impulse I^*_{sp} below, that as a rule of thumb, an increase of the specific impulse by 20 s causes the maximum possible carried payload (warheads, satellites) to be approximately doubled.

A more detailed description to derive the thrust or the specific impulse of a rocket is given below: In order to propel a rocket, a rocket engine ejects combustion gases with low molecular mass but high velocity z through the nozzle. The rocket has the mass M and initially moves with velocity u. If the rocket ejects combustion gases of mass of Δm within the time Δt and with the velocity z, a decrease in mass $M - \Delta m$ and an increase in the velocity to $u + \Delta u$ results. Due to the conservation of the impulse, the following can be derived:

$$\begin{aligned} Mu &= (M - \Delta m)(u + \Delta u) + \Delta m z \\ \cancel{Mu} &= \cancel{Mu} + M\Delta u - \Delta m u - \Delta m \Delta u + \Delta m z \\ M\Delta u &= \Delta m(\underbrace{u + \Delta u - z}_{v_e}) \end{aligned}$$

$$M\frac{\Delta u}{\Delta t} = \frac{\Delta m}{\Delta t}v_e$$

$$M\frac{du}{dt} = \frac{dm}{dt}v_e$$

Therefore $v_e = u + \Delta u - z$ and is the velocity of the ejected combustion gases relative to the rocket. Here the force $(dm/dt)v_e$ corresponds to the **thrust** $F_{impulse}$, which originates from the impulse of the system:

$$F_{impulse} = \frac{dm}{dt}v_e$$

However, this is only equal to the total thrust F of the system when the pressure p_e at the end of the nozzle is the same as the pressure of its surroundings p_a. In general, a correction term is needed which is called the pressure term $F_{pressure}$.

Figure 2.11 schematically shows the pressure profile which is one factor that influences rocket performance. The length of the arrows shows the contribution of the pressure from inside and outside the walls. While the atmospheric pressure outside is constant, the inside pressure of the combustion chamber is at its largest and decreases in the direction of the nozzle end. The pressure term is proportional to the diameter A_e:

$$F = F_{impulse} + F_{pressure} = \frac{dm}{dt}v_e + (p_e - p_a)A_e$$

If the pressure at the end of the nozzle is smaller than the pressure of the surroundings (which occurs in the case of a so-called overexpanding nozzle), the pressure term then has a negative value and reduces the total thrust. Therefore, a pressure p_e which is the same or higher (underexpanding nozzle) than the air pressure is desirable. Figure 2.11a shows schematically the difference in size between a sea-level and vacuum-optimized nozzle.

Since the air pressure decreases with increasing flight altitude, at constant nozzle diameter, the total thrust increases with increasing flight altitude. This increase can correspond to approximately 10 to 30% of the total thrust depending on the rocket. The maximum thrust is reached in vacuo. The so-called effective ejection velocity c_{eff} (of the combustion gases) is defined as the ratio between the thrust and the mass flux (dm/dt):

$$\frac{F}{\frac{dm}{dt}} = c_{eff} = v_e + \frac{(p_e - p_a)A_e}{\frac{dm}{dt}}$$

We can write:

$$F = \frac{dm}{dt}c_{eff}$$

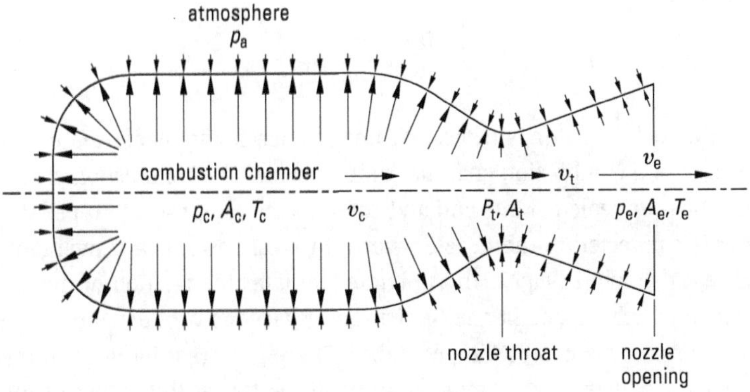

Fig. 2.11: Combustion chamber and nozzle.

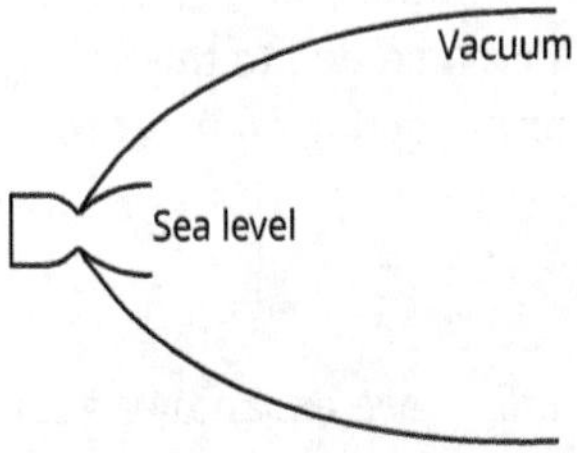

Fig. 2.11a: A sea-level and vacuum-optimized nozzle.

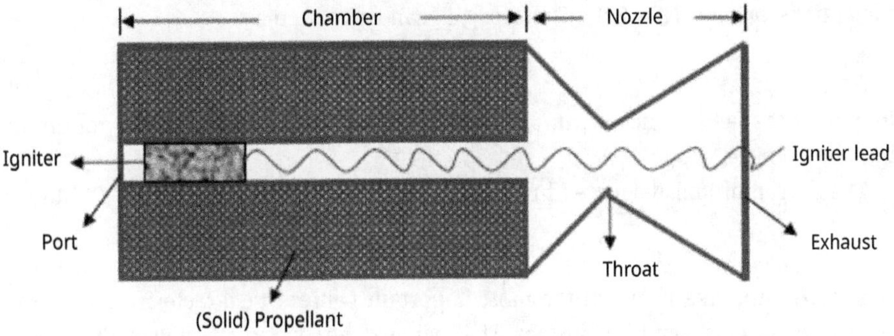

Fig. 2.11b: Schematic design of a solid rocket motor.

The effective ejection velocity is an average. In reality the velocity distribution is not constant throughout the whole nozzle diameter. Assuming a constant ejection velocity allows a one-dimensional description of the problem.

The so-called total impulse I_t is the integral of the total thrust integrated over the total combustion time t:

$$I_t = \int_0^t F dt$$

The actual design of a rocket motor is of crucial importance for its performance. A motor basically consists of a combustion chamber and a nozzle. The combustion chamber is usually a metallic tube sealed at one end and contains the solid rocket propellant. The rocket chamber is connected to the nozzle, which, in most cases, is a convergent-divergent (CD) nozzle, as shown in Fig. 2.11b. An igniter initiates the ignition of the entire propellant surface which results in the formation of gaseous products and high temperature and pressure in the combustion chamber. There is a great increase in the velocity of the gaseous products when they expand from the throat to the exit of the nozzle. As a consequence of Newton's third law of motion, the exiting gases propel the rocket into the opposite direction.

The total thrust with which a rocket is propelled has two components (see above). The first component of the thrust (F) is due to the thrust F_1 created due to the imbalance of the chamber pressure p_c and the exhaust gas pressure p_e acting on the throat, the area of which is A_t (Fig. 2.11). Therefore, F_1 can be written as:

$$F_1 = (p_c - p_e)A_t$$

Since the gases under high pressure in the combustion chamber are expanding after passing through the throat, p_c is always greater than p_e, therefore F_1 is always positive.

The second component of the thrust (F) is due to the thrust F_2 created due to the imbalance of the exhaust gas pressure (p_e) and the pressure p_a outside of the rocket acting on the exhaust (Fig. 2.11). Therefore, F_2 can be written as:

$$F_2 = (p_e - p_a)A_e$$

(Note: underexpanding nozzle, if: $p_e > p_a$; overexpanding nozzle, if: $p_e < p_a$; optimum expanding nozzle, if: $p_e = p_a$.)

The overall propulsive force (thrust F) a rocket experiences is the sum of F_1 and F_2:

$$F = F_1 + F_2 = (p_c - p_e)A_t + (p_e - p_a)A_e$$

The **specific impulse** is one of the most important values to characterize the performance of a rocket propulsion system. The higher this value is, the higher the performance will be. The specific impulse is defined as the total impulse per mass unit:

$$I_s^* = \frac{\int_0^t F dt}{g_0 \int_0^t \frac{dm}{dt} dt}$$

where $g_0 = 9.81$ m s^{-2} at sea level.

For constant thrust and mass flux, the equation above can be simplified to

$$I_{sp}^{*} = \frac{F}{\frac{dm}{dt} g_0}$$

Using the SI system, the specific impulse can be given in the units s^{-1}. However, the value g_0 is not always the same. Therefore, the specific impulse I_{sp} is often also given by the value

$$I_{sp} = \frac{F}{\frac{dm}{dt}}$$

This has the advantage that it is independent of the effect of gravity. It is given in the unit $N\,s\,kg^{-1}$ or $m\,s^{-1}$. At sea level, both expressions are linked by a factor of approximately 10.

By inserting

$$F = \frac{dm}{dt} c_{eff}$$

in

$$I_{sp}^{*} = \frac{F}{\frac{dm}{dt} g_0}$$

the mass flux is shortened, and one obtains

$$I_{sp}^{*} = \frac{c_{eff}}{g_0}$$

The effective ejection velocity c_{eff} is different from the specific impulse I_{sp} only due to g_0 (see above).

The so-called **rocket equation** describes the fundamental equations of rocket propulsion. If we consider the simplest case, in which a monostage rocket accelerates in a gravity-free vacuum (i.e. a slow-down due to gravitation and friction is not taken into consideration), if the rocket has a velocity of zero at the start and ejects the propellant with a constant ejection velocity v_e, the velocity u of the rocket after time t corresponds to:

$$u(t) = v_e \ln\left(\frac{m(0)}{m(t)}\right)$$

Where $m(0)$ is the mass of the rocket at the start and $m(t)$ is the mass of the rocket at time t.

For rocket launches from earth, the formula must be changed to include the gravitation of earth and corresponds at low altitudes with a constant earth gravitation g of $9.81\ m\,s^{-2}$:

$$u(t) = v_e \ln\left(\frac{m(0)}{m(t)}\right) - gt$$

Since g is dependent on the altitude, the following correction is needed:

$$u(t) = v_e \ln\left(\frac{m(0)}{m(t)}\right) - \int_0^t g(t')\mathrm{d}t'$$

In addition to the gravitation of earth, a rocket launched from earth must also overcome the air resistance of the atmosphere, which means that the rocket equation for such cases is only an approximation. Planes, RAM and SCRAM jets which are propelled by jet engines, transport their fuel with them, but they also suck air in and use the oxygen from the air for the combustion of the fuel. They only carry the fuel but not the oxidizer with them. The rocket equation is not valid for such vehicles, which are referred to as air-breathing engines.

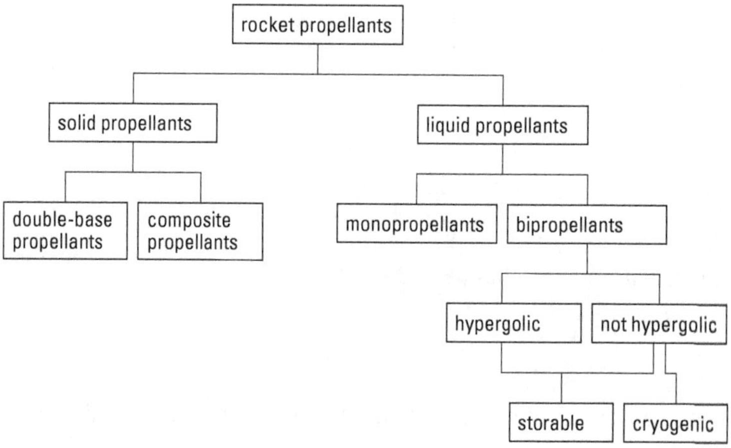

Fig. 2.12: Classification of rocket propellants.

Propellant			
Small arms propellant	**Mortar propellant**	**Gun propellant**	**Rocket propellant**
– Single-base – Double-base	– Single-base – Double-base	– Single-base – Double-base – Triple-base – Nitramine-base – Liquid	– Double-base – Composite – Composite modified double–base (CMDB) – Liquid

Generally, propellants can be classified based on either where they are going to be used as rocket or gun propellants or according to their chemical composition. A possible classification of propellants based on their use is shown in the diagram above.

Generally, one can categorize rocket propellants into those for solid propellants and those for liquid propellants (Fig. 2.12). The **solid propellants** can be categorized further into double-base (homogeneous) and composite propellants (heterogeneous). The homogeneous propellants are generally based on NC, whereas the heterogeneous propellants are generally based on AP. The double-base propellants are homogeneous NC/NG formulations. The composite propellants are mixtures of a crystalline oxidizer (e.g. ammonium perchlorate, AP) and a polymeric binder, which has been cured with isocyanate (e.g. hydroxy-terminated polybutadiene, HTPB/diisocyanate) and which contains the fuel (e.g. Al) (Fig. 2.13). The purpose of the polymer matrix (the binder) is to form a solid, elastic body of the propellant ingredients with sufficient mechanical properties. The binder is also used as a fuel because it contains mainly hydrogen and carbon. Two examples of double-base and composite propellants are given in Tab. 2.5.

Tab. 2.5: Solid propellants.

Type	Composition	I^*_{sp} / s	T_c / K
Double-base	NC (12.6% N)[a] NG Plasticizer Other additives	200 2,500	2,500
Composite	AP Al HTPB Additives	259	4,273

[a]A typical double-base composition for minimum smoke missiles would be the M36 propellant with the following composition: NC (12.6% N) 49.0%, NG 40.6%, Di-*n*-propyladipate 3.0%, 2-nitrodiphenylamine 2.0%, ballistic modifier (e.g. lead resorcinate) 5.3%, wax 0.1%.

The binder is a liquid, which is loaded with all solid components and acts as a fuel itself. It is the backbone of the polymeric system and determines the structural and mechanical properties of the final propellant.

The choice of suitable binders is limited. The most commonly used binders nowadays are epoxy-cured and polyurethane systems based on polybutadiene.

Current polymeric binders like HTPB, GAP etc. are hydroxy-terminated polymers. These polymers are cured by reaction of the hydroxyl end groups with isocyanate-based curing agents to form polyurethane (PU) bonds. PU-based polymers have been widely used in polymer-bonded explosives and solid composite rocket propellants, since PUs show acceptable compatibilities with other ingredients, provide good mixing/castability properties, as well as excellent mechanical properties and thermal stability. The disadvantage of PUs lies in the use of isocyanates, since these are potentially toxic, carcinogenic and sensitizing, causing occupational asthma, irritation of the skin and eyes, as well as other problems. Isocyanates are increasingly regulated through REACH, which calls for a move towards more sustainable alternatives.

Because of the generally excellent properties of PUs, alternative curing systems have been identified that use different precursors, but which still result in PU bonds. Four alternative, non-isocyanate polyurethane (NIPU) curing systems that have been identified recently [1] are schematically shown in Fig. 2.12a.

The most general approach to synthesizing polyurethanes without using isocyanates is through the reaction of cyclic carbonates and amines (polyaddition). However, also polycondensation, rearrangement and ring opening polymerization can be used. Table 2.5a shows an overview of the NIPU methods, including remarks or concerns for each of the pathways.

Following the conclusions from a review paper [1], the cyclic carbonate/amine system has the potential to be introduced on an industrial scale, as it does not have as many obstacles as the other NIPU methods. Other advantages of cyclic carbonates are biodegradability and low toxicity compared to isocyanates. Furthermore, biobased feedstocks (such as derivatives of plant oils), could be successfully used as precursors in the synthesis of cyclic carbonates [1].

However, challenges for the application of cyclic carbonate/amine NIPU systems as a replacement for the conventional isocyanate-based curing systems exist, namely:

- Development and synthesis of cyclic carbonates that resemble the backbone of currently used hydroxyl-terminated polybutadiene (HTPB) (Fig. 2.13)
- Identification of suitable diamines that yield polyurethane bonds when reacting with cyclic carbonates
- Investigation of reaction conditions, selectivity, purity (formation of by-products) and yield in the preparation of PUs from cyclic carbonates and diamines (reactivity, reaction time, temperature, need for catalysts, . . .)
- Compatibility of NIPUs with propellant ingredients
- Mechanical and thermal properties of pure binder

A typical basic missile design is shown in Fig. 2.13a (top), with a schematic drawing of the solid rocket motor shown in Fig. 2.13a (bottom).

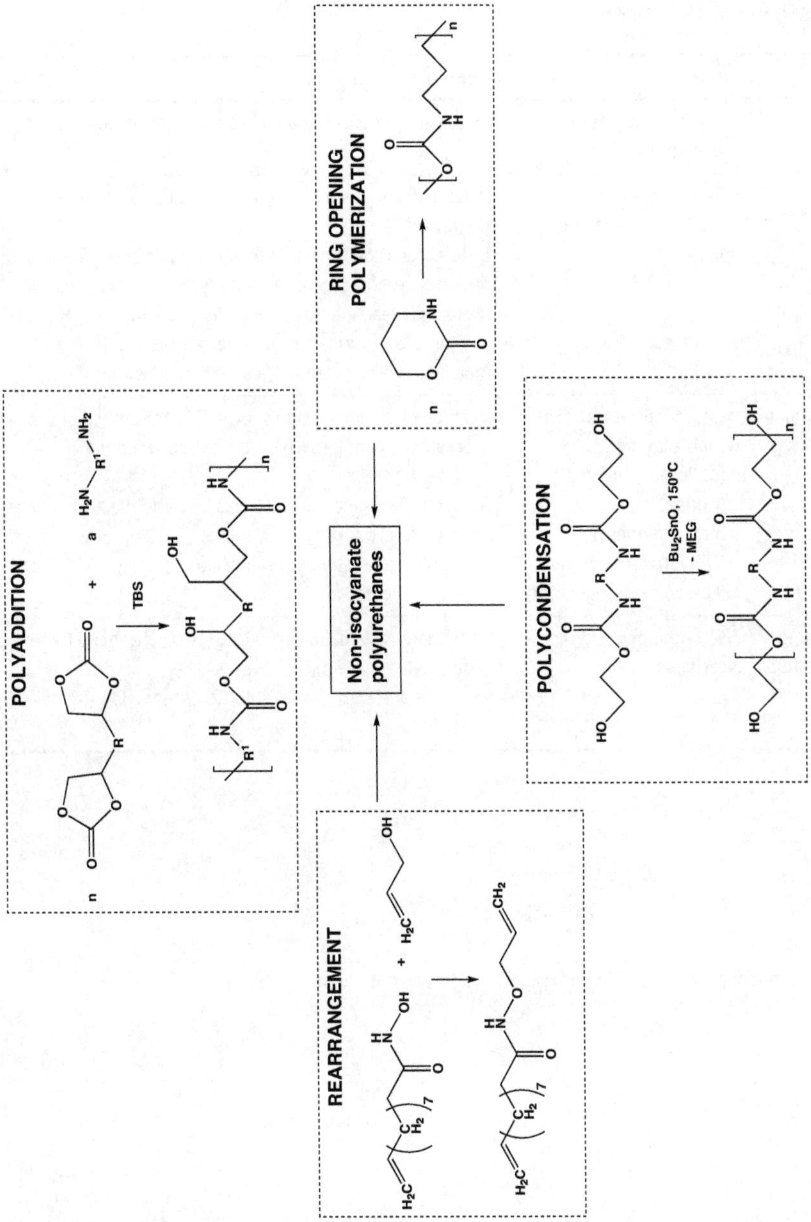

Fig. 2.12a: Possible non-isocyanate polyurethanes synthesis pathways [1] [Reproduced with friendly permission of MDPI materials and Piotr Stachak, Politechnika Krakowska].

Tab. 2.5a: Overview of alternative NIPU pathways.

Pathway	Precursors	Remarks
Polyaddition	Cyclic carbonates + diamines	Low reactivity of some amine-carbonate systems
Polycondensation	(a) Polycarbamate + diol (b) Polycarbonate + diamine	Polycarbamates are mainly produced using phosgene (highly toxic) Utilizing polycondensation on a commercial scale is limited due to several reasons: the requirement of a catalyst, prolonged reaction time, necessity of purification of final polymers, and formation via side-reactions of low molecular weight compounds that may be toxic
Rearrangement	(a) Acylazides (Curtius rearrangement) (b) Primary amides (Hofmann rearrangement) (c) Hydroxamate ester (Lossen rearrangement)	In all cases an isocyanate is formed in the course of the process, so is not less toxic than the conventional approach Halogens (bromine and chlorine) increase the amount of toxic waste produced Depending on the type of rearrangement, N_2 or CO_2 need to be removed
Ring opening polymerization	6–7 membered cyclic carbamates	Carbamates are mainly produced using phosgene (highly toxic) Synthesis preferably at 200 °C (increases yield), use of catalysts

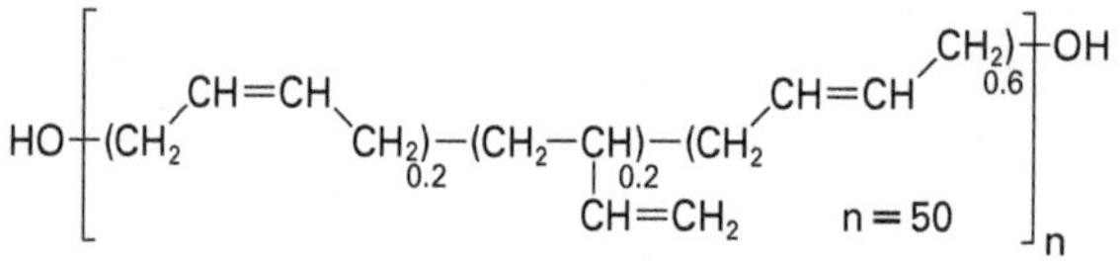

Fig. 2.13: Homopolymeric R45HT (hydroxyl terminated polybutadiene, HTPB).

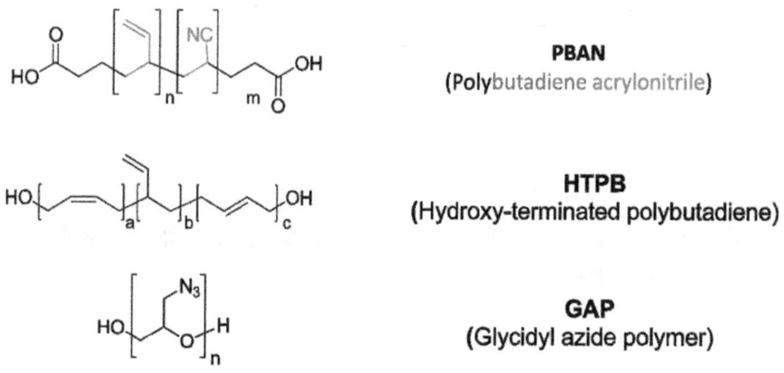

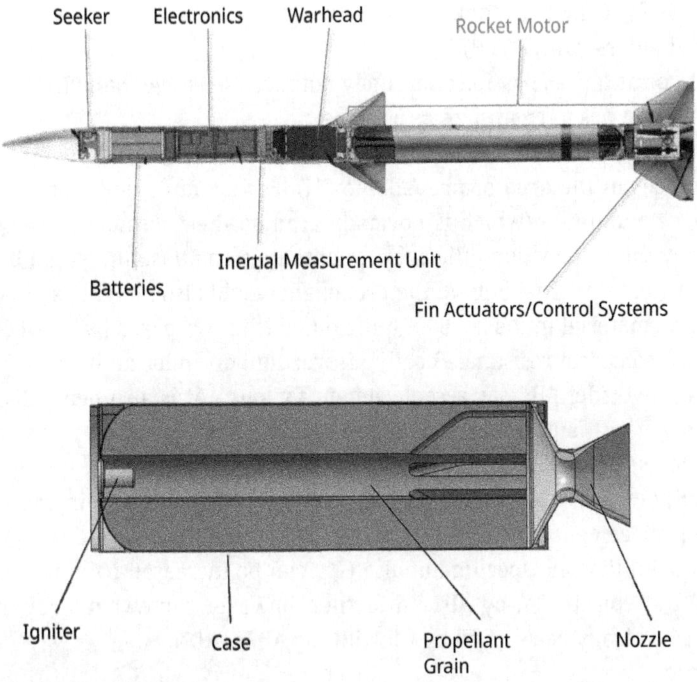

Fig. 2.13a: Basic missile design (top) with a schematic drawing of the solid rocket motor (bottom) [reproduced with kind permission of Bayern Chemie, an MBDA company].

Presently, research is on-going into trying to find alternatives to AP/Al (see also Ch. 1.2.4). The problems with the AP/Al mixtures which contain HTPB as a binder, are two-fold. On the one hand AP is toxic and should be substituted for this reason alone (see Ch. 1.2.4). On the other hand, such formulations are also problematic in slow cook-off tests (SCO test, see Ch. 6.2). It appears to be the case that here the AP slowly decomposes during the formation of acidic side-products. These acidic side-products then react with the HTPB binder, which can result in the formation of cracks and cavities in the composite, which consequently negatively affects the performance and sensitivity. Possible alternatives for AP are ADN, HNF and TAGNF. However, they cause other problems, such as, for example, the low thermal stability (ADN melts at 93 °C and already decomposes at 135 °C) and the binder compatibility is not always guaranteed either. Further research work is absolutely necessary in order to find better oxidizers for solid propellants. In this context, the following requirements must be fulfilled:

1. density has to be as high as possible, if possible over 2 g cm^{-3}
2. oxygen balance has to be good, if possible better than that of AP
3. melting point has to be above 150 °C
4. vapor pressure has to be low
5. decomposition temperature has to be above 200 °C

6. synthesis has to be facile and economic
7. compatible with binders (with HTPB)
8. sensitivity has to be as low as possible, definitely not more sensitive than PETN
9. enthalpy of formation has to be as high as possible.

There are also advances in the area of propellants (Al). For example, attempts are being made to replace aluminum (which is normally used on the micrometer scale) with nano-Al. Thereby the combustion efficiency (quantitative combustion) would be increased, but the air (oxygen) sensitivity of the propellant would also increase significantly. ALEX is a nano material in the 20–60 nm region which is prepared by an electrical explosion (quick electrical resistance heating in vacuum or under an inert gas). The use of aluminum hydride, AlH_3, as a replacement for pure Al is also being discussed currently, however it is even more sensitive to air (oxidation) when it is not stabilized through the use of wax.

Another possibility which is also currently being researched in order to increase performance is the replacement of the aluminum (Al) by aluminum hydride (AlH_3). Calculations have proven that the specific impulse (I^*_{sp}) can be increased by approximately 8% (or 20 s) by replacing Al by AlH_3. A further, however somewhat smaller improvement can theoretically be reached by substituting AP by ADN:

$$\begin{array}{lll} \text{AP / Al} & (0.70 / 0.30) & I^*_{sp} = 252\ \text{s} \\ \text{AP / AlH}_3 & (0.75 / 0.25) & I^*_{sp} = 272\ \text{s} \\ \text{ADN / AlH}_3 & (0.70 / 0.30) & I^*_{sp} = 287\ \text{s} \end{array}$$

Aluminum hydride, alane, AlH_3 is an excellent compound for propulsion applications. The low density (1.49 g/cc) of alane is the only technical weakness. However, this makes alane unsuitable for lower stages of larger launcher systems. With artillery rockets or anti-tank missiles, there would be a loss of density in comparison with Al, but this could possibly be compensated for by other properties (e.g. increased combustion rate, reduced pressure exponent, reduced flame temperature, etc.). AlH_3 is probably already used in the upper stages of Russian Intercontinental Ballistic Missiles (ICBM).

Silanes, and in particular the higher silanes such as e.g. *cyclo*-pentasilane, Si_5H_{10}, have also been discussed recently as possible propellants. These could possibly be suitable for RAM and SCRAM jets, since in addition to forming SiO_2, they could burn using nitrogen from the air to form Si_3N_4 and thereby use not only the oxygen, but also the nitrogen from the air as oxidizers (air-breathing engines).

Double-base propellants commonly consist of a binder polymer (nitrocellulose) which is plasticized with a nitric ester (nitroglycerin). In fact, double-base propellants are considered as one of the oldest propellant families, and were developed as a consequence of the development of propulsion. At the end of World War I, gun propellants were essentially powders which were based on nitrocellulose. Later, it was found that the inclusion of nitroglycerin on nitrocellulose made it possible to increase the energy level, but due to the increase of the combustion temperature, its usage as a gun propellant has been limited. However, despite this, they have found use as rocket motor propellants.

Unfortunately, propellants containing NC continuously slowly decompose. The decomposition products released in the decomposition process increase the rate of decomposition, and a self-accelerating behavior is observed. To prevent this autocatalysis, stabilizers are added to NC/NG-based propellants. The action of the stabilizers is to trap the nitrous decomposition products and form stable compounds, which prevent or delay further decomposition of the NC/NG-based propellant. The most common stabilizers are diphenylamine and its derivatives (Fig. 2.13b).

Since it was necessary to increase the energy level of the propellants in order to fulfill the special needs and requirements that are being continuously demanded for propellants, it was necessary to add some energetic additives, such as ammonium perchlorate (AP), Aluminum powder (Al), or some nitramines, such as RDX or HMX, to form so-called composite modified double-base propellants (CMDB) propellants. However, this resulted in a new set of problems such as the formation of toxic gases and an exhaust smoke plume due to using AP as an oxidizer additive or Al as a fuel in CMDB propellants respectively.

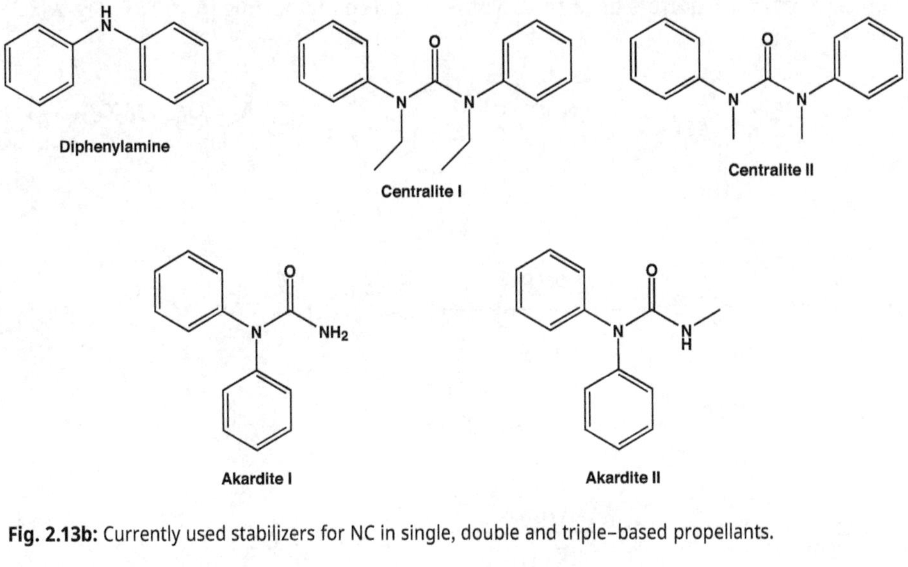

Fig. 2.13b: Currently used stabilizers for NC in single, double and triple-based propellants.

The generation of smoke has several detrimental effects such as: it may indicate the launch site of the missile and allow the missile to be located in flight, or it can result in the loss of control of the missile in cases involving optically guided missiles, and finally a negative environmental impact due to the emission of HCl gas in the exhaust plumes. The smoke produced from the combustion of composite solid rocket propellants can be classified into two general classes: primary smoke, which is a result of the gaseous products such as CO, CO_2, H_2, H_2O, Al_2O_3, and HCl, and secondary smoke which is caused by the combustion of the ammonium perchlorate-containing propellant producing HCl gas under certain atmospheric conditions (temperature and humidity). The combination with air or humidity results in the formation of a white cloud.

Composite propellants can be classified according to the composition of the smoke they produce: high smoke propellants in which the composite propellant contains Al fuel and AP oxidizer as the main ingredients, and the exhaust plume which is produced contains high primary smoke (Al_2O_3) particles and HCl gas, and reduced smoke propellants in which the composite propellant contains a small amount of Al or Mg and a reduced quantity of the AP oxidizer (by replacing AP with other materials), and the exhaust plume which is produced contains either a low amount of HCl gas or no HCl gas, as well as a low amount of primary smoke. Minimum smoke propellants are those in which the propellant does not contain Al fuel or AP oxidizer. Instead, AP is replaced by a nitramine such as HMX, RDX or ammonium nitrate (AN) with nitramine and the exhaust plume is free of HCl gas.

Possible replacements for AP are high-energy density oxidizers (HEDOs) which have been discussed recently. These compounds consist of C, H, N, and O and have a high oxygen content. 2,2,2-Trinitroethanol (TNE) is one of the most suitable starting materials with an oxygen balance (Ω_{CO}) of 30.9% and is easily synthesized through a Henry reaction. Several new oxidizers have been synthesized recently starting from TNE (Fig. 2.13c).

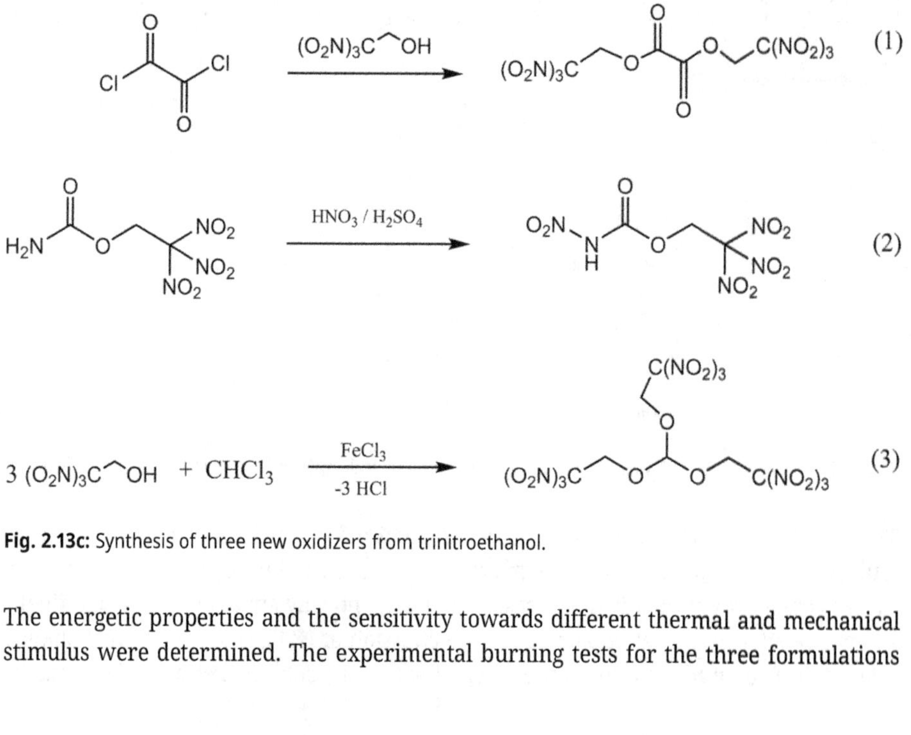

Fig. 2.13c: Synthesis of three new oxidizers from trinitroethanol.

The energetic properties and the sensitivity towards different thermal and mechanical stimulus were determined. The experimental burning tests for the three formulations of the new propellants containing the three new oxidizer compounds (**1**, **2**, and **3** from Fig. 2.13c) showed very good homogeneity with the fuel polymer (NC) which was indicated by homogeneous burning and observed using a high speed camera. Also, a high burning rate was observed in case of the second (NC/TNENC) and third (NC/TNEF) propellant formulations, while the first formulation (NC/BTNEO) showed a low burning rate and small flame in comparison with the other two compounds which produced a very bright flame on burning. This can be clearly observed in Fig. 2.13d. In addition, from the theoretical calculations, the high oxygen balance of the formulation containing NC/TNENC Ω_{CO_2} [%] ≈ 15 could be predicted as well as its high specific impulse.

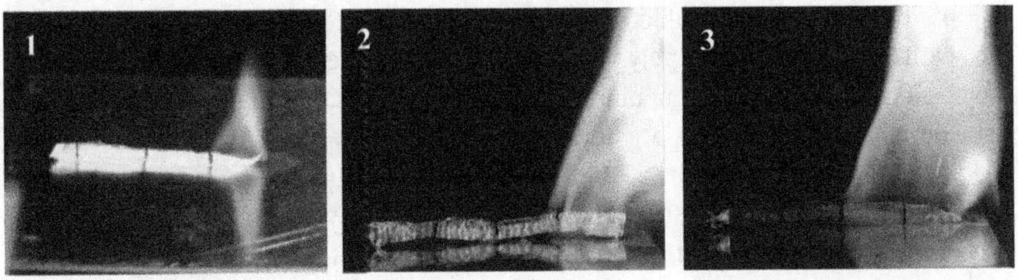

Fig. 2.13d: Burning of three different new formulations of smokeless double-base propellants (from left to right: NC/BTNEO, NC/TNENC and NC/TNEF).

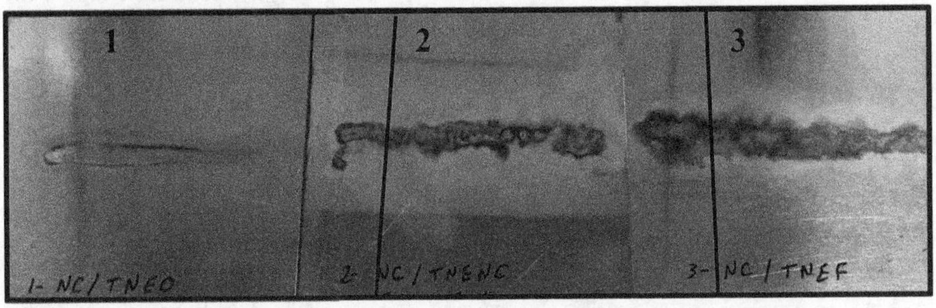

Fig. 2.13e: Effect of burning on the copper plate for the different three propellant formulations.

Furthermore, the effect of the temperature of the burning flame temperature on the copper plate (see Fig. 2.13e) shows that the first formulation (NC/BTNEO), which was based on oxalate, has a very low effect on the copper plate due to the low burning flame temperature, but the opposite is observed for the other two formulations, which are based on the carbamate and formate respectively.

The new high-performance, smokeless, double-base propellants based on the trinitroethyl unit have been prepared and studied on a laboratory scale. Three different formulations (NC/BTNEO, NC/TNENC and NC/TNEF) showed homogeneous burning with good burning rates and all were completely smokeless. The second formula with carbamate (NC/TNENC) showed the highest burning rate, followed by the third (NC/TNEF), with the lowest burning rate shown by the first formula containing oxalate (NC/BTNEO). Also, the lowest burning temperature (lowest effect on the copper plate during burning) and relatively small flame was the formula with oxalate (NC/BTNEO), while the highest burning temperature (highest effect on the copper plate during burning) and brightest flame was observed for the formula with the formate (NC/TNEF).

The search for promising substituents in the field of high energy density oxidizers is a challenging area of energetic materials. However, recently it has been found that the trinitromethyl ($-C(NO_2)_3$) and dinitromethyl ($-CH(NO_2)_2$) groups show considerable promise. Both of these groups are, of course, polynitro groups. In addition, another strategy has been adopted in combination with the use of these polynitro groups, and that is the introduction of a nitrogen-rich heterocyclic compound (Het) as the scaffold, to which these polynitro groups can be attached. The scaffolds of choice have so far been five or six-membered heterocyclic rings. In order to synthesize such heterocyclic scaffolds with polynitro groups attached, two strategies are commonly used – either the attachment of the already formed polynitro group to the scaffold (Fig. 2.13f – (A)), or the formation of the polynitro group by nitration of an appropriate precursor group which is attached to the scaffold (Fig. 2.13f – (B)). For example, the nitration of an acetonyl or ethyl acetate group attached to the heterocyclic scaffold using mixed acid has been shown recently to be effective and generally applicable (Fig. 2.13f – (B), (C)). Recently, Shreeve et al. have extended this synthetic approach to include nitration of the cyano group attached to a heterocyclic scaffold to form the corresponding trinitromethyl compound (Fig. 2.13f – (D), (E)).

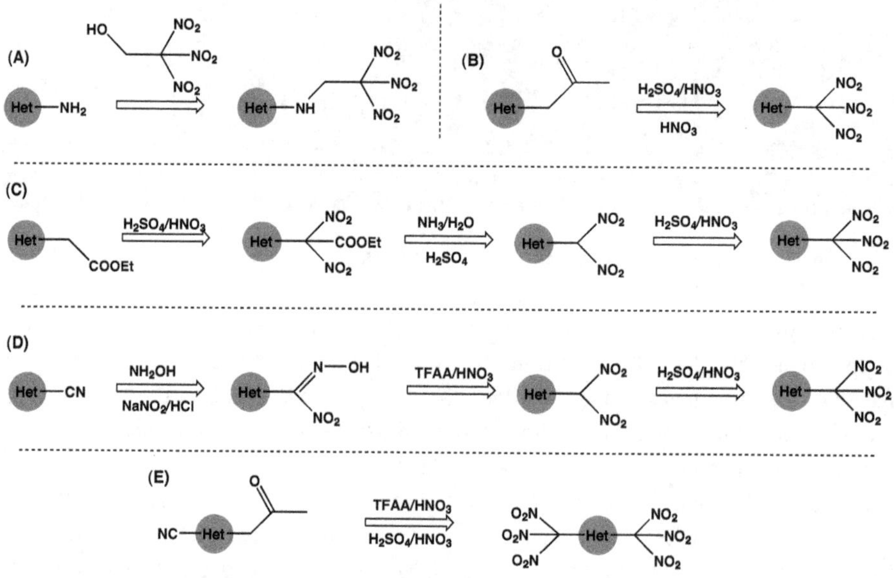

Fig. 2.13f: Synthesis of polynitro-substituted heterocyclic (Het) compounds [according to J. M. Shreeve et al., *Org. Lett.* **2019**, *21*, 1,073 – 1,077].

Figure 2.13g shows several interesting trinitromethyl and trinitroethyl organic oxidizers. The different compounds have advantages, but also severe disadvantages. For example, compounds 1, 6, 7, and 8 have decomposition temperatures which are far too low for application. Compounds 2, 4 and 5 have only oxygen balances between 5.4% and 7.8% with respect to CO_2. Compound 3 has a good oxygen balance (13.1%) and is thermally stable up to 191 °C, but has an impact sensitivity of 1 J.

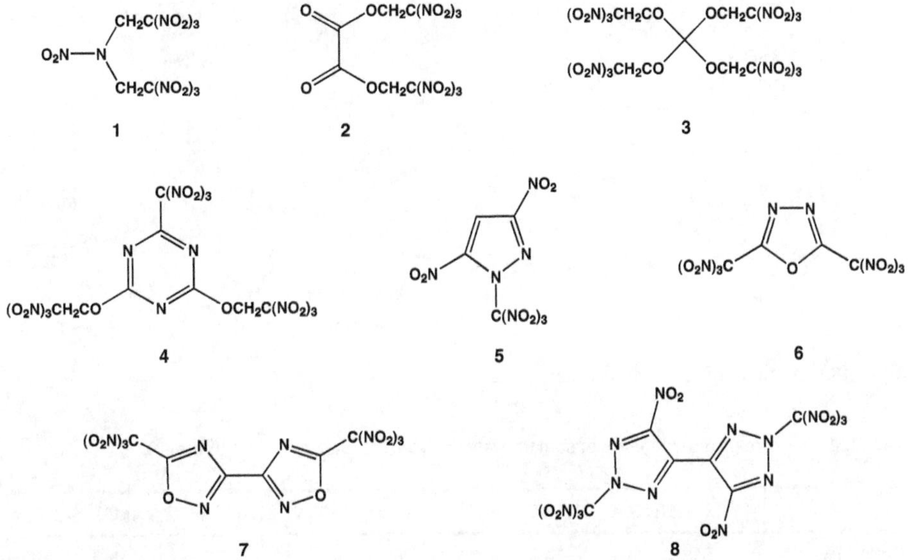

Fig. 2.13g: Typical organic oxidizers [see A. B. Sheremetev et al., *J. Mater. Chem A* **2018**, *6*, 14,780–14,786].

Figure 2.13h presents several new and advanced oxidizers. They also possess advantages and disadvantages. In Tab. 2.5b and Fig. 2.13i, the values for these compounds for the most relevant parameters for use as oxidizers are summarized.

For **liquid rocket propellants**, there is a difference between mono and bipropellants. Monopropellants are endothermic liquids (e.g. hydrazine), which decompose exothermically – mainly catalytically (e.g. Shell-405; Ir/Al_2O_3) – in the absence of oxygen:

$$N_2H_4 \xrightarrow{\text{catalyst}} N_2 + 2\ H_2 \quad \Delta H = -51\ \text{kJ mol}^{-1}$$

Monopropellants possess a relatively small energy content and small specific impulse and are therefore only used in small missiles and small satellites (for correcting orbits), where no large thrust is necessary. A summary of some monopropellants can be found in Tab. 2.6.

In a bipropellant system, the oxidizer and the fuel are transported in two tanks and are injected into the combustion chamber. The bipropellants can be separated into two different classes, either in cryogenic bipropellants, which can be handled only at very low temperatures and are therefore unsuitable for military applications (e.g. H_2/O_2) and

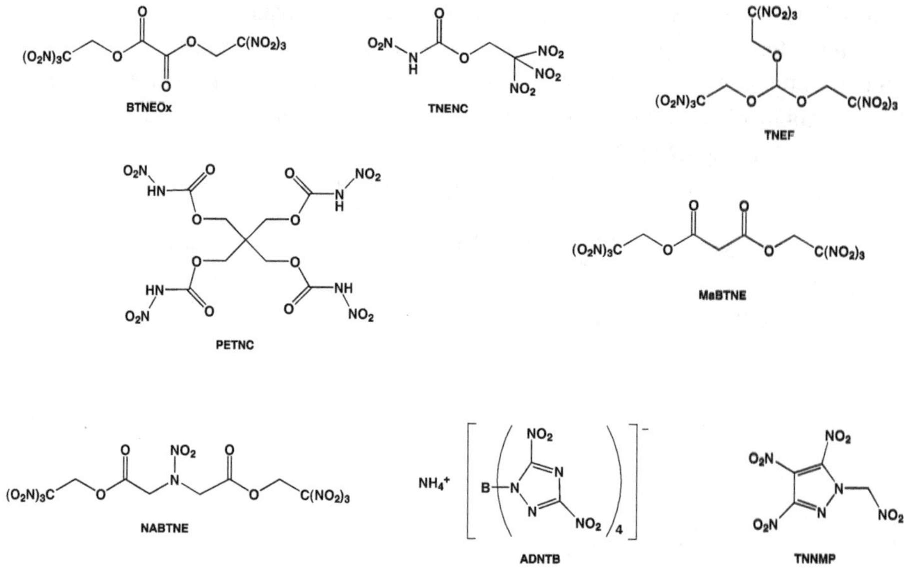

Fig. 2.13h: Examples of advanced oxidizers.

Tab. 2.5b: Performance parameters of some advanced oxidizers (desired values in green).

	BTNEOx	TNENC	TNEF[a]	PETNC	MaBTNE	NABTNE	TNNMP[b]	ADNTB[c]
$\Omega(CO_2)$ / %	+7.7	+14.9	+10.1	−26.2	−3.7	−6.3	−6.1	−8.5
$-Q_{ex}$ / kJ/kg (anaerobic)[§]	9,561	9,432	10,508	8,290	9,355	9,324	8,915	7,659
T_{ex} / K[§]	6,023	5,954	6,439	4,877	5,770	5,674	5,499	4,974
p_{CJ} / GPa[§]	14.3	14.4	17.7	12.4	13.2	13.7	14.6	15.4
VoD / m/s[§]	5,569	5,671	6,109	5,701	5,455	5,633	5,982	6,099
I_{sp} / s[#]	262	251	299	228	273	278	288	277

[§]calculated for metallized (Mg/Al) oxidizer with an energetic binder, oxidizer content 50%.
[#]isobaric combustion, equilibrium expansion through nozzle, 70 bar → vacuum.
[a]D. E. Dosch, K. Andrade, T. M. Klapötke, B. Krumm, *Propellants, Explosives, Pyrotechnics* **2021**, *46*, 895 – 898.
[b]J. Shreeve et al., *J. Mater Chem* A **2017**, *5*, 10437.
[c]R. Haiges, K. O. Christe, C. J. B. Jones, US Patent 9,309,266 (**2014**).

storable bipropellants (e.g. monomethylhydrazine/HNO_3), or in accordance with their ignition behavior in hypergolic and non-hypergolic mixtures. **Hypergolic** describes rocket propellants whose components react spontaneously with one another (in less than 20 ms) when they are mixed or somehow contact each other. The components of hypergolic propellants are mostly oxidizers or reducing agents, which react immediately upon contact, and ignite partly explosively. Since the fuel reacts immediately by burning on contact with the oxidizer at the time of injection into the combustion chamber, it is not possible for too much fuel to accumulate in the combustion chamber prior to ignition which could lead to an explosion and damage of the rocket engine. A further important advan-

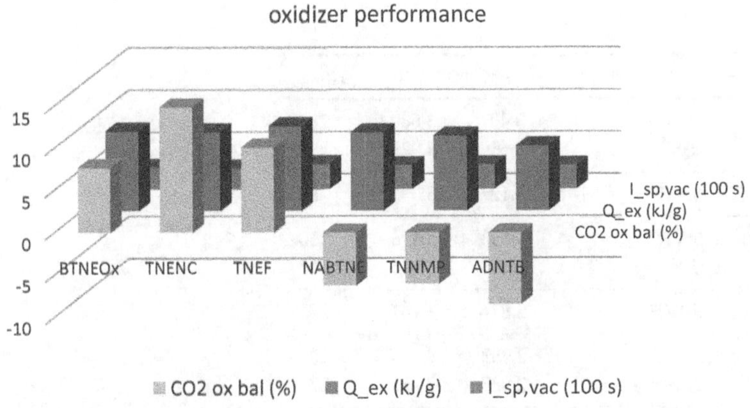

Fig. 2.13i: Performance of some advanced oxidizers.

Tab. 2.6: Monopropellants.

Monopropellant	Formula	I^*_{sp} / s	T_c / K
Hydrazine	$H_2N—NH_2$	186	1,500
Hydrogen peroxide	HO—OH	119	900
Isopropylnitrate	$C_3H_7—O—NO_2$	157	1,300
Nitromethane	$CH_3—NO_2$	213	2,400

Tab. 2.7: Bipropellants.

oxidizer	fuel	T_c / °C	T / K	I^*_{sp} / s
LOX	H_2	2,740	3,013	389
	H_2 / Be (49 : 51)	2,557	2,831	459
	CH_4	3,260	3,533	310
	C_2H_6	3,320	3,593	307
	B_2H_6	3,484	3,762	342
	N_2H_4	3,132	3,405	312
F_2	H_2	3,689	3,962	412
	MMH	4,074	4,347	348
	N_2H_4	4,461	4,734	365
OF_2	H_2	3,311	3,584	410
FLOX (30 / 70)	H_2	2,954	3,227	395
N_2F_4	CH_4	3,707	3,978	319
	N_2H_4	4,214	4,487	335
ClF_5	MMH	3,577	3,850	302
	N_2H_4	3,894	4,167	313

Tab. 2.7 (continued)

oxidizer	fuel	T_c / °C	T / K	I^*_{sp} / s
ClF_3	MMH	3,407	3,680	285
	N_2H_4	3,650	3,923	294
N_2O_4, NTO	MMH	3,122	3,395	289
MON-25 (25% NO)	MMH	3,153	3,426	290
	N_2H_4	3,023	3,296	293
IRFNA (III-A)[a]	UDMH	2,874	3,147	272
IRFNA (IV HDA)[b]	MMH	2,953	3,226	280
	UDMH	2,983	3,256	277
H_2O_2	N_2H_4	2,651	2,924	287
	MMH	2,718	2,991	285

[a]IRFNA (III-A): 83.4% HNO_3, 14% NO_2, 2% H_2O, 0.6% HF
[b]IRFNA (IV HDA): 54.3% HNO_3, 44% NO_2, 1% H_2O, 0.7% HF

tage is that the ignition definitely occurs, which is important for weapon systems such as intercontinental rockets, pulsed engines and upper stages of launch vehicles (re-ignition in space) for example. (*Note:* Hypergolic systems are also used in incendiary devices e.g. for mine clearance.) Hydrazine derivatives (hydrazine, MMH and UDMH) with HNO_3 or NTO (dinitrogen tetroxide) are practically the only hypergolic propellants used today. Table 2.7 shows a summary of different liquid bipropellant mixtures.

However, there are currently efforts to replace carcinogenic MMH and UDMH by toxicologically less damaging, but still hypergolic substances. One possible candidate is dimethylaminoethylazide (DMAZ) which has been produced by the company MACH I (PA):

DMAZ
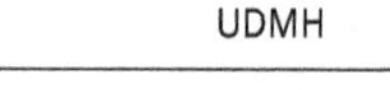
2-Dimethylaminoethylazide

The work that has been reported so far indicates that DMAZ would make a logical alternative to MMH as a candidate for a gel fuel component in a hypergolic bipropellant design (GAP may be a suitable gellant). MACH I (along with Aerojet-Rocketdyne) has demonstrated that DMAZ-based fuel gels display several advantages such as a lower toxicity, higher density, and wider liquid range as a function of temperature. Furthermore, it has been shown that gels can be formulated with DMAZ to meet the rheological requirements of certain system designs. It has also been demonstrated that the issue of longer native ignition delay of DMAZ, compared to that of MMH, can be effectively addressed using additives (soluble Co or Mn salts).

Another interesting compound is tetramethyl tetrazene, which was suggested by CNRS, Airbus Safran Launchers. 1,1,4,4-tetramethyl-2-tetrazene (TMTZ) is also considered as a prospective replacement for toxic hydrazines used in liquid rocket propulsion. The heat of formation of TMTZ was computed and measured, giving values well above those of the hydrazines commonly used in propulsion. This led to a predicted maximum I_{sp} of 337 s for TMTZ/N_2O_4 mixtures, which is a value comparable to that of MMH. TMTZ is thus an attractive liquid propellant candidate, with a performance comparable to hydrazines, but a lower vapor pressure and toxicity.

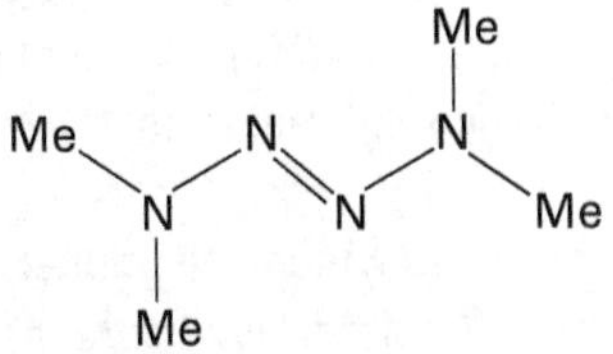

TMTZ

Out of the bipropellant mixtures shown in Tab. 2.7, only a few are used in practial applications. In particular, LOX / H_2 has proven useful in the cryogenic main engines of the civil Space Shuttle and Ariane V. The Aestus upper stage of the Ariane V relies on using NTO/MMH. The engines of the Delta (RS-27) and Atlas rockets work with LOX / HC (HC = hydrocarbons). Russian rockets in particular often use UDMH, which is very similar to MMH. As storable oxidizers, practially only NTO or RFNA (red fuming nitric acid) are used. Despite the very good specific impulse of fluorine containing (F_2 or FLOX) oxidizers and beryllium containing propellants, they have not found application due to technical (handling and pumping of liquid fluorine) and ecological (toxicity of HF and Be compounds) grounds.

At this point, the more recent development of so-called **gel propellants** should be mentioned. We have seen above that for military missiles, hypergolicity is a desired property. On the other hand, transportation of IRFNA or NTO in combination with MMH or UDMH on a ship or submarine for example also carries high risks and danger if there is an accident with unwanted contact of the oxidizer with the fuel. Gel propellants try to combine the desired and positive properties of a liquid propellant (flow behavior, pulse-mode) with the reliability of a solid propellant. The term gelled fuel is used to describe a fuel which has an additional gelling agent (approx. 5–6%) and an additive, so that it behaves like a semi-solid and only begins to liquefy under pressure. Inorganic (e.g. Aerosil 200, SiO_2 or aluminium octanoate) or organic substances (e.g. cellulose or thixatrol = modified castor oil derivative) are suitable gelling agents. The gel becomes semi-solid as a result of the presence of strong π-π- or van-der-Waals interactions and therefore it not only has a low vapor pressure but often also a higher density in comparison to pure fuel, which makes it very safe to handle. The gel only liquidizes under pressure, while the viscosity decreases with increasing shear stress. There are certain problems that still exist today in the production of oxidizer gels as a result of the high reactivity of WFNA, RFNA or NTO. Hypergolic alternatives to the strongly corrosive nitric acid derivatives are also sought after.

Hybrid rocket motors use propellants in two different states of matter, usually a solid fuel and a liquid or gaseous oxidizer. Like liquid rockets and unlike solid rockets a hybrid engine can be turned off easily, but due to the fact that fuel and oxidizer (different states of matter) would not mix in the case of an accident, hybrid rockets fail more benignly than liquid or solid motors.

The specific impulse is generally higher than for solid motors. A hybrid engine usually consists of a pressurized tank (oxidizer) and a combustion chamber containing the solid propellant (fuel). When thrust is desired, the liquid oxidizer flows into the combustion chamber where it is vaporized and then reacted with the solid fuel at chamber pressures of about 7–10 bar.

Commonly used fuels are polymers, e.g. HTPB or polyethylene which can be metallized (aluminized) in order to increase the specific impulse. Common oxidizers include gaseous or liquid oxygen or nitrous oxide.

Some of the disadvantages of the fuels presently used are a relatively low regression rate and a low combustion efficiency. Such problems could be overcome in the future by using cryogenic solid fuels like frozen pentane (@ 77 K). Such fuels would form a melt layer during the combustion process and have a higher regression rate. Alternatively, paraffin wax (not cryogenic) would also form a melt layer during combustion and could easily be aluminized. Such systems have specific impulses of up to 360 s.

The most commonly used oxidizers have the disadvantage of either being cryogenic (LOX) or toxic (N_2O_4). Newer oxidizer systems which are presently under investigation include refrigerated (approx. –40 to –80 °C) mixtures of N_2O and O_2 ("Nytox", 80% N_2O, 20% O_2). The advantage of such a mixture would be a relatively high vapor pressure (compared to pure N_2O) and a high density due to the presence of O_2, but without being cryogenic. In combination with a metallized paraffin (wax), Nytox as a fuel could help to overcome performance as well as toxicity issues.

Other developments go into the direction of adding up to approx. 10% aluminum hydride (AlH_3) or $LiAlH_4$ to the HTPB fuel in order to increase the regression rate and deliver a specific impulse of up to 370 s.

Literature

[1] P. Stachak, I. Łukaszewska, E. Hebda and K. Pielichowski, *Recent advances in fabrication of non-isocyanate polyurethane-based composite materials*, Materials **2021**, *14*, 3497.

2.4.1 Chemical thermal propulsion (CTP)

A further special area of propulsion systems is **Chemical Thermal Propulsion** (CTP). CTP is defined in contrast to STP (solar thermal propulsion) and NTP (nuclear thermal propulsion). In CTP, in a very exothermic chemical reaction in a closed system, heat but no pressure is generated since the products of the reaction are solid or liquid. The heat energy is then transferred to a liquid medium (the propellant) using a heat exchanger, which is responsible for the propulsion of for example, the torpedo. Suitable propellants are e.g. water (the torpedo can suck it in directly from its surroundsings) or H_2 or He, due to their very low molecular or atomic masses. The basic principles of CTP can also be used in special heat generators. A good example for a chemical reaction which is suitable for CTP is the reaction of (non-toxic) SF_6 (sulfur hexafluoride) with easily liquified lithium (m.p. 180 °C):

$$SF_6 + 8\,Li \rightarrow 6\,LiF + Li_2S \qquad \Delta H = -14727\,kJ\,kg^{-1}(\text{mixture})$$

In comparison to this, the reaction of MMH with NTO generates only 6,515 kJ per kilogramm of a stoichiometric mixture. Table 2.8 clearly shows the influence of the average molecular mass of the "combustion products" (in this case the propellants) on the specific impulse.

Tab. 2.8: Specific impulse I^*_{sp} of the CTP reaction Li/SF_6 occurring at 10 bar and 2,500 K, depending on the propellant used.

Propellant	M (propellant) / g mol^{-1}	I^*_{sp} / s
H_2	2	900
He	4	500
H_2O	18	320
N_2	28	230

Further chemical reactions which are principally suitable for CTP are outlined in the following equations:

$8\,Li + SF_6$	$\rightarrow$	$Li_2S + 6\,LiF$	$\Delta_r H =$ −3,520 kcal kg^{-1}
$4\,Li + OF_2$	$\rightarrow$	$Li_2O + 2\,LiF$	−5,415
$6\,Li + NF_3$	$\rightarrow$	$Li_3N + 3\,LiF$	−3,999
$8\,Li + SeF_6$	$\rightarrow$	$Li_2Se + 6\,LiF$	−2,974
$6\,Li + ClF_5$	$\rightarrow$	$LiCl + 5\,LiF$	−4,513
$6\,Li + BrF_5$	$\rightarrow$	$LiBr + 5\,LiF$	−3,212
$6\,Li + IF_5$	$\rightarrow$	$LiI + 5\,LiF$	−2,222
$4\,Li + ClF_3$	$\rightarrow$	$3\,LiF + LiCl$	−4,160
$8\,Li + ClO_3F$	$\rightarrow$	$3\,Li_2O + LiCl + LiF$	−4,869

It is obvious that the Li / SF_6 system (although it is energetically lower than others) is preferred over other systems because it is easy to handle and has low health and environmental hazards in comparison to other mixtures involving different oxidizers.

An excellent and up-to-date introduction into rocket propellants can be found in:

H. K. Ciezki, Rocket Propellants – An introduction and handbook for students, propulsion engineers and scientists, Wiley-VCH, Weinheim, **2025**.

2.5 Pyrotechnics

2.5.1 Detonators, initiators, delay compositions, and heat-generating pyrotechnics

A pyrotechnic composition is a substance or mixture of substances designed to produce an effect by heat, light, sound, gas or smoke or a combination of these, as a result of non-detonative self-sustaining exothermic chemical reactions. Pyrotechnic substances do not rely on oxygen from external sources to sustain the reaction.

The term pyrotechnic is derived from ancient Greek and translates as 'the art of fire' or 'the art of handling fire'. Similarly, as with explosives or propellants, the effects of pyrotechnics are also based on strongly exothermic reactions, whereby explosives show the largest and propellants relatively speaking the lowest speeds of reaction. Pyrotechnics are located somewhere in between. In contrast to explosives which release large quantities of gas during a reaction, pyrotechnics form mainly solid as well as gaseous products. Generally every pyrotechnic consists of an oxidizer as well as a reducing agent. Moreover, depending on the intended use, they can also contain a binder, a propellant charge and a colorant as well as smoke and noise generating additives. In contrast to explosives, which often contain the oxidizer and reducing moieties combined in one molecule (e.g. TNT, RDX), traditionally pyrotechnics are mixtures of different substances. Many pyrotechnic reactions are therefore solid state reactions, which result in a well-defined particle-size and highest possible homogenity. The energy released in a pyrotechnic reaction usually results in a flame and some smoke, light and also gas formation. Pyrotechnics have many applications, some of which are summarized in Tab. 2.9.

Tab. 2.9: Applications of pyrotechnics.

Pyrotechnic	Area of application	Example
Heat generating pyrotechnics	First fires, primers	See Fig. 2.14
	Pyrotechnic mixtures in detonators (blasting caps, detonators or primers)	Ox: KNO_3, BaO_2 Fuel: Mg, Ti, Si
	Electric "matches" in initiators ("first fires")	See Fig. 2.14
	Incendiary devices	
	Percussion primers	$KClO_3$ + TNT + PbO_2, Sb_2S_3
	Delay compositions (bombs, projectiles, grenades)	Gassy: blackpowder Gasless: metal oxide and -chromate + metal powder
	Matches	Head: Sb_2S_3 or S_8 + $KClO_3$ striking surface: P_{red} + glass
Smoke-generating pyrotechnics	Smoke munition	
Light-generating pyrotechnics	Signal flares	
	Decoy munition (IR region)	MTV
	fireworks	

Pyrotechnic mixtures which are used for ignition in blasting caps or detonators or which are used as first fires in propellant charges, are very easy to initiate. Here the initiator generates a flame if it is hit with a metal (percussion primer) or if it is initiated electrically using resistance heating (bridge wire). The first fire then initiates the propellant charge (percussion primer) or the first fire of an electrical match initiates a pyrotechnic mixture in a detonator, which then initiates a primary explosive which subsequently initiates a secondary explosive in the detonator. The shockwave of that detonator then initiates the main charge (not a component of the detonator). Figure 2.14 schematically shows the construction of a detonator which is initiated using an electrical match (initiator). For safety reasons, the initiator is introduced into the detonator only shortly before use. Typical examples for primary explosives used in detonators are lead azide and lead styphnate, and PETN for secondary explosives.

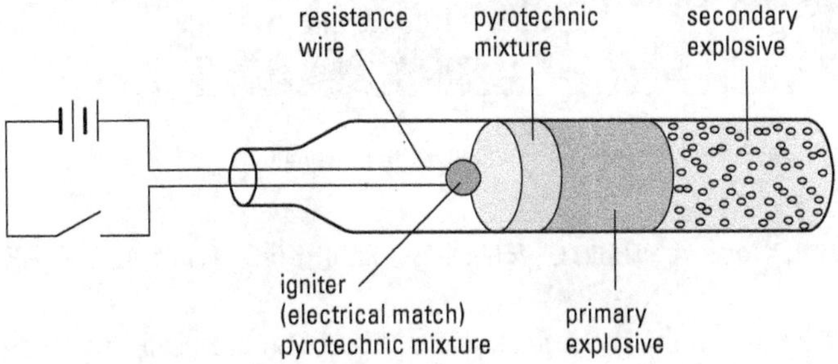

Fig. 2.14: Schematic representation of a detonator with an electric initiator.

SINOXID (= without rust) is the trademark used for the traditional primer compositions of the company Dynamit Nobel. In contrast to the mixture of mercury fulminate and potassium chlorate which was used previously, SINOXID is made-up of lead styphnate (Fig. 1.17), tetrazene (Fig. 2.2), barium nitrate, lead dioxide, antimony sulfide and calcium silicide. SINOXID compositions feature very good chemical stability and storage life, they are abrasion-, erosion- and corrosion-free and ignite propellants with precision. **SINTOX** (non-toxic) is a recently developed primer composition of Dynamit Nobel. It was developed specifically for ambient air in indoor firing ranges, which should not to be polluted with combustion products containing lead, antimony or barium. SINTOX consists essentially of diazodinitrophenol (or strontium diazodinitroresorcinate) and tetrazene as initial explosives and zinc peroxide as the oxidizer.

In fireworks, blackpowder is still used as the first fire.

Delay compositions are divided into two types: gasless (e.g. metal oxides or metal chromates with an elemental fuel) and gassy (e.g. blackpowder). Gasless delays are used in conditions of confinement (e.g. bombs, grenades, projectiles) or at high altitudes, where it is important that there are no variations of normal, ambient pressure. The desired burn rate (mm ms^{-1} to mm s^{-1}) is dependent on the purpose of use. Delay compositions with high burning rates for example are used in projectiles and bombs which should explode on impact, whereas those with low burning rates are primarily used in ground chemical munitions such as smoke pots, smoke grenades and tear pots. The following mixtures are examples of gasless delay compositions:

1. Fuel: B, Si oxidizer: $K_2Cr_2O_7$
2. Fuel: W oxidizer: $BaCrO_4$, $KClO_4$
3. Fuel: Mn oxidizer: $PbCrO_4$, $BaCrO_4$
4. Fuel: Cr oxidizer: $BaCrO_4$, $KClO_4$.

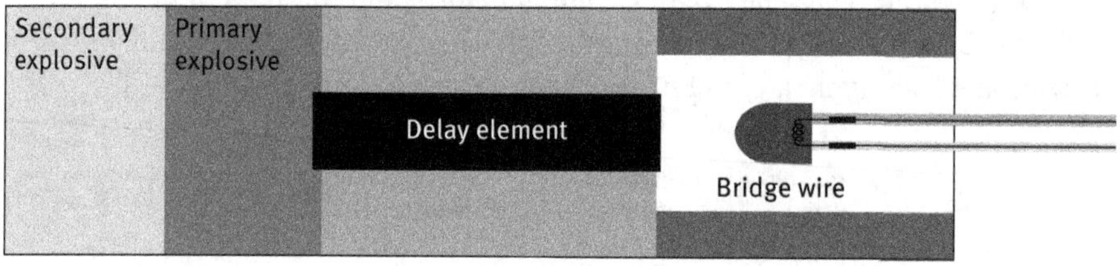

Fig. 2.14a: Schematic drawing of an electric detonator containing a delay element.

A typical delay composition would be: $BaCrO_4$ (56%), W (32%), $KClO_4$ (11%) and VAAR (~ 1%).

Gasless (producing less than 10 mL g^{-1} of gas) delays (Fig. 2.14a) are used in a large number of items every year in a range of US military applications, for example in grenades of the fuse type M201A1. Up until now, they have often been based on thermite reactions such as that shown in the following equation (eq. 1).

$$Pb_3O_4 + 2\,Si \rightarrow 2\,SiO_2 + 3\,Pb \qquad (1)$$

Different additives are used in order to increase the ignitability and combustion time. One of the most prominent mixtures with a burn rate (BR) between 0.6 and 150 mm s^{-1} is the W/$BaCrO_4$/$KClO_4$/diatomaceous earth (dilutant) composition.

A more complex reaction equation for this system is given in the following equation (eq. 2).

$$W + {}^{3}/_{8}\,KClO_4 + BaCrO_4 \rightarrow WO_3 + {}^{3}/_{8}\,KCl + {}^{1}/_{2}\,Cr_2O_3 + BaO + 508\,kJ \qquad (2)$$

Further examples of prominent gasless delay compositions are:
- barium chromate, potassium perchlorate and zirconium/nickel alloy (BR = 1.7–25 mm/s)
- boron and barium chromate (BR = 7–50 mm/s)
- potassium dichromate, boron and silicon (BR = 1.7–25 mm/s).

The time range of the delay element which is required varies strongly depending on the application. Very recently, progress on two condensed phase reactions e.g Ti/C ($Ti + C \rightarrow TiC$, $\Delta H_R = -3{,}079\ J\ g^{-1}$) and 3 Ni/Al ($3\ Ni + Al \rightarrow Ni_3Al$, $\Delta H_R = -753\ J\ g^{-1}$), as potential time delay compositions has been published.

Pyrotechnic time delays are reactive systems that burn for a desired period of time at a fixed rate, and are commonly used in military applications such as grenades and hand-held signals. More recently, there has been considerable advocacy for the development and use of environmentally friendly pyrotechnic ignition delays, by the removal of barium chromate ($BaCrO_4$). In a recent study by Groven et al., the viability of two possible replacements – strontium molybdate ($SrMoO_4$) and barium molybdate ($BaMoO_4$) – for the harmful component, $BaCrO_4$, in traditional delay compositions was evaluated. The molybdates of strontium and barium have been shown to be viable options as replacements for $BaCrO_4$ in traditional ignition delay formulations.

In general, the current trend is to favor 'green' compositions which are targeted to replace compositions containing perchlorates, chromates, lead, and barium. Thermite-based reactions are still the most promising alternatives.

Table 2.9a summarizes the burn rates for slow and fast delay compositions and thermites.

Tab. 2.9a: Burn rates for slow and fast delay compositions and thermites (Focke, Tichapondwa et al., PEP 2019, 44, 55–93).

Fuel/oxidant	Tube material	Burn rate/mm s^{-1}
Si / Pb_3O_4	Cast metal	2.2–3.8
Si / $CaSO_4$	Al	7–30
Si / Sb_2O_3	Pb	7–14
Si / $BaSO_4$	Al	8.4–16
Si / $PbCrO_4$	Pb	10–30
Si / Bi_2O_3	Pb	15–155
Al / WO_3	–	0.1–4.1
Al / MoO_3	–	4–12
Al / MoO_3	Acrylic	600–1,000
Mn / Sb_2O_3	Pb	4–9
Mn / CuO	Al	5–10

Figure 2.15 shows a schematic representation of the construction of a 12.7 mm multi-purpose projectile with a pyrotechnic mixture (incendiary) and secondary explosive. Figure 2.16 schematically shows the construction of a hand grenade with a pyrotechnic delay composition. Bombs such as, for example, the MK 80 series, where approx. 45% of the total mass is the explosive, usually contain pyrotechnic ("nose" or "tail") fuses.

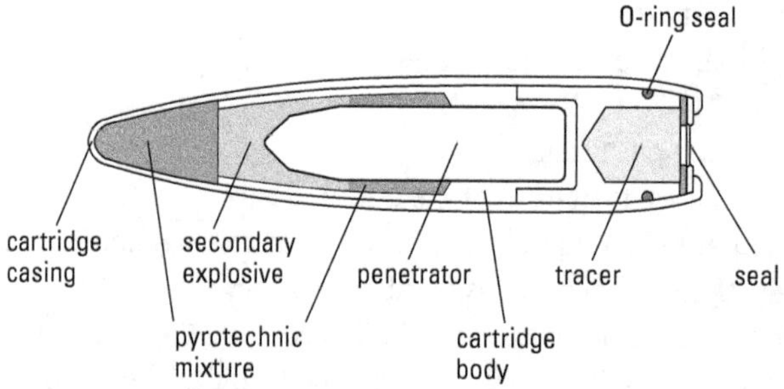

Fig. 2.15: Schematic representation of the construction of a 12.7 mm multipurpose projectile.

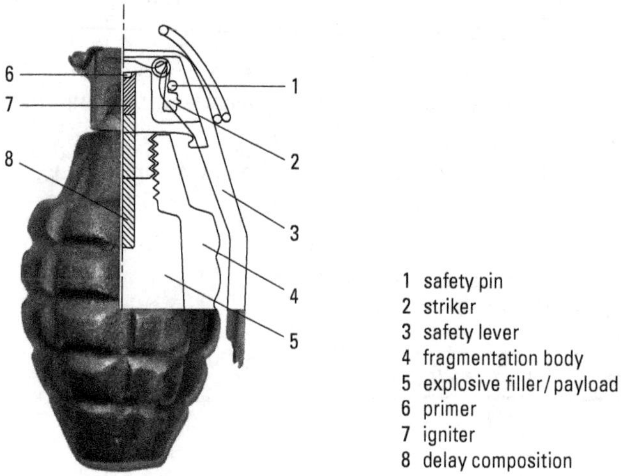

Fig. 2.16: Schematic representation of the construction of a hand grenade with a pyrotechnical delay device.

2.5.2 Light-generating pyrotechnics

Pyrotechnic formulations which emit electromagnetic radiation in the visible region are used as parachutes or hand-held signal flares for large-area illumination or to mark specific positions (targets, landing areas for planes or paratroopers etc.) [13–16]. For example, red is often used by aircraft personnel, to indicate their position in the event of an emergency landing or accident. Green and yellow are used by ground troops to mark their positions and white ($Mg + Ba(NO_3)_2 + KNO_3$; Al, Ti and Zr are also used as fuels for producing white lights) allows large-area illumination at night. The intensity and wavelength (color of the emitted light) depend on the components in the burning pyrotechnic mixture. Pyrotechnics which burn at approximately 2,200 °C, usually contain perchlorates as the oxidizer as well as organic propellants (e.g. nitrocellulose). In order to increase the flame temperature to 2,500–3,000 °C, metal powders such as magnesium are added. The emission of specific spectral colors occurs as a result of the addition of other metals and other metal salts. Sodium is therefore used for yellow, strontium for red and barium is usually used for green. The main emitters of the corresponding spectral bands are atomic Na for yellow, $Sr^{(I)}OH$ and $Sr^{(I)}Cl$ for red, as well as $Ba^{(I)}Cl$ and $Ba^{(I)}OH$ for green (Tab. 2.10). BaO is also a strong emitter in barium illuminants (Tab. 2.10). The hydrogen necessary for the formation of the monohydroxides SrOH and BaOH comes from the decomposition products of the binder or the PVC (Tab. 2.10). The Cl atoms come from the perchlorate oxidizing agent but also from the PVC. Table 2.11 shows the composition of some important illuminants in the visible range.

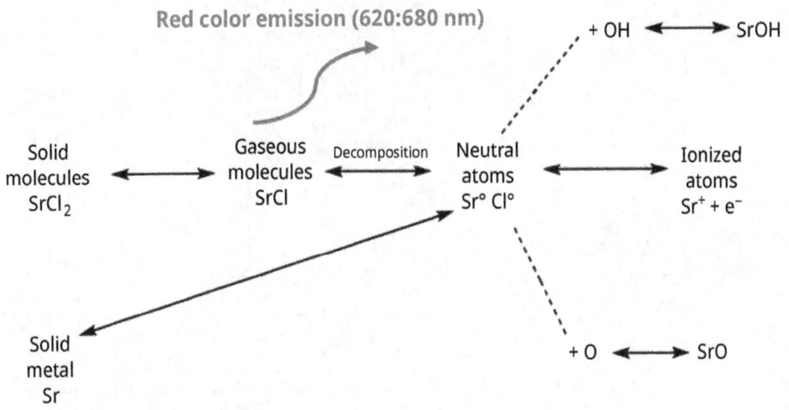

Fig. 2.16a: Sequence of emission phenomena for the SrCl molecule.

The emission in pyrotechnic compositions generally originates from one of the four following sources:
- Atomic (elemental) emission (e.g. Na) → line spectrum
- Molecular emission (e.g. BaCl, SrOH) → band spectrum
- Incandescence → continuous spectrum
- Luminescence → band spectrum

Colored flames are usually produced by the excitation of species which emit metal atoms (atomic emission). Other metal atoms may combine with other radicals in the flame, forming a color emitting molecule (molecular emission). The vapors of the emitting species are then excited by the thermal energy of the flame. The atoms or molecules in an excited state then relax to the ground state with the emission of visible light. For example, for the color red, strontium chloride is the main emitting species (Fig. 2.16a). The sequence of emission phenomena occurs in rapid succession. Figure 2.16a demonstrates the entire sequence of emission phenomena for strontium chloride.

In order to design colored illuminants, the properties of the human eye have to be considered (sample rate about 20 Hz). An eye can distinguish in the wavelength range of approximately 400 to 700 nm at an accuracy of about ± 2.5 nm, although an eye measures over very wide spectral bands in only three channels (Fig. 2.16b). The high resolution is achieved by a highly advanced, real time signal processing of overlapping bands, each of which has a different response over the entire VIS range (visual light range). The red color cells also respond to blue light, which enables the perception of purple hues. Without this property, an eye would not be able to perceive a large sector of the chromaticity diagram, because no single wavelength can represent a purple color (Fig. 2.18).

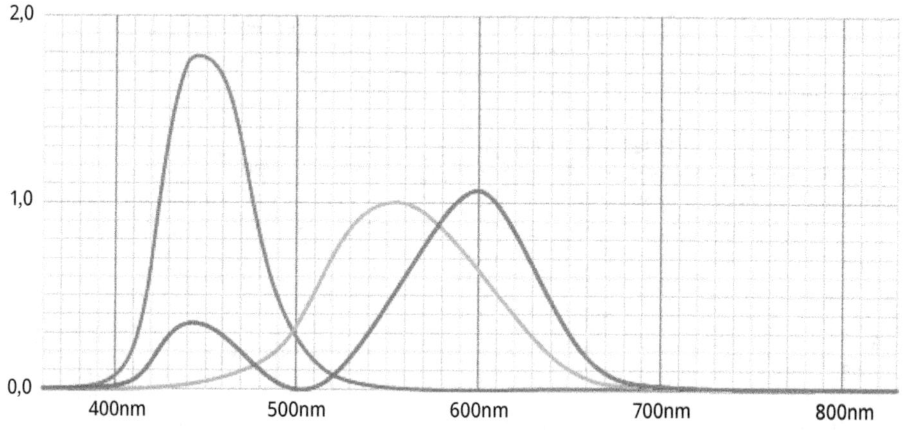

Fig. 2.16b: CIE 1931 tristimulus functions (sensitivity curves of each color cell type in a human eye). x-axis: wavelength (nm), y-axis: sensitivity (arbitrary units).

All three channels in an eye can record light over an even wider range than 400 to 700 nm (Fig. 2.16b), but the sensitivity outside of this region is very low. Hence, it is possible to see even near infrared radiation, if the source is strong enough, such as a laser. Since such sources are rare, the range of visible wavelengths in the standard observer response functions has been cut to include the above VIS range only. These cut-off gain functions are called the color matching functions. The response of these functions was measured for the first time in 1928 and published in 1931, which is why these functions are called CIE 1931 standard observer. A revised version is called CIE 1964. In this work, the CIE 1931 color matching functions with a white point at the coordinates (1/3,1/3) were used, since most emission measurements in the literature are based on these. In addition, conversion functions are available to convert from one color space to another. The white point was set into the center of the chromaticity diagram (cf. Fig. 2.18), because the measurements were not convolved with another light source. The broad bands of the tristimulus functions (Fig. 2.16b) enable creating relatively broad band light sources with high color saturation. It is, however, important to keep all emissions within a narrow enough band to excite only one of the detectors in the eye. For red light, this means that the emissions should be kept above 600 nm wavelength, to make sure the green color detector in the eye is not appreciably excited.

Tab. 2.10: Spectra of the most important atoms and molecules for pyrotechnics.

Element	Emitter	Wavelength / nm	Color
Lithium	Atomic Li	670.8	Red
		610.4	Red
		460	Blue
		413	Violet
		497	Blue/green
		427	Violet
Sodium	Atomic Na	589.0, 589.6	Yellow
Copper	CuCl	420–460	Blue/violet
		510–550	Green
Strontium	SrCl	661.4, 662.0, 674.5, 675.6	Red
	SrCl	623.9, 636.2, 648.5	Orange
	SrCl	393.7, 396.1, 400.9	Violet
	SrOH	605.0, 646.0	Orange
		659.0, 667.5, 682.0	Red
	Atomic Sr	460.7	Blue
Barium	BaCl	507, 513.8, 516.2, 524.1, 532.1,	Green
		649	Red
	BaOH	487,	Blue/green
		512	Green
	BaO	604, 610, 617, 622, 629	Orange
	Atomic Ba	553.5,	Green
		660	Red

Tab. 2.11: Composition of selected illuminants in the visible region (%).

Ingredient	Red navy	Red highway	Green navy	Yellow navy	White navy
Mg	24.4		21.0	30.3	30
$KClO_4$	20.5	6.0	32.5	21.0	
$Sr(NO_3)_2$	34.7	74.0			varied
$Ba(NO_3)_2$			22.5	20.0	53
PVC	11.4		12.0		
$Na_2(COO)_2$				19.8	
Cu powder			7.0		
Bitumen	9.0			3.9	
Binder (VAAR)		10.0	5.0	5.0	5
S_8		10.0			

From the ecological and toxicological point of view the relative high perchlorate content of all of the illuminating compositions (Tab. 2.11) as well as the use of the heavy metal barium in green illuminants is problematic. During the production and burning of the light-generating compositions, toxic perchlorate and soluble barium salts can enter into ground water. The ClO_4^- anion is a possible teratogen and has shown negative effects on thyroid function. The up-take of iodine into the thyroid is suppressed by chronic perchlorate up-take, which in consequence causes thyroid subfunction. Moreover, the binders and other ingredients (PVC, Bitumen, sulfur) often result in burning with strong smoke. Current research is therefore concerned with the development of:
- Perchlorate-free red and green colored illuminants
- Barium in the form of sparingly soluble Ba salts (not soluble $Ba(NO_3)_2$)
- Barium (heavy metal)-free green illuminants
- weakly / smoke-free red, green and yellow illuminants.

Many of the newly developed pyrotechnics are based on metal complexes with very nitrogen-rich ligands, mainly tetrazole derivatives (Fig. 2.17). Unlike conventional energetic substances they do not derive their energy through the oxidation of a carbon back-bone, but through their high (positive) heat of formation. These compounds are suited for application as propellants, reducing agents or colorants – preferably with less toxic metal salts such as Cu^{II} instead of Ba^{II}.

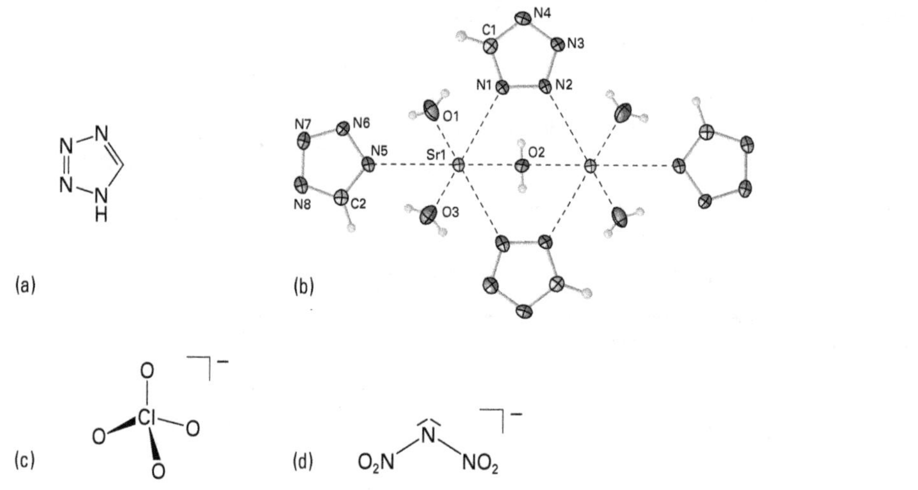

Fig. 2.17: Molecular structure of 1*H*-tetrazole (a), the strontium compound as colorant in the environmentally friendly red pyrotechnic mixture (b), the perchlorate anion (c) and the dinitramide anion (d).

Since environmentally compatible pyrotechnics should not contain perchlorates or heavy metals in the first instance, new red and green formulations are being developed which contain Cu instead of Ba and which contain other oxidizers (e.g. nitrate or dinitramide, Fig. 2.17) instead of perchlorates. The high nitrogen content also guarantees a significant reduction of the smoke and particulate matter released and consequently results in considerably-higher color brilliance.

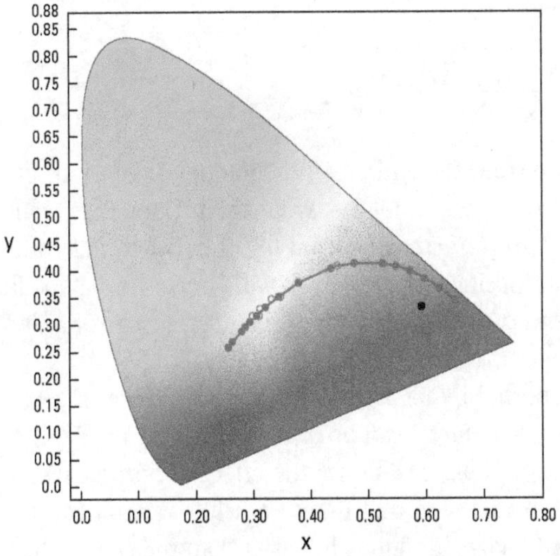

Fig. 2.18: Chromaticity diagram for the high-N strontium-based flare.

Furthermore, high-N red and green formulations show an increase in burn time, an increase in luminous intensity and an increase in spectral purity. Two examples of new red and green formulations developed by ARDEC/LMU are:

Red:	$Sr(NO_3)_2$	39%
	Mg	29%
	PVC	15%
	high-N Sr-tetrazolate salt	10%
	binder	7%
Green:	$Ba(NO_3)_2$	46%
	Mg	22%
	PVC	15%
	high-N Ba-tetrazolate salt	10%
	Binder	7%

In the above mentioned red and green formulations, it is possible to replace the high-N metal tetrazolate salt with more metal nitrate and free aminotetrazole. This causes longer burn times but lower intensities.

The strontium formulation shows a reasonably high red color purity as can be seen from the chromaticity diagram in Fig. 2.18. The color purity (p_e) of a visible flare with the chromaticity (x, y) is its difference from the illuminant's white point relative to the furthest point on the chromaticity diagram with the same hue (dominant wavelength for monochromatic sources): where (x_n, y_n) is the chromaticity of the white point and (x_I, y_I) is the point on the perimeter whose line segment to the white point contains the chromaticity of the stimulus.

$$p_e \sqrt{\frac{(x-x_n)^2+(y-y_n)^2}{(x_I-x_n)^2+(y_I-y_n)^2}}$$

The color purity is the relative distance from the white point. Contours of constant purity can be found by shrinking the spectral locus about the white point. The points along the line segment have the same hue, with p_e increasing from 0 to 1 between the white point and position on the spectral locus (position of the color on the horseshoe shape in the diagram) or (as at the saturated end of the line shown in the diagram) position on the line of purples.

A breakthrough discovery was reported by Sabatini (ARDEC) in the area of green illuminants. Formulations without any heavy metal can be based on boron carbide (B_4C, fuel) with a suitable oxidizer (e.g. KNO_3). Table 2.12 shows the effect of boron carbide (B_4C) in pyrotechnical compositions. It can be seen that flares with 100% boron carbide as the fuel show longer burn times and higher luminous intensity than the control barium nitrate based flare (M125 A1), while the spectral (color) purity is slightly lower.

Tab. 2.12: Effect of boron carbide in pyrotechnics.

Fuel ($B : B_4C$) in formulation[a]	Burn time [s]	Luminous intensity [cd]	Dominant wave-length [nm]	Spectral purity [%]
control [b]	8.15	1357	562	61
100 : 0	2.29	1707	559	55
50 : 50	5.89	2,545	563	54
40 : 60	6.45	2,169	563	54
30 : 70	8.67	1914	562	53
20 : 80	8.10	1819	563	53
10 : 90	8.92	1458	562	52
0 : 100	9.69	1403	563	52

[a]Formulation: KNO_3 83%, fuel 10%, epoxy binder 7%
[b]$Ba(NO_3)_2$ 46%, Mg 33%, PVC 16%, Laminac 4,116/Lupersol binder 5%

Due to issues such as the ratio of ingredients employed, the molecular behavior of light-emitting species, and the combustion temperatures typically associated with the generation of colored pyrotechnic flames, the production of blue-light-emitting pyrotechnic illuminants is believed to be a most challenging field for chemists. Typically, blue flames are generated by the use of copper or copper-containing compounds in conjunction with a chlorine source (Tab. 2.12a). Chlorine sources used in blue-light-emitting pyrotechnics include ammonium perchlorate, potassium perchlorate, or chlorine donors such as poly(vinyl) chloride. When copper combines with a chlorine source at elevated temperature, copper(I) chloride (CuCl) is formed, which has always been believed to be the lone molecular emitter toward achieving high quality blue flames.

However, the use of perchlorates in pyrotechnic formulations is discouraged due to the supposed toxicity of this chemical which has been linked to causing thyroid disorders. Furthermore, the combustion of polychlorinated organic materials such as poly(vinyl) chloride has been shown to produce polychlorinated biphenyls (PCBs), polychlorinated dibenzo-*p*-dioxins (PCDDs), and polychlorinated dibenzofurans (PCDFs). These polychlorinated chemicals are highly toxic and are potent carcinogens. Therefore, the removal of perchlorates and chlorinated organic materials in pyrotechnic formulations would eliminate the formation of these aforementioned pollutants.

In a joint research effort between LMU Munich (Klapötke and Rusan) and ARDEC (Sabatini), the development of the first known perchlorate-free and chlorine-free blue-light-emitting pyrotechnic formulation has been developed. The optimal formulation (Tab. 2.12b) relies on the generation of copper(I) iodide (CuI), which serves as a strong blue-light-emitting species (Tab. 2.12c, Fig. 2.18a). In addition to the environmentally friendly goal of eliminating perchlorates, the formation of polyiodated biphenyls – which are likely to form during the combustion process – is not perceived to be an environmental hazard. Polyiodated biphenyls are used as contrast agents for radiological purposes in medicinal applications.

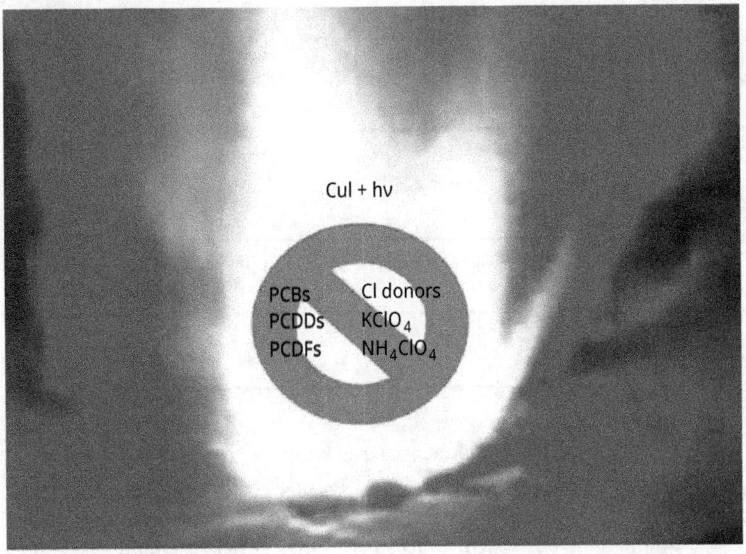

Fig. 2.18a: Blue flame originating from formulation **A.**

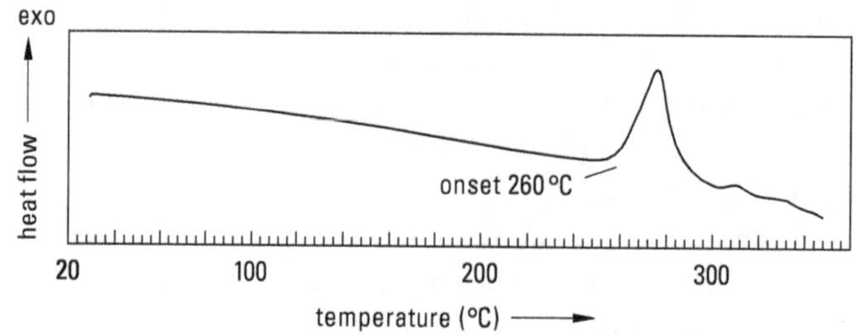

Fig. 2.18b: DSC thermogram of the new environmentally friendly red mixture.

The US military has embarked on many initiatives over the years to mitigate environmental concerns associated with their munitions. Since the US military has an interest in blue-light-emitting signaling flares, the new formulations may provide a significant benefit to the Department of Defense. In addition, civilian firework companies have been under increasing scrutiny from the Environmental Protection Agency (EPA) to "green" their firework displays. The removal of perchlorates and chlorinated organic compounds while generating blue light represents an important step toward that goal, since blue is believed to be the most difficult of all colors to generate in an environmentally friendly fashion.

Tab. 2.12a: Shimizus' chlorine-containing blue-light-emitting formulation (reference formulation).

	$KClO_4$ [wt%]	Cu [wt%]	PVC [wt%]	Starch [wt%]
Control	68	15	17	5

Tab. 2.12b: Formulations A and B.

	$Cu(IO_3)_2$ [wt%]	Guanidine nitrate [wt%]	Mg [wt%]	Urea [wt%]	Cu [wt%]	Epon 828/Epikure 3140 [wt%]
A	20	50	10	–	15	5
B	30	35	9	21	–	5

Tab. 2.12c: Performance and sensitivity data of formulations A and B.

	burn time [s]	Dw [nm]	Sp [%]	LI [cd]	LE [cdsg^{-1}]	IS [J]	FS [N]	grain size [µm]	T_{dec} [°C]
Control	4	475 (552)	61	54	360	8	324	307	307
A	6	477 (555)	64	80	1,067	> 40	> 360	< 100	198
B	6	476 (525)	63	78	780	> 40	> 360	< 100	180

Extremely important for the safe handling of such pyrotechnic mixtures are, in addition to the chemical stability (compatability with the binder and other additional compounds) particularly the thermal stability and the lowest possible impact, friction and electrostatic sensitivities. Figure 2.18b shows an example of a DSC thermogramm (Differential Scanning Calorimetry), which shows that the new environmentally friendly red mixture (colorant components from Fig. 2.17) is thermally stable up to approx. 260 °C. But the sensitivity of the mixture towards electrostatic discharge (ESD) was lowered to 1 J (cf. typical values for the human body are within the range 0.005–0.02 J).

Traditional red-light emitting pyrotechnic formulations are based on strontium and chlorinated organic materials. Usually, the main component is strontium nitrate, which is combined with a chlorine donor, such as polyvinyl chloride (PVC), to generate metastable strontium(I) chloride, which acts as the red-light emitting species. However, one drawback of using such mixtures is that highly carcinogenic compounds such as polychlorinated biphenyl, dibenzofuran, and dibenzo-p-dioxin derivatives are formed during the combustion of chlorine-containing pyrotechnic formulations. A recent report from the United States Environmental Protection Agency (U. S. EPA) concluded that strontium is also a potential health hazard. This resulted in regulation of the allowed concentration of strontium in drinking water from 2014 onwards. Strontium has been reported to adversely affect skeletal development in children and adolescents by replacing calcium in bones and interfering with bone strength. Strontium has been found to be present in 99% of all public water systems in the USA, and 7% of these have been classified as containing concerning levels of strontium. Although these studies did not include U. S. military training grounds, it is likely that raised strontium levels are also present at these facilities, since strontium is not only used in current red-light illuminating signaling pyrotechnic compositions, but also in older formulations as well, for example, in the red star

Tab. 2.12d: Comparison of the compositions of Sr-based formulations which act as red-light illuminants.

Components	**Weight %**		
Formulation	**Sabatini chlorine-free (exp.) = control**	**In-service M158 red star cluster**	**M126A1 red star parachute illuminant**
$Sr(NO_3)_2$	48	48	39.3
Mg 30/50	–	33	14.7
PVC	–	15	14.7
Mg 50/100 mesh	33	–	14.7
Nitrocellulose	–	–	–
Hexamine	–	–	–
5–AT	12	–	–
Epoxy binder	7	–	–
Laminac 4116/Lupersol	–	4	6.8
$KClO_4$	–	–	6.8

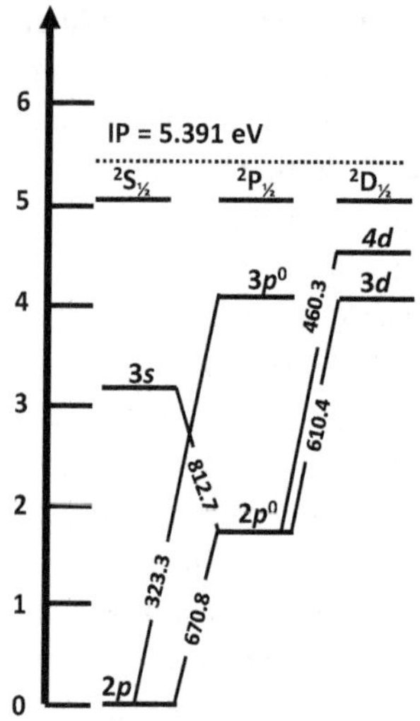

Fig. 2.18c: Atomic term diagram for lithium. The ordinate gives the corresponding energy in units of eV; IP is the ionization potential.

cluster, as well as in the red star parachute illuminant (Tab. 2.12d). A further use of strontium nitrate is in formulations used in emergency highway signals (civilian sector).

All of the above-mentioned issues have initiated research to find alternative pyrotechnic formulations which are environmentally benign and emit red light, but which do not contain any chlorine source or strontium in the formulation. It has been found that strontium nitrate can be replaced by lithium salts as the red coloring agent. This has the considerable advantage that the use of halogenated materials (e.g. PVC) can be avoided, since atomic lithium acts as the metastable light-emitting species instead of Sr(I)Cl – which is the case in strontium nitrate/PVC-based formulations.

When a metal wire loop that has been dipped into an aqueous solution containing a lithium salt is held in the flame of a laboratory Bunsen burner, a red flame is observed, which is due to the emission lines at 610 and 670 (see Tab. 2.10). These emissions originate from radiative transitions of gaseous atomic lithium. The main line at 670.8 nm corresponds to a $2s^2S_{0½}$ (ground level) $-2p^2P^0_{0½,1½}$ (two energetically similar excited levels) transition as shown in Fig. 2.18c.

In common pyrotechnic flames, these lines are superimposed on a continuum starting at 280 nm with a maximum at 400 nm. This continuum affects the color purity, and is due to the emission of LiOH (g) formed according to the following equation:

$$\mathrm{Li\,(g) + H_2O\,(g) \rightarrow LiOH\,(g) + H\,(g)}$$

Lithium atoms are scavenged by ever-present H_2O molecules which results in the formation of thermodynamically stable LiOH (g). Therefore, to obtain maximum atomic emission, the scavenging of Li (g) atoms by water forming LiOH (g) in pyrotechnic flames must be prevented.

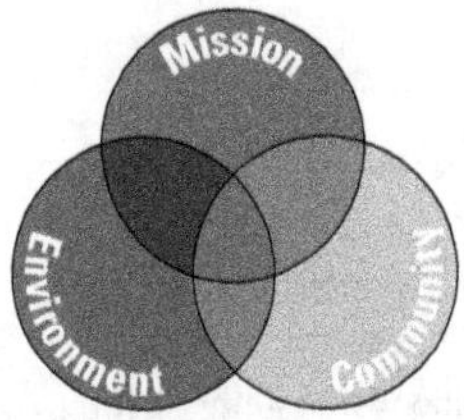

"Sustain the Mission – Secure the Future"

Fig. 2.19: Triple bottom line plus.

It is also undesirable for halogens – especially chlorine – to be present, since this has a negative effect on the red flame color in lithium flames. Although lithium perchlorate ($LiClO_4$) is superior to any other oxidizer (60.15% oxygen by weight is available for oxidation) based on its oxygen content, it shows only a pale pink flame color, due to the presence of chlorine which results in the formation of lithium chloride according to the equation below (X = Cl or any other halogen)

$$\mathrm{Li}\,(g) + \mathrm{X}\,(g) \rightarrow \mathrm{LiX}\,(g)$$

Lithium chloride (as well as the other halides of lithium) exhibits no lines in the visible range, but shows strong continuum emission with superimposed bands at 281, 285 and 290 nm.

Interestingly, although the presence of halogens suppresses the flame spectrum of lithium, this is not the case for barium, strontium or calcium. These elements produce relatively stable monohalide molecules in the presence of chlorine or bromine, which emit bands in either the green, red or orange-red regions of the spectrum respectively.

An excellent review article on the evaluation of Li compounds as color agents for pyrotechnic flames has been published by Koch (see ref. 7, below).

The protection of the environment through the use of sustainable technologies is one of the biggest and most important tasks of our time. Within the concept "triple bottom line plus" (mission, environment, community + economy) (Fig. 2.19), the US Army in co-operation with leading scientists world-wide, is trying to implement the sustainability of their operations and missions. A good example of this is the research into new, less environmentally hazardous pyrotechnics, which contribute to a successful operation as a result of their properties, but don't damage the environment due to their "green" characteristics. Furthermore, they are of general economic interest (new innovative technology, reduce the cost of purifying ground water) and of particular interest for local communities (reduced exposure to particulate matter, heavy metals and perchlorate in the vicinity of troop training camps).

Literature

[1] U. S. Environmental Protection Agency (EPA) Makes Preliminary Determination to Regulate Strontium in Drinking Water, https://yosemite.epa.gov/opa/admpress.nsf/6427a6b7538955c585257359003f0230/327f339e63facb5a85257d77005f4bf9!OpenDocument, Accessed Oct 2017.

[2] J. Roberts, Wisconsin strontium levels among highest in U. S. drinking water supplies, http://wisconsinwatch.org/2016/03/wisconsin-strontium-levels-among-highest-in-u-s-drinking-water-supplies/, Accessed Oct 2017.

[3] A. J. O'Donnell, D. A. Lytle, S. Harmon, K. Vu, H. Chait and D. D. Dionysiou, Water Res., 2016, 103, 319–333.

[4] J. J. Sabatini, E.-C. Koch, J. C. Poret, J. D. Moretti and S. M. Harbol, Angew. Chem. Int. Ed., 2015, 54, 10968–10970.

[5] J. J. Sabatini and J. D. Moretti, Chem. – Eur. J., 2013, 19, 12839–12845.

[6] J. J. Sabatini, in Green Energetic Materials, John Wiley & Sons, Ltd, 2014, pp. 63–102.

[7] E-C. Koch, Journal of Pyrotechnics 2001, 13, 1–8.

2.5.3 Decoy flares

The terms decoy flare or countermeasure munition are used to denote a system, which imitates the IR signal of a plane and therefore ground-air, water-air or air-air rockets, so-called heat-seeking missiles, get lured away from their targets. One of the first and best-known IR seekers – "Sidewinder" – was developed in China Lake (Fig. 2.20).

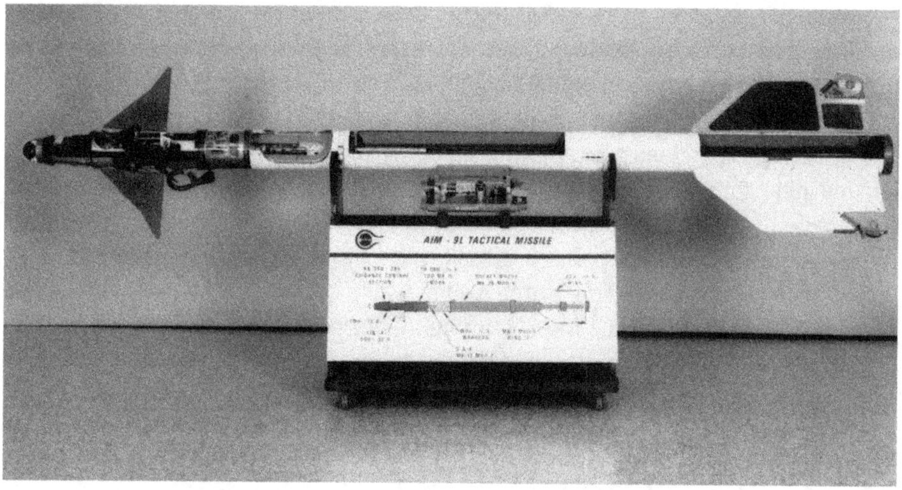

Fig. 2.20: The AIM-9 Sidewinder (AIM = Aerial Intercept Missile) is a heat-seeking guided missile for short-distance combat, which can be carried in most western fighter planes. In its 50 year history, more planes have been shot down using the Sidewinder than with any other device.

The Sidewinder has a high explosive warhead and an IR-based heat-seeking guidance system. Its other components are an optical target detector, the fins and the rocket engine. The guidance-system directs the missile straight into the hot engine of the enemy's plane (or whatever the missile considers this to be). An infra-red unit costs less than any other guidance-system. After it has been launched, the missile does not need anymore support from the launching platform and steers itself towards the target.

In order to be protected against such heat-seeking guided missiles, pyrotechnic decoy flares, which are based on blackbody radiation (or better greybody radiation), have been and are still being developed since the late 1950's. The specific radiant emittance of a radiating body (W in W cm^{-2} µm^{-1}) can be described using Planck's rule:

$$W_\lambda = \frac{2\,\pi\,h\,c^2}{\lambda^5}\,\frac{1}{ehcl/\lambda kT - 1}$$

where λ (in µm) is the wavelength, h is Planck's constant (6.626 10^{-34} W s^2), T is the absolute temperature of the radiator (in K), c is the speed of light (2.998 10^{10} cm s^{-1}) and k is the Boltzmann constant (1.38 10^{-23} W s K^{-1}).

In accordance with the rule of Wien, the maximum wavelength of the blackbody radiation λ_{max} (µm) shifts towards shorter wavelengths (higher energy) with increasing temperature:

$$\lambda_{max} = 2897.756 \mu\text{m K}\,T^{-1}$$

Since in reality there is no actual blackbody radiator (for which the Planck law applies, W), but a more realistic so-called greybody radiator (W'), the value ε has been defined as the quotient W'/ W, where ε can be given values between 0 and 1 (1 = genuine blackbody). For example, soot ($\varepsilon > 0.95$) behaves almost like a blackbody radiator, whereas MgO behaves more like a greybody radiator.

Heat-seeking missiles mainly work with photo-conductive IR detectors in the so-called α band (PbS, 2–3 µm) or in the β band (PbSe, 3–5 µm). While the hot tail pipes of a plane turbine typically emit a greybody radiation with λ_{max} between 2 and 2.5 µm, this is superimposed by the selective radiation of the hot exhaust gases (CO_2, CO, H_2O) in the region of λ_{max} = 3–5 µm. Figure 2.21 shows the typical IR signiature of a plane in the α and β bands. In order to imitate this signiature, initially Al / WO_3 thermites in graphite-balls were used (diameter 10 cm, thickness 2 mm), since graphite in the region < 2.8 µm emits a very good blackbody radiation with $\varepsilon = 0.95$. Nowadays, most of the pyrotechnic mixtures used are mixtures of Mg and perfluorinated polymers (Mg / Teflon / Viton, MTV), since the carbon soot ($\varepsilon \approx 0.85$) which is produced at very high temperatures (approx. 2,200 K) is a good IR emitter. The schematic construction of a IR decoy flare is shown in Fig. 2.22. The main reaction between Mg and Teflon can be formulated as follows, where $m \geq 2$:

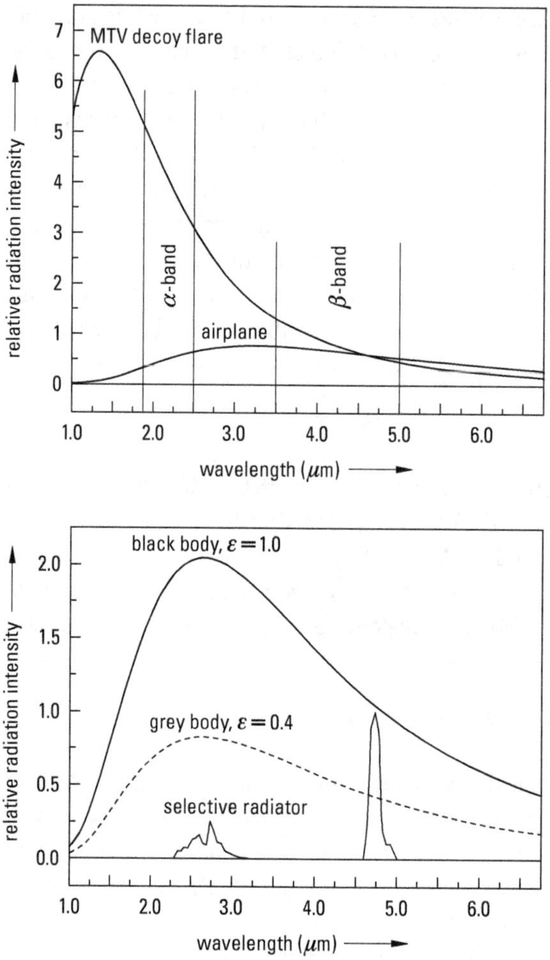

Fig. 2.21: Comparison of the relative radiation intensity of a plane and a MTV decoy flare.

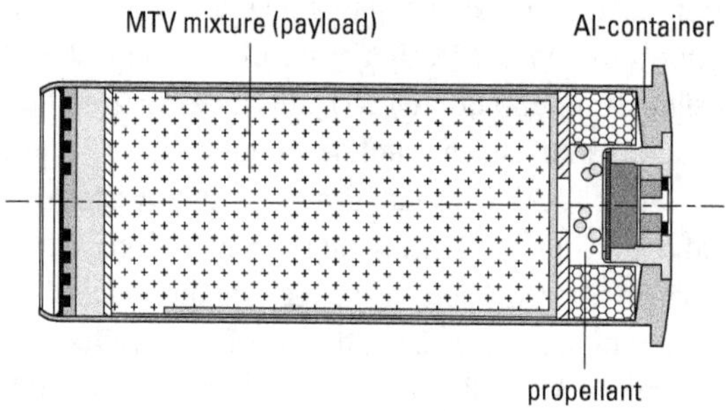

Fig. 2.22: Schematic construction of an IR decoy flare.

$$m\,Mg + —(C_2F_4)— \rightarrow 2\,MgF_2\,(l) + m\text{-}2\,Mg\,(g) + 2\,C + h\nu$$

The highly exothermic reaction for the formation of MgF_2 heats up the carbon soot formed to approx. 2,200 K, which then emits the IR radiation. Moreover, in Mg rich formulations (m ≥ 2), the evaporating Mg is oxidized in the gas phase (3,100 K). In addition, the carbon which is formed from the reductive elimination of fluorine from Teflon can be oxidized further to CO or CO_2 by atmospheric oxygen:

$$m\,Mg + —(C_2F_4)— + O_2 \rightarrow 2\,MgF_2\,(l) + m\text{-}2\,MgO\,(s) + 2\,CO_2 + h\nu$$

Therefore, during the radiation of a MTV decoy flare, the greybody radiation of the carbon soot (intensive) is superimposed with the less intensive selective radiation of the CO or CO_2, respectively. In addition to the typical wavelength dependent intensity of the radiation of a plane, Fig. 2.21 also shows that of a MTV decoy flare.

Typical pyrotechnic compositions for the generation of blackbody radiation contain magnesium as the fuel and a fluorine-containing oxidizer such as e.g. polytetrafluorethylene, PTFE, $(C_2F_4)_n$. During the reaction, (1) large quantities of carbon form (which is a strong emitter in the IR region) and become thermally excited to radiation as a result of the heat of reaction. Decoy compositions are always rich in magnesium (50–70 weight-% Mg) and use the oxygen in the air as a complimentary oxidizer (2). This will now be described in greater detail by using a MTV countermeasure as an example, consisting of 63 weight-% Mg and 37 weight-% PTFE:

$$\begin{aligned} 2.592\,Mg\,(s) + 0.3699\,(C_2F_4)_n &\rightarrow 0.7398\,MgF_2\,(s)\,0.7398 + 1.8522\,Mg\,(s) \\ &\quad + 0.7398\,C\,(gr) + 532\,kJ \end{aligned} \qquad (1)$$

$$1.8522\,Mg\,(s) + 0.7398\,C\,(gr) + xs.\,O_2 \rightarrow 1.8522\,MgO\,(s) + 0.7398\,CO_2 + 1403\,kJ \qquad (2)$$

The primary reaction (1) contributes only about a third of the total heat of the reaction. Despite this, the primary reaction significantly influences the mass flow and the radiation characteristics. Therefore, it is sensible to increase the exothermicity of the primary reaction, in order to reach the highest possible spectral efficiency, E, [$J\,g^{-1}\,sr^{-1}$]. For this reason, it has been suggested to use oxidizers which would result in a higher primary enthalpies of reaction than PTFE. Suitable substances with a higher molar enthalpy of reaction with magnesium than PTFE are graphite fluoride, $(CF_x)_n$, perfluorinated carbocyclene $(CF)_n$ or difluoroamine substituted compounds $\{C(NF_2)\}_n$.

The flame of a MTV composition consists of an inner zone of higher emission intensity, which contains a high concentration of carbon particles, which are formed through a primary reaction (1) (Fig. 2.22a). The primary reaction zone is surrounded by a second layer of lower emission intensity in which the oxidation of the magnesium and the carbon with oxygen from the air occurs (eq. 2). This intermediate zone is finally surrounded by an outer zone in which the cooling of the reaction products CO_2 and MgO occurs through radiation transfer and conduction of heat to the surrounding atmosphere. The high combustion temperature of the second layer heats up the inner zone

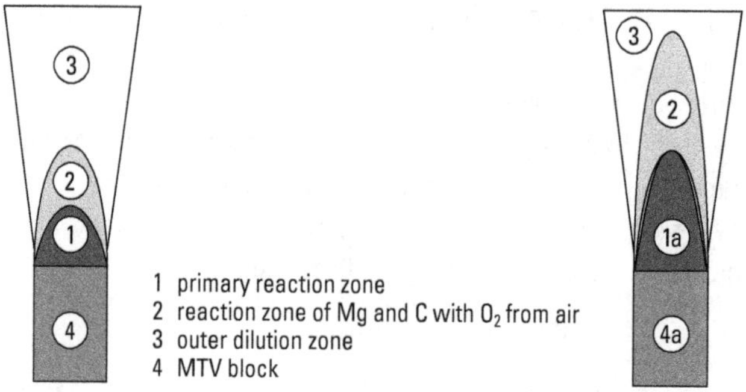

Fig. 2.22a: Typical reaction zones of a conventional decoy flare (left) and a high-nitrogen flare based on perfluoroalkylated tetrazoles (right).

even further and thereby influences the radiation characteristics considerably. Furthermore, the emission is also affected by the spatial expansion of the inner zone as well as the concentration of the carbon particles in it. Often, the concentration of the carbon particles in the inner zone is so high, that they cannot emit directly to the surroundings and therefore do not contribute to the radiation intensity. By dilution of the inner zone using an optically transparent gas such as e.g. nitrogen, the inner zone can be expanded and the concentration can also be reduced, so that a larger proportion of carbon particles can emit to the surroundings. Therefore, a radiation optimized decoy composition should deliver a sufficiently high proportion of nitrogen during combustion (Fig. 2.22a).

Recently it has been discovered that mixtures of magnesium and perfluoralkylated tetrazolates show a higher radiation intensity than compositions which are currently state-of-the-art and based on magnesium and e.g. PTFE, although the molar enthalpies of reaction for the reactions analogous to eq. (1) are lower than e.g. MTV.

It is obvious that the intensity of the radiant intensity of MTV flares increases with the mass flow (kg s^{-1}, larger flares) and the burning rate. Current research is amongst other problems concerned with finding additives (e.g. 10% Zr powder), which raise the burning rate by a factor of up to 1.5. The largest influence on the burning rate, r, is the pressure p (see Ch. 1.3), while the coefficient β describes the influence of the temperature ($\beta = f(T)$) on the linear burn rate and the index α, the pressure dependence:

$$r = \beta p a$$

To increase the temperature of reaction, components which have less negative heats of formation than PTFE can be used as well. Therefore, in some mixtures, graphite fluoride or polycarbonmonofluoride (PMF) are used instead of PTFE. This achieves higher combustion enthalpies. Furthermore, PMF shows higher thermal stability than PTFE resulting in a better (longer) storability. A typical PMF mixture has the following approximate composition:

- 55–65% Mg (fuel)
- 30–40% PMF (oxidizer)
- 5% Viton (binder)

As a result of the pressure dependence, the linear burn rate also decreases considerably with increasing altitude (Fig. 2.23).

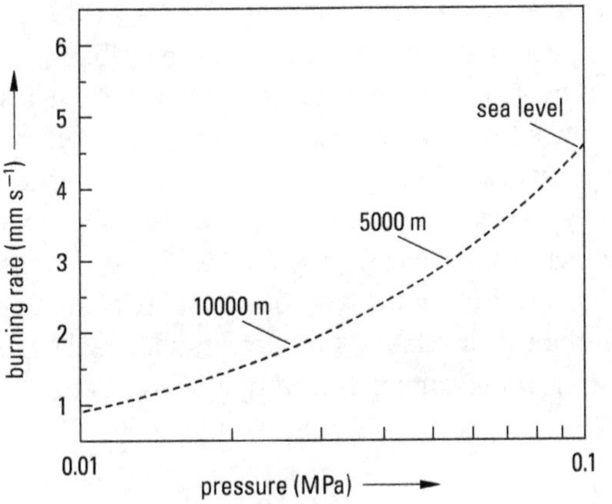

Fig. 2.23: Pressure dependence of the linear burning rate of a MTV mixture (50% Mg, 45% PTFE, 5% Viton).

The biggest problem and therefore also the biggest challenge for research is however, the intensity ratio Θ of the relative intensities in the α (I_α) and β bands (I_β). Whereas the values for this relationship lie approximately between 1.3–1.4 for MTV mixtures Θ (MTV) = (I_α) / (I_β), planes show Θ (plane) = (I_α) / (I_β) values of between approx. 0.5 to 0.8 (see Fig. 2.24). This means that modern heat-seeking guided missiles can differentiate between a real target (plane) and a MTV decoy flare. New decoy flares which show a strong emission (in β band) as a result of CO_2 or HBO_2 could solve this problem. One possibility would be to add organic substances (organic fuels) or boron compounds (formation of

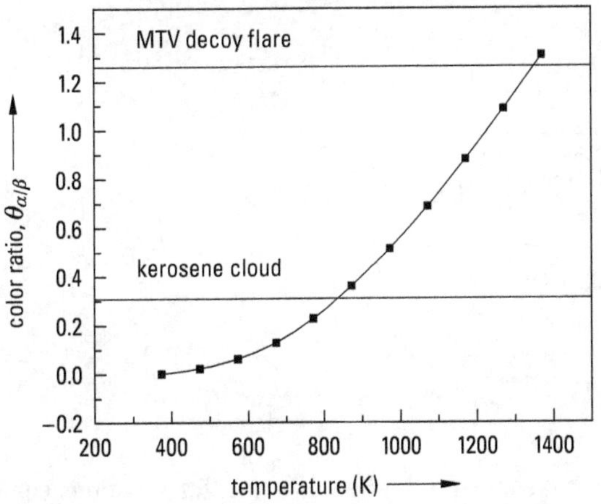

Fig. 2.24: Color ratio $\Theta_{\alpha/\beta}$ for a blackbody as a function of the temperature.

HBO_2) as well as additional oxidizers (AP) to the decoy flare compositions, but this often results in a too low intensity of the radiation, which is problematic. This is a current area of research and cannot be answered conclusively at this time.

Figure 2.21 (see above) displays the relative radiant intensity for classical MTV flares and a target signature. It can be seen that the radiant intensity distribution is totally different for MTV compared to the actual target signature. While MTV yields ratios $\Theta_{\alpha/\beta} = 1.33$, true targets exhibit ratios between $0.5 < \Theta_{\alpha/\beta} < 0.8$ depending on the type of engine. Figure 2.24 shows the intensity ratio for both α and β bands as a function of the temperature. It is obvious that only a body with a temperature of less than 900 K will provide a color ratio $\Theta_{\alpha/\beta}$ smaller than 0.8. The lower graph of Fig. 2.21 shows the total flame temperature of a plane taking into account the large amount of selective CO_2 radiation. In practice, this could mean the use of relatively cool continuous radiators, such as e.g. pyrophoric metal foils, or even the combustion of hydrocarbons to CO_2 and H_2O.

2.5.4 Smoke munitions

Smoke-generating munition is used predominantly as obscurants for self-protection (camouflage in visible and IR regions), in smoke signals and for blinding of the enemy (Fig. 2.25). An obscurant smoke is an aerosol cloud brought into the line of sight between an observer and a target (Fig. 2.26). When considering an obscurant aerosol one has to distinguish between hygroscopic (P_4O_{10}) and non-hygroscopic aerosols. Almost all aerosols which are currently used are based on red phosphorus and are therefore hygroscopic, meaning that their effectiveness is also dependent on the relative humidity (Fig. 2.27).

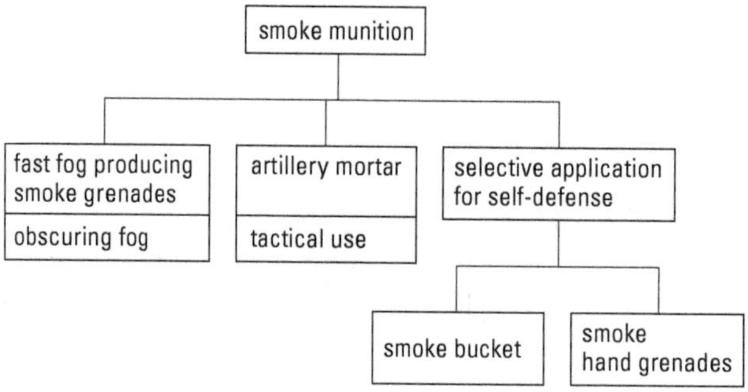

Fig. 2.25: Use of smoke munition.

In order to increase the combustion efficiency and the burning rate, formulations containing sodium nitrate or potassium nitrate, an organic binder as well as an aerosol

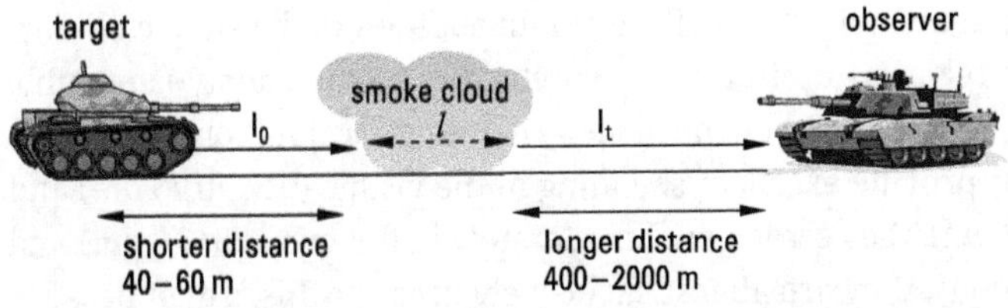

Fig. 2.26: Typical obscurant scenario (I_0 = intensity of the incoming light; I_t = intensity of the transmitted light; l = length of the fog cloud).

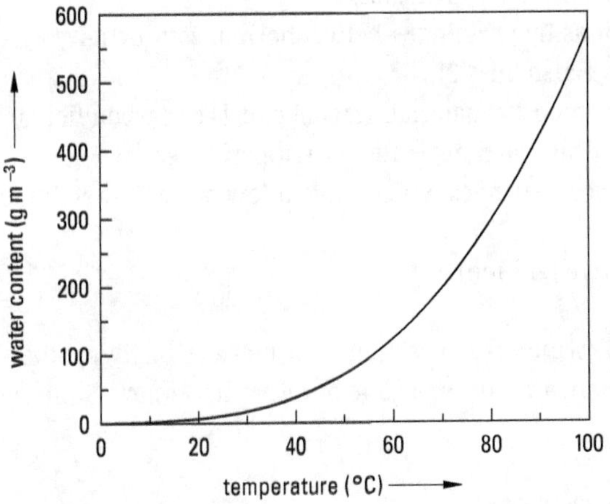

Fig. 2.27: Maximum water content of the atmosphere as a function of temperature.

agent (NH_4Cl) are used in addition to red phosphorus. Typical, simple smoke generating compositions, which are based on red phosphorus and contain approx. 10–75% P are:

Composition 1	red phosphorus	smoke generator	10%
	KNO_3	oxidizer	30%
	NH_4Cl	aerosol additive	60%
Composition 2	red phosphorus	smoke generator	50%
	$CaSO_4$	oxidizer	37%
	boron	fuel & burn rate modifier	10%
	Viton	binder	3%
Composition 3	red phosphorus	smoke generator	75%
	CuO	oxidizer	10%
	Mg	fuel & burn rate modifier	11%
	various binders	binder	4%

Increasingly, a shielding in the mid and far IR region in addition to the simple obscurent effects in the visible region is desirable, since detection by the enemy is possible using electro-optical heat sensors. Although smoke generating formulations which are based on red phosphorus provide excellent shielding in the visible (0.36–0.76 μm) and near IR regions (0.76–1.3 μm), they screen only moderately in the mid (1.5–5.5 μm) and far IR (5.5–1,000 μm) regions; they are almost exclusively used for this region because of their unsurpassed properties (aerosol formation, visible region, low toxicity etc.). Therefore, the only way to guarantee good shielding in both the mid and far IR region is by achieving higher mass flow rates (per time) of red phosphorus. This can be achieved using, amongst others, the following strategies:

1. Higher density of the formulations by pressing → better thermal conductivity
2. Higher combustion pressure (see also Ch. 1.3)
3. Higher thermal conductivity of composite materials (metallic fuels or carbon fibers)
4. More strongly exothermic chemical reactions → higher temperatures
5. Larger surface areas (no spherical particles which have a low surface to volume ratio)
6. Better ignition process (detonative ignition)

Particularly interesting is aspect (4) which is concerned with more strongly exothermic reactions. This makes the presence of an energetic additive necessary. Suitable additives are, for example:

a) Metals / Metal nitrates
b) Metals / Fluorocarbons (see also Ch. 2.5.3)
c) Metals / Oxides
d) Other homogeneous highly energetic materials.

Examples of such "advanced" smoke-generating compositions are the following two formulations:

Composition 4	Mg	7–12%
	Red phosphorus	65–70%
	PTFE	17%
	Polychloroprene binder	6%
Composition 5	Mg	12%
	Red phosphorus	66%
	KNO_3	16%
	Polyvinylalcohol (PVA)	6%

One fundamental problem with formulations containing red phosphorus and magnesium, in particular in formulations which contain a substoichiometric quantity of oxidizer, is that a thermodynamically favorable side-reaction between the phosphorus and magnesium can occur:

$$2\,P + 3\,Mg \rightarrow Mg_3P_2$$

The magnesium phosphide that is subsequently formed can react with atmospheric moisture to form phosphine, PH_3 which is toxic:

$$Mg_3P_2 + 6\,H_2O \rightarrow 3\,Mg(OH)_2 + 2\,PH_3$$

This unwanted reaction can also occur very slowly if smoke munition is stored for several years. This can result in considerable contamination of the munition depot (mainly bunkers or tunnel-shaped caves) with gaseous PH_3. For this reason, in modern formulations Mg is often replaced by Ti or Zr. These metals could in principal also form the corresponding metal phosphides, however, these metal phosphides do not react with water under normal conditions to form PH_3. Moreover, boron and/or silicon additives increase the shielding in the mid and far IR regions as a result of the formation of B_2O_3 (4–5 μm) or SiO_2 emmitters (9–10 μm). A typical Mg-free smoke formulation, which also shields good in the mid and far IR regions is:

Composition 6	Red phosphorus	Smoke generator	58.5%
	KNO_3	Oxidizer	21.1%
	Zr	Energetic fuel	4.7%
	Si	Energetic fuel	4.7%
	B	Energetic fuel	4.7%
	Chloroprene	Binder	6.3%

The potential of **phosphorus(V) nitride, P_3N_5,** as a replacement for red phosphorus (PR) in pyrotechnic obscurants has been theoretically and experimentally investigated by Koch and Cudzilo [Angew. Chem. Int. Ed. **2016**, 55, 15439–15422]. P_3N_5 can be safely mixed with KNO_3, and even $KClO_3$ and $KClO_4$. The corresponding formulations are surprisingly insensitive to friction and only mildly impact-sensitive. P_3N_5 / KNO_3 pyrolants with 20–80 wt % P_3N_5 burn 200 times faster than the corresponding mixtures based on PR and generate dense smoke. An oxygen-balanced formulation would have the following approximate composition: 35% P_3N_5, 65% KNO_3. Furthermore, unlike PR (which slowly degrades in moist air, forming phosphoric acids and phosphine (PH_3)), P_3N_5 is stable under these conditions and does not produce any acids or PH_3.

The size of the smoke cloud and its duration are of course strongly dependent on the amounts of smoke munition used (small hand grenades or 18 kg floating smoke pot) as well as the weather. While no good fog is formed (see Fig. 2.27) in extremely dry conditions, the duration of the cloud is significantly shorter in strong winds than in wind-free conditions. Typical values e.g. for a 155 mm artillary smoke greande are summarized in Tab. 2.13.

Previously, ZnO / Al / hexachloroethane (HC) mixtures were used both as smoke- and fog-generating munitions as well. In those mixtures, the intermediate $ZnCl_2$ (and C, graphite) is formed, which then quickly hydrolyses forming HCl and ZnO. Such non-

Tab. 2.13: Typical values using 155 mm artillery smoke grenade.

	Visible region (0.4–0.7 μm)	Mid IR region (3–5 μm)	Far IR region (8–14 μm)
Duration of smoke	260 s	180 s	180 s
Effective length	400 m	180 m	180 m
Effective height	12–16 m	8–12 m	8–12 m
Effective depth	ca. 40 m	ca. 40 m	ca. 40 m

hygroscopic formulations are only used in handheld flares where the formation of acidic products (H_3PO_4) can't be tolerated. However, some of the side-products of the reactions have been found to be toxicologically and ecologically critical (e.g. hexachlorobenzene, hexachlorbutadiene, chlorinated dibenzofurane, dibenzodioxine), therefore research is underway to find suitable replacements for HC. Other munitions, which have to the knowledge of the author of this book, never been fielded by NATO countries contain TiO_2 / Al / HC and also Ti / ZnO / HC.

$$C_2Cl_6 + 2\,Al \longrightarrow 2\,C + 2\,AlCl_3$$

$$2\,AlCl_3 + 3\,ZnO \longrightarrow Al_2O_3 + 3\,ZnCl_2$$

$$3\,ZnCl_2 + 3\,H_2O \longrightarrow 3\,ZnO + 6\,HCl$$

In addition to phosphorus-based obscurant smokes for self-protection there are also colored smoke compositions for signaling purposes, often in a military context. These can be produced by smoke grenades, or by various other pyrotechnical devices.

Smokes, in general, are cooler burning pyrotechnic mixtures. The reason is twofold. 1. A cooler burning smoke will rise gradually from the ground when deployed, which will serve as an effective screening/obscuration tool for military personnel. A smoke mixture which is too hot in temperature will result in a smoke rising too rapidly. When this happens, the screening/obscuration purpose of the smoke is lost. 2. A smoke which reaches high temperatures will result in combustion of the smoke dye (*i.e.* oxidation to carbon monoxide and carbon dioxide). Smoke dyes are added into smoke formulations for sublimation purposes. Not only does the sublimation event keep smokes cool (*i.e.* sublimation is an endothermic process), but it is the sublimation of the smoke dye which gives the smoke its distinct color (*i.e.* blue, red, yellow, green, black, etc.).

As has been commonly taught in smokes for years, the temperatures at which oxidizers and fuels melt is largely responsible for the temperature which a smoke will ultimately reach. Potassium chlorate has been identified as the "only oxidizer" to be used in smoke mixes because it has an exothermic decomposition, thus contributing energy to the pyrotechnic system, but also has a relatively low melting point. Scientists at ARDEC are currently looking at ways to replace potassium chlorate oxidizer in smoke mixes with other oxidizing agents. The EPA is looking at potassium chlorate now and questioning whether potassium chlorate is also bad like potassium perchlorate. Another problem with potassium chlorate is its sensitivity. Potassium chlorate used to be used in many of the early fireworks, but it was replaced by potassium perchlorate because potassium perchlorate was found to be inherently safer to work with.

As pointed out above, $KClO_3$ is much less stable than perchlorate and therefore more hazardous (especially mixtures with sulfur and phosphorus). On the other hand, it shows high burning rates and easy ignition. $KClO_3$ is slightly more hygroscopic than potassium nitrate and produces smoke of KCl. Furthermore, it can act as a chlorine donor.

The mixture used for producing colored smoke is usually a cooler-burning formula based on potassium chlorate oxidizer, $KClO_3$ (ca. 35%) sugar as a fuel (ca. 20%), and one or more dyes, with about 40–50% content of the dye. About 2% sodium bicarbonate may be added as a coolant, to lower the burning temperature. Its coolant properties arise because 2 moles of sodium bicarbonate will endothermically decompose into one mole each of sodium carbonate, carbon dioxide and water. In addition to reducing flame temperatures of smokes, coolants are also used to reduce the sensitivity of smoke mixes, given the large amount of potassium chlorate used in these mixes. The presence of the coolants minimizes undesired acid formation and helps to prolong storage of the smoke mixes.

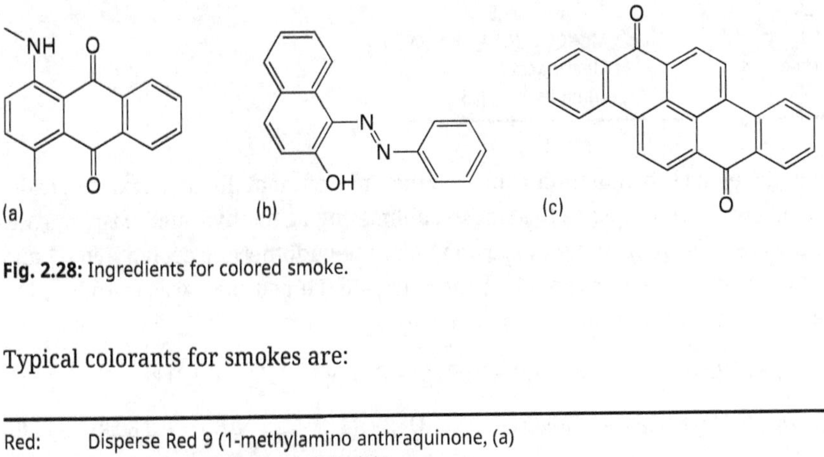

Fig. 2.28: Ingredients for colored smoke.

Typical colorants for smokes are:

Red:	Disperse Red 9 (1-methylamino anthraquinone, (a)
Orange:	Sudan I, 1-phenylazonaphth-2-ol (b)
Yellow:	Vat Yellow 4 (c) together with Benzanthrone (BZA)
Violet:	Disperse Red 9 (a) with 1,4-Diamino-2,3-dihydroanthraquinone(1,4-Diaminoanthracene-9,10-dione)
Green:	Vat Yellow 4 (c) together with with benzanthrone and Solvent Green 3
Black:	Naphthalene and anthracene

In the US, Nation Ford Chemical is the main supplier to the US military of the complete line of domestically produced dyes for smoke signaling and cover: http://www.nationfordchem.com/products/smoke-dyes/.

Sabatini, Moretti et al. from ARDEC located at Picatinny Arsenal, N.J. have now developed a prototype of a hand-held signal flare with an environmentally benign yellow smoke formulation to replace the environmentally and toxicologically hazardous mixture currently specified for the U. S. Army's M194 yellow smoke hand-held signal (*ACS Sustainable Chem. Eng.* **2013**, *1*, 673–678). This new formulation meets the burn time parameters outlined in military specifications and is composed entirely of dry, powdered, solid ingredients without the need of solvent-based binders. In addition, this formulation was found to have relatively low sensitivity towards impact, friction and electric discharge.

The main goal was to replace toxic dyes such as Benzanthrone and Vat Yellow 4 – which are presently used in the in-service M194 formulation – with environmentally benign dyes such as Solvent Yellow 33 (Fig. 2.28a, Tab. 2.13a).

Tab. 2.13a: In-service and new formulations for yellow smoke.

In-serviceM194 Ingredients / wt%	New formulation Ingredients / wt%
Vat Yellow 4 / 13	Solvent Yellow 33 / 37
$KClO_3$ / 35	$KClO_3$ / 34.5
Sucrose / 20	Sucrose / 21.5
$NaHCO_3$ / 3	$Mg_5(CO_3)_4(OH)_2 \cdot 4\,H_2O$ / 5.5
Benzanthrone / 28	Stearic acid / 1
VAAR / 1	Fumed silica / 0.5

The newly developed formulation contains: the environmentally and toxicologically benign dye Solvent Yellow 33 as the smoke sublimating agent, hydrated basic magnesium carbonate instead of sodium bicarbonate as the endothermic coolant (eqs. 1 and 2) and stearic acid as the lubricant and processing aid. Fumed silica was introduced in the absence of a binder to promote homogeneity

$$2\,NaHCO_3(s) \rightarrow Na_2CO_3(s) + H_2O\,(g) + CO_2\,(g) \quad \Delta H_r = +0.81\,kJ/g \tag{1}$$

$$Mg_5(CO_3)_4(OH)_2 \cdot 4H_2O \rightarrow 5\,MgO\,(s) + 5\,H_2O\,(g) + 4\,CO_2(g) \quad \Delta H_r = +0.80\,kJ/g \tag{2}$$

Despite the fact that both reactions (1) and (2) are endothermic and have comparable per gram enthalpies, reaction (1) occurs completely within the cool-burning temperature range of sucrose/$KClO_3$ smoke compositions, whereas reaction (2) is incomplete. Sodium bicarbonate is therefore a more aggressive coolant than hydromagnesite, as it effectively removes more energy from the smoke composition in the temperature range in which they burn. The use of $NaHCO_3$ in place of $Mg_5(CO_3)_4(OH)_2 \cdot 4H_2O$ results in longer burn times, i.e. slower burn rates. Therefore, it is essential to employ $Mg_5(CO_3)_4(OH)_2 \cdot 4\,H_2O$

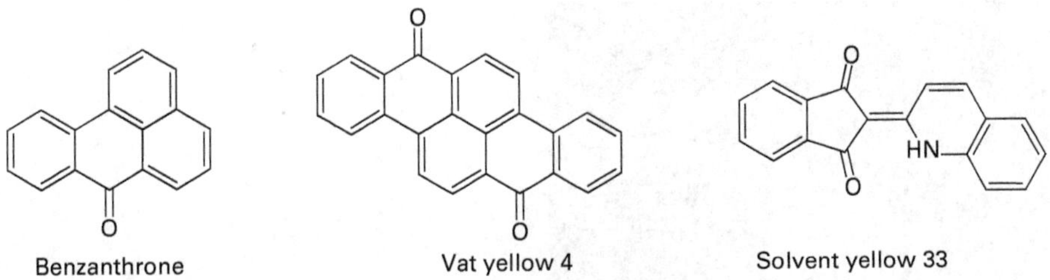

Fig. 2.28a: Chemical structures of yellow dyes.

in order to achieve a yellow smoke formulation that meets the short burn time requirements of M194.

Table 2.13b summarizes the performance and sensitivity data of the new smoke formulation in comparison with the military requirements and Fig. 2.28b shows an image of a smoke fountain generated by the new yellow smoke formulation.

Tab. 2.13b: Comparison of the performance and sensitivity data of the newly developed smoke formulation with those of the military requirements.

	Military requirement for M-194	**New formulation**
Burn time / s	9–18	15
Burn rate / g s^{-1}	3.89–7.78	4.64
IS / J		17
FS / N		> 360
ESD / J		> 0.25

It should also be noted that white smokes are also of interest as well. There are several ways to produce a white smoke. In the old days, sublimation of sulfur did the trick (again, a low melting fuel), but the formation of sulfur dioxide always was seen as a problem. The hexachloroethane (HC) smoke was around for years (see above), which consisted of hexachloroethane, zinc oxide and aluminum. The trick in using the HC smoke was that it led to the formation of the Lewis acids zinc chloride and aluminum trichloride, and these smokes readily reacted with moisture in the air to produce aluminum hydroxide, zinc hydroxide, and most importantly from a smoke perspective, large amounts of hydrochloric acid. It was the hot HCl droplets that formed which produced a thick smoke upon cooling and pulling additional moisture from the air. Metal and metal oxide particles also contributed to the smoke production due to these finer particles scattering over a rather wide area. However, HC is now deemed toxic, and in fact, HC smokes are no longer being produced by the military in the USA. The military has been complaining about the lack of an HC replacement. Right now, the candidate to re-

Fig. 2.28b: Image of a smoke fountain generated by the new yellow smoke formulation.

place it is the terephthalic acid (TA) smoke mix. But this mix is not as good as HC in terms of efficiency (40–50% of the mix is simply residue). It takes 2–3 TA grenades to equal the obscuration power of one HC grenade, and the TA smoke does not burn nearly as long. The smoke is less efficient for the TA smoke because instead of pulling moisture from the air and producing smoke with acid droplets, one is relying on sublimation and condensation of TA to produce the smoke. TA does not pull moisture from the air, which hampers its ability as a smoke obscuration agent. Nonetheless, efforts are underway at ARDEC to find a quality replacement for the HC smoke mix. Another method of producing white smoke would be the volatilization of oil, but this has not been proven to work to any appreciable extent, even after much research.

For white (camouflage) smokes, the yield factor Y, is defined as the mass of aerosol (m_a) divided by the mass of the pyrotechnic payload (m_c).

Yield factor = mass of aerosol / mass of pyrotechnic payload:

$$Y = \frac{m_a}{m_c}$$

The figure of merit (FM_m) is a quality criteria which allows different compositions to be compared with each other. It is defined as the product of the yield factor multiplied with the extinction coefficient:

$$\mathrm{FM_m} = \alpha_\lambda \cdot Y = \left(\frac{-V \cdot \ln T}{m_a \cdot T}\right) \cdot \left(\frac{m_a}{m_c}\right) = \frac{-V \cdot \ln T}{m_c \cdot L}$$

The extinction coefficient α_λ is defined according to the Lambert-Beer law,

$$\alpha_\lambda = -\frac{\ln T(obsc)}{cL} \qquad T(obsc) = \frac{I}{I(0)}$$

with:
α_λ = extinction coefficient
T = degree of transmission
$I(0)$ = intensity without smoke
c = conc. of aerosol in obscurant cloud [kg/m^3]
L = path length [m]
V = chamber volume [m^3].

It should also be mentioned that stearic acid or graphite is sometimes added in small amounts as an additive to smoke compositions. These materials serve as a lubricant, which really helps reduce sensitivities (especially to friction and impact) of smoke mixes. As for pyrotechnic binders in smokes, vinyl alcohol acetate resin (VAAR) has been used in the past. The most commonly used binder system now in smokes is polyvinyl alcohol (PVA), and it is applied wet.

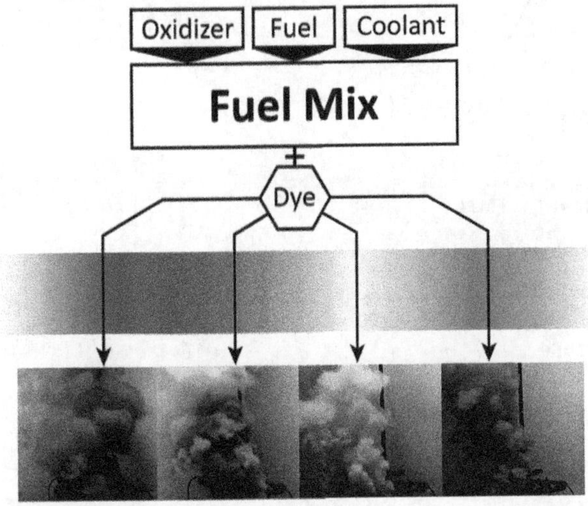

Fig. 2.28c: The concept of fuel mixes.

The development of sugar-free multi-colored smoke formulations – so-called fuel mixes – has recently been reported (Küblböck et al., *New J. Chem.*, **2018**, *42*, 10670-10675.:

DOI: 10.1039/c8nj01786g). These simple four ingredient-based mixtures of dye, potassium chlorate, 5-amino-1*H*-tetrazole, and a magnesium carbonate derivative are able to produce a variety of colors by applying the same pyrotechnical system (Fig. 2.28c). All components except the dye are pre-mixed; the dye is added in the final step. Based on previous results which indicated an overall higher smoke performance in terms of efficiency and persistence by applying 5-amino-1*H*-tetrazole as the fuel in smoke formulations, new colored smoke formulations were developed. For large producers as well as consumers, the development of fuel mixes is an effective way to reduce costs and provide a higher degree of safety. The main focus was on dyes applied in the U. S. M18 colored smoke grenades.

The most promising fuel mixes are illustrated within the ternary diagram shown in Fig. 2.28d. The first fuel mix, FM1, consisted of 50 wt% 5-AT, 30 wt% $KClO_3$, and 20 wt% MCPH and therefore, contained the highest amount of oxidizer. FM2 had a composition of 50 wt% 5-AT, 20 wt% $KClO_3$, and 30 wt% MCPH and finally fuel mix FM3, contained the highest amount of fuel with 60 wt% 5-AT and equal wt.% of $KClO_3$ and MCPH (20 wt%). The reference fuel mix Ref-FM, contained 40 wt% sucrose, 40 wt% $KClO_3$, and 20 wt% MCPH.

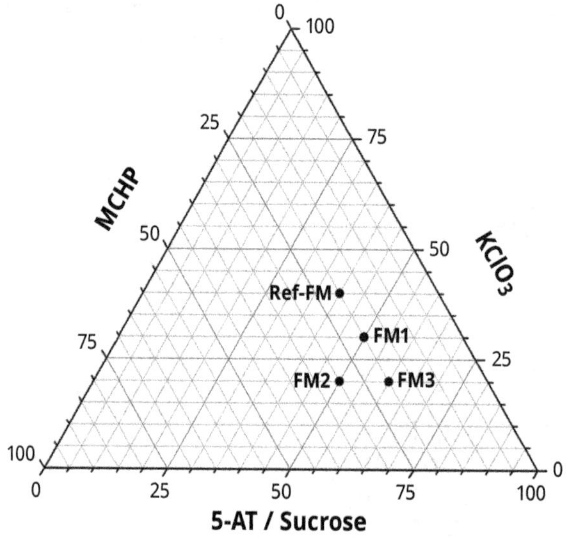

Fig. 2.28d: Developed fuel mixes. FM1 = 5-AT (50 wt%), $KClO_3$ (30 wt%), MCHP (20 wt%); FM2 = 5-AT (50 wt%), $KClO_3$ (20 wt%), MCHP (30 wt%); FM3 = 5-AT (60 wt%), $KClO_3$ (20 wt%), MCHP (20 wt%); Ref-FM = sucrose (40 wt%), $KClO_3$ (40 wt%), MCHP (20 wt%).

It was therefore successfully demonstrated that 5-AT-based fuel mixes can be used to produce green, yellow, red and violet smoke. The formulations that have been developed revealed that non-traditional high-nitrogen fuels can produce smoke of high color quality. More precisely, whereas FM1-based mixtures result in rapid, strong smoke generation in a short time, FM3-based compositions are characterized by a slow, continuous smoke generation over a much longer period of time. 5-AT is only one of many potential candidates which will be investigated in the future. All the formulations are insensitive towards friction and in addition, all the 5-AT-based colored smoke formulations are completely insensitive towards impact. The fuel mixes were more sensitive towards mechanical stimuli. A comparison of the collected aerosol revealed similar yields for FM1-based and sugar-based formulations. The superiority of the sugar-based reference formulations was illustrated by the measured transfer rates. Future investigations will no doubt focus on providing even more colored smoke formulations, e.g. blue and black, by applying the same fuel mixes. However, in order to achieve a proper evaluation, new strategies to characterize the aerosol produced by a mixture of two dyes (e.g. green dye mix) have to be established.

The illustration employing a triangle diagram (Fig. 2.28e) is a powerful tool to summarize the results of all kinds of pyrotechnical formulations. Such a diagram has three axes representing three different components with values ranging from 0 to 100%. Every point corresponds to a unique ratio of components. Since the reading direction is unintuitive for diagrams with three axes, it is marked with red lines in Fig. 2.28e for a mixture containing 50% of component 1, 30% of component 2 and 20% of component 3. It is mandatory that the percentages of all three components result in 100%. The artificial lines at the scale of every axis additionally support the given reading direction. Triangle diagrams provide assistance in discovering hidden trends and relationships between different ratios of components, and for this reason, enable an optimum pyrotechnical formula-

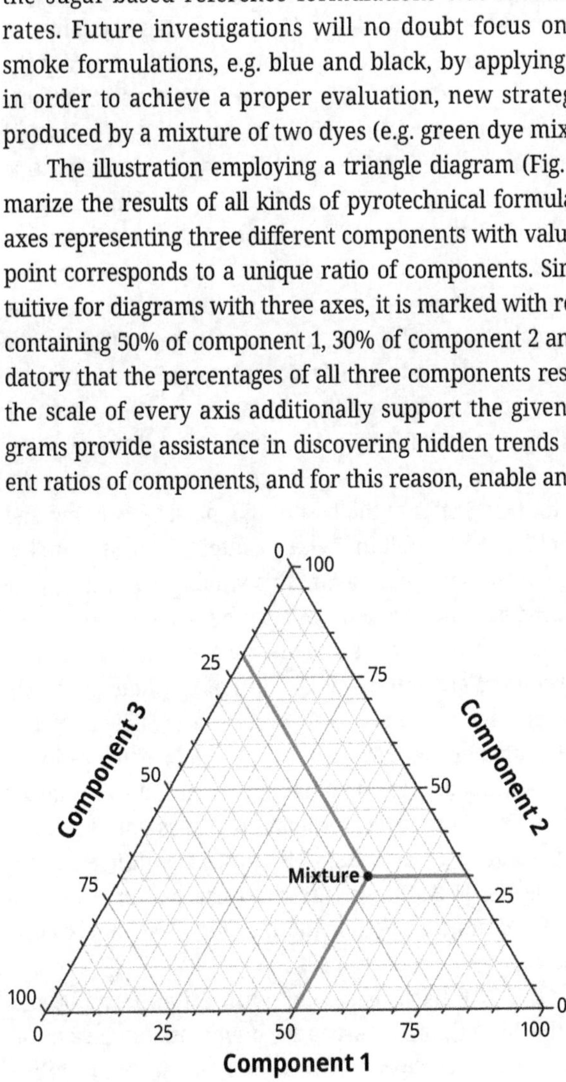

Fig. 2.28e: Triangle diagram and how to read it.

tion to be found. A more detailed explanation and study exercises are given by Kosanke (K. L. Kosanke, B. J. Kosanke, Selected pyrotechnic publications of K. L. and B. J. Kosanke, Journal of Pyrotechnics, Whitewater, CO, USA, **1995**).

2.5.5 Near-infrared (NIR) compositions

Since the mid-twentieth century, electro-optical systems such as night vision goggles and near infrared vision devices have become available. In general, these systems enable vision at night and make use of either non-visible photons and/or intensify the scarcely visible ones. The NIR range is considered to be between 700 nm and 2,000 nm. However, for detection purposes, the range between 700–1,000 nm is used extensively by several common detector materials in the UV, VIS and NIR range. For clandestine signalling and illuminating purposes in this range (700–1,000 nm) pyrotechnic compositions based on either potassium (K) and/or cesium (Cs) compounds that are based on the intense emission lines of K and Cs in the infrared, are used. These flare compositions contain silicon as a high-energy non-luminous fuel, which is known to burn in the condensed phase. A typical composition is: KNO_3 70%, hexamine 16%, Si 10%, binder 4%. Furthermore, these compositions contain nitrogen-rich compounds like e.g. hexamine or azodicarbonamide as blowing agents to peel off the condensed reaction products. While hexamine is certainly a historically used fuel in pyrotechnic IR applications, especially given its relatively low light output due to its high-nitrogen content, one of the drawbacks of hexamine (at least in the USA), is a) the cost of the material is rising and b) hexamine is getting harder and harder to obtain. In this context, lactose monohydrate has potential to serve as a hexamine replacement. This would solve the supply issue and reduce cost if it can be used in IR applications, as lactose monohydrate is very cheap, and is a fraction of the cost of hexamine. It has a higher heat of combustion than hexamine, but still gives off a substantial amount of gas (i.e. lots of CO_2 and some CO). While materials with higher heats of combustion are likely to increase visible light output, this can be minimized or eliminated altogether by adjusting oxidizer/fuel ratios in the formulations. Presently research is focussing on compounds with an even higher nitrogen content such as potassium or cesium azotetrazolate or potassium or cesium bis(tetrazolyl)amine. Typical oxidizers are the peroxides of barium, strontium and zinc, and stannic oxide. Besides NIR tracers, NIR illuminants comprise parachute signal flares as well as illuminating rounds for mortars. For clandestine aerial reconnaissance, NIR illumination rockets can be used.

From Fig. 2.29 (top) it is clear that NIR flares based on pure cesium formulations ($CsNO_3$) are superior emitters in the NIR region, whereas cesium/potassium based formulations ($CsNO_3 / KNO_3$) (Fig. 2.29 bottom) show a NIR color impurity with tailing into the visible red region.

The radiant intensity (I) is a measure of radiometric power per unit solid angle Ω ($\Omega = A / r^2$), expressed in watts per steradian ($W\ sr^{-1}$). The value of the solid angle Ω is

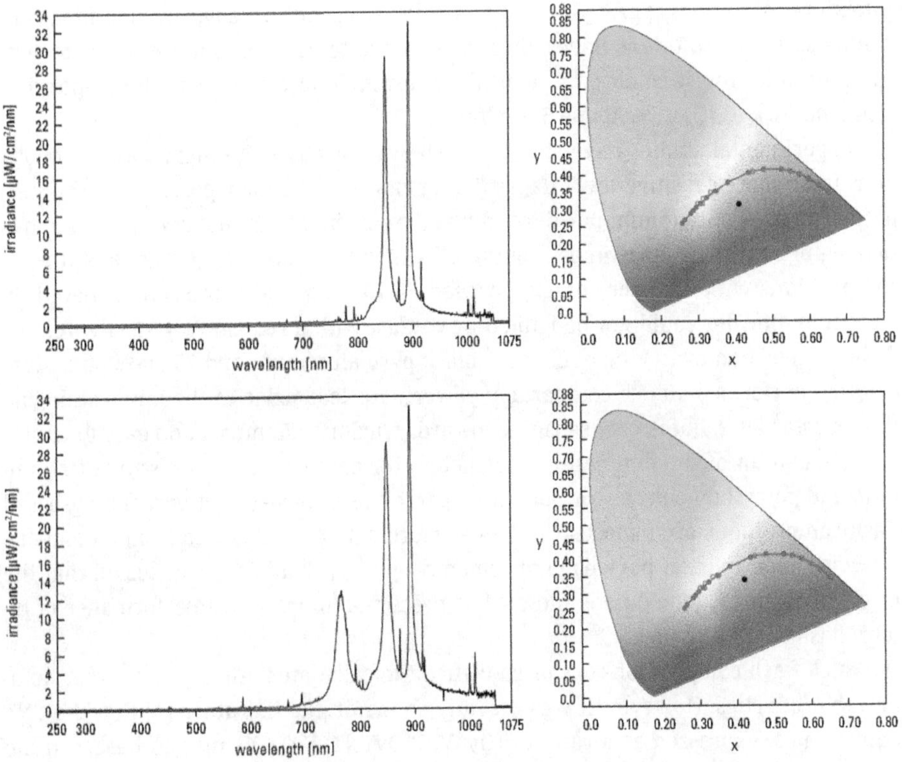

Fig. 2.29: NIR flare emission spectra (left) and chromaticity diagrams (right) for $CsNO_3$ based flares (top) and KNO_3/$CsNO_3$ based formulations (bottom).

numerically equal to the size of that area divided by the square of the radius of the sphere. For hand-held NIR flares one wants a reasonable burn time (ca. 45 s), a high concealment index X which is defined as the ratio of the radiant intensity in the near IR region (I_{NIR}, 700–1,000 nm) over the radiant intensity in the visible region (I_{VIS}, 400–700 nm) and high radiant intensities in the NIR regions I_1(600–900 nm) and I_2(695–1,050):

$$X = I_{NIR} / I_{VIS} > 25;\ I_1 > 25\ \mathrm{W\ sr^{-1}};\ I_2 > 30\ \mathrm{W\ sr^{-1}};\ \text{luminous intensity (visible)} < 350\ \mathrm{cd}$$

2.5.6 Flash compositions

Flash compositions are widely recognized as extremely dangerous pyrotechnic compositions. Despite this, they are used in civil fireworks as well as in military and law enforcement applications, e.g. effect simulators or flash grenades. Flash compositions comprise a wide field of pyrotechnics, containing simple binary compositions of metals (e.g. Al, Mg, Fe), oxidizers [e.g. $KClO_3$, $KClO_4$, KNO_3, $Ba(NO_3)_2$], and compositions with admixtures of sulfides (e.g. Sb_2S_3), sulfur, or coloring agents. Usually, a composition containing sulfur or sulfide is believed to be easier to ignite, but is more prone to self-ignition under moist conditions, as well. An example of a typical flash composition would be: 70% $KClO_4$, 25% Al and 5% sulfur.

Experimental studies have shown the chlorate-containing binary flash compositions to be more sensitive towards friction or impact than their perchlorate counterparts. The use of aluminum powder reduces the sensitivities of the resulting composition significantly but does impact ignitability as well. Binary aluminum-based flash compositions with oxidizer contents exceeding 70 wt-% and compositions based on aluminum powder could not be ignited by contact with a red glowing steel wire. The flash composition consisting of 30 wt-% black pyro aluminum and 70 wt-% potassium chlorate performed almost comparably powerful as lead azide in the micro-scale ballistic mortar tests, but is very sensitive towards friction and impact and easy to ignite.

Calculation of the detonation parameters for compositions consisting of aluminum and potassium chlorate or potassium perchlorate showed that perchlorate-based flash compositions are expected to deliver more energy in their explosion. Comparison with experimental power measurements however, led to the conclusion that the calculated values alone do not allow a fair prediction of explosive performance of binary flash compositions.

Such flash compositions are often difficult to calculate with regard to their detonation properties. However, it was recently shown in the literature [4] that the TNT equivalence, expressed as a ratio of (Qv*Vo)/(QvTNT*VoTNT), or also based on the C-J pressure (pC-J), provides the best Trauzl test results (in the Trauzl test and in mining applications, since the amount of gas produced and, above all, the detonation heat are crucial here). In the following we refer to the TNT equivalents (pC-J), since the pressure built up is of crucial importance in the present case.
The calculation results of the typical mixture of 70% $KClO_4$, 25% Al and 5% sulfur compared to TNT are listed in the table below (Tab. 2.13c).

Tab. 2.13c: Computational results (EXPLO5) for a flash composition of the following nature: 70% $KClO_4$, 25% Al and 5% sulfur.

Explosive	Density / g cm^{-3}	VoD / m s^{-1}	$p_{C\text{-}J}$ / GPa	Q_{ex} / kJ kg^{-1}	V_0 / L kg^{-1}	TNT-Eq. ($p_{C\text{-}J}$)
TNT	1.64	6,773	18.2	4,351	596	100
70% $KClO_4$, 25% Al, 5% sulfur	2.54	4,889	8.4	8,219	47	45

Below, the calculation of pressure wave parameters (pressure, momentum, etc.) from a hemispherical surface shock was carried out using the Kingery-Bulmash equation. The parameters were calculated as a function of the explosive type, the mass (W) and a distance (R) or scaled distance (Z = r/W0.33) from the detonation point. The calculations were carried out for a mass of 1 kg.

For a distance of 4 m from the center of the explosion, with a TNT equivalent of 0.45 and an explosive mass of 1 kg, the following expected effects on people result:

Injuries are ubiquitous, deaths are common (Fig. 2.30).

At a distance of 6 m under otherwise identical conditions, the expected effect is:

Serious injuries are common and deaths can occur.

At a distance of 8 m under otherwise identical conditions, the expected effect is:

People are injured by flying glass and debris.

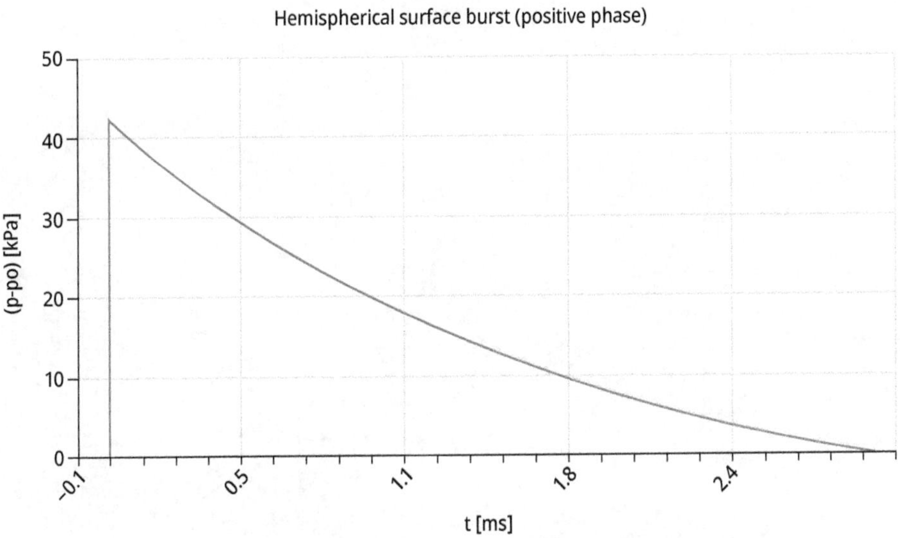

Fig. 2.30: Overpressure over time diagram for a 1 kg flash composition at a distance of 4 m from the explosion.

Literature

[1] TM 43-0001-37, Army Munition Data Sheets for Military Pyrotechnics, Dept. of the Army, Washington, DC, USA, 1994, p. 5–5 (M117 Simulator).

[2] a) J. Donner, A Professional's Guide to Pyrotechnics, Paladin Press, USA, 1997; b) H. Ellern, Military and Civilian Pyrotechnics, Chemical Publishing Company Inc., USA, 1968.

[3] T. M. Klapötke, F. X. Steemann, M. Suceska, Propellants Explos. Pyrotech. 2013, 38, 29–34.

[4] I. Dobrilovic, M. Dobrilovic, M. Suceska, Defense Technology, 2024, 36, 163–174.: https://doi.org/10.1016/j.dt.2023.08.013.

[5] EXPLO5, version 7.01.01, M. Suceska, Zagreb, Croatia, 1991–2023.
[6] W. G. Proud, Explosives and Explosive Effects, in: CBRNE: Challenges in the 21st Century, Edited by P. D. E. Biggins and D. Chana Springer, Switzerland, 2022, pp 101–136.

Check your knowledge: see chapter 14 for study questions.

3 Detonation, detonation velocity and detonation pressure

In Chapter 1.3 we defined a detonation as the propagation of a chemical reaction through an energetic material under the influence of a shock-wave at speeds faster than the speed of sound in the material. The velocity at which the energetic material decomposes is therefore only dependent on the velocity of the shock-wave. It is not determined by a heat-transfer process as it is the case for deflagration or combustion.

A detonation can result from either a continuously accelerating combustion (DDT, Deflagration-to-Detonation Transition) or from a shock (use of a primary explosive in a detonator to initiate a secondary explosive).

In the first case, we can assume that the linear burning rate increases proportionally to the pressure on the surface of the explosive:

$$r = \beta p^{\alpha}$$

Where β is the temperature dependent coefficient ($\beta = f(T)$) and α is the index of the burning rate, which describes the pressure dependence. For deflagrations $\alpha < 1$, however, this value increases to $\alpha > 1$ for detonations. The DDT transition can occur when an explosive is ignited in a confined tube, where the gases formed cannot fully escape. This results in a sharp increase of the pressure and reaction velocity. Therefore, in a detonating explosive, the reaction velocity can increase above the speed of sound, turning the deflagration into a detonation.

In the second case we assume that an explosive is subjected to a shock-wave. This method is used in detonators (see Ch. 2.5.1); the shock-wave of a detonating primary explosive initiates a secondary explosive. This shock-wave compresses the secondary explosive enough to cause the temperature to increase to higher than the decomposition temperature through adiabatic heating and the explosive material that is directly behind the shock-wave front reacts. During this reaction the shock-wave accelerates as a result of the strongly exothermic reaction of the secondary explosive. As a result of the influence of the shock-wave, the density of the explosive shortly before the reaction zone increases to 1.3–1.5 times the maximum density in the crystal (TMD), while in the thin (up to ca. 0.2 mm) chemical reaction zone directly behind the shock-wave front, temperatures of up to and over 3,000 K and at its end pressures of over 330 kbar can result. If the propagation of the shock-wave through the explosive occurs at speeds faster than the speed of sound, the process is described as a detonation. The shock-wave moves through the explosive under constant acceleration until it reaches a stationary state. The **stationary state** is reached, when the **free energy released** through this exothermic chemical reaction **is the same** as the **energy which is released to the surroundings as heat + the energy which is necessary to compress and move the crystal**. This means that when these chemical reactions occur with the

https://doi.org/10.1515/9783111447087-003

release of heat at constant pressure and temperature, the propagation of the shock-wave will become a self-sustaining process.

A critical characteristic of energetic materials is the Chapman–Jouguet (CJ) state, which describes the chemical equilibrium of the products at the end of the reaction zone of the detonation wave before the isentropic expansion. In the classical Zel'do-vich-Neumann–Döring (ZND) detonation model, the detonation wave propagates at a constant velocity, for which the CJ point characterizes the state of reaction products in which the local speed of sound decreases to the detonation velocity as the product gases expand. Although a simple concept, the time and duration of the CJ state have thwarted its experimental characterization. The experimental challenges have stimulated theoretical developments, such as variational perturbation theory and integral equation theory, to provide numerical models. Unfortunately, these approaches have been hampered by a lack of knowledge about the atomistic chemical and physical processes and by inaccuracy in the assumed equations of state (EOS).

According to the classical ZND model, the thermodynamic quantities of material in an initial unshocked state and final shocked state are related by the equations of mass, momentum, and energy conservation across the shock front as follows:

$$\rho_0 D = \rho(D - u) \tag{3.1}$$

$$p - p_0 = \rho_0 u D \tag{3.2}$$

$$e + pv + \frac{1}{2}(D - u)^2 = e_0 + p_0 v_0 + \frac{1}{2}D^2 \tag{3.3}$$

where ρ is the density, D is the velocity of the detonation wave propagating through the material, u is the velocity of the products behind the detonation wave, p is the pressure, e is the specific internal energy, and V = 1/ρ is the specific volume. The term "'specific"' refers to the quantity per unit mass, while the subscript "0" refers to the quantity in the initial unshocked state.

Using eqs. (3.1) and (3.2), eq. (3.3) can then be written, which satisfies the Hugoniot function as:

$$H_g = 0 = e - e_0 - \frac{1}{2}(P + P_0)(V_0 - V) \tag{3.4}$$

The detonation velocity D can then be calculated by solving eqs. (3.1) and (3.2), which results in

$$P - P_0 = \rho_0 D^2\left(1 - \frac{V}{V_0}\right) \tag{3.5}$$

The set of parameters for which $H_g = 0$ determines the Hugoniot curve of shocked states in the unreacted material, and the states along the Rayleigh line in the reacting material.

The Rayleigh line is a straight line connecting points corresponding to the initial and final states plotted on a graph of pressure versus volume for a substance which is subjected to a shock wave. The CJ state is the tangent point between the fully reacted Hugoniot curve and the Rayleigh line.

Under the influence of the dynamic behavior of the shock-wave, a thin layer of still unreacted explosive gets compressed alongside the shock adiabat of the corresponding explosive (or Hugoniot adiabat) from the original specific volume V_0 ($V_0 = 1/\rho_0$) to volume V_1 (Fig. 3.1). As a consequence of the dynamic compression, the pressure increases from p_0 to p_1, which in turn causes an increase in the temperature in the thin compressed layer of the explosive (Fig. 3.1), which causes an initiation of the chemical reaction. At the end of the chemical reaction, the specific volume and pressure have the values V_2 and p_2. This state corresponds to the point of the detonation product on the shock adiabat (Fig. 3.1). At this point it is important to emphasize again that in a deflagration, the propagation of the reaction occurs through thermal processes, while for the considerably quicker occurring detonation a shockwave mechanism occurs.

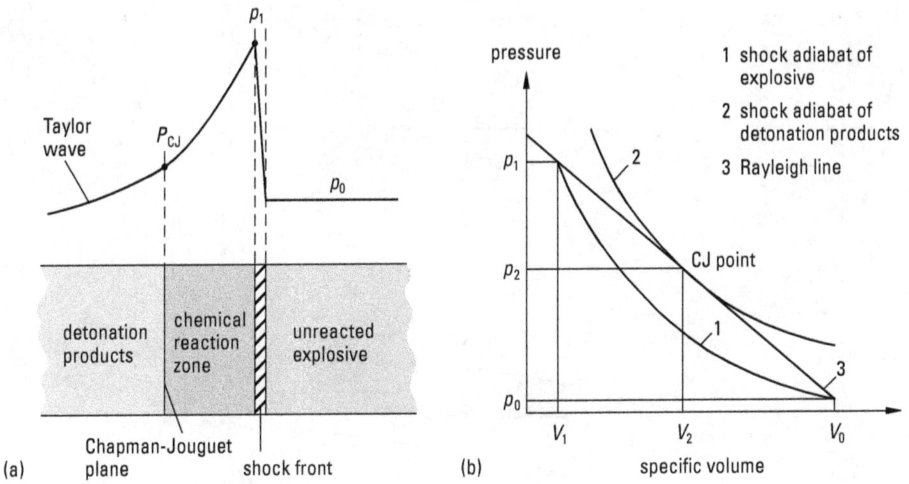

Fig. 3.1: Schematic representation of the detonation process and the shock-wave structure (a) as well as the shock adiabat for an explosive and the detonation products (detonation in a stationary state) (b).

The desirable situation for the occurrence and linear propagation of a detonation for homogeneous explosives exists in confined tubes or cylindrical shaped explosives, in which the system should not fall below the critical diameter (characteristic for every individual explosive) as it otherwise causes the wave front to be disturbed ("loss" of energy to outside) and therefore the detonation velocity will be reduced. While for many secondary explosives one inch is a "good" tube diameter, for primary explosives often considerably smaller diameters of 5 mm are sufficient. Figure 3.2 schematically shows the propagation of a shockwave in a cylindrical shaped explosive. Since the wave front is convex and not level, the linear detonation velocity is highest in the centre of the cylindrical explosive and decreases towards the surface. For large diameters this effect plays a negligible role, but for very small diameters the surface effects can become dominant, which causes the wave front to be unstable. This so-called critical diameter must be taken into account for measurements of the detonation velocity, so that one stays above the critical diameter.

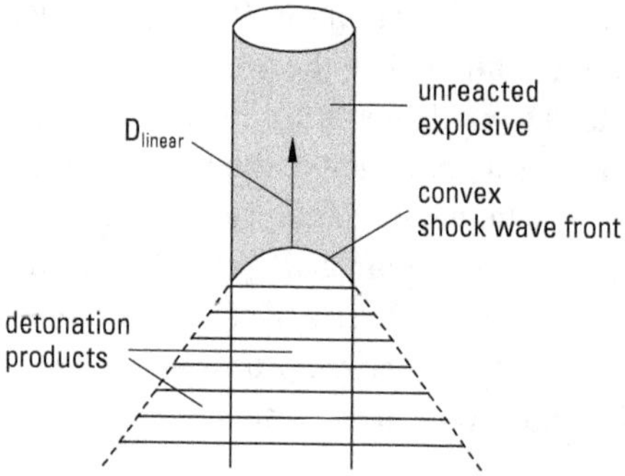

Fig. 3.2: Schematic propagation of a shock wave in a cylindrical shaped explosive.

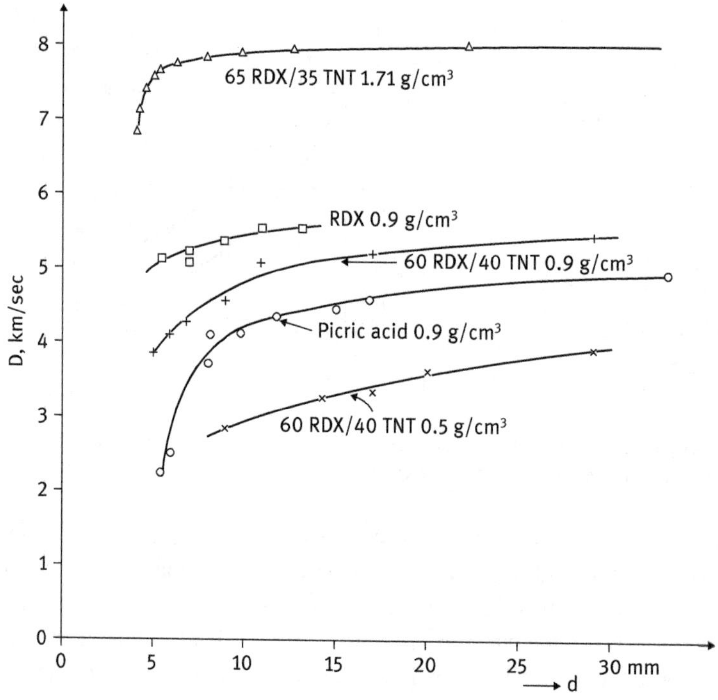

Fig. 3.2a: The dependence of the detonation velocity on the charge diameter of various high (secondary) explosives at low densities.

The critical diameter is defined as the diameter of a cylindrical charge, below which stable detonation cannot propagate, no matter how high the velocity of the initiating shock wave is. Primary explosives have usually very small critical diameters (sometimes in the μm region), whereas secondary explosives typically have larger critical diameters (in the mm to cm range). The critical diameter is also usually larger in cast than in pressed charges. Moreover, the critical diameter is of importance in considering the safety of proposed designs for detonators containing detonatable materials.

Above the critical diameter, the dependence of the detonation velocity (*D*) on the charge diameter (*d*) is characteristic for many explosives, particularly at lower densities and for explosives which exhibit non-ideal behavior. The reason for this is due to the fact that radial loses of energy are higher at lower charge diameters. Figure 3.2a shows the dependence of the detonation velocity on the charge diameter of various high (secondary) explosives at low densities.

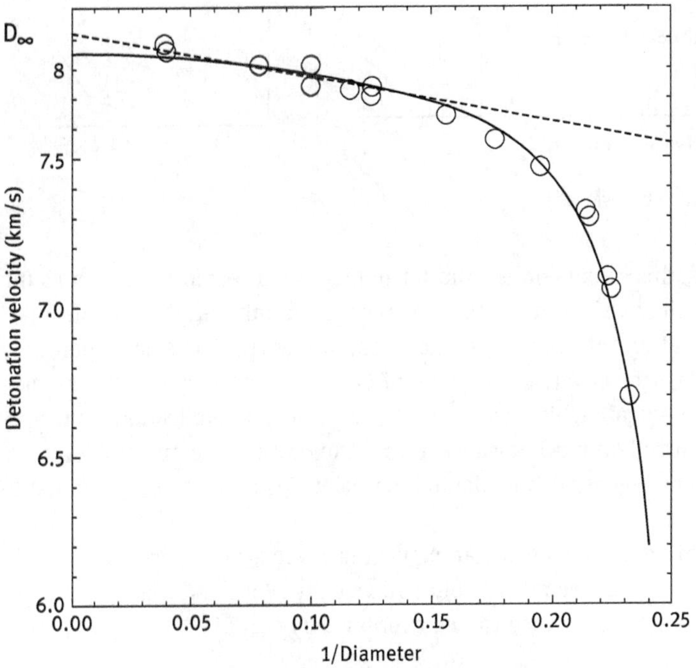

Fig. 3.2b: Extrapolation of the detonation velocity D for an infinitely large diameter of a cylindrical charge.

If the measured detonation velocity (*D*) is plotted against the reciprocal diameter of the cylindrical charge used (Fig. 3.2b), the detonation velocity for an infinitely large diameter (D_∞) can be extrapolated according to the following equation, where A_L is a constant characteristic for the explosive used:

$$D = D_{\infty}\left(1 - A_L \frac{1}{d}\right)$$

The general detonation behavior of an explosive is schematically shown in Fig. 3.2c.

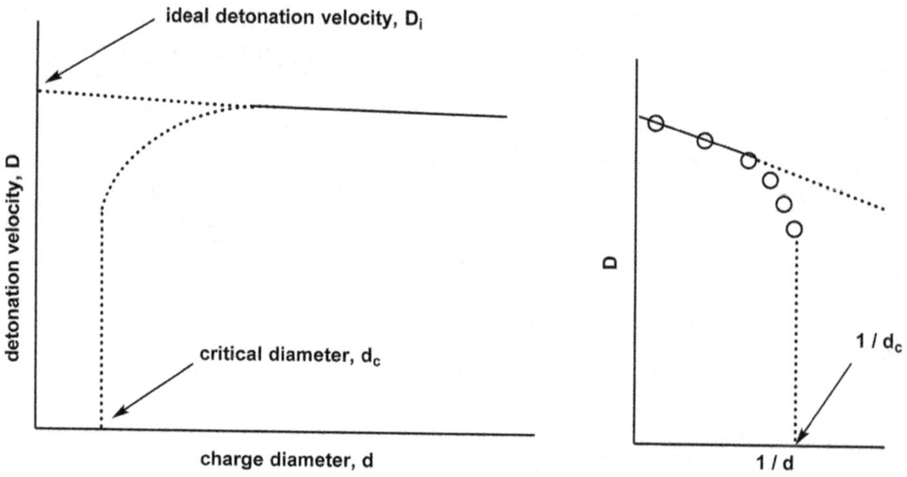

Fig.3.2c: General detonation behavior of an explosive.

As previously stated, this discussion is valid for homogeneous explosives, such as the ones used in the military, since their reactions are predominantly intramolecular. Such explosives are often referred to as "ideal" explosives, in particular when they can be described using the steady state model of Chapman and Jouguet. In heterogeneous explosives (non-ideal), which are currently used in civil applications, intermolecular (diffusion controlled) mechanisms are predominant for the air bubbles, cavities or cracks (etc.). As a general rule, detonation velocities increase proportional to the diameter.

For many unconfined, monomolecular explosives, there is an experimentally validated linear relationship between the detonation velocity (VoD) and the reciprocal of the diameter of the charge, 1/d, which can be expressed by eq. (3.16):

$$V(d) = D\,(1 - [a/d]) \tag{3.16}$$

where V(d) is the VoD of a detonating unconfined cylindrical charge of diameter d, and D and a are curve-fitting constants. D is the limiting value of the VoD for a charge of infinite diameter, and a can be considered a constant which is characteristic of the explosive formulation.

However, it was found that unconfined RDX-driven PBXs and composite explosives usually fit better with an elliptical relationship between V(d) and 1/d of the form (eq. 3.17):

$$V(d)^2 = D^2\left(1 - [a^*/d]^2\right) \tag{3.17}$$

where D and a* are again curve-fitting constants, the values of which can be obtained from a linear plot of $V(d)^2$ against $(1/d)^2$.

Based on experimentally obtained data, it was found that the best fit can be obtained with eq. 3.18:

$$d_c = 2.208\ a^* \tag{3.18}$$

By combining eqs. (3.17) and (3.18), eq. (3.19) is obtained:

$$[V(d)/D]^2 = 1 - [d_c/2.208\ d]^2 \tag{3.19}$$

This enables an estimation of the "lower threshold velocity" or "cut off velocity", $V(d_c)$, for an RDX-driven formulation (the V(d) profile of which follows eq.(3.17)) to be estimated according to eq. (3.20):

$$V(d_c) = 0.892\ D \tag{3.20}$$

From the column on the right-hand side in Tab. 3.1, it can be seen that there is very good agreement between the experimentally determined d_c values and the calculated critical diameters for these RDX-driven composite formulations.

Tab. 3.1: Comparison of the experimentally determined and calculated critical diameters for various unconfined heterogeneous RDX-driven explosives.

Formulation #	**V(d) parameter, a*, mm (from eq. 3.17)**	**Critical diameter, d_c, mm (exptl.)**	**Critical diameter, d_c, mm (calcd.) (d_c = 2.208 a*) (eq. 3.18)**
PBXW-115	35.25	80	77.8
Cyclotol	2.0	6.0	4.4
Comp. B	1.66	4.3	3.4
Amatex	8.74	20	19.3

#D. J. Whelan, G. Bocksteiner, *J. Energ. Mater.*, **1995**, *13*, 15–34.

A novel method has been introduced which allows the simple estimation of the critical diameter of CHNO pure and composite explosives (M. H. Keshavarz et al., *J. Energ. Mat.*, **2019**, *37*, 331–339). It uses the number of oxygen and nitrogen atoms as well as an additional value for C_{Shock} which is based on shock sensitivity and impact sensitivity for desired explosives under certain situations. Equation (3.21) enables good predicted values to be achieved for a wide range of CHNO pure and composite high explosives containing nitro, nitramine and nitrate ester functional groups.

$$d_c = -3.19\ n_N + 5.38\ n_O + 21.80\ C_{Shock} \tag{3.21}$$

$$d_{c,core} = -11.97\ n_N + 14.11\ n_O \tag{3.22}$$

Different values of C_{Shock} are given as:

a) For pure explosives, the following statements have to be taken into account:
   (i) The value of C_{Shock} equals −0.4 if $P_{98\%TMD} > 38$ kbar and $E_{IS} > 28$ J.
   (ii) For $E_{IS} < 22$ J there are not sufficient experimental values to determine values for C_{Shock}. Thus, d_c is taken to be equal to 2.5.
b) For composite explosives, the value for C_{Shock} can be estimated as follows:
   (iii) For composite explosives containing PETN, the value of C_{Shock} is −0.3.
   (iv) The value of C_{Shock} equals −0.1 if $P_{98\%TMD} < 24.4$ kbar.
c) For $d_{c,core} < 0$, eq. (3.22): The value of C_{Shock} is 0.8.

Equation (3.21) provides the easiest pathway for predicting critical diameters using the conditions of $P_{98\%TMD}$ of pure and composite explosives as well as the calculated E_{IS} and the negative values of $d_{c,core}$.

In accordance with the detonation model for the stationary state, the points (V_0, p_0), (V_1, p_1) and (V_2, p_2) lie on one line (Fig. 1.22) which is called the **Rayleigh line**. The gradient of the Rayleigh line is determined by the detonation velocity of the explosive. In agreement with the postulate of Chapman and Jouguet, this Rayleigh line is also a tangent to the shock adiabat of the detonation products at exactly the point which corresponds to the end of the chemical reaction (V_2, p_2). This point is therefore also referred to as the **Chapman-Jouguet Point** (C-J point).

Generally, for homogeneous explosives, the **detonation velocity** increases proportional to the density of the energetic formulation. This means that reaching the maximum loading density is essential in order to obtain good performance. Different technical processes can be used in order to achieve the highest possible density including pressing, melt casting or extruding processes. For pure substances, the limiting, theoretical maximum density (TMD) is that of the single crystal (obtained from X-ray data at room temperature) (Tab. 3.2).

Usually the experimental densities are determined by single crystal X-ray diffraction at low temperature. Therefore it is necessary to convert the low-temperature densities into room-temperature values, in order to obtain the theoretical maximum density (TMD) (Tab. 3.3):

$$d_{298\ K} = \frac{d_T}{1 + \alpha_V\,(298 - T_0)} \rightarrow \alpha = \frac{d_T}{d_{298}} \div (298 - T)\ - 1 \div (298 - \mathrm{T})$$

As a good approximation the following formula can also be used:

$$d_{298\ K} = \frac{d_T}{1 + \alpha_V(298 - T_0)} \qquad \text{with} \qquad \alpha_V = 1.5 \cdot 10^{-4}\mathrm{K}^{-1}$$

Tab. 3.2: Comparison of the predicted critical diameter (mm) using eqs. (3.22) and (3.21) with the experimental data (the values of $P_{98\%\ TMD}$ in kbar and E_{IS} in J were calculated).

Name	Density (g cm^{-3})	Formula	$P_{98\%\ TMD}$/kbar	E_{IS}/J	Exp./mm	Eq. (3.22)/mm	Dev/mm	C_{Shock}	Eq. (3.21)/mm	Dev/mm
PBX-9502 (TATB/Kel-F 800 95/5)	1.895	$C_{2.30}H_{2.23}N_{2.21}O_{2.21}Cl_{0.038}F_{0.13}$	28.11		9.0	4.74	−4.26	0	4.84	−4.16
PBX-9404 (HMX/NC/CEF 94/3/3)	1.846	$C_{1.40}H_{2.75}N_{2.57}O_{2.69}Cl_{0.03}P_{0.01}$	21.47		1.18	7.20	6.02	−0.1	4.09	2.91
XTX 8003 (PETN/Silicon rubber 80/20)	1.53	$C_{1.80}H_{3.64}N_{1.01}O_{3.31}Si_{0.27}$	26.13		0.36	34.62	34.26	−0.3	8.04	7.68
XTX 8004 (RDX/Sylgard 80/20)	1.53	$C_{1.62}H_{3.78}N_{2.16}O_{2.43}Si_{0.27}$	23.67		1.4	8.44	7.04	−0.1	4.00	2.60
DAAF (3,3′-Diamino-4,4′-azoxyfurazan)	1.60	$C_4H_4N_8O_3$	32.08	26.95	1.25	−53.40	−54.65	0.8	8.05	6.80
BPTAP (2,4,8,10-Tetranitrobenzopyrido-1,3a,6,6a-tetraazapentalene)		$C_{11}H_3N_9O_8$	35.40	23.03	3.0 (<3)	5.18	2.18	0	2.50	−0.50
RDX/Wax 95/5	1.05	$C_{1.64}H_{3.28}N_{2.57}O_{2.57}$	27.03		4.5 (4.0–5.0)	5.50	1.00	−0.1	3.44	−1.06
RDX/Wax 90/10	1.1	$C_{1.93}H_{3.86}N_{2.43}O_{2.43}$	28.61		4.5 (4.0–5.0)	5.21	0.71	−0.1	3.14	−1.36
RDX/Wax 80/20	1.25	$C_{2.51}H_{5.02}N_{2.16}O_{2.16}$	31.77		4.4 (3.8–5.0)	4.63	0.23	0	4.73	0.33

(continued)

Tab. 3.2 (continued)

Name	Density (g cm^{-3})	Formula	$P_{98\% TMD}$/kbar	E_{IS}/J	Exp./mm	Eq. (3.22)/mm	Dev/mm	C_{Shock}	Eq. (3.21)/mm	Dev/mm
RDX/Wax 72/28	1.39	$C_{2.97}H_{5.95}N_{1.95}O_{1.95}$	34.29		4.4 (3.8–5.0)	4.17	−0.23	0	4.26	−0.14
B 2214 (NTO/HMX/HTPB 72/12/16)		$C_{2.44}H_{3.23}N_{2.54}O_{1.99}$	25.85		65	−2.32	−67.32	1	24.40	−40.60
B 2248 (NTO/HMX/HTPB 46/42/12)		$C_{2.15}H_{3.19}N_{2.55}O_{2.20}$	24.70		11	0.53	−8.97	0	3.70	−5.80
NTO (3-nitro-1,2,4-triazol-5-one)		$C_2H_2N_4O_3$	25.45	21.56	25.4	−5.54	−30.94	0.8	20.82	−4.58
RMSD							21.77 (All data except d_c=2.5 in eq.(3.21))			3.96 (All data except d_c=2.5 in eq. (3.21))

Tab. 3.3: The coefficients of volume expansion.

CL-20	$1.5 \cdot 10^{-4}\ K^{-1}$
HMX*	$1.6 \cdot 10^{-4}\ K^{-1}$
TNT	$2.26 \cdot 10^{-4}\ K^{-1}$
BTF	$3.57 \cdot 10^{-4}\ K^{-1}$
TKX-50	$1.10 \cdot 10^{-4}\ K^{-1}$

*C. Xu, et al., *Propellants, Explos. Pyrotech.* **2010**, *35*, 333–338

Kamlet and Jacobs suggested an empirical relationship between the detonation velocity and the **detonation pressure**. In this, the detonation velocity *D* is linear and the detonation pressure $p_{C\text{-}J}$ to the power of two dependent on the loading density ρ_0 (in g cm^{-3}) [17–19]:

$$p_{C\text{-}J}[\text{kbar}] = K \rho_0^2 \Phi$$

$$D\left[\text{mm}\,\mu\,\text{s}^{-1}\right] = A\,\Phi^{0.5}\,(1 + B\,\rho_0)$$

The constants K, A and B are defined as follows:
K = 15.88
A = 1.01
B = 1.30

The value Φ is therefore

$$\Phi = N(M)^{0.5}\,(Q)^{0.5}$$

where *N* is the number of moles of gas released per gram of explosive, *M* is the mass of gas in gram per mole of gas and *Q* is the heat of explosion in cal per gram.

The Kamlet-Jacobs equations for the Chapman-Jouguet pressure ($p_{C\text{-}J}$) and the detonation velocity D (VoD), show that detonation velocity and detonation pressure are strongly dependent on (a) the loading density and (b) the parameter φ. In the past, the importance of the role of the density was greatly emphasized, since it appears to a higher power than φ in the Kamlet-Jacobs equations. This resulted in the density being referred to as "the primary physical parameter in detonation performance" (P. Politzer, J. S. Murray, *High Performance, Low Sensitivity: Conflicting or Compatible?*, PEP, **2016**, *41*, 414–425). In subsequent years, however, it became more and more clear, that not only is the parameter φ as important in determining the detonation velocity as the density is, it is also needed in order to reliably predict the detonation pressure. It can be stated, however, that an increase in density by 0.02 g/cc causes an increase in the detonation velocity of approximately 100 m/s.

The Kamlet-Jacobs parameter φ reflects the specific inherent properties of an individual explosive compound. The loading density, on the other hand, depends not only on the crystal structure of the explosive compound, but also on external factors such as the method of preparation used to prepare the explosive. As a result of this difference, Politzer and Murray suggested that φ should be considered to be a measure of the intrinsic detonation potential of the explosive compound (P. Politzer, J. Murray, *PEP*, **2019**, *44*, 844–849). The intrinsic detonation potential is highly interesting, since it can provide useful insights into factors which promote a high detonation performance. An explosive compound that has a high φ value–even if coupled with a low density which will result in an unsatisfactory detonation performance–may draw attention to desirable molecular features that the compound possesses. It is worthwhile to point out that nitramines and nitrates usually show the highest intrinsic detonation potentials, with all having $\varphi \geq 6.48$. In contrast, compounds which contain C—NO_2 linkages usually have $\varphi < 5.2$. One exception to the aforementioned is FOX-7, which has a φ value of 6.10. Generally, it can be therefore be concluded that explosives containing nitramine (N—NO_2) and nitrate (O—NO_2) linkages will generally show a much higher level of detonation performance than for explosive compounds containing only the C—NO_2 linkage. Interestingly, even attaching four nitro groups to a single carbon atom is not sufficient to achieve a high intrinsic detonation potential, since tetranitromethane (TNM) has the lowest φ of 4.02. Unfortunately, it has been found that the desirable detonation behavior of nitramines and nitrates is usually accompanied by a high level of sensitivity (impact and friction) which is undesirable. At this point it is worthwhile to point out that RDX and HMX have exactly the same φ value, which means that in the case of these two compounds, the additional —CH_2—$N(NO_2)$ group in HMX does not increase the intrinsic detonation potential. The considerably better detonation performance of HMX can be entirely and exclusively attributed to its higher density.

Conceptually, the intrinsic detonation potential can be considered to be similar to the property known as the explosive power, which was discussed in chapter 4.1.

It has been demonstrated that if the intrinsic detonation potential of explosives is known, it can provide useful insight with respect to the detonation performances of explosives. Even if the density of an explosive compound is low–which will result in an unsatisfactory detonation performance for that explosive–if the compound shows a high value of φ, it may reveal desirable molecular features which could then be target features to include in hitherto unknown explosive compounds. The elucidation of a high φ value for an explosive compound may also act as a motivational force to increase efforts aimed at modifying the crystallization procedure to increase the density of the explosive in order to take advantage of the high φ.

Similarly, empirical relationships have also been published to correlate the specific impulse (I_{sp} / N s g^{-1}) with the detonation velocity (D / km s^{-1}) and the detonation pressure ($p_{C\text{-}J}$ / kbar) [see also: M. H. Keshavarz, *Chem. of Heterocyclic Comp.*, **2017**, *53*, 797–801]:

$$D\left[\mathrm{km\,s^{-1}}\right] = 1.453\, I_{sp}\left[\mathrm{N\,s\,g^{-1}}\right]\,\rho_0\left[\mathrm{g\,cm^{-3}}\right] + 1.98$$

$$p_{C\text{-}J}[\mathrm{kbar}] = 44.4\, I_{sp}\,[\mathrm{N\,s\,g^{-1}}]\,\rho_0^2[\mathrm{g^2\,cm^{-6}}] - 21$$

Table 3.4 shows how the detonation velocity and the detonation pressure are dependent on the density for selected explosives.

Tab. 3.4: Loading density dependency of the detonation velocity D and the detonation pressure $p_{C\text{-}J}$ (experimental data) [20].

Explosive	Density, ρ/g cm^{-3}	Detonation velocity, D/m s^{-1}	Detonation pressure, $p_{C\text{-}J}$/kbar
TNT	1.64	6,950	210
	1.53	6,810	171
	1.00	5,000	67
RDX	1.80	8,750	347
	1.66	8,240	293
	1.20	6,770	152
	1.00	6,100	107
PETN	1.76	8,270	315
	1.60	7,750	266
	1.50	7,480	240
	1.26	6,590	160

Non-ideal explosives have Chapman-Jouguet (C-J) detonation parameters (pressure/velocity) significantly different from that expected from a thermochemical / hydrodynamic code for equilibrium and steady state calculations. Therefore, common computer codes such as CHEETAH, EXPLO5 and others often do not correctly predict the detonation parameters of non-ideal explosives such as ANFO or metalized explosives.

For a wide range of CHNOFClAl(AN) explosives a very useful and simple method for the calculation of the detonation parameters of CHNOFCl, aluminized and ammonium nitrate explosives was recently suggested by M. H. Keshavarz et al.:

$$D = 5.468\,\alpha^{0.5}\left(\overline{Mw}_g Q_d\right)^{0.25}\rho_0 + 2.045$$

D, Q_d and ρ_0 are detonation velocity (in km s^{-1}), heat of detonation (in kJ g^{-1}), and initial density (in g cm^{-3}), respectively. α is the number of moles of gaseous products per gram of explosive (α) and $\overline{Mw}_g$ the average molecular weight of gaseous products.

Quite recently, Keshavarz also provided a general and simple method for predicting the detonation pressure (DP) of different kinds of ideal and non-ideal explosives containing aluminum (Al) and ammonium nitrate (AN). The new model can be applied for CHNO and CHNOFCl explosives in pure form or in mixtures, as well as non-ideal mixed explosives containing Al and AN. It can also be used for different plastic bonded explosives (PBXs). The detonation pressure can be calculated according to the following equation:

$$DP = p_{C-J} = 24.436\,\alpha\,(\overline{Mw}_g\,Q_d)^{0.5}\rho_0{}^2 - 0.874$$

Whereby:

α: moles of gaseous products per gram of explosive (mol g^{-1})
$\overline{Mw}_g$: average molecular mass of gaseous products (g mol^{-1})
Q_d: heat of detonation (kJ g^{-1})
ρ_0: density (g cm^{-3})

Empirical methods for the calculation of the detonation velocity, detonation pressure and density

Detonation velocity and detonation pressure

The use of empirical approaches to calculate the detonation velocity (VoD) and detonation pressure (P_{C-J}), has, of course, a long history in energetic materials. While many different approaches have been used over the years, perhaps the most well-known and widely used is the Kamlet-Jacobs method. A detailed discussion of the Kamlet-Jacobs method is outlined in the previous chapter, as well as various empirical equations proposed by Keshavarz.

The empirical approaches to calculating the detonation velocity and detonation pressure (i.e. performance prediction) can be largely grouped into three main categories. The first of these involves solving hydrodynamic equations obtained from the conservation of mass, momentum and energy conditions (which require that the equation of state of the products is known). The equation of state for each product was developed by Kistiakowsky and Wilson, and implemented in the early BKW, Ruby and Tiger computer codes from the 1960s. The second group contains the Kamlet-Jacobs approach and is based on simple parametric fits to results calculated by the methods implemented in the Ruby etc. codes. This group has the advantage of not requiring complicated computer codes. The third group is for methods that involve parametric fits to experimental data. The parameter which should be fitted to the experimental data can range from the oxygen balance to the specific impulse or impact sensitivities. In this section, this third group of methods is looked at in more detail, since although empirical in nature, many require little input data and can be a useful alternative for compounds for which neither density nor heat of formation are known – both of which are required for EXPLO5 and CHEETAH_9.0 calculations.

Rothstein-Petersen[1]: This method proposes that there is a simple, empirical linear relationship between the detonation velocity at TMD and a factor (F) that is only dependent on the chemical composition and connectivity for ideal C,H,N,O explosives.

It does not require knowledge of the heat of formation or the density. The only input this method requires is the number of C, H (n(H)), N (n(N)) and O (n(O)) atoms in the explosive, as well as the number of oxygen atoms in excess of those required to convert all C and H atoms in the explosive into CO_2 and H_2O (n(B)), the number of O atoms doubly bonded to C (n(C)), the number of O atoms singly bonded to C (n(D)), as well as the number of nitrate ester groups or salts of nitric acid (n(E)). All of these values can be obtained from a drawing of the structure[1]. Since the VoD (D') is dependent on the density, a linear regression plot of factor F vs. VoD can be generated using experimentally obtained VoD values at TMD for various known explosives.

The VoD calculated by the Rothstein-Petersen method[1] is given by the following equation (D' in km/s):

$$D' = \frac{F - 0.26}{0.55}$$

Where D' is the calculated VoD at TMD obtained using the Rothstein-Petersen method and *F* is the factor obtained from the following equation based on the composition and connectivity of the explosive[1]:

$$F = \left[100\,x \frac{n(O) + n(N) - \frac{n(H)}{2n(O)} + \frac{A}{3} - \frac{n(B)}{1.75} - \frac{n(D)}{4} - \frac{n(E)}{5}}{MWt.} \right] - G$$

The values n(O), n(N), n(H), n(B), n(D) and n(E) are explained above, a value of G = 0.4 is used for liquid explosives and G = 0 for solid explosives, and finally, A = 1 for aromatic explosives and A = 0 for all non-aromatic explosives.

The calculated (R-P method) D' value for the VoD at TMD is approximately related to the experimentally determined VoD (D_0) values for known explosives at TMD (ρ_0) by the following equation[1] (D' and D_0 in km/s):

$$D' = D_0 + (\rho_{TMD} - \rho_0)\,x\,3.0$$

For the detonation pressure ($P'_{C\text{-}J}$), it was found that a linear regression plot of experimentally determined detonation pressures vs. experimentally determined detonation velocities fitted the following relationship (with a 0.99 correlation coefficient)[1] ($P'_{C\text{-}J}$ in kbar):

$$P'_{C-J} = 93.3\,D - 456$$

An interesting aspect of this method, is that in contrast to most other methods (e.g. Kamlet-Jacobs) and programs (EXPLO5, CHEETAH_9.0) which are used to estimate VoD, the Rothstein-Petersen method doesn't require a value for the heat of formation as an input. However, one drawback of this method is that it isn't suitable for formulations.

Keshavarz[2],[3]: Another method that avoids the use of the heat of formation is an empirical approach developed by Keshavarz, in which the specific impulse is calculated empirically based solely on the knowledge of the number of C, H, N and O atoms present in the explosive, as well as the number of >NH or $-NH_2$ groups, and number of aromatic rings (including triazole, tetrazole, etc. rings as well as carbocyclic aromatic rings) it contains. This means that only a sketch on paper of the compound is required to generate the input values needed in order to estimate the specific impulse (I_{sp})[2] (I_{sp} in Ns/g):

$$I_{sp} = 2.4205 - 0.074\text{a} - 0.0036\text{b} + 0.0237\text{c} + 0.04\text{d} - 0.1001\text{n-NH}_x + 0.1466\,(\text{nAr} - 1)$$

The value estimated for the specific impulse (I_{sp}) can then be used to estimate both the VoD and $P_{C\text{-}J}$, with only the density being required as an additional input[3] (VoD in km/s, $P_{C\text{-}J}$ in kbar):

$$\text{VoD} = 1.453\, I_{sp}\, \rho_0 + 1.98$$

$$P_{C\text{-}J} = 44.4\, I_{sp}\, \rho_0^2 - 21$$

Again, this approach by Keshavarz requires only minimal knowledge of the energetic compound to estimate the VoD and $P_{C\text{-}J}$ and is a facile method, since it considers the VoD and $P_{C\text{-}J}$ as a function of the compound's composition and density.

Stine[4]: In contrast to the method of Keshavarz outlined above, the method of Stine assumes that the VoD for organic explosives of the type $C_aH_bN_cO_d$ is dependent on the chemical composition, (bulk) density and the heat of formation. Stine proposed the following general equation to describe the VoD (in km/s) which again incorporates several empirical factors (α, β, γ, δ and h) which were assigned the following values based on obtaining the best fit between the experimental and calculated detonation values for a chosen test group of explosives: α = −13.85, β = 3.95, γ = 37.74, δ = 68.11, h = 0.6917, D_0 = 3.69 km/s using ΔH_f given in kcal/mol[4] (D and D_0 in km/s):

$$\text{D} = D_0 + \rho\,\frac{(\alpha\text{a} + \beta\text{b} + \gamma c + \delta\text{d} + h\Delta H_f)}{M}$$

A comparison between the results calculated by these various methods with the experimental data has been reported in the literature[5],[6] and the results for several well-known explosives are briefly outlined in Tab. 3.5 below. However, it should be pointed out that a comparison of the Rothstein-Petersen values with the experimental values is only sensible when the VoD at TMD has been determined since the Rothstein-Petersen value calculates the VoD at TMD.

Tab. 3.5: A comparison of the experimentally determined VoD with values calculated using different methods at the same density.

Method/compound	Density (g/ cm³)/heat of formation (kJ/mol)	Exptl. value (m/s)	Kamlet-Jacobs (m/s)	Rothstein-Petersen (m/s) (@ TMD)	Keshavarz (based on I_{sp}) (m/s)	Stine (m/s)
PETN	1.76/−538.5	8,260	8,652	8,078	8,620	8,379
Tetryl	1.70/29	7,860	7,692	7,772	7,718	7,695
Trinitroaniline	1.76/−72.8	7,420	7,588	7,497	7,598	7,396
HMX	1.89/47.3	9,110	9,003	9,042	9,135	8,992

Density

Holden's method is an old method for calculating the density of energetic materials. It uses the smallest number of parameters to calculate the density compared to other, newer group additivity methods. It may however, give large deviations for new energetic compounds because it cannot consider intermolecular and intramolecular interactions. The first chapter of the book (The Properties of Energetic Materials: Sensitivity, Physical and Thermodynamic Properties, M. H. Keshavarz, T. M. Klapötke, deGruyter, 2018) discusses other group additivity methods. The simple computer code EMDB-1.0 can calculate the density for many types of energetic compounds with good reliability. For example, EMDB 1.0 calculates the crystal density of octanitrocubane (ONC) as being 1.99 g cm^{-3}, however, using Holden's method, Willer obtains a value of 2.16 g cm^{-3}.[7] The experimental value for the density of octanitrocubane is 1.96 g cm^{-3}.[8] The method to predict densities by Ammon[9] gives a density for ONC of 2.02 g cm^{-3}.

Literature

[1] L. R. Rothstein, R. Petersen, Predicting High Explosive Detonation Velocities from their Composition and Structure, Propellants, Explosives, Pyrotech., 1979, 4, 56–60.

[2] M. H. Keshavarz, Prediction Method for Specific Impulse Used as Performance Quantity for Explosives, Propellants, Explosives, Pyrotechnics, 2008, 33, 360–364.

[3] M. Jafari, M. Kamalvand, M. H. Keshavarz, A. Zamani, H. Fazeli, A Simple Approach for Prediction of the Volume of Explosion Gases of Energetic Compounds, Ind. J. Eng. Mater. Sci., 2004, 11, 429–432.

[4] J. R. Stine, On Predicting Properties of Explosives – Detonation Velocity, J. Energet. Mater., 1990, 8, 41–73.

[5] H. Shekhar, Studies on Empirical Approaches for Estimation of Detonation Velocity of High Explosives, Centr. Eur. J. Energet. Mater., 2012, 9, 39–48.

[6] Energetic Materials Designing Bench (EMDB), Version 1.0, M. H. Keshavarz, T. M. Klapötke, M. Suceska, Propellants Explosives Pyrotechnics 2017, 42, 854–856.

[7] R. L. Willer, Journal of the Mexican Chemical Society 2009, 53(3), pp.108–119.

[8] P. E. Eaton, M. X. Zhang, R. Gilardi, N. Gelber, S. Iyer, R. Surapaneni, Propellants, Explosives, Pyrotechnics 2002 27(1), 1–6.
[9] H. L. Ammon, Propellants, Explosives, Pyrotechnics, 2008, 33, 92–102.

Check your knowledge: See Chapter 14 for Study Questions.

4 Thermodynamics

4.1 Theoretical basis

As we have already discussed above, the main performance criteria for secondary explosives are:
1. heat of explosion Q (in kJ kg^{-1})
2. detonation velocity D (m s^{-1})
3. detonation pressure p (in kbar)

and, slightly less importantly,
4. explosion temperature T (K) and
5. volume of gas released V per kg explosive (in l kg^{-1})

To calculate the detonation velocity and the detonation pressure we require (see Ch. 3) thermodynamic values such as e.g. the heat of detonation, from which the detonation temperature can be obtained.

Before we are concerned with detailed thermodynamic calculations, it is important not to forget that for the good design of a secondary explosive, a balanced oxygen content (**oxygen balance** Ω) is desirable. Generally, oxygen balance describes the relative amount of oxygen excess or defecit (with negative sign "–"), to achieve a balanced ratio between the oxidizer and the combustable components (fuel). A mixture of substances with a balanced oxygen balance ($\Omega = 0$) can be converted into fully oxidized products by heating them in a closed container, without any added external oxygen and without any excess oxidizer or fuel. In accordance with this definition, a CHNO-containing explosive is completely converted to CO_2, H_2O and N_2. For a compound with the general formula $C_aH_bN_cO_d$, the oxygen balance Ω (in %) is defined as follows:

$$\Omega = \Omega_{CO_2} = \frac{\left[d - (2a) - \left(\frac{b}{2}\right)\right] \times 1600}{M}; \; \Omega_{CO} = \frac{\left[d - a - \left(\frac{b}{2}\right)\right] \times 1600}{M}$$

where M is the molecular mass of the explosive. For example, TNT has a molecular mass of M (TNT) = 227 g mol^{-1} and a composition with the formula $C_7H_5N_3O_6$. Therefore the oxygen balance for TNT is Ω (TNT) = –74%:

$$C_7H_5N_3O_6 \rightarrow \; 7\,CO_2 + 2.5\ H_2O + 1.5\ N_2 - 10.5\,O$$

Table 4.1 shows the oxygen balances of some secondary explosives.

Figure 4.1 shows the influence of the oxygen balance on conventional CHNO-explosives. Usually (this is the case for very nitrogen-rich compounds as well) a good oxygen balance results in a greater (more negative) heat of explosion and therefore leads to better performance of the explosive.

https://doi.org/10.1515/9783111447087-004

Tab. 4.1: Summary of the oxygen balances of important secondary explosives.

Secondary explosive	Formula	Oxygen balance, Ω_{CO_2} / %
Ammonium nitrate, AN	NH_4NO_3	+20.0
Nitroglycerin, NG	$C_3H_5N_3O_9$	+3.5
Nitropenta, PETN	$C_5H_8N_4O_{12}$	-10.1
RDX	$C_3H_6N_6O_6$	-21.6
HMX	$C_4H_8N_8O_8$	-21.6
Nitroguanidine, NQ	$CH_4N_4O_2$	-30.7
Picric acid, PA	$C_6H_3N_3O_7$	-45.4
Hexanitrostilbene, HNS	$C_{14}H_6N_6O_{12}$	-67.6
Trinitrotoluene, TNT	$C_7H_5N_3O_6$	-74.0

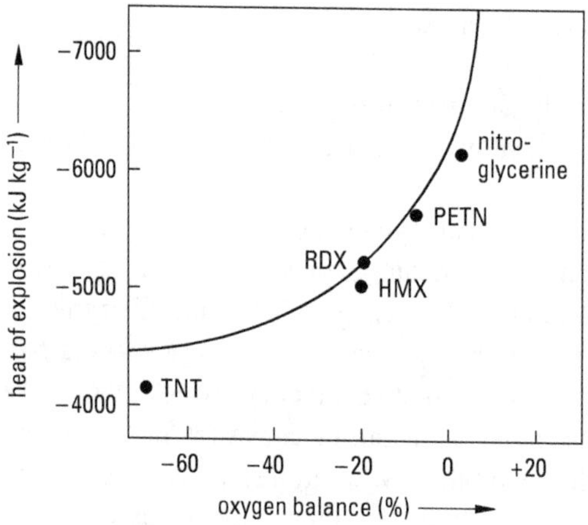

Fig. 4.1: Influence of the oxygen balance Ω on the heat of explosion Q.

The oxygen balance of an explosive alone does not help when the thermodynamics of the reaction are to be estimated. In this case, it is (particularly for compounds with too-low oxygen balances) necessary to estimate how much of the oxygen gets converted into CO, CO_2 and H_2O. Additionally, since all explosions occur at high temperatures (approx. 3,000 K), even with oxygen balances of $\Omega = 0$, not only CO_2 but also CO will form as stated in the Boudouard equilibrium. An approximate, but very simple scheme for the estimation of the detonation products is given by the so-called modified **Springall-Roberts rules**. These are given below and must be applied in order from 1 to 6:

1. C atoms are converted into CO
2. if O atoms remain, they oxidize hydrogen to H_2O
3. if there are remaining O atoms, they oxidize the already formed CO to CO_2
4. all nitrogen is converted to N_2

5. one third of the CO formed is converted into C and CO_2
6. one sixth of the CO formed originally is converted into C and water

Tab. 4.2: Determination of the detonation products of TNT ($C_7H_5N_3O_6$) according to the Springall-Roberts-Rules.

C atoms are converted into CO	$6\,C \rightarrow 6\,CO$
If O atoms remain, they oxidize hydrogen to H_2O	All oxygen is already used-up
If there are remaining O atoms, they oxidize the already formed CO to CO_2	All oxygen is already used-up
All nitrogen is converted to N_2	$3\,N \rightarrow 1.5\,N_2$
One third of the CO formed is converted into C and CO_2	$2\,CO \rightarrow C + CO_2$
One sixth of the CO formed originally is converted into C and water	$CO + H_2 \rightarrow C + H_2O$
Total reaction	$C_7H_5N_3O_6 \rightarrow 3\,CO + CO_2 + 3\,C + 1.5\,H_2 + H_2O + 1.5\,N_2$

As an example, in Tab. 4.2, the Springall-Roberts-Rules have been applied to work out the detonation products of TNT.

Therefore, the enthalpy of explosion (ΔH_{ex}) for TNT corresponds to the reaction enthalpy for the following reaction:

$$C_7H_5N_3O_6(s) \rightarrow 3CO(g) + CO_2(g) + 3C(s) + 1.5\ H_2(g) + H_2O(g) + 1.5\ N_2(g)\ \Delta H_{ex}(TNT)$$

In contrast, the **enthalpy of combustion** for TNT (ΔH_{comb}) corresponds to the change in enthalpy for the following reaction and is more negative (more exothermic) than the explosion:

$$C_7H_5N_3O_6(s) + 5.25\ O_2(g) \rightarrow 7CO_2(g) + 2.5H_2O(g) + 1.5\ N_2(g)\ \Delta H_{comb}(TNT)$$

For the definition of the enthalpy of combustion (ΔH_{comb}), it is important to check if the values are given for H_2O (g) or H_2O (l).

Since the exact standard enthalpies of formation for all possible detonation products are known (see e.g. http://webbook.nist.gov/chemistry/), the enthalpy of detonation can easily be calculated, as long as the enthalpy of combustion is exactly known from the experimental data, which unfortunately is not always the case.

According to the first fundamental law of thermodynamics, energy is conserved in any process involving a thermodynamic system and its surroundings. It is convenient to focus on changes in the assumed internal energy (U) and to regard them as due to a combination of heat (Q) added to the system and work done by the system (W). Taking ΔU as an incremental change in internal energy, we can write:

$$\Delta U = W + Q$$

Since work, W is defined as:

$$W = -\int_{V_1}^{V_2} p\mathrm{d}V = -p\Delta V$$

for V = const.: $\Delta U = Q_V$
and for p = const.: $\Delta U = Q_p - p\,\Delta V$

For propellant charges which burn in a gun barrel and for secondary explosives, a good approximation is V = const. (isochoric) and therefore $\Delta U = Q_V$, while for rocket fuels (free expansion of the gases in the atmosphere) p = const. (isobaric) and therefore $\Delta U = Q_p - p\,\Delta V$ is a good approximation.

Since the equation of state H (enthalpy) is defined as $H = U + pV$, the following generalization can be made:

$$\Delta H = \Delta U + p\,\Delta V + V\,\Delta p$$

Therefore, for p = const.: $\Delta H = Q_p$.

This means that we can generalize as follows:

$$Q_V = \sum \Delta_f U^o_{(\text{detonation products})} - \sum \Delta_f U^o_{f\,(\text{explosive})}$$

and

$$Q_p = \sum \Delta_f H^o_{(\text{detonation products})} - \sum \Delta_f H^o_{f\,(\text{explosive})}$$

In a first approximation, the energy of an explosion can be taken to be equal to that of the Helmholtz Free Energy:

$$\int p\mathrm{d}V = -\Delta A = -\Delta U + T\Delta S$$

The molar energies of formation U and enthalpies H for a given explosive with the composition $C_aH_bN_cO_d$ can be connected as follows:

$$H = U + \Delta n\,R\,T$$

here, n is the change of moles of the *gaseous* substances and R is the general gas constant. If we consider the hypothetical "equation of formation (from the elements)" for TNT for example, the change of moles is negative and Δn corresponds to –7:

$$7\,C(s) + 2.5\,H_2(g) + 1.5\,N_2(g) + 3\,O_2(g) \rightarrow C_7H_5N_3O_6(s)$$

In practice, in a good approximation H and U can often be taken to be equal.

The liberated volume of detonation gases (V_0, usually back-calculated for standard conditions of 273 K and 1 bar) is also an important parameter when assessing the performance of an explosive. V_0 can easily be calculated using the reaction equation derived from the Springall-Roberts rules and applying the ideal gas law and it is usually given in l kg^{-1}.

Table 4.3 shows a summary of typical values for the volumes of gas released during an explosion (under STP).

Tab. 4.3: Calculated volumes of gas released in an explosion (under STP).

Explosive	V_0 / l kg^{-1}
NG	740
PETN	780
RDX	908
HMX	908
NQ	1,077
PA	831
HNS	747
TNT	740

The **Explosive Power** has been defined as the product of the volume of gas released V_0 (STP) and the heat of explosion Q. Traditionally Q is given in kJ kg^{-1}, and V_0 in l kg^{-1} (Tab. 4.3). The value of the explosive power is usually compared with the explosive power of a standard explosive (picric acid) resulting in power indices, as shown below (Tab. 4.4).

$$\text{Explosive power} = Q\,(\text{kJ kg}^{-1}) \times V_0\,(\text{l kg}^{-1}) \times 10^{-4}\,(\text{kg}^2\,\text{kJ}^{-1}\,\text{l}^{-1})$$

$$\text{Power index} = \frac{Q \times V_0}{Q_{\text{PA}} \times V_{\text{PA}}} \times 100$$

Tab. 4.4: Explosive power and power index.

Explosive	$-Q$ / kJ kg^{-1}	V_0 / l kg^{-1}	Explosive power	Power index
$Pb(N_3)_2$	1,610	218	35	13
NG	6,195	740	458	170
PETN	5,794	780	452	167
RDX	5,036	908	457	169
HMX	5,010	908	455	169
NQ	2,471	1,077	266	99
PA	**3,249**	**831**	**270**	**100**
HNS	3,942	747	294	109
TNT	4,247	740	314	116

Tab. 4.5: Average heat capacities C_V (in J K^{-1} mol^{-1}).

T_{ex} / K	CO_2	CO	H_2O	H_2	N_2
2000	45.371	25.037	34.459	22.782	24.698
2100	45.744	25.204	34.945	22.966	24.866
2,200	46.087	25.359	35.413	23.146	25.025
2,300	46.409	25.506	35.865	23.322	25.175
2,400	46.710	25.640	36.292	23.493	25.317
2,500	46.991	25.769	36.706	23.665	25.451
2,600	47.258	25.895	37.104	23.832	25.581
2,700	47.509	26.012	37.485	23.995	25.703
2,800	47.744	26.121	37.849	24.154	25.820
2,900	47.965	26.221	38.200	24.309	25.928
3,000	48.175	26.317	38.535	24.260	26.029
3,100	48.375	26.409	38.861	24.606	26.129
3,200	48.568	26.502	39.171	24.748	26.225
3,300	48.748	26.589	39.472	24.886	26.317
3,400	48.924	26.669	39.761	25.025	26.401
3,500	49.091	26.744	40.037	25.158	26.481
3,600	49.250	26.819	40.305	25.248	26.560
3,700	49.401	26.891	40.560	25.405	26.635
3,800	49.546	26.962	40.974	25.527	26.707
3,900	49.690	27.029	41.045	25.644	26.778
4,000	49.823	27.091	41.271	25.757	26.845
4,500	50.430	27.372	42.300	26.296	27.154
5,000	50.949	27.623	43.137	26.769	27.397

The detonation or explosion temperature T_{ex} is the theoretical temperature of the detonation products, assuming that the explosion occurs in a confined and undestroyable adiabatic environment (adiabatic conditions). The detonation temperature can be calculated with the assumption that the heat content of the detonation products must be the same as the calculated heat of detonation (Q). We can assume, that the detonation products of the initial temperature T_i (normally 298 K) will be brought up to a temperature of T_{ex} due to the heat of detonation, i. e. T_{ex} is dependent on Q. The connection between Q and T is given in the following equation, whereby C_V is the molar heat capacity of the detonation products:

$$Q = \sum \int_{T_i}^{T_{ex}} C_V \mathrm{d}T \quad \text{with:} \quad C_V = \left(\frac{\partial Q}{\partial T}\right)_V = \left(\frac{\partial U}{\partial T}\right)_V$$

$$\text{and} \quad C_P = \left(\frac{\partial Q}{\partial T}\right)_P = \left(\frac{\partial H}{\partial T}\right)_P$$

Therefore, the detonation temperature T_{ex} can be estimated as follows:

$$T_{ex} = \frac{Q}{\sum C_V} + T_i$$

If a value which is slightly too low and a value which is slightly too high are used for the detonation temperature and the value for Q is calculated using the values given in Tab. 4.5, the "correct" detonation temperature T_{ex} can be estimated iteratively.

The **"energy content"** of an energetic material is the chemical energy released by the detonation (or combustion) of an explosive. The energy is released in the form of heat energy, and is measured in per unit mass or per unit of volume. This means that the heat of detonation corresponds to the energy content. For example, the heat of detonation of RDX (ρ = 1.8 g cm^{-3}) is 5,729 J g^{-1}, and when this value is multiplied by the density, the heat of detonation per unit volume results (i.e. 5,729 J g^{-1} · 1.8 g cm^{-3} = 10.31 kJ cm^{-3} or 10,310 kJ L^{-1}). For HMX (ρ = 1.9 g cm^{-3}), the heat of detonation is 5,692 J g^{-1} (i.e. 10.82 kJ cm^{-3} or 10,815 kJ L^{-1}). Thus, the heat of detonation (energy content) per volume is higher for HMX (since its density is higher) than for RDX.

The detonation energy is obtained by integration of the pressure – volume along the isentropic expansion of the detonation products (it represents work that can be done by detonation products). The detonation energy is very similar to the calorimetrically determined heat of detonation, since the work done by the detonation products should be equal to the heat energy.

For RDX, the detonation energy is 10.46 kJ cm^{-3} and for HMX it is 11.02 kJ cm^{-3}, which is slightly higher (as is also the case for the heat of detonation).

Since detonation occurs under constant volume conditions, the heat of detonation is equal to the change of internal energy (not enthalpy) of the detonation products (at the detonation temperature) and reactants (initial explosive):

$$Q_V = \Delta U$$

In the case of isobaric combustion, the heat of detonation equals the value for the change in enthalpy:

$$Q_p = \Delta U + p \quad \Delta V = \Delta H$$

Literatur

M. Suceska, *Prop. Explos. Pyrotech.* **1999**, *24*, 280–285.

4.2 Computational methods

4.2.1 Thermodynamics

The thermodynamic data as well as the detonation parameters can nowadays be very reliably obtained by using quantum-mechanical computer calculations. On the one hand it is important to check experimental results, and on the other hand – and even more importantly – it is important to predict the properties of potential new energetic materials without any prior experimental parameters, for example during the planning of synthetic work. Moreover, such computational methods are ideal for the estimation of the detonation parameters of newly synthesized compounds, which have not been obtained in the 50 – 100 g quantities which are necessary for the experimental determination of such detonation parameters (e.g. detonation velocity).

In order to be able to calculate the detonation parameters of a particular neutral or ionic compound, it is advisable to calculate the enthalpy (H) and free energy (G) quantum chemically with very exact methods (e.g. G2MP2, G3 or CBS-4M). To achieve this, the Gaussian (G09W or G16W) programme is suitable. In the following section, we will concentrate on the CBS-4M method developed by Petersson and co-workers. In the CBS method, (complete basis set) the asymptotic convergence behavior of natural orbitals is used, in order to extrapolate the energy limit for an infinitely large basis set. The CBS method begins with a HF/3–21G(d) calculation in order to optimize the structure and for the calculations of the zero-point energy. Then, using a larger basis set, the so-called base-energy is calculated. A MP2 /6–31+G calculation with a CBS extrapolation gives the perturbation-theory corrected energy, which takes the electron correlation into account. A MP4 (SDQ) /6–31+G(d,p) calculation is used, to estimate the correlation contributions of a higher order. The most widely used CBS-4M version today is a re-parametrization of the original CBS-4 version, which contains additional empirical correction terms (M stands for "minimal population localization" here).

The enthalpies of the gaseous species M can now be calculated using the method of the atomization energies [21–23]:

$$\Delta_f H^\circ(g,m) = H^\circ_{(\text{molecule})} - \sum H^\circ_{(\text{atoms})} + \sum \Delta_f H^\circ_{(\text{atoms})}$$

Two examples should be considered for further discussion: the covalent nitroglycerin (NG) and the ionic ammonium dinitramide (ADN). The calculated enthalpies (H) and free energies (G) for NG, the ions NH_4^+ and $N(NO_2)_2^-$ as well as the relevant atoms H, C, N and O using the CBS-4M method, are summarized in Tab. 4.6.

Therefore in Tab. 4.6 we already have the $H^\circ_{(\text{molecules})}$ and $H^\circ_{(\text{atom})}$ values (given in a. u. = atomic units; 1 a. u. = 1 H = 627.089 kcal mol^{-1}). The values for $\Delta_f H^\circ_{(\text{atoms})}$ are easily obtained from the literature and are summarized in Tab. 4.7.

Tab. 4.6: CBS-4M values for NG, the NH_4^+ ions and $N(NO_2)_2^-$ and the relevant atoms H, C, N and O.

	$-H^{298}$ / a. u.	$-G^{298}$ / a. u.
NG	957.089607	957.149231
NH_4^+	56.796608	56.817694
$N(NO_2)_2^-$	464.499549	464.536783
H	0.500991	0.514005
C	37.786156	37.803062
N	54.522462	54.539858
O	74.991202	75.008515

Tab. 4.7: Literature values for $\Delta_f H^\circ_{(\text{atoms})}$ (in kcal mol^{-1}).

	Ref. [21]	NIST [24]
H	52.6	52.1
C	170.2	171.3
N	113.5	113.0
O	60.0	59.6

Tab. 4.8: Calculated standard enthalpies of formation $\Delta_f H^\circ$ (g) for the gas-phase.

Gas-phase species	Formula	$\Delta_f H^\circ$(g) / kcal mol^{-1}	$\Delta_f H^\circ$(g) / kJ mol^{-1}
NG	$C_3H_5N_3O_9$	- 67.2	-281.1
ammonium ion	NH_4^+	+151.9	+635.5
dinitramide ion	$N(NO_2)_2^-$	- 29.6	-123.8

According to the equation given above, we can now easily calculate the standard enthalpies of formation $\Delta_f H^\circ$(g) for the gas-phase species NG, NH_4^+ and $N(NO_2)_2^-$ (Tab. 4.8).

In order to be able to convert the standard enthalpies of formation $\Delta_f H^\circ$(g) for the gas-phase into values for the condensed phase, for covalent molecules (NG) we additionally require the enthalpy of sublimation $\Delta H_{\text{sub.}}$ (for solids) or the enthalpy of vaporization $\Delta H_{\text{vap.}}$ (for liquids). Both values can be estimated using **Trouton's rule**, in which T_m is the melting point of the solid and T_b is the boiling point of the liquid [25]:

$$\Delta H_{sub.}\left[\text{J mol}^{-1}\right] = 188\ T_m\ [\text{K}]$$

$$\Delta H_{vap.}\left[\text{J mol}^{-1}\right] = 90\ T_b\ [\text{K}]$$

NG is a liquid and its extrapolated boiling point is 302 °C (= 575 K). Therefore, the enthalpy of vaporization is calculated to be $\Delta H_{vap.}$(NG) = 51.8 kJ mol^{-1} (12.3 kcal mol^{-1}).

Recently, **Muravyev** published a new equation to calculate the sublimation enthalpy of energetic CHNO materials [*PCCP* **2021**, *23*, 15,522]:

$$\Delta H_f{}^\circ{}_m(s) = \Delta H_f{}^\circ{}_m(g) - \Delta H_{sub}{}^\circ{}_m$$

with:

$$\Delta H_{sub}{}^\circ{}_m = 0.15T_m(K) + 3.27[H] + 5.30[N] + 3.30[O]$$

However, presently the best way to determine the enthalpy of sublimation is based on a machine learning model which has been implemented into the **RoseBoom** computational code [Research output software for energetic materials based on observational modelling 2.1 (RoseBoom2.1©), S. Wahler, T. M. Klapötke, *Materials Advances*, **2022**, *3*, 7,976 – 7,986, DOI: 10.1039/D2MA00502F; S. Wahler, P. Chung, T. M. Klapötke, *J. Energ. Mat.*, **2023**, https://doi.org/10.1080/07370652.2023.2219678]. Figure 4.1a shows enthalpies of sublimation predicted using Trouton's rule (Fig. 4.1a, top), based on the Muravyev equation (Fig. 4.1a, middle) and predicted using the RoseBoom ML code (Fig. 4.1a, bottom).

In the case of ionic solids of the type AB, AB_2 or A_2B, the lattice energy (ΔU_L) and lattice enthalpy (ΔH_L) can be calculated by using the Jenkin's method [26–29]. Only the molecular volumes of the ions are required. These can be most easily obtained from single crystal X-ray diffraction data:

$$\Delta U_L = |z_+||z_-|\nu\left[\frac{a}{\sqrt[3]{V_M}} + \beta\right]$$

Here $|z_+|$ and $|z_-|$ are the dimensionless charges of the cations and anions and ν is the number of ions per 'molecule' (2 for ADN, 3 for Ba $(DN)_2$). V_M is the volume in nm^3 of a formula unit (e.g. V_M(ADN) = $V_M(NH_4^+)$ + $V_M(DN^-)$; $V_M(Ba(DN)_2)$ = $V_M(Ba^{2+})$ + 2 $V_M(DN^-)$). The constants α and β are dependent on the composition of the salt, and are summarized in Tab. 4.9.

In terms of density (ρ_m in g cm^{-3}), the above equation for ΔU_L can also be written in the following form:

$$\Delta U_L\left(\text{kJ mol}^{-1}\right) = \gamma\left(\rho_m/M_m\right)^{1/3} + \delta$$

where M_m is the chemical formula mass of the ionic material (in g), and the coefficients γ (in kJ mol^{-1} cm) and δ (kJ mol^{-1}) are assigned the values listed in Tab. 4.9 for the corresponding stoichiometry.

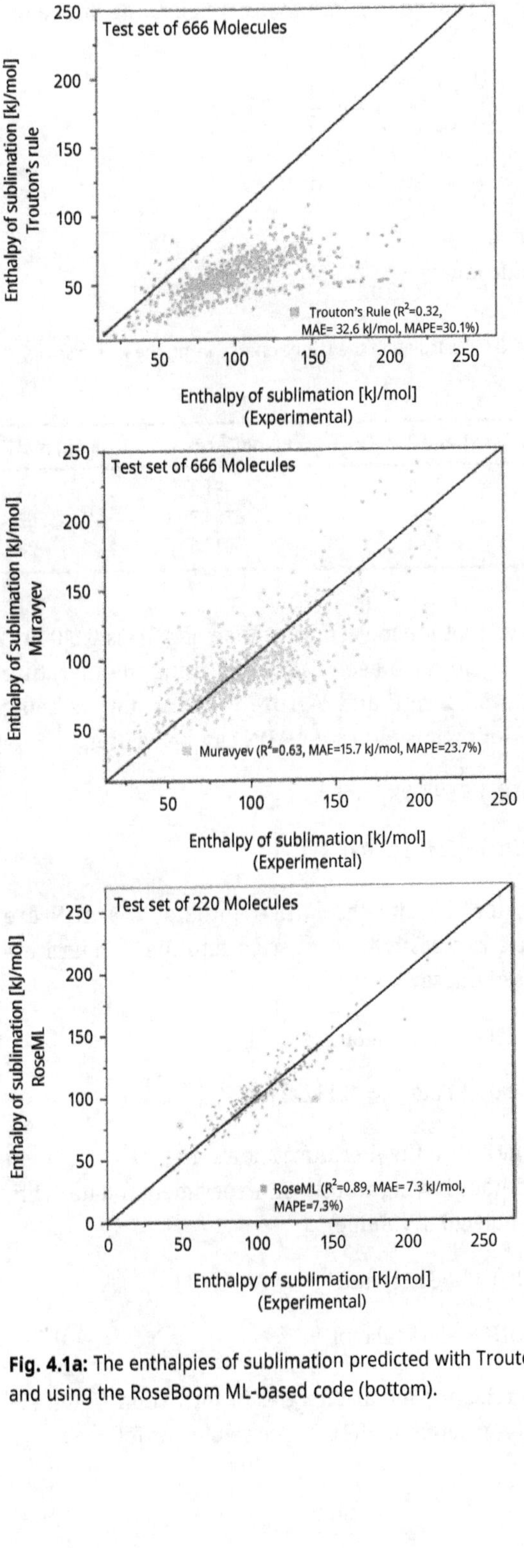

Fig. 4.1a: The enthalpies of sublimation predicted with Trouton's rule (top), Muravyev's equation (center) and using the RoseBoom ML-based code (bottom).

The lattice energy ΔU_L can easily be converted into the corresponding lattice enthalpy ΔH_L:

$$\Delta H_L(A_pB_q) = \Delta U_L + \left[p\left(\frac{n_A}{2} - 2\right) + q\left(\frac{n_B}{2} - 2\right)\right]RT$$

n_A, n_B: = 3 for monoatomic ions
5 for linear, polyatomic ions
6 for non-linear, polyatomic ions

Tab. 4.9: The values for the coefficients α and β for calculating the lattice energy using the Jenkin's method.

Salt type	α / kJ nm mol^{-1}	β / kJ mol^{-1}	γ / kJ mol^{-1} cm	δ / kJ mol^{-1}
AB	117.3	51.9	1981.2	103.8
AB_2	133.5	60.9	6764.3	365.4
A_2B	165.3	-29.8	8375.6	-178.8

From X-ray diffraction data it is known that the molecular volume of ADN is 0.110 nm^3 [30]. If this value was not known, it could have been calculated using the literature values for the volumes of $V_M(NH_4^+)$ = 0.021 nm^3 and $V_M(DN^-)$ = 0.089 nm^3 as being V_M(ADN) = 0.110 nm^3. Therefore, the following values for ADN can be written:

$$\Delta U_L(\text{ADN}) = 593.4\,\text{kJ mol}^{-1}$$

$$\Delta H_L(\text{ADN}) = 598.4\,\text{kJ mol}^{-1}$$

Since the enthalpy of vaporization for NG and the lattice enthalpy for ADN are known, the gas-phase enthalpy values can easily be converted into the standard enthalpies of formation for the condensed phase:

$$\Delta_f H°(\text{NG}) = -332.9\,\text{kJ mol}^{-1} = -80\,\text{kcal mol}^{-1}$$

$$\Delta_f H°(\text{ADN, s}) = -86.7\,\text{kJ mol}^{-1} = -21\,\text{kcal mol}^{-1}$$

A comparison with experimental values from the literature shows that the calculation outlined here gives a good result, without having to rely on experimental data (with the exception of the density and the molecular volume):

$$\Delta_f H°(\text{NG}) = -88\,\text{kcal mol}^{-1}$$

$$\Delta_f H°(\text{ADN}) = -36\,\text{kcal mol}^{-1}$$

As we have already shown in Chapter 1.4, the molar energies of formation U and enthalpy H for a given explosive with composition $C_aH_bN_cO_d$ are related as follows:

$$H = U + \Delta n RT$$

Here n is the change in the number of moles of gaseous substances and R is the ideal gas constant. Therefore, for NG (Δn = −8.5) and ADN (Δn = −6) the following values can be written:

$$\Delta_f U°(\text{NG}) = -311.8 \text{ kJ mol}^{-1} = -1373.0 \text{ kJ kg}^{-1}$$

$$\Delta_f U°(\text{ADN}) = -71.8 \text{ kJ mol}^{-1} = -579.0 \text{ kJ kg}^{-1}$$

While relatively-standard methods are used by almost all synthetic energetic materials groups world-wide for the calculation of the enthalpies of formation for neutral compounds and salts, it has recently become apparent that there are severe limitations and errors that result from using Jenkins equation to estimate the lattice enthalpies of salts.

For neutral compounds and ions in the ***gas***-phase the best way to obtain reliable enthalpies of formation is still based on composite methods such as CBS-4M, CBS-QB3 or CBS-APNO. These increase in the given order in accuracy but also in cpu time. However, the accuracy gained from one method to the next higher one can almost be neglected in comparison with the inaccuracy that is made by calculating the lattice energies for salts.

Therefore, for ***neutral compounds*** it is recommended to calculate the enthalpy of formation in the gas-phase at CBS-QB3 level of theory then convert this value into the standard state (solid or occasionally liquid) using the enthalpies of sublimation or vaporization calculated based on ML methods using the RoseBoom© code (see above). The CBS-4M has been modified by the inclusion of diffuse functions in the geometry optimization step to give CBS-QB3. The five-step CBS-QB3 series of calculations starts with a geometry optimization at the B3LYP level, followed by a frequency calculation to obtain thermal corrections, zero-point vibrational energy, and entropic information. The next three computations are single-point calculations (SPCs) at the CCSD(T), MP4SDQ, and MP2 levels. The CBS extrapolation then computes the final energies.

For ***salts*** the situation is more complicated, and it appears that the problem lies in the estimation of lattice energies/enthalpies of salts. While it has been shown recently that the Jenkins equation is not good enough to estimate the lattice energies and enthalpies for 1:1 salts, the improved equation by Gutowski (which is a reparametrization of the Jenkins equation for 1:1 salts and which is more suitable for energetic compounds), may be used in combination with the CBS-QB3 calculated values for the enthalpy of formation for the individual ions in the gas phase:

$$\Delta U_L = |z_+||z_-|\nu\left[\frac{\alpha}{\sqrt[3]{V_M}} + \beta\right]$$

with the constants: α = 19.9 (kcal nm)/mol and β = 37.6 kcal/mol

The conversion from density (ρ) to molecular volume (V_M) of a substance is straightforward and is carried out using the following equations, depending on whether the molecular volume is required in Å³ or nm³:

$$V_M\left[nm^3\right] = \frac{M\left[\frac{g}{mol}\right]10^{21}\left[\frac{nm}{cm^3}\right]}{\rho\left[\frac{g}{cm^3}\right]N_L} \qquad V_M\left[\text{Å}^3\right] = \frac{M\left[\frac{g}{mol}\right]10^{24}\left[\frac{\text{Å}}{cm^3}\right]}{\rho\left[\frac{g}{cm^3}\right]N_L}$$

For 1:2 and 2:1 salts the situation is more complicated since there is no re-parameterized version of the Jenkins equation available and for salts of the type A_2B such as TKX-50, the Jenkins equation provides a very unsatisfactory value for the lattice enthalpy.

At the moment, one option to estimate the enthalpy of formation of 2:1 (A_2B) and 1:2 (AB_2) salts which avoids the necessity of estimating the lattice enthalpy is to undertake the following steps:

1. The enthalpy for the "salt reaction" between the neutral base in its standard state and the acid in its standard state is calculated. The value of $\Delta H_{\text{salt reaction}}$ is a value which is specific for every individual base (Tab. 4.9a). It is obtained according to Sinditskii's method by subtracting from the experimentally determined value of the enthalpy of formation of the Base-$H^+NO_3^-$ (s) salt, the enthalpies of formation of the base and acid in their standard states. It is then assumed that this value for the $\Delta H_{\text{salt reaction}}$ is the same for a specific base, regardless of which acid is involved in protonating the base. In order to calculate the enthalpy of formation of the ionic salt, the enthalpies of formation of the neutral base, and acids in their standard states as well as the $\Delta H_{\text{salt reaction}}$ for the given base are added together. An example is given below using TKX-50:

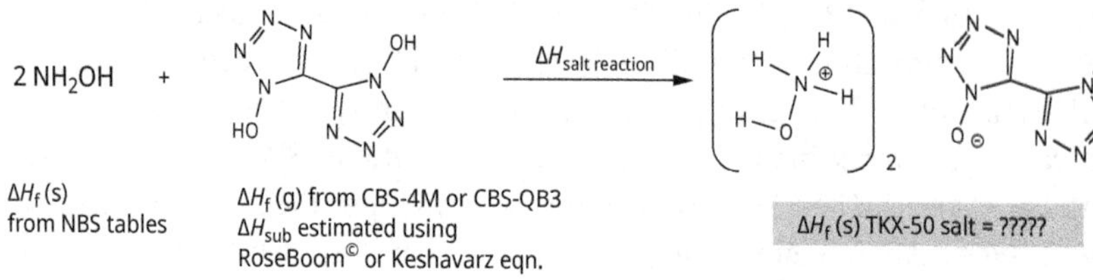

ΔH_f (s) TKX-50 salt = 2 × ΔH_f NH_2OH (s) + ΔH_f diol (s) + 2 × $\Delta H_{\text{salt reaction}}$

2. In the second step, the enthalpy of formation of a (hypothetical) neutral adduct between the acid and base(s) in the gas-phase is calculated at e.g. CBS-QB3 level of theory. The enthalpies of sublimation/vaporization for the component acid and base are then subtracted from the gas-phase enthalpy of formation of the adduct. An example is given below using TKX-50:

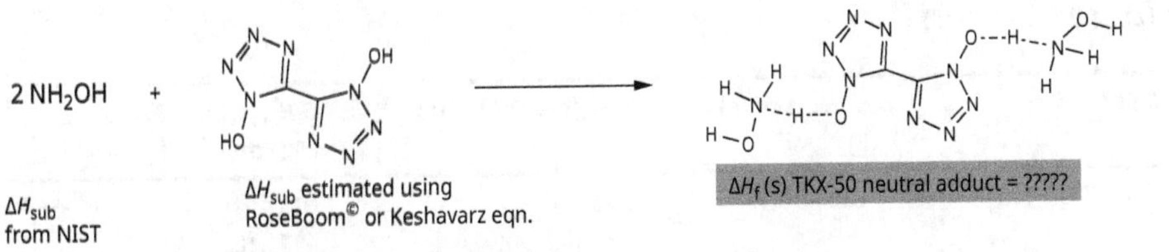

ΔH_f (s) TKX-50 neutral adduct = ΔH_f TKX-50 neutral adduct (g) – 2 × ΔH_{sub} NH_2OH (s) – ΔH_{sub} diol (s)

3. In the final step, the values from steps 1 and 2 are weighted 80:20 to obtain an estimated value for the enthalpy of formation for the salt:

ΔH_f (s) TKX-50 (est.) = 0.8 × ΔH_f TKX-50 salt + 0.2 × ΔH_f TKX-50 neutral adduct

Table 4.9a gives a summary of the estimated values for the $\Delta H_{\text{salt reaction}}$ for a range of neutral bases. The experimentally determined values for the enthalpy of formation ($\Delta_f H^\circ_m$(s)) of the corresponding base-$H^+NO_3^-$ salts were used and not of other salts such as the corresponding perchlorate salt.

Tab. 4.9a: Estimated values for the $\Delta H_{\text{salt reaction}}$ for a range of bases, estimated using the values of $\Delta_f H^\circ_m$(s) of only the corresponding nitrate salts.

Salt	$\Delta_f H^\circ_m$(salt, s) / kJ · mol^{-1}	$\Delta_f H^\circ_m$(base) / kJ · mol^{-1}	$\Delta_f H^\circ_m$(acid) / kJ · mol^{-1}	$\Delta H_{\text{salt reaction}}$ / kJ · mol^{-1}
$NH_3OH^+NO_3^-$	−366.5	NH_2OH(s) = −114.2	HNO_3(l) = −174.10	−78.2
$NH_4^+NO_3^-$	−365.56	NH_3(g) = −46.11	HNO_3(l) = −174.10	−145.35
$C(NH_2)_3^+NO_3^-$	−386.94	guanidine(s) = − 56.1	HNO_3(l) = −174.10	−156.8
$AG^+NO_3^-$	−278.7	AG(s) = + 58.5	HNO_3(l) = −174.10	−163.1
$DAG^+NO_3^-$	−157.3	DAG(s) = + 167.4	HNO_3(l) = −174.10	−150.6
$TAG^+NO_3^-$	−50.2	TAG(s) = + 287.7	HNO_3(l) = −174.10	−163.8
$5\text{-}AT^+NO_3^-$	−27.6	5-AT(s) = + 207.9	HNO_3(l) = −174.10	−61.4
$N_2H_5^+NO_3^-$	−251.58	N_2H_4(l) = + 50.63	HNO_3 (l) = −174.10	−128.1

Tab. 4.9a (continued)

Salt	$\Delta_f H^\circ_m$(salt, s) / kJ · mol^{-1}	$\Delta_f H^\circ_m$(base) / kJ · mol^{-1}	$\Delta_f H^\circ_m$(acid) / kJ · mol^{-1}	$\Delta H_{salt\ reaction}$ / kJ · mol^{-1}
$(H_2N)_2COH^+NO_3^-$	−564.0	urea(cr) = − 333.51	HNO_3(l) = −174.10	−56.39
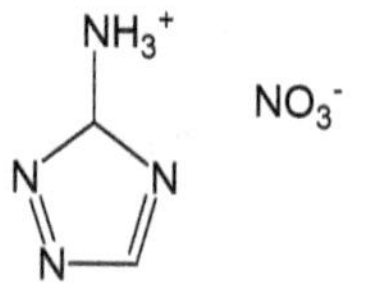 3-amino-1,2,4-triazolium$^+NO_3^-$	−171.1	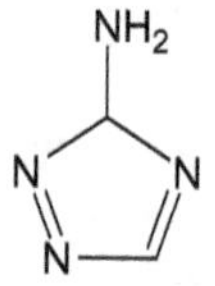 3-amino-1,2,4-triazole (s) = + 77.0	HNO_3(l) = −174.10	−74
Anilinium$^+NO_3^-$	−182	aniline (l) = + 31.3	HNO_3(l) = −174.10	−40
$EtNH_3^+NO_3^-$	−366.9	$EtNH_2$(l) = − 74.1	HNO_3(l) = −174.10	−118.7
$Me_2NH_2^+NO_3^-$	−352.0	Me_2NH(l) = − 43.9	HNO_3(l) = −174.10	−134.0
$Me_3NH^+NO_3^-$	−343.9	Me_3N(l) = − 46.0	HNO_3(l) = −174.10	−123.8
$MeNH_3^+NO_3^-$	−354.4	$MeNH_2$(l) = − 47.3	HNO_3(l) = −174.10	−132.6
$Et_2NH_2^+NO_3^-$	−418.8	Et_2NH(l) = − 103.3	HNO_3(l) = −174.10	−141.7
$Et_3NH^+NO_3^-$	−447.7	Et_3N(l) = − 169.0	HNO_3(l) = −174.10	−104.6
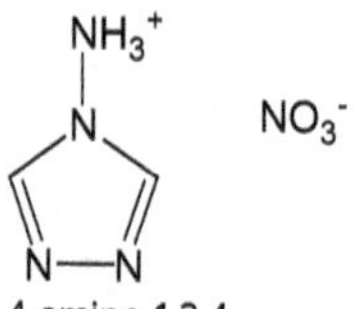 4-amino-1,2,4-triazolium nitrate	+2	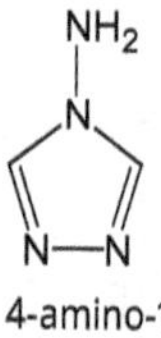 4-amino-1,2,4-triazole(s) = + 223.13	HNO_3(l) = −174.10	−47.03

Table 4.9b shows a summary for the calculated and measured enthalpies of formation of various 2:1 salts using the Jenkins equation and the 80:20 method as described above.

Tab. 4.9b: Estimated values for the $\Delta_f H^\circ_m$ (s) for 2:1 salts using values calculated at CBS-QB3 level of theory where necessary, as well as the experimentally determined $\Delta_f H^\circ_m$ (s) values previously reported in the literature.

Compound	**$\Delta_f H^\circ_m$ / kJ · mol^{-1} CBS-QB3: Cation** *Anion* **ΔH_{latt} / kJ · mol^{-1} (Jenkins)**	**$\Delta_f H^\circ_m$ kJ · mol^{-1} (CBS-QB3 Jenkins)**	**$\Delta_f H^\circ_m$ kJ · mol^{-1} (Salt)**	**$\Delta_f H^\circ_m$ / kJ · mol^{-1} (Neutral adduct)**	**Average of salt + adduct / kJ. mol^{-1}**	**80:20 / kJ. mol^{-1}**	**Exptl. value/ kJ. mol^{-1}**	**Δ / kJ.mol^{-1} (exptl. and 80:20 values)**
TKX-50	**2 × + 669.2** *1 × + 541.05* −1488.95	+390.5	+168.3	+240.49	+204.4	+182.74	+193 (average)	−10.26
TKX-50	**2 × + 669.2** *1 × + 541.05* −1488.95	+390.5	+168.3	+252.39	+210.3	+185.12	+193 (average)	−7.88
GZT	**2 × 570.31** *1 × + 790.5* −1287.98	+643.14	+374.2	+532.37	+453.3	+405.83	+410, +452, +387	−4.17, +18.83, −46.17
AG_2AzT	**2 × 668.9** *1 × + 790.5* −1246.01	+882.29	−509.8	+737.9	+664.4	+620.22	+462, +434, +782	+158.22, +186.22, −161.78
DAG_2AzT	**2 × 813.1** *1 × + 790.5* −1215.11	+1201.59	+833.6	+885.56	+859.6	+843.99	+709	+134.99

(continued)

Tab. 4.9b (continued)

Compound	**$\Delta_f H^\circ_m$ / kJ · mol^{-1} CBS-QB3: Cation** *Anion* **ΔH_{latt} / kJ · mol^{-1} (Jenkins)**	**$\Delta_f H^\circ_m$ kJ · mol^{-1} (CBS-QB3 Jenkins)**	**$\Delta_f H^\circ_m$ kJ · mol^{-1} (Salt)**	**$\Delta_f H^\circ_m$ / kJ · mol^{-1} (Neutral adduct)**	**Average of salt + adduct / kJ. mol^{-1}**	**80:20 / kJ. mol^{-1}**	**Exptl. value/ kJ. mol^{-1}**	**Δ / kJ.mol^{-1} (exptl. and 80:20 values)**
TAG_2AzT	**2 × 920.8** *1 × + 790.5* −1177.40	+1454.7	+1047.8	+1167.04	+1107.42	+1071.65	+1,075, +1,065	−3.35, +6.65
$(NH_4)_2AzT$	**2 × 632.12** *1 × + 790.5* −1467.43	+587.31	+417.08	+623.03	+520.1	+458.27	+443.9, +452, +551	+14.37, +6.27, −92.73
$(N_2H_5)_2AzT$	**2 × 765.52** *1 × + 790.5* –	density unknown	+645.06	+784.88	+714.97	+673.02	+659 +858	+14.02, −184.98
G_2CO_3	**2 × + 570.31** *1 × − 232.2* −1415.0	−505.68	−1044.6	−902.36	−973.48	−1016.15	−971.1	−45.05
$(NH_4)_2SO_4$	**2 × + 632.12** *1 × − 607.3* −1806.74	−1149.8	−1196.9	−989.22	−1093.1	−1155.36	−1180.9	+25.54
$(NH_3OH)_2SO_4$	**2 × + 669.2** *1 × − 607.3* −1712.2	−981.1	−1200.0	−1143.6	−1171.8	−1188.72	−181.98	−6.74

Literature

A. L. R. Silva, A. R. R. P. Almeida, M. D. M. C. Ribeiro da Silva, J. Reinhardt, T. M. Klapötke, *Propellants, Explosives, Pyrotechnics*, **2023**, *48*, e202200361, http://doi.org/10.1002/prep.202200361.

A. L. R. Silva, G. P. León, M. D. M. C. Ribeiro da Silva, T. M. Klapötke, J. Reinhardt, *Thermo*, **2023**, *3*(4), 549–565; https://doi.org/10.3390/thermo3040033.

V.P. Sinditskii, S. A. Filatov, V. I. Kolesov, K O. Kapranov, A. F. Asachenko, M. S. Nechaev, V. V. Lunin, N. I. Shishov, *Thermochim. Acta*, **2015**, *614*, 85–92.

H. D. B. Jenkins, H. K. Roobottom, J. Passmore, L. Glasser, *Inorg. Chem.*, **1999**, *38*, 3609–3620.

K. E. Gutowski, R. D. Rogers, D. A. Dixon, *J. Phys. Chem. B*, **2007**, *111*, 4788–4800.

E. F. C. Byrd, B. M. Rice, *J. Phys. Chem. A*, **2009**, *113*, 345–352.

4.2.2 Detonation parameters

There now are various different codes available for computing detonation parameters (e.g. TIGER, CHEETAH, EXPLO5 etc.). In this discussion, we will concentrate on the application of one such code, namely the program EXPLO5. This program is based on the chemical equilibrium, a steady-state model of detonation. It uses the Becker-Kistiakowsky-Wilson's equation of state (BKW-EOS) for gaseous detonation products and Cowan-Fickett's equation of state for solid carbon [31–34]. The calculation of the equilibrium composition of the detonation products is done by applying the modified White, Johnson and Dantzig's free energy minimization technique. The program is designed to enable the calculation of detonation parameters at the C-J point (see Ch. 3).

The ideal gas law of the form

$$pV = nRT$$

with the ideal gas constant R enables us to calculate for an ideal gas the pressure p for a certain temperature T and also the volume V for a known molar amount n.

To calculate the detonation pressure, this equation is however not good enough, because there are strong deviations from the ideal gas behavior. The Becker-Kistiakowsky-Wilson equation of state (BKW-EOS)

$$\frac{pV}{RT} = 1 + \mathrm{x}\,\mathrm{e}^{\beta \mathrm{x}} \qquad \mathrm{x} = \frac{\mathrm{k}}{VT^{\alpha}}$$

which contains the covolume constant k takes into account the residual volume of the molecules of the gaseous components. The α and β parameters are obtained by comparisons with the experimental data (fitted empirically). For extremely low temperatures, however, the pressure becomes infinitely large. Cowan and Fickett developed the Becker-Kistiakowsky-Wilson equation of state to:

$$\frac{pV}{RT} = 1 + \mathrm{x}\, e^{\beta \mathrm{x}} \qquad \mathrm{X} = \frac{\kappa \sum X_i k_i}{V(T+\theta)^a}$$

whereby also $(\delta p / \delta T)_V$ remains positive. The covolume constant is replaced by the weighted sum of the products of the mole fractions X_i and the geometric covolumes k_i of the gaseous components i. The Becker-Kistiakowsky-Wilson-Neumann parameters α, β, κ and θ were obtained empirically by adapting the computational results to the experimental data.

However, this equation can not be used for solids. For such cases Cowan and Fickett suggest an equation of the following form with the factors $p_1(V)$, $a(V)$ and $b(V)$ as polynomial functions of the compression of the material η relative to the crystal density of the solid in the standard state, which is known as the Cowan-Fickett equation of state for solids:

$$p = p_1(V) + a(V)T + b(V)T^2$$

$$\eta = \frac{V^o(T^o)}{V} = \frac{\rho}{\rho_o}$$

The $p(V)$ diagram shows the shock adiabat of the explosive, which is also called the Hugoniot curve or Hugoniot adiabat (see Fig. 4.2). The shock adiabat can be calculated for both the non-reacted explosive as well as for the reaction products (see Fig. 4.2). The C-J point, which represents the point where the C-J conditions are fulfilled, is therefore the point where the shock adiabat of the reaction products touches the Rayleigh line (tangent), which is described by the following equation:

$$p - p_0 = \rho_0^2 U^2 (V_0 - V)$$

Here, ρ_0 is the material density of the unreacted explosive, U is the impact velocity and V and V_0 are the specific volumes. The Rayleigh line is a straight line with the slope D^2 / V^2 connecting (V_0, p_0), (V_1, p_1) and (V_2, p_2) in the $p(V)$ diagram.

At the point where the Rayleigh line touches the schock adiabat of the reaction products the slope of both functions is the same, and the following relationship, where U is the velocity of the products, is valid:

$$\frac{\partial p}{\partial V} = \frac{D^2}{V^2} = \frac{U^2}{V^2}$$

Under the assumption of a steady-state model of detonation, the EXPLO5 program allows calculation of the detonation parameters and also the chemical composition of the system at the C-J point. For the calculations, the BKW equation of state is applied for gases, where X_i is the mole fraction of the i-th gaseous component and k_i is the molar covolume of the i-th gaseous detonation products:

$$\frac{pV}{RT} = 1 + x\,e^{\beta x} \qquad x = \frac{\kappa \sum X_i k_i}{V(T+\theta)^a}$$

$\alpha = 0.5,\quad \beta = 0.176,\quad \kappa = 14.71,\quad \theta = 6{,}620$ (EXPLO5 V5.03)
$\alpha = 0.5,\quad \beta = 0.096,\quad \kappa = 17.56,\quad \theta = 4{,}950$ (EXPLO5 V5.04)
$\alpha = 0.5,\quad \beta = 0.38,\quad \kappa = 9.41,\quad \theta = 4{,}250$ (EXPLO5 V6.02)

For solid carbon, the Cowan-Fickett equation of state is applied in the following form, with the presumption that carbon is present in the graphite modification:

$$p = p_1(V) + a(V)T' + b(V)T'^2$$

where:
$T' = T / 11605.6$ K,
$p_1(V) = -2.467 + 6.769\,\eta - 6.956\,\eta^2 + 3.040\,\eta^3 - 0.3869\,\eta^4$,
$a(V) = -0.2267 + 0.2712\,\eta$,
$b(V) = 0.08316 - 0.07804\,\eta^{-1} + 0.03068\,\eta^{-2}$,

$$\text{compression of the material } \eta = \frac{V^\circ(T^\circ)}{V} = \frac{\rho}{\rho^\circ}$$

The minimization of the free energy in accordance with White-Johnson-Dantzig allows the determination of the composition of the detonation products in the equilibrium state. The thermodynamic parameters (enthalpy, entropy, free enthalpy, free energy) of the detonation products can be calculated using functions based on the enthalpy $H_T^\circ - H_0^\circ$. The coefficients $c_0, \ldots c_4$ can be found in the literature [*J. Chem. Phys* **1958**, *28*, 751–755; *Prop. Expl. Pyrotech.* **1985**, *10*, 47–52]:

$$(H_T^\circ - H_0^\circ) = c_0 + c_1 T + c_2 T^2 + c_3 T^3 + c_4 T^4$$

The EXPLO5 program calculates the parameters of state of the products along the shock adiabat, starting from a density of a given explosive (ρ_0) and then increasing it in an arbitrary chosen step up to the density of about $1.5 \cdot \rho_0$. Then it determines the C-J point as a point on the shock adiabat at which the detonation velocity has a minimum value (this minimum, D_{min}, is determined by the minimum of the first derivative of the Hugoniot adiabat, see Fig. 4.2). Once the C-J point is determined, the detonation parameters can be calculated by applying the well-known relationships between them.

The hydrodynamic detonation theory allows correlation of the detonation parameters on the basis of the laws of mass, impulse and energy conservation and to calculate them independently of p and V:

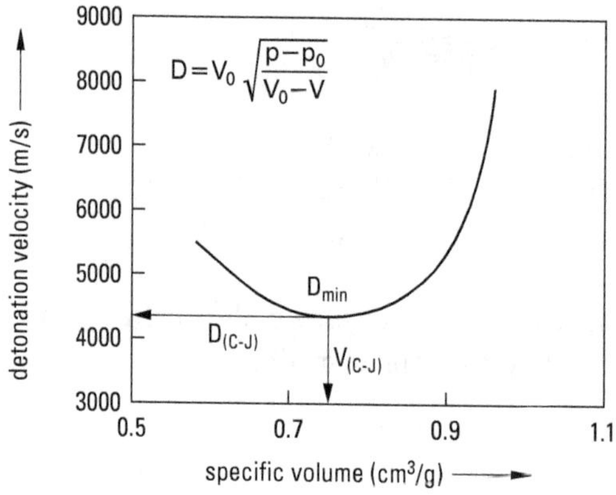

Fig. 4.2: Change of detonation velocity with specific volume of detonation products.

$$\frac{D}{V_0} = \frac{D-U}{V}$$

$$\frac{D^2}{V_0} + p_0 = \frac{(D-U)^2}{V} + p$$

$$U_0 + D^{2} + p_0 V_0 = U + 1/2\,(D-U)^{2} + pV$$

From this, for the detonation velocity and the velocity of the shock front, the following equations can be written:

$$D = V_0\sqrt{\frac{p-p_0}{V_0-V}}$$

$$U = (V_0 - V)\sqrt{\frac{p-p_0}{V_0-V}}$$

The detonation parameters for NG, TNT and RDX calculated using the EXPLO5-program are summarized in Tab. 4.10, as well as the experimental data for comparison.

Sometimes, the terms “heat of explosion” and “heat of detonation” get mixed-up and are often (wrongly!) used as if they were equivalent. The heat of explosion and the heat of detonation are both heats of reaction but are obtained under different conditions (by applying Hess’ law). In the detonation run, the heat of reaction corresponds to the “heat of reaction at the C-J point”, and is called heat of detonation. The heat of reaction obtained under “constant volume condition” combustion is usually called the “heat of combustion” or “heat of explosion”. There is one additional term which is worth mentioning here and that is the “heat of complete combustion”. This term is usually used to denote constant volume combustion in an oxygen atmosphere.

Tab. 4.10: Calculated and experimentally determined detonation parameters.

explosive	density	method	D / m s^{-1}	$p_{C\text{-}J}$ / kbar	T_{ex} / K	$Q_{C\text{-}J}$ / kJ kg^{-1}
NG	1.60	experimental	7,700	253	4,260	
		EXPLO5	7,819	242	4,727	-6,229
TNT	1.64	experimental	6,950	210		
		EXPLO5	7,150	202	3,744	-5,087
RDX	1.80	experimental	8,750	347		
		EXPLO5	8,920	345	4,354	6,033
HNS	1.65	experimental	7,030	215		
		EXPLO5	7,230	212	4,079	-5,239
PETN	1.76	experimental	8,270	315		
		EXPLO5	8,660	311	4,349	5,889

In a shock wave, the Rankine-Hugoniot jump equations apply across the detonation front. The shock pressure is replaced by the C-J pressure p_{CJ}, shock particle velocity u_{CJ}, and the shock velocity by the detonation velocity D.

mass balance: $\rho_{CJ} / \rho_0 = D / (D - u_{CJ})$

Momentum balance: $P_{CJ} = \rho_0 u_{CJ} D$,

Where P_{CJ} is the C-J pressure (GPa), ρ_{CJ} is the density in the C-J state (g cm^{-3}), ρ_0 is the density of the unreacted explosive (g cm^{-3}), u_{CJ} the particle velocity at the C-J state (km s^{-1}), and D the detonation velocity (km s^{-1}).

The momentum equation can be further simplified by applying the following approximation, which predicts the C-J pressure within an accuracy of 7% for most explosives.

$$P_{CJ} = \rho_0 D^2 / 4$$

4.2.3 Combustion parameters

Using the calculated thermodynamic data outlined in 4.2.1, not only the detonation parameters (4.2.2) but also the most important performance parameters for rocket propellants can be calculated. For this there are various programs which are available as well, which we are going to refer back to but we will concentrate again on the EXPLO5 code.

The combustion of an energetic material is an irreversible process in which mainly gaseous – and to a lesser extent solid – combustion products are formed. One can differentiate between two borderline cases of combustion:

1. isobaric
2. isochoric

In an **isobaric combustion**, the combustion occurs without any loss of heat to the surroundings (adiabatic) and at a constant pressure (isobar), while the combustion products are found in a chemical equilibrium (e.g. rocket propellants).

In an **isochoric combustion** it is also assumed that the combustion occurs without any loss of heat to the surroundings (adiabatic) but at a constant volume (isochoric) and the combustion products are found in chemical equilibrium (e.g. gun propellants).

For **rocket propellants**, as we have already seen in Chapter 4.1, the assumption of free expansion of the gases in the atmosphere or in space, p = const. (and thereby $\Delta U = Q_p - p\Delta V$) is a good approximation and therefore the process is best described as **isobaric**. For the theoretical calculation of the performance of rocket propellants, the following assumptions are made:
1. the pressure in the combustion chamber and at the nozzle throat is constant
2. energy and momentum conservation equations are applicable
3. the velocity of the products of combustion in the combustion chamber is equal to zero
4. there is no temperature and velocity lag between condensed and gaseous species
5. the expansion in the nozzle (nozzle, Fig. 4.3) occurs isentropically (a change of state is isentropic, when the entropy S remains constant, i. e. S = const. or $dS = 0$).

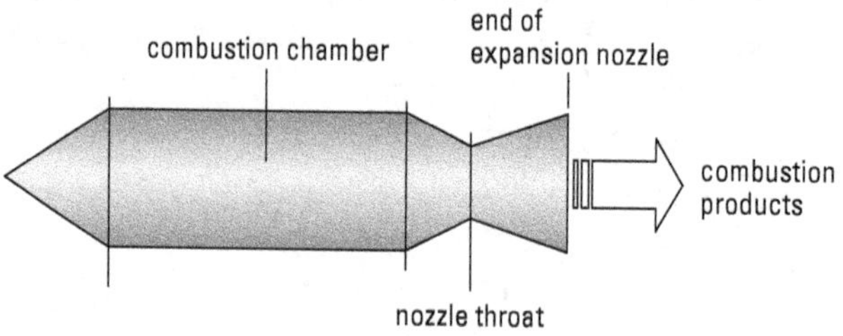

Fig. 4.3: Schematic representation of a rocket combustion chamber with expansion nozzle.

The important performance parameters for the performance of a rocket propellant are the thrust F and the specific impulse I^*_{sp}

As we have already seen above (Ch. 2.4), the average thrust and the specific impulse I_{sp} are related to each other as follows ($I^*_{sp} = I_{sp} / g$):

$$\bar{F} = I_{sp} \frac{\Delta m}{\Delta t}$$

For the thrust we can finally write:

$$F = \frac{dm}{dt} v_e + (p_e - p_a) A_e$$

Here, v_e is the velocity of the combustion gases at the end of the expansion nozzle (see Fig. 2.20), p_e and p_a are the pressure at the end of the expansion nozzle and atmospheric pressure respectively and A_e is the cross-section at the end of the expansion nozzle.

Therefore, we can write the following for the specific impulse:

$$I_{sp} = \frac{F}{\frac{dm}{dt}} = v_e + \frac{(p_e - p_a) A_e}{\frac{dm}{dt}}$$

Using the EXPLO5 programme, we can calculate the following performance parameters (amongst others) for rocket propellants under isobaric conditions for different pressures in the combustion chamber:
- isobaric heat of combustion Q_p (kJ kg^{-1})
- isobaric combustion temperature T_c (K)
- composition of the products of combustion
- temperature and pressure at the nozzle throat
- flow velocity at the nozzle throat
- temperature at the end of the expansion nozzle (p_e = 1 bar)
- specific impulse

For isochoric combustion processes, the following parameters (amongst others) can be calculated:
- isochoric heat of combustion Q_v (kJ kg^{-1})
- total pressure in closed systems (bar)
- composition of the combustion products
- specific energy: $F = n\,RT_c$ (J kg^{-1}), T_c = isochoric combustion temperature.

Currently, a new propellant is being researched. This propellant is hydrazinium aminotetrazolate (**2**), which is obtained in a facile route from hydrazine and aminotetrazole (**1**) (Fig. 4.4) [35a].

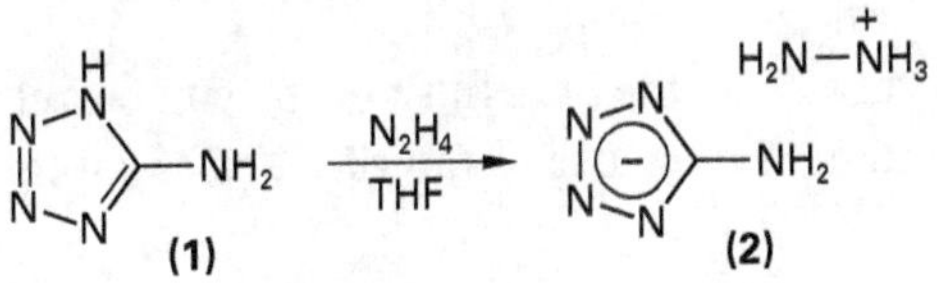

Fig. 4.4: Synthesis of hydrazinium aminotetrazolate (**2**, Hy-At).

Tab. 4.11: Calculated combustion parameters for hydrazinium aminotetrazolate – Hy-At /ADN formulations with p = 45 bar.

Oxidizer ADN[a]	Fuel Hy-At	ρ / g cm^{-3}	Ω / %	T_c / K	I_{sp} / s
10	90	1.573	−65.0	1863	227
20	80	1.599	−55.0	1922	229
30	70	1.625	−44.9	2,110	236
40	60	1.651	−34.8	2,377	246
50	50	1.678	−24.7	2,653	254
60	40	1.704	−14.6	2,916	260
70	30	1.730	−4.5	3,091	261
80	20	1.756	+5.6	2,954	250
90	10	1.782	+15.7	2,570	229
AP[b]	Al[c]				
70	30	2.178	−2.9	4,273	255

[a]ADN, ammonium dinitramide, [b]ammonium perchlorate, [c]aluminum.

The programme EXPLO5 was used to calculate the rocket propellant parameters for this compound in combination with ADN as the oxidizer while presuming a combustion chamber pressure of 45 bar (Tab. 4.11).

We can see that the specific impulse of a 70 : 30 mixture of Hy-At is approximately 5 s higher than that of a stoichiometric mixture of AP and Al. Therefore, the new fuel Hy-At in combination with the environmentally friendly ADN (in contrast to AP) could potentially be of interest.

For **gun propellants** we can assume isochoric combustion with the specific energy f_E or force or impetus ($f_E = n\,RT$), the combustion temperature T_c (K), the co-volume b_E (cm^3 g^{-1}) and the pressure p (bar; 3,000–4,000 bar) being the most important parameters. Moreover, a large N_2 / CO ratio is desirable in order to avoid erosion problems. The loading densities are by far not as important as for high explosives. Comparing M1 and EX-99 with the newly developed high-nitrogen propellant (HNP) which is based on triaminoguanidinium azotetrazolate (TAGzT) and with NICO (based on TAG_2BT) (Tab. 4.11a), very similar performance values for M1, HNP and NICO, but a greatly reduced errosion coefficient (N_2/CO) can be found (Tab. 4.11b, Fig. 4.4a). It is also worth mentioning that the HNP and NICO are similar in their performance to NILE (Navy Insensitive Low Erosion Propellant: 40% RDX, 32% GUDN, 7% acetyl triethyl citrate, 14% cellulose acetate butyrate, 5% hydroxyl propylcellulose, 2% plasticizer).

The term gun erosion refers to the loss of steel in the bore (inside) of a gun barrel. Consequently, if gun erosion occurs, it results in the wear of the rifle and/or an increase in the bore diameter.

Since as far back as even World War II, it has been common knowledge that the erosion of gun barrels leads to two main types of problems:

Tab. 4.11a: Propellant charge compositions.

Propellant charge formulation	Ingredient	Amount / % (weight)
M1	NC (13.25)	86
	2,4-DNT	10
	dibutyl phthalate (DBP)	3
	diphenyl amine (DPA)	1
EX-99	RDX	76
	cellulose acetate	12
	BDNPA/F[a]	8
	NC (13.25)	4
High-N-1	RDX	56
	TAGzT	20
	cellulose acetate	12
	BDNPA/F[a]	8
	NC (13.25)	4
High-N-2 (HNP)	RDX	40
	TAGzT	20
	FOX-12/GUDN	16
	cellulose acetate	12
	BDNPA/F[a]	8
	NC (13.25)	4
TAG_2-BT-High-N-2 (NICO)	RDX	40
	TAG_2-BT[b]	20
	FOX-12/GUDN	16
	cellulose acetate	12
	BDNPA/F[a]	8
	NC (13.25)	4

[a]Bis(dinitropropyl)acetal (BDNPA) : Bis(dinitropropyl)formal (BDNPF) = 50 : 50
[b]TAG_2-BT = $(CN_6H_9)_2$5,5′-Bistetrazolate

Tab. 4.11b: Calculated performance of various gun propellants[a].

	T_c / K	$p_{max.}$ / bar	f_E / kJ g^{-1}	b_E / cm^3 g^{-1}	N_2 / CO (w/w)
M1	2,834	2,591	1.005	1.125	0.23
EX-99	3,406	3,249	1.257	1.129	0.71
High-N-1	2,922	3,042	1.161	1.185	0.95
High-N-2 (HNP)	2,735	2,848	1.088	1.181	1.05
TAG_2-BT-High-N-2 (NICO)	2,756	2,896	1.105	1.185	1.03

[a]loading density = 0.2 g cm^{-3}; virial equations of state (real gas)

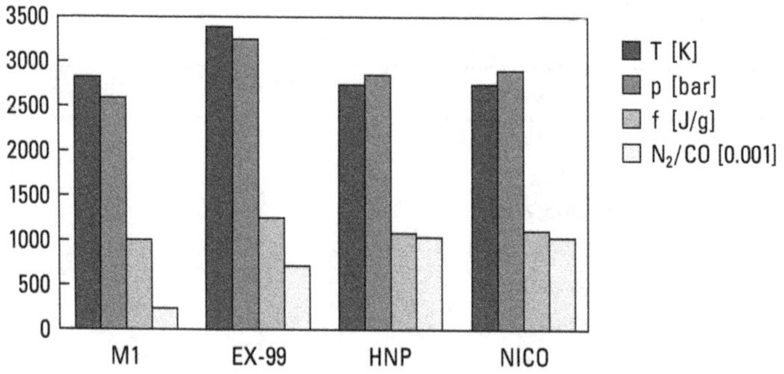

Fig. 4.4a: Calculated performance of various gun propellants (HNP = High-N-2, NICO = TAG_2-BT-High-N-2).

(i) financial: due to barrel replacement costs over the lifespan of the weapon system and
(ii) reduced operational effectiveness: due to inconsistent gun performance and availability.

These problems are consequences of the effects of repeated firings on gun barrels. Even under normal firing conditions and usage, damage to the bore surface and therefore a continuous and progressive increase in the bore diameter result as consequences. Financial costs ensue not only due to the necessity of replacing the barrel itself, but also for the transportation of the gun barrel from its production facility to the location of the gun system. The latter is a logistical cost which depends on the distance and fuel prices, but is not trivial in its sum.

Three main approaches are employed to attempt to reduce the amount of gun barrel erosion which occurs, namely: the development of propellants which cause less erosion, the use of coatings and the production of treated barrel materials and liners, as well as the employment of additives and lubricants which reduce erosion. Out of these three approaches, the use of protective coatings, or treating the surface of the gun barrel bore have found the most wide-spread use. One of the most commonly employed methods used to achieve this includes the deposition of a thin layer of hard chromium inside the gun barrel or by nitriding (hardening) the inside of the barrel. Such coatings/treatments are designed to combat all of the mechanisms of erosion which are discussed below. Although gun barrel erosion is complex and no one single mechanism is usually exclusively responsible, it can be very approximately divided into three categories:
(i) mechanical action
(ii) heat transfer effects and
(iii) changes in chemical composition

These three categories can be classified as mechanical, thermal and chemical erosion respectively, and are highly interdependent in that they act in combination to erode the barrels. Mechanical erosion results from friction due to a projectile (or its driving band) moving through the gun barrel. Thermal erosion, on the other hand, is due to morphological changes of the gun barrel steel, (*i.e.* austenite and martensite phase transformation of steel) as well as melting of the barrel material. That the latter occurs is unsurprising when it is considered that depending on the propellant which is used and the rate of fire of the gun system, the steel temperature of the barrel can reach 1800 K in a matter of only milliseconds, which is high enough to cause partial melting of the gun steel, since typical gun steel melts at 1723 K. Such temperatures are, however, only achieved if very hot propellants are used and usually the temperature of the gun barrel will be significantly lower. A good illustrative example here is the barrel surface temperature for a Navy 5"/54 gun which has been evaluated at 1450 K for a certain propellant which has a flame temperature of 2654 K.

Chemical erosion is the term used to describe the effect of chemical reactions that occur between the gun steel and the combustion gases generated by the propellant. The main combustion gases generated by a gun propellant are carbon monoxide (CO), carbon dioxide (CO_2), water (H_2O), hydrogen (H_2) and nitrogen (N_2). At elevated temperatures, each of the above-mentioned gases can react with gun steel to form various compounds. Some of these reactions are shown in eqs. (1 to 4).

$$Fe + CO_2 \rightarrow FeO + CO \tag{1}$$

$$3\,Fe + 2\,CO \rightarrow Fe_3C + CO_2 \tag{2}$$

$$4\,Fe + CO \rightarrow FeO + Fe_3C \tag{3}$$

$$5\,Fe + CO_2 \rightarrow 2\,FeO + Fe_3C \tag{4}$$

Iron carbide, Fe_3C is formed due to the presence of CO and CO_2 in the gun barrel. In order to minimize erosion of the gun barrel, the formation of iron carbide (with its low melting point of approximately 1420 K) and other low melting point ferrous compounds should ideally be avoided in order to minimize erosion. It is interesting and worthwhile to note that there is currently no consensus on which combustion gas is the most erosive out of the combustion gases that are produced. Despite this, the following order has been proposed which lists the most (CO_2) to least (N_2) chemically erosive combustion gas. The four compounds to the left of "0" correspond to gases which contribute to gun erosion, whereas N_2 to the right of "0" with a chemical erosion value of less than zero, contributes to the stability of a gun barrel due its formation of iron nitride on reaction with the barrel.

$$CO_2 > CO > H_2O > H_2 > 0 > N_2$$

The formation of iron nitride (mostly ε-Fe_3N at the surface and Fe_4N in lower layers) as a consequence of nitrogen gas being present in the combustion gases has been established, and it is generally accepted that in contrast to CO_2, CO, H_2O and H_2, nitrogen gas has a lowering effect on gun barrel wear. Consequently, it has been proposed that combustion gases which possess a high-nitrogen content could, in fact, contribute to the re-nitridation of the gun barrel and thereby increase the service lifetime by a significant amount (up to a factor of four).

The interdependence between the propellant and the thermal and chemical erosion can be more easily understood since combustion of the propellant is responsible for generating the heat in the gun, and the composition of the combustion gases depends not only on the molecular composition of the propellant, but also on its flame temperature. And of course, the reactions of the steel and the combustion gases are not only dependent on what combustion gases are present, but also on the temperature of the gases.

Therefore, it is not surprising and also to be expected, that changes to either the flame temperature and/or the composition of the combustion gases will have a direct impact on the erosion capability of a propellant. The amount of CO that is produced can be decreased by increasing the oxygen balance of the propellant, resulting in combustion gases with a higher CO_2 and H_2O content, but without affecting the nitrogen content. Since nitrogen is a combustion gas which is not considered to be erosive, the generation of nitrogen should be prioritized. Importantly, since the N_2/CO ratio of combustion gases can be predicted using thermochemical calculations which negate the use of experiments, this ratio can be used to estimate how erosive a particular propellant will be.

If all of the above-mentioned points are taken into consideration, then it can be concluded that in designing a gun propellant which shows low erosion properties, it is of particular importance to achieve a lowering of the flame temperature, as well as to increase the nitrogen content of the combustion gases. Consequently, nitrogen-rich materials are particularly promising since being nitrogen-rich they show very high (often > 80% by weight) nitrogen-contents. In addition, the advantageous effect of nitrogen-rich materials on the burning rates of propellants has also been shown. In fact, it has been demonstrated that for certain nitrogen-rich propellants, even although the flame temperature is lower, the experimentally determined maximum pressure in a closed vessel was the same as those of formulations that did not contain nitrogen-rich components, which indicates that the same level of performance is achievable.

The most common equation of state for interior ballistics is that of Nobel-Abel:

$$p\,(v - b_E) = n\,RT$$

(with: b_E = co-volume, n = mol number, R = gas constant).

The co-volume is a parameter which takes the physical size of the molecules and any intermolecular forces created by their proximity to one another into account.

If experimental test results (manometric pressure bomb) for the maximum pressure using different loading densities are available, the force f_E and the co-volumes b_E can be calculated as follows:

$$f_E = \frac{p_2}{d_2} \times \frac{p_1}{d_1} \times \frac{d_2 - d_1}{p_2 - p_1}$$

$$b_E = \frac{\frac{p_2}{d_2} - \frac{p_1}{d_1}}{p_2 p_1}$$

where:
f_E force
b_E co-volume
p_1 maximum pressure at the lower density d_1
p_2 maximum pressure at the higher density d_2
d_1 lower loading density
d_2 higher loading density.

Assuming that the powder combustion occurs at a temperature equal to the explosion temperature T_{ex} and under consideration of the fact, that after combustion, the specific volume V and the charge density (*d*) are equal, the following equation applies:

$$\frac{p_{max}}{d} = b_E\,p_{max} + f_E$$

The EXPLO5 code can also be used for calculation of the thermodynamic properties of propellant gases and shows an error of usually less than 5% (loading densities ca. 0.2 g cm^{-3}) when the maximum pressure, the specific energy and the co-volume are considered.

Once a suitable propellant formulation has been identified, the design of the solid grain becomes an important issue (and goes beyond the scope of this book). Fig. 4.4b shows some typical solid grain designs for regressive, neutral, and progressive burning.

The pressure-time profile inside the gun chamber and the actual value of the peak pressure are very important. The propellant grain configuration should, therefore, be designed so that the required peak pressure is achieved within a few milliseconds to propel the projectile with the desired velocity. If the surface area of the grain starts decreasing as the burning proceeds, it is called regressive burning (e.g. a solid cylindrical grain) (Fig. 4.4b). If the surface area of the grain remains constant during burning, we talk about neutral burning (e.g. tubular grains) (Fig. 4.4b). If a multi-perforated grain is used, e.g. a hepta-tubular grain, the rate of increase of the surface area originating from the seven holes overcompensates the decrease of the surface area due to burning from the periphery. As a result, the chamber pressure increases during burning (progressive burning) (Fig. 4.4b). The multi-tubular geometry of the grain is common among gun propellants for high performance guns such as tank guns.

The investigation of propellant charge powders in a manometric bomb or pressure bomb provides values for the burning behavior, e.g. the vivacity. The schematic structure of a manometric bomb is shown in Fig. 4.4c.

A manometric bomb (closed vessel) is a steel cylinder with a defined internal volume (1). The propellant powder is ignited by an electric igniter (2). The pressure development is measured by a pressure sensor (3) (usually a piezo element) and recorded. After complete burn-up, the excess pressure can be released via the outlet valve (4).

The increase in pressure over time is measured until the propellant charge is completely burned. The maximum achievable pressure depends on the amount of

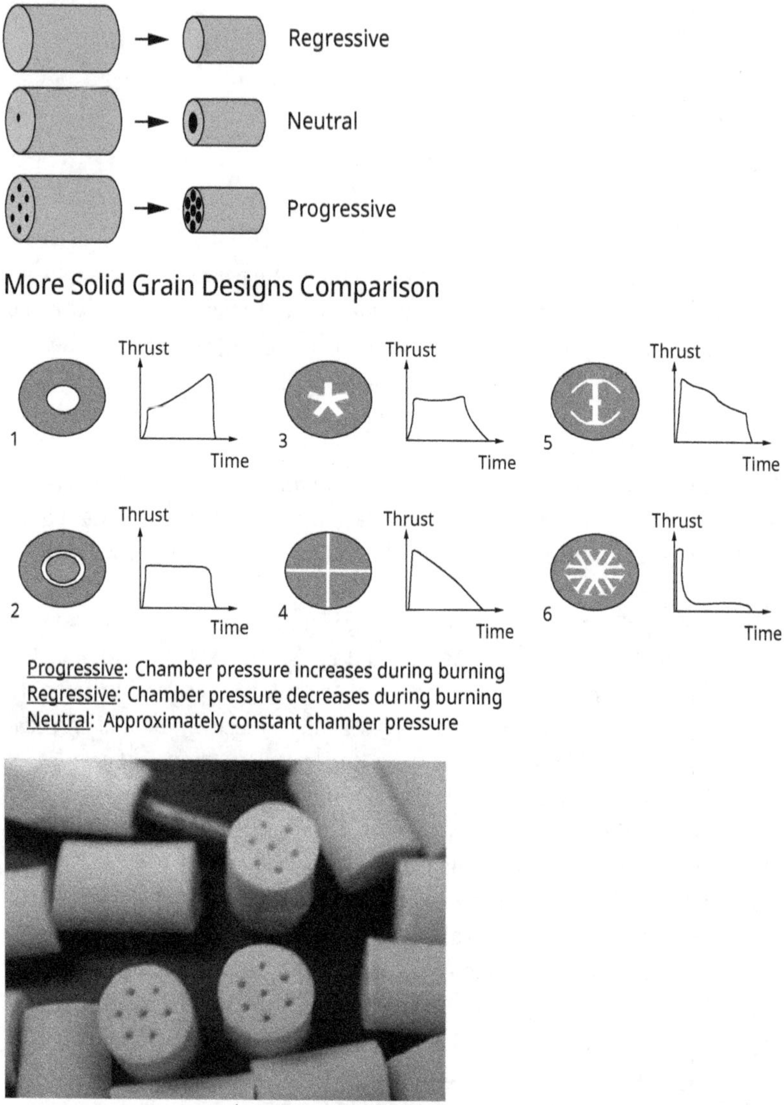

Fig. 4.4b: Solid grain design (top two figures) and actual grains of manufactured HNP (see Tab. 4.11a) with the following parameters: l = 0.055 inches, outer diameter = 0.06 inches, inner diameter = 0.03 inches, web = 0.016 inches.

powder, temperature and the released gas molecules, and thus, indirectly on the composition. A comparison of the pressure curves of single-, double- and triple-base propellants with the same filling quantity is shown in Fig. 4.4d.

Single-base propellants have the lowest energy content and generate the lowest gas pressure for the same propellant mass. Double-base propellants generate a significantly

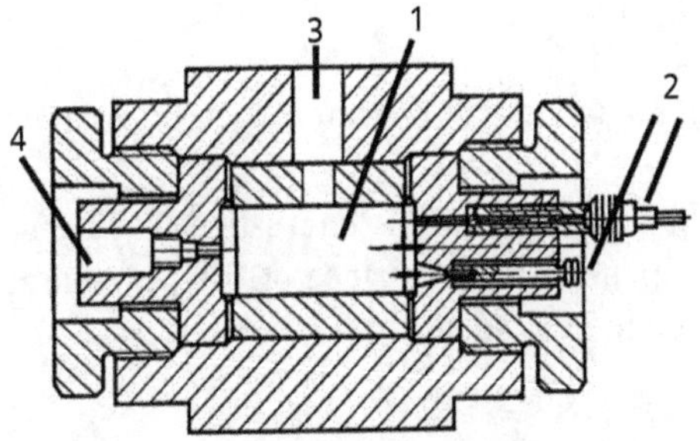

Fig. 4.4c: Schematic representation of a manometric bomb. 1: inner volume, 2: electrical igniter, 3: pressure sensor, 4: pressure relieve valve.

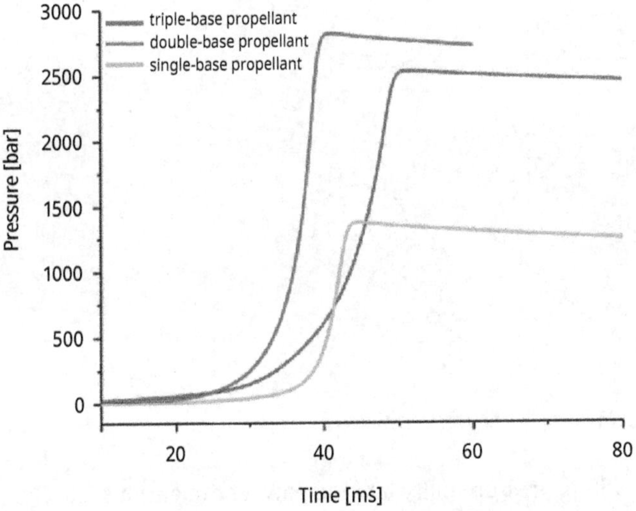

Fig. 4.4d: Pressure-time function of single-, double- and triple-base propellants using the same amounts of powder.

larger gas volume by the addition of nitroglycerin. With the same filling quantity, they show a higher maximum pressure than a triple-base propellant. This is due to the higher combustion temperature, not a higher energy content. The crystalline energy source (*e.g.* NQ) increases the energy content, while at the same time lowering the combustion temperature. Since the gas volume is directly proportional to the temperature even under pressure according to Noble-Abel (see ideal gas law), a higher maximum gas pressure is achieved with a double-base compared to a triple-base propellant.

The resulting pressure per mass unit of powder burned is referred to as the (powder) force F. The explosion temperature T_{ex} can be determined according to eq. (1). The force can be read from the positive ordinate value after reaching the maximum pressure.

$$F = n \cdot R \cdot T_{ex} \tag{1}$$

F: force, *n:* moles of combustion gases per mass unit, *R:* gas constant, T_{ex}: explosion or combustion temperature.

Furthermore, the type of burning and how it propagates is interesting. The burning and thus the volume of gas released depends on the surface of the propellant charge. The burn rate can be determined from the Vieille law (eq. 2).

$$r = \beta \cdot p^{\alpha} \tag{2}$$

r: linear burn rate, *β*: powder-specific constant, *p:* pressure, *α:* pressure exponent (for propellant charges typically 0.8 to 1.0).

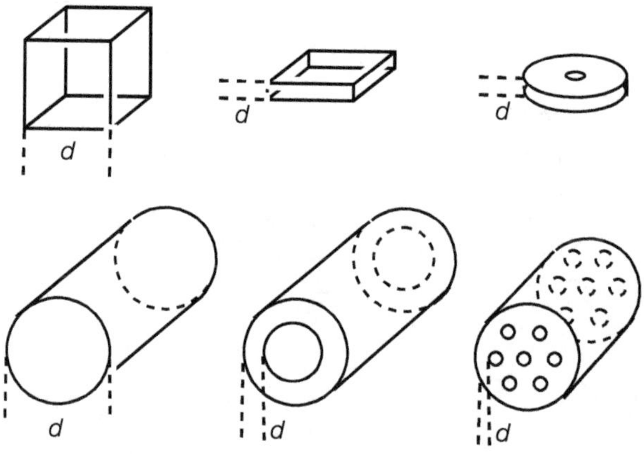

Fig. 4.4e: Smallest distance *d* in different powder geometries.

Here, *β* depends on the resulting pressure pulse and the powder dimensions (eq. 3):

$$\beta = \frac{1}{I} \cdot \frac{d}{2} \tag{3}$$

I: pressure pulse, *d:* powder dimensions

The pressure pulse *I* is the integral of the pressure-time curve in Fig. 4.4d between the start of combustion and the maximum pressure. The powder dimension *d*, is the smallest distance in the powder geometry, as shown in Fig. 4.4e.

Since the dependency of the powder geometry cannot be seen directly from the pressure-time curve, the dependency of the vivacity versus the powder turnover is considered. The vivacity is defined according to eq. (4):

$$L = \frac{dp}{dt} \cdot \frac{1}{p} \cdot \frac{1}{p_m} \tag{4}$$

L: vivacity, *p:* pressure at time t, *t:* time and p_m: maximum pressure

By multiplying by $1/p_m$, the dependence on the maximum gas pressure is minimized and the dependence on the geometry of the propellant charge is eliminated. The vivacity is determined for each point on the pressure-time curve and plotted against p/p_m, as shown in Fig. 4.4f. Under ideal conditions, the numerical value 0 corresponds to ignition, the value 0.5 to half of the powder conversion and 1 to complete combustion.

The shape function of the powder, which takes into account the geometric relationships (ball, flake, cylinder, hole, powder) during combustion is shown in eq. (5):

$$\rho(z) = \text{actual surface area / surface area at beginning of combustion} \tag{5}$$

Figure 4.4f shows a comparison of the different combustion curves depending on the powder geometry. The green and black curves show the burning of propellant charges without perforation. The surface gradually decreases during the burning process. Accordingly, gas production is greatest at the beginning and decreases as it progresses. Combustion is called regressive and is particularly desirable for ammunition for small-caliber weapons (mostly single-base propellant charges).

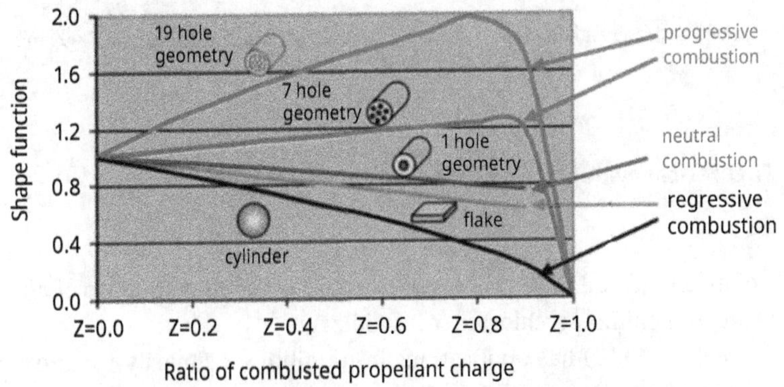

Fig. 4.4f: Schematic presentation of the vivacity of different propellant charge geometries.

The blue curve (see Fig. 4.4f) shows the vivacity curve of a cylindrical single-hole powder. The surface remains constant during the idealized burning, since the front advances both from the outer surface inwards and from the hole surface outwards. Gas production is constant during burning, which is why the burn is referred to as neutral. Powders with this geometry are used in ammunition for small and medium-caliber weapons (single-base or double-base propellants).

The red and pink curves show the vivacity curves of propellants with multiple perforations. With these geometries, the surface becomes larger with increasing combustion. The erosion front runs inwards from the outer surface, and outwards from all holes at the same time. With increasing burning, the gas volume production increases, which is referred to as progressive. Propellants with such a geometry are therefore preferably used for ammunition in large-caliber weapons, in order to be able to compensate for the increasing space in the barrel when the projectile is propelled forward (mostly double- and triple-base propellants).

Figure 4.4g shows the typical design of a cartridge (filled with propellant grains) and projectile used in light weapons, e.g. .5 cartridge (12.7 × 99 mm NATO).

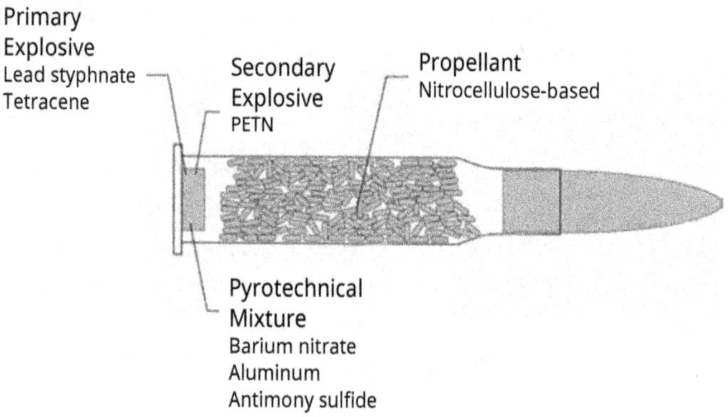

Fig. 4.4g: Typical design of a cartrdge and projectile used in light weapons, e.g. .5 cartridge (12.7 × 99 mm NATO).

4.2.4 Example: Theoretical evaluation of new solid rocket propellants

A. Oxalates

Solid propellants of essentially all solid rocket boosters are based on a mixture of aluminum (Al, fuel) and ammonium perchlorate (AP, oxidizer).

Ammonium perchlorate (AP) has applications in munitions, primarily as an oxidizer for solid rocket and missile propellants. It is also used as an air-bag inflator in the automotive industry, in fireworks, and is a contaminant in agricultural fertilizers. Because of these uses and ammonium perchlorate's high solubility, chemical stability, and persistence, it has become distributed widely in surface and ground water systems. There is little information about the effects of perchlorate in these systems or on the aquatic life that inhabits them. However, it is known that perchlorate is an endocrine disrupting chemical that interferes with normal thyroid function and that thyroid dysfunction impacts both growth and development in vertebrates. Because perchlorate competes for iodine binding sites in the thyroid, the addition of iodine to culture water has been examined in order to determine if perchlorate effects can be mitigated. Finally, perchlorate is known to affect normal pigmentation of amphibian embryos. In the US alone the cost for remediation is estimated to be several billion dollars, money that is needed in other defense areas.

In the course of the emerging global interest in high-energetic, dense materials (HEDM) the LMU group is currently developing new energetic materials, which preferably have a positive oxygen balance value (see Ch. 4.1) [35b].

The objective of the ongoing work is to explore the chemical synthesis of possible replacements for AP as an oxidizer in tactical missile rocket motors. The synthesis, sensitivities, thermal stability, binder compatibility and decomposition pathways of these new high-oxygen materials are currently being researched. In the following example, we theoretically want to evaluate, the suitability of nitrosyl (NO^+) and nitronium (NO_2^+) oxalate as a potential ingredient for solid rocket propellants.

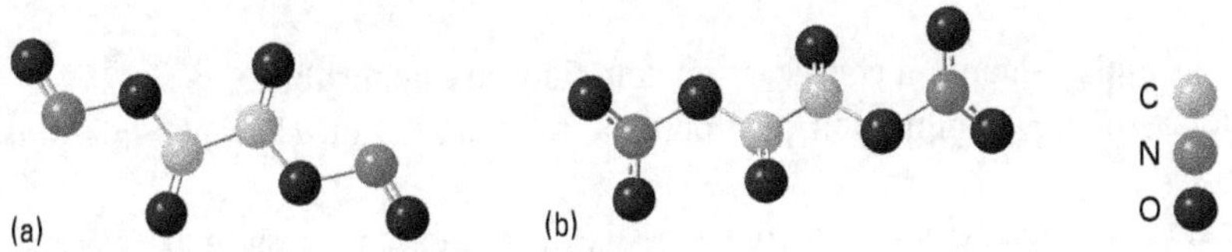

Fig. 4.5: Optimized molecular structures of $(ON)O_2C—CO_2(NO)$ (a) and $(O_2N)O_2C—CO_2(NO_2)$ (b).

Tab. 4.12: Solid state energies of formation ($\Delta_f U°$).

	$\Delta_f H°(s)$ / kcal mol^{-1}	Δn	$\Delta_f U°(s)$ / kcal mol^{-1}	M / g mol^{-1}	$\Delta_f U°(s)$ / kJ kg^{-1}
$[NO_2]_2[O_2C—CO_2]$	−86.6	−5	−83.6	180.0	−1943.2
$[NO]_2[O_2C—CO_2]$	−107.0	−4	−104.6	148.0	−2957.1
$O_2N—O_2C—CO_2—NO_2$	−113.5	−5	−110.5	180.0	−2568.5
$ON—O_2C—CO_2—NO$	−96.5	−4	−94.1	148.0	−2660.2

The molecular structures of neutral $(ON)O_2C—CO_2(NO)$ and $(O_2N)O_2C—CO_2(NO_2)$ were fully optimized without symmetry constraints in both cases to C_i symmetry (Fig. 4.5).

Table 4.12 presents the calculated energies of formation for the solid neutral species and salts based on the CBS-4M method (see Ch. 4.2.1). Furthermore we see from Tab. 4.12 that for the nitronium ($[NO_2]^+$) species the covalently bound form is favored over the ionic salt by 26.9 kcal mol^{-1} while for the nitrosonium species ($[NO]^+$) the salt is favored over the covalent isomer by 10.5 kcal mol^{-1}. This change from the preferred covalent form of – NO_2 compound (actually a nitrato ester) to the ionic nitronium salt can be attributed almost exclusively to the increased lattice enthalpy of the (smaller!) NO^+ species (ΔH_L (NO^+ – NO_2^+ salt) = 31.4 kcal mol^{-1}) (*N.B.* The difference in the ionization potentials of NO (215 kcal mol^{-1}) and NO_2 (221 kcal mol^{-1}) is only marginal).

For a solid rocket propellant, free expansion of the combustion products into space (or atmosphere) with p = const. can be assumed and therefore the below equation is a good approximation, i.e. the combustion process can be considered isobaric:

$$\Delta U = Q_p - p\,\Delta V$$

In this study, we assume firing the rocket motor against an ambient atmosphere (p = 1 bar) as it is commonly the case for tactical missiles.

The following combustion calculations were carried out under isobaric conditions, based on the assumption that the combustion of fuel proceeds without any heat loss to the surroundings (i.e. adiabatically) and that the state of chemical equilibrium establishes in the combustion products. The calculation of the theoretical rocket performances was based on the following assumptions:

1. the pressure in the combustion chamber and the chamber cross-section area are constant
2. the energy and momentum conservation equations are applicable
3. the velocity of the combustion products at the combustion chamber is equal to zero
4. there is no temperature and velocity lag between condensed and gaseous species
5. the expansion in the nozzle is isentropic (*N.B.* In thermodynamics, an isentropic process or isoentropic process is one during which the entropy of the system remains constant)

The theoretical characteristics of the rocket motor propellant may be derived from the analysis of the expansion of the combustion products through the nozzle. The first step in the calculation of the theoretical rocket performance is to calculate the parameters in the combustion chamber, and the next step is to calculate the expansion through the nozzle (see Fig. 4.3). The expansion through the nozzle is assumed to be isentropic ($\Delta S = 0$). The EXPLO5 and ICT program code provide the following options:

- *frozen* flow (composition of combustion products remains unchanged, frozen, during the expansion through the nozzle, i.e. equal to composition in the combustion chamber),
- *equilibrium* flow (composition of combustion products at any station in the nozzle is defined by chemical equilibrium).

The frozen performance is based on the assumption that the composition of combustion products remains constant ("frozen"), while equilibrium performance is based on the assumption of instantaneous chemical equilibrium during the expansion in the nozzle.

The specific impulse I_{sp} is the change of the impulse (impulse = mass × velocity, or force × time) per propellant mass unit. The specific impulse is an important parameter for the characterization of rocket propellants and can be interpreted as the effective exhaust velocity of the combustion gases when exiting the expansion nozzle:

$$I_{sp} = \frac{F \times t_b}{m} = \frac{1}{m}\int_0^{t_b} F(t)\,dt$$

The force F is the time-dependent thrust, $F(t)$, or the average thrust, F, t_b is the burning time of the motor and m the mass of the propellant. Therefore, the unit of the specific impulse I_{sp} is N s kg^{-1} or m s^{-1}.

Tab. 4.13: Combustion properties (solid rocket motor) of Al formulations with essentially zero oxygen balance (frozen expansion).

	$O_2N—O_2C—CO_2—NO_2$: Al = 0.70 : 0.30	**$[NO]_2[O_2C—CO_2]$: Al = 80 : 20**	**AP : Al = 0.70 : 0.30**
Condition	Isobaric	Isobaric	Isobaric
p / bar	70	70	70
ρ / g cm^{-3}	1.93	1.82	2.18
Ω / %	-2.0	-0.6	-2.8
Q_p / kJ kg^{-1}	-6,473	-5,347	-6,787
$T_{comb.}$ / K	4,642	4,039	4,290
I^*_{sp} / s	223	220	243

It is convention to divide the specific impulse I_{sp} by g_0 (standard gravity, $g_0 = 9.81$ m s^{-2}) so that the resulting specific impulse I^*_{sp} has the unit s (seconds):

$$I^*_{sp} = \frac{I_{sp}}{g_0}$$

The specific impulse I^*_{sp} can also be defined according to the following equation with $\gamma = C_p / C_v$

$$I^*_{sp} = \frac{1}{g}\sqrt{\frac{1\gamma RT_C}{(\gamma-1)M}} \qquad \gamma = \frac{C_p}{C_v}$$

Table 4.13 summarizes the calculated propulsion parameters for aluminized formulations in which the Al content has been varied in order to achieve an oxygen balance that is close to zero (with respect to CO_2, see eq. 2). Table 4.13 contains the corresponding values for a AP /Al formulation for comparison as well. Finally, Tab. 4.14 shows the calculated specific impulses for equilibrium expansion for the three optimized formulations (covalent $O_2N—O_2C—CO_2—NO_2$ /Al; ionic $[NO]_2[O_2C—CO_2]$ /Al and AP / Al). The results of Tab. 4.14 are graphically summarized in Fig. 4.6.

From Tab. 4.13 and Fig. 4.6 we can conclude that generally the agreement between the EXPLO5 and ICT calculated equilibrium specific impulses is reasonably good with the ICT code always predicting a slightly better performance. It is further apparent that the formulation with covalent $O_2N—O_2C—CO_2—NO_2$ and Al results in better performance than the formulation using ionic $[NO]_2[O_2C—CO_2]$ and Al. The specific impulse of the chlorine and perchlorate-free formulation with covalent $O_2N—O_2C—CO_2—NO_2$ and Al (I) is just slightly lower than that of the AP/Al (**III**) formulation. Therefore it can be concluded that bis(nitronium) oxalate (or oxalic acid dinitrato ester), $O_2N—O_2C—CO_2—NO_2$, may be a promising new perchlorate-free and environmentally benign oxidizer for formulations to be used in solid rocket motors.

Tab. 4.14: Specific impulses (solid rocket motor) of Al formulations calculated using different codes (EXPLO5 and ICT).

	$O_2N—O_2C—CO_2—NO_2$: Al = 0.70 : 0.30		$[NO]_2[O_2C—CO_2]$: Al = 80 : 20		AP : Al = 0.70 : 0.30	
Formulation	I		II		III	
	EXPLO5	ICT	EXPLO5	ICT	EXPLO5	ICT
Condition	Isobaric		Isobaric		Isobaric	
p / bar	70		70		70	
ρ / g cm^{-3}	1.82		1.74		2.13	
Ω / %	−1.8		−0.5		−2.85	
I^*_{sp} / s	223	206	220	215	243	229
Frozen equilibrium	226	245	225	230	247	257

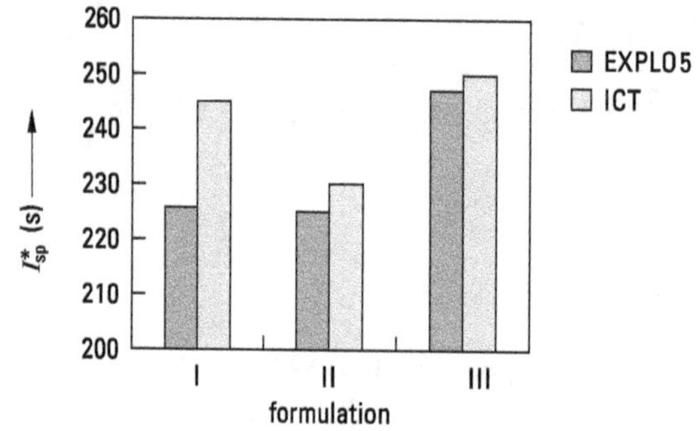

Fig. 4.6: Graphical representation of the calculated equilibrium specific impulses for the optimized formulations **I**: covalent $O_2N—O_2C—CO_2—NO_2$ / Al, **II**: ionic $[NO]_2[O_2C—CO_2]$ / Al and **III**: AP / Al (see Tab. 4.14).

The above discussion clearly revealed that the covalently bound oxalic acid dinitrate ester is the most promising candidate as a high oxidizer and potential replacement for AP in this series. In order to evaluate its thermodynamic and kinetic stability we calculated the decomposition into CO_2 and NO_2. The reaction enthalpy of $\Delta H = -56.5$ kcal mol^{-1} clearly indicates that the oxalic acid dinitrate ester is (as expected) thermodynamically unstable with respect to its decomposition into CO_2 and NO_2.

$$O_2N—O_2C—CO_2—NO_2\ (s) \xrightarrow[\Delta H\ =\ -56.5\ \text{kcal mol}^{-1}]{} 2\,CO_2\ (g) + 2\,NO_2\ (g)$$

In order to evaluate the kinetic stability of covalently bound $O_2N—O_2C—CO_2—NO_2$ it was decided to compute a two dimensional potential energy hypersurface (Fig. 4.7) at B3LYP/6–31G* level of theory. As one can see, the simultaneous dissociation of $O_2N—O_2C—CO_2—NO_2$ into CO_2 and NO_2 has a relatively high activation barrier (Fig. 4.7). The transition state (Fig. 4.8) was located 37.1 kcal mol^{-1} (CBS-4M) above the dinitrate ester. It is interesting that in agreement with Hammond's postulate, the transition state lies more towards the higher-energy starting material.

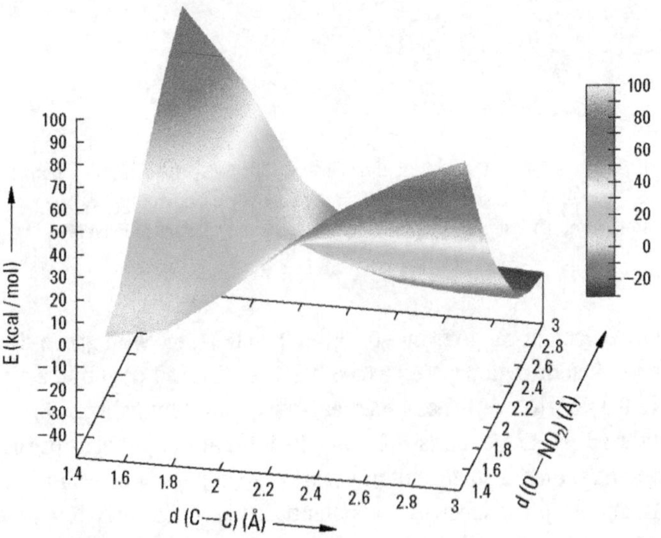

Fig. 4.7: Potential energy hypersurface for the simultaneous dissociation of $O_2N—O_2C—CO_2—NO_2$ into CO_2 and NO_2 at B3LYP/6–31G* level of theory.

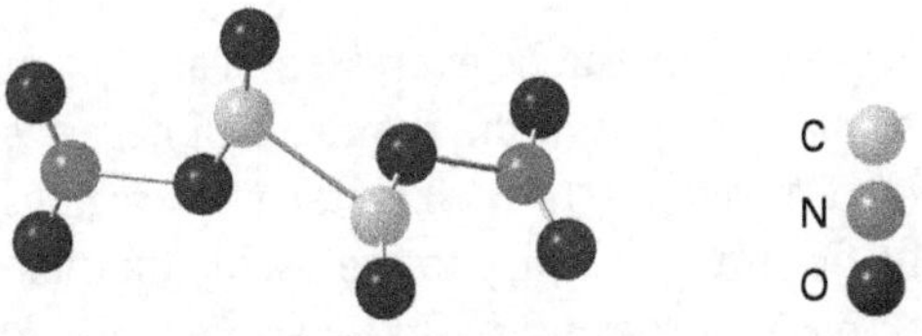

Fig. 4.8: Transition state structure for the simultaneous dissociation of $O_2N—O_2C—CO_2—NO_2$ into CO_2 and NO_2 at CBS-4M level of theory (NIMAG = 1, ν_1 = −817 cm^{-1}, d(C—C) = 2.33 Å, d(O—NO_2) = 1.91 Å).

The electrostatic potential (ESP) of covalent $O_2N—O_2C—CO_2—NO_2$ was computed at the optimized structure at the B3LYP/6–31G(d) level of theory. Figure 4.9 shows the electrostatic potential for the 0.001 electron/bohr^3 isosurface of electron density evaluated at

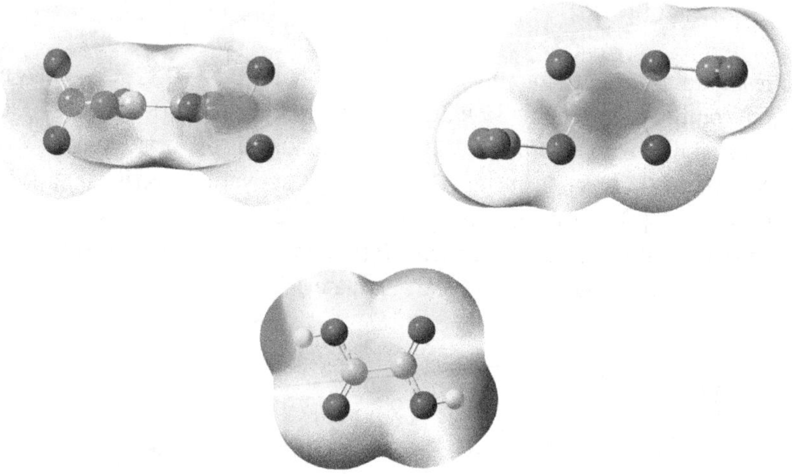

Fig. 4.9: Electrostatic potentials of $O_2N—O_2C—CO_2—NO_2$ and oxalic acid (B3LYP/6-31G(d), 0.001 e bohr^{-3} isosurface, energy values −0.06 H to + 0.06 H); color coding: red (very negative), orange (negative), yellow (slightly negative), green (neutral), turquoise (slightly positive), light blue (positive), dark blue (very positive).

the B3LYP level of theory. The colors range from −0.06 to + 0.06 hartrees with green denoting extremely electron-deficient regions ($V(r) > 0.06$ hartree) and red denoting electron-rich regions ($V(r) < -0.06$ hartrees). It has recently been proven by Politzer, Murray et al. and extensively used by Rice et al. (see Ch. 8) [39–44] that the patterns of the computed electrostatic potential on the surface of molecules can generally be related to the sensitivity of the bulk material. The electrostatic potential at any point r is given by the following equation in which Z_A is the charge on nucleus A, located at R_A.

$$V(r) = \sum \frac{Z_A}{|R_A - r|} - \int \frac{\rho(r')}{|r' - r|} dr'$$

Politzer et al. were able to show that impact sensitivity can be expressed as a function of the extent of this anomalous reversal of the strengths of the positive and negative surface potentials. In most nitro ($—NO_2$) and nitrato ($—O—NO_2$) systems the regions of positive potential are stronger than the negative, contrary to the usual situation. This atypical imbalance between stronger positive regions and weaker negative regions can be related to the impact sensitivities. The calculated electrostatic potential of $O_2N—O_2C—CO_2—NO_2$ (Fig. 4.9) shows strong positive regions over the nitro ($—NO_2$) groups with the positive areas extending into the $O—NO_2$ region (oxygen—NO_2 bond). Furthermore, there is also a strong positive region over the relatively weak C—C bond. This is in good accord with the labile $O—NO_2$ and C—C bonds and also accounts for the easy bond cleavage. In comparison, free oxalic acid (Fig. 4.9) does not show any positive regions over the bonds in the molecule.

From this computational study the following conclusions can be drawn:

1. Covalently bound and ionic nitronium and nitrosonium oxalate were researched with respect to their potential use as energetic materials or oxidizers for solid rocket motors. None of these compounds can be expected to be a good high explosive. However, the covalent molecule oxalic acid dinitrate ester, $O_2N—O_2C—CO_2—NO_2$, was identified to be a potentially interesting oxidizer.
2. The computed specific impulse of a $O_2N—O_2C—CO_2—NO_2$ /Al formulation (80 : 20) is comparable to that of a conventional AP /Al (70 : 30) formulation, however, it is free of toxic perchlorate or any halogen.
3. Oxalic acid dinitrate ester, $O_2N—O_2C—CO_2—NO_2$ is metastable with respect to decomposition into CO_2 and NO_2. The reaction barrier (transition state) for the monomolecular dissociation was calculated to be 37 kcal mol^{-1} at CBS-4M level of theory.
4. The computed electrostatic potential of $O_2N—O_2C—CO_2—NO_2$ shows strong positive areas at the 0.001 s $bohr^{-3}$ isosurface over the $O—NO_2$ and C—C bonds indicating the relative weakness of the particular bonds.

The results obtained in this study should encourage synthetic work in order to prepare oxalic acid dinitrate ester, $O_2N—O_2C—CO_2—NO_2$ on a laboratory scale and to experimentally evaluate its properties, first and foremost its thermal stability.

Literature

J. Lavoie, C.-F. Petre, C. Dubois, *Propellants Explos. Pyrotech.* **2018**, *43*, 879–892.

B. Evaluation of TKX-50 as an ingredient in rocket propellants

Generally speaking, composite rocket propellants usually consist of a powdered oxidizer (e.g. ammonium perchlorate, AP), powdered metal fuel (e.g. aluminum, Al) and a plasticized binder (e.g. HTPB). The category of propellants known as high-energy composite (HEC) propellants, contain in addition, a high-energy explosive (e.g. RDX, HMX or CL-20) in the formulation, and AP is partly replaced by these high-energy materials. Although this results in an increase in the specific impulse (I_{sp}), implementation of HECs is limited due to the increased hazard that the high-energy explosive additives bring with them, since they are impact and friction sensitive additives.

Computational results for TKX-50-based solid rocket propellant formulations (22% energetic binder, 20% oxidizer, 18% Al, 40% TKX-50) in the literature indicate that TKX-50 shows a higher density impulse than HMX, but is lower than that of CL-20. The NASA SP-273 computer code was used to predict the parameters of rocket propellant formulations, and showed that TKX-50 containing formulations exhibit a superior specific impulse value compared to corresponding formulations in which TKX-50 was replaced by other common explosives. Based on this work, the authors concluded that the theoretically predicted data indicated that TKX-50 has potential for application in rocket and gun propellants.

The properties of composite propellants containing TKX-50 have been investigated experimentally with different mass fractions of TKX-50 particles – specifically on TKX-50/HTPB binder slurries and AP/HTPB/Al propellant slurries. It is possible with the inclusion of specially selected energetic materials to increase the gravimetric specific impulse or the density specific impulse of propellants, however, the energetic materials chosen for this purpose must show low sensitivity to external stimuli such as impact and friction. Since TKX-50 shows a high detonation velocity, but at the same time a low sensitivity to friction (comparable to or lower than those of RDX, HMX and CL-20) and impact (lower than those of HMX, RDX and CL-20), it is an interesting candidate not only as an explosive, but also as an ingredient in propulsion compositions. In addition, the effects of TKX-50 on the properties of HTPB-based composite solid propellants have been studied computationally. It has been concluded, that TKX-50 particles could be used as the energetic components which would improve the combustion properties of the composite propellant.

A typical composite propellant formulation (70% AP, 16% Al, 12% HTPB and 2% epoxy) with real-life applications has a calculated specific impulse of 264 s. The calculated specific impulses of the pure (100%) individual energetic ingredients are: NG 259 s, HMX 266 s, CL-20 272 s and TKX-50 268 s.

High-energy composite (HEC) propellants with the following composition have been evaluated:

Glycidyl azide polymer	(GAP)	10%
Nitroglycerin	(NG)	15%
Aluminum	(Al)	18%
Energetic filler	(HMX, CL-20, TKX-50)	X %
Ammonium perchlorate	(AP)	Y %

The percentage ratios X + Y = 57% were optimized, in order to obtain an oxygen balance with respect to CO_2 of −33%, which corresponds to that of the typical composite propellant formulation (70% AP, 16% Al, 12% HTPB and 2% epoxy) with real-life applications mentioned above. The results are summarized in Tab. 4.14a.

It is shown in Tab. 4.14a, that the high-energy composite (HEC) propellant formulation based on TKX-50 is calculated to have the highest specific impulse, exceeding those of the corresponding HMX and CL-20-based formulations. In addition, the typical composite propellant formulation of 70% AP, 16% Al, 12% HTPB and 2% epoxy has a calculated specific impulse of only 264 s. This means that all of the three HECs shown in Tab. 4.14a were calculated to possess higher specific impulses than that of the secondary explosive-free standard composite propellant formulation. However, it is important to point out that importantly, TKX-50 is less impact and friction sensitive than

CL-20. This is important since it should result in a significant reduction in the hazard of the propellant formulation in comparison with the same propellant composition containing CL-20 instead of TKX-50.

Tab. 4.14a: EXPLO5 calculation results for various high-energy composite (HEC) propellant formulations, as well as sensitivity data for the individual secondary explosives (HMX, CL-20 and TKX-50) taken from the literature.

High-energy explosive component	HMX	CL-20	TKX-50
Impact sensitivity [J]	7.4	3–4	20
Friction sensitivity [N]	120	ca. 96	120
ESD [J][11]	0.2	ca. 0.1	0.1
Formulation	10% GAP 15% NG 18% Al 45% HMX 12% AP	10% GAP 15% NG 18% Al 54% CL-20 3% AP	10% GAP 15% NG 18% Al 41% TKX-50 16% AP
Oxygen balance [%]	−33%	−33%	−33%
Specific impulse [s]	274	275	278
Isochoric combustion temperature [K] Heat of combustion [kJ/kg]	3,806 6,094	3,964 5,971	3,701 5,996
Exhaust velocity [m/s]	2,690	2,694	2,723

Literature

[1] Babuk V, Glebov A, Dolotkazin I, Conti A, Galfetti L, DeLuca LT, Vorozhtsov A. Condensed Combustion Products from Burning of Nanoaluminum-Based Propellants: Properties and Formation Mechanism, Progress in Propulsion Physics, 2009; 1: 3–16.

[2] Chaturvedi S, Dave PN. Solid Propellants: AP/HTPB Composite Propellants, Arabian J. Chem. 2019; 12(8): 2061–2068.

[3] DeLuca LT. Innovative Solid Formulations for Rocket Propulsion, Eurasian Chemico-Technological J. 2016; 18(3): 181–196.

[4] DeLuca LT, Shimada T, Sindinskii V, Calabro M. Chemical Rocket Propulsion, Springer, Switzerland, 2017.

[5] Pang, W, DeLuca LT, Gromov AA, Cumming AS. Innovative Energetic Materials: Properties, Combustion Performance and Application, Springer, Singapore, 2020.

[6] Yang V, Brill TB, Ren W-Z. Solid Propellant Chemistry, Combustion, and Motor Interior Ballistics, Zarchan P (ed.), American Institute of Aeronautics and Astronautics Inc., USA, 2000; 185.

[7] Yan X-T, Xia Z-X, Huang L-Y, Feng Y-C, Na X-D. Experimental Study on Combustion Process of NEPE Propellant, 7th Europ. Conf. for Aeronautics and Space Sciences (EUCASS), 2015; DOI:10.13009/EUCASS2017–324.

[8] Talawar MB, Nandagopal S, Singh S, Majahan AP, Badgujar DM, Gupta M, Khan MAS. ChemistrySelect, 2018; 3(43): 12175–12182.

[9] Pang W, Li J, Wang K, Fan X, De Luca LT, Bi F, Li H. Effects of Dihydroxylammonium 5,5'-Bistetrazole-1,1'-Diolate on the Properties of HTPB Based Composite Propellant. Propellants, Explos Pyrotech. 2018; 43(10): 1013–1022.
[10] Pang WQ, DeLuca LT. Effects of TKX-50 on the Properties of HTPB-Based Composite Solid Propellant, 7th Europ. Conf. for Aeronautics and Space Sciences (EUCASS), 2015; DOI:10.13009/EUCASS2017–544.
[11] Energetic Materials Encyclopedia, Volumes 1–3, 2nd edn., T. M. Klapötke, De Gruyter, Berlin/Boston, 2021.
[12] Theoretical evaluation of TKX-50 as an ingredient in rocket propellants, T. M. Klapötke, M. Suceska, Z. Anorg. Allg. Chem, 2021; 647(5): 572–574.

4.2.5 Example: EXPLO5 calculation of the gun propellant properties of single, double and triple-base propellants

In order to elucidate the different properties of single, double and triple-base gun propellants and to compare these values with the ones obtained for a new high-nitrogen propellant (e.g. NILE, see above), the relevant combustion parameters were calculated using the EXPLO5 code under the assumption of isochoric combustion. The results are summarized in Tab. 4.15.

The general trend in terms of performance can clearly be seen in the calculated results shown in Tab. 4.15. While a double-base propellant performs much better than a single-base formulation (much higher force and pressure), the combustion temperature is also considerably higher (over 360 K) which causes increasing erosion problems. The triple-base propellant on the other hand shows performance in between the single- and double-base propellant while having a combustion temperature only slightly higher (137 K) than that of the single-base propellant.

Tab. 4.15: Computed (EXPLO5) gun propellant parameters for a single, double and triple-base propellants in comparison with a new high-nitrogen formulation (isochoric conditions).

propellant	ρ / g cm^{-3}	loading density / g cm^{-3}	f_E / kJ g^{-1}	b_E / cm^3 g^{-1}	$T_{comb.}$ / K	p_{max} / bar	N_2 / CO
NC (12.5)	1.66	0.2	1.05	1.08	3,119	2,674	0.26
NC : NG (50 : 50)	1.63	0.2	1.21	1.02	3,987	3,035	0.59
NC : NG : NQ (25 : 25 : 50)	1.70	0.2	1.13	1.06	3,256	2,875	1.25
TAGzT : NC (85 : 15)	1.63	0.2	1.09	1.22	2,525	2,875	6.07

The experimental formulation of a new high-nitrogen propellant with 85% TAGzT (triaminoguanidinium azotetrazolate) and 15% NC has the lowest combustion temperature of all shown formulations (nearly 600 K below $T_{comb.}$ of the single-base propellant). Furthermore, the performance of the new high-nitrogen formulation can be expected to be similar (e.g. same maximum pressure as a triple-base propellant) to a triple-base propellant. Equally important, the N_2 / CO ratio of a triple-base propellant increases from 1.25 to 6.07. This, together with the significantly reduced combustion temperature should help to drastically reduce erosion problems. Initial field studies using TAGzT-based propellants for large caliber (105 Howitzer) guns have indicated that the life-time of a gun barrel could possibly be increased by a factor of up to 4. This would justify the (still) higher cost for high-nitrogen formulations because of the increased life-time and therefore drastically reduced cost of the gun system.

A large number of empirical, numerical, semi-empirical and quantum-chemical methods for the prediction of various explosive properties such as:
- Detonation performances
- Impact, shock, friction and electrostatic discharge sensitivities
- Thermal stabilities
- Densities
- Melting points
- Reaction enthalpies
- . . .

have been published by M. H. Keshavarz et al. from the Malek-ashtar University of Technology, Shahin-shahr, Iran.

See for example (and the cited literature therein):

(a) Keshavarz, Mohammad Hossein, Explosive Materials (2011), 179–201.
(b) Keshavarz, Mohammad Hossein, Explosive Materials (2011), 103–123.
(c) Keshavarz, Mohammad Hossein; Shokrolahi, Arash; Esmailpoor, Karim; Zali, Abbas; Hafizi, Hamid Reza; Azamiamehraban, Jamshid, Hanneng Cailiao (2008), 16(1), 113–120.
(d) Keshavarz, Mohammad Hossein; Esmailpour, Karim; Zamani, Mehdi; Roknabadi, Akbar Gholami, Propellants, Explosives, Pyrotechnics (2015), 40, 886–891.
(e) Rahmani, Mehdi; Vahedi, Mohamad Kazem; Ahmadi-Rudi, Behzad; Abasi, Saeed; Keshavarz, Mohammad Hossein, International Journal of Energetic Materials and Chemical Propulsion (2014), 13(3), 229–250.
(f) Keshavarz, Mohammad Hossein; Soury, Hossein; Motamedoshariati, Hadi; Dashtizadeh, Ahmad, Structural Chemistry (2015), 26(2), 455–466.
(g) Oftadeh, Mohsen; Keshavarz, Mohammad Hossein; Khodadadi, Razieh, Central European Journal of Energetic Materials (2014), 11(1), 143–156.
(h) Keshavarz, Mohammad Hossein, Propellants, Explosives, Pyrotechnics (2015), 40(1), 150–155.
(i) Keshavarz, Mohammad Hossein; Zamani, Ahmad; Shafiee, Mehdi Propellants, Explosives, Pyrotechnics (2014), 39(5), 749–754.
(j) Keshavarz, Mohammad Hossein; Pouretedal, Hamid Reza; Ghaedsharafi, Ali Reza; Taghizadeh, Seyed Ehsan, Propellants, Explosives, Pyrotechnics (2014), 39(6), 815–818.
(k) Keshavarz, Mohammad Hossein; Pouretedal, Hamid Reza; Ghaedsharafi, Ali Reza; Taghizadeh, Seyed Ehsan, Propellants, Explosives, Pyrotechnics (2014), 39(6), 815–818.

(l) Keshavarz, Mohammad Hossein; Motamedoshariati, Hadi; Moghayadnia, Reza; Ghanbarzadeh, Majid; Azarniamehraban, Jamshid, Propellants, Explosives, Pyrotechnics (2014), 39(1), 95–101.
(m) Keshavarz, Mohammad Hossein; Seif, Farhad; Soury, Hossein, Propellants, Explosives, Pyrotechnics (2014), 39(2), 284–288.
(n) Rahmani, Mehdi; Ahmadi-rudi, Behzad; Mahmoodnejad, Mahmood Reza; Senokesh, Akbar Jafari; Keshavarz, Mohammad Hossein, International Journal of Energetic Materials and Chemical Propulsion (2013), 12(1), 41–60.
(o) Keshavarz, Mohammad Hossein, Propellants, Explosives, Pyrotechnics (2013), 38(6), 754–760.
(p) Keshavarz, Mohammad Hossein; Seif, Farhad, Propellants, Explosives, Pyrotechnics (2013), 38(5), 709–714.
(q) Keshavarz, Mohammad H.; Motamedoshariati, Hadi; Moghayadnia, Reza; Ghanbarzadeh, Majid; Azarniamehraban, Jamshid, Propellants, Explosives, Pyrotechnics (2013), 38(1), 95–102.
(r) Keshavarz, Mohammad Hossein, Propellants, Explosives, Pyrotechnics (2012), 37(4), 489–497.
(s) Rahimi, R.; Keshavarz, M. H.; Akbarzadeh, A. R., Central Europ. J. of Energ. Mat., (2016), 13, 73.
(t) M. Jafari, M. H. Keshavarz, A. Zamani, S. Zakinejad, I. Alekaram, Propellants Explos. Pyrotech., (2018), 43, 342–347.

4.2.6 Semiempirical calculations (EMDB)

The ability to reliably predict the performance and thermochemical properties of a new energetic material from its molecular structure is highly desirable. Several thermochemical equilibrium codes, such as CHEETAH, ICT, and EXPLO5 (which utilize suitable equations of state (EOS)), as well as different empirical packages such as EDPHT (which use calibrated empirical methods), have been introduced in order to determine the performance and thermochemical properties of different EMs.

EMDB_1 is a new professional package in the area of EMs, which calculates more than thirty physicochemical and detonation parameters for different pure explosives or energetic formulations (for C-H-N-O-F-Cl-Al-Br-I-S compounds). The unique property of the EMDB code is that, in contrast to other codes, it does not require the density or enthalpy of formation as an input but estimates these values on the basis of the molecular structure.

The predecessor of EMDB is the improved version of the EDPHT [82] program package. In the viewpoint of capabilities, methods, and appearance, EMDB is therefore the improved, updated, and advanced successor of the EDPHT code. Many of correlations have been modified and upgraded based on the latest literature, which has resulted in a significant improvement in the accuracy and precision of the results which are obtained using EMDB in comparison with the values obtained from the EDPHT code.

Here we want to look at the results of performance (VoD) calculations which have each been calculated using the EMDB, EXPLO5, and CHEETAH codes with the experimentally determined values. The impact and friction sensitivities calculated using the EMDB code are compared with the measured values.

All calculations were carried out on a desktop PC using the EXPLO5_6.03 and EMDM_1.0 codes [83]. The results of CHEETAH_8.0 calculations were taken from the literature [84].

Table 4.16 summarizes the calculated and measured detonation velocities of ten well established and new explosives. The molecular structures of the newer explosives are shown in Fig. 4.10.

Tab. 4.16: Calculated and measured detonation velocities (in m s^{-1}) [a] [85].

	density / g cm^{-3}	EMDB_1.0	EXPLO5_6.03	CHEETAH_8.0	LASEM	measured @ TMD
TNT	1.65	7,230 (+3%)	6,878 (−2%)	7,192 (+2%)	6,990 ± 230	7,026 ± 119
HNS	1.74	7,620 (+6%)	7,209 (±0%)	7,499 (+4%)	7,200 ± 210	7,200 ± 71
NTO	1.93	8,080 (−3%)	8,601 (+3%)	8,656 (+4%)	8,300 ± 250	8,335 ± 120
RDX	1.80	8,670 (−2%)	8,919 (+1%)	8,803 (±0%)	8,850 ± 190	8,833 ± 64
ε-CL-20	2.04	9,600 (±0%)	9,882 (+3%)	9,833 (+3%)	9,560 ± 240	9,570
TKX-50	1.88	9,140 (−3%)	9,995 (+6%)	9,735 (+3%)	9,560 ± 280	9,432
BDNAPM	1.80	8,090	8,220	8,171	8,630 ± 210	
BTNPM	1.93	9,020	9,348	9,276	9,910 ± 310	
TKX-55	1.84	7,860	7,666	7,548	8,230 ± 260	
DAAF	1.75	7,960 (−2%)	8,163 (+1%)	8,124 (±0%)	8,050 ± 260	8,110 ± 30

[a]The percentage values in parentheses refer to the deviation of the calculated from the measured VoD values.

Inspection of the calculated and measured VoD values reveals that there is generally reasonably good agreement. Both, the thermodynamic programs (CHEETAH and EXPLO5) as well as the empirical code (EMDB) predict VoD values within a few percent of the experimental values.

Table 4.17 shows the calculated (EMDB) and measured sensitivity parameters for impact and friction. These values can only be predicted using the EMDB code, but not using one of the thermodynamic programs. The EMDB predicted values clearly show the correct trend in terms of the sensitivity categories: very sensitive (BTNPM), mod-

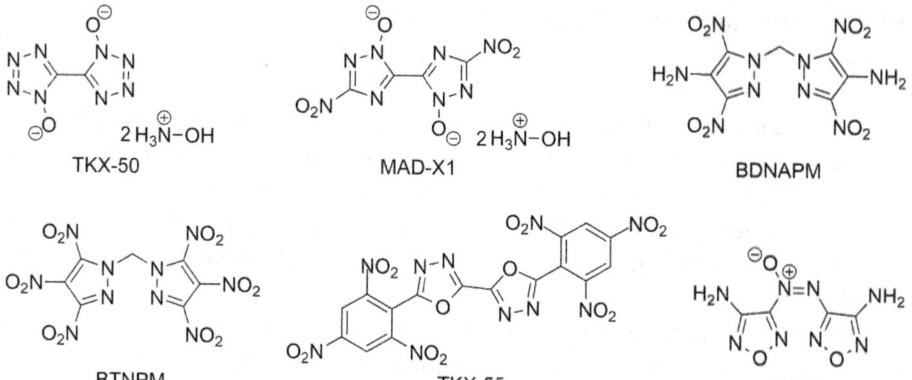

Fig. 4.10: Molecular structures of some more recent explosives.

Tab. 4.17: Measured and calculated (EMDB) impact (IS) and friction (FS) sensitivities.

	IS / J (exptl)	IS / J (EMDB_1.0)	FS / N (exptl)	FS / N (EMDB_1.0)
TNT	15	33	> 353	210
HNS	5	20	> 240	227
NTO	> 120	39	> 353	144
RDX	7.5	5	120	156
ε-CL-20	4	3	48	128
TKX-50	20	15	120	82
BDNAPM	11	39	> 360	166
BTNPM	4	3	144	160
TKX-55	5	12.5	> 360	229
DAAF	7	17	> 360	175

erately sensitive (RDX), and insensitive (NTO) for impact sensitivity. The correct trend is also predicted for the friction sensitivities using the EMDB code.

Both the thermodynamic programs (EXPLO5 and CHEETAH) and the empirical code (EMDB) predict VoD values which agree well with the experimental data. In addition, the EMDB code can predict many more properties. For example, safety parameters, phase transition thermochemical data, as well as performance and thermodynamic values.

4.2.7 EXPLO5: Capabilities and background

EXPLO5 is probably the most popular and comprehensive thermodynamic equilibrium code for the calculation of the properties of energetic materials. The code can predict the performance of ideal and non-ideal high explosives and various explosive formulations, propellants and pyrotechnic mixtures. It predicts the performance under constant volume combustion (explosives, gun propellants), constant pressure combustion (rocket propellants) and detonation.

The main features of EXPLO5, Version V7.02 (2025), can be summarized as follows:

- EXPLO5 is a chemical equilibrium computer code which solves thermodynamic equations between reaction products to find the equilibrium composition.
- Calculation of the equilibrium composition of the reaction products is achieved by applying the modified White, Johnson, and Dantzig's free energy minimization technique.
- The state of gaseous detonation products is described by the Becker-Kistiakowsky-Wilson equation of state (BKW EOS) and Exp-6 EOS.
- The state of gaseous combustion products is described by the virial EOS.
- The state of condensed products (such as compressible solids and liquids) is described by the Murnaghan equations of state.
- The thermodynamic functions of gaseous products are derived from thermodynamic equations using the BKW EOS and the Exp-6 EOS (for detonation products) and the virial EOS (for combustion products).
- The thermodynamic functions of compressible condensed products are derived using Murnaghan equations of state.
- The equation system describing mathematically the state of equilibrium is solved by applying the modified Newton-Raphson method.
- The thermodynamic functions of the reaction products in their standard state are calculated from the enthalpy, which is expressed in a fourth degree polynomial form as a function of temperature.
- The C-J point is determined as a point on the shock adiabat of the detonation products at which the detonation velocity has the minimum value (see below and Figs. 4.14 and 4.15).
- EXPLO5 has a built-in fitting algorithm to estimate the coefficients in the Jones-Wilkins-Lee ((JWL) EOS), which are used to calculate the detonation energy of explosives.
- The Wood-Kirkwood (slightly divergent) detonation theory enables calculation of the self-propagating detonation velocity of non-ideal explosives, as well as other detonation parameters as a function of the charge diameter.
- EXPLO5 can calculate the combustion or detonation parameters of a wide variety of different molecules (containing 52 different chemical elements: C, H, N, O, Al, Cl, Si, F, B, Ba, Ca, Na, P, Li, K, S, Mg, Mn, Zr, Mo, Cu, Fe, Ni, Pb, Sb, Hg, Be, Ti, I, Xe, U, W, Sr, Cr, Br, Co, Ag, Zn, Sn, Bi, Cs, Hf, Ge, Nb, Ta, Yb, Y, V, Ar, He, Cd and Ce).
- EXPLO5 can handle multiple condensed phases (e.g. graphite, diamond, and liquid carbon), and the method of calculation of the chemical activity of reaction products can be specified by the user.
- The reaction product composition is estimated automatically by the program based on the formula of the reactant, the products available from the products' database, and running mode, but can also be customized by the user. The user

can specify the concentration threshold of the reaction products, so that the composition of the products can dynamically change depending on the state of equilibrium at a given reaction condition.
- EXPLO5 has a reactants and products database which contains the most frequently used explosives, binders, and additives. The database contains around 550 reactants, while the products database contains more than 950 products (including different phases of products).
- The user can add/delete new reactants and products, as well as modify data in the reactants and products database.
- EXPLO5 produces a summary of the calculation which contains the most important information. Detailed calculation results may be stored by the user for future analysis.

EXPLO5 is a thermochemical computer code that can predict the performance of ideal high explosives and various explosive formulations, propellants, and pyrotechnic mixtures. It predicts the performance under constant volume combustion, constant pressure combustion, and ideal detonation:

A. ***Ideal detonation***

I – Calculation of the equilibrium composition and thermodynamic parameters of the state of detonation products along the shock adiabat of detonation products

II – Calculation of the C-J point and detonation parameters (D, p, T, v, Q, E, etc.) according to the C-J detonation model

III – Isentropic expansion of the detonation products from the C-J point down to atmospheric pressure as well as room temperature, as well as thermodynamic functions of state along the expansion isentrope

IV – Evaluation of the coefficients in the JWL equation of state along the expansion isentrope, and calculation of the detonation energy applying the JWL model

B. ***Non-ideal detonation***

I – Integration of ordinary differential Wood and Kirkwood flow equations supplemented by several reaction rate models, in order to determine the self-propagating detonation velocity and other detonation parameters (p, T, v, Q, E, etc.) as a function of charge radius.

II – Calculation of the equilibrium composition and thermodynamic parameters of the unreacted explosive and at detonation at a given p, v, λ.

III – Evaluation of the coefficients in the JWL equation of state along the expansion isentrope and calculation of the detonation energy by applying the JWL model

C. ***Isobaric combustion***

I – Calculation of the equilibrium composition of combustion products and thermodynamic parameters of state under constant pressure conditions

II – Calculation of the adiabatic combustion temperature and heat
III – Calculation of the equilibrium composition of combustion products and thermodynamic parameters of state during expansion through the nozzle
IV – Calculation of theoretical rocket performance (e.g. pressure and flow velocity at the nozzle throat, exhaust, and sound velocity at the nozzle exit, thrust coefficient, nozzle expansion ratio, specific impulse, etc.)

D. ***Isochoric combustion***
I – Calculation of the equilibrium composition of combustion products and thermodynamic parameters of state under constant volume conditions
II – Calculation of the adiabatic combustion temperature and heat under constant volume conditions
III – Calculation of pressure, co-volume, and compressibility factor of gaseous combustion products as a function of loading density
IV – Calculation of the force (energy) of energetic materials
V – Modeling of the isochoric cooling of combustion products from the combustion temperature to room temperature, and calculation of the calorimetric heat of combustion

E. ***Database of reactants and products***
I – EXPLO5 has a default database which includes about 950 of the most frequently used reactants, and more than 1,000 products, as well as the co-volumes and constants in the BKW and the virial EOS for gaseous products, and the Murnaghan EOS for condensed products.
II – EXPLO5's database contains the necessary input parameters for thirty explosives and ingredients of non-ideal composite explosives (parameters in Murnaghan EOS, thermodynamic parameters, reaction rate parameters).
III – EXPLO5 enables database access to be customized for multiple users on one computer. It means each user can use her/his own database and configuration files.
IV – EXPLO5 can handle a wide variety of different molecules and their mixtures (containing up to 52 different chemical elements: C, H, N, O, Al, Cl, Si, F, B, Ba, Ca, Na, P, Li, K, S, Mg, Mn, Zr, Mo, Cu, Fe, Ni, Pb, Sb, Hg, Be, Ti, I, Xe, U, W, Sr, Cr, Br, Co, Ag, Zn, Sn, Bi, Cs, Hf, Ge, Nb, Ta, Yb, Y, V, Ar, He, Cd and Ce).
V – Users can manipulate the database (add and remove compounds, modify data, and create the user's own database file).
VI – Users can customize the composition of the combustion and detonation products.

Ideal detonation

Detonation is a process in which the layer-by-layer supersonic propagation of chemical reactions through an explosive occurs. According to the generally accepted Zeldovichvon Neumann-Doering (ZND) model of detonation, chemical reactions occur at a definite rate in the chemical reaction zone, under the action of a shock wave (Fig. 4.11).

Under the influence of the dynamic action of the shock wave, a thin layer of the explosive is compressed from the initial specific volume V_0 ($V_0 = 1/\rho_0$; ρ_0 is the initial density of explosive) to volume V_1 in accordance with the shock adiabat (or also called Hugoniot) for a given explosive (Fig. 4.12). The equation defines the relationship between the density (or volume) and the pressure during the shock compression of the explosive. As a consequence of shock compression, the pressure is increased to the value p_1, which results in a significant temperature increase in the compressed explosive layer, where the initiation of chemical reactions then occurs. When the chemical reactions are finished, the volume and pressure of the reaction products correspond to the values V_2 and p_2. This state corresponds to the point lying on the shock adiabat for detonation products. From that point (the C-J point), the products expand isentropically (isentropic = adiabatic expansion, no heat transfer between the system and its surroundings: dQ = TdS, if dQ = 0 → dS = 0) (Taylor wave) into the surrounding medium.

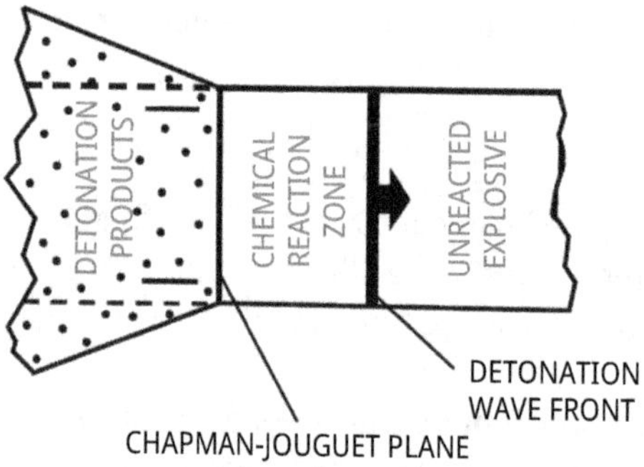

Fig. 4.11: Schematic representation of the detonation process for an ideal explosive.

According to the steady-state model of detonation, the points (V_0, p_0), (V_1, p_1) and (V_2, p_2) lie on one straight line. This line is called the Rayleigh line. The slope of the Rayleigh line is determined by the detonation velocity of a given explosive (Fig. 4.13). According to the Chapman and Jouguet hypothesis, the Rayleigh line is tangent to the shock adiabat of the detonation products at the point that corresponds to the end of the chemical reaction. That point is assigned as the Chapman-Jouguet point (C-J point).

The detonation process may be described mathematically by applying thermodynamic and hydrodynamic laws. The state and motion of the matter in the detonation wave may be expressed by means of the laws of conservation of mass, momentum, and energy. These laws can be written in the form:

$$\rho_0 D = \rho(D - U_P)$$

$$p - p_0 = \rho_0 D U_p$$

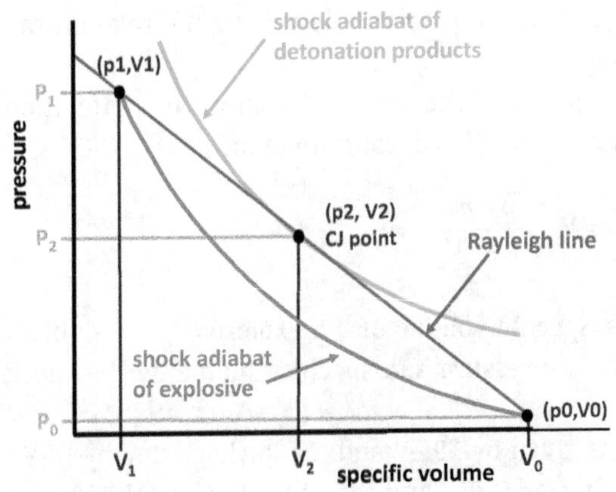

Fig. 4.12: Shock adiabat of an unreacted explosive and its detonation products in the case of steady state detonation.

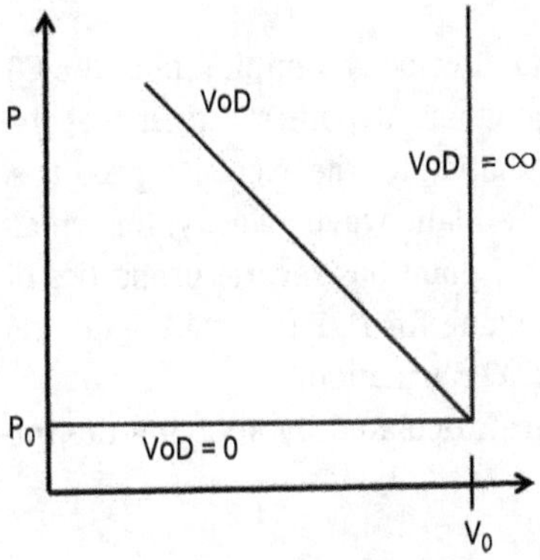

Fig. 4.13: The slope of the Rayleigh line is determined by the detonation velocity of a given explosive.

$$E - E_0 = (1/2)(p + p_0)(V_0 - V)$$

where D is the detonation velocity, U_p is the particle velocity, E is the internal energy of the detonation products, and V is the specific volume (subscript "0" indicates values for the unreacted explosive).

Combining the above equations and considering the equation resulting from the Chapman-Jouguet hypothesis results in:

$$\gamma = -\left(\frac{\partial \ln p}{\partial \ln V}\right)_s = -\frac{V}{p}\left(\frac{\partial p}{\partial V}\right)_s = -\frac{V}{p}\left(\frac{p - p_0}{V - V_0}\right)$$

where γ is the polytropic exponent. It is also possible to determine the relationship between other detonation parameters.

The C-J point is determined as a point on the shock adiabat of the detonation products at which the detonation velocity, calculated using equation:

$$D = V_0 \sqrt{\frac{p - p_0}{V_0 - V}}$$

has its minimum ($V_0 = 1/\rho_0$ and is the specific volume of the explosive, p_0 is ambient pressure, and p and V are the values of pressure and specific volume on the shock adiabat).

If the slope of the shock adiabat is given by δP/δV and the Rayleigh line by D^2/V^2, then we can write for the C-J point $\delta P/\delta V = D^2/V^2$, and therefore $D^2 = V^2\ (\delta P/\delta V)$, and consequently

$$D = V_0 \sqrt{\frac{p - p_0}{V_0 - V}} \text{(see above)}$$

An interesting question to ask is why has the detonation velocity a minimum at the C-J point? For a plane detonation wave, Chapman and Jouguet's hypothesis states that the line $(P_0,V_0) \rightarrow (P,V)$ is tangent to the Hugoniot (shock adiabat) for the explosion products. In this case, the point (P,V) is called the C-J point. The plane-wave velocity, for which the C-J condition is satisfied, is denoted by $D_{C\text{-}J}$. The C-J point has several properties of interest. For example, $D_{C\text{-}J}$ is the minimum possible value for D. Figures 4.14 and 4.15 show the minimum detonation velocity from an EXPLO5 calculation.

Other detonation parameters at the C-J point are calculated by applying the following equations:

- Particle velocity (U_p):

$$U_p = \sqrt{(p - p_0) \cdot (V_0 - V)}$$

- Polytrope exponent (γ):

$$\gamma = \frac{\rho_0 D^2}{p} - 1$$

Upon detonation, the explosive turns into hot, highly compressed gases whose density in the detonation wave front is greater than the original density of the explosive (at the C-J state). This compressed hot gas is now available to push on its surroundings. This pushing is the work it is doing, or the energy it is transferring.

The work process is considered isentropic ($dS = 0$), and the expansion is represented by the expansion isentrope on the p-V diagram (Fig. 4.16). An isentropic process is an idealization of the expansion process, which assumes there is no heat transfer between the system and its surroundings (adiabatic process). Because entropy is

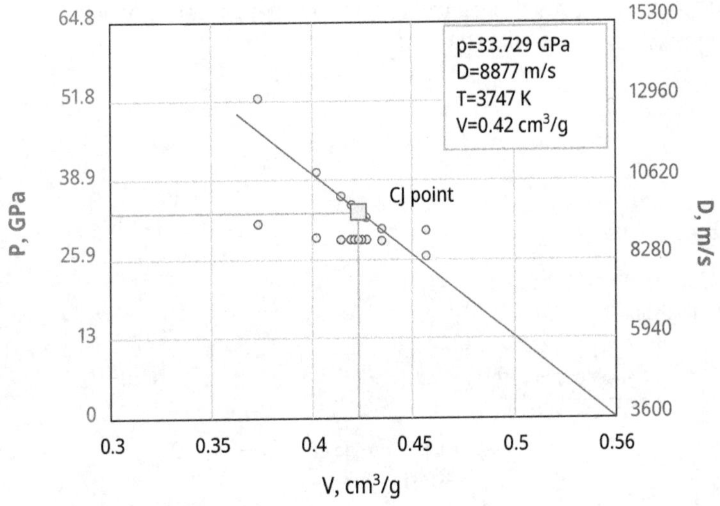

Fig. 4.14: Minimum detonation velocity at the C-J point. Line = Rayleigh line, blue circles = VoD calcd. by eq. $D = V_0 (p - p_0 / V_0 - V)^{0.5}$, orange circles = pressure.

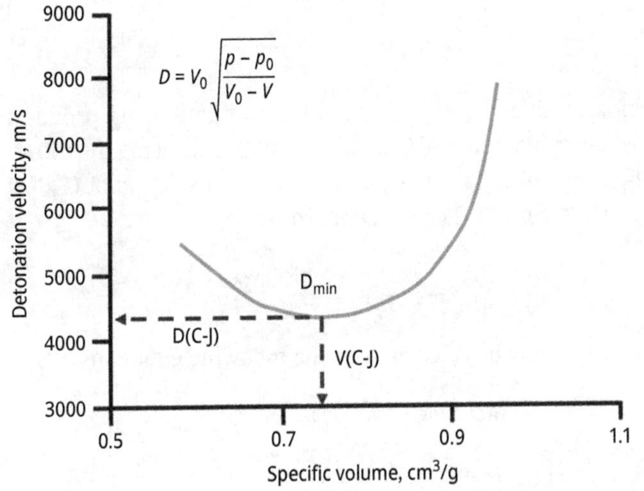

Fig. 4.15: Change of detonation velocity with specific volume of detonation products.

directly related to heat transfer by the equation $dQ = T\,dS$, it means that if $dQ = 0$ then $dS = 0$.

Assuming that the detonating explosive is instantly compressed from the room temperature and atmospheric pressure value up the Rayleigh line to the C-J point, and then it expands down the isentrope, the energy available for work on the surroundings can be calculated as being the difference between energy on the expansion isentrope (*ES*) and compression energy (*EC*) (Fig. 4.16)

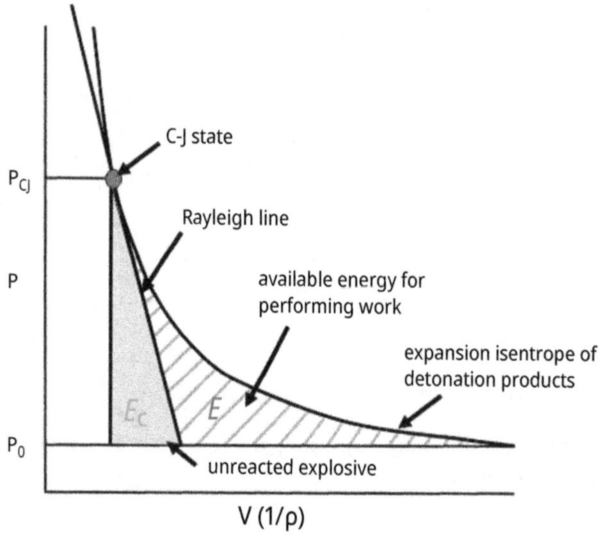

Fig. 4.16: Expansion isentrope of detonation products.

The energy of the detonation products at an infinite volume of detonation products ($E_d(v \to \infty)$ is defined as the detonation energy (E_0). It is calculated as being the difference between the internal energy of the detonation products at the C-J point ($E_s(CJ)$) and the energy of shock compression of the detonation products up to the C-J point (so-called shock energy, E_c),

$$E_0 = -[E_s(CJ) - E_c]$$

The energy of shock compression may be calculated by the following equations:

$$E_c = (1/2)(p_{CJ} + p_0)(V_0 - V_{CJ})$$

or

$$E_C = \frac{\rho_0 D^2 (1 - v_{CJ})^2}{2}$$

where: V_0 is the specific volume of initial explosive, V_{CJ} is the specific volume of detonation products at the C-J point, v_{CJ} is the relative volume of detonation products at the C-J point ($v_{CJ} = V_{CJ}/V_0$), p_0 is atmospheric pressure, and p_{CJ} is pressure of detonation products at the C-J point.

Non-ideal detonation

It is important to stress that the above description is a simplified one, and is only valid for ideal explosives. In an ideal explosive, the shock front is a plane and not curved (Fig. 4.11), whereas in a non-ideal detonation, the shock front is somewhat curved (Figs. 4.17 and 4.18), and the flow of the reaction products significantly diverges. A consequence of this is that reactions in a non-ideal detonation are never complete in the detonation driving zone (DDZ) – i.e. between the shock front and the sonic line. The detonation (a detonation is a reactive shock – shock followed by chemical reactions) driving reaction terminates at the "sonic line" and contributes to supporting the detonation process.

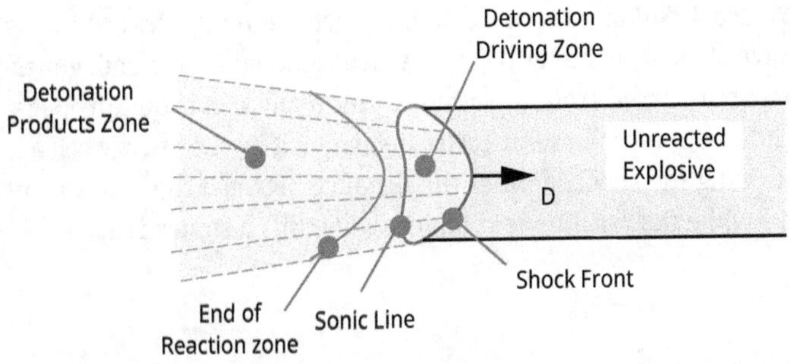

Fig. 4.17: Schematic representation of the detonation process for a non-ideal explosive.

In the case of non-ideal detonation, the shock front is notably curved, causing the flow of reaction products to spread out significantly. As a result, reactions in the detonation driving zone (DDZ) – the region between the shock front and the sonic line – remain incomplete. Within the DDZ, the flow is subsonic, allowing perturbations such as compression and rarefaction waves, which travel at the local speed of sound, to help sustain the shock front. However, beyond the sonic line, the flow becomes supersonic, meaning these perturbations cannot reach the DDZ to influence the shock front (see Figs. 4.17 and 4.18).

When a cylindrical column of explosive is detonated, the velocity of detonation (VoD) tends to decrease as the diameter of the column becomes smaller. This phenomenon occurs due to a reduction in pressure along the sides of the column. For larger diameters, energy losses are minimal compared to the energy produced at the wave front. However, as the diameter decreases, these losses become more significant relative to the energy generated, leading to a continued decrease in velocity. This reduction persists until the column reaches a critical diameter, known as the failure diameter, where energy losses outweigh production, causing the detonation to cease.

Moreover, it is well-established that increased confinement can have an effect similar to enlarging the charge diameter. In non-ideal conditions, as depicted in Fig. 4.18, the shock front remains curved, the flow of reaction products diverges, and the reaction remains incomplete within the detonation zone. The detonation driving zone (DDZ) ends at the sonic line and plays a key role in sustaining the detonation process. In this scenario, the detonation velocity may approach, but never exceed, the ideal detonation velocity. The degree of non-ideality in an explosive is often gauged by the difference between its ideal and actual detonation velocities.

Beyond the sonic line, the flow becomes supersonic, meaning that perturbations like compression and rarefaction waves which travel at the local speed of sound, cannot catch up with the DDZ and, thus, cannot affect or reduce the speed of the detonation wave. The rarefaction (Taylor wave) occurring in the supersonic, still-reactive flow between the sonic line and the end of the reaction zone is particularly significant in commercial explosives, which may have slow energy-releasing reactions that contribute significantly to the blast. Figure 4.18 illustrates the non-ideal detonation process, showing how a cylindrical explosive stick of diameter d (mm), when initiated at one end, generates a detonation wave with a curved front of radius R_s (mm) and a constant propagation velocity D (km/s). The position of the sonic locus, determined by the energy release in the detonation zone driving the shock, defines the distance between the shock front and the sonic locus as x_{CJ}, while the reaction is complete when this distance is x_{rz}.

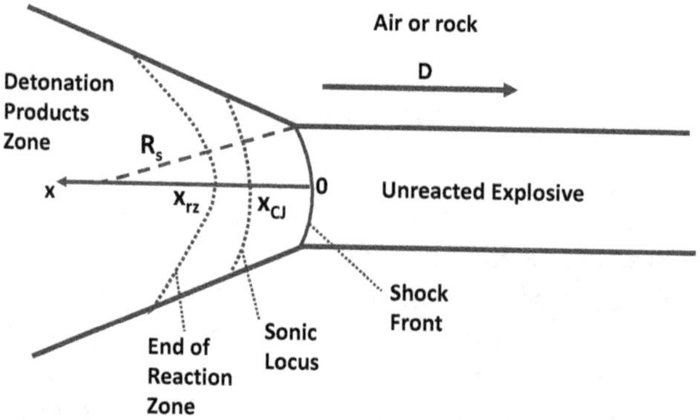

Fig. 4.18: Detonation front for a non-ideal cylindrical explosive.

For an ideal explosive at the C-J point (at the end of the chemical reaction zone), conversion of the explosive into the products corresponding to the thermochemical equilibrium is complete (1.0), whereas for a non-ideal explosive the conversion is < 1. Thus, for a non-ideal explosive, the C-J point is also called the "sonic point" (end of DDZ) (Fig. 4.18), and where the Rayleigh line has the same slope as the Hugoniot (shock adiabat) of the detonation products formed at the sonic point.

For both ideal and non-ideal detonations, at the C-J point the "flow" (detonation velocity, D – particle velocity, U_p) corresponds to the speed of sound, C0:

$$D - U_p = C0$$

Literature

EXPLO5, Version 7.02, Muhamed Sućeska, Prilaz baruna Filipovića 28, 10000 Zagreb, Croatia.

M. Sućeska, Calculation of detonation properties of C-H-N-O explosives, *Propellants, Explos. Pyrotechnics*, **1991**, *16*, 197–202.

M. Sućeska, EXPLO5 computer program for calculation of detonation parameters, *Proc. of 32nd Int. Annual Conference of ICT, Karlsruhe*, **2001**, pp. 110/1–110/13.

M. Sućeska, Calculation of detonation heat by EXPLO5 computer code results, *Proc. of 30th Int. Annual Conference of ICT, Karlsruhe*, **1999**, pp. 50/1–50/14.

M. Sućeska, Evaluation of detonation energy from EXPLO5 computer code results, *Propellants, Explos. Pyrotechnics*, **1999**, *28*, 280–285.

M. Sućeska, calculation of thermodynamic parameters of combustion products of propellants under constant volume conditions using virial equation of state. Influence of values of virial coefficients, *J. Energetic Mater.*, **1999**, 17, 253–277.

M. Sućeska, Calculation of detonation parameters by EXPLO5 computer program, *Materials Science Forum*, **2004**, *465–466*, 325–330.

M. Sućeska, C.H.Y. Serene, A. How-Ghee, Can the accuracy of BKW EOS be improved, *15th International Detonation Symposium, San Francisco, USA, July 13–18*, **2014**, pp. 1247–1256.

S. Esen, P. C. Souers, P. Vitello, Prediction of the non-ideal detonation performance of commercial explosives using the DeNE and JWL++ codes, *Int. J. Numerical Methods Engineer.*, **2005**, *64*, 188–1914. https://doi.org/10.1002/nme.1424.

G. Vasilescu, A. Kovacs, E. Gheorghiosu, B. Garaliu, G. Ilcea, Numerical simulation for determining detonation parameters of explosive substances using EXPLO5 thermo-chemical prediction software, *MATEC Web conf.*, **2020**, *305*, 00049. https://doi.org/10.1051/matecconf/202030500049.

4.2.8 Gun barrel design

The problem with the barrels of large caliber weapons is that at maximum power high temperatures are reached, and the CO from the propellant powder reacts with the barrel (steel) to form iron carbide, which leads to erosion problems. The normal approach used to address this problem is to lower the temperature and use N-rich powders, which form more N_2 than CO, so more iron nitride is formed (i.e. less iron carbide), which tends to harden the tubes and make them more resilient. Of course, the question then arises as to which alloy components must be present in steel, that will react more strongly with nitrogen than with carbon and thereby encourage formation of iron nitride rather than iron carbide. Basically, these questions have been clarified for the high-temperature apparatus used for the construction for gas turbines, nuclear reactors and the like, where high strength and high corrosion resistance at high temperatures must be guaranteed. In addition, these alloys must be producible on an industrial scale, workable with machine tools and not too expensive. Three technical alloys from the USA are important in this context: Hastelloy, Waspaloy and Inconel.

HASTELLOY belongs to the group of highly corrosion-resistant, nickel-chromium-molybdenum-tungsten alloys. It is characterized by showing high resistance to crevice corrosion, pitting corrosion and stress corrosion cracking in oxidizing and reducing media.

WASPALOY is an age-hardened, nickel-based superalloy which offers outstanding resistance at high temperature, as well as excellent oxidation resistance.

INCONEL alloys have good resistance to corrosion, making them suitable for use in extreme environments. When heated, a thick, stable oxide layer forms that protects the surface. In addition, its strength is maintained over a wide temperature range.

Another approach is to electrolytically deposit 3d, 4d and 5d metals on the inner tube, whereby the enthalpy of formation of the corresponding metal nitride should be higher than that of the carbide. The metals which come into consideration are:

Ti, V, Cr, Mn, Fe, Co, Ni
Zr, Nb, Mo
Hf, Ta, W

The enthalpies of formation per g-atom of the metal carbides and nitrides should be known. Metals which have a higher enthalpy of formation with respect to the formation of the nitride compared to the carbide, can then be electrolytically applied to the inner tube.

The Binnewies thermochemical data appear are good, and without exception, for all of the 3d, 4d, and 5d metals which are listed, the nitrides have the more negative enthalpies of formation. Therefore, the idea of providing a lot of nitrogen in the "mortar tube", in order to reduce the amount of carbide formation which occurs, does not seem to be so bad.

There are already patents for mortar tubes made of Inconel.

The following values for the carbides and nitrides were taken from the reference book: M. Binnewies, Thermochemical data of elements and compounds, Wiley-VCH, **ISBN/ISSN/ISMN** 3527297758 T:

Ti V Cr Mn Fe Co Ni
Zr Nb Mo
Hf Ta W

The metals which are not listed (e.g Sc, Y, Rh, Pd) have been excluded due to their very high cost.

The data have been converted to the values of the sum of 2 stoichiometries for better comparison, with the original data being given in square brackets, e.g. BW 541 Fe_4N $\Delta H^\circ_{298} = -11.1$ kJ/mol results for a sum total of 2:

$Fe_{1.6}N_{0.4}$ $\Delta H^\circ_{298} = (-11.1/5)\cdot 2 = -4.44$ kJ/mol

TiC	$\Delta H^\circ_{298} = -184.1 +/- 4$ kJ /mol (BW255)
TiN	$\Delta H^\circ_{298} = -337.7 +/- 4.2$ kJ /mol (BW697)
ZrC	$\Delta H^\circ_{298} = -207.1$ kJ /mol (BW256)
ZrN	$\Delta H^\circ_{298} = -365.3 +/- 8.3$ kJ /mol (BW697)
HfC	$\Delta H^\circ_{298} = -226$ kJ /mol (BW242)
HfN	$\Delta H^\circ_{298} = -373.6$ kJ /mol (BW586)

$V_{1.33}C_{0.67}$	$\Delta H°_{298}$ = –78.1 kJ /mol [V_2C $\Delta H°_{298}$ = –117.2 kJ /mol (BW256)]
VN	$\Delta H°_{298}$ = –217.2 kJ /mol (BW697)
NbC	$\Delta H°_{298}$ = –138.9 kJ /mol (BW250)
$Nb_{1.33}C_{0.67}$	$\Delta H°_{298}$ = –123.9 kJ /mol [Nb_2C $\Delta H°_{298}$ = –185.8 kJ /mol (BW250)]
NbN	$\Delta H°_{298}$ = –235.1 kJ /mol (BW694)
$Nb_{1.5}N_{0.5}$	$\Delta H°_{298}$ = –164.6 kJ /mol [Nb_2N $\Delta H°_{298}$ = –246.9 kJ /mol (BW694)]
TaC	$\Delta H°_{298}$ = –144.1 kJ /mol (BW255)
$Ta_{1.33}C_{0.67}$	$\Delta H°_{298}$ = –138.9 kJ /mol [Ta_2C $\Delta H°_{298}$ = –208.4 kJ /mol (BW255)]
TaN	$\Delta H°_{298}$ = –252.3 kJ /mol (BW696)
$Cr_{1.2}C_{0.8}$	$\Delta H°_{298}$ = –34.0 +/–4.8 kJ /mol [Cr_3C_2 $\Delta H°_{298}$= –85.1 +/–12.1 kJ /mol (BW25)]
$Cr_{1.4}C_{0.6}$	$\Delta H°_{298}$ = –32.1 +/–4.8 kJ /mol [Cr_7C_3 $\Delta H°_{298}$= –160.7+/–16.7 kJ /mol (BW266)]
CrN	$\Delta H°_{298}$ = –117.2 kJ /mol (BW419)
MoC	$\Delta H°_{298}$ = –28.5 kJ /mol (BW247)
$Mo_{1.33}C_{0.67}$	$\Delta H°_{298}$ = –33.0 kJ /mol [Mo_2C $\Delta H°_{298}$ = –49.5 kJ /mol (BW247)]
$Mo_{1.33}N_{0.67}$	$\Delta H°_{298}$ = –54.4 kJ /mol [Mo_2N $\Delta H°_{298}$ = –81.6 kJ /mol (BW660)]
WC	$\Delta H°_{298}$ = –40.6 kJ /mol (BW256)
$W_{1.33}C_{0.67}$	$\Delta H°_{298}$ = –17.6 kJ /mol [W_2C $\Delta H°_{298}$ = –26.4 kJ /mol (BW256)]
WN	missing

The following metals which were considered are: Mn, Fe, Co, Ni

$Mn_{1.5}C_{0.5}$	$\Delta H°_{298}$ = –7.6 kJ /mol [Mn_3C $\Delta H°_{298}$ = –15.1 kJ /mol (BW246)]
$Mn_{1.58}C_{0.42}$	$\Delta H°_{298}$ = –18.5 kJ /mol [$Mn_{15}C_4$ $\Delta H°_{298}$ = –175.7 kJ /mol (BW269)]
$Mn_{1.43}N_{0.57}$	$\Delta H°_{298}$ = –58.3 kJ /mol [Mn_5N_2 $\Delta H°_{298}$ = –204.2 kJ /mol (BW684)]
$Mn_{1.6}N_{0.4}$	$\Delta H°_{298}$ = –51.5 kJ /mol [Mn_4N $\Delta H°_{298}$ = –128.7 kJ /mol (BW684)]
$Fe_{1.5}C_{0.5}$	$\Delta H°_{298}$ = +12.6 kJ /mol (BW238) [Fe_3C $\Delta H°_{298}$ = +25.1 kJ /mol (BW238)]
$Fe_{1.6}N_{0.4}$	$\Delta H°_{298}$ = –4.4 kJ /mol (BW238) [Fe_4N $\Delta H°_{298}$ = –11.1 kJ /mol (BW238)]
Co-C	missing
$Co_{1.5}N_{0.5}$	$\Delta H°_{298}$ = +4.2 kJ /mol [Co_3N $\Delta H°_{298}$ = +8.4 kJ /mol (BW415)]
$Ni_{1.5}C_{0.5}$	$\Delta H°_{298}$ = +33.7 kJ /mol [Ni_3C $\Delta H°_{298}$ = +67.4 kJ /mol (BW250)]
Ni-N	missing

The thermochemical data fit well, since the nitrides all have the more negative enthalpies of formation – without exception!

In conclusion, if there is “a lot” of nitrogen in the “mortar tube”, the amount of carbide formation can be reduced!

Super alloys

Some of these materials for mortar barrels are so hard, that they can only be processed by grinding. For Inconel, there is already a patent for pipe diameters of up to 81 mm.

The compositions also include, for example, the low-melting Al. This should not be underestimated if titanium is present. TiAl is a high-melting, very stable compound.

The very latest (2022) example is NASA alloy GRX-1810, in which yttrium oxide (Y_2O_3) nanoparticles are embedded between the granules. This blocks the diffusion and migration of dislocations at high temperatures, which increases stability.

Ni-based Superalloys:
A **superalloy**, or **high-performance alloy**, is an alloy with the ability to operate at a high fraction of its melting point. Key characteristics of a superalloy include mechanical strength, thermal creep deformation resistance, surface stability, as well as corrosion and oxidation resistance.

The crystal structure of a superalloy is typically face-centered-cubic (fcc). Examples of such alloys are Hastelloy, Inconel, Waspaloy, Rene alloys, Incoloy, MP98T, TMS alloys, and CMSX single crystal alloys.

Alloying elements: Al, Ni, Co, Cr, Ta, Ti, Re, W . . .

These superalloy materials (such as Inconcel) are becoming more common in defense applications because of their high strength, corrosion resistance and high survivability in extreme environments.

Current processes to manufacture 60 mm mortar tubes rely on traditional mechanical cutting and grinding of the material to achieve the required geometry. However, tough materials that are good for weapons are inherently difficult to machine. Also in aeronautics, General Electric (GE) has developed and patented an electro-erosion process called Blue Arc for the production of aircraft engines.

Check your knowledge: see chapter 14 for study questions.

---

5 Initiation

5.1 Introduction

An explosive can be initiated using different stimuli (e.g. heat or shock) and can then either ignite, deflagrate and turn into a detonation, or directly detonate, if it is initiated using a strong shock (Fig. 5.1). The ignition occurs at an ignition temperature which is characteristic for individual substances, if the linear loss of heat of the surroundings is smaller than the heat generated through the exothermic reaction.

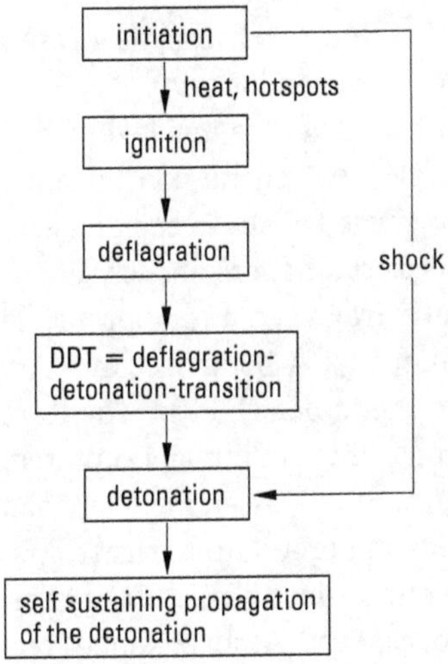

Fig. 5.1: Transition from initiation to detonation.

Generally, it can be said that the **initiation** of explosives is mostly a **thermal process.** However, the ignition (initiation) of an explosive can also occur by impact linking via the low frequency vibrational modes (doorway modes) between 200 and 600 cm^{-1}. In cases where the initiation occurs through a shock-wave, we also observe strong warming-up through adiabatic compression. If mechanical (impact, friction) or electrostatic (ESD) mechanisms are the cause, it can be assumed, that the mechanical or electric energy is also first converted into heat. This occurs, for example, during the formation of so-called **hotspots**. Here, there are small gas bubbles (0.1–10 μm), which are then strongly heated (up to 900 °C) as well through adiabatic compression and therefore initiate the explosive. These can either be gas bubbles in liquids or in solids. The larger the difference in pressure between the original pressure (p_1) and the final pressure on compression p_2, the higher the jump in temperature:

$$T_2 = T_1 \left(\frac{p_2}{p_1}\right)^{\frac{\gamma-1}{\gamma}}$$

https://doi.org/10.1515/9783111447087-005

Another type of possible hotspots are small, very hard crystals or crystal needles. Prior to breaking the crystal, energy must be applied either through friction or through pressure, in order to bring charges of the same sign closer together. This energy is released again on breaking of the crystals and can result in the formation of a hotspot. However, not every hotspot results in an ignition and finally in detonation. For example, when the energy is released to the surroundings without resulting in an initiation. Generally, it can be said that the temperature in a hotspot must be at least 430 °C, in order to be sufficient enough to cause the initiation of a secondary explosive. On the other hand, liquids (e.g. NG) can be initiated through hotspots (gas bubbles of dissolved gases) as well through adiabatic compression. On average hotspots exist only for 10^{-3}–10^{-5} s.

When crystals rub together, hotspots can form only through **friction**, as a result of the friction heat and the (in comparison to metals) low heat conductivity.

Usually it is assumed, that on **impact** most primary explosives are initiated by hotspots from inter-crystalline friction. For secondary explosives, as a general rule, initiation by impact results in the formation of hotspots, which originate (adiabatic compression) from gas bubbles between the crystals. Such hotspots only exist for approximately 10^{-6} s.

The thermal energy of hot spots must be efficiently transferred to appropriate molecular vibrational modes, a process called "up-pumping", if bond-breaking and subsequent exothermic decomposition and detonation are to be achieved. The term "trigger linkage" has been applied to the key bond or bonds that are initially ruptured. Any dissipation of hot spot energy, for instance by diffusion, will lessen the likelihood of these processes. Thus, Kamlet suggested that free rotation around the trigger linkage (e.g. C-NO_2, N-NO_2, O-NO_2) can have a desensitizing effect, since it uses energy that could otherwise go into bond-breaking vibrational modes. Furthermore, Kamlet recognized the significance of decomposition steps that follow the initial bond rupture, for example the autocatalysis (by NO_2 radicals) for the decomposition of nitramines.

The fact that solids composed of very small particles are less sensitive can now be explained on the grounds that these will have smaller hot spots and thus require higher temperatures for ignition.

It should also be noted that initiation of detonation can occur even in a homogeneous (defect-free) solid (which doesn't really exist). This can occur, for example, if there is efficient anharmonic coupling to channel energy from lattice into the critical molecular vibrations.

Finally, let us consider the situation in TATB which is well known for its remarkable insensitivity. An outstanding feature of TATB is the strong and extensive hydrogen bonding, both inter- and intramolecular, between the NO_2 and NH_2 groups. This would be anticipated from its molecular structure, and has been confirmed by X-ray diffraction. The intermolecular H-bonding gives rise, in the solid, to a two-dimensional network, to which is attributed the relatively high thermal conductivity of TATB. In terms of the hot spot concept, this is a desensitizing factor, since it results in a more rapid dissipation of energy through thermal diffusion. This is also a reason why TATB requires a high ignition temperature. On the other hand, the strong intermolecular H-bonding should increase the NO_2 rotational barrier, an expectation for which DFT calculations provide support. By Kamlet's reasoning (see above), this should have a sensitizing effect, since it diminishes the loss of energy through rotation and makes more of it available for the key NO_2 vibrational modes. It is believed that C-NO_2 homolysis is the first or one of the first steps in TATB initiation.

5.2 Ignition and initiation of energetic materials

As was mentioned above, the ignition of energetic materials and in particular the initiation of explosives can occur in manifold ways and depending on the output of the energetic material, the initiating system is called a detonator (shockwave) or an igniter (flame) (Fig. 5.2). The simplest method is the direct ignition by flame. This was the method which was used in the past to ignite black powder by a burning fuse. Flame as a simple initiating impulse is still used to initiate energetic materials, for example, in fuse-type blasting caps (Fig. 5.3, A).

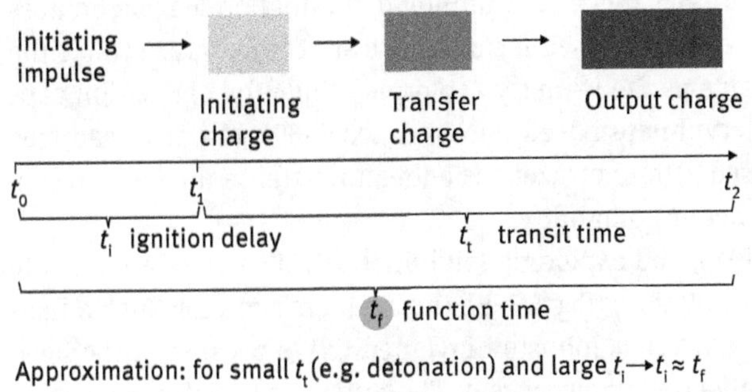

Fig. 5.2: Schematic presentation of an initiation chain.

Another method similar to flame as an impulse is to ignite the explosive by heat (Fig. 5.3, B + C). A bridgewire which is heated by an electric current is either in direct contact with the primary explosive (hot-wire initiator), or first it is in contact with a pyrotechnical composition which then initiates the primary explosive by flame. This second type, called an electric match, is the most common initiation method in blasting caps worldwide. However, there is also a type of detonator known which achieves detonation without a primary explosive. This type of detonator is called NPED (Non Primary Explosives Detonator) and uses the ability of PETN to undergo DDT after ignition by a pyrotechnic composition. NPEDs also represent a form of electric match type detonators, but provide a higher degree of safety due to the absence of the sensitive primary explosive.

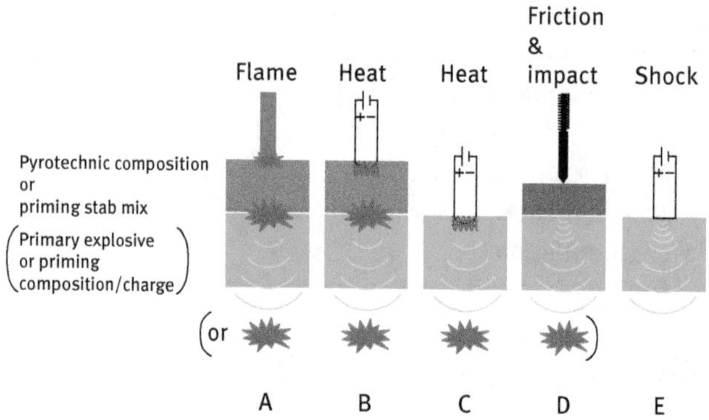

Fig. 5.3: Schematic representation of common initiation and ignition methods of energetic materials. A) Fuse type B) Electric match type C) Hot-wire type D) Percussion / Stab type E) Exploding-bridgewire type.

In contrast, the initiation of bombs and grenades, or the ignition of gun propellants is generally achieved by an impact or friction impulse (Fig. 5.3, D). Independent of the desired output signal, the primary explosive or priming charge must exhibit a high sensitivity towards friction to ensure reliable initiation. However, it hinders the handling of the explosive and increases the risk of unintended initiation. For detonators like the M55 stab detonator, lead azide-based formulations are commonly used for the stab mix, followed by lead azide as the primary explosive. While for percussion caps which only deflagrate, lead styphnate based (SINOXID, NOL-60) and also lead-free (SINTOX) formulations are used. Often tetrazene is added to increase the friction sensitivity, in order to achieve a better ignitability.

Exploding-bridgewire (EBW), and exploding foil initiators (EFI) are relatively safe and very reliable electric type initiators (Fig. 5.3, E). During their work for the Manhattan Project, Luis Alvarez and Lawrence Johnston invented EBWs because of the need for a precise and highly reliable detonator system. The initiating impulse is a shockwave which is formed by the explosion of a thin bridgewire. The gold or platinum wire is heated very quickly by the passing high current (such high currents are supplied by low-inductance, high voltage capacitors). While the metal vaporizes, the electrical resistance of the bridgewire increases rapidly and the current decreases. However, a lot of current still passes through the high density metal vapor, and so heating is continued up to the point where inertia is overcome. This results in explosive expansion of the metal vapor. Further effects amplify the shock. This shockwave finally initiates the secondary explosive which – due to practical reasons – is PETN in applications. The advantage of EBWs over classic-type detonators (e.g. hot-wire, stab) is their precise and consistent function time (variations of $\leq 0.1\ \mu s$).

However, all of the above-mentioned initiating systems have some disadvantages. Stab detonators or percussion caps contain materials which are extremely friction and impact sensitive. Therefore, there exists a high risk of unintended initiation, and the handling of these materials is problematic. Furthermore, the electric type detonators are susceptible towards electrostatic discharge, electromagnetic interference and corrosion. In the case of the safe EBWs and EFIs, stray currents would not initiate the detonator but may damage it. A further disadvantage of EBWs and EFIs is the bulky size of the power sources which are required.

5.3 Laser ignition and initiation

Due to the disadvantages of conventional initiators, a new ignition and initiation method was introduced at the end of the 60s and beginning of the 70s. The first published laser initiation experiments were reported by the Russian research group of Brish *et al.* and by the Americans Menichelli and Yang. Brish *et al.* successfully initiated samples of PETN and LA by laser irradiation. As the radiation source a neodymium glass laser (λ = 1,060 nm) operating in the Q-switched mode was used. LA was initiated by a laser pulse with an output power of 10 MW, a pulse length τ of 0.1 µs and energy of 0.5 J. The pulse had a power density at the surface of the explosive sample of up to 8 MW cm^{-2}. For the initiation of PETN, the laser beam had to be focused to increase the power density at the surface. Further investigations with a neodymium glass laser and an additional ruby laser (λ = 694 nm) were made by Brish *et al.*, and four possible initiation mechanisms were considered:

(i) Light impact. This could be excluded because the exerted light pressure was several orders of magnitude lower than the pressures required for shock explosion.
(ii) Electrical breakdown could not fully be excluded but did not fit to all observations during the laser initiation experiments. For example, no differences in the initiation of explosives were observed for dielectrics and metal powders.
(iii) Photochemical initiation could be explained by a multiquantum photoelectric effect, but the probability for initiation was low and a dependency between the laser wavelength (1,064 or 694 nm) and the initiation threshold was not observed. Thus, photochemical initiation does not seem to be the appropriate mechanism for laser initiation.
(iv) Photothermal (or simply called thermal) initiation explains all of the experimental results. The laser radiation heats the explosive and leads to an increase of the pressure inside the sample. The rapid heating of the sample results in an explosive decomposition described as a conversion of light energy into shockwave energy.

While Brish *et al.* directly irradiated the explosive sample with the laser beam, Menichelli and Yang coated a glass window with a 1,000 Å thick aluminum film. Irradiation of this film generated shockwaves by vaporization of the metal, similar to the functionality of EBWs. [15c, 15d]. The generated shockwave is able to directly initiate secondary explosives. In the work of Menichelli and Yang, a Q-switched ruby laser was successfully used (E = 0.8 to 4.0 J; τ = 25 ns) to initiate PETN, RDX and tetryl.

However, the practical use of these solid-state lasers (ruby and neodymium glass) was limited because of their large size. The advent of commercially available laser diodes (electrically pumped semiconductor laser) afforded new possibilities in the initiation and ignition of energetic materials, although solid-state lasers were and are still used for investigations. One of the first reports on diode laser ignition of explosives and pyrotechnics was published by Kunz and Salas in 1988. A GaAlAs diode laser with a power of 1 W, a wavelength of 830 nm, and a pulse duration of 2–10 ms was used to thermally ignite pentaammine(5-cyano-2*H*-tetrazolate)cobalt(III) perchlorate (CP, Figure 9) and a pyrotechnic mixture of titanium and potassium perchlorate ($Ti/KClO_4$). In contrast to the experiments of Brish *et al.* and Menichelli and Yang, the laser beam was directly coupled to the explosive's surface by an optical fiber (core diameter of 100 µm) and not focused. CP and $Ti/KClO_4$ were tested upon laser irradiation in pure and in doped (carbon black, graphite and laser dye) form. Additionally, various particle sizes and compressed densities were investigated. In the work of Kunz and Salas, it was shown that the laser ignition threshold is strongly influenced by the absorbance of the energetic materials. Thus, CP exhibited a considerably lower absorbance at 830 nm than the $Ti/KClO_4$ mixture and therefore a higher energy was necessary for ignition ($Ti/KClO_4$: $E_{crit.}$ = 3 mJ; CP: $E_{crit.}$ = 6–7 mJ). Doping CP with 0.8% carbon black resulted in a similar absorbance ($\alpha \approx 0.8$) to $Ti/KClO_4$, and also the critical energy could be decreased to a value of 3 mJ similar to $Ti/KClO_4$. Additionally, absorbance measurements demonstrated that the particle size and pressed powder density influences the absorbance (α increases with decreasing particle size and an increase in the density) although the effect on the ignition threshold is low. One reason could be that an increasing density also increases the thermal conductivity and – related to this – the ignition threshold. Although the diode laser systems are relatively cheap, small in size and offer significantly improved safety over electric type ones, laser diodes provide a considerably lower output power than the solid-state lasers. Consequentially, in the past, laser diode ignition exhibited long function times and was not suitable as a replacement for fast-functioning electric type initiators like EBWs.

Various further experiments with laser radiation as an ignition and initiation source have been conducted since the seventies. For example, for the ignition of propellants, various pyrotechnic mixtures and metastable intermolecular (or intermixed) composites (MICs), the effects of the laser wavelength, gas pressure or dopants on the initiation thresholds, the development of laser initiated detonators, as well as a fundamental understanding of the initiation mechanism were investigated. A remarkable development of a NPED which was successfully initiated by laser irradiation has been achieved. The NPED uses the principle of flying plate detonators and consists of a PETN/graphite/Al mixture as a laser sensitive ignition charge. The ignition charge drives a thin metal plate (called flying plate) which initiates low-density PETN (ρ = 0.7 g cm^{-3}) by shock. This charge is followed by two further charges with higher densities (PETN with a density of

1.0 g cm^{-3} and RDX with a density of 1.6 g cm^{-3}). A schematic representation of the set-up is shown in Fig. 5.4. Although no values were given for this detonator, it can be expected that the function times are in the range of milliseconds.

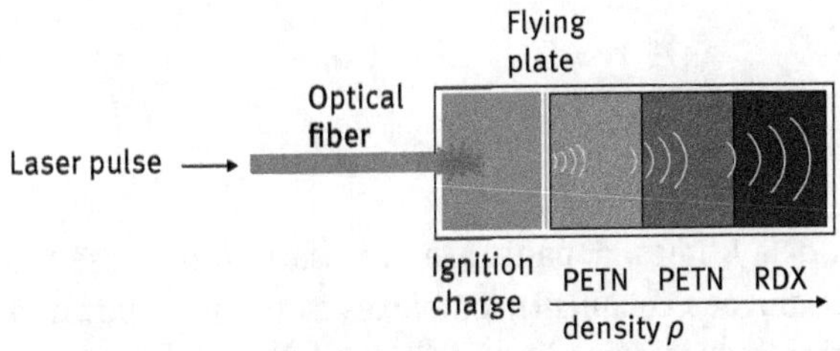

Fig. 5.4: Schematic representation of the laser initiated NPED.

Other recent investigations deal with energetic coordination compounds (ECCs) which exhibit a high sensitivity towards laser irradiation. The investigation of ECCs with respect to their behavior towards laser irradiation started with CP and its promising replacement tetraammine-*cis*-bis(5-nitro-2*H*-tetrazolato-N^2)cobalt(III) perchlorate (BNCP). BNCP (Figs. 5.5 and Fig. 5.6) was already reported to exhibit initiating capabilities by Bates in 1986. Since that time, BNCP has been extensively investigated as an initiating explosive which is able to undergo fast DDT after ignition by electric devices or laser irradiation. BNCP exhibits excellent energetic properties, e.g. a detonation velocity of 8.1 km s^{-1} was calculated for BNCP at a density of 1.97 g cm^{-3}. With the advent of more powerful diode lasers, the ignition delay times of BNCP could be decreased considerably and excellent values of 4–5 μs were possible.

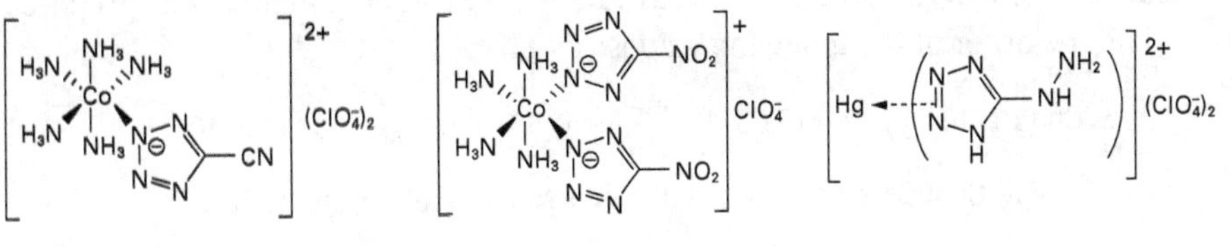

Fig. 5.5: Molecular structures of three prominent laser ignitable explosives CP, BNCP, and HTMP.

At the beginning of the twenty-first century, Zhilin *et al.* investigated BNCP and some BNCP-analogous tetraamminecobalt(III) perchlorate complexes with 1*H*-tetrazolate, 5-methyl-1*H*-tetrazolate, 5-amino-1*H*-tetrazolate, 5-nitriminotetrazolate, 1,5-diamino-1*H*-tetrazole, and 5-amino-1-methyl-1*H*-tetrazole as ligands with respect to laser irradiation ($\tau = 2$ ms, $E = 1.5$ J, $\lambda = 1{,}064$ nm). Only the complexes with 5-nitro-2*H*-tetrazolate, 5-amino-1*H*-tetrazolate and 1,5-diamino-1*H*-tetrazole as ligands detonated after initiation by laser irradiation. The others only combusted or were unable to be ignited.

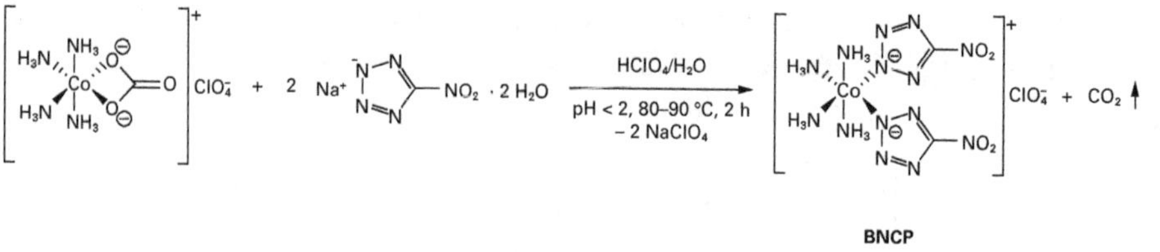

Fig. 5.6: Synthesis of BNCP according to the literature.

In order to obtain short ignition delay times, dopants are necessary to increase the absorptance of CP, BNCP or its analogous cobalt(III) complexes in the near infrared region. Therefore, explosives with a higher own absorbance have been investigated. One class of explosives which exhibit a high sensitivity towards laser irradiation are metal complexes with various hydrazinoazoles as ligands. The advantage of this ligand type is the high positive enthalpy of formation for the azole, combined with the low ionization potential of the hydrazine moiety. The following hydrazinoazoles have been investigated as ligands in mercury(II) perchlorate complexes: 3-hydrazino-5-aminopyrazole, 3-hydrazino-5-aminopyrazolone, 3-hydrazino-4-amino-1,2,4-triazole, 3-hydrazino-4-amino-5-methyl-1,2,4-triazole, 3-hydrazino-4-amino-5-mercapto-1,2,4-triazole and 5-hydrazino-1*H*-tetrazole.[26] The 5-hydrazino-1*H*-tetrazolemercury(II) perchlorate complex (HTMP, Fig. 5.5) exhibited the highest sensitivity towards laser irradiation ($\tau = 30$ ns, $E = 1 \cdot 10^{-5}$ J) of the investigated azoles (e.g. 3-hydrazino-4-aminopyrazole: $\tau = 30$ ns, $E = 2.8 \cdot 10^{-4}$ J). Based on these results, four metal(II) perchlorate complexes (M = Co, Ni, Co, Cd) of the type $[M(HATr)_2](ClO_4)_2$ with 3-hydrazino-4-amino-1,2,4-triazole (HATr) as the ligand were synthesized and tested towards laser irradiation. All complexes were successfully initiated and the critical initiation energies determined. A correlation between the initiation energy and the ionization potential was proposed to explain the order of the following values:

$$E\text{: Cu}\left(1.1 \cdot 10^{-5}\,\text{J}\right) < \text{Cd}\left(5.03 \cdot 10^{-4}\,\text{J}\right) < \text{Ni}\left(5.75 \cdot 10^{-4}\,\text{J}\right) < \text{Co}\left(1.36 \cdot 10^{-3}\,\text{J}\right)$$

$$I_1 + I_2\text{: Cu (28.02 eV)} > \text{Cd (25.90 eV)} > \text{Ni (25.78 eV)} > \text{Co (24.92 eV)}$$

A mechanism (eqs. 5.1 and 5.2) for the initial stages in the decomposition of the abovementioned complexes was claimed by Ilyushin *et al.*:

$$2ClO_4^- + M^{2+} \rightarrow ClO_4^\bullet + M^+ + ClO_4^- \tag{5.1}$$

$$ClO_4^- + M^+ \rightarrow ClO_4^\bullet + M^0 \tag{5.2}$$

A highly active perchlorate radical is formed by a laser induced redox reaction between the metal cation and the anion. This radical subsequently oxidizes the organic ligand. The more noble metal cation (higher ionization potential) can be easier reduced. Although the suggested mechanism explains, in agreement with the experimental data, how the ionization potential influences the initiation energy threshold, it doesn't fit with the thermal mechanism proposed by numerous researchers and is therefore doubtful. For example, the mechanism by Ilyushin *et al.* cannot explain the influence of light absorbing particles like carbon black on the initiation threshold.

To conclude, the results of the investigations which have been carried out so far suggest that laser initiation seems to be a thermal mechanism, although the mechanism itself and the parameters which influence it are still not completely understood. For example, Ilyushin *et al.* state that there seems to be no dependence between optical properties and the laser initiation threshold for various cobalt(III) complexes which is not plausible for a photothermal process. A thermal initiation mechanism means that laser light is absorbed and converted into heat by internal conversion. At lower laser powers (10^{-1}–10^{1} W range), the explosives undergo a DDT after being heated up to the auto-ignition temperature. According to the literature, the explosive is shock initiated at higher power (approximately 10^{2} W and more) and DDT no longer takes place. Hafenrichter *et al.* described four regimes for the laser initiation depending on the power densities:

Region I: Power density is low and the steady state temperature is reached before auto-ignition → no ignition occurs.

Region II: Power density is higher (kW cm^{-2}) and ignition occurs, although a significant amount of energy is conducted away due to thermal conductivity of the materials → slow ignition delay times in the order of milliseconds.

Region III: At even higher power densities (MW cm^{-2}) the auto-ignition temperature is reached quickly before large amounts of heat are conducted away → decreasing ignition delay times in the order of microseconds.

Region IV: At extremely high power densities (GW cm^{-2}), the explosive is shock initiated by laser ablation of the surface → ignition delay times in the order of nanoseconds.

While Region II can be compared with the hot-wire initiation of primary explosives or pyrotechnic compositions, the laser power densities in region IV also make it possible to directly shock initiate secondary explosives by laser irradiation. The laser power densities of Region IV are achieved by solid-state lasers with laser powers of at least 100 W. In contrast, laser diodes (~ 1–10 W) only provide power densities which fall into the regions II and III. However, more powerful laser diodes have been gradually developed and therefore, laser diode initiators (LDI) have become more and more attractive for applications due to their small size and the low cost of the diodes.

All in all, laser initiation provides several advantages compared to the conventional initiation by electric devices. However, it has to be differentiated between hot-wire, exploding-bridgewire and exploding foil initiators. A comparison of the electric type and typical laser type (e.g. Nd:YAG [$Y_{2.97}Nd_{0.03}Al_5O_{12}$] solid-state laser and GaAlAs or InGaAs diode laser) initiators is given in Tab. 5.1. The advantages and disadvantages are discussed based on the literature.

Tab. 5.1: Comparison of electric and laser-type detonators.

Parameter	Hot-wire	EBW	EFI	Nd:YAG laser	Diode laser
Current [A]	5	500	3,000	10^1–10^2 [a]	10
Voltage [V]	20	500	1500	10^0–10^2 [a]	2
Power [W]	1	10^5	$3 \cdot 10^6$	$\geq 10^2$	1–10
Function Time [s]	10^{-3}	10^{-6}	10^{-7}	10^{-7}–10^{-6}	10^{-6}–10^{-3}
Initiating Charge	LA, LS, PETN	PETN	PETN, HNS	PETN, RDX, HMX, Tetryl	BNCP, CP, Pyrotechnic Mixtures

[a]Current and voltage depends on the laser pumping source used (e.g. flash lamp, arc lamp, laser diode).

Hot-wire initiators are cheap to produce and are one of the most manufactured initiator types. Hot-wires are operated with low current (~ 5 A) and voltage (~ 20 V) and therefore only a small current source is necessary. On the other hand, hot-wires are susceptible towards unintended initiation by electric impulses, e.g. electrostatic discharge, because of their low threshold. In addition, primary explosives (sensitive toward thermal and mechanical stimuli) are required for the generation of a shockwave. Depending on the application, the slow and less consistent function times in the range of milliseconds can be neglected. EBWs and EFIs exhibit considerably faster function times with very minor standard deviations (down to 25 ns). Furthermore, these two types of detonators do not consist of a primary explosive, which greatly increases the safety. However, a main disadvantage of the EBWs and EFIs is the bulky size of their power sources and the heavy cables which are required, which makes them inapplicable for many applications. Therefore, laser ignition and initiation provides some advantages, such as the high isolation of the energetic material which cannot be achieved with electric type detonators. Unintended initiation by electric impulses is nearly impossible because the induced currents are not able to produce a laser pulse exceeding the initiation energy threshold. In addition, a bridgewire which is susceptible to corrosion is not present, and therefore considerably reduces the risk of failures in initiation. Furthermore, the currents which are required for laser initiation are in the range of hot-wire detonators, but the function times can achieve the values of EBWs. In addition, the direct initiation of secondary explosives (e.g. RDX) by solid-state lasers offers new possibilities and considerably increases the safety.

Although solid-state lasers provide a considerably higher output power and can therefore be used similar to the functionality of EBWs and EFIs, the configuration is relatively large compared to diode lasers. Standard diode lasers are only several millimeters in size and can be produced with relatively low cost making them suitable for applications.

Concluding, based on these aspects, the laser initiation of explosives by laser diode radiation seems to be a very attractive initiation method of the future. However, due to the limited power of laser diodes, the explosives might be thermally ignited and undergo a DDT, in contrast to the shock initiation which results by high-power solid-state laser radiation. The operating mode of laser diode initiators can be compared to that of electric hot-wires but with faster function times as has been demonstrated in the literature.

5.4 Electric detonators

In general, detonators (or blasting caps) are used to initiate high explosives. The output of a detonator is a strong shock. There are four basic types of detonators:

– electric detonators:	EBW, EFI (also known as "slapper detonator"), hot-wire detonators
– non-electric detonators:	percussion devices, shock tube (90% HMX, 10% Al)
– instantaneous:	
– delay:	pyrotechnic, electronic

In contrast to a detonator, an igniter has a flame (= hot flash) as the output. Igniters are used for the initiation of low explosives.

Electric bridge wire detonators (EBW) and Exploding Foil Initiators (EFI) were originally developed for military applications involving nuclear weapons. They have now also found use in numerous non-military applications (e.g. explosive welding, oil exploration, mining, etc.), although they still are not as common as the more popular hot wire initiators. EBWs were invented in the 1940s for the Manhattan project in order to be able to obtain the desired simultaneity for a nuclear device. Later, EFIs were invented by LLNL in 1965 (see below).

EBWs and EFIs are both secondary explosive detonators that require a high amplitude, but short duration electrical impulse to function. The difference is that the secondary explosive in an EBW is directly in contact with the bridgewire, and it is shock-initiated by the exploding wire. In an EFI, the exploding foil accelerates a disc across a gap, and the explosive (not directly in contact with the exploding foil) is initiated by the kinetic energy of the flying disc.

Although electric detonators appear to be much more complicated than simple hot wire devices, the advantages of electric detonators are obvious:
- safety, insensitivity
- reliability
- precision
- repeatability
- simultaneity
- shot to shot reliability (under 5 microseconds)

In Fig. 5.7, a comparison between a typical EBW detonator and a common hot-wire detonator is shown. The major differences are in the bridgewires. EBWs generally use gold or platinum wires (inertness), while hot-wire devices use high resistance materials such as Cr-Ni-alloys. The secondary explosive against the bridgewire in an EBW is usually PETN, or less commonly, RDX or a thermite. Hot-wire detonators usually contain LA or LS in contact with the bridgewire. In EBWs, the explosion of the wire starts a detonation without any deflagration (no DDT), as is also the case for a hot-wire detonator

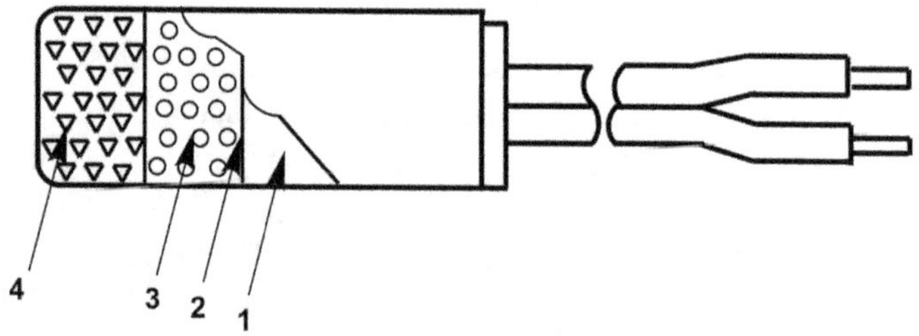

	hot-wire	EBW
1 head	plastic	plastic
2 bridgewire	hi-resist	lo-resist
3 initial pressing	lead azide	PETN
4 output pellet	PETN / RDX	PETN / RDX

Fig. 5.7: Comparison between a common hot-wire and an EBW detonator [V. Varosh, *Propellants Explosives Pyrotech* **1996**, *21*, 150–154].

Exploding foil initiators (EFI) have been used since the early 1980s as a safe and reliable method of initiating insensitive secondary explosives in the first stage of an explosive chain. Figure 5.8 illustrates the major components of an EFI. Tampers can be any rigid dielectric material such as plastic, sapphire etc., and the bridge foil can be made from any conductor (e.g. Cu or Al). The dielectric flyer is usually polyimide. The barrel is usually made of a dielectric material, but a conductor can also be used instead. The diameter of the barrel hole equals the length of the bridge foil, approx. 0.2 mm. The above mentioned four components are laminated into one subassembly and clamped against a high density explosive pellet, usually HNS. In operation, a high current explodes the narrow section of the bridge foil, which shears out a disc of a dielectric material which accelerates down the barrel and initiates the high explosive pellet (HNS) by means of kinetic energy.

In summary, both the EBW and EFI detonators, explode because an electric current is heating the conductor which tries to expand, but the conductor is heated faster than it can expand. The clear advantage of EFIs is that they are able to initiate HNS, whereas EBWs usually contain thermally less stable PETN or RDX.

Table 5.2 summarizes the most important electrical characteristics of EBW and EFI detonators in comparison with conventional hot-wire detonators.

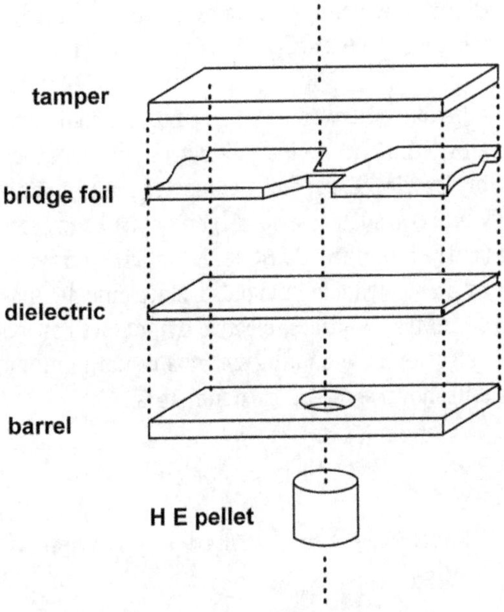

Fig. 5.8: Schematic presentation of an EFI detonator [V. Varosh, *Propellants Explosives Pyrotech* **1996**, *21*, 150–154].

Tab. 5.2: Comparison of the electric characteristics of electric detonators.

	Hot-wire	EBW	EFI
Operating current[a] / A	5	500	3,000
Voltage[a] / V	20	500	1,500
Energy / J	0.2	0.2	0.2
Power / W	1	100,000	3,000,000
Function time / µs	1,000	1	0.1

[a]Current and voltage depends on the used laser pumping source (e.g. flash lamp, arc lamp, laser diode)

Apart from use in nuclear devices, EBWs and EFIs are used in the oil industry for initiating perforating guns, because they are inherently radio frequency (RF) safe. However, they are expensive, have temperature limitations and are therefore only used in a small percentage of perforating jobs.

The biggest non-nuclear use of EBWs/EFIs is as precise initiation timing for scientific studies into detonation physics involving high-speed photography and flash X-ray. EBW's and EFI's are also safer for use in laboratory applications and require less safety management in terms of vetting allowable electrical devices/practices in the testing laboratory.

That said, EBWs and EFIs also introduce complexity into commercial systems. For nuclear applications they are highly reliable, but for commercial applications, they are much, much less reliable than traditional low-voltage detonators, because of the high degree of technical expertise and stringent electrical system requirements necessary for them to function properly.

Strictly speaking, a distinction should be made between electric and electronic detonators (Fig. 5.9). Unlike electric detonators, the wire leads of electronic detonators don't attach directly to a match head or bridgewire. Electronic detonators feature additional protection from extraneous energy sources, such as a spark gap device to protect against static discharge events, as well as current limiting resistors. Although both types of detonators are susceptible to damage by an electromagnetic pulse, electronic detonators typically have built-in protection from EMPs. Whereas, electric detonators can be used with any appropriate firing device, electronic detonating systems remain unique and must never be interchanged. Figure 5.10 shows some in-use detonators.

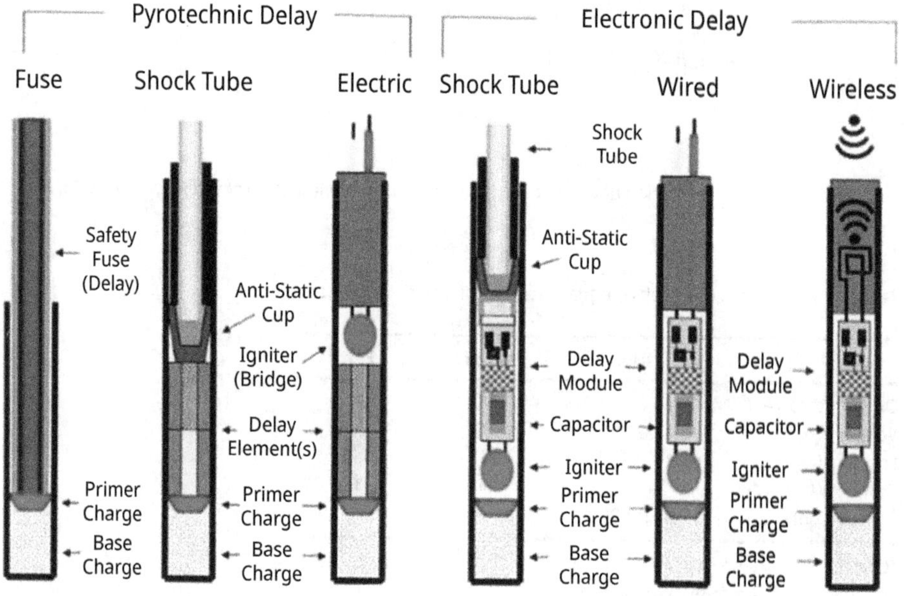

Fig. 5.9: Basic detonator construction (photo: Mine Safety and Health Administration). [https://www.safetyandhealthmagazine.com/articles/20789-electronic-vs-electric-detonators-msha-safety-alert-highlights-the-differences].

Check your knowledge: see chapter 14 for study questions.

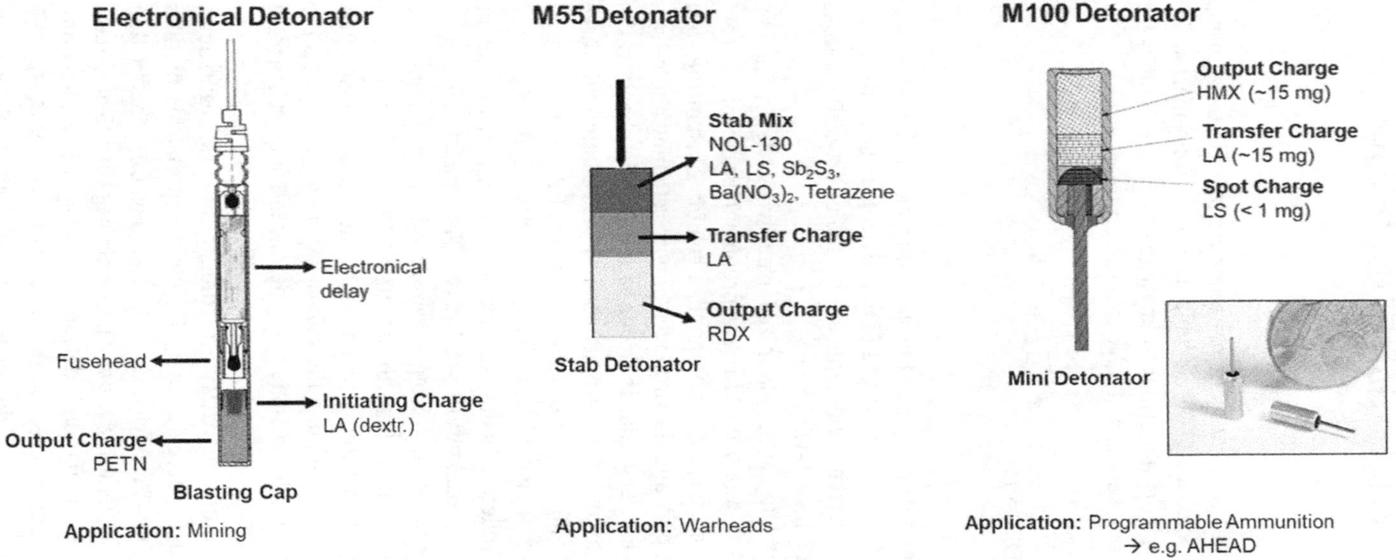

Fig. 5.10: Different types of detonators.

6 Experimental characterization of explosives

6.1 Sensitivities

As we have seen in the previous chapter, energetic materials are often initiated using thermal processes. However, the explosion stimuli could also come from mechanical or electrostatic sources. Therefore, it is important to know the exact sensitivities for explosive compounds. The important values that have to be determined are:

1. the impact sensitivity
2. the friction sensitivity
3. the electrostatic sensitivity (ESD) and
4. the thermal sensitivity

Testing the response of solid, liquid or pasty substances to impact, friction and thermal stimuli is required in various standards such as EEC, Official Journal of the European Communities as well as UN Recommendations on the Transport of Dangerous Goods, 13.4.2 Test 3(a)(ii) BAM drop hammer.

The **impact sensitivity** of solid, liquid or gelatinized explosives is determined by using the drophammer method (Tab. 6.1a, Tab. 6.1b). The drophammer essentially consists of a cast steel block with a cast base, a round anvil, a column fixed at the steel block, hardened, smoothed guide bars and the drop weight with retaining and releasing device. The heavy iron block is essential in order to adsorb the shock waves caused by the falling weight. Both guide bars are attached to the column with three brackets. An adjustable metre rule allows an exact measurement of the drop height.

In this test the sample (approx. 40 mg) to be investigated is placed in the plunger assembly, consisting of two steel rollers, a hollow steel collar and a centering ring for fixation (Fig. 6.1). The assembly is placed onto a small anvil. The impact energy (energy = work × distance = mass × acceleration × distance) can be varied by changing the drop-height (approx. 0.1–1 m) and the mass (approx. 0.1–10 kg). The minimum impact energy is determined by looking at which one had at least one out of five samples explode. It is important that while determining the impact sensitivity of crystals the size of the crystals must be given; smaller crystals are generally less impact sensitive than larger crystals of the same substance.

In order to determine the **friction sensitivity** according to the BAM regulations, the sample is placed onto a rough porcelain plate (25 × 25 × 5 mm). This plate is clamped onto the moving platform of the friction apparatus (Fig. 6.2). The friction force between the moving porcelain plate and a static porcelain peg (10 × 15 mm) (curvature radius 10 mm) causing sample initiation is determined. A set of 13 weights supplied with the instrument allow friction forces in the range of 0.5–360 N (*N.B*) to be achieved. The force applied (at end points of the lever) is proportional to the ratio of the length of the lever arm measured between the fulcrum and application point of

https://doi.org/10.1515/9783111447087-006

Tab. 6.1a: Impact sensitivity values (if more than one value is reported, the literature disagrees on the value for this particular compound).

Compound	IS (J)	Compound	IS (J)
5-ADP, 5-amino-3,4-dinitropyrazole 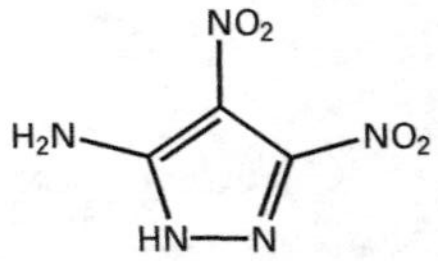	23	3,6-Bis(Tetrazol-5-yl)-1,2,4,5-tetrazine 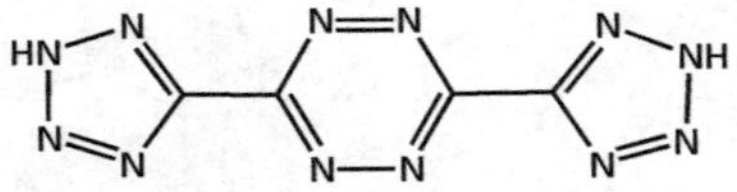	6
ADNQ, Ammonium dinitroguanidine 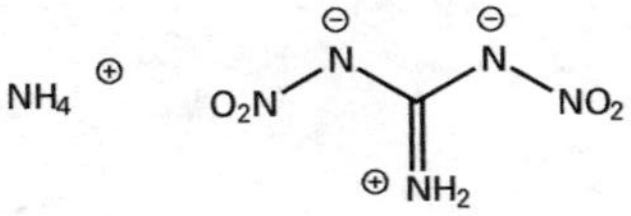	10	5,5′-Bis(1*H*-tetrazolyl)amine 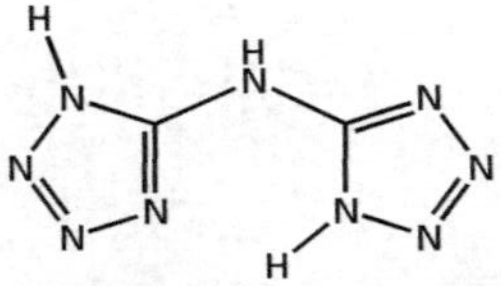	> 30
Silver nitrotetrazolate 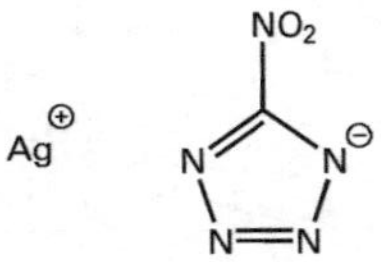	1	BNFF-1, 3,4-Bis(4-nitro-1,2,5-oxadiazol-3-yl)-1,2,5-oxadiazole 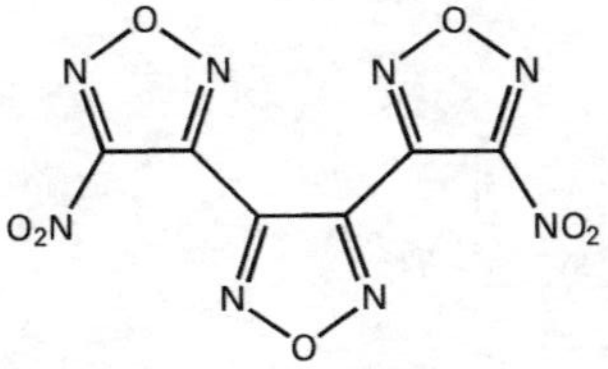	11.5

(continued)

Tab. 6.1a (continued)

Compound	IS (J)	Compound	IS (J)
1-Amino-3,6-dinitropyrazolo[4,3-*c*]pyrazole 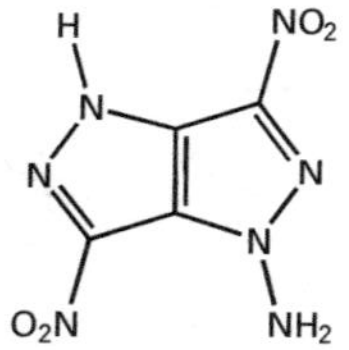	14	BTATz, 3,6-Bis(1*H*-1,2,3,4-tetrazol-5-yl-amino)-*s*-tetrazine 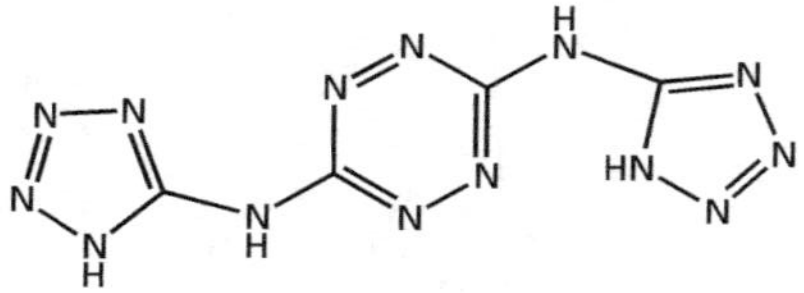	7
3-Amino-4-nitrofurazan 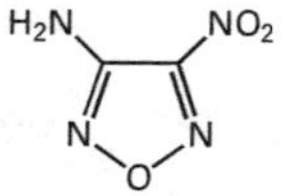	10	1,5-Diamino-1*H*-tetrazolium perchlorate 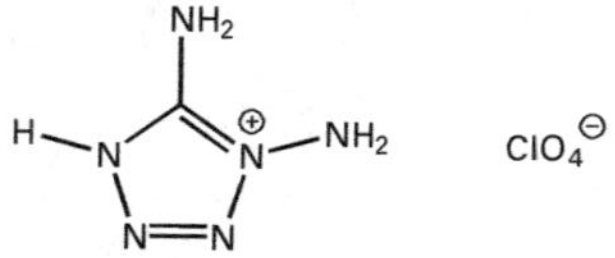	7
Ammonium nitroformate $NH_4^{\oplus}$ $C(NO_2)_3^{\ominus}$	3	1,5-Diaminotetrazolium dinitramide 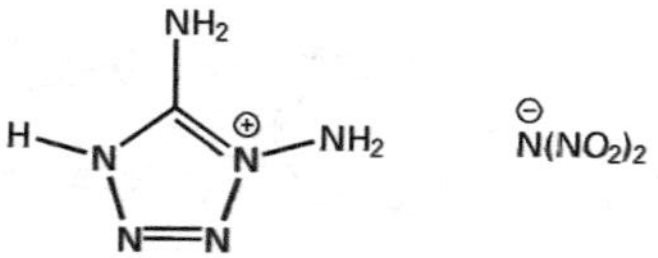	2

ANTTO, 1-Azido-8-nitrato-2,4,6-trinitro-2,4,6-triazaoctane 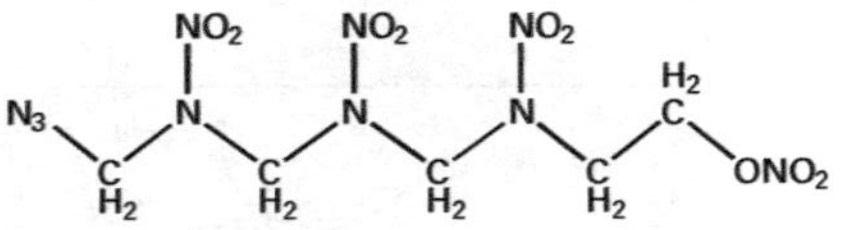	4	1,5-Diamino-1*H*-tetrazolium nitrate 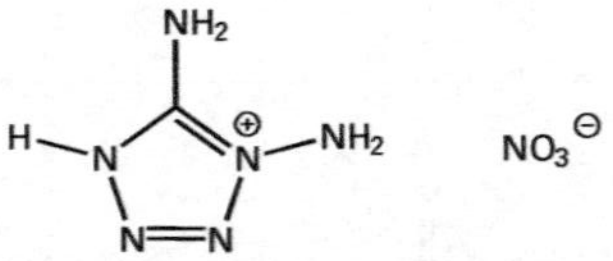	9
APX, 1,7-Diamino-1,7-dinitrimino-2,4,6-trinitro-2,4,6-triazaheptane 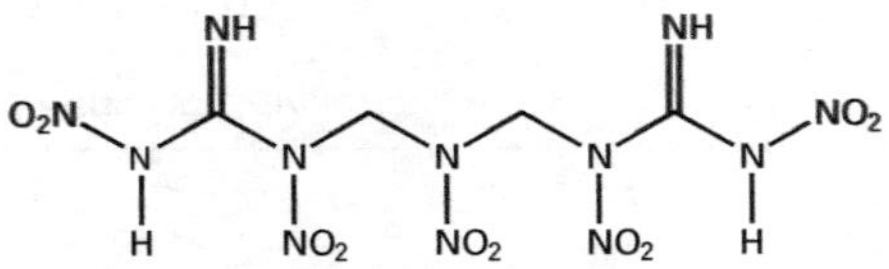	≥3	Dinitrourea 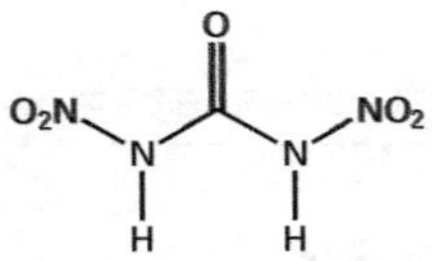	5
5-Amino-1*H*-tetrazolium perchlorate 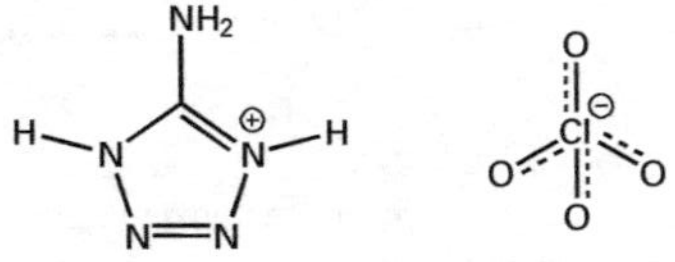	1.5	3(5),4-Dinitropyrazole 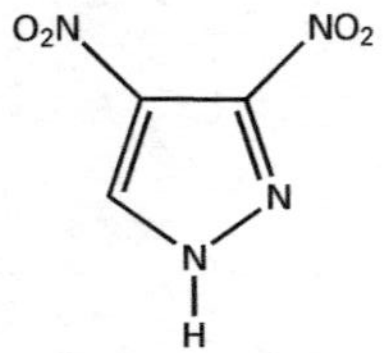	20
5-Aminotetrazolium dinitramide 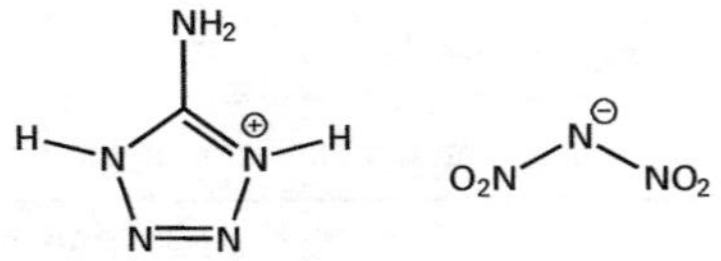	2	Potassium 5-Nitrotetrazolate 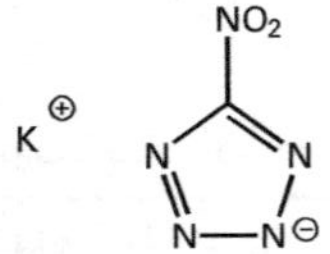	10

(continued)

Tab. 6.1a (continued)

Compound	IS (J)	Compound	IS (J)
5-Aminotetrazolium nitrate	10	Potassium tetrazolate	> 100
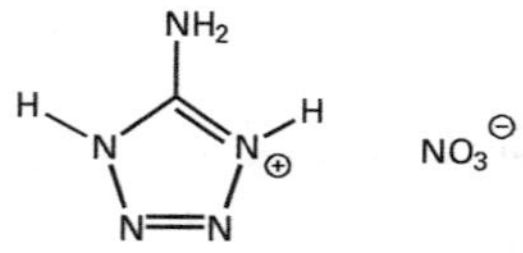		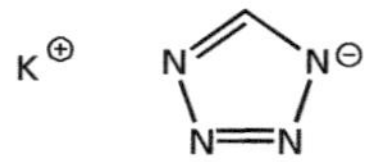	
Diaminoazobistetrazine	5	Keto-RDX	1.5
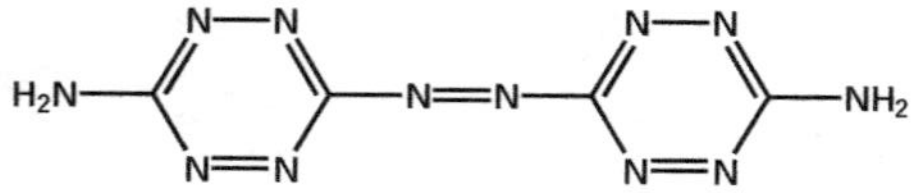		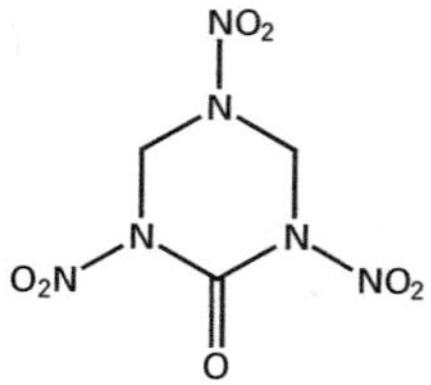	
5-Nitriminotetrazole	1.5	BTOx, 2,2,2-Bis(trinitroethyl) Oxalate	10
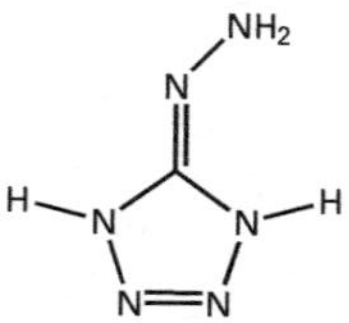		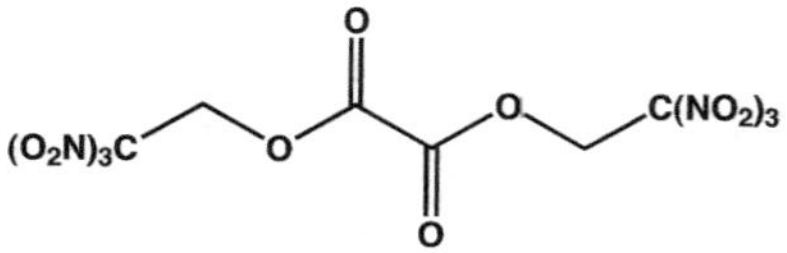	

Pentanitrobenzene 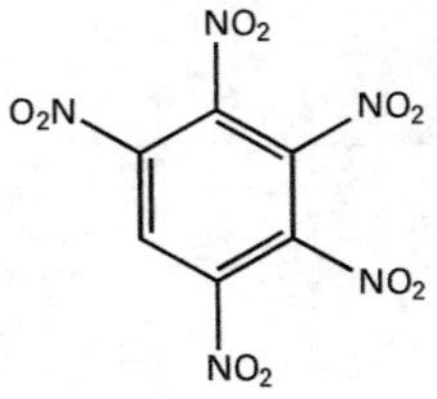	5	TKX-55, 5,5′-Bis(2,4,6-trinitrophenyl)-2,2′-bi(1,3,4-oxadiazole) 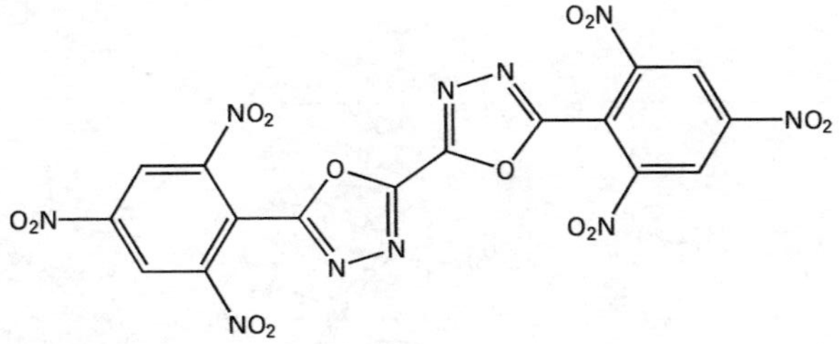	5
Sodium 5-Nitrotetrazolate dihydrate 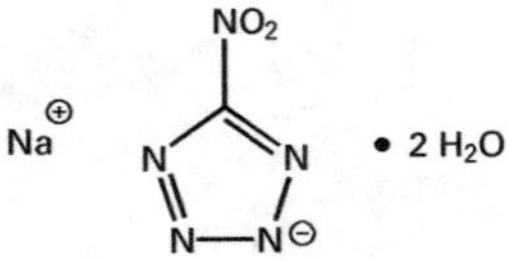	> 30	*Bis*(3,4,5-trinitropyrazolyl)methane 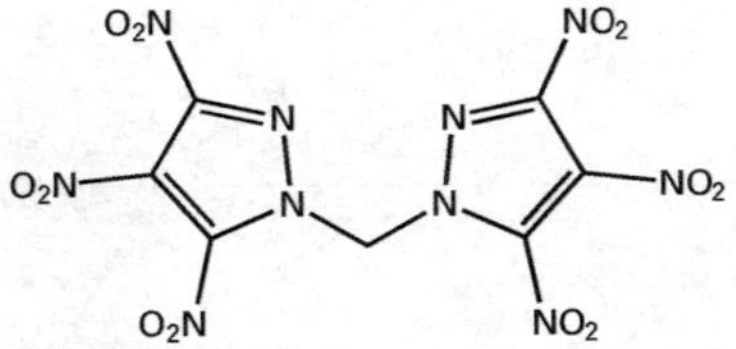	4
Triaminoguanidinium nitroformate 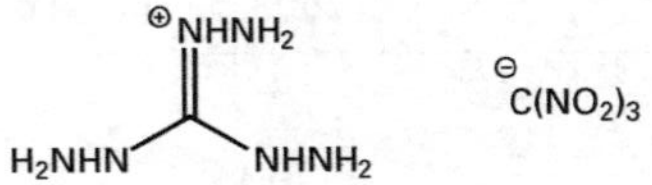	2	Butanetriol Trinitrate 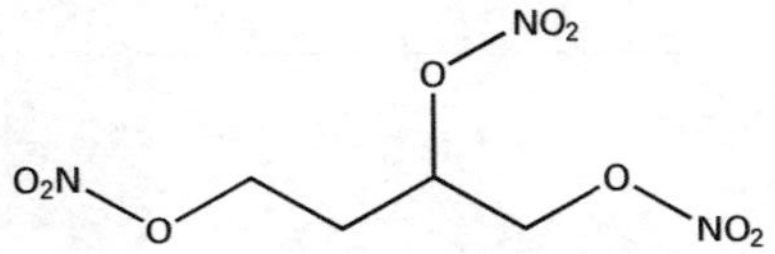	1

(continued)

Tab. 6.1a (continued)

Compound	IS (J)	Compound	IS (J)
Ammonium dinitramide 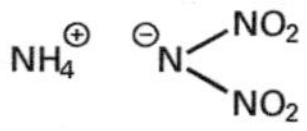	3–5	BuNENA, *N*-Butyl-*N*-(2-nitroxyethyl)nitramine 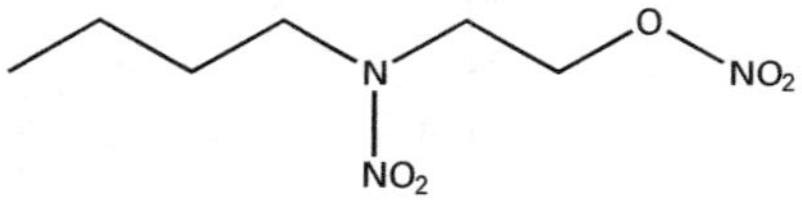	6
Ammonium nitrate	> 40	ε-CL-20 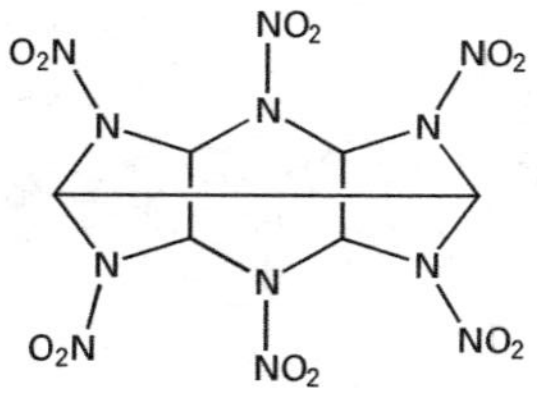	3
Ammonium perchlorate	15	DBX-1, Copper(I) 5-Nitrotetrazolate 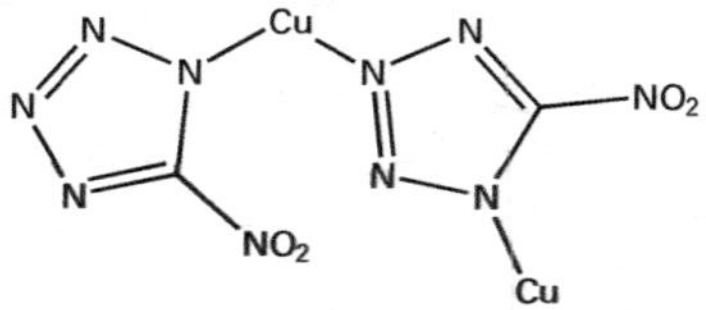	0.036

Ammnoum picrate	33	Cyanuric azide 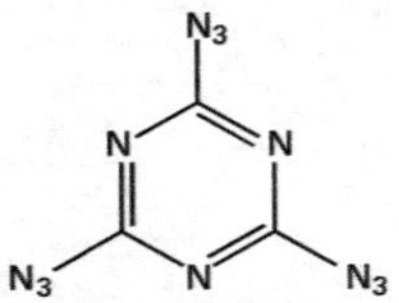	0.15
Azotriazolone 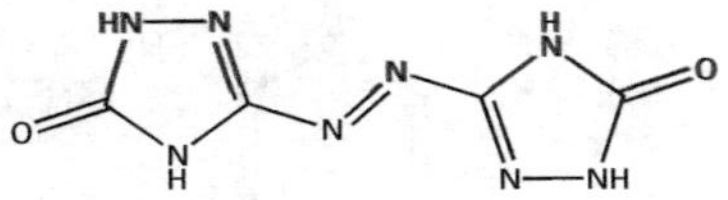	15	DADP, Diacetone diperoxide 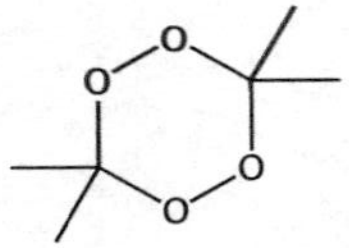	5
Benzoyl peroxide 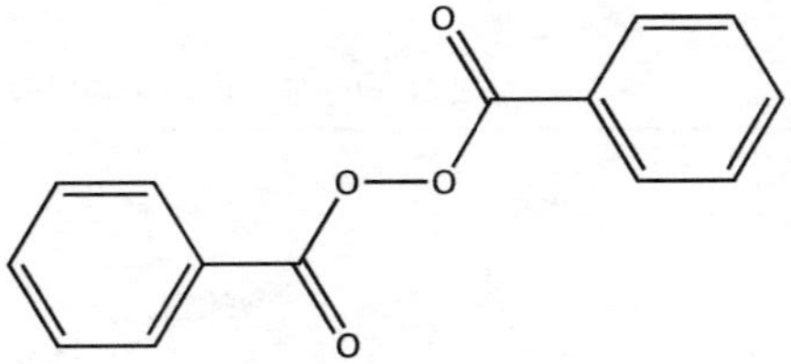	5	DAAF, 3,3′-Diamino-4,4′-azoxyfurazan 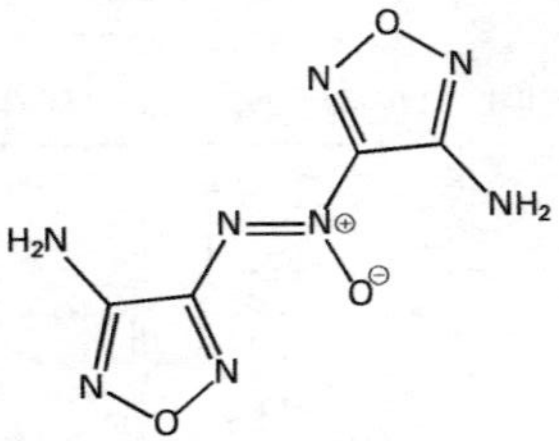	7

(continued)

Tab. 6.1a (continued)

Compound	IS (J)	Compound	IS (J)
Bis(3,5- dinitro-4-aminopyrazolyl)methane 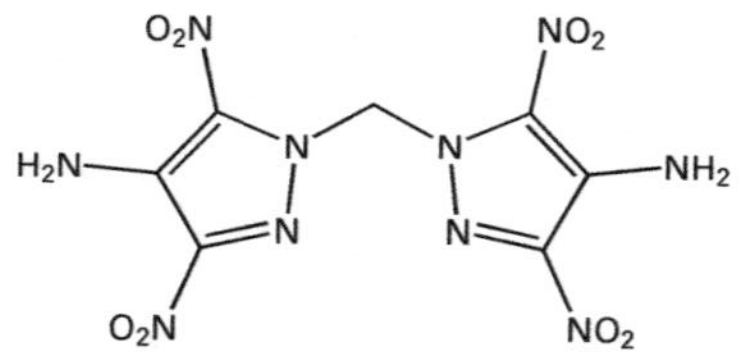	11	DDNP, 2-Diazonium-4,6-dinitrophenolate 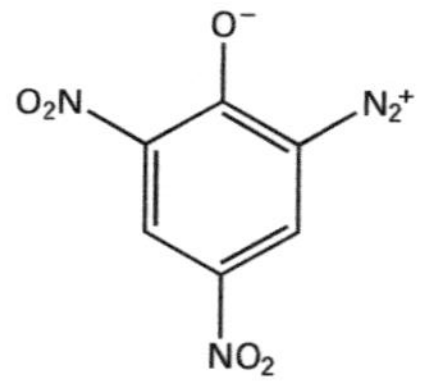	1
Bis(nitramino)triazinone 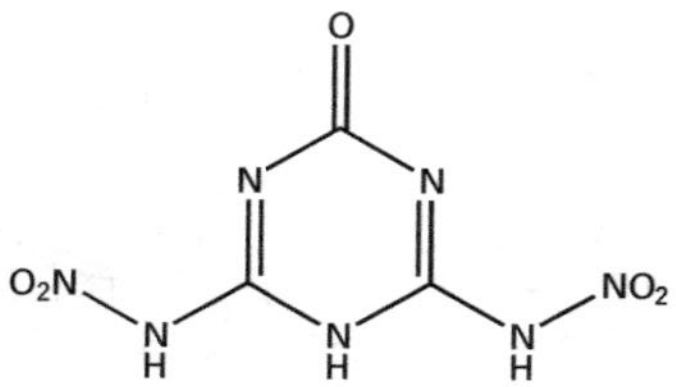	> 50.5	DEGN, Diethyleneglycol Dinitrate 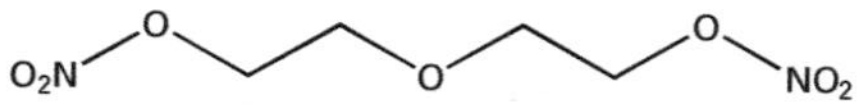	0.1

MAD-X1, Dihydroxylammonium-3,3′-dinitro-5,5′-bis(1,2,4-triazole)-1,1′-diolate 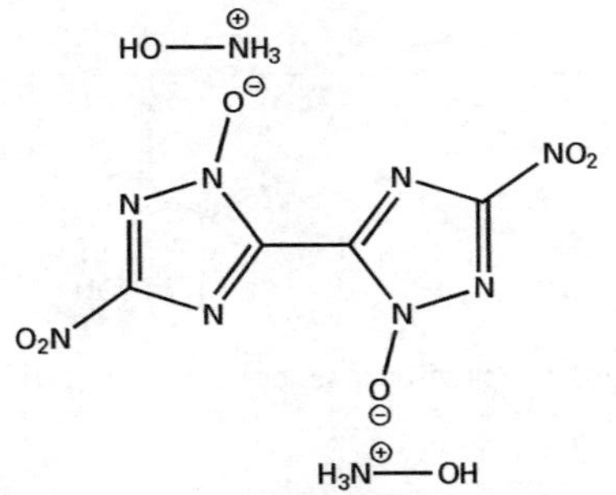	> 40	Diglycerol tetranitrate 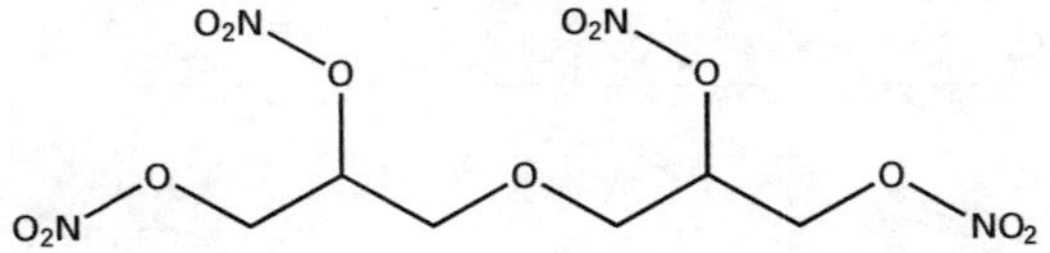	1.5
DMDNB, 2,3-Dimethyl-2,3-dinitrobutane 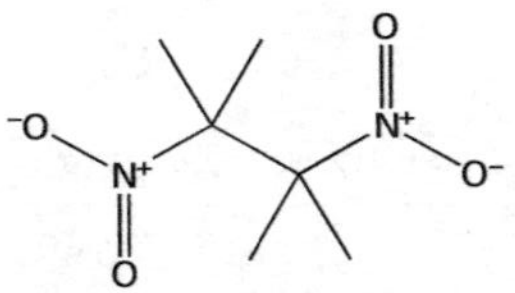	40	TKX-50, Dihydroxylammonium 5,5′-bitetrazole-1,1′-dioxide 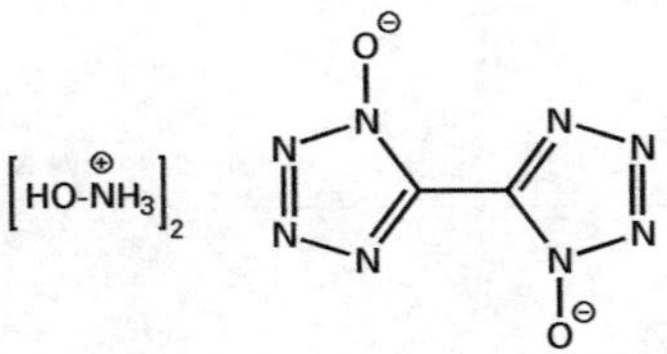	20

(continued)

Tab. 6.1a (continued)

Compound	IS (J)	Compound	IS (J)
DNAN, 2,4-Dinitroanisole 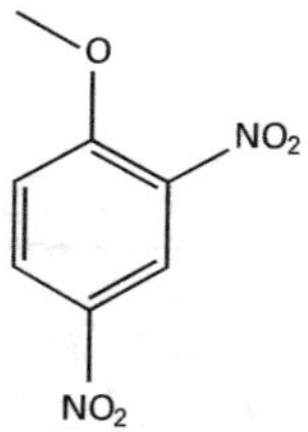	> 50	LA	2.5–4
DNDA-5, 2,4-Dinitro-2,4-diazapentane 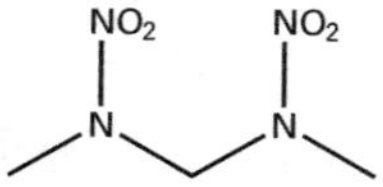	> 29.43	D-Mannitol hexanitrate 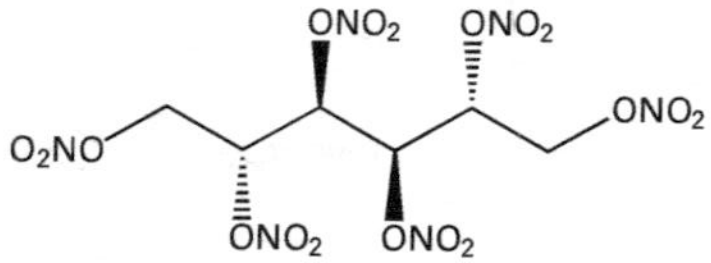	1
DNDMOA, Dinitrodimethyloxamide 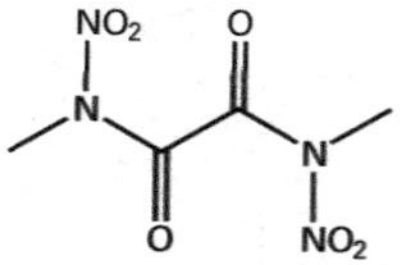	6	Nitroaminoguanidine 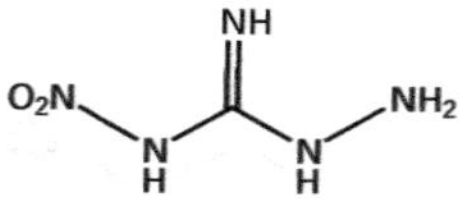	20

DINGU, 1,4-Dinitroglycolurile 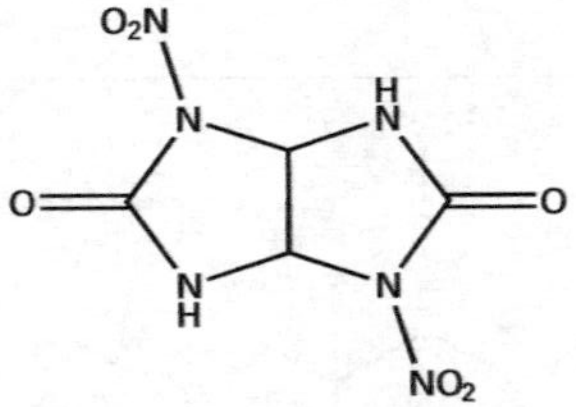	5.5	NG, Nitroglycerine 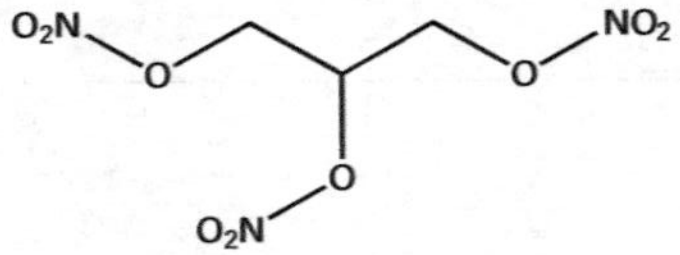	0.2
Dinitroorthocresol 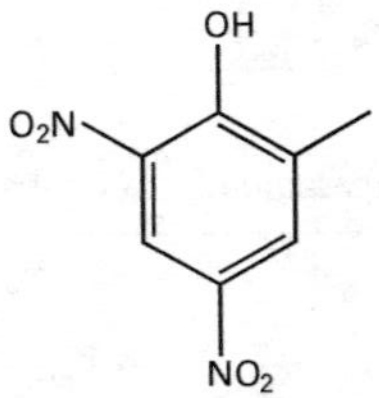	> 50	NQ, Nitroguanidine 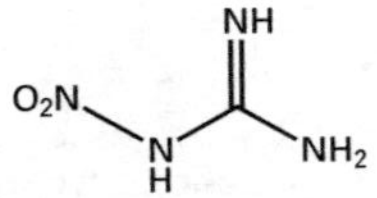	> 50
TEX, 4,10-Dinitro-2,6,8,12-tetraoxa-4,10-diazaisowurtzitane 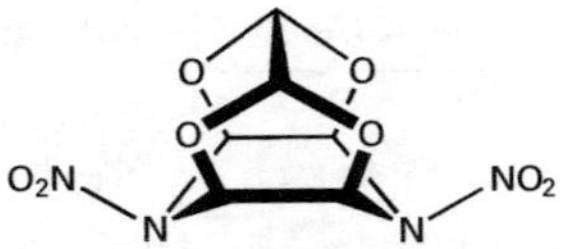	32	Nitromethane	40

(continued)

Tab. 6.1a (continued)

Compound	IS (J)	Compound	IS (J)
2,6-DNT, 2,6-Dinitrotoluene	> 40	2-Nitrotoluene	> 40
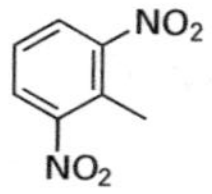		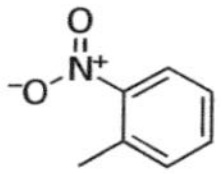	
DINA, Dioxyethylnitramine Dinitrate	6	NTO, 3-Nitro-1,2,4-triazole-5-one	
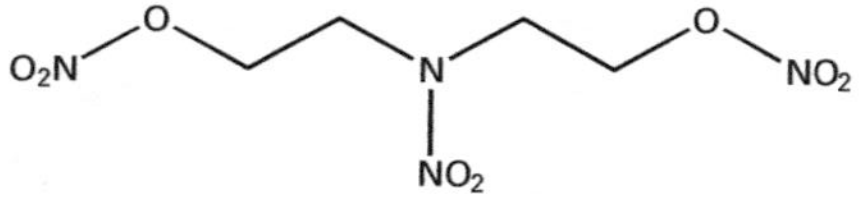		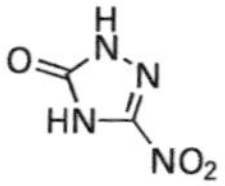	
DIPEHN, Dipentaerythritol Hexanitrate	4	HMX, Octogen	6.35
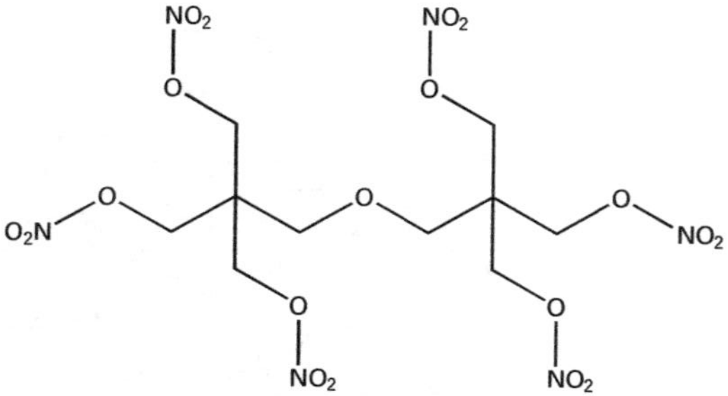		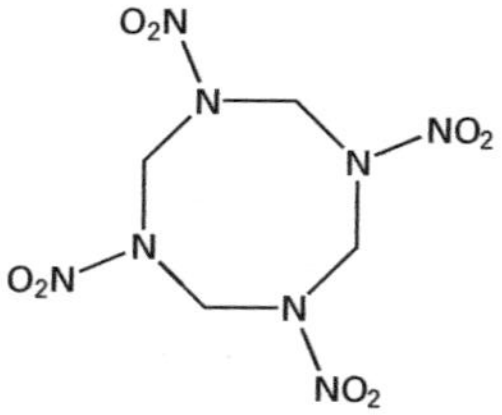	

ETN, Erythritol Tetranitrate 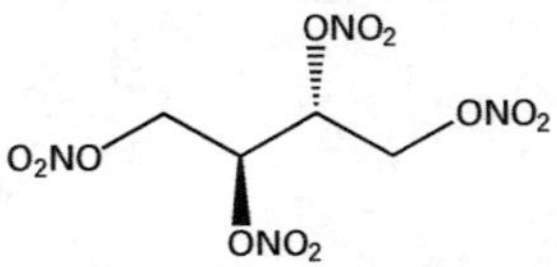	3	PETN, Pentaerythritol Tetranitrate 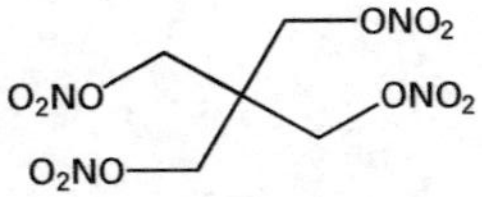	2.90
EDD, Ethylenediamine Dinitrate 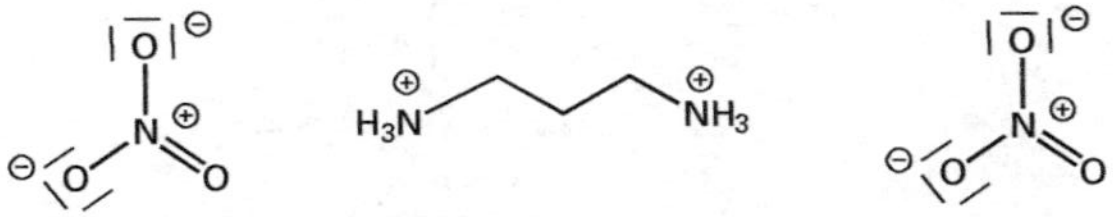	10	Picramic acid 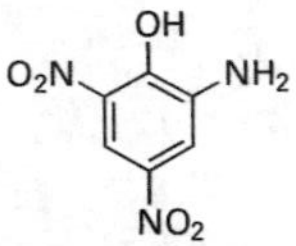	34
EDNA, Ethylene dinitramine 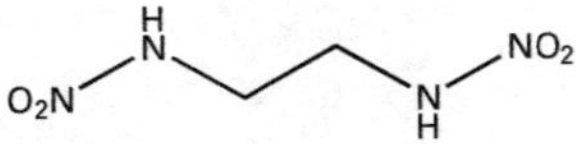	8	Picric acid 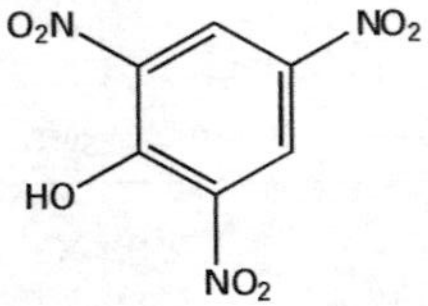	7.4

(continued)

Tab. 6.1a (continued)

Compound	IS (J)	Compound	IS (J)
Ethylene glycol dinitrate 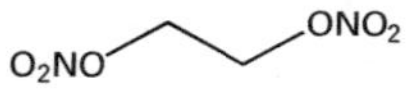	0.2	K2DNABT, Potassium 1,1′-Dinitramino-5,5′-bistetrazolate 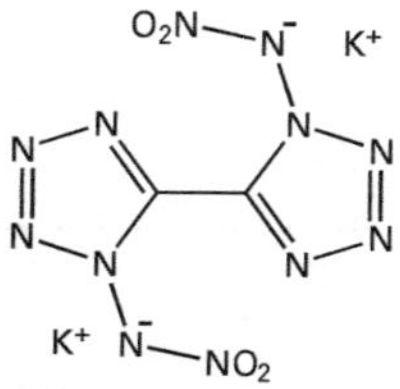	1
Ethyltetryl 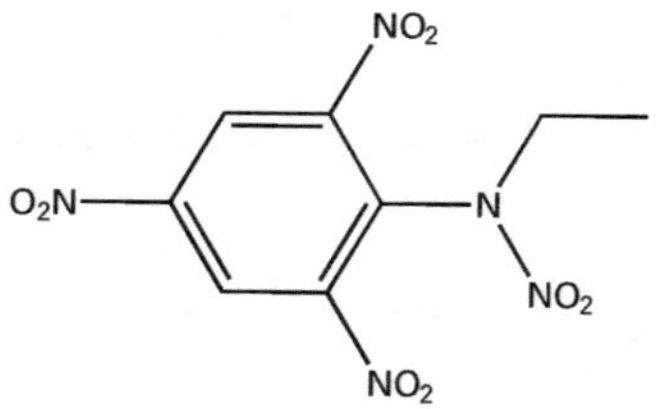	5	KDNP, Potassium 5,7-dinitro-[2,1,3]-benzoxadiazol-4-olate 3-oxide 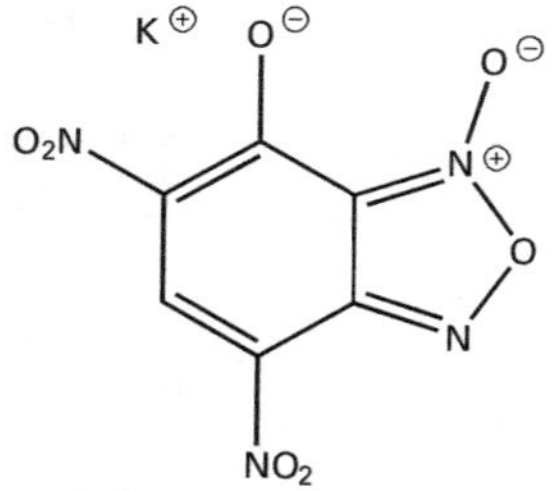	10

FOX-7, 1,1-diamino-2,2-dinitroethene 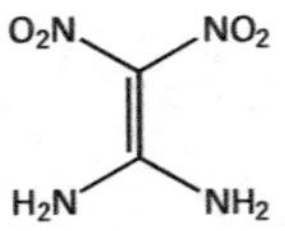	15 – 40	PYX, 3,5-dinitro2,6-bispicrylmino pyridine 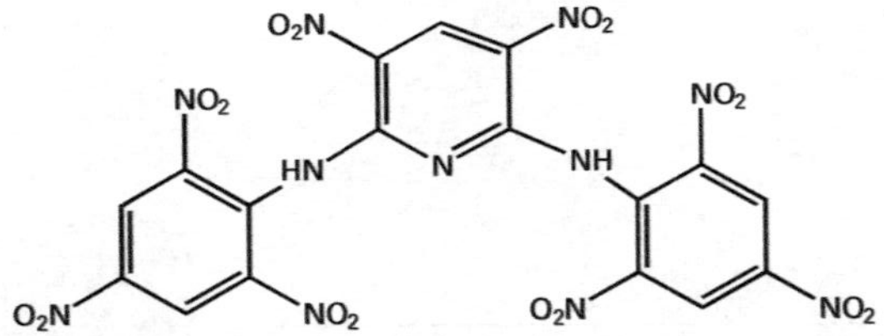	8
FOX-12, Guanylurea Dinitramide 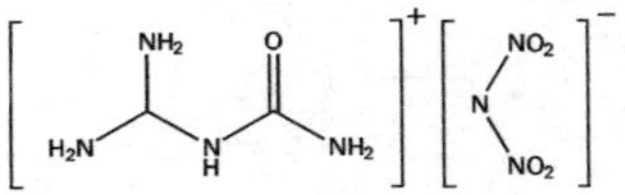	31	Silver azide	1.18
Glycerol 1,3-Dinitrate 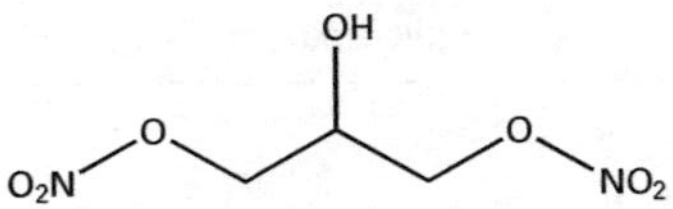	1.5	Silver fulminate	~0.8
Hexamethylenetetramine Dinitrate 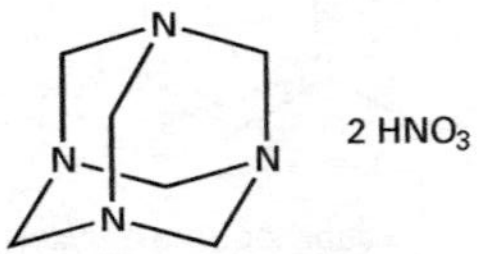	15	Styphnic acid 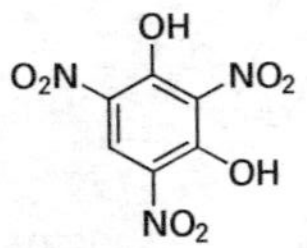	7.4

(continued)

Tab. 6.1a (continued)

Compound	IS (J)	Compound	IS (J)
Hexanitroazobenzene 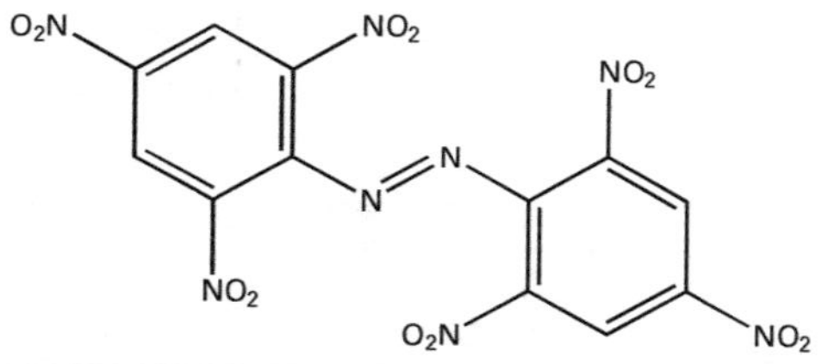	8.57	Tacot 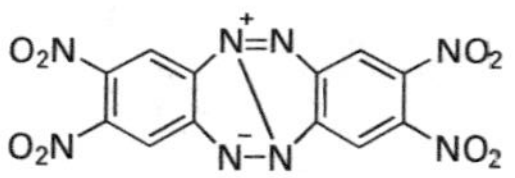	69
2,4,6,2′,4′,6′-Hexanitrobiphenyl 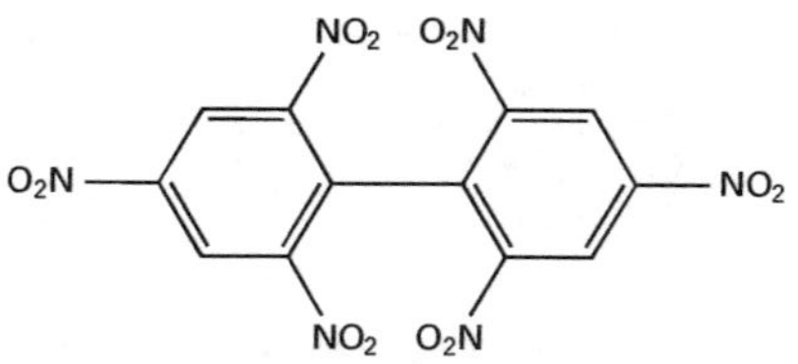	2 – 20	TATP, Triacetonetriperoxide 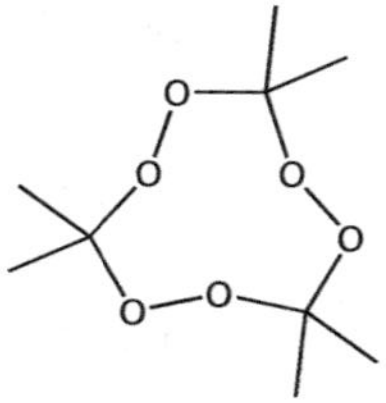	0.3

HDNP, 2,4,6,2′,4′,6′-Hexanitrodiphenylamine 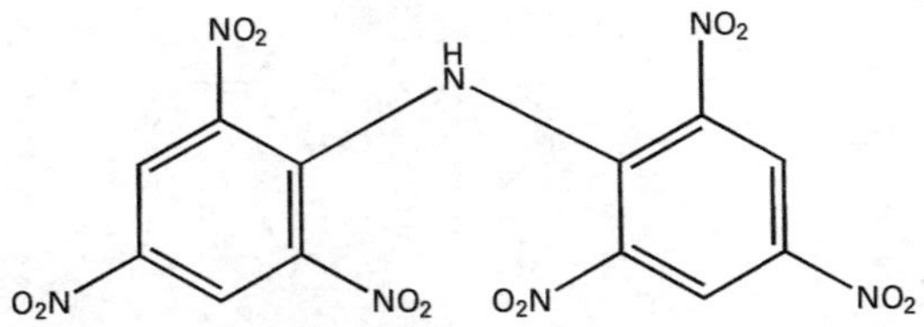	7.5	2,3,4,6-Tetranitroaniline 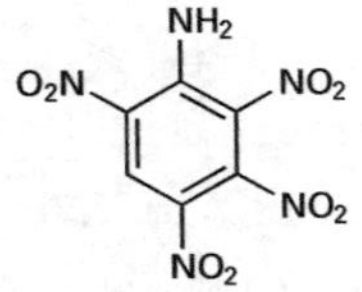	6
2,4,6,2′,4′,6′-Hexanitrodiphenylsulfide 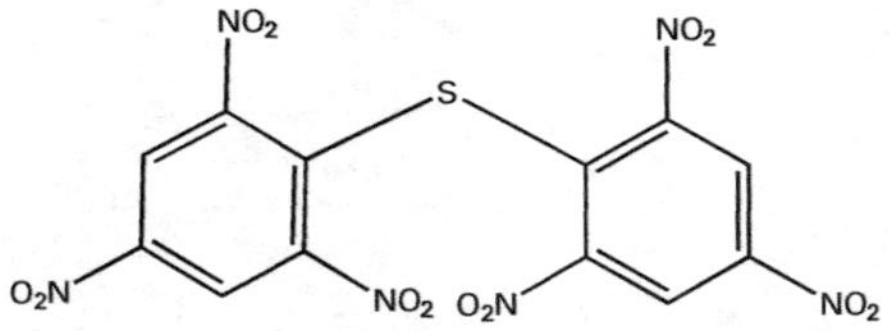	6	Sorguyl 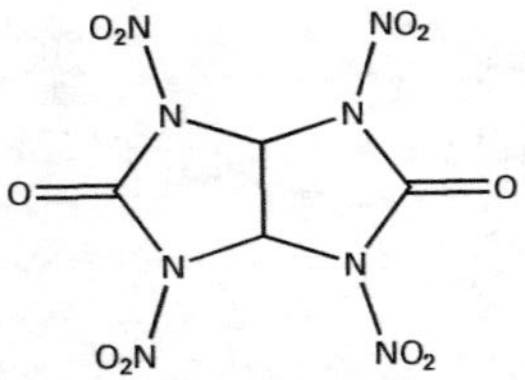	2.04

(continued)

Tab. 6.1a (continued)

Compound	IS (J)	Compound	IS (J)
2,4,6,2′,4′,6′-Hexanitrodiphenylsulfone 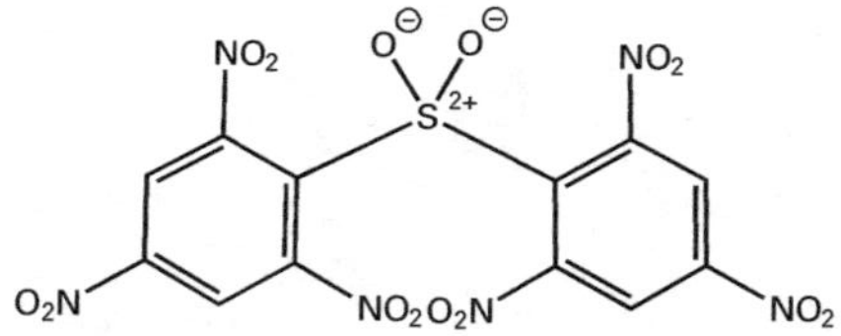	3.86	Tetrazene 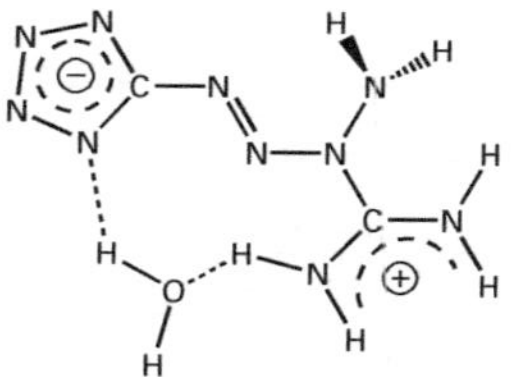 X-ray structure 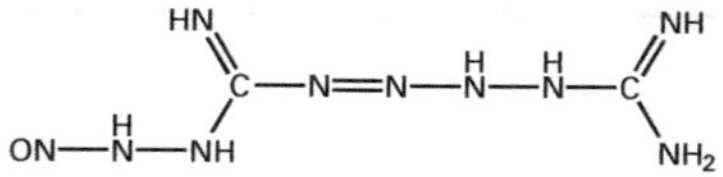 historical formula I 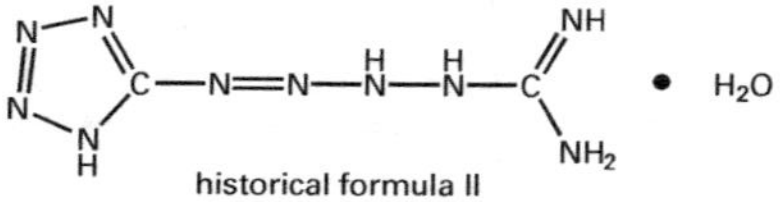 historical formula II	1
HNO, Hexanitrooxanilide 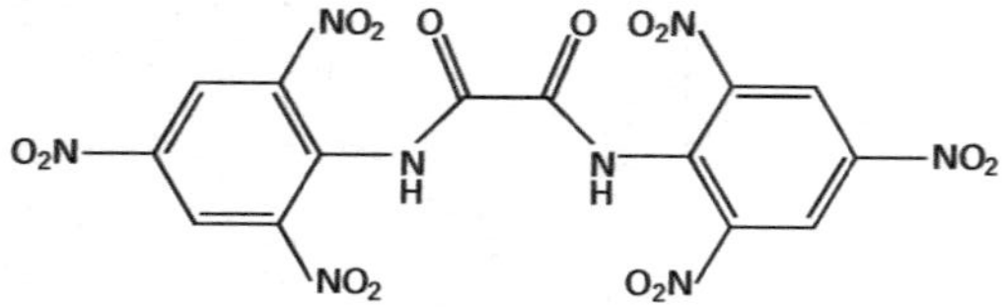	7.50–14.2	Tetryl 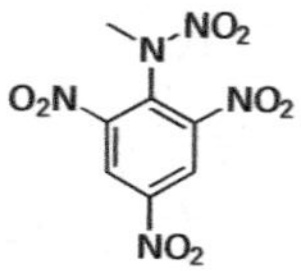	3

HNS, Hexanitrostilbene 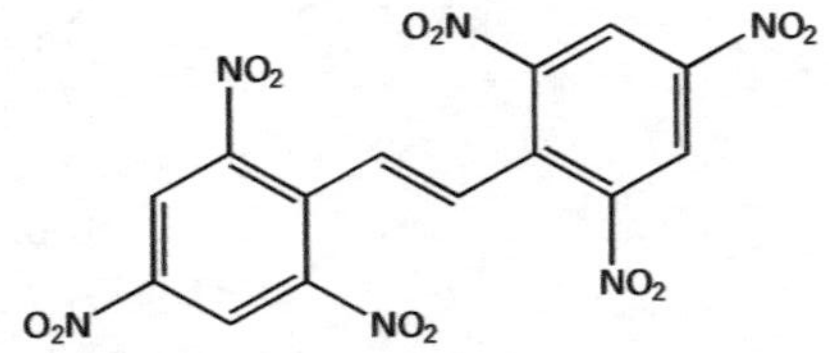	5	TAGN, triaminoguanidinium nitrate 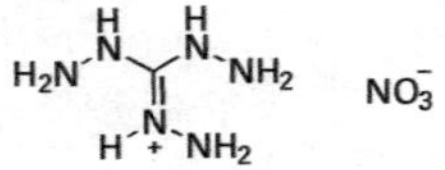	4
RDX, Hexogen 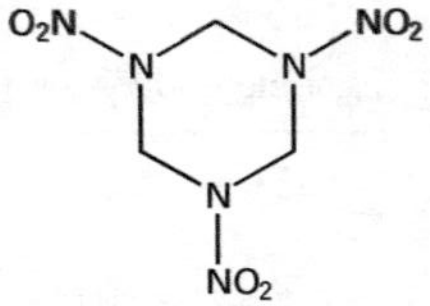	7.5	TATB, 1,3,5-Triamino-2,4,6-trinitrobenzene 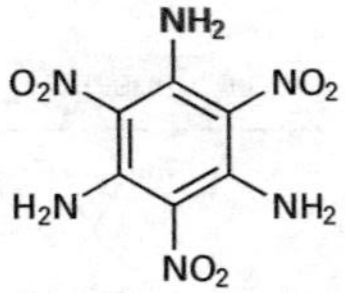	50
HMTD 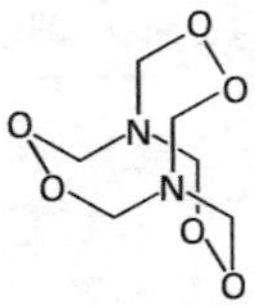	2	TATNB, 1,3,5-Triazido-2,4,6-trinitrobenzene 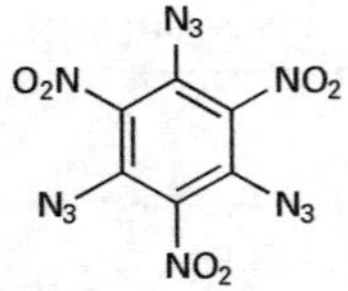	< 4.9
Hydrazinium Nitrate $N_2H_5^+ \quad NO_3^-$	7.4	Trinitroaniline 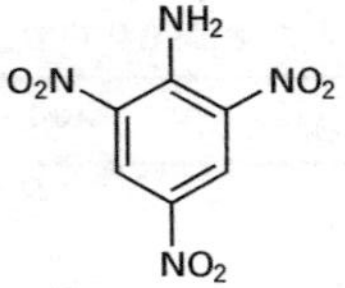	15

(continued)

Tab. 6.1a (continued)

Compound	IS (J)	Compound	IS (J)
Hydrazinium Nitroformate $N_2H_5^+\ C(NO_2)_3^-$	4	Trinitroanisole 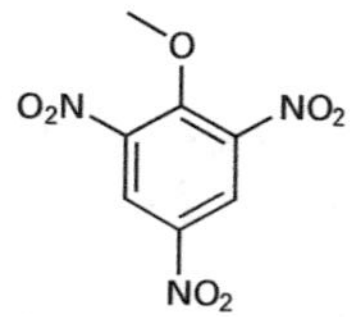	20
Hydrazinium Perchlorate $N_2H_5^+\ ClO_4^-$	2	TNAZ, Trinitroazetidine 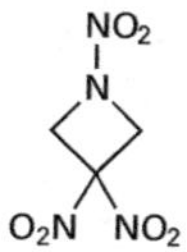	6
Isosorbitol Dinitrate 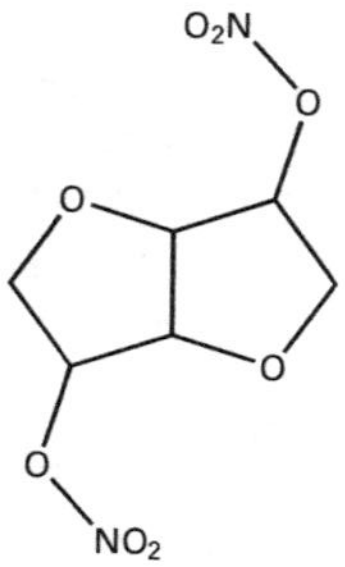	15	Trinitrobenzoic acid 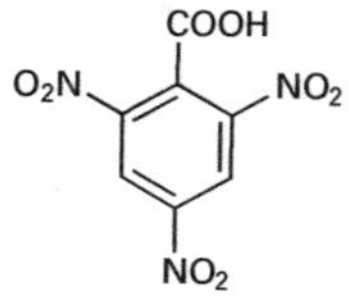	10

		2,4,6-Trinitrocresol 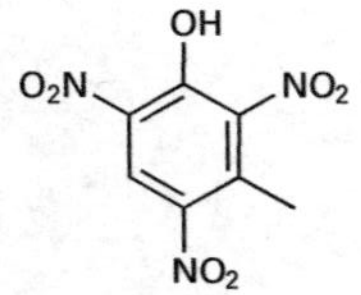	12
Pentryl 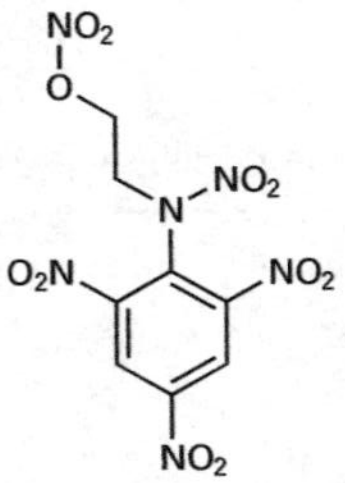	4	TNX, trinitroxylene 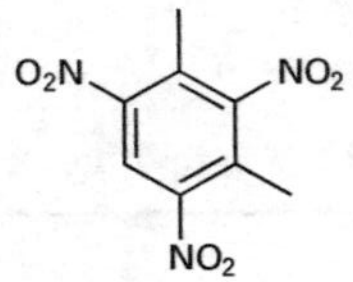	10
Trinitropyridine 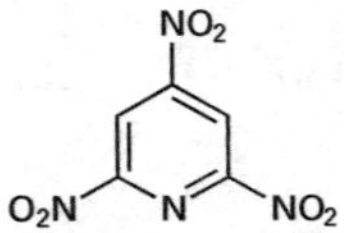	4.5–6.5	Uronium Nitrate 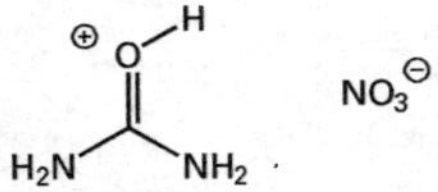	> 40

(continued)

Tab. 6.1a (continued)

Compound	IS (J)	Compound	IS (J)
TNPyOx, Trinitropyridine-*N*-oxide 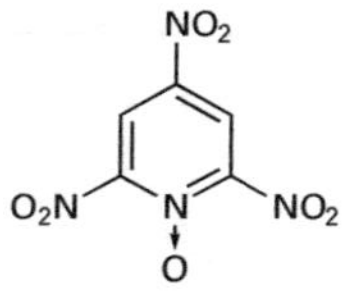	1.5–3.0		
TNT 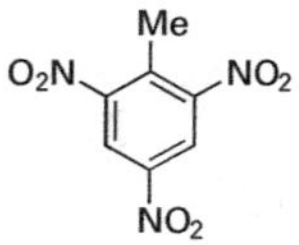	36.6 (15 J corresponds to the value for a fast-refrigerated TNT melt)		

(Values taken from: T. M. Klapötke, Energetic Materials Encyclopedia, de Gruyter, Berlin, 2nd edn., **2021**).

(a)

(b)

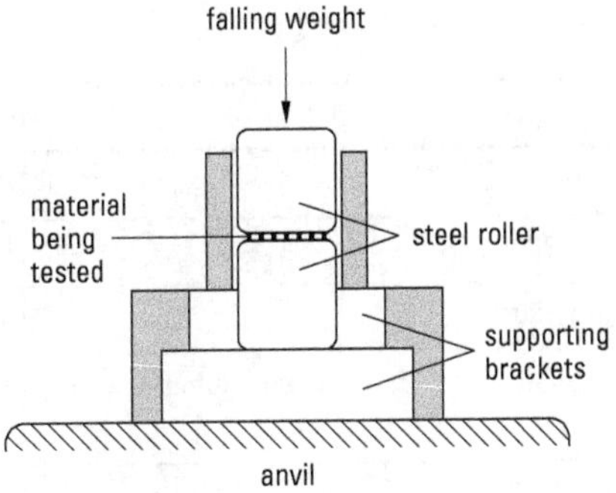

Fig. 6.1: Drophammer in accordance with the BAM (a) and detailed sketch of the drop-hammer test with the two steel rollers, a hollow steel collar and a centring ring for fixation (b).

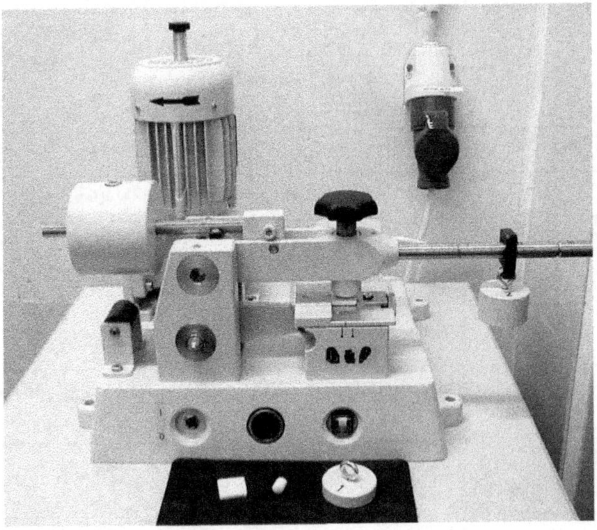

Fig. 6.2: Friction tester according to BAM regulations.

the force applied at each end of the lever. This test method is applicable both for sensitive primary explosives and less sensitive high explosives. The porcelain plate moves back and forth under the porcelain peg. The minimum force of friction is determined. During this process at least one sample out of six ignites, turns black, makes a cracking sound or explodes. As discussed above for the impact sensitivity, before determining the friction sensitivity, the sample should be sieved in order to perform the test on a sample with a uniform and defined crystal size.

Electrostatic discharge is one of the most frequent and the least characterized cause of accidental explosions of energetic materials. It is therefore imperative in

Tab. 6.1b: Friction sensitivity values (if more than one value is reported, the literature disagrees on the value for this particular compound).

compound	FS (N)	compound	FS (N)
ADN	64 → > 350	EDD	> 353
AN	> 360	EDNA	47.4
AP	> 100 → > 363	EGDN	> 360
Ammonium picrate	unaffected in friction pendulum test	Ethyltetryl	> 353
Benzoyl peroxide	120, 240	FOX-7	216, > 350
DNAM	216, > 360	FOX-12	> 352
BTOx	> 360	GAP	> 352
TKX-55	> 360	Guanidinium nitrate	> 353
BTNPM	144	Hexamethylenetetramine dinitrate	240
CL-20	28–96	HNDP	> 353
DBX-1	0.1	Hexanitroethane	240
Cyanuric azide	0.1, < 0.5	HNS	192, 240, > 360
DADP	5, 1.75, 2.99, 49 (crude)	RDX	120, 13.73, 250, > 360
DAAF	> 360	HMTD	< 5, 0.63, 0.1, 0.01
LLM-105/ ANPZ-O	360	Hydrazinium nitroformate	16 → 36, 12 – → 14, 25
DDNP5 – 22		Hydrazinium perchlorate	> 10
TKX-50	120, 50, 40	Isosorbitol dinitrate	> 160
MAD-X1	> 360	LA	0.1 → 1.0, 10
DNAM	170, > 360	LS	< 1, 0.1, 1.45
BTNEO	> 360	Basic LS	< 1
DINGU	20 → 300	Mannitol hexanitrate	30
TEX	161.3, > 360	MF	5.3, 6.48
DNT	> 360	Nitroaminoguanidine	144, 240

Tab. 6.1b (continued)

compound	FS (N)	compound	FS (N)
DIPEHN	less sensitive than PETN	NC	353
ETN	~50	NG	360, > 353, 112
NQ	> 360	Picryl chloride	353
NM	> 360	Trinitrocresol	353
NTO	> 353, > 360	Trinitropyridine	> 353
Octogen/HMX	120 → 240	TNT	353, > 360
PETN	80, 73, 60	Urea nitrate	353, > 360
Picramic acid	> 353	$Ni(N_2H_4)_3(NO_3)_2$	same as silver azide
PA	> 363, explodes if exposed to friction between plates of iron, steel or stone	Ammonium nitroformate	96
AMMO/ polyAMMO	40 (monomer)	Aminotetrazolium perchlorate	8
PVN	196	Aminotetrazolium dinitramide	20
KDN	> 360	Aminotetrazolium nitrate	> 324 → > 360
KDNBF	same as LS, 38	H_2BTA	360
KDNP	~10	Diaminotetrazolium perchlorate	60
PYX	280, 360	Diaminotetrazolium dinitramide	5
TNENC	96	Diaminotetrazolium nitrate	192
Styphnic acid	> 353	Dinitrourea	76
TATP	< 5, 1.6, 0.2 (crude)		
Sorguyl	54		
Tetrazene	8.57		
Tetryl	360, 353		
$TAG^+NO_3^-$	120, > 120, 216		
TATB	353, 360		
TNEF	96		

Tab. 6.1b (continued)

compound	FS (N)	compound	FS (N)
Trinitroaniline	353		
Trinitroanisole	353		
TNAZ	110, 160, 324		
Trinitrobenzene	353		
Trinitrobenzoic acid	353		

(Values taken from: T. M. Klapötke, Energetic Materials Encyclopedia, de Gruyter, Berlin, 2nd edn., **2021**).

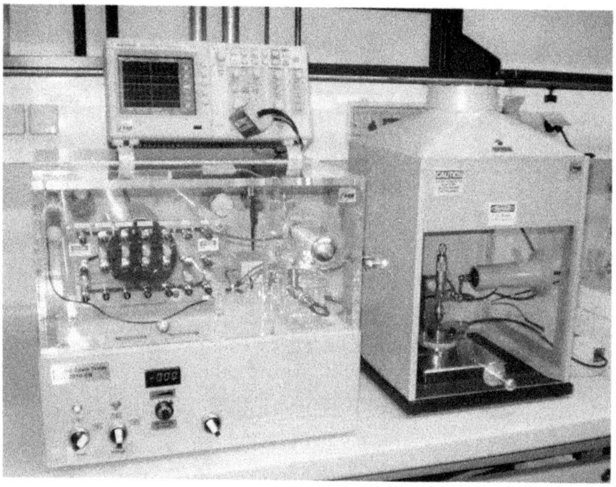

Fig. 6.3: Electrostatic test equipment.

R&D, manufacture, processing, loading or demilitarization to have reliable data on electrostatic spark sensitivities of energetic materials. The **electrostatic sensitivity** (electrostatic discharge sensitivity, ESD) is determined by using an ESD test apparatus. Different spark energies (usually between 0.001 and 20 J (Fig. 6.3)) can be applied, by using variable capacitive resistances C (in Farad, F) and loading voltages V (in Volt, V):

$$E = \frac{1}{2} C U^2$$

This is a particularly important test for the safe handling of explosives since the human body can be electrically charged (depending on the type of clothing worn and humidity etc.) which on discharge can cause spark formation. Typical values for the human body are (Fig. 6.3a):

C = 0.0001–0.0004 μF
U = 10,000 V
E = 0.005–0.02 J.

For capacitive discharge calculations with C[F] and Q[C], V[V] and E[J], the following can be written:

$$C = \frac{Q}{V} \text{ and } E = \frac{1}{2}CV^2 = \frac{1}{2}QV = \frac{Q^2}{2C}$$

As is shown in Fig. 6.4, the ESD values are greatly influenced by the particle size; in these tests the sample must be carefully sieved prior to measuring. The finer the powder of a particular sample is, the higher the ESD sensitivity values are. Usually, compounds with ESD < 0.1 J are classified as sensitive, those with ESD > 0.1 J as insensitive.

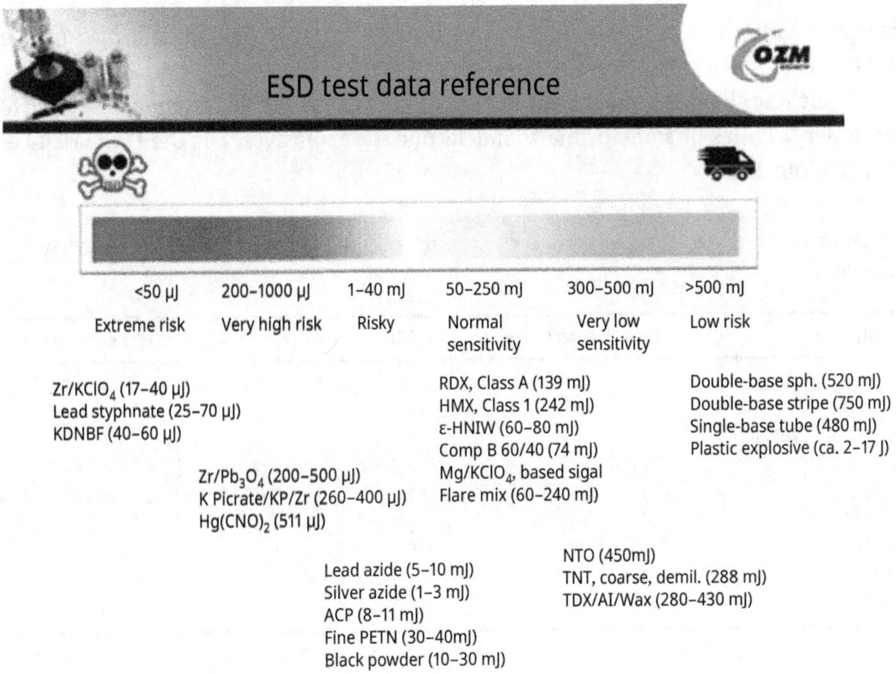

Fig. 6.3a: ESD test data reference (according to OZM).

In a similar way as the ESD sensitivity depends on the grain size, the impact sensitivity often also depends on the crystal size. One explanation for measuring a lower sensitivity of larger crystals is as follows: small crystals have a very short DDT length, while larger crystals have a much longer DDT length. The thickness of the test sample

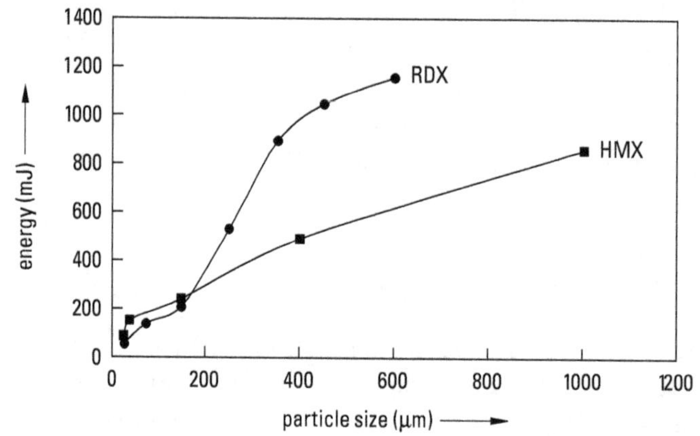

Fig. 6.4: Dependence of the ESD sensitivity on the particle size for hexogen (RDX) and octogen (HMX).

is very small and is probably in the same order as the DDT length, and may not be long enough for the larger crystal sizes.

Table 6.1c shows a summary of typical values for the impact, friction and electrostatic sensitivities of some primary and secondary explosives. For the UN classification see Tab. 6.2.

Tab. 6.1c: Values for the impact, friction and electrostatic sensitivities of some primary and secondary explosives.

explosive	impact sensitivity / J	friction sensitivity / N	ESD sensitivity / J
primary explosive			
$Pb(N_3)_2$	2.5–4	0.1–1	0.005
secondary explosive			
TNT	15	> 353	0.46 – 0.57
RDX	7.5	120	0.15 – 0.20
β-HMX	7.4	120	0.21 – 0.23
PETN	3	60	0.19
NQ	> 49	> 353	0.60
TATB	50	> 353	2.5 – 4.24

Table 6.1d presents the ESD sensitivities of common primary explosives.

Tab. 6.1d: ESD sensitivities of metal-containing energetic materials.

Compound	Structure	ESD	Lit.
DBX-1, copper(I) 5-nitrotetrazolate	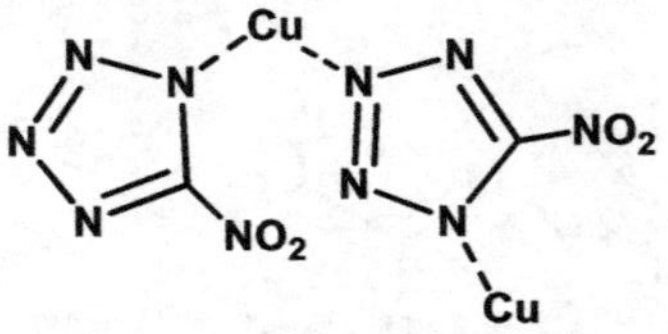	3 mJ	1
copper(II) azide	$Cu(N_3)_2$	0.1 – 0.2 mJ	1
cuprous azide, copper(I) azide	$Cu_2(N_3)_2$	0.1 – 0.2 μJ	1
silver azide	AgN_3	ca. 0.1 mJ	2
silver fulminate	AgCNO	0.04 – 0.3 μ J	1
silver nitrotetrazolate	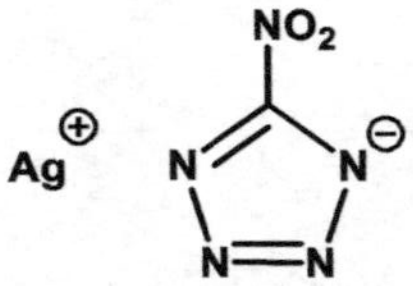	12 μ J - < 50 mJ	1
lead azide	$Pb(N_3)_2$	ca. 5 mJ	1

(continued)

Tab. 6.1d (continued)

Compound	Structure	ESD	Lit.
lead 2,4-dinitroresorcinol	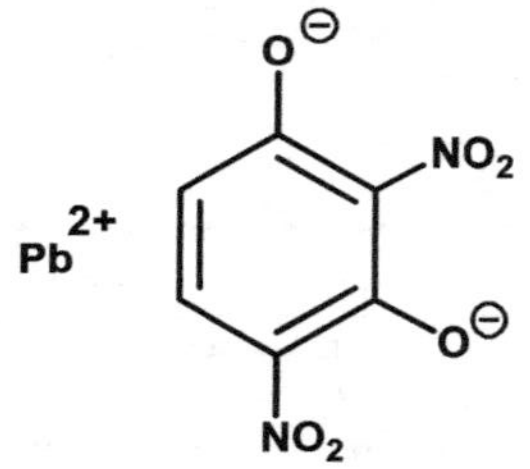		
lead picrate	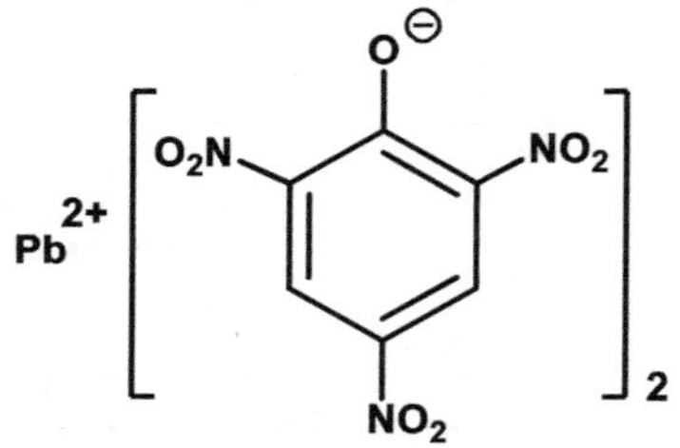		
lead styphnate	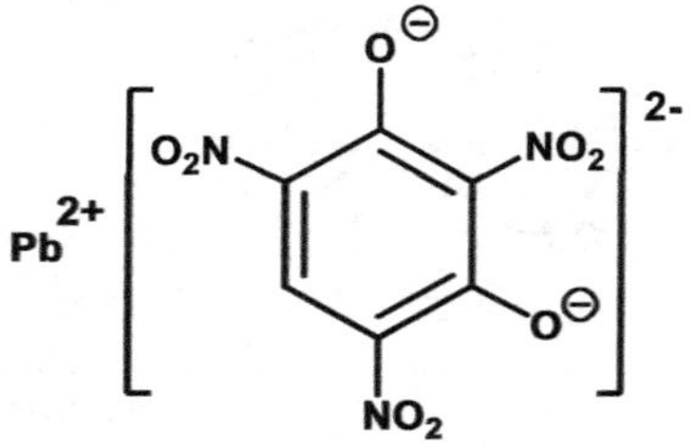	ca. 0.1 mJ	1

mercury fulminate	$Hg(CNO)_2$	ca. 0.5 mJ	2
mercury nitrotetrazolate	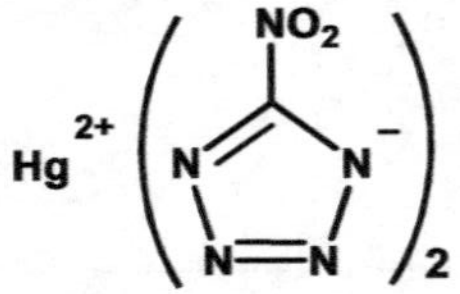	0.27 mJ	1
KDNBF	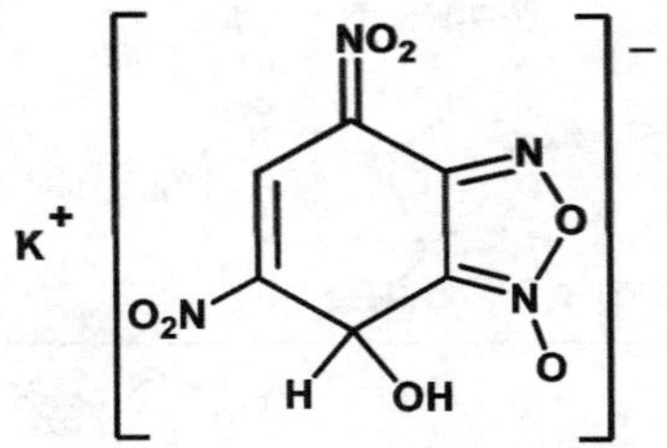	20 µJ	3
KDNP	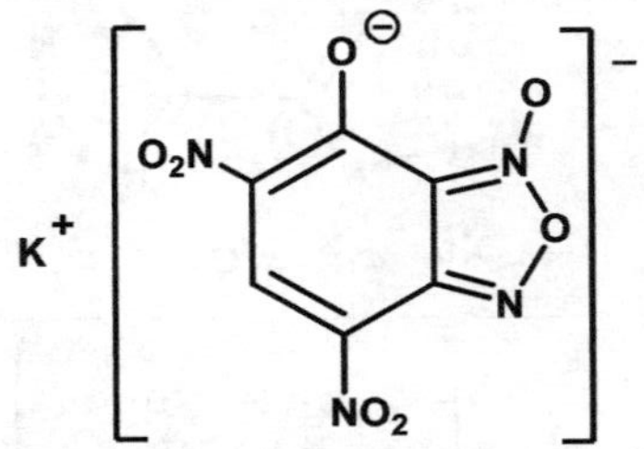	675 µJ	4

(continued)

Tab. 6.1d (continued)

Compound	Structure	ESD	Lit.
NHN	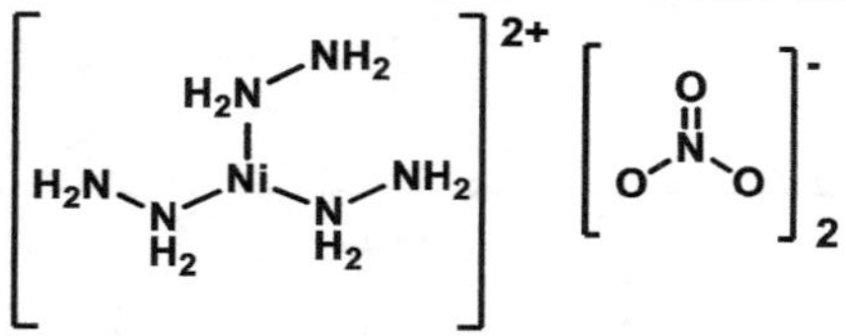	20 mJ	5
K_2DNABT	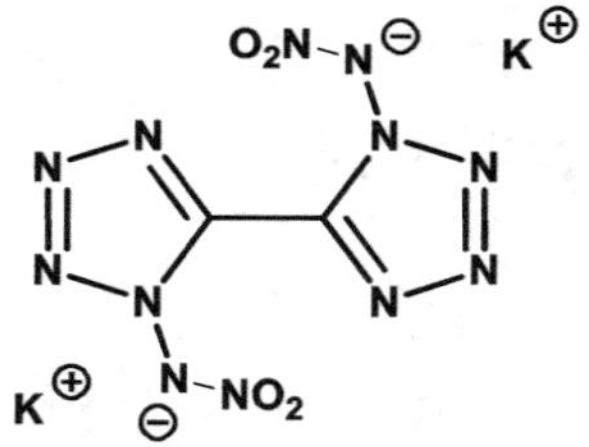	3 mJ	1
$[Cu(N_3)_2(1\text{-}MTZ)]$	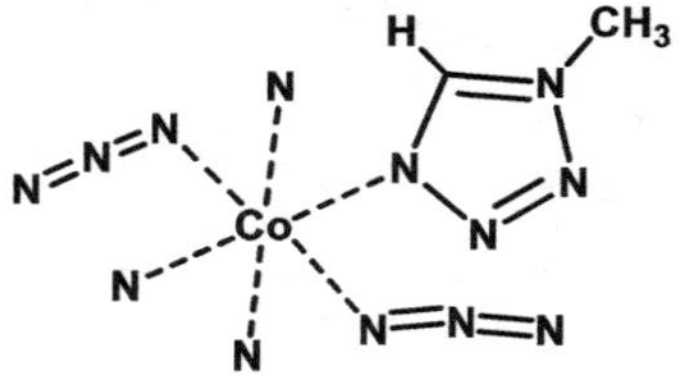	0.79 mJ	6

$Ag_4NT_4 \cdot NH_3$	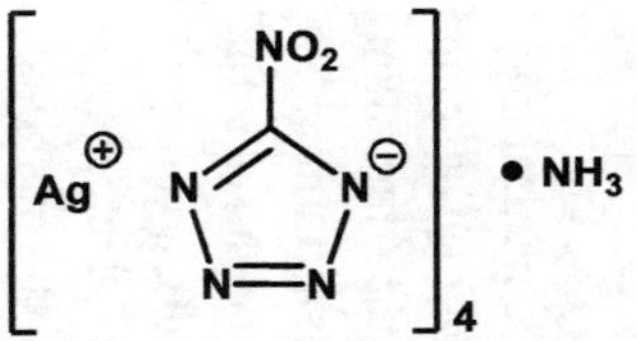	8.3 mJ	7
BNCP	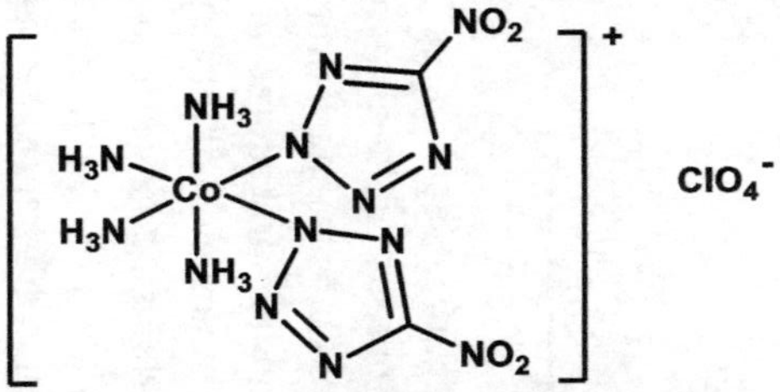	690 mJ	8

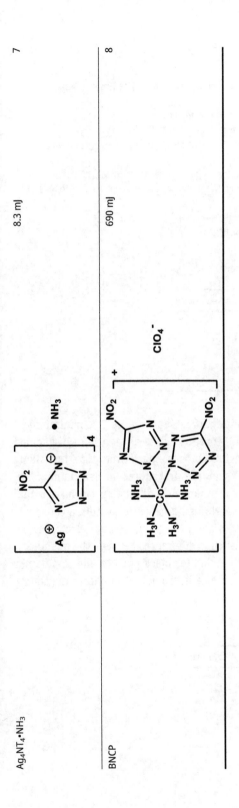

Literature

1. *Energetic Materials Encyclopedia*, Volumes 1 – 3, 2nd edn ., T. M. Klapötke, De Gruyter, Berlin / Boston, **2021.**
2. *Primary Explosives, R. Matyas, J. Pachman, Springer, Heidelberg,* **2013**.
3. M. Zahálka, V. Pelikán, R. Matyáš, *Sci. Technol. Energ. Mater.*, **2019**, *80*, 28-30.
4. J. W. Fronabarger, M. D. Williams, W. B. Sanborn, D. A. Parrish, M. Bichay, *Prop. Explos. Pyrotechn.*, **2011**, *36*, 459-470.
5. Z. Shunguan, W. Youchen, Z. Wenyi, M. Jingyan, *Prop. Explos. Pyrotechn.*, **1997**, *22*, 317-320.
6. M. H. H. Wurzenberger, M. Lommel, M. S. Gruhne, N. Szimhardt, *Angew. Chem. Int. Ed.*, **2020**, *59*, 12367-12370.
7. M. Lommel, R. Schirra, J. Stierstorfer, T. M. Klapötke, *Prop. Explos. Pyrotechn.*, **2024.**
8. M. Lommel, Bachelor Thesis, LMU Munich, Munich, **2018**.

Tab. 6.2: UN classification for the transport of dangerous goods.

	impact sensitivity / J	friction sensitivity / N
insensitive	> 40	> 360
less sensitive	35–40	ca. 360
sensitive	4–35	80–360
very sensitive	< 4	10–80
extremely sensitive		< 10

Related to the ESD but **not** identical is the minimum ignition energy. The minimum energy is the energy (MIE, usually given in mJ) that can ignite a mixture of a specified flammable material with air or oxygen, measured by a standard procedure. Depending on the specific application, there are several standard procedures for determining MIE of dust clouds, solvent vapors and gases. The common element in all procedures is that the energy is generated by an electrostatic spark discharge released from a capacitive electrical circuit. The exact circuit components and the arrangement of electrodes between which sparks are generated are the principle differences between the methods.

In the following table (Tab. 6.2a) MIE is quoted for black powder and some other flammable substances mixed with air. A reference is provided to indicate the source of the data.

Tab. 6.2a: MIE values for flammable substances.

Substance	MIE / mJ	Reference
acetone	1.15	1
acetylene	0.017	1
aluminum	50	1
black powder	**320**	**3**
boron	60	3

Tab. 6.2a (continued)

Substance	MIE / mJ	Reference
butane	0.25	1
carbon disulfide	0.01	1, 3
cellulose	35	1
cellulose acetate	20 – 50	3
charcoal	20	3
coal	40	1
cotton (filler)	25	3
cotton linters	1920	3
dextrine	40	1
diethyl ether	0.2	1, 3
dimethyl sulfide	0.5	2, 3
epoxy resin	15	1
gasoline	0.8	3
hydrogen	0.01	1
hydrogen sulfide	0.07	1, 3
isopropyl alcohol	0.65	1
lycopodium	50	3
methane	0.3	1, 3
PMMA	15 – 20	3
PVA	160	3
potato starch	20 – 25	1,3
sulfur	15	1
THF	0.54	1
thorium	5	1
TNT	**75**	**3**
toluene	0.24	3
uranium	45	1
zirconium	5 – 15	1, 3

1. Haase, H. (1977) Electrostatic Hazards, Their Evaluation and Control, Verlag Chemie, Weinheim.
2. Berufsgenossenschaften, Richtlinien Statische Elektrizität, ZH1/200 (1980), Bonn. Buschman, C.H. (1962) De Veiligheid 38: 20–28.
3. Babrauskak, V. (2003) Ignition Handbook, Fire Science Publishers, Issaquah WA.

The most important indicator for assessing the safety of explosives is the sensitivity of an explosive towards external stimuli. The external stimuli sources which have to be considered include impact (impact sensitivity (IS)) and friction (friction sensitivity (FS)) among others. However, the sensitivity of an explosive towards – for example – impact and friction are not absolute values, and are affected by several factors, one of which, the morphology, will be discussed in more detail below, using the secondary explosive TKX-50 as an example.

TKX-50 has been prepared as samples with a range of different morphologies – from needle crystals to block-shaped crystals. As shown in Tab. 6.2b, small variations in the crystal density of TKX-50 samples with different morphologies is observed, and in fact, the crystal quality of raw TKX-50 material can be improved and defects decreased by crystallization (Tab. 6.2b). The results of impact sensitivity tests on samples of each morphology indicate that the impact sensitivity is affected by the crystal morphology of the TKX-50 sample. The short rod crystals show the lowest impact sensitivity with a value of 19.6 J, which is lower than that of the block-shaped crystalline TKX-50 sample which was found to have an impact sensitivity of 13.2 J. From the re-crystallized samples, the TKX-50 sample consisting of needle-like crystals was found to be the most impact sensitive (6.3 J), but still less sensitive than the raw TKX-50 sample (3.2 J) which was found to be the most impact sensitive of all. The friction sensitivity of the sample varied over a wide range from 0 – 60%. The H_{50} values for the impact sensitivity show good correlation with the values for the friction sensitivities, which indicates that, for these TKX-50 samples, the lower impact sensitivity is, the lower friction sensitivity is. These results demonstrate that the sensitivity of the secondary explosive TKX-50 can be changed – either to higher or lower sensitivity – by modifying the crystal morphology and decreasing crystal defects.

Tab. 6.2b: Mechanical sensitivity of different TKX-50 samples.

sample	morphology	density / g cm^{-3}	impact sensitivity / J	friction sensitivity / %
TKX-50	flake-like	1.8772	4	60
TKX-50	needle-like	1.8818	8.9	24
TKX-50	block-like	1.8807	16.5	0
TKX-50	short stick-like	1.8887	24.5	0

Crystal engineering is a promising emerging strategy to achieve a balance between the energy and safety of energetic crystals. Using crystal engineering, the morphology and crystal structure of a given explosive can be adjusted by fine-tuning the crystallization conditions. In particular, crystallization can be used to result in a significant decrease in the number of crystal defects in an explosive substance, which is then reflected in the lower probability of hot-spot formation in the crystalline explosive sample when it is exposed to external stimuli. In addition, the thermal stability of an explosive can also be enhanced (Fig. 6.4a) in addition to decreasing the sensitivity using this approach, since crystallized EMs generally feature smaller particle sizes and fewer inclusions and defects.

In general, energetic crystals with smaller particle sizes, fewer crystal defects, smoother surfaces, and a narrower particle size distribution are less sensitive to external stimuli. Nano- and micro-sized energetic crystals can be obtained by evaporation or precipitation crystallization methods, of which, the method employing solvent/antisolvent crystallization is the most cost-effective and most commonly used. However, the solvent/antisolvent method is less suited for the production of nano-sized energetic

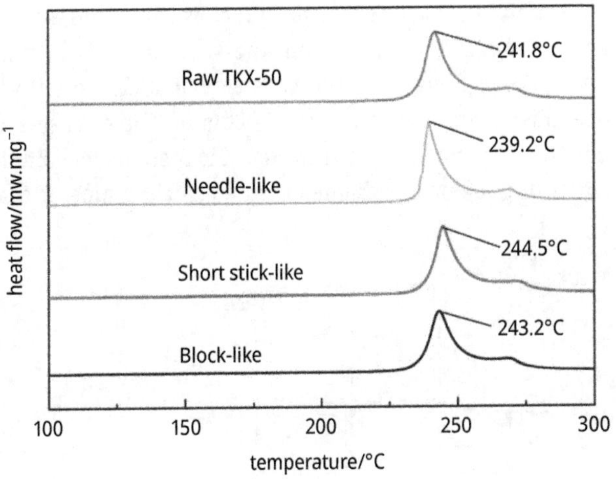

Fig. 6.4a: DSC curves for the raw material and three different crystal forms of the secondary explosive TKX-50.

crystals with a narrower particle size distribution. To overcome this issue, ultrasound and spray-assisted techniques were introduced to the solvent/antisolvent process.

It has recently been shown that energetic materials which show a packing structure which enables shear-sliding to easily occur, tend to possess a low impact sensitivity, in combination with a high molecular stability. This can be found in structures in which a perfect face-to-face π-stacking is present, which is constructed by large π-bonded molecules showing strong intra- and intermolecular hydrogen bonding. If the intermolecular interactions and the anisotropy of these interactions in an energetic crystal can be enhanced, the impact sensitivity of the energetic material is expected to be lower.

Literature

X. Xu, D. Chen, H. Li, R. Xu, H. Zhao, Crystal Morphology Modification of 5, 5′-Bisthiazole-1,1′-dioxyhydroxyammonium Salt, *ChemistrySelect*, **2020**, *5*, 1919–1924.

B. Tian, Y. Xiong, L. Chen, C. Zhang, Relationship Between the Crystal Packing and Impact Sensitivity of Energetic Materials, CrystEngComm, **2018**, *20*, 837–848.

In accordance with the UN guidelines for the transport of dangerous/hazardous goods, the substances in Tab. 6.2 have been classified based on their impact and friction sensitivities.

For the thorough classification of a substance (which is necessary for safe handling), the **thermal stability** must also be determined. The first indication of the thermal stability of energetic compounds or formulations can be obtained from DSC data (see Fig. 2.18b, Ch. 2.5.2). The so-called steel-sleeve test (or Koenen test) is used particularly for assessing the transport safety. In this test, the substance is filled into a steel sleeve (internal diameter: 24 mm, length: 75 mm, wall thickness: 0.5 mm, $V = 25$ mL) up to a height of 15 mm beneath the top edge and then the sleeve is closed with a nozzle plate. Nozzle plates are available with an orifice of 1.0–20 mm. The steel sleeve is attached to the nozzle plate with a two-part threaded socket. It is then heated simultaneously by four burners. By varying the orifice diameters, the critical diameter can

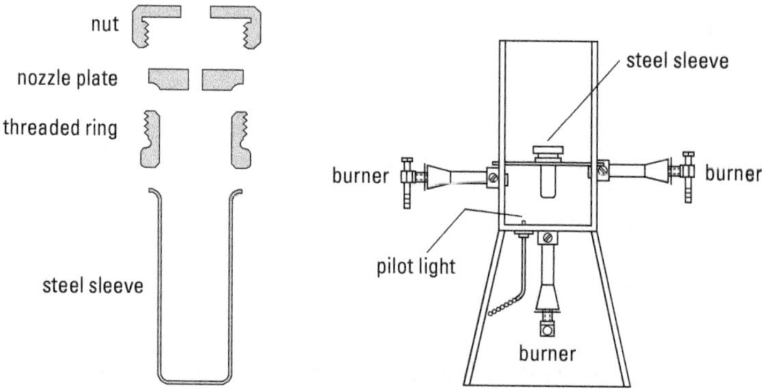

Fig. 6.5: Schematic construction of a steel sleeve test.

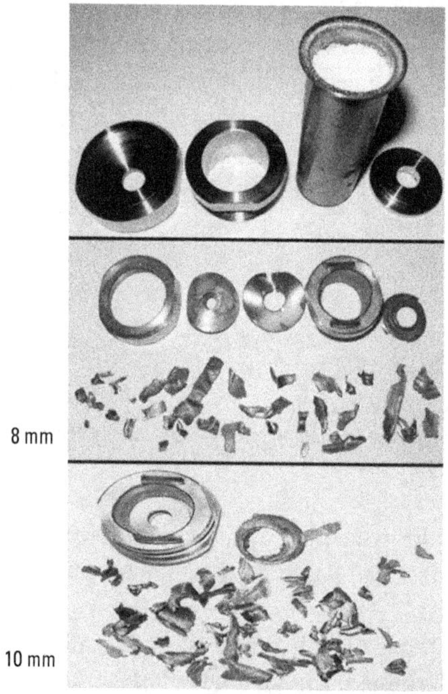

Fig. 6.6: Results of steel sleeve tests for a new explosive with 8 and 10 mm holes in the nozzle plate.

be determined. It is the diameter at which the pressure increase on burning and the subsequent explosion destroys the sleeve into at least four small splinters. Figure 6.5 schematically shows the construction of a steel sleeve test, and Fig. 6.6 shows the results of some tests.

According to UN guidelines, all energetic materials can be divided into four main categories based on their potential danger/hazard. These UN classifications are summarized in Tab. 6.3.

Tab. 6.3: UN hazard classification.

UN classification	hazard potential
1.1	mass detonation with possible formation of fragments
1.2 (18)	no mass detonation, whereby most fragments land within the stated distance (in m)
1.2 (12)	
1.2 (08)	
1.2 (04)	
1.3	mass fire
1.4	moderate fire

6.2 Long-term stabilities

Simple **DSC measurements** (Differential Scanning Calorimetry) are an ideal method to quickly investigate the thermal stability of a substance while only using very small quantities of substance, which is particularly important when dealing with dangerous samples. During DSC measurements, heating rates of 5 °C min^{-1} are commonly used to give an initial indication of the thermal stability and decomposition temperature (see Ch. 2.5.2). However, it is necessary to research the long-term stability by using additional measurements. In this context, the so-called isotherm safety calorimeter can be used for example. This can investigate the thermal stability of a substance at a certain temperature (e.g. 30 °C below the decomposition temperature) for 48 h or longer. In isoperibolic long-term evaluations (approximately isothermal; heat exchange), the stability of substances or mixtures of substances in a sample cell (RADEX cell) can be investigated, with or without an additional pressure sensor.

Figure 6.7 shows an example of DSC plots of a highly energetic material (ammonium 1-methyl-5-nitriminotetrazolate) using different heat rates, while Fig. 6.8 shows the long-term stability of various other compounds in a **FlexyTSC** safety calorimeter.

For the unambiguous determination of the thermal stability of a substance or mixture of substances, it is generally necessary to determine the heat flow rate P (in μW g^{-1}). The substance can be considered to be thermally stable at a certain temperature, if the heat flow corresponds to a value not higher than 300 μW g^{-1} over a time period of 7 days. Figure 6.9 shows an example of such an investigation for diaminote-

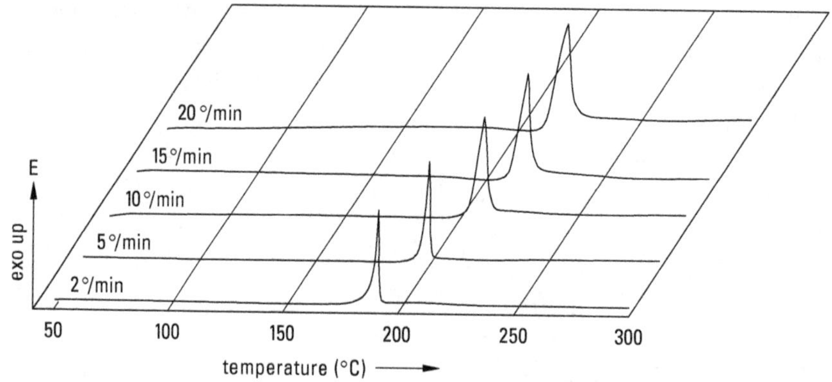

Fig. 6.7: DSC plot for ammonium 1-methyl-5-nitriminotetrazolate using different heat rates (up = exo).

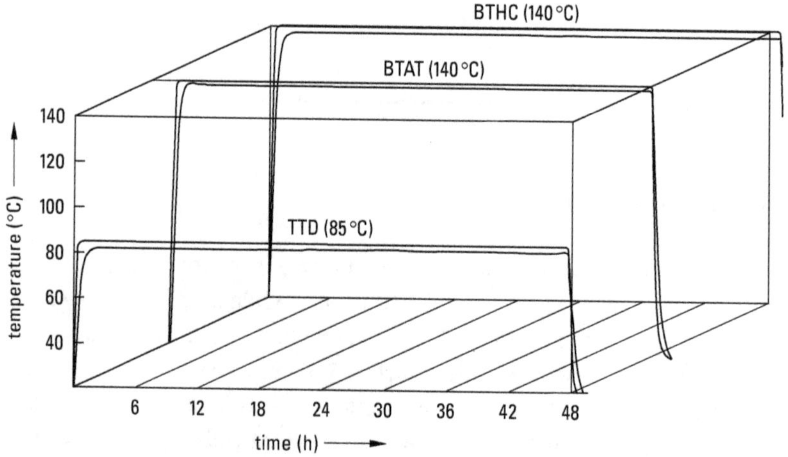

Fig. 6.8: Isothermal long-term measurement (48 h) for TTD, BTAT and BTHC without decomposition.

trazolium nitrate (HDAT nitrate, Fig. 6.10) at a temperature of 89 °C. Here it can be observed that the substance is thermally stable at this temperature.

The so-called **Kneisl-Test** is a simple method which can be used for the estimation of the long-term thermal stability, particularly for primary explosives. A defined quantity (e.g. 100 mg) of the substance under investigation is air sealed in a glass ampoule. This ampoule is then placed in an oven for 100 h at the temperature for which thermal stability should be determined. The ampoule is then opened, the remaining mass of substance weighed and any gaseous decomposed products which may have been formed are analyzed (IR, MS, GC-MS). In order to "pass" the test, the mass lost due to decomposition should not exceed 2% of the original sample mass.

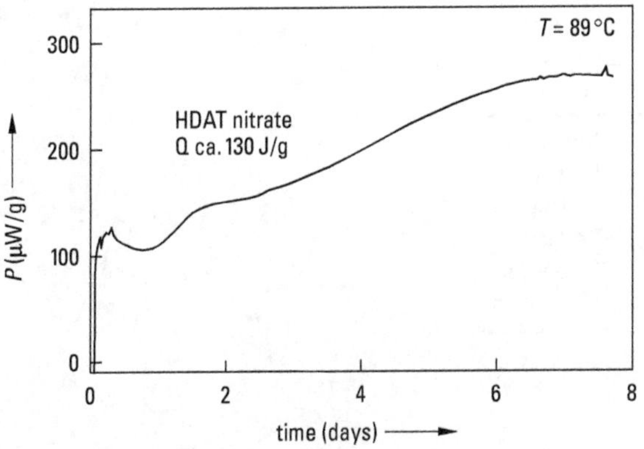

Fig. 6.9: Heat flow for HDAT nitrate at 89 °C.

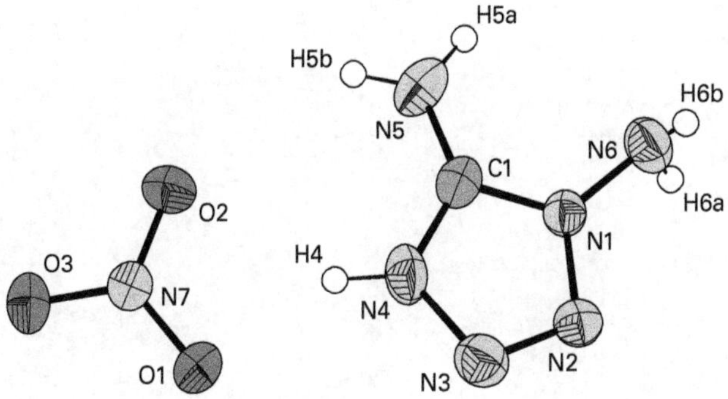

Fig. 6.10: Molecular structure of HDAT nitrate.

6.3 Insensitive munitions

The term insensitive munition (IM, see STANAG 4,439) is used to describe munition which is particularly safe to handle and which is difficult to initiate accidentally, but at the same time has the power and reliability to fulfil the requirements necessary to complete the mission. The insensitivity of munition can be tested and classified into six categories (Tab. 6.4).

Six tests are performed on each substance in order to classify them into one of these categories. The results of these tests can range from “no reaction” to “complete detonation”. The tests are briefly summarized in Tab. 6.5.

Since this area is so important, NATO has founded a Munitions Safety Information Analysis Center (MSIAC).

Tab. 6.4: Categories for insensitive munition.

	German	English
NR	Keine Reaktion	No reaction
V	Abbrand	Burning
IV	Deflagration	Deflagration
III	Explosion	Explosion
II	Teilweise Detonation	Partial detonation
I	Vollständige Detonation	Detonation

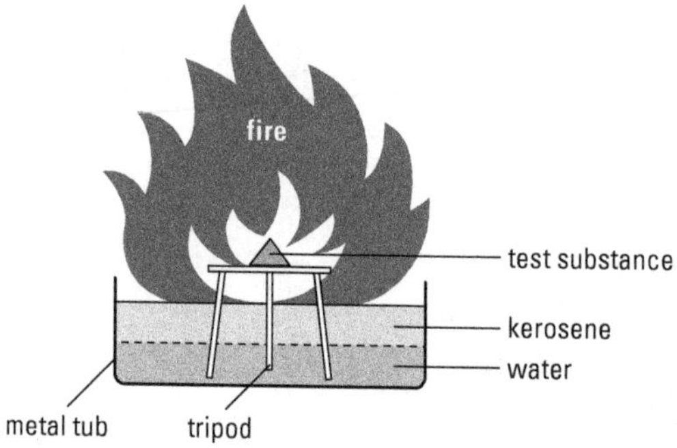

Fig. 6.11: Simple construction of an FCO test (fast cook-off test).

Tab. 6.5: IM tests.

Scenario	IM test	Test procedure	Allowed response for IM (see Tab. 6.4)
small arms attacks	BI	attack (bullet impact) with three 12.7 mm bullets with 850 m s^{-1}	V – burning
fragmenting munitions attacks	FI	attack with five 16–250 g-fragments with 2,530 m s^{-1}	V – burning
magazine, store, plane, ship or truck in fuel fire	FCO	fast cook-off-test, simulation of a fuel fire (see Fig. 6.11)	V – burning
fire in adjacent magazine, store or vehicle	SCO	slow cook-off-test (SCO) with a heat rate of 3.3 °C / h	V – burning
shaped charge weapon attack	SCI	impact of a shaped charge between 50 and 62 mm	III – no detonation
detonation in magazine, plane, truck etc.	SR	reaction to detonation of a neighboring charge (sympathetic reaction)	III – no detonation

6.4 Gap test

In the so-called gap test the sensitivity towards shockwaves is determined [36]. The apparatus for the gap test (Fig. 6.12) is used to measure the ability of a material to propagate a detonation by subjecting it to a detonating booster charge under confinement in a polycarbonate tube. The gap test enables the determination of the minimum shock wave pressure that can cause complete detonation of the tested explosive. The explosive under investigation is subjected to the action of the shock wave of a known pressure. Such a wave is generated by a booster and a shock wave pressure attenuator. Whether or not the shock wave causes the complete detonation of the explosive can be concluded on the basis of the mechanical effects produced after the detonation of the explosive: hole cutting in a steel plate, dent depth in a witness steel block or compression of a copper cylinder. In this test, the gap medium (usually water) stops flying particles and direct heat transmission, serving as a heat filter. Consequently, the shock wave is the only energy transmitted to the explosive. If other factors are constant, the possibility of the transmission of the detonation is determined mostly by both the sensitivity of the acceptor charge and by the initiating strength of the donor charge. The gap test method is based on the determination of the distance between the donor and the acceptor charge of given masses at which transmission or failure of detonation occurs.

Figure 6.12 schematically shows the set-up of the gap test. Typical values for the donor charge are 10 g RDX (with 0.6 g PETN detonator). The medium is water (or alternatively air), and the confinement is a Plexiglass tube of approx. 20 mm diameter. With such an experimental set-up, values for the initiation limits are approx. 25–27 mm for PA and 21–23 mm for Tetryl. In contrast, the value for TNT (less impact sensitive) is

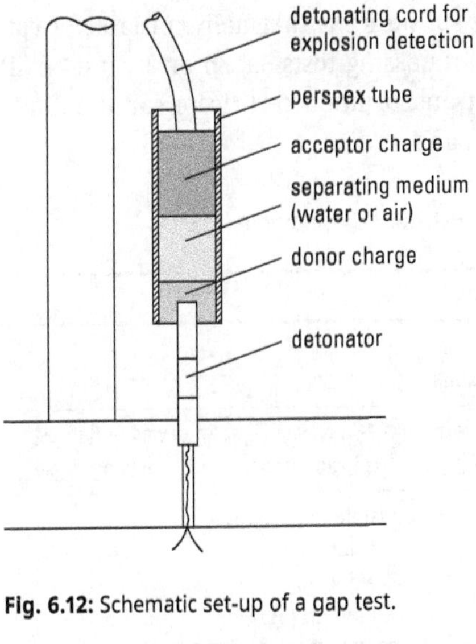

Fig. 6.12: Schematic set-up of a gap test.

only 7–8 mm. The extremely insensitive compound FOX-12 only shows a response at 2 mm under similar conditions.

In Fig. 6.12a, a typical set-up for a water-gap test according to WIWEB is shown.

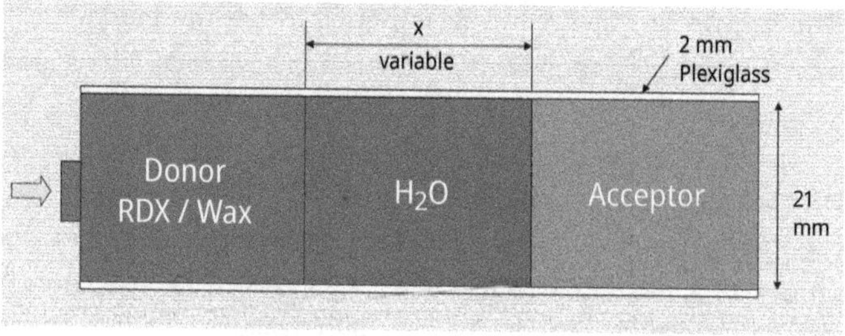

Fig. 6.12a: Water-Gap test according to WIWEB (with friendly permission of Dr. Paul Wanninger).

6.5 Classification

If new energetic materials are synthesized in a research laboratory and are sent to other institutes for further analysis, various stability tests have to be carried out, so that the materials can initially be classified (IHC, **interim hazard classification)** as being at least hazard class "1.1 D" (see Ch. 6.1). The requirements to "pass" the necessary tests (UN 3a–UN 3d) are summarized in Tab. 6.6. A positive test result (+) means that the substance did *not* pass the test.

If a sample does not pass test 3c, but is shown to be thermally stable using a DSC, test 4a must additionally be performed. If the substance is thermally unstable, it can not be transported. In the case of a sample not passing tests 3a, 3b or 3c, test 4b (ii) must be performed (Tab. 6.7). A preliminary transportation certificate can then only be issued if the sample passes tests 3a–3d or 4a and 4b (ii).

Tab. 6.6: Required UN tests 3a–3d to obtain an IHC (interim hazard classification).

UN test	Test	Conditions	+
UN 3a	Impact sensitivity	Five tests; positive, if 50% of the tests give positive results	≤ 3.5 J
UN 3b	Friction sensitivity	Five tests; positive, if one out of five tests is positive	≤ 184 N, visible sparks, visible flames, explosion heard, clear crackling noise
UN 3c	Thermal stability	75 °C, 48 h	visible examination: – color change – explosion – ignition (burning)

Tab. 6.6 (continued)

UN test	Test	Conditions	+
			weight loss (except adsorbed moisture)
			RADEX: self-heating max. 3 °C (RADEX, see Ch. 6.2)
UN 3d	Small-scale-burning test (see Fig. 6.13)	sawdust soaked in kerosene, not confined	Explosion or detonation

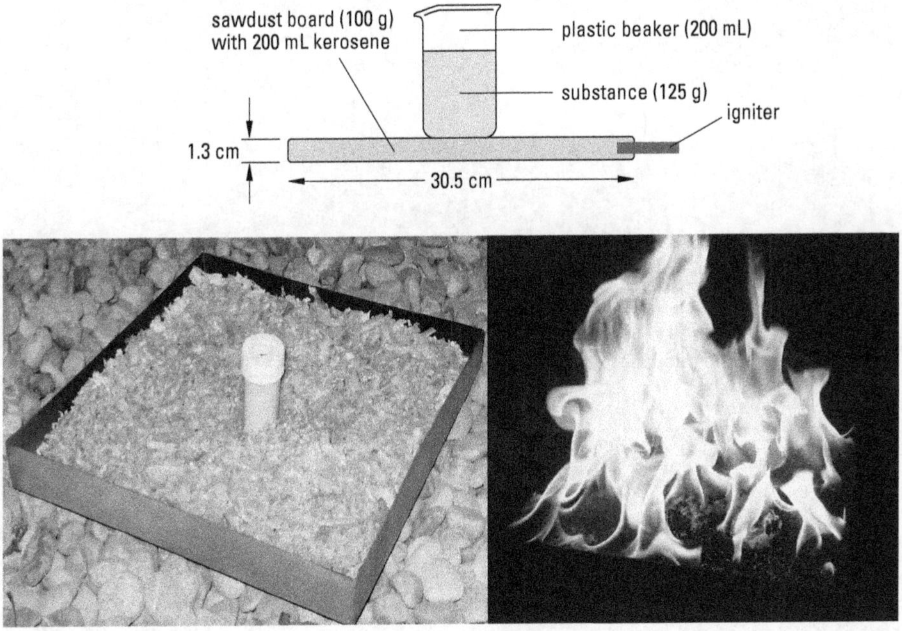

Fig. 6.13: Experimental set-up for a small-scale-burning test (UN 3d).

Tab. 6.7: UN tests 4a and 4b (ii) to establish an IHC (interim hazard classification).

UN test	test	conditions	+
UN 4a	Thermal stability of packed and unpacked energetic materials	75 °C, 48 h	Visible examination: – color change – explosion – ignition (burning)

Tab. 6.7 (continued)

UN test	test	conditions	+
			weight loss (excluding adsorbed moisture)
			RADEX: self-heating max. 3 °C (RADEX, see Ch. 6.2)
UN 4b (ii)	Free fall from 12 m height	12 m free fall (see Fig. 6.14)	Fire, explosion or detonation (damage of the packaging does not count as a "fail")

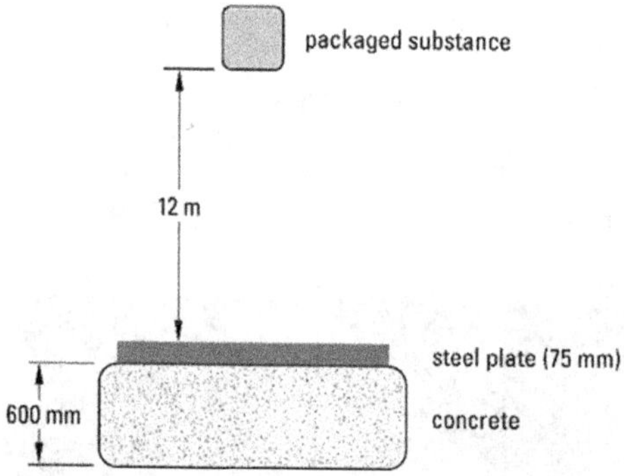

Fig. 6.14: Set-up for the free fall test (UN 4b(ii)).

6.6 Trauzl test

The lead block test (or Trauzl test) is a method which can be used to determine the strength of an explosive. To perform this test, 10 g of the explosive under investigation is wrapped in tin foil and placed into the central borehole (125 mm deep, 25 mm diameter) of a massive lead cylinder (20 cm in diameter, 20 cm long). A blasting cap with an electrical primer is introduced into the center of the explosive charge, and the remaining free space is filled with quartz sand (Fig. 6.15 and 6.16). After the explosion, the volume of the resulting cavity is determined by filling with water. A volume of 61 cm^3 (which is the original volume of the cavity) is deducted from the result which is obtained. Typical lead block excavation values are (per 10 g of explosive, given in cm^3):

PETN	520
RDX	480
TNT	300
AP	190
AN	180

Quite recently, Keshavarz et al. introduced a new model for the prediction of the strength – and therefore power index – of pure explosives and mixtures containing $C_aH_bN_cO_d$ high energy materials, on the basis of volume expansions in the Trauzl lead block [M. Kamalvand, M. H. Keshavarz, M. Jafari, *Propellants, Explos. Pyrotech.* **2015**, *40*, 551–557; M. Jafari, M. Kamalvand, M. H. Keshavarz, S. Farrashi, *Z. Anorg. Allg. Chem.*, **2015**, *641*, 2446–2451].

The optimized correlation has the following form:

$$\Delta V_{\text{Trauzl}} = 578.8 - 175.4(a/d) - 88.88(b/d) + 140.3(V+) - 122.4(V-) \quad (1)$$

The $V+$ and $V-$ are raising and lowering correction functions, respectively.

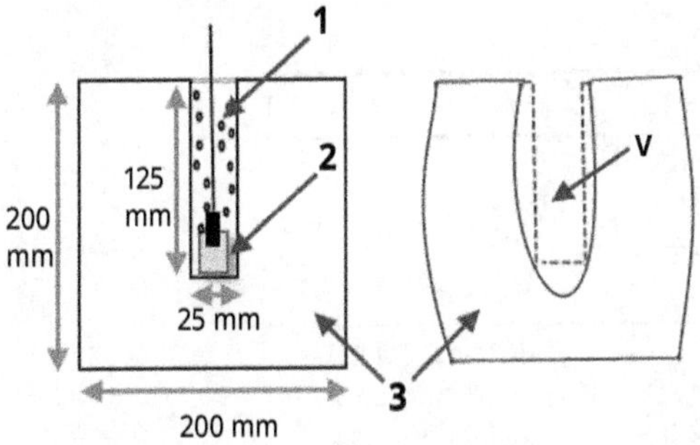

Fig. 6.15: Trauzl (lead block) test: 1 sand stemming, 2 test sample, 3 lead block, V original volume of the cavity (61 cm³).

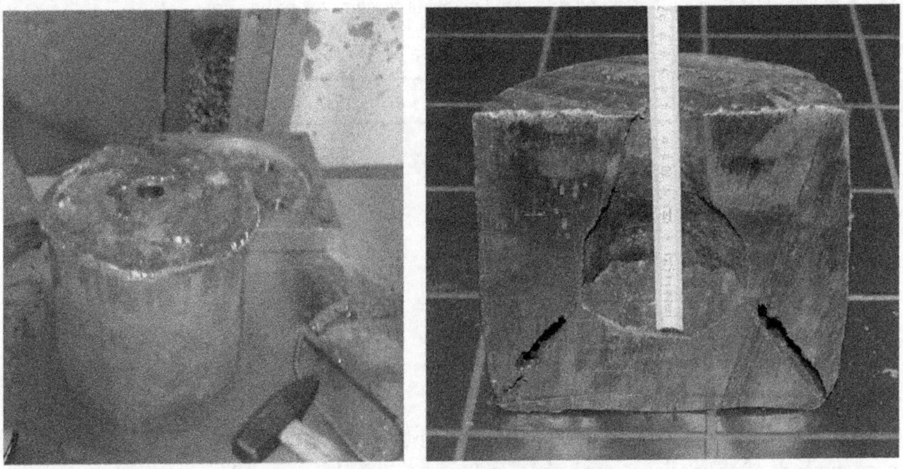

Fig. 6.16: Lead block for Trauzl test before (left) and after (right) the test.

Nitrate and nitramine groups are two important substituents in energetic materials. The energy released during the detonation of common explosives is mainly derived from the oxidation of fuel atoms, i.e. carbon and hydrogen, by oxygen atoms which can be provided by $-NO_2$ groups during the initial stage of a detonation reaction. In contrast, the oxygen of carbonyl and hydroxyl groups cannot lead to a net increase in the energy, because the energies of their cleavage and recombination are equivalent. In other words, nitrate and nitramine groups can increase the strength of a high energy molecule by increasing the value of Q_{det}. The predicted ΔV_{Trauzl} on the basis of *a*/*d* and *b*/*d* was also underestimated in the case of aliphatic energetic compounds, therefore the raising parameter, i.e. *V*+, for correcting the predicted values of ΔV_{Trauzl} must be introduced. Molecular moieties as well as their corresponding values for *V*+ are given in Tab. 6.8.

Tab. 6.8: Values of *V*+ and *V*− which are used to account for the presence of certain molecular moieties.

Molecular moiety	*V*+ (cm^{-3})	*V*− (cm^{-3})
R-(ONO_2)x x = 1, 2	1.0	
R-(ONO_2)x x ≥ 3	0.5	
R-(NNO_2)x x = 1, 2	0.5	
Ph-(NO_2)x x ≤ 2	0.8	
H_2N-C(=O)-NH-R		1.0
phenyl-(OH)x or phenyl-(ONH_4)x		0.5 x
phenyl-(NH_2)x or phenyl-(NHR)x		0.4 x
phenyl-(OR)x		0.2 x
phenyl-(COOH)x		0.9 x

Tab. 6.9: Experimentally obtained and calculated excavation values for some pure energetic compounds.

Explosive	Exptl. ΔV_{Trauzl} (cm^3 / 10 g)	Calcd. ΔV_{Trauzl} (cm^3 / 10 g)
NQ	302	313
NG	520	541
HMX	428	472
PETN	520	517
TATB	175	168
TNT	300	300
HNS	301	330
RDX	480	472

Strong intermolecular attractions – especially hydrogen bonding – can decrease the energy content of energetic compounds and increase their thermodynamic stability. Some polar groups, such as $-COOH$, $-OH$, and NH_2, in aromatic energetic molecules decrease the value of Q_{det}. Similarly, the strength of these energetic materials is overestimated on the basis of *a*/*d* and *b*/*d*. Thus, a lowering parameter, *V*−, should be included in eq. (1) for correcting the predicted ΔV_{Trauzl}. Another molecular fragment, which can lead to the formation of significant intermolecular hydrogen bonding, is the (NH_2)-CO-(NH)-

fragment. The lowering parameter, *V*–, is assigned to account for the presence of this fragment in energetic molecules. Tab. 6.8 contains the functional groups and molecular moieties, which can be used to determine the values of *V*– .

Table 6.9 gives the experimentally obtained and calculated excavation values for some pure energetic compounds.

The Trauzl lead block test allows the determination of the approximate performance of explosives in blasting applications by measuring the volume increase (expansion) that is produced by the detonation of an explosive charge in the cavity of a lead block. The increase in the volume of the lead block cavity was found to correlate best with the product of the heat of detonation and the root of the volume of detonation products:

$$\Delta V_T = a_0 \cdot Q\ V_0^{0.5}$$

ΔV_T net volume increase of the cavity (cm^3)
Q heat of detonation ($MJ \cdot kg^{-1}$)
V_0 volume of detonation products at standard state ($L \cdot kg^{-1}$)
a_0 constant

Dobrilovic I, Dobrilovic M, Suceska M, Revisiting the theoretical prediction of the explosive performance found by the Trauzl test, *Defence* Technology, **2024**, 36, 163 – 174. doi: https://doi.org/10.1016/j.dt.2023.08.013.

The Trauzl Lead Block Test has been in use for many years, as an established experimental measure of the relative performance of an explosive. More recently, the Trauzl Lead Block Test was examined against various detonation parameters, including TNT Equivalence (TNTe) methods and the Berthelot theory in an attempt to identify any strong relationships.

The following equations are used:

Power Index (PI)

$$\text{TNTe } PI_X = (Q_X \cdot V_X)/(Q_{TNT} \cdot V_{TNT}) \quad (1)$$

Hydrodynamic Work Function (HW)

$$\text{TNTe } HW_X = (P_{CJ\,X}/P_{CJ\,TNT})/(\rho_{0\,X}/\rho_{0\,TNT})^{0.96} \quad (2)$$

Heat of Detonation (Q)

$$\text{TNTe } Q_X = Q_X/Q^{TNT} \quad (3)$$

Experimental Trauzl Test (T)

For experimentally determined Trauzl test values, the following can be defined:

$$\text{TNTe } T_X = VT_X/VT_{TNT} \quad (4)$$

Where the following definitions apply:

D_{TNT}	Detonation Velocity for TNT (m/s)
D_X	Detonation Velocity for explosive X (m/s)
EP_X	Explosive Power (J/kg)
$P_{CJ\ TNT}$	Detonation Pressure at the Chapman-Jouget (CJ) point for TNT (Pa)
$P_{CJ\ X}$	Detonation Pressure at the Chapman-Jouget (CJ) point for explosive X (Pa)
Q_{TNT}	Heat of Detonation for TNT (J/kg)
Q_X	Heat of Detonation for explosive X (J/kg)
$\rho_{0\ TNT}$	Density of unreacted TNT (kg/m^3)
$\rho_{0\ X}$	Density of unreacted explosive X (kg/m^3)
TNTe HW_X	TNT equivalence from the Hydrodynamic Work Function for explosive X
TNTe PI_X	TNT equivalence from the Power Index for explosive X
TNTe Q_X	TNT equivalence from the Heat of Detonation for explosive X
TNTe T_X	TNT equivalence from the Trauzl test for explosive X
V_{TNT}	Volume of expansion of the gases from P_{CJ} to SATP for TNT (m^3/kg)
V_X	Volume of expansion of the gases from P_{CJ} to SATP for explosive X (m^3/kg)
VT_{TNT}	Volume of void increase in the Trauzl test for TNT (cm^3/10 g)
VT_X	Volume of void increase in the Trauzl test for explosive X (cm^3/10 g)

Allowing for experimental variation in the Trauzl test and detonation parameters, it can be concluded that the hypothesis is correct, that both the TNTe Power Index and TNTe Berthelot are linearly related to the TNTe Trauzl. The TNTe Power Index is particularly closely linked with the TNTe Trauzl values, so that the Trauzl Test can be said to be a good measure of EP, or that the Power Index (PI) is a good theoretical predictor of EP. Within experimental error it can be said that:

$$\text{TNTe } PI_X = \text{TNTe } T_X = VT_X / VT_{TNT} \tag{5}$$

Lit.: P. M. Locking, TNT Equivalence, Berthelot Theory and the Trauzl Lead Block Test, 32nd International Symposium on Ballistics, Reno, Nevada, USA, May 9th – 13th, **2022.**

There are different definitions for the term "TNT-equivalent" in terms of calculated values.

From the C-J pressure: $TNT_{eq} = [P_{C\text{-}J}(\text{explosive}) / P_{C\text{-}J}(\text{TNT})] \times 100\%$

From the heat of detonation: $TNT_{eq} = [Q_d(\text{explosive}) / Q_d(\text{TNT})] \times 100\%$

From the combined detonation energy and gas volume:

$$TNT_{eq} = [(Q_v(\text{explosive})^*V_0(\text{explosive})) / (Q_v(\text{TNT})^*V_0(\text{TNT}))] \times 100\%$$

The ratio of $(Q_v{}^*V_0)/(Q_v(\text{TNT})^*V_0(\text{TNT}))$ best describes the Trauzl test results, and therefore, works best for mining applications.

6.7 Sand test

The sand test evaluates the explosive power by measuring the amount of sand crushed by an explosive within a heavy, thick-walled bomb. The set-up is illustrated in Fig. 6.17. In this test, a 0.4 g sample of the explosive is pressed into a No. 6 blasting cap cup at 3,000 psi. Following this, 0.25 g of tetryl and 0.2 g of lead azide are pressed at 3,000 psi on top of the test sample to serve as the initiating charge, which is ignited using a 9 inch safety fuse. The complete sample is then placed in a test bomb containing 200 g of sand, which is coarse enough not to pass through a 30-mesh sieve. After detonation, the sand is again sieved, and the amount that now passes through the 30-mesh sieve is weighed. The results – reported in grams of sand that pass through the 30-mesh sieve – indicate the brisance of the explosive or shattering power. The larger the mass of sand that passes through the 30-mesh sieve, the greater the impact of the explosive.

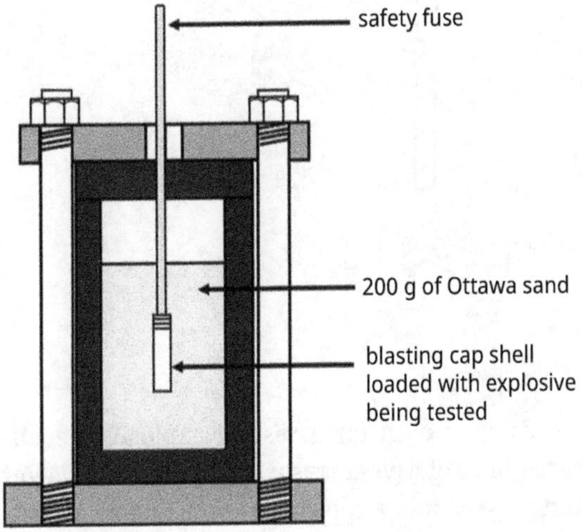

Fig. 6.17: Schematic set-up for the sand test.

6.8 Ballistic mortar test

In the ballistic mortar test, a sample of explosive is placed into a cylindrical hole within a heavy steel mortar (see Fig. 6.18). This hole is then sealed with a steel cylinder, acting as the projectile. The mortar itself is suspended from an arm, functioning as the "bob" at the end of a pendulum. When the explosive is detonated, it propels the projectile into a sand pile, while simultaneously causing the mortar to swing backward. The maximum height that the mortar reaches along a quadrant is recorded. The test is then repeated using different masses of samples of the explosive being tested. The amount of explosive (W in grams) that causes the mortar to reach the same height that is achieved when 10 grams of TNT is used is determined. The result is then reported as "percent TNT equivalent" using the following formula:

$$\%\,\text{TNT equivalent} = \left(\frac{10\,\text{g of TNT}}{\text{W}}\right) \times 100\%$$

Where W is the weight in grams of the test explosive (see also Tab. 6.10).

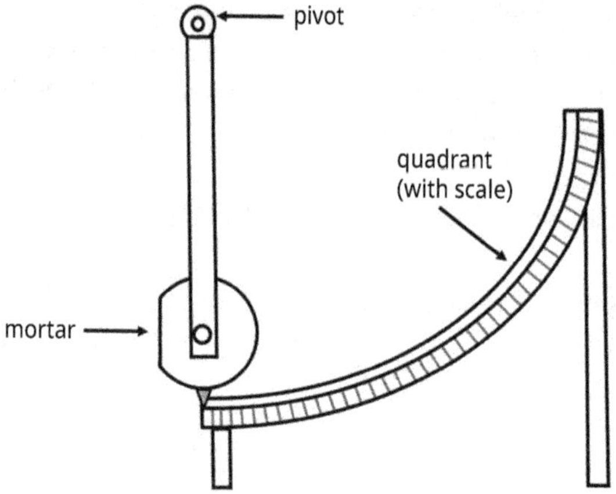

Fig. 6.18: Schematic set-up for the ballistic mortar test.

6.9 Cylinder test

The cylinder test is another method which is used for assessing explosive performance. In this test, a cylindrical charge of explosive, measuring 1 inch in diameter and 12 inches in length, is inserted and pressed into a copper sleeve with a wall thickness of 0.10 inch. A high-speed streak camera is employed to capture the radial velocity of the copper sleeve as it expands after detonation has occurred (see Fig. 6.19).

The camera records the expansion at a point 4 inches from the open end of the charge (8 inches from the detonator). The velocity of expansion is measured at two points: after 5 to 6 mm of wall displacement and after 19 mm. These measurements are used to analyze the energy output in both the "head-on" and "tangential" directions relative to the detonation. The results are expressed as relative energy in these two modes and are compared with data from known explosives.

An excellent introduction into test methods can be found in: P. W. Cooper, S. R. Kurowski, *Introduction to the Technology of Explosives*, Wiley-VCH, New York, Weinheim, **1996**.

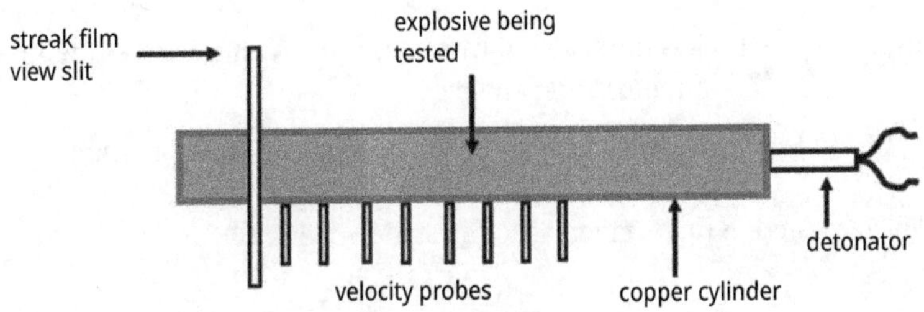

Fig. 6.19: Schematic set-up for the cylinder test.

Tab. 6.10: Overview of explosive output test data obtained from the sand, Ballistic mortar, Trauzl and plate dent tests.

Explosive	Sand test (g)	Ballistic mortar test (% TNT)	Trauzl test (% TNT)	Plate dent test (% TNT)
Composition-B	54	133	130	132
RDX	60	150	157	135
HMX	60	150	145	
Lead azide	19		39	
Nitroglycerine (NG)	51	140	181	
Nitroguanidine (NQ)	36	104	101	95
PETN	63	145	173	129
Picric acid (PA)	49	112	101	107
Tetryl	54	130	125	115
TNT	48	100	100	100
Tritonal (80/20)		124	125	93

6.10 Aging of energetic materials

The aging phenomenon of energetic materials is a very important aspect, since aging can result in the energetic material exhibiting different performance properties such as the heat of reaction and reaction kinetic parameters. Obviously, to be of practical importance and usage, energetic materials must show an extremely slow decomposition rate under ambient conditions. However, this stability has the consequence of making it essentially impossible to investigate the effects of aging on an energetic material using experimental methods over a timespan of decades. In order to overcome this problem, theoretical prediction methods such as the Arrhenius approach can be used. The Arrhenius approach predicts the shelf-life of a chosen energetic material based on the ratio of reaction rates between the storage temperature (T_1) and the elevated temperature (T_2) which is selected as the aging temperature and at which the accelerated aging experiment is performed. The Kissinger evaluation uses the Arrhenius approach as shown below:

$k(T) = Z \cdot \exp\left(-\frac{Ea}{R \cdot T}\right)$ k is the reaction rate constant which describes the Arrhenius T dependence

$\frac{d\alpha(T)}{dt} = k(T) \cdot f(\alpha)$ which corresponds to the general rate expression

in which the nth order reaction f(α) is given as follows:

$$f(\alpha) = (1 - \alpha)^n$$

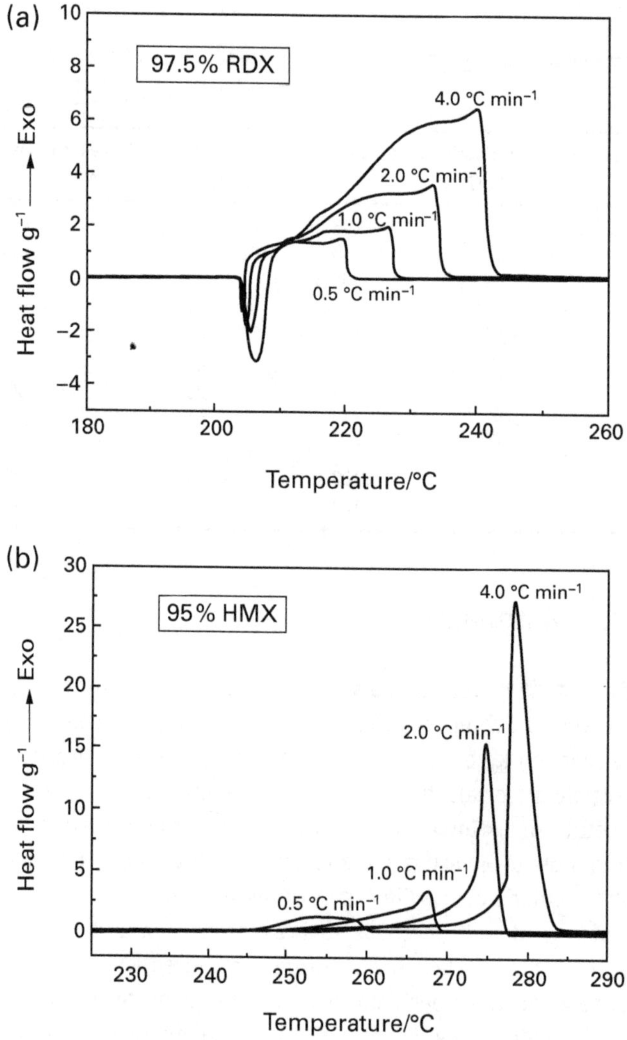

Fig. 6.20: Non-isothermal DSC thermogram for (a) 97.5% RDX and (b) 95% HMX.

The degree of reaction (conversion α) is a constant value at the peak maximum in a set of DSC curves and it is independent of the linear heating rate h (see Fig. 6.20). Therefore, if the value of conversion α is obtained and the second derivative thereof calculated, then an equation can be generated which relates the heating rate h, peak maximum temperature T_P, conversion at peak maximum α_P and the Arrhenius activation parameters Z and E_a.

$$\frac{d\left(\frac{d\alpha(T)}{dt}\right)}{dt} = \frac{dk(T)}{dt} \cdot f(\alpha) + k(T) \cdot \frac{df(\alpha)}{dt} \cdot \frac{d\alpha(T)}{dt}$$

$$\frac{d\left(\frac{d\alpha(T_P)}{dt}\right)}{dt} = \frac{dk(T_P)}{dt} \cdot f(\alpha(T_P)) + k(T_P) \cdot \frac{df(\alpha(T_P))}{dt} \cdot \frac{d\alpha(T_P)}{dt} = 0$$

$$\frac{dk(T_P)}{dt} \cdot f(\alpha(T_P)) = -k(T_P) \cdot \frac{df(\alpha(T_P))}{dt} \cdot \frac{d\alpha(T_P)}{dt}$$

$$Z \cdot \exp\left(-\frac{Ea}{R \cdot T_P}\right) \cdot \frac{Ea}{R \cdot T_P^2} \cdot \frac{dT_P}{dt} \cdot f(\alpha(T_P)) = -Z \cdot \exp\left(-\frac{Ea}{R \cdot T_P}\right) \cdot \frac{df(\alpha(T_P))}{dt} \cdot \frac{d\alpha(T_P)}{dt}$$

$$k(T_P) \cdot \frac{Ea}{R \cdot T_P^2} \cdot \frac{dT_P}{dt} \cdot f(\alpha(T_P)) = -Z \cdot \exp\left(-\frac{Ea}{R \cdot T_P}\right) \cdot \frac{df(\alpha(T_P))}{dt} \cdot k(T_P) \cdot f(\alpha(T_P))$$

$$\frac{Ea}{R \cdot T_P^2} \cdot h = -Z \cdot \exp\left(-\frac{Ea}{R \cdot T_P}\right) \cdot \frac{df(\alpha(T_P))}{dt}$$

$$\frac{h}{T_P^2} = -Z \cdot \frac{Ea}{R} \cdot \frac{df(\alpha(T_P))}{dt} \cdot \exp\left(-\frac{Ea}{R \cdot T_P}\right)$$

$$\ln\left(\frac{h}{T_P^2}\right) = \ln\left(-Z \cdot \frac{Ea}{R} \cdot \frac{df(\alpha(T_P))}{dt}\right) - \frac{Ea}{R \cdot T_P}$$

$$\ln\left(\frac{h}{T_P^2}\right) = \ln(Ak) - \frac{Ea}{R \cdot T_P} \quad \text{general Kissinger equation}$$

$$Ak = -Z \cdot \frac{Ea}{R} \cdot \frac{df(\alpha(T_P))}{dt}$$

For a reaction of nth order:

$$\ln\left(\frac{h}{T_P^2}\right) = \ln\left(n \cdot (1-\alpha_P)^{(n-1)} \cdot \frac{R}{Ea} \cdot Z\right) - \frac{Ea}{R \cdot T_P}$$

$$Ak = n \cdot (1-\alpha_P)^{(n-1)} \cdot \frac{R}{Ea} \cdot Z$$

$$Zk = n \cdot (1-\alpha_P)^{(n-1)} \cdot Z$$

For a reaction of first order, n = 1:

$$\ln\left(\frac{h}{T_P^2}\right) = \ln\left(\frac{R}{Ea}\cdot Z\right) - \frac{Ea}{R\cdot T_P}$$

$$Ak = \frac{R}{Ea}\cdot Z$$

$$Zk = Z$$

Procedure for prediction with the case first order decomposition of the substance:

If the values of $\ln\left(\frac{h}{T_P^2}\right)$ are plotted against $1/T_p$ in 1/Kelvin, the activation energy E_a can be obtained from the slope of the plot and the ordinate section (where the slope of the graph intersects the y axis) gives $\ln\left(\frac{R}{Ea}\cdot Z\right)$, from which Z can be calculated.

Via Arrhenius the reaction rate constants at specific temperatures are available.

Via the ratios of rate constants at different temperatures, the times can be calculated, since the reaction rate constant and time are reciprocal connected.

From the principle of determination this is valid for 50% conversion of the substance.

$$\frac{k(T_1)}{k(T_2)} = \frac{Z\cdot\exp\left(-\frac{Ea}{R\cdot T_1}\right)}{Z\cdot\exp\left(-\frac{Ea}{R\cdot T_2}\right)} = \frac{t(T_2)}{t(T_1)}$$

$$t(T_2) = t(T_1)\cdot\exp\left(-\frac{Ea}{R}\cdot\left(\frac{1}{T_1}-\frac{1}{T_2}\right)\right)$$

Whereby t_1 and t_2 are the shelf-lives at temperatures T_1 and T_2, respectively, and k_1 and k_2 are the reaction rates at temperatures T_1 and T_2, respectively.

A word of caution has to be given though, since it has been reported in the literature that this Arrhenius approach generally over-predicts the shelf-life of an energetic material.

A general strategy which can be used in order to predict the shelf-life of an energetic material (or a formulation of energetic materials) involves the following steps:

- Artificial aging of the energetic material at elevated temperatures for defined periods of time
- Determination of various energetic material properties after the artificial aging process
- Extrapolation of these results to estimate the effect of exposure for longer periods of time at lower temperatures based on kinetic models, e.g.

$$k = A\cdot e^{-\frac{E_A}{kT}}$$

$$F = e^{\left(\frac{E_A}{R}\cdot\frac{\Delta T_F}{T_T^2}\right)}$$

$$t_E = t_T \cdot F^{\frac{T_T - T_E}{dT_F}} \cdot \frac{1}{365.25}$$

where: T_E = service temperature
T_T = test temperature
F = factor by which the rate changes for every ΔT_F change in temperature
t_E = estimated lifetime (in years)
t_T = duration of aging test (in days)

The most important test methods used for assessing the chemical stability of high explosives (HE) and double-base propellants, as well as for determining compatibility are summarized below:

Chemical stability
Main test methods for HEs and primary explosives:
- Vacuum stability test (VST)
- Mass loss
- Heat flow calorimetry (HFC)

Main test methods for double base propellants:
- Stabilizer depletion
- Bergman-Junk test
- Abel test

Compatibility
- Vacuum stability test (VST)
- Heat flow calorimetry (HFC)
- DSC and TGA
- Stabilizer depletion

In the **VST**, usually 2.5 g of sample are heated to 100 °C in vacuum and kept at this temperature for 40 h. The volume of gas evolved by 1 g of the sample is calculated at the end of the test. The thresholds are usually defined as ≤ 2 mL/g for HE, propellants and pyrotechnics. Typical values for RDX are 0.2 mL/g and for Comp-B 0.04 mL/g.

In the **HFC**, the heat flow from a decomposing explosive at a constant temperature is determined over a given period of time. The total heat generation is determined and must be below a specific threshold value in order for storage safety to be ensured, e.g. for 10 years at 25 °C:

In the **Mass Loss** test, 2–5 g of a sample are heated for a specific time. The sample mass is measured in intervals until autocatalytic reaction or 3% mass loss is reached. The typical criterion which is used to predict that an energetic material will be stable during storage for 15 years at 25 °C is a mass loss of < 3% over a period of 16.2 days at 80 °C.

Temperature / °C	Test time / days	Threshold / µW g^{-1}
60	123	9.8
70	34.8	34.5
80	10.6	114
90	3.43	350

Figure 6.21 demonstrates the behavior of ADN in the mass loss test.

In the **stabilizer depletion** test, the stabilizer gets extracted and analysed by GC or HPLC.

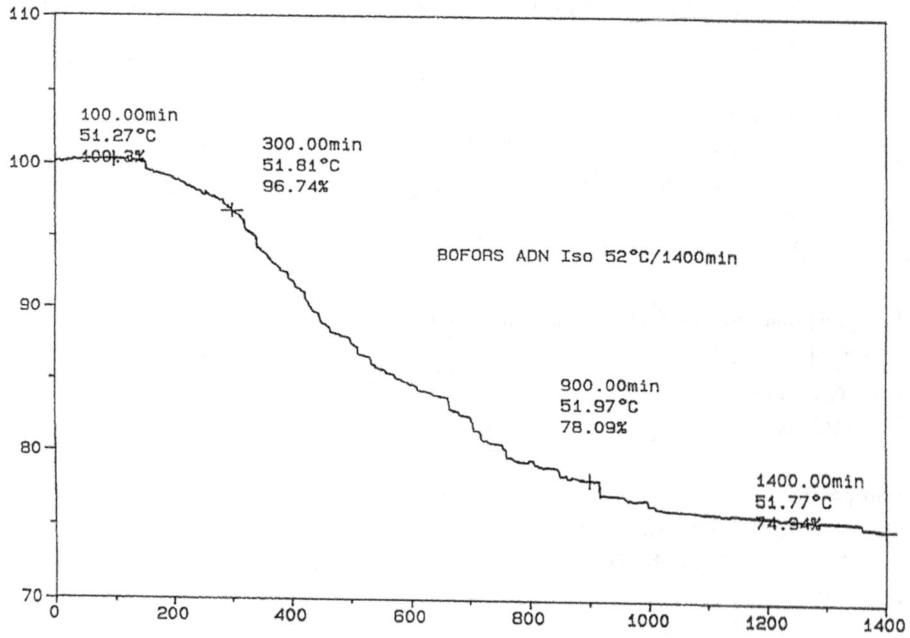

Fig. 6.21: Mass loss of ADN (isothermal TG curve, vacuum, 52 °C), y-axis weight in %, x-axis: time in s.

In the **Bergman-Junk** (BeJu™ by OZM) test, the test apparatus is used to determine the chemical stability of firearm and rocket propellants, and does so by the qualitative analysis of gaseous products evolved during the thermal decomposition of a sample. The sample is heated in a heavy, aluminum heating block at 120 or 132 °C. Gaseous products evolved during the thermal decomposition are absorbed into a glass adaptor filled with water. The amount of gaseous products is then determined by quantitative volumetric analysis of the resulting aqueous solution. The typical sample mass is 5 g, and the temperature ranges of the test are between 120–132 °C with a testing period of up to 6 h.

The **Abel test** (ABT™ by OZM) apparatus is traditionally used to determine the thermal stability of firearm and rocket propellants by measuring the time interval needed to evolve nitrogen oxide gasses from the heated sample. Iodide Starch Indicator Papers are used to detect the presence of decomposition gases. The test temperature is typically set between 60–82 °C. Nitrate esters decompose by evolving nitrogen oxide gases which are detected by the standardized color change of the Iodide Starch Indicator Paper. The period of time between inserting the sample into the heating block and the color change of the indicator paper occurring is the test time. The typical sample mass is 1–3 g and the temperature used ranges between 60–82 °C with a testing period of up to 1 h.

The **chemical compatibility** of an energetic material can be tested using the VST, HFC or DSC and/or TGA methods. These measurements are performed on 1 : 1 mixtures of the energetic material and material against which the compatibility is tested, and a heating rate of 2 K/min is used. The following table summarizes the compatibility criteria:

	Sample mass	HR, T_{max}	Change in	Compatible	Doubtful	Incompatible
DSC	1–10 mg	2 K/min, 700 °C	peak max.	Δ < 4 °C	4 °C < Δ < 20 °C	Δ > 20 °C
TGA	1–10 mg	2 K/min, 1100 °C	weight loss	Δ < 4 °C	4 °C < Δ < 20 °C	Δ > 20 °C

The **critical temperature** of a thermal explosion is also an important safety parameter, which is useful to estimate the start of combustion (K. Wang et al., J. Loss Prevention Process Industries **2015**, *38*, 199 – 203) and is calculated according to:

$$T_b = \frac{E_{ao} - \sqrt{E_{ao}^2 - 4E_{ao}\,RT_{so}}}{2R}$$

In the above formula, R is the gas constant (8.3143 J K^{-1} mol^{-1}), E_{ao} is the value of E_a by Ozawa's method and T_{so} is the onset temperature of the decomposition reaction corresponding to $\beta \rightarrow 0$ (Z. Tonglai et al., *Thermochim. Acta* **1994**, *244*, 171 – 176).

Organic explosives

When considering only pure organic explosives, it can be generalized that the chemical aging frequently begins with the homolysis of the weakest bond within the molecule (unimolecular decay), which is then subsequently followed by and accompanied by self-accelerating parallel reactions which involve the decomposition products which are formed. Therefore, the activation energy corresponds approximately to the bond energy of the weakest bond in the molecule and is consequently a good indicator of the thermal stability of a specific explosive. It has been shown that explosives of this type can be very approximately categorized into two groups: those with an activation energy to decomposition of above 170 kJ/mol and those with a value below

155 kJ/mol. The former is found to be stable for thousands of years under ambient conditions, whereas the latter show limited chemical stability, and the stability of these compounds must be more closely investigated. Table 6.11 gives an overview of many commonly used explosives including aromatic and aliphatic nitro compounds, secondary nitramines, and organic azides, which can be categorized as being very stable, whereas aliphatic nitrate esters show a much lower chemical stability, as do organic peroxides, perchlorates, and nitrogen-rich heterocycles.

Tab. 6.11: Activation energies for decomposition and the corresponding bond which undergoes homolytic cleavage for various classes of organic explosives.

Substance class	Typical representative	Functional group	Homolytic bond cleavage	E_A/kJ mol^{-1}	Chemical stability
Aromatic nitro compund	TNT, PA, TATB, HNS	AR—NO_2	C—NO_2	190 – 290	Excellent
Aliphatic nitro	Nitromethane	R—NO_2	C—NO_2	230 – 260	Excellent
Aliphatic nitramines	RDX, HMX, NQ	R_2CN—NO_2	N—NO_2	170–200	Excellent
Aliphatic azides	GAP, Et—N_3	R_3C—N_3	N—N_2	165–170	Mostly good
Aliphatic nitrate esters	NC, NG, PETN	R—O—NO_2	O—NO_2	155–190	Poor: NC, NG Good: PETN

Inorganic explosives

There are also many explosive compounds which are inorganic compounds (*e.g.* azides, acetylides, and fulminates of lead, silver, and mercury) and the chemical stabilities of these classes of compounds range from being chemically very stable (lead azide) to very unstable (mercury fulminate reacts with metals in a moist atmosphere). TNT is one of the rare examples of explosives which is used as a pure compound and also in composite explosives (blends). These blends may consist of the explosive compound as well as other explosive compound, or with inert materials such as binders. Even in such composite explosives, the chemical aging behavior of the final product is still largely dominated by the aging behavior of the main explosive compound.

High explosives retain the excellent chemical stability of their major components (aromatic nitro compounds and/or secondary nitramines).

Modern and currently used polymer-bonded rocket propellants only contain stable explosive components (*e.g.* nitramines and/or ammonium nitrate and aluminum), however, most gun propellants and also many rocket propellants, show a considerably lower chemical stability since they are based on aliphatic nitric acid esters (nitrate esters) such as nitrocellulose (cellulose nitrates with a nitrogen content between

12.2 and 13.4 mass %). In contrast, pyrotechnics – which are mixtures of oxidizers and fuels – show significantly different chemical aging behavior, in which the chemical degradation is mainly the result of oxidation and hydrolysis of the fuel (*e.g.* hydrolysis/corrosion of Mg by moisture, or oxidation of Al, Mg and Ti by oxygen).

Propellants

In this section, the problem of chemical aging with respect to propellants will be considered in more detail, since it is of particular importance for this class of compounds. As was mentioned above, chemical aging of propellants begins with homolytic breaking of the weak O—NO_2 bond of the aliphatic nitrate esters (*e.g.* nitrocellulose NC and nitroglycerine NG), which results in the formation of nitrogen dioxide and the corresponding alkoxyl radical:

$$R—O—NO_2 \rightarrow R—O\cdot + \cdot NO_2 \qquad (1)$$

The free radicals which are generated in this process are highly reactive and immediately undergo consecutive reactions with nearby nitrate ester molecules. In the case of NC, the backbone alkoxyl radical R-O· also performs internal destabilization reactions by splitting off small stable molecules:

$$\begin{array}{lcl} R—O—NO_2 + R—O\cdot & \rightarrow & N_2, N_2O, \cdot NO, \cdot NO_2 \\ & & H_2O, H_2, CO_2, CO \\ R—O—NO_2 + \cdot NO_2 & \rightarrow & C_2H_2O_4, \text{other fragments} \end{array} \qquad (2)$$

A further main decomposition pathway is the neutral to acidic hydrolysis of the nitrate esters. This reaction is catalyzed by moisture and residual acids, H_3O^+. Residual acids may be present if they were not completely removed after the nitrate ester synthesis, or as a result of water and acids formed during decomposition:

$$R—O—NO_2 + H_2O\,(H_3O^+) \rightarrow R—OH + HNO_3 \qquad (3)$$

Furthermore, another possible decomposition reaction which must be mentioned is the enhanced hydrolysis caused by the interaction between nitrate groups and N_2O_4. This proceeds by the initial conversion of the nitrate group to the nitrite group R—O—NO, followed by the subsequent hydrolysis of the O—NO bond. This reaction was found to occur with nitrogylcerin, and a significantly lower activation energy of 71 kJ/mol was found for hydrolysis of the O—NO bond in the nitrite group compared to 100 kJ/mol for the hydrolysis of the O—NO_2 bond of the nitrate group. Therefore, this means that at lower temperature, this pathway can be a dominant decomposition route.

Some reaction products of eqs. (1) and (2) undergo further reaction in presence of moisture and oxygen as is shown by the following equations:

$$2\ \cdot NO + O_2 \rightarrow 2\ \cdot NO_2 \rightarrow N_2O_4 \quad (4)$$

$$\cdot NO + \cdot NO_2 + H_2O \rightarrow 2\,HNO_2 \quad (5)$$

$$3\ \cdot NO_2 + H_2O \rightarrow 2\,HNO_3 + \cdot NO \quad (6)$$

The radicals and acids formed by reactions (1)–(6) strongly self-accelerate both radical (2) and hydrolytic (3) decomposition of the nitrate esters. This accelerating action is called autocatalysis. Although the primary homolytic reaction (1) cannot be suppressed, the consecutive reactions (2)–(3) can be slowed down nearly to zero by the removal or binding of any acids, nitric oxides, and water from the system. This strategy is used for the stabilization of nitric ester-based explosives. Most commonly, aromatic amines (e.g. diphenylamine, 2-nitro-diphenylamine, *p*-nitro-*N*-ethylaniline, *p*-nitro-*N*-methylaniline) and urea derivatives (e.g. Akardite-II (1-methyl-3,3-diphenylurea) and ethyl centralite (1,3-diethyl-1,3-diphenylurea)) are used as stabilizers, since both are capable of binding nitric acid and nitric oxides. A simplified and more phenomenological picture of the chemical aging of propellants is given in Fig. 6.22.

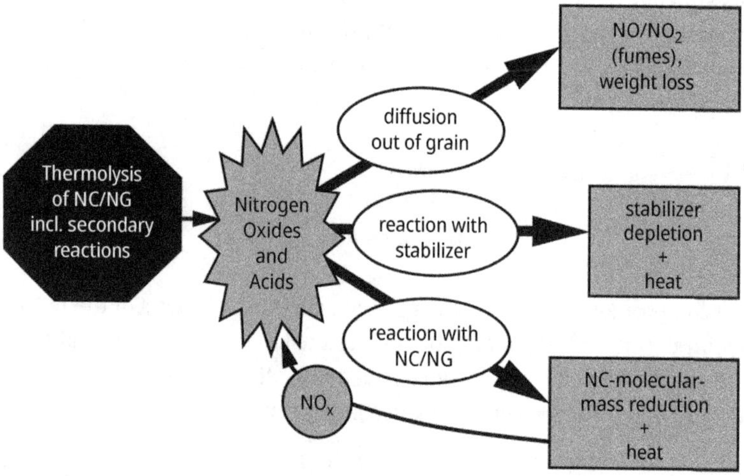

Fig. 6.22: Simplified scheme to illustrate the chemical aging of propellants. In the first step of aging, homolytic breaking of the nitrate ester bonds of nitrocellulose (NC) and nitroglycerine (NG) occurs. This produces highly reactive nitrogen oxides and acids which should be trapped by the stabilizer during consecutive reactions with moisture and oxygen, resulting in stabilizer consumption and evolution of heat. Any nitrogen oxides and acids which are not captured quickly enough can undergo the following fates: either they react further with the nitrate ester component(s) of the propellant (which reduces the molecular mass of the NC and produces more heat and nitrogen oxides, which then accelerates aging of the propellant further), or diffuse out of the propellant (which results in the appearance of 'red fumes'; the leaving nitrogen oxides, together with the other escaping backbone decomposition products such as H_2O, CO_2, CO, and N_2, also cause weight loss). (Reproduced with kind permission of CHIMIA from *Chimia*, **2004**, *58*, 401–408.).

Literature

Y. Kim, A. Ambekar, J. J. Yoh, *J Therm Anal Calorim.*, **2018**, *133*, 737–744.
B. Vogelsanger, *Chimia*, **2004**, *58*, 401–408.
M. A. Bohn, *International Annual Conference of ICT*, **1999**, *30*, 16–1–16–45.
M. A. Bohn, *Proceedings of the Workshop on the Microcalorimetry of Energetic Materials*, Leeds, UK, April 7th–9th **1997**, G1–G31.
M. A. Bohn, *International Annual Conference of ICT*, **1997**, *28*, 109–1–109–46.

Check your knowledge: see chapter 14 for study questions.

7 Special aspects of explosives

7.1 Shaped charges

A **shaped charge** is an explosive charge which is shaped to focus the effect of the explosive's energy [37]. In contrast to an explosion of a solid cylindrical explosive charge which creates a relatively wide but only moderately deep crater (Fig. 7.1), a shaped charge consisting of a solid cylinder of explosive (with rotational symmetry) with a conical hollow in one end and a central detonator causes a much deeper though narrower crater (Fig. 7.1). A shaped charge with a metal-lined conical cavity (e.g. Cu or W liner) creates an even deeper and narrower crater (Fig. 7.1). The extremely high pressure generated by the detonation of the explosive drives the liner contained within the hollow cavity inward whereby it collapses upon its central axis. The resulting collision forms and projects a high-velocity jet of metal forward along the axis (see Fig. 7.2). Most of the jet material originates from the innermost layer of the liner and remains below the melting point as solid in the jet. A typical modern lined shaped charge can penetrate armor steel to a depth of seven or more times the diameter of the charge's cone.

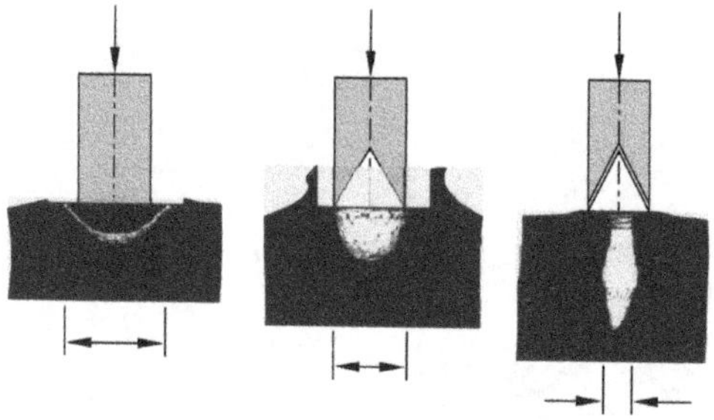

Fig. 7.1: Conventional cylindrical explosive charge, shaped charge, and shaped charge with metal liner. (This diagram is reproduced with slight modification from the original of Prof. Dr. Manfred Held, who is herewith thanked for his permission to reproduce this.)

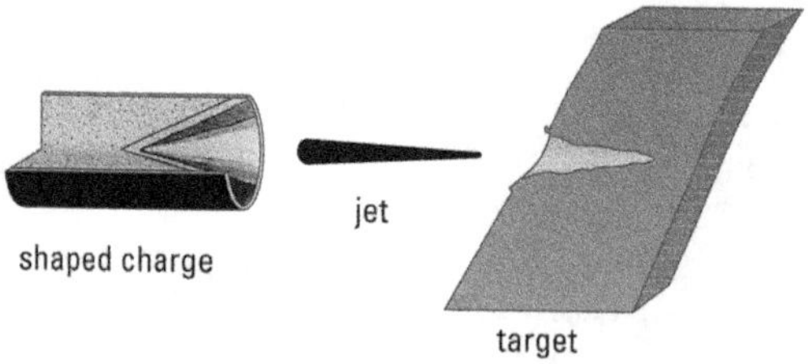

Fig. 7.2: Jet formation from a metal-lined shaped charge.
(This diagram is reproduced with slight modification from the original of Prof. Dr. Manfred Held, who is herewith thanked for his permission to reproduce this.)

https://doi.org/10.1515/9783111447087-007

Ideally, the liner consists of a malleable metal of high density. For this reason, copper is most commonly used. Tungsten and tantalum are also in use because of their extremely high densities, while depleted uranium, though in use for kinetic energy munition (see below) has not found widespread application as liner material for shaped charges. The pyrophoric effects of tantalum (and DU) further enhance the damage caused by the lined shaped charge after penetrating the armor steel.

The velocity of the jet depends on the brisance of the high explosive used and the internal apex angle of the metal liner. The smaller the angle is, the higher is the velocity of the jet. But very small angles can result in jet bifurcation or even failure of the jet to form at all. For this reason, a "good" compromise needs to be found (usually 40–90°). In any case, most of the jet formed moves at hypersonic speed with its tip reaching speeds of up to 7–12 km s^{-1}, the tail at a lower velocity of 1–3 km s^{-1} and the slug at still lower velocity (< 1 km s^{-1}). The high velocity of the jet combined with the high density of the material of the metal liner gives the jet a very high kinetic energy. When the jet hits the target, a very high pressure is formed as a result. A typical jet velocity of 10 km s^{-1} results in a pressure of about 200 GPa. The penetration process therefore generates enormous pressures, and therefore it may be considered to be hydrodynamic (Fig. 7.3). At such a pressure the jet and armor may be treated like an incompressible fluid (if their material strengths are ignored) in order to get a good approximation so that the jet penetrates the target (armor steel plate) according to the laws of hydrodynamics like a liquid (Fig. 7.3). The general penetration behavior of a projectile as a function of its velocity is shown in Tab. 7.1.

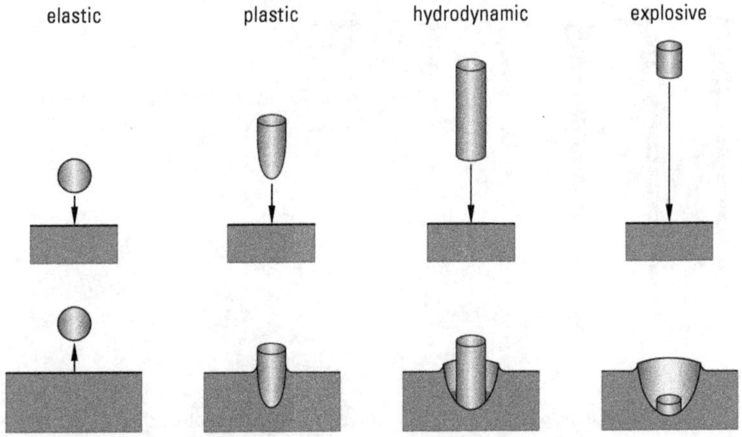

Fig. 7.3: Penetration behavior: elastic, plastic, hydrodynamic, explosive.
(This diagram is reproduced with slight modification from the original of Prof. Dr. Manfred Held, who is herewith thanked for his permission to reproduce this.)

Tab. 7.1: Penetration behavior of a projectile as a function of its velocity.

Velocity of projectile (km s^{-1})	Effect	Launch method
< 50	Elastic, plastic	Mechanical, air rifle
50–500	Plastic	Mechanical, air rifle
500–1,000	Plastic or hydrodynamic, target material appears to be very viscos	Powder gun
3,000–12,000	Hydrodynamic, target material appears to be liquid	Explosive acceleration
>12,000	Explosive, evaporation of the colliding solids	Explosive acceleration

Since the jet needs some time and space to fully form, most shaped charges have a long ballistic cap to guarantee the correct standoff distance. When the ballistic cap hits the target, the main charge gets detonated while still being at sufficient (for the jet formation) distance from the target surface. If, however, the standoff distance is too large, the jet stretches (due to different tip and tail velocities) and eventually breaks up into particles which dramatically lowers the penetration depth. Fig. 7.4 shows the influence of the standoff distance on the penetration depth for a typical metal-lined shaped charge (diameter: 10 cm, length: 18 cm).

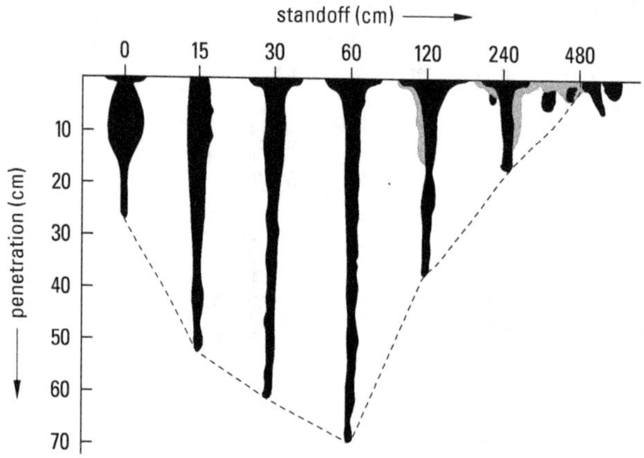

Fig. 7.4: Influence of the standoff distance on the penetration depth of a typical shaped charge with metal liner (diameter: 10 cm, length: 18 cm).
(This diagram is reproduced with slight modification from the original of Prof. Dr. Manfred Held, who is herewith thanked for his permission to reproduce this.)

Because of the extremely high jet velocity the (low) speed with which the shaped charge travels towards its target is relatively unimportant. For this reason, relatively slow and repercussion-free projectiles are often used to carry the shaped charge. The advantage of such systems is that they do not require a big and heavy launch platform which makes the entire system more mobile. Shaped charges are frequently used as warheads in anti-tank missiles and also as gun-fired projectiles, rifle grenades, mines, bomblets, torpedos and as various types of air-, land- or sea-launched guided missiles. Fig. 7.5 shows the schematic design of a shaped charge with a metal liner. Figure 7.5a shows schematically a conical-shaped charge with jet formation.

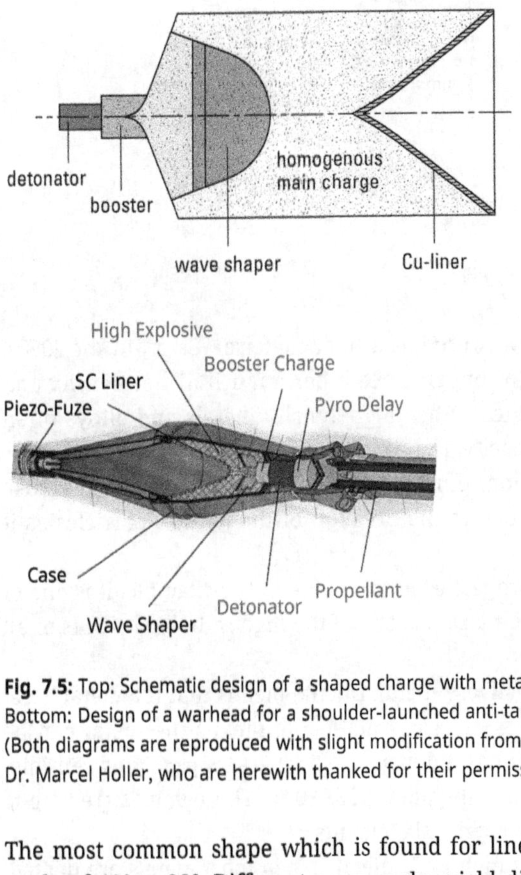

Fig. 7.5: Top: Schematic design of a shaped charge with metal liner and wave shaper.
Bottom: Design of a warhead for a shoulder-launched anti-tank missile (shaped-charge).
(Both diagrams are reproduced with slight modification from the original of Prof. Dr. Manfred Held and Dr. Marcel Holler, who are herewith thanked for their permission to reproduce the figures.)

The most common shape which is found for liners is conical, with an internal apex angle of 40° to 90°. Different apex angles yield different distributions of jet mass and velocity. Small apex angles can result in jet bifurcation, or even result in the failure of the jet to form at all. This is attributed to the collapse velocity being above a certain threshold – normally slightly higher than the liner material's bulk sound speed.

Liners have been made from many materials, including glass and various metals. The deepest penetrations are achieved with liners constructed from a dense, ductile metal, and a very common choice is copper. For some modern anti-armor weapons, molybdenum has been used.

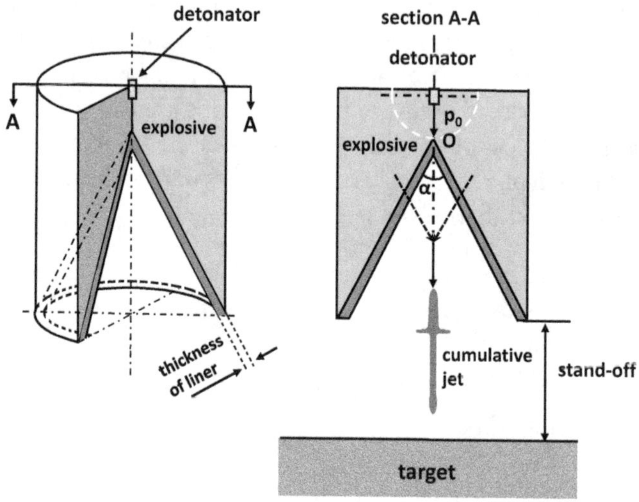

Fig. 7.5a: Conical-shaped charge with jet formation.

In early anti-tank weapons, copper was used as a liner material. Later, in the 1970s, tantalum was found to be superior to copper, since it has a much higher density and very high ductility at high strain rates. Other high-density metals and alloys often have drawbacks in terms of price, toxicity, radioactivity (DU), or lack of ductility.

To achieve maximum penetration, pure metals yield the best results, because they show greatest ductility, which delays the breakup of the jet into particles as it stretches.

Rhenium has some interesting properties in this context. For example, it is one of the most dense elements known, and it also has one of the highest boiling points of all the elements.

Rhenium is a ductile, malleable, silvery metal. Ductile means that the metal is capable of being drawn into thin wires, whereas malleable means the metal can be hammered into thin sheets. Rhenium has a density of 21.02 grams per cubic centimeter, a melting point of 3,180 °C, and a boiling point of 5,630 °C. These values are among the highest to be found for any of the chemical elements.

Rhenium is quite dense, which is high for a metal. When most metals are heated, they reach a point where they change from being ductile to being brittle. This means that the metals can be worked with at temperatures below that point, but not above it, since above this transition temperature they become brittle, and on attempts to bend or shape them, they break apart. The unusual behavior of rhenium (high ductility) means that it can be heated and recycled many times without breaking apart.

Quite recently, experiments involving reactive liner materials (RMs) have also been performed (*Propellants Explos. Pyrotech.* **2018**, *43*, 955–961). RMs liners were prepared using a mixture of Al and PTFE powder with a mass ratio of approximately 26%

Al and 74% PTFE. The liner samples were prepared by the following process: mixing, cold pressing, and sintering. The results are promising, but further work is necessary to reach a conclusion with respect to the suitability of RMs liners.

Wide angle cones and other liner shapes such as plates or dishes with apex angles greater than 100° do not jet, but instead give an **explosively formed projectile or explosively formed penetrator (EFP)** (Fig. 7.6). The projectile forms through dynamic plastic flow and has a velocity of 1–3 km s^{-1}. Target penetration is much smaller than that of a jet, but the hole diameter is larger with more armor backspall.

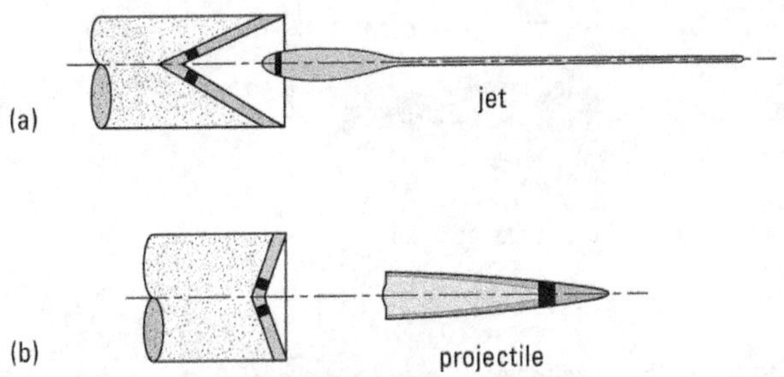

Fig. 7.6: Comparison between a shaped charge (a) and an explosively formed projectile (b). (This diagram is reproduced with slight modification from the original of Prof. Dr. Manfred Held, who is herewith thanked for his permission to reproduce this.)

In general, Fig. 7.6a shows the jet formation as a function of the liner geometry (Fig. 7.6.a).

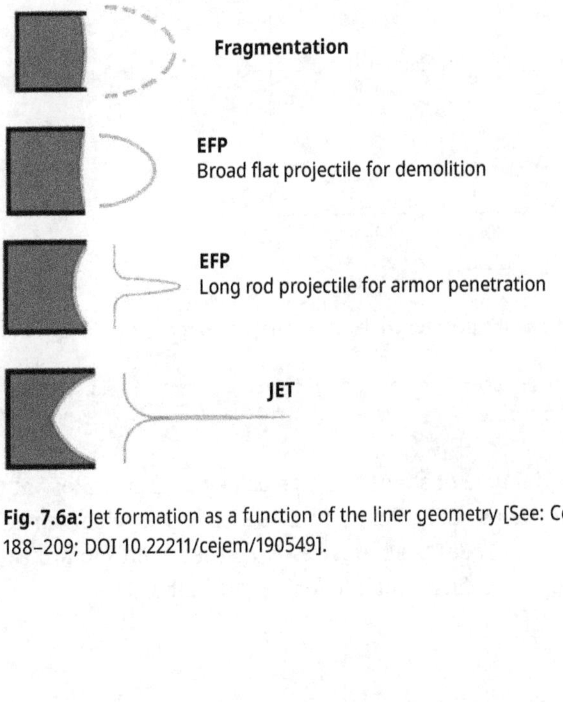

Fig. 7.6a: Jet formation as a function of the liner geometry [See: Cent. Eur. J. Energ. Mater. 2024, 21(2): 188–209; DOI 10.22211/cejem/190549].

An EFP uses the action of the explosive's detonation wave (and to a lesser extent the propulsive effect of its detonation products) to project and deform a plate or dish of ductile metal (such as copper or tantalum) into a compact, high-velocity projectile, commonly called the slug (Fig. 7.7). This slug (one projectile with homogeneous velocity) is projected towards the target at about two kilometers per second. The main advantage of the EFP over a conventional shaped charge is its effectiveness at very large standoff distances, which is equal to hundreds of times the charge's diameter (perhaps a hundred meters for a practical device) (Fig. 7.8).

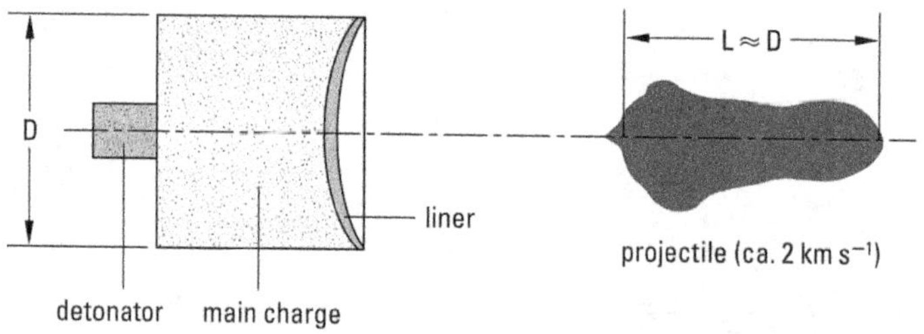

Fig. 7.7: Schematic design of a charge to generate an explosively formed projectile (EFP).
(This diagram is reproduced with slight modification from the original of Prof. Dr. Manfred Held, who is herewith thanked for his permission to reproduce this.)

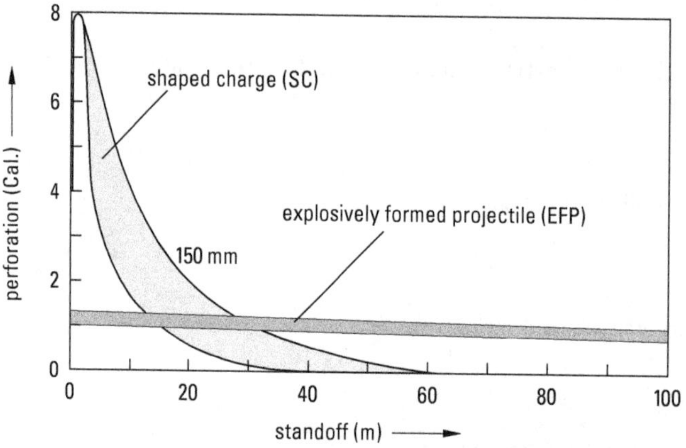

Fig. 7.8: Influence of the standoff distance on the perforation depth for shaped charges (SC) and explosively formed projectiles (EFP).
(This diagram is reproduced with slight modification from the original of Prof. Dr. Manfred Held, who is herewith thanked for his permission to reproduce this.)

A **fragmenting warhead** is a special type of shaped charge or device that explosively forms projectiles. In such a device many EFPs based on cones or cups are combined into one multi-projectile warhead (Fig. 7.8a). Such devices which send out the projectiles in a cylindrical geometry are particularly suitable for targets in the air.

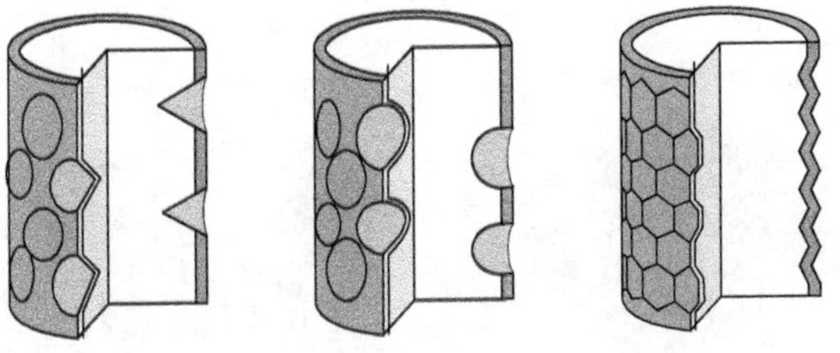

Fig. 7.8a: Schematic presentation of a multi-projectile warhead.
(This diagram is reproduced with slight modification from the original of Prof. Dr. Manfred Held, who is herewith thanked for his permission to reproduce this.)

An **explosive lens** is a very specialized type of shaped charge, which affects the shape of the detonation wave which passes through it in a similar way as light is affected when it is passed through an optical lens. The general design of an explosive lens necessitates the use of several explosive charges. These explosive charges are shaped in a manner which results in a change in the shape of the detonation wave which passes through the shaped charge. In order for this to be achieved, the different explosive charges which the explosive lens is constructed of must have different detonation velocities.

For example, in order to convert a spherically expanding wavefront (Fig. 7.9, top) into a spherically converging wavefront (Fig. 7.9, bottom), in which there is only a single boundary between the explosive with fast detonation velocity and explosive with slower detonation velocity, the boundary shape must be a hyperboloid. Instead of containing a wave shaper, many – particularly older – explosive lenses use two explosives instead which have significantly different detonation velocities (VoD) – but which are in the range from 5,000–9,000 m s^{-1}. Again, the combination of a correctly chosen pair of explosives – in which one has a high detonation velocity and the other has a low detonation velocity – results in a spherical converging detonation wave.

An explosion usually expands outward (spherically expanding) through the explosives in every direction from the point of its detonation. Therefore, for a spherically converging wavefront, the shape of the shockwave has to be manipulated so that it strikes every point on the surface of the pit (inner blue area, surrounded by a tamper, which is a container that surrounds the pit) at the same time. This can be accomplished using an explosive lens.

The detonator is located on the outside of the lens, and is attached to the explosive which possesses a high detonation velocity. A second explosive – with a lower detonation velocity – is contained inside the explosive which has a high detonation velocity. The shape of these two explosives is important since they must be shaped to similarly refract the shock wave. The inner surface of the low velocity explosive is spherical and rests against the spherical pit. The surface between the two explosives

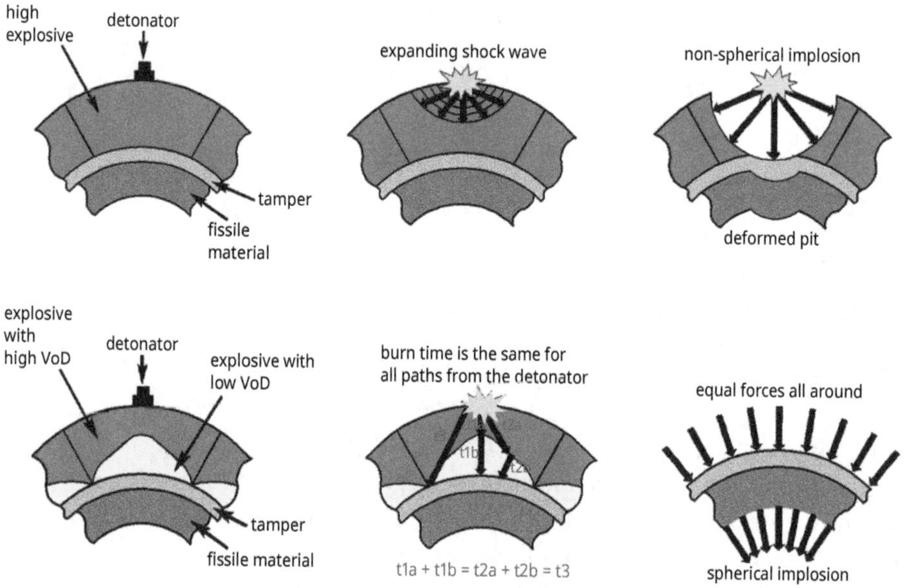

Fig. 7.9: Schematic presentation of a spherically expanding wavefront (top) and of a spherically converging wavefront (bottom), using an explosive lens consisting of two high explosives. One of the explosives has a high detonation velocity, and the other explosive a lower detonation velocity. (Reproduced with kind permission of Phillip R. Hays PhD LT USNR-R, Nuclear/Special Weapons Officer, USS *Oklahoma City* CLG-5 1970-1972 from: http://www.okieboat.com/How%20nuclear%20weapons%20work.html).

is essentially conical, but with a rounded point. Using the values of the speeds of detonation of the two explosives, the shape which the surface should adopt can be calculated. Ultimately, the explosive lens is used in order to focus the energy of the chemical explosion onto the surface of the pit.

The shock wave of an explosive quickly passes through the explosive with a high detonation velocity until it comes into contact with the explosive with low detonation velocity. At this point, the shock wave then detonates the explosive with the lower detonation velocity. The detonation wave then passes more slowly through the explosive with low detonation velocity. The shock wave reaches different points on the boundary between the two explosives at different times. It arrives first at the point which is closest to the detonator, whereas the last point it reaches is that which is furthest away from the detonator. Since there is more of the explosive with low detonation velocity along the shorter path which the shock wave takes from the detonator to the surface of the pit, this means that the shock wave requires a longer period of time to expand inward. Conversely, there is a larger quantity of the explosive with higher detonation velocity along the longer paths from the detonator to the pit, and this means that the shock wave passes through the longer distance faster. The amounts of the two explosives are calculated so that the shape of the shock wave means that it will travel along all of the paths originating from the detonator and finally reach the surface of the pit in the same amount of time (about 10 to 15 microseconds). This causes the shape of the shock wave to be that of a spherical implosion, which reaches all of the points on the surface of the pit at exactly the same time. If all of the surrounding forces are equal, the pit will be uniformly compressed.

One of the biggest problems of any implosion device using high explosives is to ensure a symmetric collapse occurs, and to avoid any jet formation and instabilities that, provoking a predetonation, would result in a low energy release (yield).

Figure 7.9a shows a nice example of the two-dimensional fairly symmetric initiation from the edges.

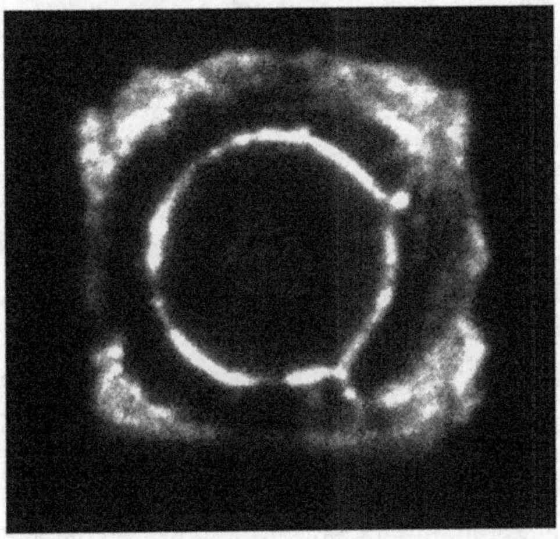

Fig. 7.9a: Two-dimensional wave-shaping with fairly symmetric initiation from the edges.

A further strategy to penetrate or perforate a target is the use of so-called **kinetic energy (KE) munition**. In this context the term "kinetic energy" indicates that the destructive energy of the munition originates from its high kinetic energy $\left(\int T = \frac{1}{2}mv^2\right)$. For this reason, a high mass (m) and a high velocity (v) of the penetrator are desirable. The Bernoulli equation shows the relationship between the penetration depth (P) and the length of the penetrator (L) with the densities of the penetrator (ρ_P) and the target (ρ_T):

$$P \sim L\sqrt{\frac{\rho_P}{\rho_T}}$$

The constant η depends on the velocity of the penetrator (v_P). For a typical penetrator velocity of 1,600 m s^{-1} the constant η is about 0.66:

$$P = \eta L\sqrt{\frac{\rho_P}{\rho_T}} = 0.66 \times L\sqrt{\frac{\rho_P}{\rho_T}} \qquad (\text{for } v_P = 1{,}600 \text{ m s}^{-1})$$

Since the density of the target material is usually 7.85 g cm^{-3} (armor grade steel) one can only influence the penetration depth by the velocity, density and length of the penetrator. Table 7.2 summarizes the penetration depths depending on the penetrator material used (penetrator density). It is apparent why from a strategical point of view penetrators made of highly dense (and pyrophoric) depleted uranium are very suitable.

Tab. 7.2: Penetration depth as a function of the penetrator material used (density) for a typical penetrator of 80 cm length with a velocity of 1,600 m s^{-1}.

Penetrator material	Penetrator density / g cm^{-3}	Penetration depth / cm
Armor steel	7.85	53
Tungsten	19.3	83
Depleted uranium (DU)	19.0	82
Tantalum	16.7	77

It should also be noted that increasing the length of the penetrator is not always easy because long penetrators need to be stabilized by tail fins to avoid spinning and to guarantee a hitting angle of exactly 0° (Fig. 7.10). (*N.B.* The NATO hitting angle is defined as the angle between the penetrator's trajectory and a line perpendicular to the target surface. This means that with a NATO angle of 0° the penetrator hits the target directly perpendicular to the target surface, as desired.) While a relatively small shaped charge (see above) travels with relatively low velocity and therefore requires only a small launch platform, kinetic energy munitions (KE) usually need large and therefore less mobile cannons to achieve the high velocities required. This makes shaped charges particularly useful for launching from highly mobile platforms such as helicopters. But the lower velocity of a shaped charge device compared to kinetic energy munition makes the shaped charge device more vulnerable during the duration of its flight.

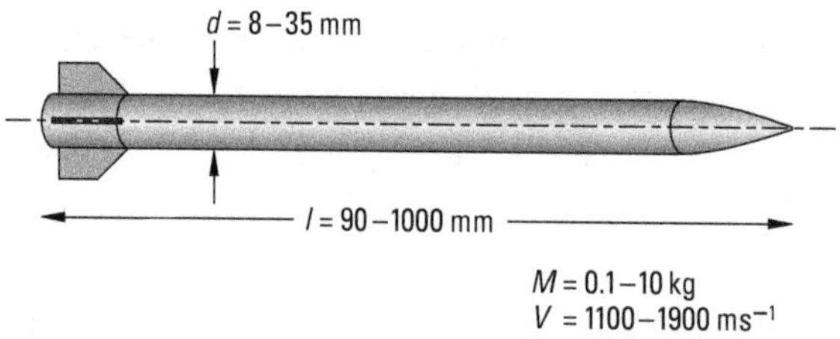

Fig. 7.10: Typical kinetic energy munition (KE) which is stabilized by tail fins.
(This diagram is reproduced with slight modification from the original of Prof. Dr. Manfred Held, who is herewith thanked for his permission to reproduce this.)

7.2 Detonation velocities

In the discussion of shaped charges (see above) it was shown that the brisance (see Ch. 2.2) of the main charge used is of utmost importance. The brisance value (*B*) is defined as the product of the loading density (ρ), the specific energy (*F* for "force of an explosive") and the detonation velocity *D*:

$$\text{Brisance: } B = \rho\, F\, D$$

The specific energy ("force") of an explosive (*F*) on the other hand can be calculated according to the general equation of state for gases:

$$\text{Specific energy: } F = p_e V = n\, R\, T$$

Therefore, we can conclude that for shaped charges the loading density of the high explosive and its detonation velocity (or to be more precise the Gurney velocity, see Ch. 7.3) are relevant performance parameters.

In Chapter 4.2.2 we already discussed methods for the theoretical calculation of the detonation velocity and detonation pressure. In this chapter we now want to focus on the experimental determination of the detonation velocity. Bearing in mind that detonation velocities of known high explosives may reach up to 10,000 m s^{-1}, the experimental determination of the detonation velocity is not easily achieved. There are several methods which are suitable to measure the detonation velocity [38]. Most of these methods are based on the fact that the detonation process is accompanied by the emission of light (autoluminous process). Depending on the measuring equipment selected, the methods for the detonation velocity determination can be divided into,

1. optical methods, which are based on the use of different types of high-speed cameras (distance-time curves) and
2. electrical methods, which are based on the use of different types of velocity probes combined with an electronic counter or an oscilloscope.

In recent years, the fiberoptic technique has widely been used for the determination of the detonation velocity. Optical fibers are capable of detecting and transmitting a light signal accompanying the detonation wave front. This light signal may be recorded by optical methods (using a high-speed streak camera), or it may be transformed into an electric signal (by a fast photodiode) which is then recorded by a suitable ultrafast signal recording technique (fast-storage oscilloscope or multi-channel analyzer). The optical fibers also serve as a convenient means of transporting the signal from the experimental assembly in the firing area to the recording shed. The length of the optical fiber may exceed 20 m. Usually highly flexible optical fibers with a relatively low attenuation factor, 1 mm core diameter and 2.2 mm outside diameter of the black plastic jacket may be used. If the explosive charge is unconfined, the optical fibers are placed directly into the explosive charge to a depth of 2/3 of the charge diameter. If the explosive charge is confined, the fiber is placed through a hole in the wall of the metal or plastic confinement right to the inner surface of the tube (Figs. 7.11 and 7.12).

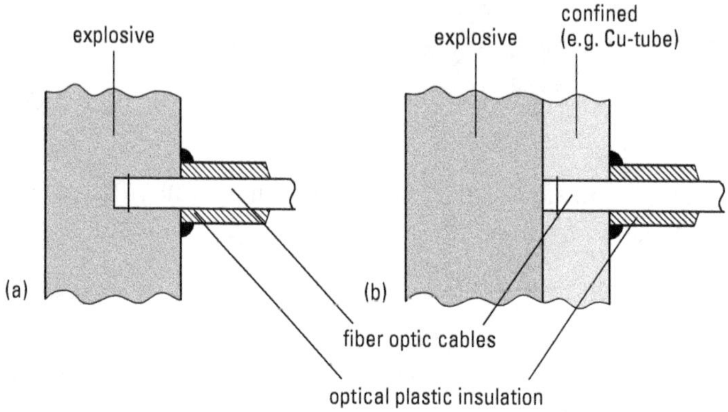

Fig. 7.11: Position of an optical fiber in case of an unconfined (a) and a confined (b) explosive charge.

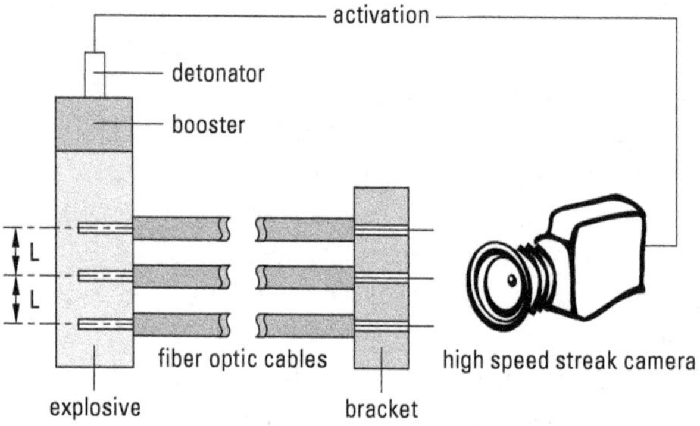

Fig. 7.12: Schematic representation of the determination of the detonation velocity using the optical fiber/streak camera technique.

Figures 7.12 and 7.13 show the experimental set-up for using a high-speed streak camera or a fast photodiode/oscilloscope for the determination of the detonation velocity.

As discussed above, a convenient way of measuring the detonation velocity is to convert the light signal using a fast photodiode that has a rising time of about 10 ns into an electric signal that may be recorded by either a fast oscilloscope (Fig.7.13) or a multi-channel analyzer (Fig. 7.14).

For the experimental measurement of the detonation velocity in a chemical laboratory (indoors), it is advisable to carry out the detonation experiment in a so-called detonation chamber (Fig. 7.15). Usually, the explosive is filled (pressed or melt-cast) into a plastic or metal tube (sealed on one end) which is wider than the critical diame-

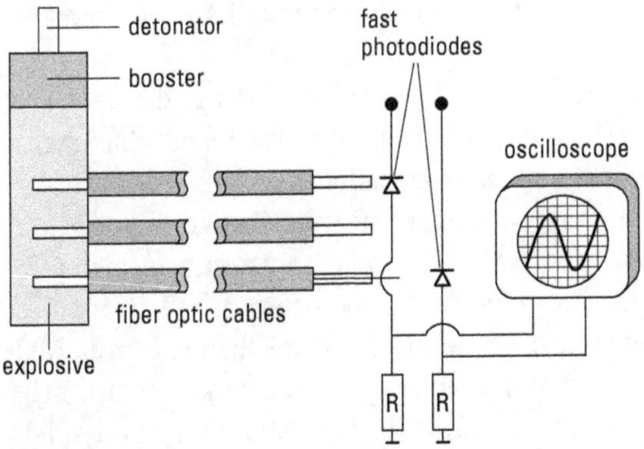

Fig. 7.13: Schematic representation of the test arrangement for the determination of the detonation velocity using the optical fiber/fast photodiode/oscilloscope technique.

Fig. 7.14: Recording the detonation velocity (VoD, velocity of detonation) using a multichannel analyzer.

ter. This tube contains at least two, but preferably more (for mean value calculation) holes for the optical fibers a distance of approx. 1 in. An experimental set-up for measuring the detonation velocity using the optical fiber method is shown in Figs. 7.16 and 7.17. For the measurement it is important to address the following:

1. The density of the explosive under investigation in the tube has to be constant and known exactly.
2. The diameter of the confinement tube has to be above the critical diameter (see Ch. 3, Fig. 3.2). While for many secondary explosives 1 inch is a "good" diameter to start with, for primary explosives usually much smaller diameters (ca. 5 mm) are sufficient. In any case, measurements with different tube diameters are recommended in order to ensure convergence of the detonation velocity on increasing tube diameter (to make sure to be above the critical diameter).
3. The closest optical fiber should not be closer to the detonator than one, preferably two calibers, to ensure that the detonation wave can stabilize and that it originates exclusively from the explosive under investigation (and not from the detonating device).

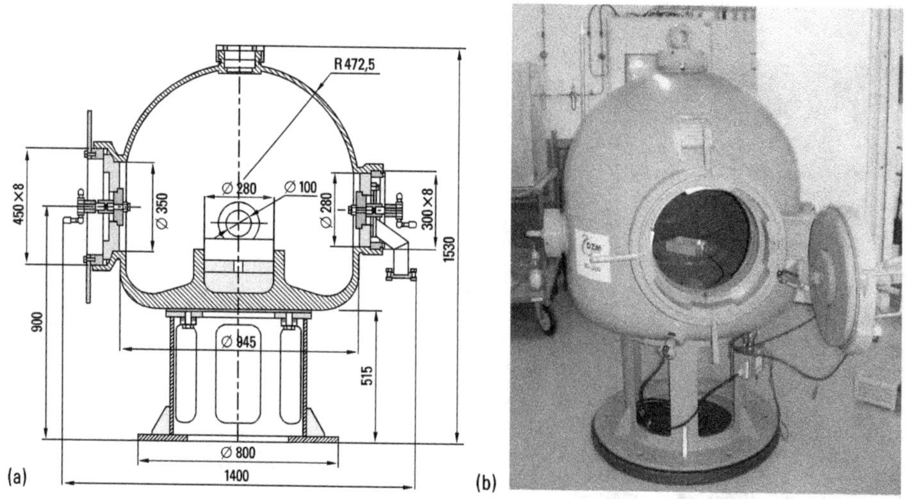

Fig. 7.15: Schematic representation (left) and photo of a KV-250 detonation chamber (up to 250 g TNT or equivalent).

Table 7.3 and Fig. 7.18 show the dependency of the calculated (EXPLO5, see Ch. 4.2.2) and experimentally measured (optical fiber method, multi channel analyzer) detonation velocities of a new high explosive (NG-A) on the loading density. The good agreement between the calculated and measured detonation velocity values reflects the quality of the measurement (time accuracy: ± 0.1 μs, accuracy of the measured detonation velocity ± 0.2%, highest measurable velocity 10,000 m s^{-1}) and gives credence to the theoretically calculated parameters.

Another, perhaps simpler, method to measure the detonation velocity is the so-called Dautriche method. This simple method can be used to obtain an approximate value for the detonation velocity, and since this method does not require any specialized or costly instruments, it is suitable for cases in which high accuracy is not re-

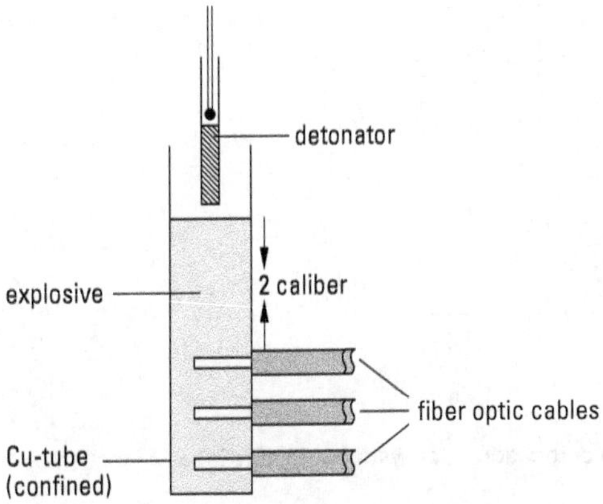

Fig. 7.16: Schematic set-up of a test for the determination of the detonation velocity using the optical fiber method.

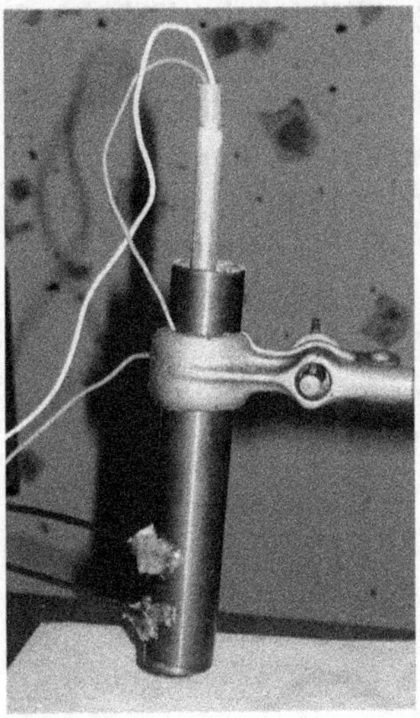

Fig. 7.17: Typical set-up for a simple determination of the detonation velocity using the optical fiber method.

Tab. 7.3: Measured and calculated detonation velocities of NG-A as a function of the loading density.

Loading density / g cm^{-3}	D / m s^{-1} (measured)	D / m s^{-1} (calculated)
0.61	4,181	4,812
1.00	6,250	6,257

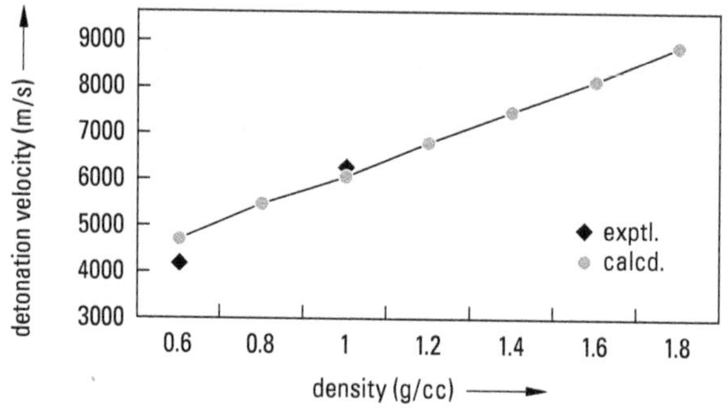

Fig. 7.18: Measured (exptl.) and calculated (calcd.) detonation velocity (det. vel.) of NG-A as a function of the loading density (density).

quired. The basis for this method is the fact that two processes that propagate at different linear velocities travel different distances in the same time interval. The difference in the length between the two distances is a simple function of the velocities of these two processes. If the distances which are travelled, as well as the velocity of one process are known, then the velocity of the other process, can be calculated. The explosive test sample has a cylindrical charge form, and is either unconfined (*i.e.* in the open) or, as is more usually the case, confined. For confined samples, a steel tube with 30 mm diameter and 3 mm wall thickness is the preferred option, however, a cardboard tube may also be used. Two holes are made in the explosive charge (indicated as points A and B in Fig. 7.18a), where the ends of the detonating cord are inserted. The ends may be capped with standard detonators. The distance between point A and point B is usually 300 mm. A distance of 100 mm or more (the distance needed for the stabilization of the detonation wave) should be present between point A and the point where the explosive charge is initiated. The middle section of the detonating cord (usually 900 mm in length) passes over a lead or aluminum plate, in such a way that the midpoint of the cord (assigned as M) is situated towards one of the ends of the metal plate. After initiation of the explosive charge, a detonation wave begins to propagate through the explosive charge at unknown velocity, D_X. It branches out at point A, progresses into the detonating cord, and travels through the explosive charge being tested before entering the other end of the detonating cord at point B. The two waves that travel through the detonating cord at known velocity (D_F), meet at a point indicated as *S*. The longer the time that is needed for the detonation to travel between points A and B in the explosive charge (*i.e.* the lower the detonation velocity of the tested explosive is), the greater the distance between the midpoint of the cord (M) and the collision point *(S)* will be. At the point in the metal plate where the waves meet, the collision path will be sharply indicated. Since the time intervals required for the detonation waves to travel the A-M-S and A-B-S paths are the same, the detonation velocity is calculated using the following equation:

$$t_1 = t_2$$

$$\frac{(L_F/2)+d}{D_F} = \frac{L}{D_X} + \frac{(L_F/2)-d}{D_F}$$

$$D_X = \frac{D_F L}{2d}$$

where: L is the distance between points A and B, L_F *is* the length of the detonating cord, and d is the distance between the detonating cord midpoint (*M*) and the point at which the detonation waves meet (*S*).

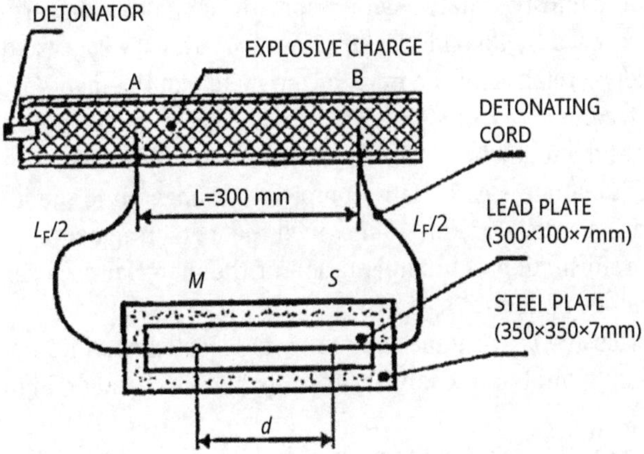

Fig. 7.18a: Schematic representation of the experimental set-up used to determine the VoD according to the Dautriche method.

While it is relatively easy to measure the detonation velocity, it is more difficult to determine the C-J pressure experimentally. However, in a shock wave, the Rankine-Hugoniot jump equations apply across the detonation front. The shock pressure is replaced by the C-J pressure P_{CJ}, the shock particle velocity u_{CJ}, and the shock velocity by the detonation velocity D.

Mass balance: $\rho_{CJ} / \rho_0 = D / (D - u_{CJ})$

Momentum balance: $P_{CJ} = \rho_0 u_{CJ} D$,

Where P_{CJ} is the C-J pressure (GPa), ρ_{CJ} the density in the C-J state (g cm^{-3}), ρ_0 the density of the unreacted explosive (g cm^{-3}), u_{CJ} the particle velocity at the C-J state (km s^{-1}), and D the detonation velocity (km s^{-1}).

The momentum equation can be further simplified by the following approximation, which predicts the C-J pressure within an accuracy of 7% for most explosives.

$$P_{CJ} = \rho_0 \, D^2/4.$$

7.3 Gurney model

As was already discussed in Chapter 7.1 (shaped charges), it is not only the detonation velocity, but also the so-called Gurney velocity, that determines how quickly metal fragments are ejected from an explosive charge with a specific shape (bombs, grenades). This question was researched by Ronald W. Gurney in 1943. Gurney suggested that there is a simple dependence relating to the mass of the metal confinement (M) and the explosive (C) on the fragment velocity (V). The simple Gurney model developed in Aberdeen (MD) assumes the following:

- The energy released on detonation is essentially completely changed into the kinetic energy of the detonation gases *and* kinetic energy of the metal fragments.
- The energy used for the deformation or fragmentation of the material used for confinement can essentially be ignored.
- During detonation, the explosive is spontaneously ($\Delta t = 0$) transformed into homogeneous and chemically completely changed gaseous products under high pressure.
- The gaseous detonation products expand with uniform density and a linear velocity gradient.
- The chemical detonation energy of the explosive is changed into kinetic energy, until the fragments have a steady-state velocity (Fig. 7.19), from which the Gurney velocity can be calculated.

The fragment velocity is largely dependent on the shape of the charge. For cylindrical charges (which are a good approximation for most bomb and missile (rocket) warheads) (Fig. 7.20):

$$\frac{V}{\sqrt{2E}} = \left(\frac{M}{C} + \frac{1}{2}\right)^{-0.5}$$

The constant $\sqrt{2E}$ is the so-called Gurney velocity (in km s^{-1}), which is dependent on the nature of the explosive.

Figure 7.20a shows the terminal fragment velocity depending on the C/M ratio, where M is the mass of the metal confinement and C is the mass of explosive.

Spherical charges (Fig. 7.20) which have initiating charges in the centre and are used as an approximation for grenades (and many cluster bombs), have a similar relationship:

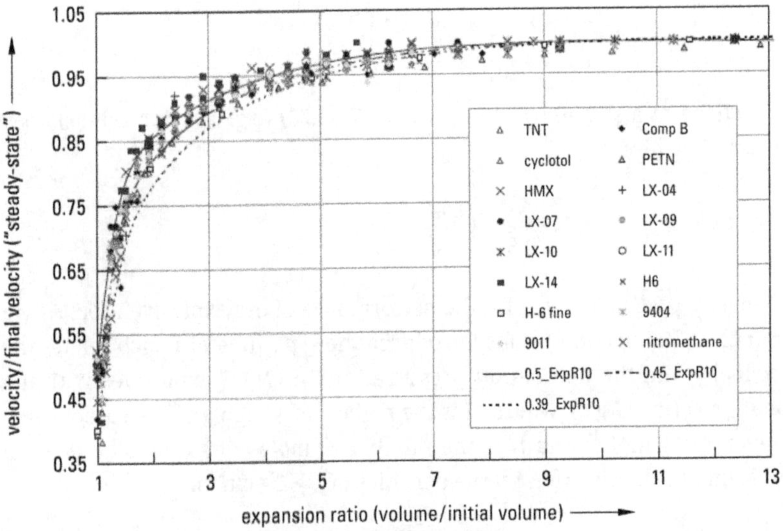

Fig. 7.19: Dependence of the normalized fragment velocity of 16 explosives (Cu cylinder) on the expansion volume. Steady-state at velocity/final velocity = 1.

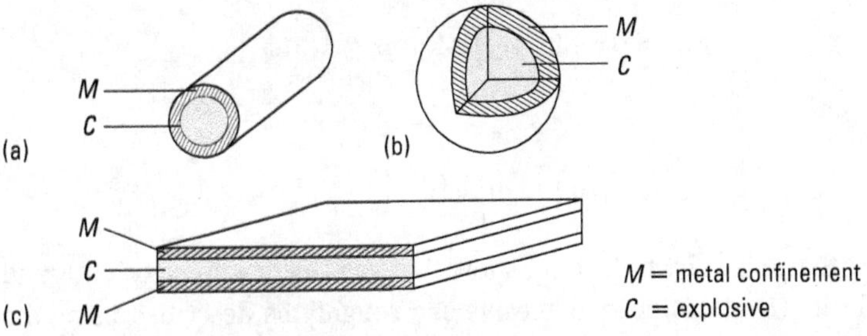

Fig. 7.20: Construction of a cylindrical (a), spherical (b) and sandwich configuration (c).

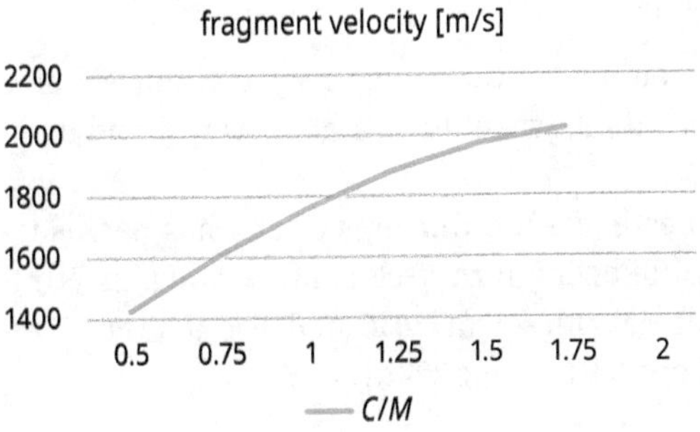

Fig. 7.20a: Fragment velocity curve of a cylindrical warhead with different *C/M* ratios.

$$\frac{V}{\sqrt{2E}} = \left(2\frac{M}{C} + \frac{3}{5}\right)^{-0.5}$$

Charges in the shape of a symmetrical sandwich (Fig. 7.20) (e.g. reactive armor) obey the following relationship:

$$\frac{V}{\sqrt{2E}} = \left(2\frac{M}{C} + \frac{1}{3}\right)^{-0.5}$$

The Gurney velocity $\sqrt{2E}$ is decisive for the performance of the explosive used. As we already saw in Ch. 3, Kamlet and Jacobs introduced the Φ parameter which relates the detonation velocity and the detonation pressure for $C_aH_bN_cO_d$ explosives with the heat of detonation Q (in cal g^{-1}), where M is the molecular weights of the gaseous detonation products (in g mol^{-1}) and N is the number of moles of gaseous detonation products per gram of explosive (for charge densities of $\rho > 1$ g cm^{-3}):

$$\Phi = N\sqrt{MQ}$$

To approximate the Gurney velocity, Hardesty, Kamlet et al. suggested the following relationship:

$$\sqrt{2E} = 0.6 + 0.54\sqrt{1.44\ \Phi\ \rho}$$

therefore,

$$\sqrt{2E} = 0.877\ \Phi^{0.5}\rho^{0.4}$$

It was only in 2002 that A. Koch et al. could show that the Gurney velocity $\sqrt{2E}$ and the detonation velocity (D = VoD) of an explosive can roughly be described using the following simple relationship:

$$\sqrt{2E} = \frac{3\sqrt{3}}{16}D \approx \frac{D}{3.08}$$

Therefore, there is a simple approximation for the Gurney velocity $\sqrt{2E}$ for either a pure explosive or for an explosive formulation, when the charge density ρ and detonation velocity are known (Tab. 7.4).

The most recent approximation to estimate the Gurney velocity has been suggested by Locking (BAE) which has the following form (see P. M. Locking, 29th Int. Symp. on Ballistics, Edinburgh, Scotland, May 9th – 13th, **2016**, and also D. Frem, *FirePhysChem.*, **2023**, *3*, 281 – 291, https://doi.org/10.1016/j.fpc.2022.11.002):

$$EG = \sqrt{2E} = \frac{D}{f_x} \quad \text{with} \quad f_x = 18.0467\,(1 + 1.3\,\rho_x/1000)/\rho_x^{0.4}$$

Tab. 7.4: Dependence of the Gurney velocity $\sqrt{2E}$ on the detonation velocity (D = VoD).

Explosive	Charge density ρ / g cm^{-3}	D / km s^{-1}	$\sqrt{2E}$ / km s^{-1}
Composition A-3[a]	1.59	8.14	2.64
Composition B[b]	1.71	7.89	2.56
HMX	1.835	8.83	2.87
Octol (75/25)[c]	1.81	8.48	2.75
PETN	1.76	8.26	2.68
RDX	1.77	8.70	2.82
Tetryl	1.62	7.57	2.46
TNT	1.63	6.86	2.23
Tritonal[d]	1.72	6.70	2.18

[a] 88% RDX, 12% binder and plasticizer
[b] 60% RDX, 39% TNT, 1% binder
[c] 75% HMX, 25% RDX
[d] 80% TNT, 20% Al

A very simple approach to approximate the Gurney velocity for $C_aH_bN_cO_d$ explosives (density ρ, in g cm^{-3}) was introduced in 2008, and is based on the following equation:

$$\sqrt{2E}\ \left[\mathrm{km\,s^{-1}}\right] = 0.404 + 1.020\,\rho - 0.021\,\mathrm{c} + 0.184\,(\mathrm{b/d}) + 0.303\,(\mathrm{d/a})$$

Some results are summarized in Tab. 7.5. However, this method fails for explosives which do not contain hydrogen atoms and is also only recommended for preliminary estimations.

Tab. 7.5: Gurney velocity $\sqrt{2E}$ obtained according to method A: $\sqrt{2E} = D / 3.08$ (see Tab. 1.1) and Method B: $\sqrt{2E}$ = 0.404 + 1.020 ρ – 0.021 c + 0.184 (b/d) + 0.303 (d/a).

Explosive	Charge density ρ / g cm^{-3}	D / km s^{-1}	$\sqrt{2E}$/ km s^{-1} method A	$\sqrt{2E}$/ km s^{-1} method B
HMX	1.835	8.83	2.87	2.90
PETN	1.76	8.26	2.68	2.97
RDX	1.77	8.70	2.82	2.87
Tetryl	1.62	7.57	2.46	2.41
TNT	1.63	6.86	2.23	2.42

Of course, the Gurney velocity can also be determined experimentally by using the measured fragment velocity V. It can be concluded that the relationship discussed above provides a good approximation, but that the Gurney velocity is also dependent on other factors including, for example, the material of the confinement used (in particular its density). This can be seen in Tab. 7.6.

Tab. 7.6: Dependence of the experimentally determined Gurney velocities $\sqrt{2E}$ on the cylinder material used.

Explosive	$\sqrt{2E}$ (steel) / m s^{-1}	$\sqrt{2E}$ (copper) / m s^{-1}
Composition A-3	2,416	2,630
Composition B	2,320	2,790
Octol (75/25)	2,310	2,700
TNT	2,040	2,370
Tetryl	2,209	2,500

Figure 7.21 also shows that the Gurney velocity for a specific explosive with a particular charge density, is only constant when the surface density (in g cm^{-2}) of the cylinder walls is above a critical value of approx. 1 g cm^{-2}.

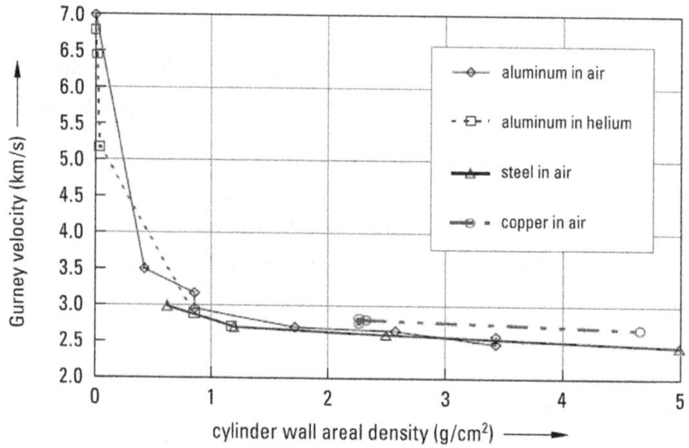

Fig. 7.21: Dependence of the Gurney velocity $\sqrt{2E}$ on the surface density (in g cm^{-2}) of the cylinder walls for Composition B (ρ = 1.71 g cm^{-3}).

The most recent work, which has mainly been reported by Joe Backofen, shows that the Gurney velocity is not only dependent on the explosive, the mass of the confinement, and the shape of the charge, but that it is also dependent on the properties of the material (see Tab. 1.2 and Fig. 7.20). Subsequently he proposed a two-step model to describe this: a brisant shock-dominated first step, followed by a gas-dynamic second (propulsion) step:

1) The 1st stage of detonation-driven-propulsion has a → vector property. The detonation-zone has a finite-width with a Poisson-Ratio effect of about 0.5, and a velocity-vector moving in the direction of its particle-velocity that is driving the shock-front. A supersonic airplane compresses the air in front of it because the airplane has kinetic energy (sustained by expending fuel-based chemical energy) that is expended while

trying to push the air in front of it to the sides. The airplane is the equivalent of the particle-velocity in a detonation. The increase in explosive composition density is what happens when it is in the process of being accelerated to the particle-velocity, while locally the crystals are being subjected to head-on-pressure and shear that "crack" many bonds due to local differences in density / geometry / acceleration. The shock-wave is where this compression begins.

2) The 2nd stage of detonation-driven propulsion is a gas-dynamic stage – an expansion of a mixture of solid and gaseous products from the explosive composition mixture which may contain materials such as aluminum particles that can absorb compression-work while within the 1st stage by decreasing their volume via a phase-transition to a denser-form and then release this via expansion (unfortunately, their volume expansion is trivial in comparison to detonation-product expansion during the 2nd stage) and if reactive-with-the-detonation-products can release chemical-energy to heat via radiation the detonation-products-gaseous-mixture. The "gaseous" expansion is from a "hydro-static-like" field that has a moving locus – the field centroid has a vector while the expansion does not.

The motion during the two stages can be used to aim the effects of the detonation-process – in the "near-field". (The examples that best describe this are the air-blast measurements by Dr. Held and many others, as well as those of explosions in the water – namely, the difference is blast and shock waves off the end of a charge versus the sides of the charge.)

In water, there is also the effect of the collapse of the crater made in the water during the detonation processes. There are the same processes in the air, but the effects are so small that most people do not notice them. In water, this is called the "bubble collapse". It is often also called the "bubble energy" and other similar terms, while relating the energy used to make the bubble to that which can be directed via shaping the bubble.

Short et al. [*Combustion and Flame* **1981**, *43*, 99–109] used these detonation products to derive wall velocities in cylinder tests of $C_aH_bN_cO_d$ explosives by least squares fitting of the experimental data as follows:

$$V_{\text{cylinder wall}} = 0.368 \times \varphi^{0.54} \times \rho^{0.84}\ (R - R_0)^{(0.212 - 0.065 \times \rho)}$$

$$\varphi = N \times M^{1/2} \times Q^{1/2}$$

where $V_{\text{cylinder wall}}$ is the cylinder wall velocity in km/s, $(R–R_0)$ is the actual radial expansion in cm, ρ is the loading density in g/cm^3, N is the number of moles of gaseous detonation products per gram of explosive, M is the average molecular weight of these gases, and Q is the heat of detonation in calories per gram.

In general, there are five types of warheads:

– Fragmentation warheads
  Approximately 30% of the energy released by the explosive detonation is used to fragment the case and impart kinetic energy to the fragments. Fragmentation warheads are designed to cause target damage through the creation of high-velocity

fragments. The weight of individual fragments varies depending on the purpose of the munition, and commonly ranges from tenths of grams up to around 16 grams. Typically, fragmentation warheads utilize approximately 30% of the energy released by the explosive detonation to separate and disperse these fragments, with the rest of the energy causing blast effects. Generally speaking, the radius of effective fragment damage exceeds the radius of blast damage, as air friction slows the blast wave much more effectively than the dispersed fragments.
- Blast warheads
  Blast warheads are designed to achieve target damage primarily through the effects of a shock wave, producing overpressure and high temperatures resulting from the detonation of high explosives. This detonation causes a compression wave to propagate outwards, causing a near-immediate rise from normal atmospheric pressure to peak overpressure, followed by a slower drop back to, and below, normal atmospheric pressure, in objects it passes through. See Table 2.2a for details on the effects of blast overpressure on the human body and structures. Blast warheads are common amongst explosive weapons of all types, including rockets and artillery projectiles.
- Shaped charge warheads
  Shaped charge warheads (SC) are characterized by a hollow metal liner (often copper or aluminum) which is formed in a conical or hemispherical shape, and typically bonded to the explosive fill on the convex side.
- Continuous-Rod Warheads (kinetic energy rounds)
- Special-purpose warheads (thermal, radiation, biological, pyrotechnic: smoke, illumination, . . .).

Figure 7.21a shows the variation of the fragment velocity depending on the explosive filler in the warhead.

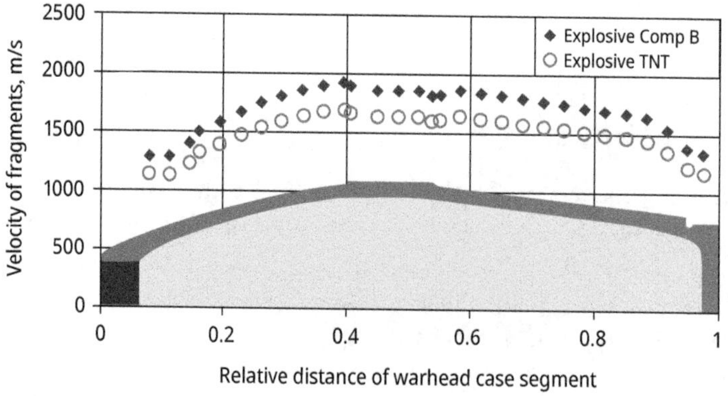

Fig. 7.21a: Variation of fragment velocity depending on the explosive [B. Zecevic, J. Terzic, A. Catovic, 15th DAAAM Intl. Symposium, 2004, Vienna].

Explosives are used in mining to produce two effects. First is the use of "brisance" to shatter the rock into smaller pieces so that it can more easily be handled and processed. (The shattering also is controlled by the use of "brisance" — the shock-wave — by using the "reflection" of shock-waves to cause the tearing-apart of the material (rock) by means of what some call "spallation". Such "spallation" was studied long ago and, as an example, has been used to break concrete using "ear-muff" "counter-charges" as well as to "cut" metals using "fracture-tape" (a "ladder" of C-4 or PE4 on a plastic frame). Therefore, brisance is very important.

The second effect is the heaving or throwing of the fractured material by means of the gas-dynamic-action of the detonation-products pushing against the material-pieces. This too is very important, especially as a hazard during mining and other blasting operations.

The second-step "gas-generation" has dominated for (too) long the development of solid, condensed explosive materials. For some applications the above mentioned brisant properties (of the first stage) can be more important such as for very small implosion charges (oil well perforators), mining to fracture rock, and for driving the detonation in the mixture of the energetic material itself with binders. Therefore, the energy-density and power-density in the crystal (solid state) is very important.

In conclusion, the energetic material needs to be "matched" to its application while remembering that there are two stages involved during detonation-driven propulsion.

The take-home message is that the first-stage does not involve "gases". It is a solid-state to solid-state process. (For example, this is how diamond is formed from the carbon chains via compression into the metastable-state which can "stay together" if the lattice is not over-compressed.)

The Gurney formulae for simple, symmetric geometries and a relationship to approximate the Gurney Velocity (Vg or $(2Eg)^{1/2}$) using the detonation rate (D) in km/sec and adiabatic expansion constant (Γ) for an explosive's gaseous detonation products are given in eq. 1 – 4:

$$Vf_{cyl} = (2Eg)^{1/2}\left[M/C + {}^{1}/_{2}\right]^{-1/2} \quad \text{(cylinder)} \tag{1}$$

$$Vf_{plate} = (2Eg)^{1/2}\left[M/C + {}^{1}/_{3}\right]^{-1/2} \quad \text{(symmetrical sandwich)} \tag{2}$$

$$Vf_{sph} = (2Eg)^{1/2}\left[M/C + {}^{3}/_{5}\right]^{-1/2} \quad \text{(sphere)} \tag{3}$$

$$(2Eg)^{1/2} \approx 0.605D/[\Gamma - 1] \quad \text{(Roth's formula)} \tag{4}$$

Where M and C represent the masses of the inert boundary material and explosive, respectively. (Cooper provides an alternative approximation that the Gurney Velocity is about 0.337 times an explosive's detonation rate).

However, it has been shown that experimental data reveal that geometry and material properties affect an explosive's performance beyond the mass-to-charge ratio used in Gurney modeling. As a result, these factors affect the measurement of an explosive's Gurney Velocity and its corresponding Gurney Energy – parameters used not only in analytical formula Gurney modeling but also in the formulation of gas-dynamic equations-of-state.

Therefore, a two-stage detonation propulsion model has been developed by Backofen.

In the following section we briefly explore a detonation propulsion model that separates propulsion into two parts:

1. initial motion imparted by a brisant shock-dominated process that depends upon intimate contact of an explosive with the propelled material, and
2. subsequent acceleration by a gas-push (gas-dynamic) process. Initial motion is envisioned as being caused by the higher-pressure region of a detonation front (i.e. envision the von Neumann spike or reaction zone region as being a finite thickness of solid materials squeezed at high pressure). The gas-push process is envisioned similar to that assumed by Gurney modeling and generally does the majority of the explosive's work.

It has been shown by Backofen that the 1st propulsion stage produced significant effects affecting overall propulsion:

- About half the final achievable velocity is provided by the 1st stage and depends upon both the propelled material and its thickness relative to the explosive,
- Energy can be wastefully absorbed by a phase transition in the propelled material during the 1st propulsion stage,
- The initial velocity is significantly affected by the angle at which the detonation wave interacts with the propelled material, and
- Charge geometry affects both the initial 1st stage and total propulsion more than is described by the Gurney formulae which only use charge and propelled material masses in their formulation.

Backofen could show that the two separate stages of the two-stage detonation propulsion model can be modeled using equations that describe trends through experimentally derived data and that these equations also can be combined to model the entire propulsion process. For example, the initial free-surface velocity (Vi) imparted to a cylinder's wall by an in-contact detonation running through a cylinder can be described by the following formula (eq. 5):

$$\mathrm{Vi} = 0.2085\,\mathrm{D}\left(\rho_{\mathrm{ex}} / \rho_{\mathrm{cyl}}\right)^{1/2}\left(\mathrm{t_{cyl}} / \mathrm{R_{ex}}\right)^{-3/40} \tag{5}$$

Where Vi and the explosive's detonation rate (D) are described using comparable units, such as km/s; the explosive density (ρ_{ex}) and cylinder density (ρ_{cyl}) are measured in g/cm^3; and the cylinder wall thickness (t_{cyl}) and the cylinder inner radius (R_{ex}) filled with explosive are measured using comparable units such as mm or cm (ρ_{plate}, t_{plate}, and T_{ex} are used when this equation is used to model the effect of a grazing wave on a plate; and 0.2085 becomes 0.417 when the detonation wave impacts “head-on” against the plate’s surface).

Backofen also showed that the “final” “steady-state” velocity (Vf) achievable during an experiment involving a cylinder can be described by using a Gurney equation and a relationship for the Gurney Velocity (V_g in km/s or $(2Eg)^{1/2}$ where Eg is the Gurney Energy in kJ/g) (eq. 6, 7):

$$Vf_{cyl} = (2Eg)^{1/2}\left[M/C + {}^{1}/_{2}\right]^{-1/2} \tag{6}$$

$$V_g = (2Eg)^{1/2} = A\,D\left(t_{cyl}\,\rho_{cyl} \,/\, R_{ex}\,\rho_{ex}\right)^{B} \tag{7}$$

Where M and C represent the cylinder wall and explosive masses in grams or g/cm, A = 0.302 for Comp B explosive in aluminum cylinders (or for $\Gamma = 3$ explosives), $A \approx 0.605 / [\Gamma - 1]$ for explosives in copper cylinders, Γ is the adiabatic expansion coefficient for the gaseous detonation products, and B = (− 5 / 30).

Equations (5) and (6) were combined with Roth’s formula for the Gurney Velocity ($V_g \approx 0.605\ D / [\Gamma - 1]$) using $\Gamma = 2.75$ to provide the following equation (eq. 8) which roughly passed through Vi/Vf data points measured during experiments as well as points approximated by using equation (5) to predict Vi values. (see Figure 7.21b)

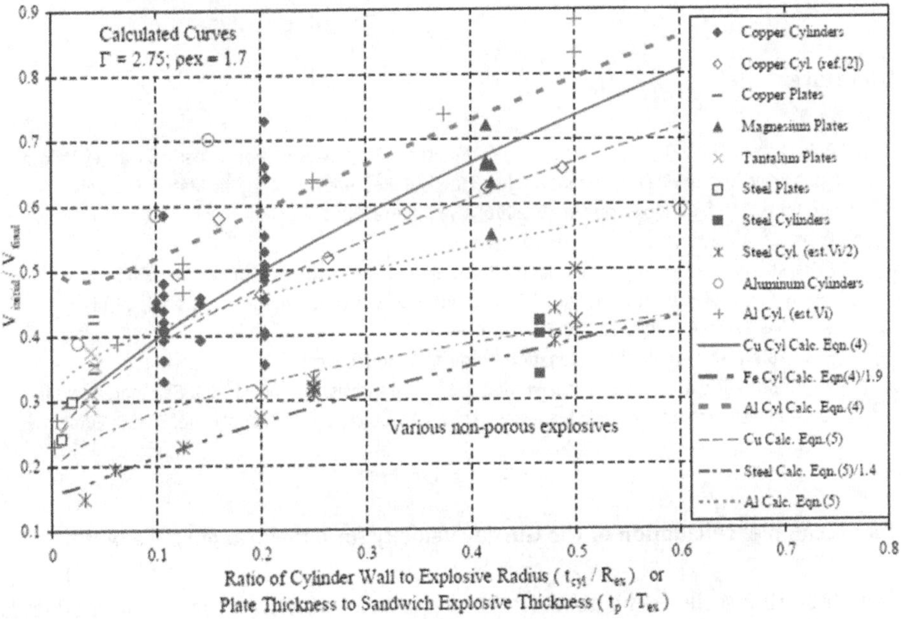

Fig 7.21b: $V_{initial}/V_{final}$ data for cylinders and plates of various inert materials subjected to grazing detonation by various non-porous explosives.

$$(Vi/Vf)_{cyl} = (0.2085/0.3457)\ (t_{cyl}/\ R_{ex})^{-3/40}[(t_{cyl}/\ R_{ex})^2 + 2\ (t_{cyl}/\ R_{ex}) + 0.5\ (\rho_{ex}/\rho_{cyl})]^{1/2} \quad (8)$$

Equation (5) can be used to describe the initial motion imparted to a shaped charge liner during the 1st propulsion stage when rewritten as (eq. 9):

$$Vi = 0.2095\,D\,(\rho_{ex}/\ \rho_{liner})^{1/2}(t_{liner}/\ C_{ex})^{-3/40} \quad (9)$$

Where $C_{ex} = [(CD/2) - (t_{liner} + R_{inner})]$ with R_{inner} as the radius from the charge axis to the liner's inner surface.

1st stage – *Initial* motion (explosives-to-plate coupling)):
- imparted by brisant shock-dominated processes that depend upon intimate contact of explosive with propelled material
- envisioned as caused by higher-pressure region of detonation front (envision the von Neumann spike or reaction zone region as a finite thickness of solid materials squeezed at high pressure while undergoing endothermic and exothermic processes)

2nd stage – Gas-push (*gas-dynamic*) propulsion:
- envisioned similar to that assumed by Gurney modeling (gaseous product expansion from a homogeneous "all burnt" condition while pushing confining boundaries to a final "steady-state" velocity as the pressure drops)

Literature

[1] J.E. Backofen, The Influence of Geometry and Material Properties on an Explosive's Gurney Velocity and Energy, in proc. 9th Int. Seminar, "New Trends in Research of Energetic Materials", Univ. Pardubice, Czech Republic, pp. 76-89, **2006**; also (Central European Journal of Energetic Materials, 3 (4), pp. 23–40, 2006).

[2] J.E. Backofen, The Two-Stage Detonation Propulsion Model: Issues to Ponder / Possibilities for Research, in proc. 10th Int. Seminar, "New Trends in Research of Energetic Materials", Univ. Pardubice, Czech Republic, pp. 80–95, **2007**.

[3] J.E. Backofen, The Two-Stage Detonation Propulsion Model: Further Exploring 1st Stage Performance Effects by a Look Backward at the Model's Basis, in proc. 11th Int. Seminar, "New Trends in Research of Energetic Materials", Univ. Pardubice, Czech Republic, pp. 78–90, **2008**.

7.3.1 Example: calculation of the Gurney velocity for a general purpose bomb

The MK80 GBP series is the abbreviation for a series of flying bombs whose effect is based on pressure and splitter effect. All variations of this series show an optimized aerodynamic shape so that they can be used in quick-flying fighter planes. The MK80 series consists of bombs which are predominantly composed of the explosive payload (TNT, Composition B, Tritonal . . .), one or more fuses (nose and tail) and the tail unit with stabilizing fins or breaking facility. Typically, the mass of the explosive corresponds to approximately half of the total mass.

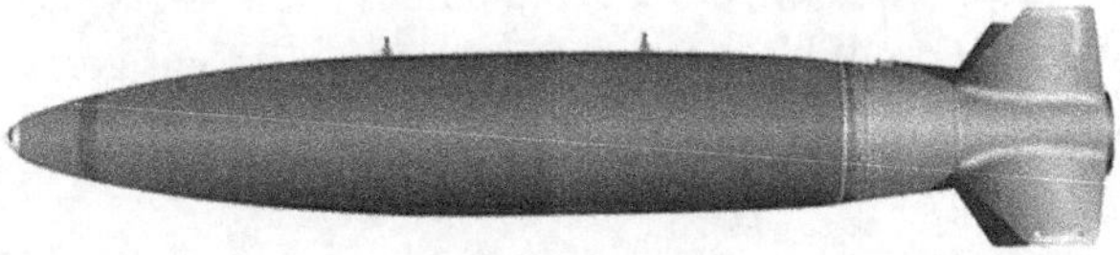

Fig. 7.22: MK84 general purpose bomb.

The MK84 (Fig. 7.22) is the largest bomb in the MK80 series. With a length of 3.8 m and a diameter of 46 cm they can be considered for the calculation of the Gurney velocity as being cylindrical. The total mass of the MK84 is 907 kg and the mass of the HE is approximately 430 kg.

Assuming that the payload of 430 kg of HE is only made up of Tritonal (80% TNT, 20% Al), the detonation velocity can be calculated to be $D = 6.70\ \text{km s}^{-1}$, based on a density of $\rho = 1.72\ \text{g cm}^{-3}$. Following the relationship discussed above:

$$\sqrt{2E} = \frac{3\sqrt{3}}{16} D \approx \frac{D}{3.08}$$

The Gurney velocity can be calculated to be

$$\sqrt{2E} = 2.175\ \text{km s}^{-1}$$

Therefore, if it is assumed to be a cylindrical charge, the fragment velocity can be approximated as follows:

$$\frac{V}{\sqrt{2E}} = \left(\frac{M}{C} + \frac{1}{2}\right)^{-0.5}$$

and

$$V = \sqrt{2E}\left(\frac{M}{C} + \frac{1}{2}\right)^{-0.5}$$

$$V = 2.175\ \text{km s}^{-1}\left(\frac{477\ \text{kg}}{430\ \text{kg}} + \frac{1}{2}\right)^{-0.5} = 1.72\ \text{km s}^{-1}$$

There are four basic types of air-delivered explosive bombs:

Penetrating: These weapons are designed to pierce armor with a directional charge and/or kinetic energy. They typically have a 25 to 30% charge-to-weight ratio.

Fragmentation: These weapons are designed to destroy personnel and soft-skinned targets such as vehicles, aircraft, and equipment primarily by fragmentation. They typically feature a thicker hardened metal case, often treated to optimize fragmentation. They typically have a 10 to 20% charge-to-weight ratio.

General Purpose: These are the most common air-delivered bombs and are designed to destroy targets through a combination of blast and fragmentation. There is generally some penetrating effect. General purpose bombs typically have a charge-to-weight ratio of approximately 50%.

High Capacity: These are weapons designed to destroy targets primarily through blast. They tend to be among the largest aerial bombs. They typically have thin cases to maximize blast energy. They typically have a 65 to 80% charge-to-weight ratio.

7.4 Plate dent tests vs. fragment velocities

A standard test procedure for determining the output of a detonator is the plate dent test. In the plate dent test, a detonator is assembled on a metallic plate of specified hardness and thickness. The depth of the dent produced after firing of the detonator is used as a measure of the strength of the detonator.

An important parameter of an explosive is its CJ pressure. Unfortunately, this is also one of the most difficult explosive properties to measure accurately. The available methods require measurement of the free-surface velocities of metal plates driven by explosive charges using a smear camera technique. Historically, CJ pressure measurement has been associated with attempts to measure the ill-defined property called brisance. Brisance is vaguely described as being the local shattering effect of an explosive.

Brisance – shattering – is what would be observed if a "pot" made from rock, ceramic, cast iron, or other "brittle" (non-energetic) material were filled with a mixture that could detonate (in other words, produce the 1st stage of detonation-driven propulsion).

Because the historical meaning of the word "brisance" implied more than mere CJ pressure (in varying and unknown degrees, effects resulting from charge size, impedance, detonation product isentropes, etc.), the older, intuitive ideas persisted. Gradually, Becker's definition was accepted.

In the early 1940s, a brisance test was developed that gave reasonable correlation with the CJ pressure data that was available. This has become known as the **Plate Dent Test**.

Described briefly, in the plate dent test, an explosive cylinder is mounted vertically on a cold-rolled steel plate of given hardness, and a detonator is fixed to the top of the explosive (Fig. 7.22a). The cylinder should be tall enough to allow an approximately steady detonation wave to develop in the cylinder after initiation. The steel plate should be sufficiently strong and massive that damage as a result of the explosion is confined to a dent, the depth of which is not dependent upon distortions of the entire plate. Plate dent tests are also conducted using aluminum alloy plates. In assessing explosives, steel or aluminum plate dent tests are used as references because these are the metals with adequately known material properties.

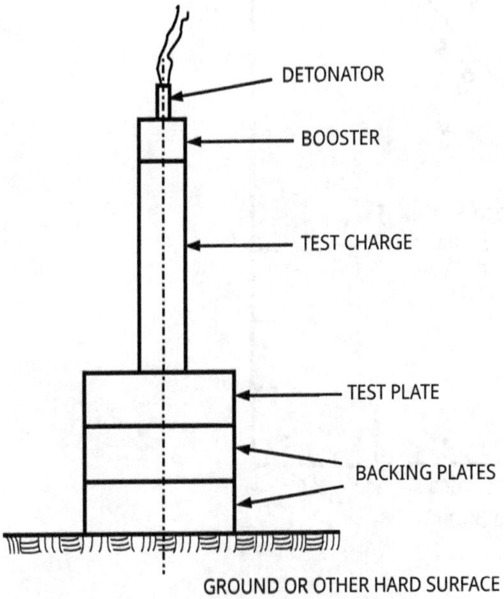

Fig. 7.22a: Plate Dent test configuration.

Figure 7.22b shows that there is a clear correlation between the CJ pressure and the plate dent. Using a familiar plate material such as steel, the plate dent test is a well-established procedure for assessing CJ pressures of new explosives. The test is fast, reliable, and inexpensive.
Lit.: G. H. Pimbley, A. L. Bowman, W. P. Fox, J. D. Kershner, C. L. Mader, M. J. Urizar, Investigating Explosive and Material Properties by Use of the Plate Dent Test, Los Alamos Scientific Laboratory Technical Report, LA-8591 MS, UC-45, Issued: November **1980**.

The Small Scale Shock Reactivity Test (SSRT test) follows a similar principle. In this test, the explosive under investigation ("sample" in Fig 7.23) is initiated by a commercial detonator and the depth in a solid aluminum block is measured for example by filling it with fine sand (SiO_2) and correlating the dent volume with the power of the explosive (Fig. 7.23, Tab. 7.7).

The depth of the dent is somewhat proportional to $\int p\, dt$ [N m^{-2} s = kg m^{-1} s^{-1}] (Figs. 7.24 and 7.25), which corresponds to the impulse density I_D (defined as: impulse I [kg m s^{-1}] per area (A [m^2]):

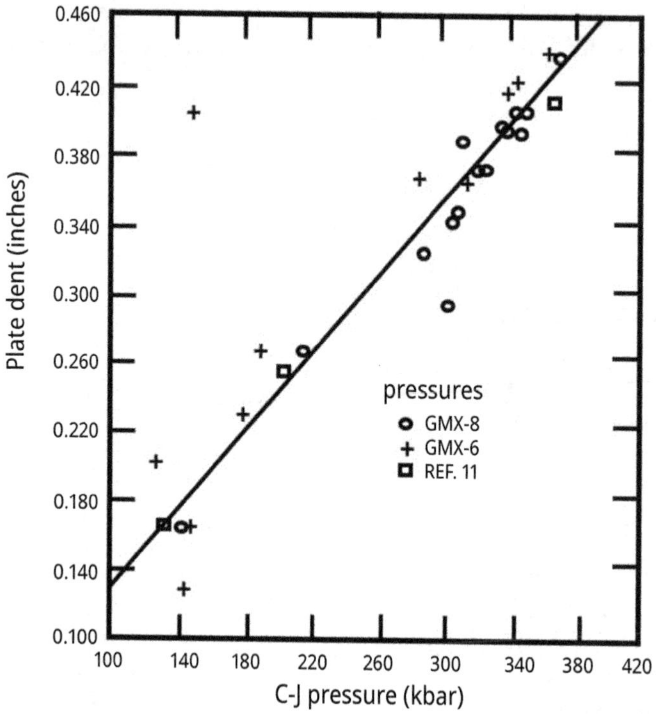

Fig. 7.22b: Plate dent vs. CJ pressure.

depth of dent	$\propto$	$\int p\, dt$	$[\mathrm{N\,m^{-2}\,s = kg\,m^{-1}\,s^{-1}}]$
impulse:	$\propto$	$I = m\,v$	$[\mathrm{kg\,m\,s^{-1}}]$
impulse density:	$\propto$	$I_D = \frac{I}{A}$	$[\mathrm{kg\,m^{-1}\,s^{-1}}]$

However, often the detonation of the next assembly (upon which the detonator acts in the explosive train) is governed, not by the time integral of a pressure pulse (Figs. 7.24 and 7.25), but by the instantaneous shock pulse. In other words, while the depth of the dent is somewhat proportional to $\int p\, dt$ $[\mathrm{N\,m^{-2}\,s = kg\,m^{-1}\,s^{-1}}]$ (Figs. 7.24 and 7.25), the strength of a detonator often corresponds better to an instantaneous shock pulse.

N.B.: A shock pulse (shock or pressure wave) develops when two pieces of moving metal contact each other in an initial impact. This shock pulse is in the ultrasonic frequency range and typically occurs at around 36 kHz. The amplitude of the shock pulse is proportional to the velocity of the impact.

It has been shown that increasing the length of the secondary (high) explosive column above a certain point does not change the velocity of the flyer plate, but increases the depth of the dent. This is due to the fact that the dent block responds to the time integral of the pressure pulse (Figs. 7.24 and 7.25). However, a longer column of an ex-

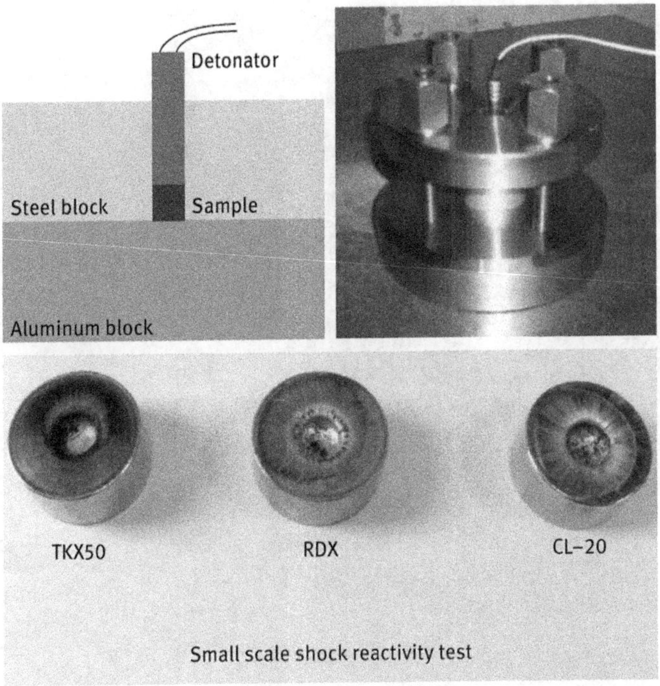

Fig. 7.23: Small Scale Shock Reactivity Test (SSRT test): schematic drawing (top left), photograph of test set-up (top right) and dents in solid aluminum discs after the test.

Tab. 7.7: SSRT test results for three high explosives (RDX, CL-20 and TKX-50).

Explosive	Mass [mg]	Dent [mg SiO_2]
RDX	504	589
CL-20	550	947
TKX-50	509	857

plosive will have the same detonation characteristics and thus throws the flyer plate with the same velocity. Therefore, a better method for determining the output of a detonator may be the calculation or measurement of fragment velocities, since the instantaneous shock pulse is proportional to the acceleration of a metal plate.

Calculated results on the acceleration of metal plates resulting from different explosive charges were obtained using the one-dimensional computer hydrocode EP [N. F. Gavrilov, G. G. Ivanova, V. N. Selin, V. N. Sofronov, UP-OK program for solving continuum mechanics problems in one-dimensional complex, VANT. Procedures and programs for numerical solution of mathematical physics problems, (3), p. 11–21, **1982**] (Figs. 7.26. 7.27 and 7.28). As a result of such calculations, a rather complete com-

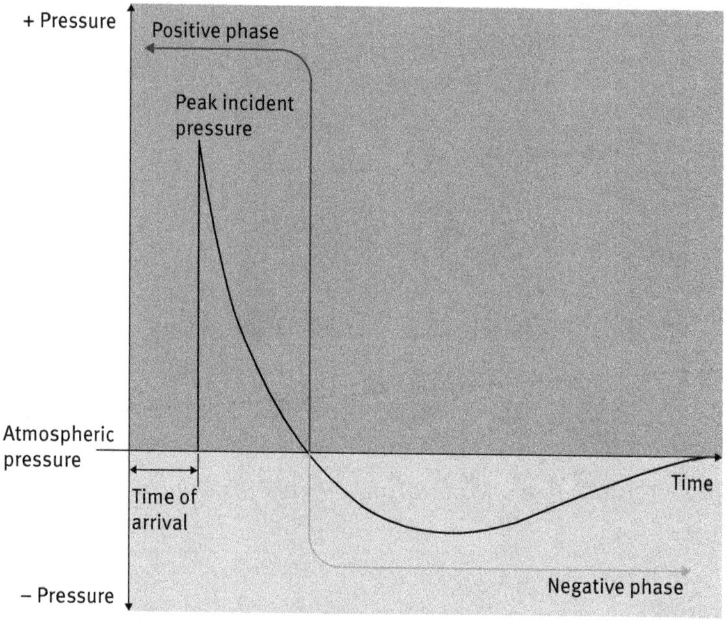

Fig. 7.24: Typical pressure-time history.

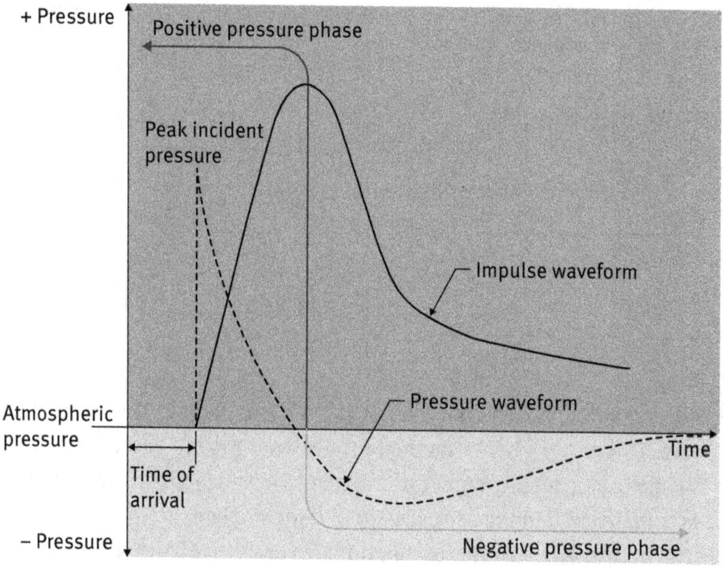

Fig. 7.25: Typical impulse waveform.

parative picture of explosion properties and characteristics of shock wave action onto barriers was obtained for several new and standard explosive materials. The results which were obtained indicate that in a wide range of their initial states (porosity,

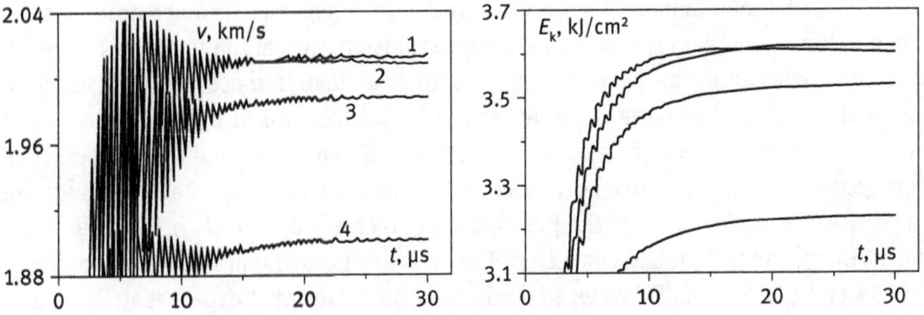

Fig. 7.26: Acceleration of plates by free charges of HMX (1), TKX-50 (2), MAD-X1 (3) and RDX (4).

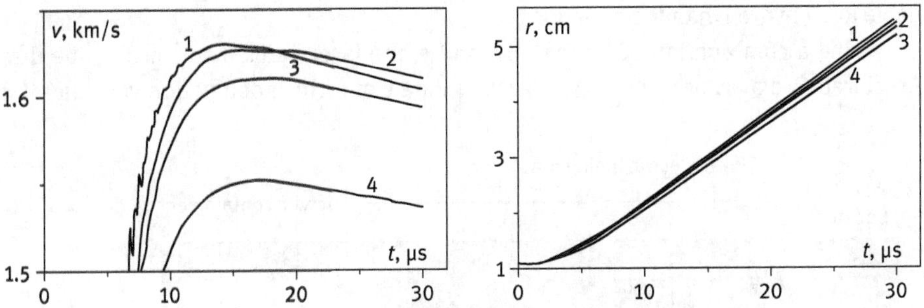

Fig. 7.27: Acceleration of Tantalum cylindrical layers by charges of HMX (1), TKX-50 (2), MAD-X1 (3) and RDX (4).

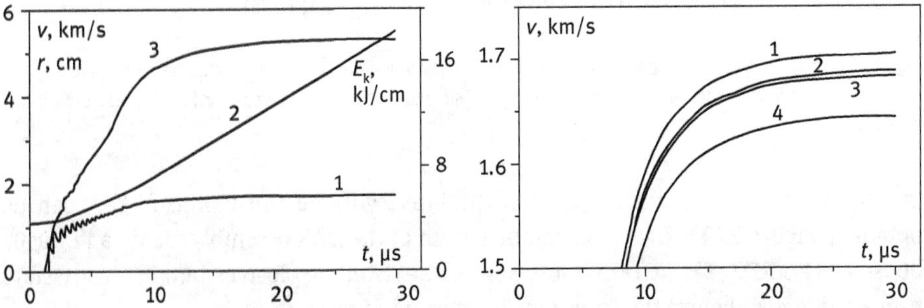

Fig. 7.28: Acceleration of a cylinder (1 v, 2 r, 3 – E_K) by a porous charge of TKX-50 (left) and cylinders by porous charges of HMX (1), TKX-50 (2), MAD-X1 (3) and RDX (4).

inert binders), the new explosive materials TKX-50 and MAD-X1 possess higher characteristics of explosion and shock wave action on practically every compact barrier in comparison with many standard explosive materials, and in particular in comparison with the most used military explosive, RDX.

The oscillations that can be seen in some figures (e.g. Fig. 7.26) are quite natural when a thin metal plate is accelerated by an explosion. In real life they are, of course, more blurred than in the calculation, which does not take into account all dissipation processes. Furthermore, the possible effect of spall fracture of inert plates was not taken into account. Nevertheless, at certain ratios of explosive/plate thicknesses, spall damage to the plates is possible. This effect can in a certain way affect the nature of the oscillation of the speed of the plates, but in no case will it really affect their energetics, that is, on the studied character of the relative performance of the explosives.

Recently, Lorenz et al. developed a relatively simple test that experimentally obtains accurate measurements of the detonation velocity, pressure, and expansion energy of new high explosives (*PEP* **2015**, *40*, 95–108). The so-called Disc Acceleration eXperiment (DAX test) provides an initial condition of steady detonation and a charge-geometry amenable to 2D hydrodynamic simulations.

Using a combination of timing pins and a one laser velocimetry probe, the disc Acceleration experiment, or DAX, generates data from which detonation velocities, C-J

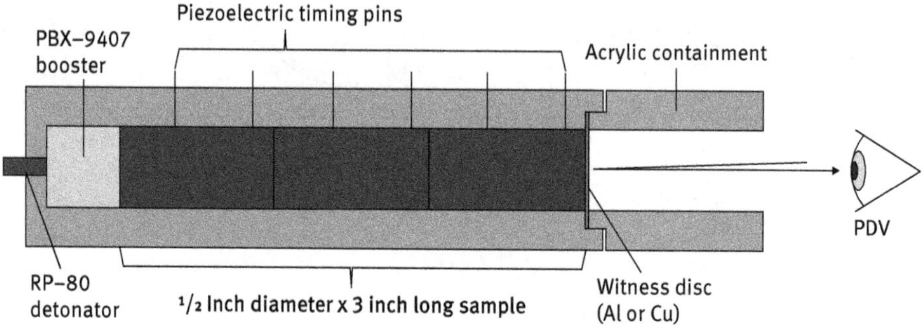

Fig. 7.29: Schematic design of the DAX assembly. The PDV velocity-probe that tracks the witness foil acceleration is fixed to the end of a 100 mm long acrylic extension and fiber coupled to the PDV detection hardware.

pressures and the detonation energy (at relative volume expansions of 2–3) can be obtained. Figure 7.29 shows a schematic design of the DAX assembly. Using a PBX9407 booster (94% RDX, 6% VCTFE), most new and existing explosives obtain steady-state propagation well before the front reaches the end of the rate stick.

The PDV diagnostic measures the accelerating witness plate, which serves as the primary measure of explosive performance for DAX. A very detailed analysis of the plate velocity allows accurate estimates of the C-J pressure, product gas energies and JWL EOSs.

Figure 7.30 shows the disc trajectory (velocity vs. time) data for a typical single DAX shot. The initial motion of the foil's rear surface (away from the sample) is a free surface "jump-off" that is approximately twice the in-material particle velocity. The disc continues to accelerate in steps, due to the reverberating pressure wave inside the disc. This is a common condition for foils and thin plates driven by explosives.

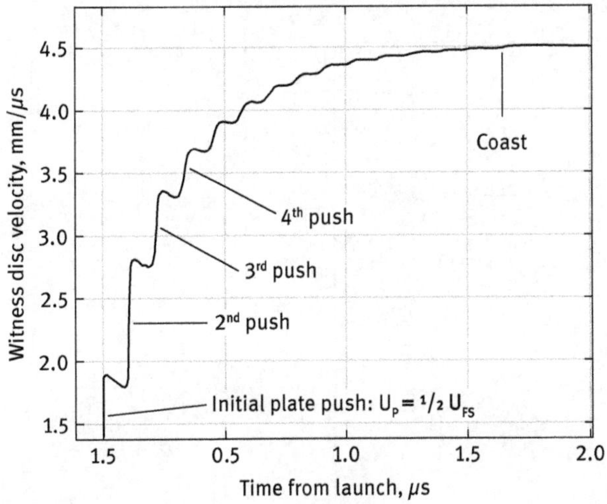

Fig. 7.30: Disc trajectory (velocity vs. time) data for a typical single DAX shot.

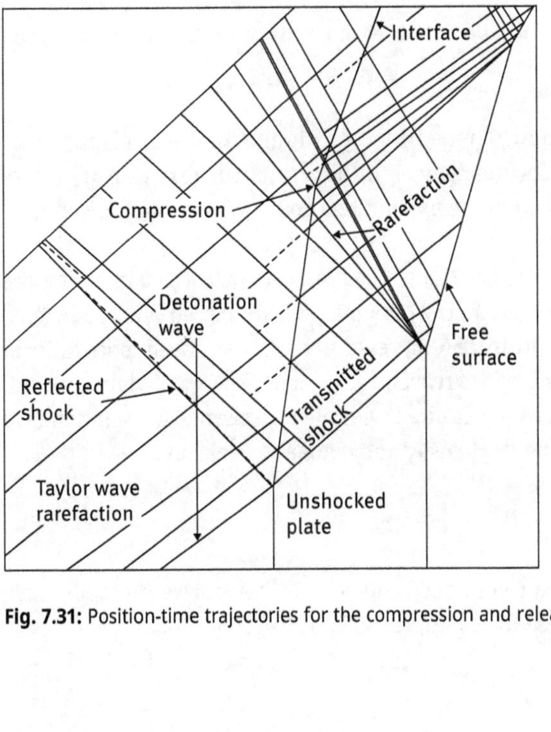

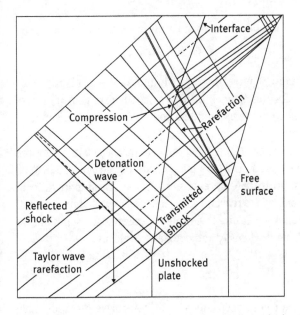

Fig. 7.31: Position-time trajectories for the compression and release waves.

The time-dependent pressure wave (Taylor wave) behind the detonation front, results in a series of compression and release waves in the disc, which are observed as accelerations and decelerations of the back surface of the witness disc. A generalized position-time description of the reverberating waves is shown in Figure 7.31.

Figure 7.32 shows velocity-time profiles of the accelerating witness disc using two different witness disc materials and thicknesses. In the "coast" region (around 2 μs),

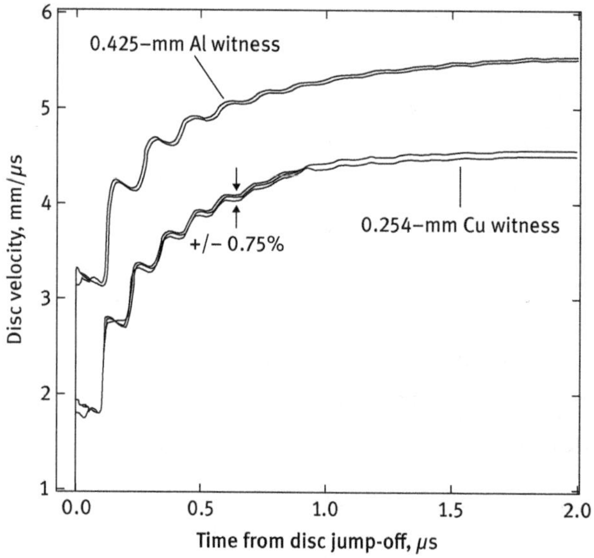

Fig. 7.32: DAX data for a PETN-based explosive using two different witness disc materials (Cu and Al) and disc thicknesses.

the copper disc has 33% more kinetic energy than the aluminum disc. This indicates the improved coupling efficiency for the heavier witness disc. The disc velocity corresponds to the internal energy change in the expanding product gases, but the coupling to the plates varies with plate mass.

The detonation velocity can be directly obtained from the piezoelectric timing pin data (cf. Fig. 7.29), and the difference (in mm) where the pins are placed into the detonation tube. However, the detonation pressure (C-J pressure) has always been more difficult to determine. This is because the end of the reaction zone is not always obvious – even if the pressure of the detonation could be measured with high precision. One solution is to simply estimate the C-J pressure from the density and detonation velocity:

$$P_{\text{C-J}} \approx \frac{1}{4}\rho_0 D_{\text{V}}^2$$

However, this approximation is not very accurate, and more precise methods have been described in the literature (see for example: *PEP* **2015**, *40*, 95–108).

7.5 Underwater explosions

Depending on the medium in which the pressure-wave generated by an explosive expands into its surroundings, one can distinguish between
(i) underwater explosions (UNDEX) and
(ii) air explosions (AIREX).

The unique properties of underwater explosions are due to the high velocity of sound in water meaning that the pressure-wave travels approximately four times faster in water than it does in air. Furthermore, due to the high density and low compressibility of water, the destructive energy (from the explosion) can be efficiently transferred over relatively large distances. The most important effects caused by an underwater explosion are the corresponding shock-wave and the gas bubble pulsations.

During detonation, an explosive undergoes rapid conversion into gaseous products with high temperature and pressure. If detonation occurs underwater, complex physicochemical processes take place. Nevertheless, taking into account the damage which is caused by the detonation, this process (underwater explosion) can be considered as resulting from a shock wave (Fig. 7.33) and the subsequent gas bubble pulsations of the detonation products (Figs. 7.34 and 7.35). The shock wave which is generated propagates radially outward from its source and acts on the surrounding water. At the front of the shock wave, the pressure increases violently to the peak pressure point and then decays exponentially to hydrostatic pressure. The pressure pulses of the detonation gas products are much slower in comparison with the shock wave. Initially, the gases which have been compressed to high pressure cause rapid expansion of the gas bubble. Due to the inertia of outwardly moving water, the bubble expands to pressures lower than the point of pressure equilibrium and this continues until the pressure difference becomes large enough to stop the outward flow of water. Subsequently, the higher pressure of the surrounded water reverses the process causing contraction of the gas bubble. Similarly, the reverse process does not stop at the point of pressure equilibrium, but recompresses the detonation products to higher pressure values. Afterwards, the second pulsation occurs (an acoustic pulse without generation of the shock wave). The bubble expansion-contraction oscillations referred to as *bubble pulses* (*bubble collapse*) take place several times. Each subsequent pulse is weaker due to loss of energy. Because of high buoyancy, the bubble of gases moves upward in a jumping manner.

The underwater explosion test is a widely used method to compare the effectiveness of various explosives and is a convenient method for measuring the brisance and heaving power. This approach is based on the following hypothesis:

- the shock energy from an underwater explosion measures the explosive's shattering action in other materials (μs time-scale)
- the bubble energy from the underwater explosion measures the heaving action of the explosive (ms time-scale)

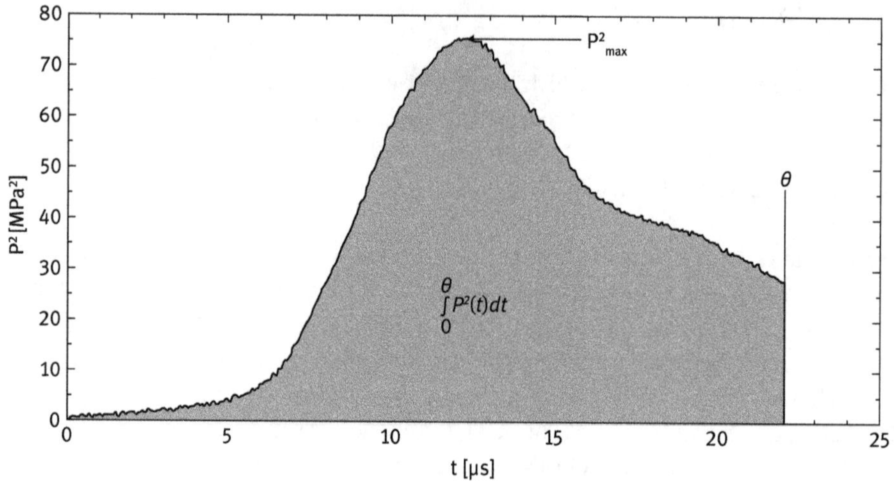

Fig. 7.33: Calculation of the shock energy equivalent E_S.

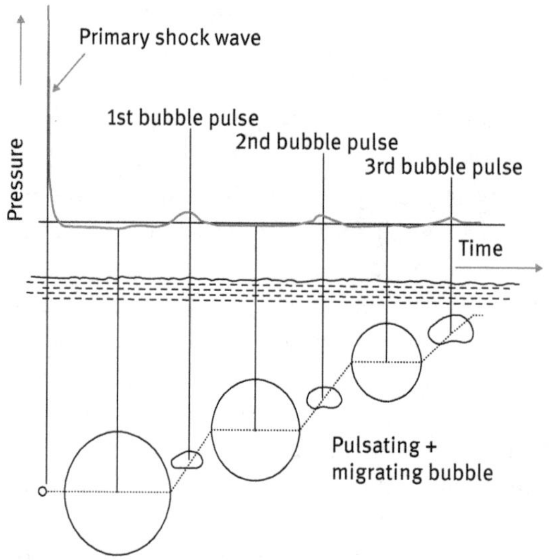

Fig. 7.34: Schematic presentation of a pulsating and migrating bubble.

The shock energy is the compressional energy radiated from an underwater detonation. In the tests, it is obtained by measuring the area under the *P2(t)* curve at a known distance. The total explosive energy is postulated to be the sum of the shock wave and bubble energies. Additionally, underwater tests also can measure the shock wave impulse – the rate of change of pressure of the shock wave – which is another indicator of explosive strength. The shock wave impulse is derived by measuring the

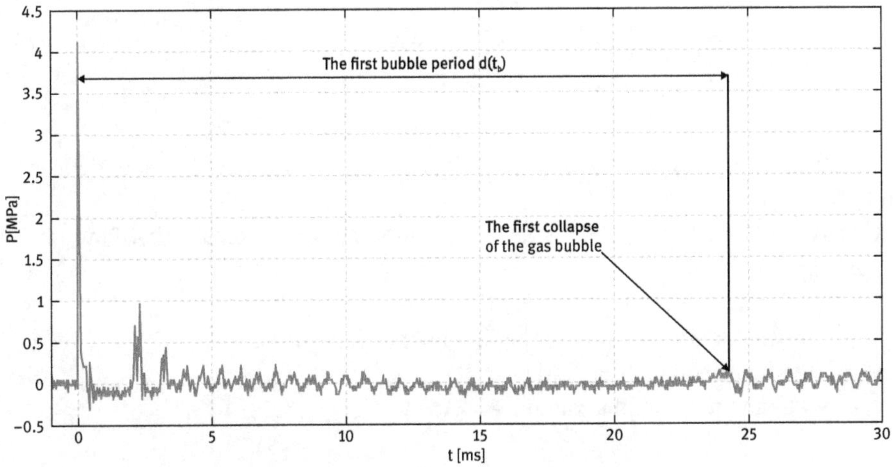

Fig. 7.35: Plot of the bubble energy equivalent $E_B = t_b^3$.

area under the *P(t)* curve for a selected integrated time interval at a known distance from the explosion.

Key characteristics of underwater explosion phenomena which are used to evaluate the performance of explosives are:

a) shock wave:
   - the peak overpressure (P_m)
   - the time constant (θ; the time required for the pressure to fall to the value $P_m \cdot e^{-1}$)
   - the overpressure (above the hydrostatic pressure) change with time (*P(t)*)
   - the impulse of the shock wave, (*I*)

The impulse, *I* is calculated using following equation:

$$I = \int_0^t P(t)dt \tag{7.1}$$

where:
P(t) – overpressure (above the hydrostatic) change with time
P_0 – hydrostatic pressure

the upper limit of the integral is usually 5 θ
- energy flux density (energy)

The energy flux density of shock wave (*E*, value of work done on a surface or the energy behind the shock front per unit area) corresponds to

$$E = \frac{1}{\rho_0 C_0}\left(1 - 2.422 \cdot 10^{-4}\, P_m - 1.031 \cdot 10^{-8} P_m{}^2 I\right) = \int_0^t P^2(t)\,dt \tag{7.2}$$

where:

$\rho_0 C_0$ – the acoustic impedance of the water (ρ_0 – density of water; C_0 – sound velocity in water)

the coefficients –2.422 and –1.031 take into account the correction for after-flow

t – usually 5 θ

b) oscillation of the first bubble:
   - the maximum pressure of the first bubble pulse (P_B)
   - the first bubble phase duration (T_1)
   - the maximum bubble radius (A_{max})

As a consequence of the many advantages (such as the huge amount of information which can be obtained during the test as well as the low costs, and the easy performance of the underwater explosion), this method has been standardized and is used to compare the initiating capability of detonators [5]. The primary shock wave energy and the bubble gas energy are determined by measuring the overpressure of the shock wave energy generated in water. The methodology which is described in the European Standard (Explosives for civil uses – Detonators and relays – Part 15: Determination of equivalent initiating capability, EN 13763-15) is based on the following equations which are used for describing the primary shock-wave energy and bubble energy [4, 6, 7].

The full description of shock wave energy generated in water

The primary shock-wave energy (E_{SW}) can be determined from the following equation

$$E_{SW} = \frac{4\pi R^2}{\rho_w C_w} \int_{t_0}^{t_0+\theta} P^2(t)\,dt \tag{7.3}$$

where:

R – distance between detonator and pressure sensor

ρ_w – water density

C_w – sound velocity in water

t_0 – moment the shock wave starts

θ – time at which the sensor output has decreased to $P_{max} \cdot e^{-1}$

The full description of the bubble energy generated in water

The time which is measured between the moment of the primary shock wave registration and the consecutive pressure wave is used to calculate the bubble gas energy (E_{BW}):

$$E_{BW} = \frac{A \cdot \sqrt{(Bh + C)^5}}{M_{HE}} t_b^3 \tag{7.4}$$

where:
A, B, C – constants which depend on the experimental conditions
h – depth the detonator is immersed
M_{HE} – mass of explosive
t_b – time interval between the shock-wave pressure peak and the first collapse of the gas bubble

Assuming there are no boundary effects, the so-called Willis formula can be used to calculate the bubble energy from the first period of pulsation of the gas bubble (t_b) [7]:

$$t_b = \frac{1.135 \cdot \sqrt[3]{E_{BW}} \cdot \sqrt{\rho_w}}{\sqrt[6]{p_h{}^5}} = \frac{1.135 \cdot \sqrt[3]{E_{BW}} \cdot \sqrt{\rho_w}}{\sqrt[6]{(\rho_w \cdot g \cdot h + 101325)^5}} \tag{7.5}$$

where:
p_h – total hydrostatic pressure at the charge depth
g – gravitational acceleration

The total explosive energy (E) is assumed to be the sum of the shock wave and bubble energies.

$$E = E_{SW} + E_{BW} \tag{7.6}$$

– Shock energy equivalent, E_s (Fig. 7.33):

$$E_s = \int_{t_0}^{t_0+\theta} P^2(t)\, dt \tag{7.7}$$

– Bubble energy equivalent, E_B (Fig. 7.35):

$$E_B = t_b^3 \tag{7.8}$$

7.6 The detonation velocity of mixtures of solid explosives with non-explosive liquids

Urbanski et al. published a paper in 1939 which considered how the detonation velocity (VoD) of a solid explosive with a density of less than the theoretical maximum density (TMD) is affected, if the air contained in the explosive gets replaced by a non-explosive liquid.[1] The authors concluded that replacing the air with a liquid that does not dissolve the explosive results in an increase in the detonation velocity. This appears to support the theory of Becker and Schmid, which states that the VoD is equal to the sum of the speed of sound (C) and the particle velocity (U) at the CJ point:[2–4]

$$\mathrm{VoD} = \mathrm{C} + \mathrm{U} \tag{1}$$

In 1968, Hikita and Fujiwara proposed a method for calculating the detonation velocity under the assumption that the inert liquid acts only as a shock transmitter during the reaction.[5] The authors showed that their proposed mechanism was in agreement with the experimental results.[5]

In order to quantify Urbanski's finding, the calculated detonation velocities of high explosives with non-explosive liquids [EXPLO5 code (V6.05.04)] can be looked at more closely.[6]

RDX has a theoretical maximum density of 1.8 g cm^{-3} and was used as the high explosive in the calculations. The results for the calculated detonation velocities at TMD and ρ = 1.35 g cm^{-3} are shown in Tab. 7.8.

Tab. 7.8: Calculated VoD for RDX with different densities.

ρ [g cm^{-3}]	VoD [m s^{-1}]
1.80 (TMD)	8,798
1.35	7,315

If a density (ρ = m/V) of 1.35 g cm^{-3} is assumed, the volume that 100 g occupy is (V = m/ρ = 100/1.35 cm^3) 74.074 cm^3. Therefore, a 100 g sample of RDX with a density of 1.35 g cm^{-3} and a total volume of 74.074 cm^3 contains 99.978 g of "net" RDX with a density of 1.800 g cm^{-3} (TMD) and occupying 55.543 cm^3, as well as 0.022 g of air occupying 18.531 cm^3. Calculation of a mixture of 99.978 mass-% RDX and 0.022 mass-% air resulted in a density of 1.35 g cm^{-3} and a VoD of 7319 m s^{-1}, which is practically identical (the difference is only 4 m s^{-1}, or 0.055%) to the calculation (ρ = 1.35 g cm^{-3}) in Tab. 7.8.

If the detonation velocities are now calculated in which the 18.531 cm^3 air is replaced by various liquids of different densities and sound velocities,[7] the mass of RDX is unchanged (99.978 g). The results are summarized in Tab. 7.9.

Tab. 7.9: Calculated VoD for RDX mixed with different non-explosive liquids.

Non-expl. liquid	Speed of sound in non-explosive liq. [m s^{-1}] [7–9]	Density of non-explosive liquid [g cm^{-3}]	Mass ratio RDX/non-expl. liquid	Density of mixture [g cm^{-3}]	VoD [m s^{-1}]
air (gas)	343	0.0012	99.978/0.022	1.351	7,314
CCl_4 (l)	926	1.59	77.238/22.762	1.600	7,714
CH_3OH (l)	1123	0.79	87.228/12.772	1.665	7,430
H_2O (l)	1497	1.00	84.363/15.637	1.555	7,861
glycerine (l)	1923	1.26	81.068/18.933	1.544	7,806
kerosene (l)	1324	0.82	86.807/13.193	1.609	7,441
acetone (l)	1161	0.79	87.228/12.772	1.572	7,509
aniline (l)	1640	1.02	84.100/15.900	1.590	7,531
benzene (l)	1298	0.88	85.977/14.023	1.612	7,365
cyclohexanol (l)	1465	0.96	84.894/15.106	1.547	7,578
ethylene glycol (l)	1660	1.20	81.805/18.195	1.600	7,762

Tables 7.9 and 7.10 show that the final detonation velocity does not correlate well with the density of the mixture, nor with the speed of sound in the added non-explosive liquid, but with the sound velocity in detonation products at the CJ point (Fig. 7.36), while the particle velocity in these mixtures with similar composition is relatively constant. This is nicely in accord with Urbanski's finding, and the theory of Becker and Schmid, which states that the VoD is equal to the sum of the speed of sound (C) and the particle velocity (U) (eq. 1). The results in Tab. 7.9 also agree nicely with the study by Hikita and Fujiwara [5] (this study: 18.9% glycerine, VoD = 7,806 m s^{-1}; Hikita and Fujiwara [5]: 20% glycerine, VoD = 7,850 m s^{-1}). The fact that the RDX/air mixture still shows a relatively high VoD can be attributed to the high mass ratio of RDX, while the other non-explosive liquids do not contribute to the energy output (but in many cases, contributed to increased sound velocities).

Tab. 7.10: Calculated sound velocities (C) and shock-wave velocities (particle velocity, U) for different non-explosive liquids.

Non-expl. liquid	Mass ratio, RDX/non-expl. liquid	Particle velocity, U [m s^{-1}]	Sound velocity, C [m s^{-1}]	VoD [m s^{-1}]
benzene (l)	85.977/14.023	1795	5,570	7,365
CH_3OH (l)	87.228/12.772	1848	5,582	7,430
kerosene (l)	86.807/13.193	1794	5,647	7,441
acetone (l)	87.228/12.772	1862	5,647	7,509
aniline (l)	84.100/15.900	1803	5,728	7,531
cyclohexanol (l)	84.894/15.106	1810	5,768	7,578
CCl_4 (l)	77.238/22.762	1892	5,822	7,714
ethylene glycol (l)	81.805/18.195	1869	5,893	7,762
glycerine (l)	81.068/18.933	1894	5,912	7,806
H_2O (l)	84.363/15.637	1914	5,947	7,861

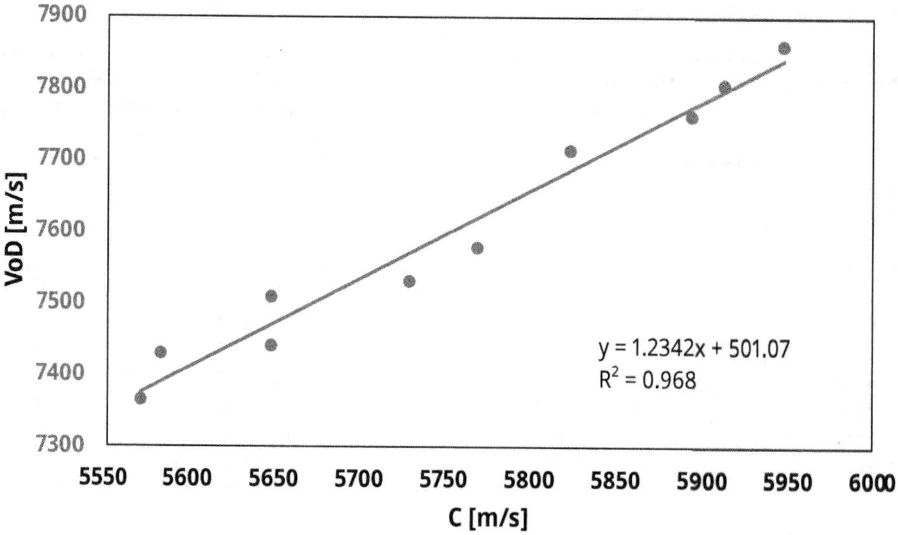

Fig. 7.36: Correlation between the detonation velocity (VoD) and the sound velocity (C) in various mixtures (see Tab. 7.10).

In summary, pressing out the air of a high-explosive powder and replacing the voids with a non-explosive liquid such as ethylene glycol, water or glycerin may help to increase the final detonation velocity.[9]

Literature

[1] T. Urbanski, T. Galas, Sur la vitesse de détonation des mélanges d'explosifs solides avec des liquids non explosives, Compt. Rend. 1939, 9th October, 558–560.

[2] T. M. Klapötke, Chemistry of High Energy Materials, 5th edn., chapter 1.3, T. M. Klapötke, Walter de Gruyter, Berlin/Boston, 2019.

[3] P. O. K. Krehl, Shock wave physics and detonation physics – a stimulus for the emergence of numerous new branches in science and engineering, Eur. Phys. J. H 2011, 36, 85–152.

[4] M. Suceska, Test Methods for Explosives, Springer, New York, 1995.

[5] T. Hikita and S. Fujiwara, A Contribution to the Theory of Detonation of Solid Explosives Containing Inert Liquids, Science and Technology of Energetic Materials, 1968, 29, 432–437.

[6] M. Suceska, EXPLO5-version V6.05.04, OZM Research, 2020, Czech Republic.

[7] http://hyperphysics.phy-astr.gsu.edu/hbase/Tables/Soundv.html (last accessed on October 14, 2020).

[8] https://www.engineeringtoolbox.com/sound-speed-liquids-d_715.html (last accessed on October 14, 2020).

[9] T. M. Klapötke, M. Suceska, Propellants, Explosives, Pyrotechnics 2021, 46, 352 – 354.

Check your knowledge: See chapter 14 for study questions.

8 Correlation between the electrostatic potential and the impact sensitivity

8.1 Electrostatic potentials

In Chapters 4.2.2 and 4.2.3 we have seen that it is often possible to predict performance parameters using quantum mechanics without requiring experimental data. There are also huge efforts being made towards being able to theoretically predict, not only the performance but also the sensitivity parameters (if possible using quantum mechanical ab initio or DFT calculations), and particularly to allow a prediction of the sensitivity of new molecules which have not been synthesized previously. This would mean that the preparation of sensitive molecules which would have little chance of ever being used as energetic materials could be avoided. In addition, such sensitivity predictions would help to increase safety and also improve financial issues related to for synthetic projects so that they could be used more effectively for more specific targets. In comparison to computer calculations, the synthetic work in chemical laboratories requires a lot more effort and is much more time consuming and therefore considerably more expensive. As discussed previously in Chapter 6.1, the impact sensitivity is one of the most important parameters for the estimation of the sensitivity of highly energetic compounds. At this point it is worthwhile to repeat that although the drop-hammer method for the determination of the impact sensitivity uses a mechanical method, the initiation most likely also occurs as a result of hot-spots. Thus, it occurs thermally. It is thanks to the work of Peter Politzer, Jane Murray and Betsy Rice [39–44] in particular, that it was realized that for covalently bonded molecules, the electrostatic potential (ESP) of an isolated molecule can be correlated with the (bulk) material properties of the compound in the condensed phase.

The electrostatic potential $V(r)$ for a particular molecular hypersurface (normally 0.001 electron $bohr^{-3}$ surface) is defined as follows:

$$V(r) = \sum_i \frac{Z_i}{|R_i - r|} - \int \frac{\rho(r')}{|r' - r|} dr'$$

Here, Z_i and R_i are the charge and coordinates (position) of atom i, and $\rho(r)$ is the electron density. The electrostatic potential can be obtained either from X-ray diffraction investigations, but more commonly it is obtained using quantum chemical calculations. Regions in which the ESP is positive are electron deficient and their electron density is low in this area. In contrast, regions which have negative potentials are electron rich and have a higher electron density. The relative magnitude and extension of positive and negative potentials on a given hypersurface of a molecule are of decisive importance for the impact sensitivity of the compound. In the following section, the 0.001 electron $bohr^{-3}$ hypersurface will always be considered. For “typical”,

https://doi.org/10.1515/9783111447087-008

non-impact sensitive organic molecules, the areas of the negative potentials are smaller but significantly stronger than the positive ones. In contrast to this, in impact sensitive compounds the regions of positive potentials are still larger, but as strong as or stronger than the negative regions (Fig. 8.1). Politzer and Rice were able to show that the impact sensitivity can be described as a function of this anomalous behavior. Figure. 8.1 shows the calculated electrostatic potentials of benzene, nitrobenzene (both insensitive) and trinitrobenzene (impact sensitive). It can also be said that the less evenly the electron density is distributed (without taking into account extreme values due to charges above atoms of the electron donating or electron withdrawing substituents) the more sensitive a compound will behave. Moreover, it could be determined that for aromatic compounds (see Fig. 8.1) the impact sensitive species show a stronger electron deficient region above the aromatic ring (more positive potential) in comparison with the less sensitive species. It appears to be the case that the more impact sensitive compounds are significantly more electron deficient in the centre of their structures (more positive potential) than their insensitive analogues. Above the C—NO_2 bond in the more impact sensitive trinitrobenzene there also is a region of positive potential.

In addition to the qualitative conclusions (see Fig. 8.1), attempts have also been made to be able to quantitatively predict the impact sensitivity, in particular for the important organo-nitro compounds. In this, the average electrostatic potential V_{Mid} (kcal mol^{-1}) at the "mid point" of the C—N bond, or the averaged potential V^*_{Mid} (kcal mol^{-1}) for all C—N bonds in the molecule, plays an important role:

$$V_{Mid} = \frac{Q_C}{0.5R} + \frac{Q_N}{0.5R}$$

$$V^*_{Mid} = \frac{1}{N}\sum_{i=1}^{N}\left(\frac{Q_C}{0.5R} + \frac{Q_N}{0.5R}\right)$$

For organo-nitro compounds, the following correlation between V^*_{Mid} and the (widely used in the USA) $h_{50\%}$ value has been proposed by Rice et al. [44]:

$$h_{50\%} = y_0 + a\,\text{exp.}\,(-b\,V^*_{Mid}) + c\,V^*_{Mid}$$

The $h_{50\%}$ value is directly correlated to the impact sensitivity and corresponds to the drop-height in cm, for which 50% of the samples explode in a drop hammer test using a 2.5 kg mass (Tab. 8.1). In the equation above, y_0, a, b and c are constants with the following values:

$a = 18922.7503$ cm
$b = 0.0879$ kcal mol^{-1}
$c = -0.3675$ cm $kcal^{-1}$ mol^{-1}
$y_0 = 63.6485$ cm

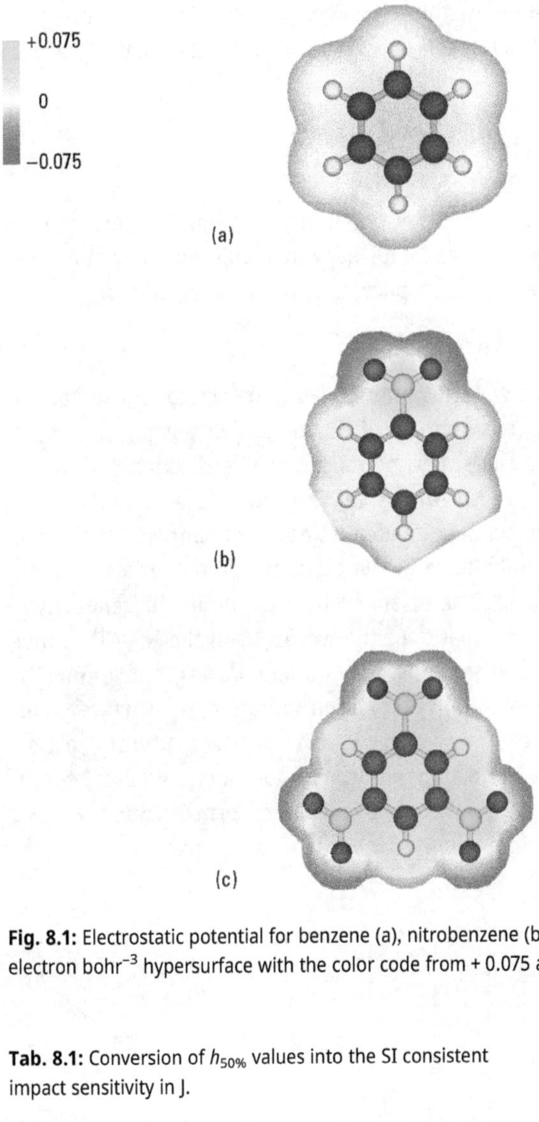

Fig. 8.1: Electrostatic potential for benzene (a), nitrobenzene (b) and trinitrobenzene (c) on the 0.001 electron bohr^{-3} hypersurface with the color code from + 0.075 a.u. (green) to −0.075 a.u. (red).

Tab. 8.1: Conversion of $h_{50\%}$ values into the SI consistent impact sensitivity in J.

compound	$h_{50\%}$ / cm	impact sensitivity / J
PETN	13	3
RDX	28	7
HMX	32	8
HNS	54	13
FOX-7	126	31
NTO	291	71

On this basis, it should also be possible for other classes of compounds to establish similar correlations, in order to predict the impact sensitivities at least semi-quantitatively.

8.2 Volume-based sensitivities

Politzer, Murray et al. have also researched the possibility of a link between the impact sensitivities of energetic compounds and the space available for their molecules in the crystal lattices [39 f, g]. To measure this space, they used the equation

$$\Delta V = V_{eff.} - V(0.002)$$

where $V_{eff.}$ is the effective molecular volume obtained from the crystal density and V (0.002) is that enclosed by the 0.002 a.u. contour of the molecule's gas phase density, determined computationally. When the experimental impact sensitivity (h_{50} values) was plotted against ΔV for a series of 20 compounds (Fig. 8.2, a), the nitramines (▴) formed a separate group showing little dependence on ΔV. Their impact sensitivities correlate well with an anomalous imbalance in the electrostatic potentials on their molecular surfaces, which is characteristic of energetic compounds in general (see above, Ch. 8.1). The imbalance is symptomatic for the weakness of the N—NO_2 bonds, caused by depletion of electronic charge. Thus, for compounds (e.g. nitramines) in which trigger-linkage-rupture is a key step in detonation initiation, the surface potential imbalance can be symptomatic of sensitivity. The impact sensitivities of non-nitramines (Fig. 8.2, a), on the other hand, depend on ΔV a lot more, and can be quite effectively related to it if an electrostatically-based correction term is added (Fig. 8.2, b). Why might this be a factor? Perhaps the availability of more space enhances the

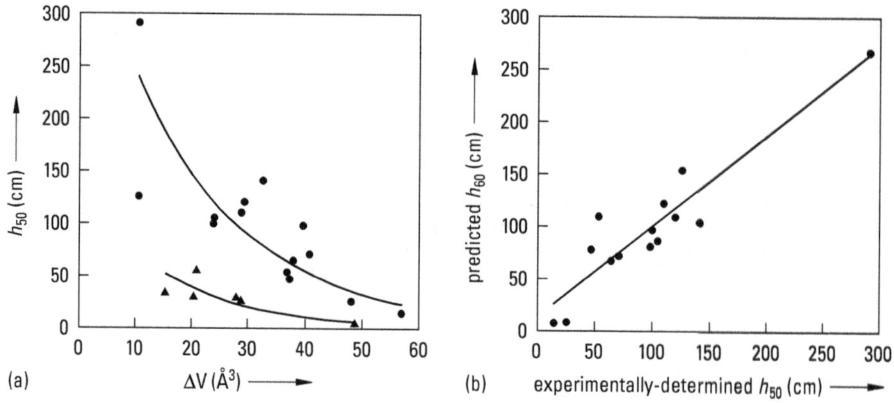

Fig. 8.2: Measured impact sensitivities (h_{50} values) plotted against computed ΔV for various energetic compounds (▴ = nitramines, • = non-nitramine energetic compounds) (a). Comparison of impact sensitivities predicted on an electrostatically corrected volume-based model [39 f,g] (b).

molecule's ability to absorb and localize the external energy coming from the impact, for example either vibrationally or translationally.

The volume-based (ΔV) h_{50} values for the non-nitramine energetic compounds were predicted using the equation

$$h_{50} = a(\Delta V)^{\frac{1}{3}} + \beta v \sigma_{tot}^2 + \gamma$$

where α (−234.83), β (−3.197) and γ (962.0) are constants, $\sigma_{tot}^2 = \sigma_+^2 + \sigma_-^2$ (with σ_+^2, σ_-^2 and σ_{tot}^2 being the positive, negative and total variances of the electrostatic potential) and ν being an electrostatic balance parameter defined as $v = (\sigma_+^2 \sigma_-^2)/(\sigma_+^2 + \sigma_-^2)^2$.

Check your knowledge: See Chapter 14 for Study Questions.

9 Design of novel energetic materials

9.1 Classification

As we have already discussed above, secondary explosives (or high explosives) are homogeneous substances which can either be covalent or ionic. However, in both cases it is important that the oxidizer and fuel are combined within *one* molecule (i.e. in one covalent molecule or in one ionic formula unit). For example, this is the case for TNT, in which the three nitro groups correspond to the oxidizer, and the C—H backbone acts as the fuel. Such molecules belong to the "**oxidation of the carbon backbone**" group of substances (Tab. 9.1).

Tab. 9.1: Overview of different approaches for the design of highly energetic compounds.

type	example	comments	structure
oxidation of the C back-bone,	TNT, PETN, RDX	compound can be exothermic or endothermic, compounds can be covalent or ionic	Fig. 1.3
ring or cage strain	CL-20, ONC	most endothermic, most covalent	Fig. 1.6
nitrogen-rich molecules	TAGzT, Hy-At	always endothermic, compounds can be covalent or ionic	Fig. 2.2

In the TNT molecule, all of the angles involving the C atoms are approximately 120° (C_6 ring) or 109° (CH_3 group) and they therefore correspond almost to the ideal angles for a sp^2 and sp^3 hybridized atom. If one now considers a strained ring or cage structure, for example in CL-20, the energy of the molecule is increased as a result of the ring or **cage strain**. Consequently, in the explosive decomposition forming CO, CO_2, H_2O and N_2, more energy is released than would be in an unstrained system (Tab. 9.1).

Principally, members of both of the above-mentioned classes of compounds (oxidation of the molecule back-bone; ring and cage strain) can be exothermic or endothermic compounds. An example of the former is TNT with $\Delta_f H° = -295.5$ kJ kg^{-1}, while RDX shows a positive standard enthalpy of formation of $\Delta_f H° = +299.7$ kJ kg^{-1}.

A third class of highly energetic materials is always endothermic and is called **nitrogen-rich molecules** (Tab. 9.1). The reasons why nitrogen-rich molecules are particularly interesting as highly energetic compounds will be considered in more detail in the next section.

If we first consider the bond energies for typical C—C, N—N and O—O single, double and triple bonds, we can see that the bond strength increases from single to double to triple bonds (which is unsurprising) (Fig. 9.1).

However even more important is the question whether an element-element double bond is twice as strong as a single bond, and a whether a triple bond is three

https://doi.org/10.1515/9783111447087-009

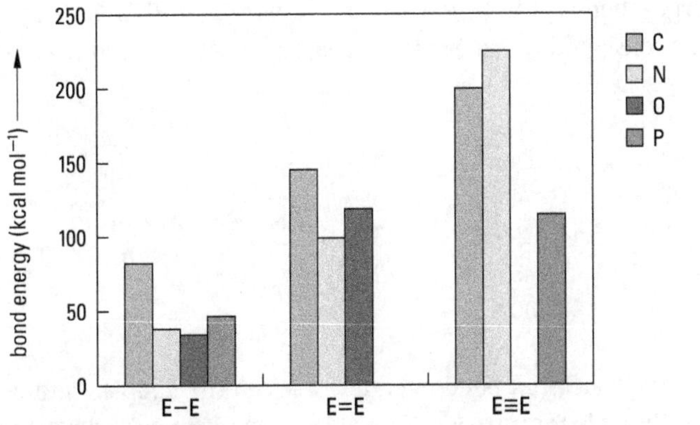

Fig. 9.1: Approximate bond energies for C—C, N—N, O—O and P—P single, double and triple bonds (in kcal mol^{-1}).

times as stable as a single bond. This means that we have to average the bond energy per 2-electron bond. This situation is illustrated in Fig. 9.2.

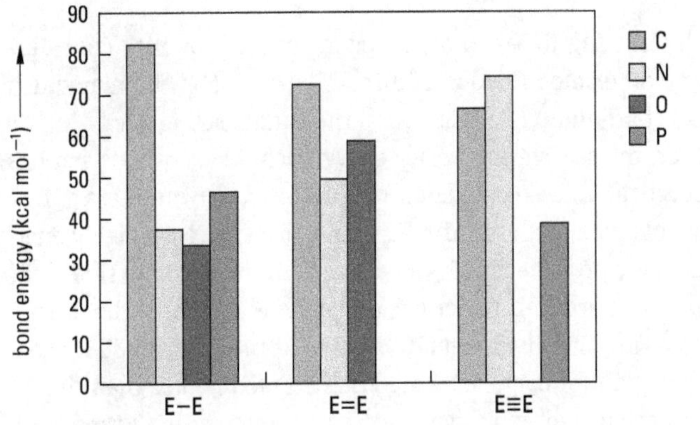

Fig. 9.2: Average element-element bond energies per 2-electron bond (in kcal mol^{-1}).

Now we can see that the element nitrogen is unique as the bond energy per 2-electron bond increases from a single to a double to a triple bond. The surprising enormously high bond energy of the N≡N triple bond is responsible for the fact that all nitrogen-rich and polynitrogen compounds show a strongly exothermic decomposition when N_2 is formed. Furthermore, we can see that unlike nitrogen, carbon prefers single bonds over double or triple bonds. This means that thermodynamically, acetylene H—C≡C—H trimerizes in an exothermic reaction forming benzene, while thermodynamically, N≡N would never oligomerize or polymerize to N_x ($x = 4, 6, 8$). The reason for the particular stability of the N≡N triple bond in comparison to carbon can be found in the small size of the N atoms (better overlapping) (covalent radii: N = 0.70 Å, C = 0.77 Å), and also as a result of the different hybridization of N in N_2 in comparison with C in HCCH, in which the N—N σ bond in N_2 has a considerably higher p-character than the C—C bond in acetylene does:

N≡N:	σ-LP (N)	64% s	36% p
	σ-(N—N)	34% s	66% p
H—C≡C—H	σ(C—C)	49% s	51% p
	σ(C—H), C	45% s	55% p
	H	100% s	

The large jump in the average 2-electron bond energies between the double and the triple bonds is particularly valuable for nitrogen. This means that during the development of new energetic substances with a very high nitrogen content (N > 60%) and a N—N bond order smaller than or equal to 2, large amounts of energy (for explosions or rocket propulsion) will be released on decomposition to molecular dinitrogen.

9.2 Polynitrogen compounds

It would be particularly interesting to take polynitrogen compounds into consideration (100% N). However, no other modification of nitrogen except N_2 is known which is metastable under normal conditions (1 atm, 298 K) in the condensed phase.

In recent years, however, the search for further energy-rich forms of nitrogen has intensified using both theoretical and experimental methods. Apart from their purely theoretical interest, the possible potential uses of metastable forms of N_x as highly energetic materials, which would only produce hot air (hot dinitrogen) on decomposition, has made N_x molecules particularly interesting. Generally, calculations show that the decomposition of highly energetic forms of poly-nitrogen to N_2 should release over 10 kJ/g energy. This energy release per gram of condensed material is much higher than that of any of the state-of-the-art propellants or explosives used today. Some of the forms of nitrogen we will consider in the following section have already been identified as being vibrationally stable, thermodynamically metastable N_x molecules. Particularly interesting in this context is the N_6 molecule, which has been most intensively investigated theoretically. The nitrogen triazide molecule $N(N_3)_3$, has also been investigated by using theoretical calculations and its synthesis has been attempted, but only the formation of five equivalents of N_2 and large quantities of energy were observed. The six most important possible structures of N_6 are summarized in Fig. 9.3, while Tab. 9.2 shows a comparison of the N_6 molecules with some known, isoelectronic organic C—H compounds. For the different N_6 structures (**1–6**) five have (**1–5**) classic analogues in organic chemistry, however only the N_6 analogues **2**, **3** and **4** represent stable minimum structures. The diazide structure **6** is the only isomer which doesn't have a classic analogue in organic chemistry.

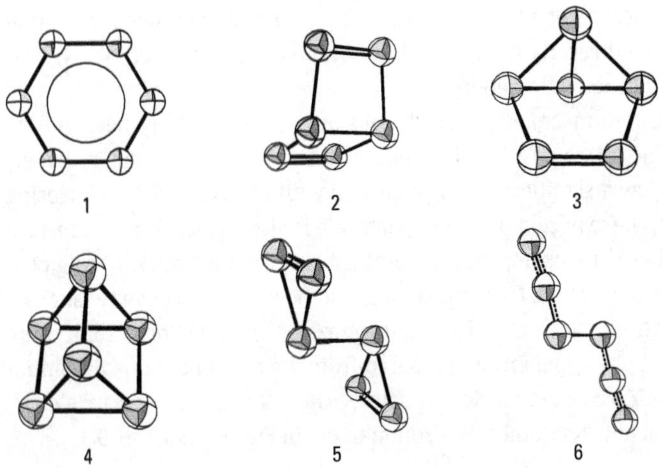

Fig. 9.3: Structures of six different N_6 isomers [45–46].

Tab. 9.2: Compilation of N_6 isomers and their isoelectronic organic CH analogues [45–46] (*Recent work suggests the diazideisomer may well be stable).

N_6 form (see Fig. 9.3)	point group	stability	rel. energy (in kJ mol^{-1})	CH analogue	point group
1	D_{6h}	unstable	899	Benzene	D_{6h}
2	C_{2v}	stable	1,037	Dewar benzene	C_{2v}
3	C_{2v}	stable	890	Benzvalene	C_{2v}
4	D_{3h}	stable	1384	Prismane	D_{3h}
5a	C_{2h}	transition state	1,020	Bicyclopropene	C_{2h}
5b	C_{2v}	transition state	1,041	Bicyclopropene	C_{2v}
6a*	C_1	transition state	769	–	–
6b*	C_2	transition state	769	–	–
N_2	$D_{\infty h}$	stable	0	Acetylene	$D_{\infty h}$

Since only isomers **2, 3** and **4** are stable minima, we will concentrate on them in the following discussion. Structure **3** contains an almost pre-formed N_2 unit (N1—N2) and in-depth studies show that this compound is therefore not kinetically stabilized enough for it to be of practical interest. The relative energies of **2** and **4** are 1,037 and 1,384 kJ mol^{-1} higher than that of N_2. These values correspond to specific energies of 14 and 19 kJ/g. These energies are noticeably high, considering that the high-performance explosives used today have typical values of 6 kJ/g.

Furthermore, both molecules **2** and **4**, have relatively high values for the lowest frequency normal vibration which is found at approximately 450 cm^{-1}. These values suggest, that the structures of these two isomers are relatively rigid and that the highly thermodynamically favorable decomposition reaction (see above) to N_2 should have a significant activation energy. Therefore, we can summarize that the N_6 analogues of Dewar benzene (**2**) and prismane (**4**) are the best candidates for N_6 for real high-energy materials. Structure **4** appears to be the most promising preparatively, since the unimolecular dissociation into three N_2 molecules is symmetry forbidden (4 + 4 + 4) and would possess considerable activation energy.

On the basis of quantum mechanical calculations, it has also recently been shown that azidopentazole (**9**, Fig. 9.4) is possibly the global minimum on the N_8 energy hypersurface. Azidopentazole has a significant energy barrier with respect to the ring closing reaction (**9** → **7**) and it can be expected that azidopentazole is also stable, also with respect to reverse cyclization. Therefore, azidopentazole probably represents not only the global minimum on the N_8 energy hypersurface, but should also be a synthetically realistic target. But how could one attempt to synthesize azidopentazole? The most promising route is probably the reaction of phenylpentazole (prepared from the corresponding diazonium salt and the azide ion) with a covalent azide compound. The structures and relative energies of the three most probable N_8 isomers are summarized in Fig. 9.4 and Tab. 9.3.

$$C_6H_5{-}NH_2 \xrightarrow{NaNO_2,\ HCl,\ 0\,°C} C_6H_5{-}N_2^+\ Cl^-$$
$$C_6H_5{-}N_2^+\ Cl^- + AgPF_6 \longrightarrow C_6H_5{-}N_2^+\ PF_6^- + AgCl$$
$$C_6H_5{-}N_2^+\ PF_6^- + NaN_3 \longrightarrow C_6H_5{-}N_5 + NaPF_6$$
$$C_6H_5{-}N_5 + R{-}N_3 \longrightarrow N_5{-}N_3\ (9) + C_6H_5{-}R$$

7 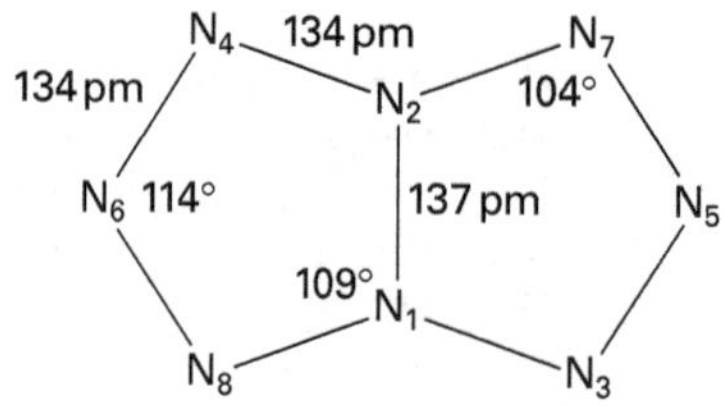

8 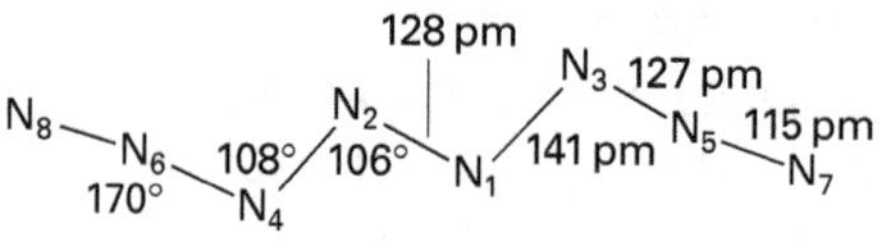

9 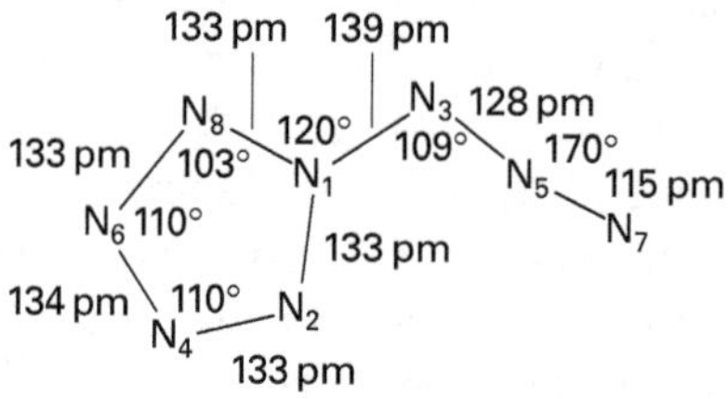

Fig. 9.4: Structures of three different N_8 isomers (**7–9**).

Tab. 9.3: Summary of N_8 isomers.

N_8 isomer	point group	stability	rel. energy (in kJ mol^{-1})
Pentalene bicycle (**7**)	D_{2h}	stable	245.1
N_8 chain (**8**)	C_s	stable	246.9
Azidopentazole (**9**)	C_s	stable	229.7
N_2	$D_{\infty h}$	stable	0.0

It is interesting to note at this point that polynitrogen compounds are not only of great interest as explosives, but also as rocket propellants. For example, a specific impulse of 408 s and a combustion chamber temperature of 7,500 K have been predicted for N_4 (T_d) (and for N_8 an I_{sp} of 500 s). From experience, since an increase in the specific impulse of 20 s results in a doubling of the payload, this would be a massive improvement. Furthermore, since polynitrogen compounds exhibit smoke-free combustion, they are also of big interest for tactical rockets.

Although currently some very nitrogen-rich compounds are known, there is no all-nitrogen/polynitrogen compound in use or even under investigation. Nonetheless, the (calculated) advantages of this class of compounds is summarized again in the following:
- only gaseous reaction products (N_2)
- large positive standard enthalpies of formation ($\Delta_f H°$)
- large propulsive or explosive power
- very high specific impulse (200% > value for hydrazine)
- very high (adiabatically calculated) reaction temperature (up to 7,500 K)
- smoke-free combustion
- minimal signature of a rocket engine
- low erosion of gun barrels (no formation of iron carbide)

In recent years it has been possible to increase the static pressure during material synthesis and investigations by more than a factor of 50. This means that it is possible to access pressures which were previously unthinkable. This breakthrough was made possible by the huge advances in the diamond-cell technique. The pressure at the centre of the earth corresponds to a value of approximately 3 500 000 bar = 3 500 kbar = 3.5 Mbar. The highest pressure that has been reached using the diamond-cell is 5.5 Mbar. It has been assumed, that the highest possible pressure accessible using this technique should be approximately 7.5 Mbar. At this pressure, the metallization of diamond begins, which is linked to a large decrease in volume, so that as a result the diamonds crack. The change of the inner energy (U) of a substance is dependent on the variables temperature (T) and pressure (p):

$$\Delta U = T\,\Delta S - p\,\Delta V$$

Due to the temperature dependent term $T\Delta S$, the state (solid, liquid, gas) can be changed, e.g. by melting or evaporation. However, the pressure-dependent term $p\Delta V$ is more dominant, because the pressure can be increased to a much greater extent than the temperature. Table. 9.4 summarizes the effects at static pressures of 5000 bar, 50 000 bar or 500 000 bar, that can be achieved for a compound which shows a decrease in volume of $\Delta V = 20\ cm^3\ mol^{-1}$.

Tab. 9.4: Correlation of the applied pressure and material changes.

p / bar	$p\Delta V$ / kcal mol^{-1}	effect
5 000	2	pressing the substance
50 000	20	deformation of bonds
500 000	200	formation of new bonds with new electronic states

The application of static pressures in the range of 5 000 to 50 000 bar has relatively speaking only the small effect of pressing the substance together and bending of the bonds. These effects can also be achieved by changing the temperature at ambient pressure. When the pressure reaches 500 000 bar, a completely new situation arises: old bonds can be broken, new bonds made and new electronic states occupied, which result in drastic changes to the physical properties of the substance.

As a highlight of such investigations, the results of the high-pressure research on nitrogen should be mentioned. As a result of the triple bond in the dinitrogen molecule, N_2, it is one of the most stable diatomic molecules. At low temperatures and pressures nitrogen can be condensed, but this solid still contains diatomic N_2 molecules and is an insulator with a large band gap.

In 1985 McMahan and LeSar predicted that the triple bond in molecular nitrogen should be breakable under very high pressures and a solid should be formed which consists of trivalent (i.e. three-coordinate) nitrogen atoms (pressure-coordination rule). Such structures already exist at normal pressures for the other group 15 elements phosphorus, arsenic, antimony and bismuth. The transformation pressure for nitrogen should lie in a range between 500 and 940 kbar. An estimation of the bond energies predicts 38 kcal mol^{-1} for a single bond and 226 kcal mol^{-1} for a triple bond. This predicts a difference of 188 kcal mol^{-1}, and corresponds to the value of 200 kcal mol^{-1} (cf. Tab. 9.4).

The preparation of trivalent nitrogen was achieved in 2004 by Eremets et al. in a diamond cell at 1 150 000 bar and 2 000 K [47–49]. The crystallographic data for the trivalent nitrogen is: cubic, lattice parameter $a = 3.4542(9)$ Å. A three-dimensional structure which consisted of trivalent nitrogen atoms (Fig. 9.5) was found. The N—N bond length at 1.1 Mbar is 1.346 Å, and the NNN angle is 108.8°. The nitrogen atoms form screws of trivalent atoms which are connected to form a three-dimensional network.

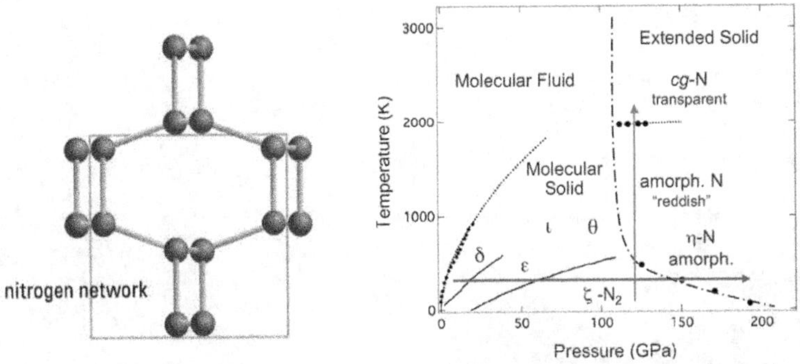

Fig. 9.5: Left: Singly bonded polymeric nitrogen (cubic modification). (Reproduced with permission from Prof. Dr. J. Evers, http://www.cup.uni-muenchen.de/ac/evers/index.html). **Right:** Conceptual phase/chemical transformation diagram of nitrogen illustrating the molecular to non-molecular transformation at high pressures and temperatures [reproduced with friendly permission of Phys Review (2007, B76, 014113)].

Unfortunately, this cubic polynitrogen is not stable under normal conditions and already decomposes back to molecular dinitrogen at a pressure of approximately 42 GPa.

The pentazolates are the final, all-nitrogen members of the azole series. Pentazolate salts have been notoriously elusive, despite enormous efforts to make them in either the gaseous or condensed phases over the last hundred years. Quite recently, Oleynik et al. [*Chem. Mater.* **2017**, *29*, 735–741] reported the successful synthesis of a solid state compound consisting of isolated pentazolate anions N_5^-, which involved compressing and laser heating cesium azide (CsN_3) mixed with cryogenic liquid N_2 in a diamond anvil cell. The experiment was suggested by theory, which predicted transformation of the mixture would occur at high pressures to form the new compound, cesium pentazolate (CsN_5). This work provides critical insight into the role of extreme conditions in probing and exploring unusual bonding types that ultimately leads to the formation of novel, high-nitrogen and also high-energy density species. Fig. 9.5a shows the X-ray structures of two CsN_5 polymorphs at 60 GPa.

In this context, it should also be mentioned that very recently Zhang et al. [**Science** *355*, 374–376.] were able to finally report the synthesis and characterization of a salt containing the pentazolate anion which is stable at room temperature and can be prepared and isolated under standard conditions: $(N_5)_6(H_3O)_3(NH_4)_4Cl$. Related to the investigations undertaken on dinitrogen (N_2) at high very high pressures and high temperatures which showed the formation of trivalent nitrogen, are investigations into whether it should be possible to prepare poly-CO as a future energetic material.

Dinitrogen, N_2 and carbon monoxide, CO both exist as diatomic molecules in the gas-phase under standard conditions, are isoelectronic (contain the same total number of electrons) and contain a very strong triple bond with a high bond dissociation energy (N_2: 945.42 kJ/mol, CO: 1076.5 kJ/mol). There is interest in establishing whether exposing the CO molecule to very high pressures will result in an extended solid, in

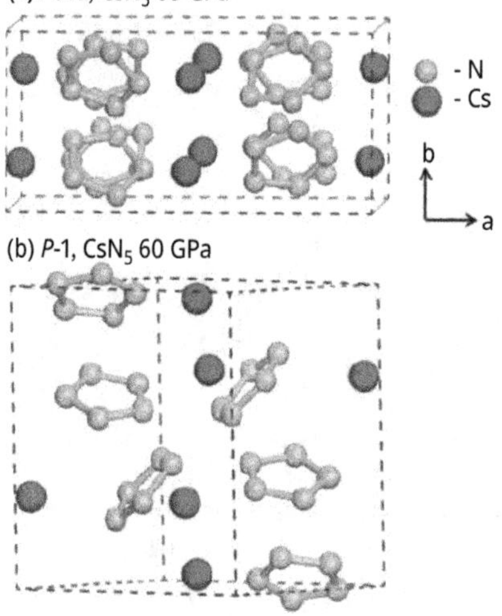

Fig. 9.5a: Crystal structures of two energetically very similar CsN_5 polymorphs at 60 GPa.

which coordination numbers are increased and bond orders lowered (Fig. 9.5b) – just as was observed for cubic, poly-nitrogen. Clearly, the potential of poly-CO to release devastating amounts of energy is clear, and in addition, the conversion of CO-δ to diatomic CO would only result in the release of CO gas and no other toxic compounds.

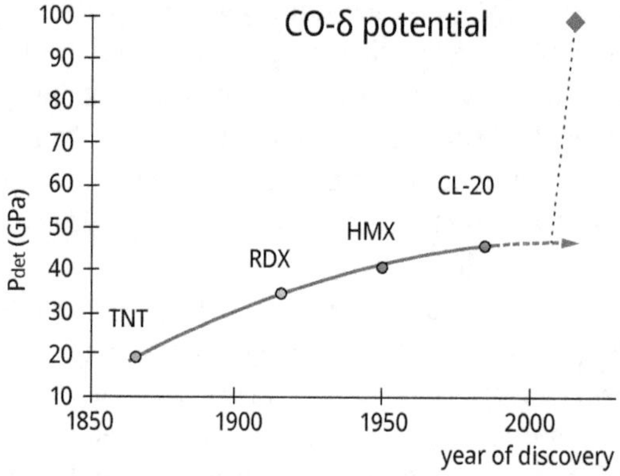

Fig. 9.5b: Calculated detonation pressure of CO-δ in which diatomic CO is produced, compared to those of state of the art secondary explosives[1].

CO-δ has been calculated to possess a heat of formation in the order of magnitude of approximately (a phenomenal!) ten times that of current state of the art secondary explosives. This, in combination with the extremely high calculated density of 2.70 g cm^{-3} results in a calculated specific impulse of almost double that of HMX or RDX, and a detonation pressure five times that of TNT and more than double that of CL-20 (Tab. 9.4a)[1].

Tab. 9.4a: Comparison of energetic performance properties of current state of the art secondary explosives, as well as the calculated values for the proposed high-pressure CO-δ[1].

compound	density (g/cm^3)	heat of formation (kJ/g)	I_{sp} (s)	P_{det} (GPa)
CO-δ	2.70	11	450	100
ε-CL-20	2.05	0.84	255	45
β-HMX	1.90	0.40	260	37
RDX	1.81	0.37	260	33
TNT	1.65	0.22	204	20

However, at this time, such a compound would be too unstable for application, since poly-CO would only be stable under extremely high pressures. Releasing the pressure would result in conversion to diatomic CO. Furthermore, the high pressures which would be required to form poly-CO require highly specialized apparatus. This means that there would be major problems to overcome, in order to be able to scale-up such ultra high-pressure energetic materials.

In a related vein, the formation of extended solids of other small molecules such as carbon dioxide is also being considered, in attempts to form future, extremely energetic materials. Again, extremely high pressures are required in order to break the multiple bonds of ambient CO_2 enabling the formation of new bonds in the covalent network (Fig. 9.5c).

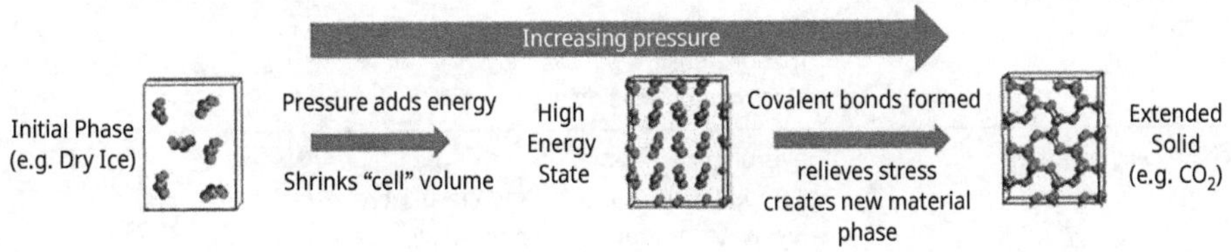

Fig.9.5c: Proposed formation of an extended solid form of CO_2 using ultra-high pressure. In contrast to CO_2 under normal conditions, in an extended solid high pressure form, higher coordination numbers as well as C-O bonds with lower bond orders would be expected, resulting in a higher heat of formation for the extended solid in comparison with that of CO_2 under normal conditions[1].

Literature

[1] J. M. Goldwasser, *Extended Solids*, DARPA Presentation, Proposer's day Workshop, 9th February **2012**.

9.3 High-nitrogen compounds

One synthetic approach is to synthesize nitrogen-rich compounds (N > 60%), which also contain oxidizing groups (e.g. nitro: $—NO_2$, nitrimino: $=N—NO_2$, Fig. 9.6), in order to achieve the best possible oxygen balance (Ω). This is important because in principal, it is always better to have the oxidizer and fuel combined in *one* molecule (in contrast to mixtures), because this generally results in superior detonation parameters such as a higher detonation velocity and detonation pressure. Of course, the new compounds synthesized also should not be worse in any of the positive properties RDX shows (e.g. thermal stability, low sensitivity etc.). Table 9.5 shows a summary of the desired properties for new, nitrogen-rich highly energetic compounds.

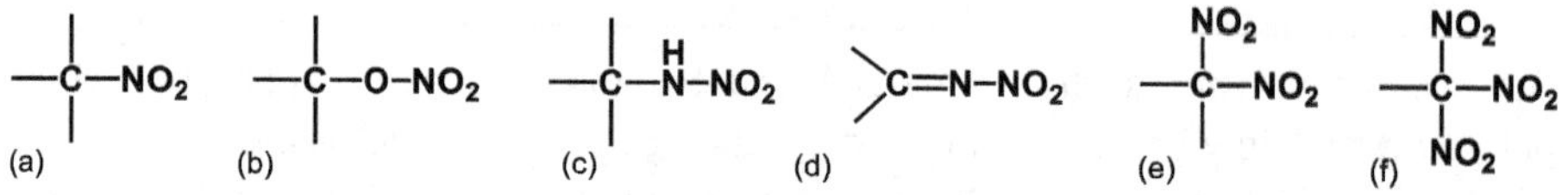

Fig. 9.6: Oxidizing groups which can facilitate a positive oxygen balance; Nitro (a), Nitrato (b), Nitramino (c), Nitrimino (d), Dinitromethyl (e) and Trinitromethyl (f).

Tab. 9.5: Desired properties of new nitrogen-rich highly energetic compounds.

performance	detonation velocity detonation pressure heat of explosion	$D > 8{,}500$ m s^{-1} $p > 340$ kbar $Q > 6{,}000$ kJ kg^{-1}
stability	thermal stability impact sensitivity friction sensitivity electrostatic sensitivity	$T_{dec.} \geq 180$ °C IS > 7 J FS > 120 N ESD > 0.2 J
chemical properties	hydrolytically stable, compatible with binder and plasticizer, low water solubility[a] (or non-toxic), smoke-free combustion, long-term stable (> 15 years under normal conditions)	

[a]low octanol-water partition coefficient.

Of course, it is difficult to fulfill all these criteria, in particular as further practical aspects also have to be taken into account, including: simplest possible synthesis, a cheap and sustainable synthesis (no chlorinated solvents), as well as being able to automate the individual synthetic steps [50, 51]. So far it has been possible to identify two classes of compounds which show many of the properties mentioned above:
- tetrazole compounds [52, 53]
- trinitroethyl compounds [54]

9.3.1 Tetrazole and dinitramide chemistry

While triazole compounds are not often energetic enough and pentazole species are kinetically too labile, the tetrazole derivatives often combine the desired endothermicity with kinetic stability (Fig. 9.7). Aminotetrazole (AT) and diaminotetrazole (DAT) are good starting materials for the synthesis of new highly energetic, nitrogen-rich compounds. AT and DAT can be synthesized in accordance with the synthetic schemes outlined in Fig. 9.8. For DAT, the synthesis starts from thiosemicarbazide (easily obtained from KSCN and hydrazine) in acidic medium.

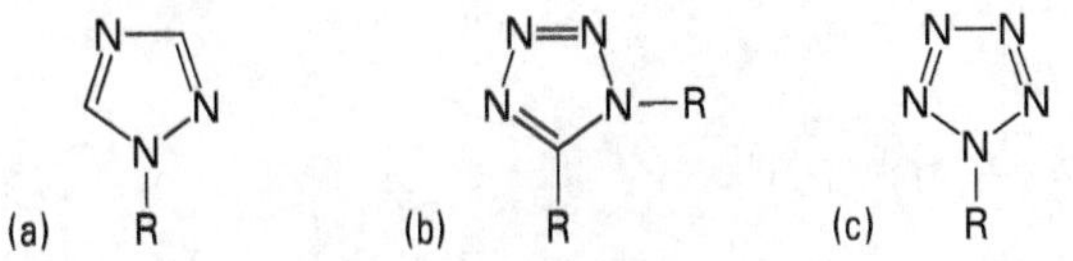

Fig. 9.7: The basic structures of 1,2,4-triazole (a), tetrazole (b) and pentazole compounds (c).

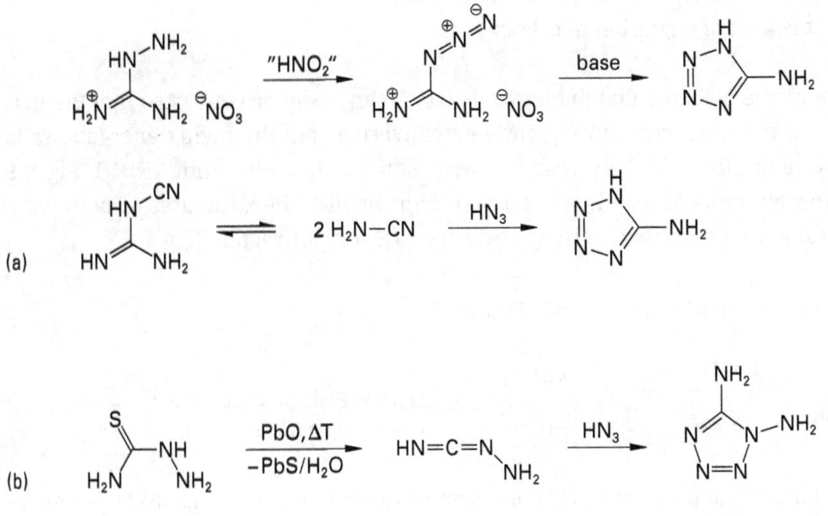

Fig. 9.8: Two synthetic routes for the synthesis of aminotetrazole (AT) (a) and the synthetic route for diaminotetrazole (DAT) (b).

Often, energetic compounds composed of salts show advantages over their covalent analogues; they show a lower vapor pressure for example, essentially eliminating the risk of inhalation of toxic compounds. Furthermore, ionic compounds usually have higher densities, higher thermal stabilities and larger critical diameters (IM) [55]. On the other hand, salts are often water soluble and hygroscopic, which is negative. The solid-state density (ρ) is a particularly important parameter since not only the detonation pressure (p) but also the detonation velocity (D) are directly dependent on the density, which we have already seen using the simple relationship of Kamlet and Jacobs which was discussed in Chapter 3 [10].

$$p_{C-J}\ [\text{kbar}] = K\rho_0^2\,\Phi$$
$$D\left[\text{mm}\,\mu s^{-1}\right] = A\,\Phi^{0.5}(1 + B\,\rho_0)$$

The constants K, A and B are defined as follows:

$$K = 15.88$$
$$A = 1.01$$
$$B = 1.30$$

The parameter Φ is therefore

$$\Phi = N(M)^{0.5}\,(Q)^{0.5}$$

where
- N is the no. of moles of gas released per gram of explosive
- M the mass of gas in gram per mole of gas and
- Q is the heat of explosion in cal per gram

Aminotetrazole (AT) and diaminotetrazole (DAT) can easily be converted into the perchlorate salts, which can subsequently be converted into the highly energetic salts HAT^+DN^- and $HDAT^+DN^-$ by reaction with potassium dinitramide (KDN). Fig. 9.9 shows the synthesis of diaminotetrazolium dinitramide. The structures of both high-energy salts were determined using single crystal X-ray diffraction (Fig. 9.10).

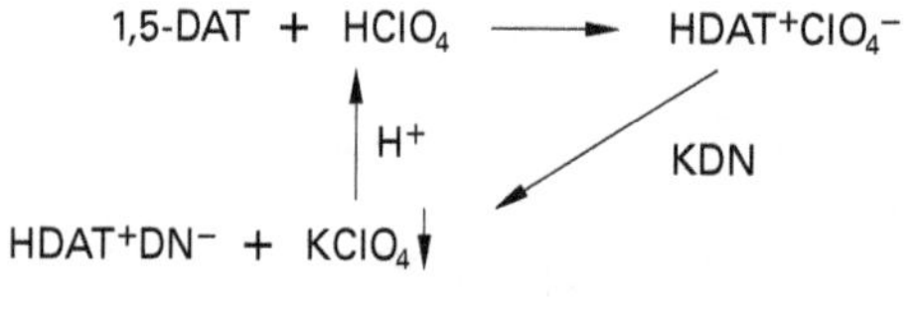

Fig. 9.9: Synthesis of diaminotetrazolium dinitramide ($HDAT^+DN^-$).

The most important parameters for the characterization of the performance of secondary explosives for use in energetic formulations in war heads and general purpose bombs (GPBs) are the heat of detonation (Q), the detonation pressure (p) and the detonation velocity (D). For comparison, Fig. 9.11 shows a summary of the important values for aminotetrazolium dinitramide (HAT^+DN^-), diaminotetrazolium dinitramide ($HDAT^+DN^-$) and RDX.

Tetrazole compounds are also good starting materials for the synthesis of energetic materials with substituted side-chains attached to the tetrazole skeleton. For example, tetrazole can be alkylated by reaction with 1-chloro-2-nitro-2-azapropane and converted into various energetic compounds (Figs. 9.12 and 9.13).

Compound **8** for example is stable up to 184 °C (melting point of 150 °C), is not impact sensitive (IS > 100 J) and is only slightly friction (FS = 120 N) and electrostatic sensitive

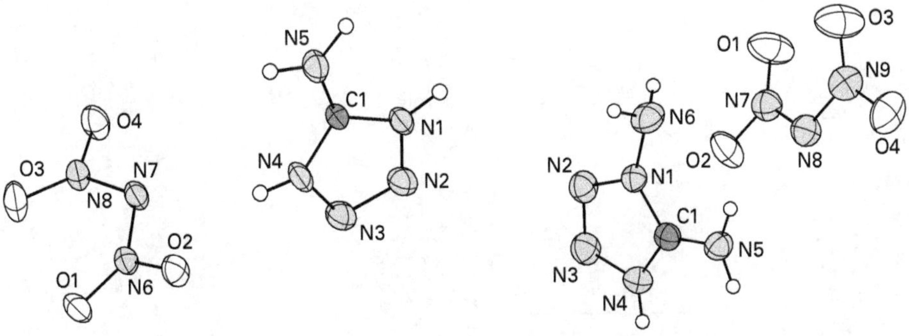

Fig. 9.10: Structures of aminotetrazolium dinitramide (HAT^+DN^-, left) and diaminotetrazolium dinitramide ($HDAT^+DN^-$, right).

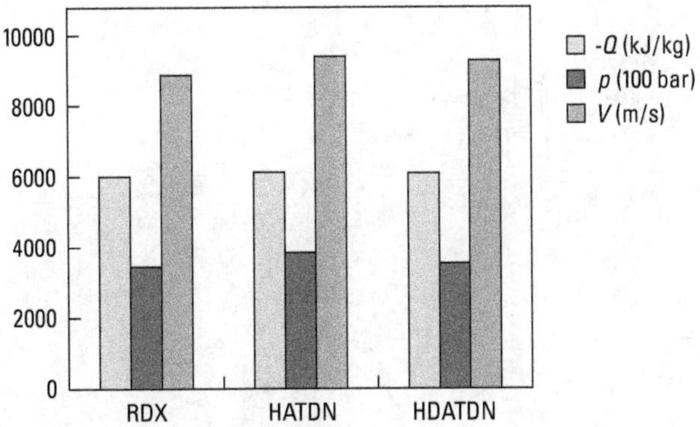

Fig. 9.11: Comparison of the performance data for aminotetrazolium dinitramide (HAT^+DN^-), diaminotetrazolium dinitramide ($HDAT^+DN^-$) and hexogen (RDX).

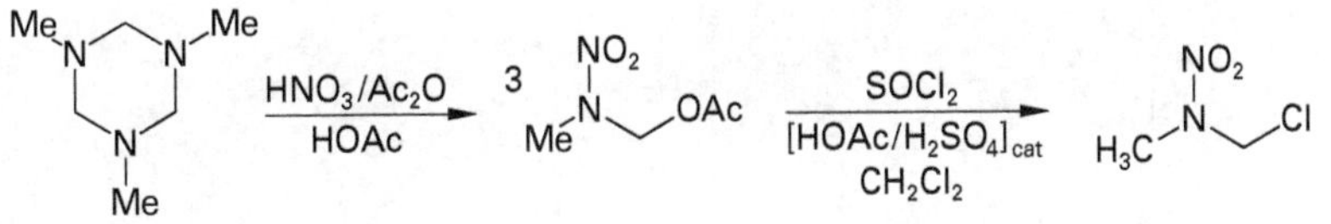

Fig. 9.12: Synthesis of 1-chloro-2-nitro-2-azapropane.

(ESD = 0.22 J). The performance parameters such as the detonation velocity (8,467 m s^{-1}), detonation pressure (273 kbar) and heat of explosion (5,368 kJ kg^{-1}) are significantly higher than the values for TNT and only marginally lower than those of RDX.

Out of the different possible derivatives of tetrazoles, we have so far only discussed (**A**) and nitriminotetrazole (**D**) (Fig. 9.14) in more detail. Although the azide-substituted tetrazole compound of type **C** is very energetic, it is still not useful for practical application since it is extremely sensitive. Additionally of interest are the neutral and ionic nitrotetrazole derivatives of type **F**, which are more stable (Fig. 9.14).

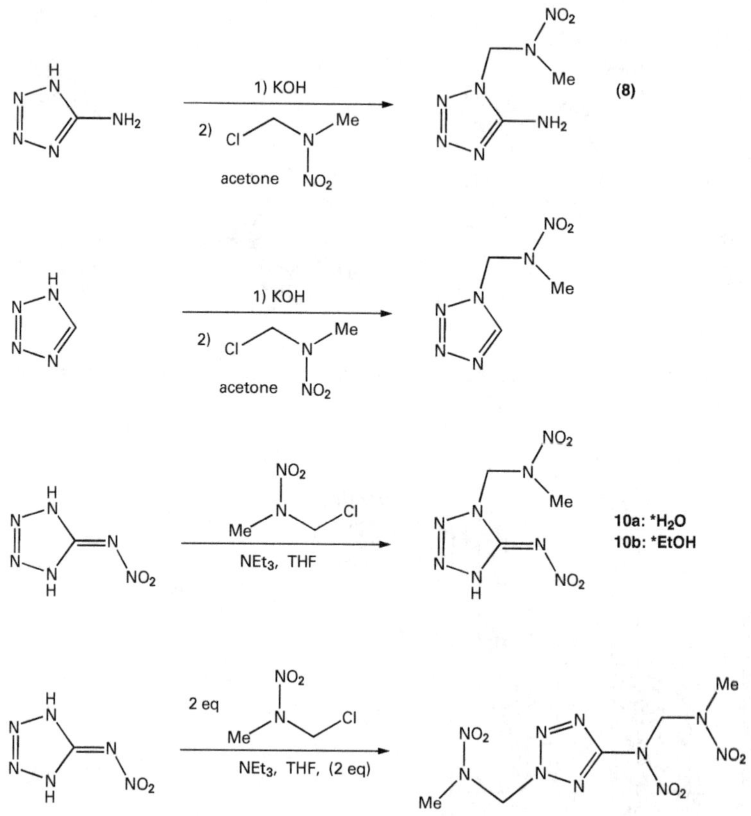

Fig. 9.13: Synthesis of tetrazole compounds with energetic side-chains.

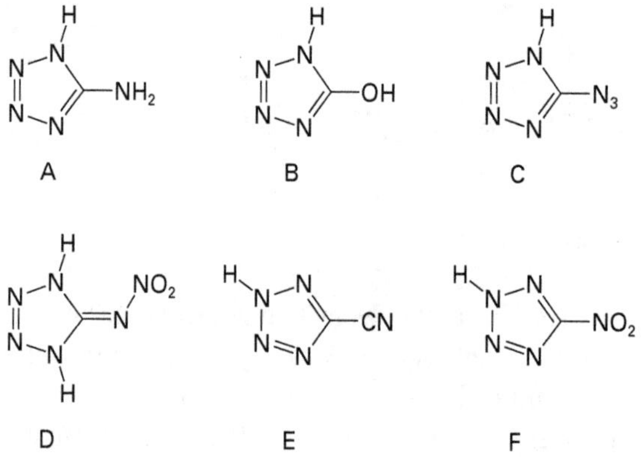

Fig. 9.14: Tetrazole derivatives with different substituents.

While free nitrotetrazole is extremely sensitive and only of limited thermal stability, the salts of nitrotetrazole are, at least thermally, often significantly more stable. The best synthetic route starts from 5-aminotetrazole via the corresponding copper salt (M = alkali metal):

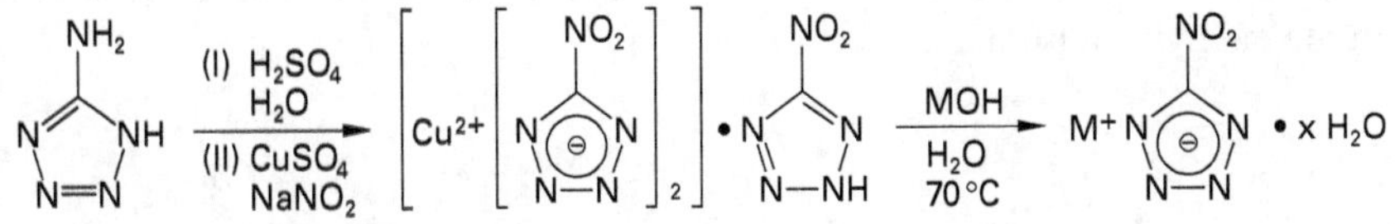

Although AgNT can be obtained from NaNT and $AgNO_3$, ammonium nitrotetrazolate is best prepared as the hemihydrate (Fig. 9.15) by the acidification of sodium nitrotetrazolate with sulfuric acid, followed by extraction and reaction with ammonia gas. This compound is particularly suited to the synthesis of further energetic nitrotetrazolate

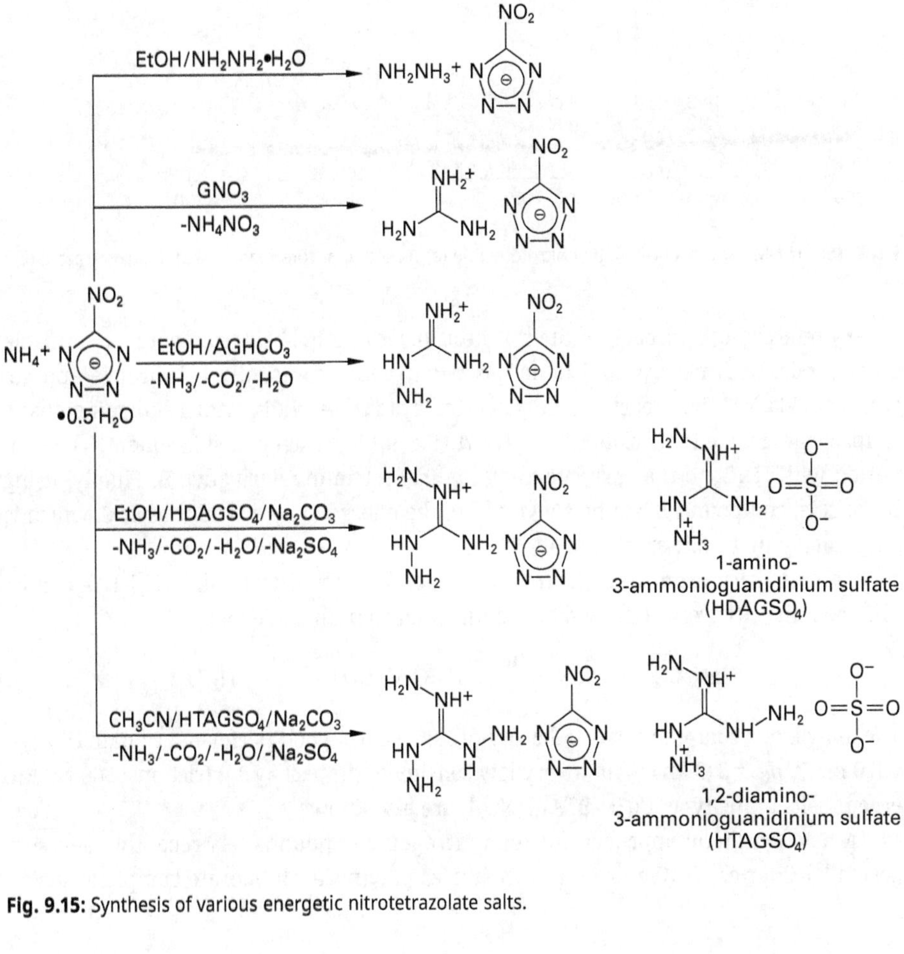

Fig. 9.15: Synthesis of various energetic nitrotetrazolate salts.

salts such as, for example the hydrazinium, guanidinium, aminoguanidinium, diaminoguanidinium and triaminoguanidinium salts (Fig. 9.15).

Even though they are mainly of academic interest, the nitrotetrazole compounds are very suitable species to study using ^{15}N NMR spectroscopy. Figure 9.16 shows an example of very nicely resolved ^{15}N NMR spectra of free nitrotetrazole, as well as of both the methylated isomers.

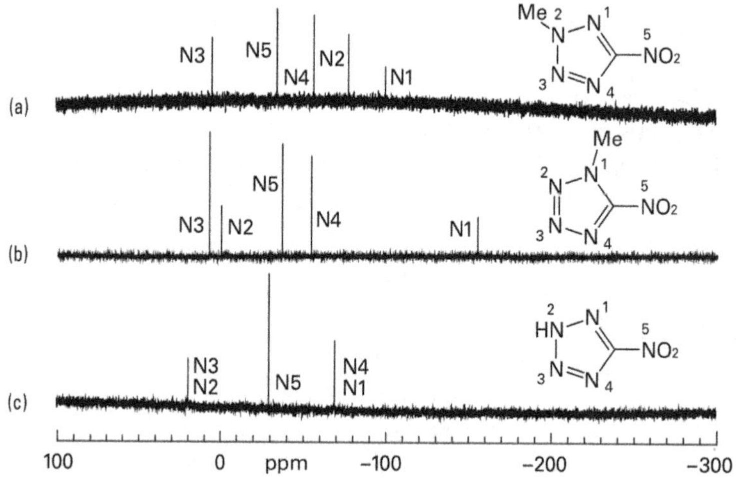

Fig. 9.16: ^{15}N NMR spectra of 2-methyl-nitrotetrazole (a), 1-methyl-nitrotetrazole (b) and nitrotetrazole (c).

A very interesting approach to the synthesis of new, ethyl-bridged, energetic tetrazole compounds, both neutral and ionic has been reported recently by Shreeve and coworkers [56a, b]. They react *in situ* generated cyanazide, N_3CN, with 1,2-diaminoethane to form the bridged tetrazole derivative **A** (Fig. 9.17), which can subsequently be converted with HNO_3 to the corresponding neutral nitrimino derivative **B**. Finally, using hydrazine, compound **B** can be converted to the energetic hydrazinium salt **C**, which is surprisingly thermally stable (m. p. = 223 °C).

Another very promising tetrazole compound, bistetrazolylamine (H_2BTA), can be obtained directly from sodium dicyanamide and sodium azide [56c]:

$$NaN(CN)_2 + 2NaN_3 \xrightarrow{HCl} H_2BTA{\bullet}H_2O \xrightarrow{-H_2O} H_2BTA$$

The anhydrous compound has a density of 1.86 g cm^{-3}, good performance data (VoD = 9,120 m s^{-1}, $p_{C\text{-}J}$ = 343 kbar) but a very low sensitivity (impact > 30 J, friction > 360 N). Energetic salts of the type $[Cu(H_2BTA)_2][NO_3]_2$ are also known.

A new synthetic approach for high-nitrogen compounds has recently been suggested by Charles H. Winter et al. which uses poly(pyrazolyl)borate complexes [56d].

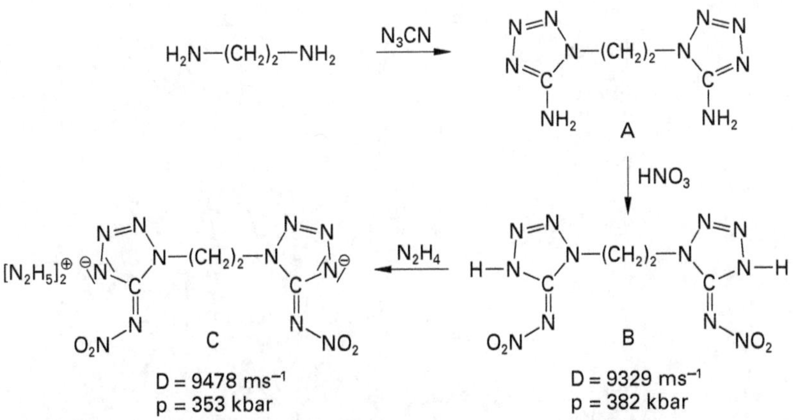

Fig. 9.17: Synthetic scheme for the preparation of bridged nitriminotetrazole compounds.

Since a borate anion $[BR]_4^-$ carries only one negative charge but can include four heterocyclic ring systems R, theoretically, up to 8 N_2 molecules can get liberated if R is a tetrazole ring system, i. e. if it is a tetrakis(tetrazolyl)borate anion.

9.3.2 Tetrazole, tetrazine and trinitroethyl chemistry

Through the combination of very oxygen-rich trinitroethyl groups with the nitrogen-rich and endothermic tetrazole and tetrazine ring systems, the syntheses of three highly energetic trinitroethyl derivatives were achieved:

1. trinitroethyl-tetrazole-1,5-diamine (TTD)
2. bis(trinitroethyl)-tetrazole-1,5-diamine (BTTD)
3. bis(trinitroethyl)-1,2,4,5-tetrazine-3,6-diamine (BTAT)

The synthesis of these compounds is relatively facile using trinitroethanol and the corresponding amines, while for the synthesis of TTD diaminotetrazole was used, for BTTD TTD was used, and for BTAT, diamino-tetrazine was used (Fig. 9.18). At pH values above 6, trinitroethanol behaves as an acid, so that for the Mannich reaction shown in Fig. 9.18, two different mechanisms can be discussed.

An important characteristic of these compounds is that due to the dipolar nitro groups, the molecules form strong intra and intermolecular interactions including hydrogen bonds, which contribute not only to an increase in the thermal stability, but also to a higher density. This is particularly clear in trinitroethanol due to its intermolecular dipolar nitro group interactions (Fig. 9.19). The molecular structures of TTD, BTTD and BTAT which were obtained using single crystal X-ray diffraction are shown in Figs. 9.20–9.22. The high densities, combined with the endothermic character of the

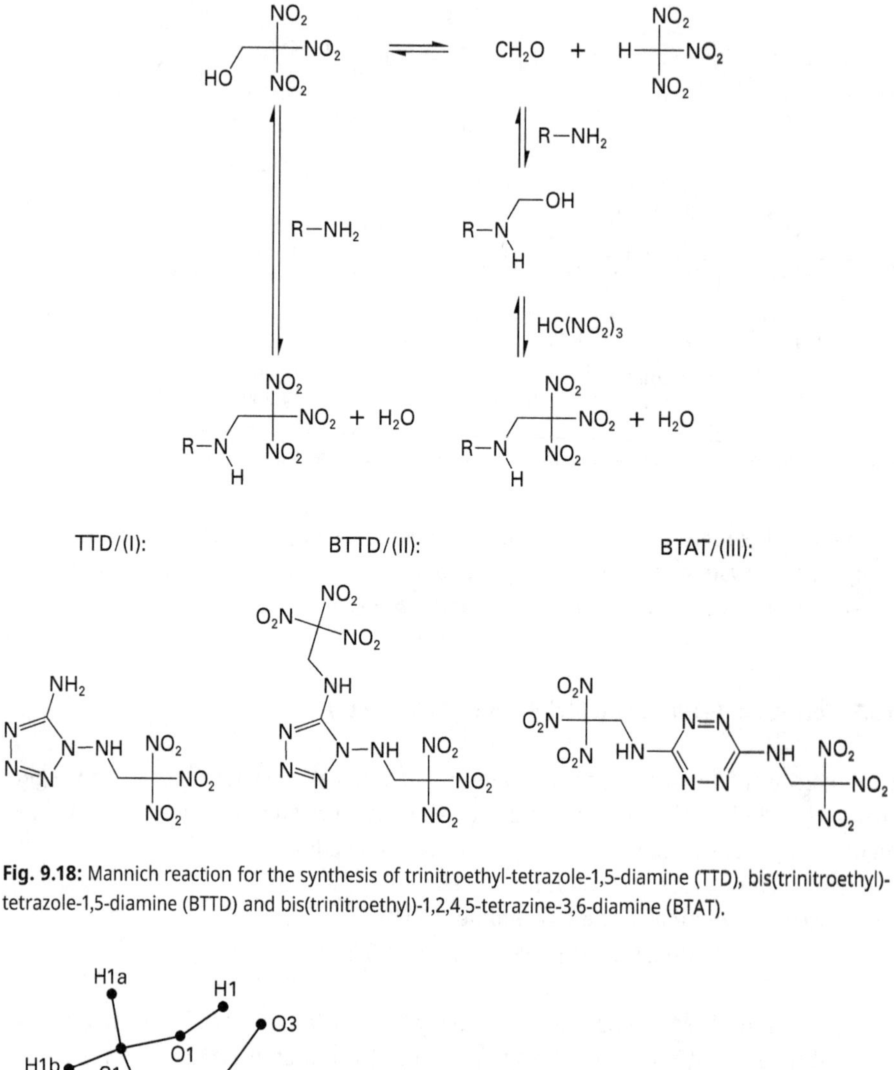

Fig. 9.18: Mannich reaction for the synthesis of trinitroethyl-tetrazole-1,5-diamine (TTD), bis(trinitroethyl)-tetrazole-1,5-diamine (BTTD) and bis(trinitroethyl)-1,2,4,5-tetrazine-3,6-diamine (BTAT).

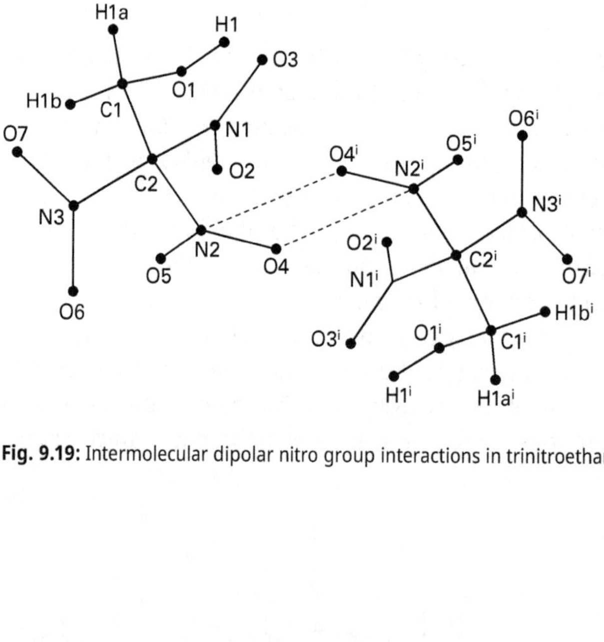

Fig. 9.19: Intermolecular dipolar nitro group interactions in trinitroethanol.

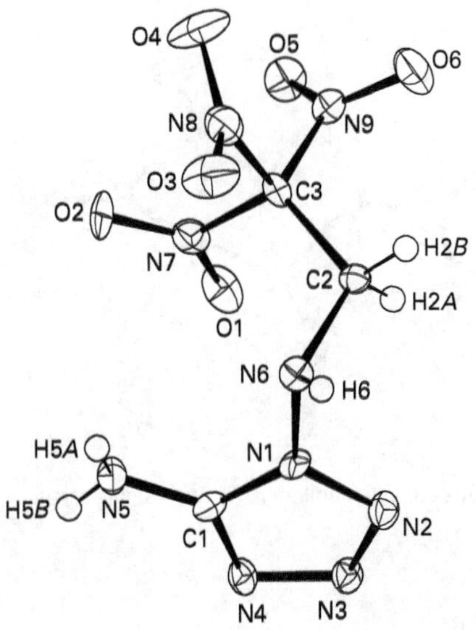

Fig. 9.20: Molecular structure of trinitroethyl-tetrazole-1,5-diamine (TTD) from single crystal X-ray diffraction data.

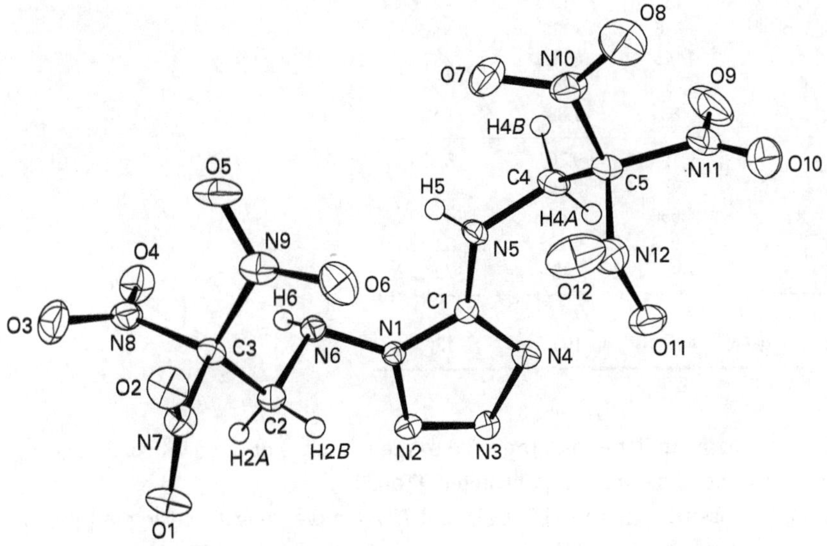

Fig. 9.21: Molecular structure of bis(trinitroethyl)-tetrazole-1,5-diamine (BTTD) from single crystal X-ray diffraction data.

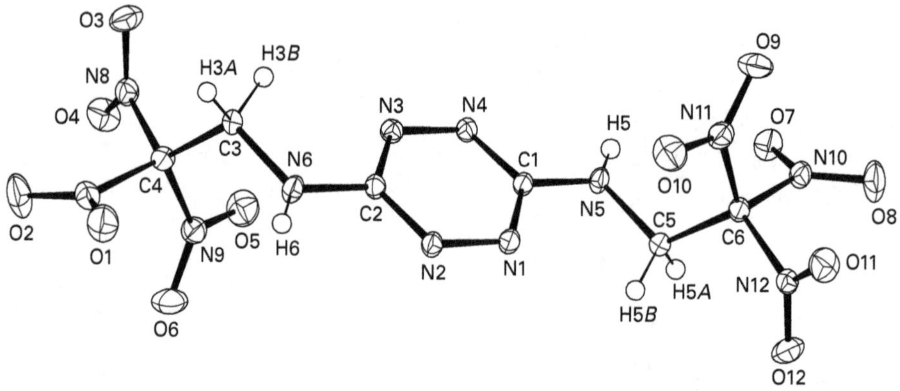

Fig. 9.22: Molecular structure of bis(trinitroethyl)-1,2,4,5-tetrazine-3,6-diamine (BTAT) from single crystal X-ray diffraction data.

Tab. 9.6: Properties of TTD and BTAT in comparison with RDX.

	TTD	BTAT	RDX
formula	$C_3H_5N_9O_6$	$C_6H_6N_{12}O_{12}$	$C_3H_6N_6O_6$
Ω / %	−15.2	−10.9	−21.6
$\Delta_f H°$ (s) / kJ mol^{-1}	+356	+336	+ 85
$\Delta_f U°$ (s) / kJ kg^{-1}	1443	852	280
$T_{dec.}$ / °C	126	184	202
ρ / g cm^{-3}	1.831	1.886	1.80
D / m s^{-1}	9,194	9,261	8,894
$p_{C\text{-}J}$ / kbar	370	389	338
T_{ex} / K	4,650	4,867	4,290
$Q_{C\text{-}J}$ / kJ kg^{-1}	6,018	6,135	5,875
V_0 / l kg^{-1}	788	743	797
impact sensitivity / J	30	7	7
friction sensitivity / N	40	160	120

tetrazole or tetrazine moieties and the good oxygen balance of the trinitroethyl group, all result in very good detonation parameters (Tab. 9.6).

As can be observed in Tab. 9.6, TTD and BTAT have good to very good performance parameters with acceptable sensitivities. However, since TTD decomposes at 126 °C, its thermal stability is too low. In contrast, BTAT is stable to 184 °C (DSC, 5 °C/min). A long-term study at 140 °C for 48 h also showed no decomposition of BTAT.

At this point it should be pointed out, that BTAT has the same CHNO composition as CL-20, but it is significantly less sensitive than the energetic ε modification of CL-20. Figure 9.23 shows a photo of a steel-sleeve test (Koenen test) performed using BTAT with a critical diameter of 10 mm (see Ch. 6.1).

$C_6H_6N_{12}O_{12}$, $T_{dec} > 180\,°C$, $\Sigma(O,N) > 82\,\%$, $D = 9261\ m\ s^{-1}$, $P_{CJ} = 389\ kbar$

Fig. 9.23: Explosion of bis(trinitroethyl)-1,2,4,5-tetrazine-3,6-diamine (BTAT) in the Koenen test.

As we have already seen, trinitromethane and trinitroethanol are important starting materials in the synthetic chemistry of highly nitrated compounds. The precursor potassium trinitromethanide can be prepared best from tetranitromethane (TNM):

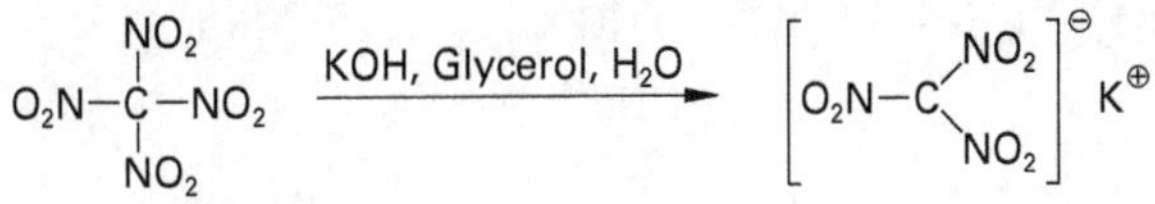

Free trinitromethane can be obtained by acidification with phosphoric acid:

$$K^+\ [C(NO_2)_3]^- \xrightarrow{H_3PO_4} H{-}C(NO_2)_3$$

Tetranitromethane can itself be prepared in good yield by the nitration of acetic anhydride or isopropanol using conc. HNO_3. The synthesis requires very exact control of the reaction conditions and is not without danger.

$$4\,(CH_3CO)_2O + 4\,HNO_3 \rightarrow C\,(NO_2)_4 + 7\,CH_3COOH + CO_2$$

Finally, trinitroethanol (TNE) can be obtained by the condensation of trinitromethane with formaldehyde (paraformaldehyde) in acidic conditions:

$$H{-}C(NO_2)_3 + H_2CO \rightarrow HO{-}CH_2{-}C(NO_2)_3$$

The corresponding chlorocarbonate can be obtained by the reaction of free TNE with phosgene, and is also an important building block in the chemistry of polynitro-compounds (Fig. 9.24).

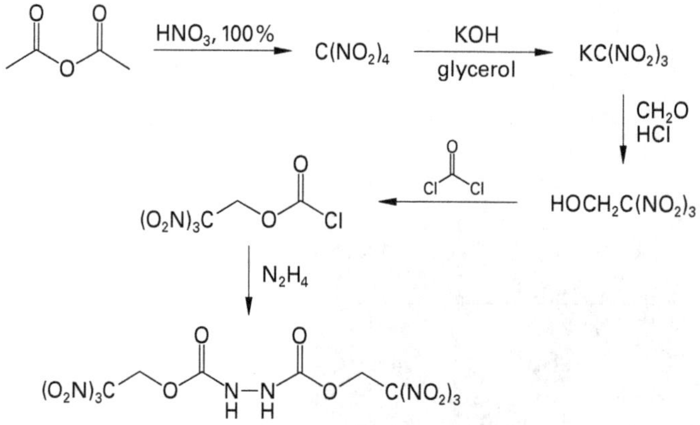

Fig. 9.24: Synthesis of trinitroethanol and trinitroethylchlorocarbonate.

A better method for obtaining the chlorocarbonate is by stirring a mixture of TNE, $FeCl_3$, CCl_4 and a few drops of water under reflux for an hour.

The synthesis of hydrazinium nitroformate (HNF) occurs best through the reaction of anhydrous hydrazine with trinitromethane (nitroform) in methanol or diethyl ether:

$$H{-}C(NO_2)_3 + N_2H_4 \rightarrow \underbrace{[N_2H_5]^+ [C(NO_2)_3]^-}_{\text{HNF}}$$

9.3.3 Ionic liquids

The term ionic liquids refers to liquids that exclusively contain ions. They are liquid salts which do not need to be dissolved in a solvent such as water. As a rule, the term ionic liquids is used when the corresponding salts are liquid at temperatures lower than 100 °C. Examples of suitable cations are alkylated imidazolium, pyridinium, ammonium or phosphonium ions. The size and the symmetry of the participating ions prevents the formation of a strong crystal lattice, meaning that only small amounts of thermal energy are needed to overcome the lattice energy and break up the solid crystal structure.

Current research is focused on finding suitable nitrogen-rich anions, so that ionic liquids could also be used as fuels in bipropellants (see Ch. 2.2 and Ch. 2.4). In comparison with the currently used fuels such as MMH and UDMH, ionic liquids would possess a negligible low vapor pressure and therefore considerably reduce the inhalation toxicity. Initial investigations have shown that the dicyanamide anion, $[N(CN)_2]^-$, in combination with the 1-propargyl-3-methyl-imidazolium cation may be a suitable ionic liquid for rocket propulsion. However, using WFNA as the oxidizer, the ignition delay is 15 ms, in comparison to 1 ms for the MMH / NTO system and 9 ms for MMH / WFNA [57].

Ionic liquids should also be very suitable for use as monopropellants, but the salts used must contain either the oxidizer and fuel combined, or salt mixtures which contain both oxidizing and reducing salts. Since these mixtures are homogeneous systems which contain both the oxidizer and fuel, they can be labeled as monopropellants, just as hydrazine is. Particularly interesting are salt mixtures which are less toxic and have a lower vapor pressure than hydrazine. Such mixtures are also known as "green propellants". Suitable anions are the nitrate or dinitramide ions [58]. A combination which has already been studied intensively as an oxidizer is the HAN, hydroxylammonium nitrate system. ADN, ammonium nitrate (AN) and hydrazinium nitrate (HN) have also been investigated. As fuels, hydroxylammonium azide (HAA), ammonium azide (AA) or hydrazinium azide (HA) may be appropriate. As a rule, these salt mixtures are not used as pure substances on safety grounds, but with 20 or 40% water added; they then decompose catalytically in an exothermic reaction. Table 9.7 shows the dependence of the calculated specific impulses on the water content for such salt mixtures.

Tab. 9.7: Influence of the water content on the specific impulse (in s) of oxidizing and reducing salt mixtures.

oxidizer	fuel	I^*_{sp} (0% water)	(20% water)	I^*_{sp} (40% water)
HAN	HAA	372	356	319
HAN	AA	366	342	303
HAN	HA	378	355	317
ADN	HAA	376	352	317
ADN	AA	362	335	303
ADN	HA	373	347	314
AN	AA	349	324	281
AN	HAA	366	342	303
AN	HA	359	334	293
HN	AA	374	349	309
HN	HAA	401	354	316
HN	HA	379	354	314
	Hydrazine	224		

Quite recently several new energetic ionic liquids (EILs) have been synthesized either consisting of imidazole salts or 1,2,4-triazole salts, but until now only few EILs from derivatives of tetrazole have been described in the literature. Tetrazoles are built of 5-membered heterocycles with four nitrogen atoms in the ring. The heats of formation ($\Delta_f H°$) become more and more positive with increasing number of ring nitrogen atoms (imidazole $\Delta_f H° = +58.5\ kJ\ mol^{-1}$, 1,2,4-triazole $\Delta_f H° = +109.0\ kJ\ mol^{-1}$, 1,2,3,4-tetrazole $\Delta_f H° = +237.2\ kJ\ mol^{-1}$). Furthermore, the use of tetrazoles is of great interest since the generation of molecular nitrogen as an end-product of propulsion is highly desirable. Also, the formation of carbon residues can be minimized. Tetrazoles can be synthesized by many different synthetic routes, which are mostly low in price and also possible on larger scales.

In the future, 1,4,5-substituted tetrazolium salts with the general formula depicted in Fig. 9.25 should be synthesized and evaluated. Due to the introduction of hydrophobic side chains (except for NH_2) one would expect these compounds to have melting points below 100 °C.

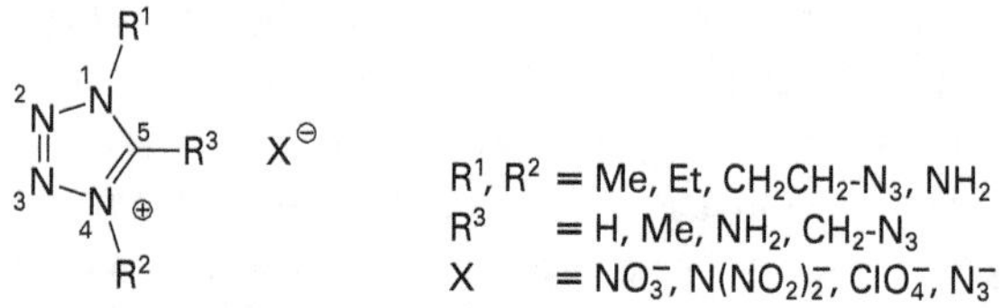

Fig. 9.25: 1,4,5-Substituted tetrazolium cation with different energetic anions X.

It should be advantageous to introduce methyl-, ethyl-, ethylazide- and amino-groups, on the 1- and 4-positions of the tetrazole ring system as well as hydrogen, methyl-, amino- and methylazido groups to the 5-position. Theoretical calculations have shown that these tetrazolium salts in combination with classic energetic and oxidizing anions like nitrate (NO_3^-), dinitramide ($N(NO_2)_2^-$), perchlorate (ClO_4^-) or azide (N_3^-) should have high heats of formation and, consequently, high specific impulses. For example, the 1,4,5-trimethyltetrazolium cation (Fig. 9.26), which is not able to form classical hydrogen bonds, has a calculated gas phase enthalpy of formation of 127.6 kcal mol^{-1}. In combination with the energetic anions X = nitrate, dinitramide, perchlorate, and azide (Fig. 9.25) four new EILs, with high heats of formation can be discussed. Table 9.8 summarize the calculated thermodynamic parameters.

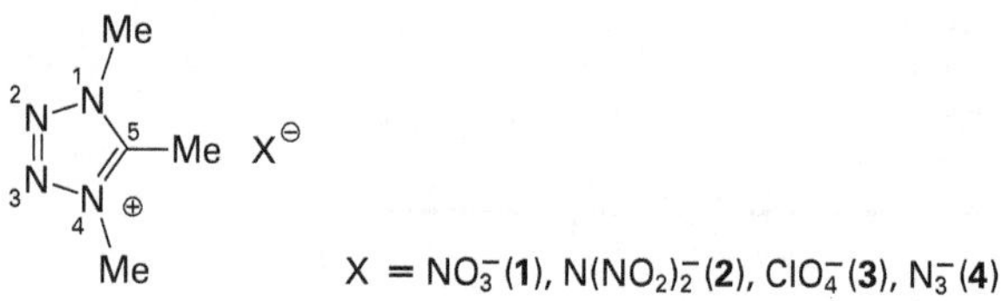

Fig. 9.26: Energetic Ionic Liquids based on the 1,4,5-trimethyltetrazolium (TMTz$^+$) cation.

Tab. 9.8: Calculated gas phase heats of formation of 1–4.

	1	2	3	4
$\Delta_f H°$ (g) / kcal mol^{-1}	128	173	136	248

A significant step to a lower-toxicity bipropulsion system would be the demonstration of hypergolicity (spontaneous ignition) between an ionic liquid, which is a paragon of low vapor toxicity, and a **safer oxidizer**. Apart from cryogens, hydrogen peroxide seems to be especially promising because of its high performance, less-toxic vapor and corrosivity, and its environmentally benign decomposition products, which make handling this oxidizer considerably less difficult than N_2O_4 or HNO_3. A high fuel performance can be fostered by light metals with large combustion energies and relatively light products. Elements with considerable performance advantages and non-toxic products are aluminum and boron. The need for light combustion products through the production of hydrogen gas and water vapor is fulfilled by a high hydrogen content. Aluminum and boron are well known for their ability to serve as hydrogen carriers in neutral and ionic molecules.

Quite recently, Schneider and other researchers from the Air Force Research Laboratory at Edwards AFB tested various ILs with 90% and 98% H_2O_2 [57b]. In view of the advantages of high hydrogen content, ILs containing $[Al(BH_4)_4]^-$ ions may be viewed as a densified form of hydrogen stabilized by metal atoms. In this recently published study, Schneider et al. investigated the new compounds trihexyltetradecylphosphonium tetrahydridoborate [THTDP][BH_4] (**1**) and [THTDP][$Al(BH_4)_4$] (**2**) (Fig. 9.27).

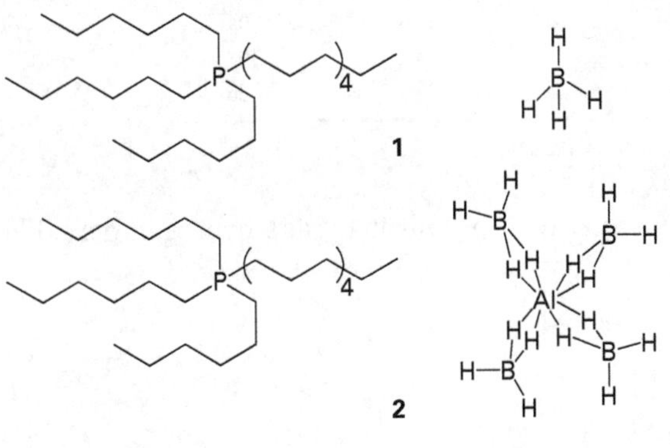

Fig. 9.27: Borohydride based ionic liquids [THTDP][BH_4] (**1**) and [THTDP][$Al(BH_4)_4$] (**2**).

Compounds **1** and **2** were subjected to drop tests to determine their reactivity with common propulsion oxidizers, including 90% and 98% H_2O_2 (Tab. 9.9a). While **1** lights only 3 s after dropping onto H_2O_2, the ignition delay of **2** was quite short. The hypergolic dicyanamide ILs display changes in delay times (with white fuming nitric acid, WFNA) from 30 ms to 1,000 ms upon reversal of the order of addition of oxidizer and fuel. In contrast, the ignition of **2** is equally fast regardless of the order of addition. These simple drop tests place only upper limits on the ignition delays because ignition may be initiated by hydrogen, which burns with an almost invisible flame. However, these tests do demonstrate that an IL with a complex $[Al(BH_4)_4]^-$ anion is universally reactive with traditional rocket oxidizers including less hazardous H_2O_2. The computed specific impulses for various H_2O_2/2 ratios in comparison with the systems NTO (N_2O_4) / MMH and N_2H_4 / H_2O_2 are summarized in Tab. 9.9b. Furthermore, this new class of ILs holds the potential for enabling high-performing, noncryogenic, green bipropulsion for the first time.

Tab. 9.9a: Drop test results of **2** on four oxidizers.

	90% H_2O_2	**98% H_2O_2**	**N_2O_4 (NTO)**	**WFNA[a]**
reaction	ignition	ignition	ignition	explosion
ignition delay / ms	< 30	< 30	ignition in vapor phase before the liquids combined	–

[a]WFNA = white fuming nitric acid

Tab. 9.9b: Computed equilibrium I_{sp} values[b].

O/F (mass %)	**NTO / MMH**	**N_2H_4 / H_2O_2**	**(THTDP) [Al (BH$_4$)$_4$] (2)**
50 / 50	287	292	218
60 / 40	311	**304**	226
70 / 30	**321**	303	257
80 / 20	286	269	291
85 / 15			**302**
90 / 10			286

[b]chamber pressure = 15 bar, expansion pressure = 0.05 bar

A very promising ionic liquid as a fuel with propulsion-grade hydrogen peroxide (H_2O_2) is 1-ethyl-3-methyl-imidazolium-thiocyanate (EMIM-SCN).

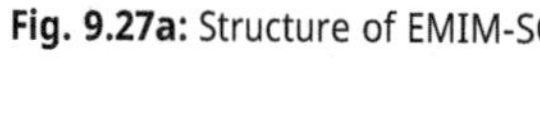

Fig. 9.27a: Structure of EMIM-SCN.

9.4 Dinitroguanidine derivatives

Exceeding the detonation performance of known explosives such as RDX (1,3,5-trinitro -1,3,5-hexahydrotriazine) is a long-term research goal in the field of new energetic compounds. The development of strained cages like CL-20 (HNIW: 2,4,6,8,10,12-hexanitro-2,4,6,8,10,12-hexaazaiso-wurzitane) or ONC (octanitrocubane) has attracted lots of interest, however both materials have several drawbacks. The existence of polymorphs as well as a cost intensive preparation makes both compounds disputable for contemporary application. As an aside, highly nitrated compounds like HNB (hexanitrobenzene) show the most promising properties, but their chemical stability towards acids or bases is often not good enough. As material density is the key for well performing high explosives, a different approach has been started with the concept of zwitterionic structures. A molecule capable of fulfilling these requirements is 1,2-dinitroguanidine (DNQ), which has been reported and well described in the literature previously. The crystallographic characterization of DNQ shows different nitro group surroundings, contrary to dissolved DNQ which shows rapid prototropic tautomerism. The presence of an acidic nitramine (pK_a = 1.11) in DNQ makes it possible to produce monoanionic dinitroguanidine salts. Nucleophilic substitution reactions involving the dinitroguanidine anion are also reported in the literature.

1,2-Dinitroguanidine (**1**) can easily be synthesized as reported by Astrat'yev. The synthetic route is shown in Fig. 9.28. Since it is known that strong nucleophiles like hydrazine or ammonia can substitute nitramine groups at elevated temperatures, **ADNQ** (ammonium dinitroguanidine) was produced from a stoichiometric amount of ammonium carbonate to avoid the free base. 1,7-Diamino-1,7-dinitrimino-2,4,6-trinitro-2,4,6-triazaheptane (**APX**) was also synthesized according to the method of Astrat'yev, starting from 1,3-bis(chloromethyl)nitramine (**2**) and potassium dinitroguanidine (**3**). The former compound is readily available from the nitrolysis of hexamine with subsequent chlorination.

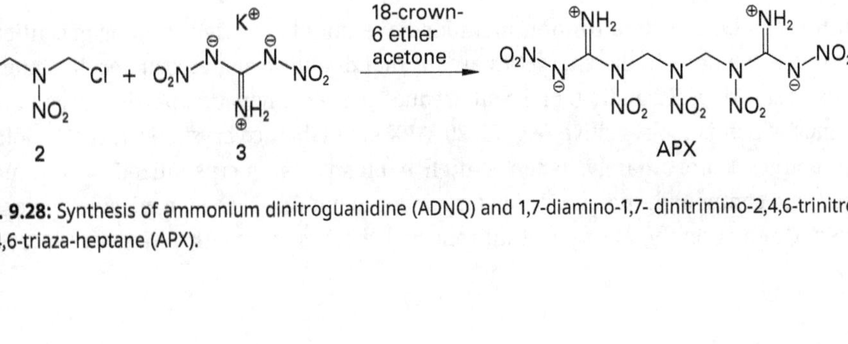

Fig. 9.28: Synthesis of ammonium dinitroguanidine (ADNQ) and 1,7-diamino-1,7- dinitrimino-2,4,6-trinitro -2,4,6-triaza-heptane (APX).

Differential scanning calorimetry (DSC) measurements were used to determine the decomposition temperatures of APX and ADNQ, and indicate that decomposition of APX starts with an onset temperature of 174 °C. In contrast to this behavior, the decomposition temperature of the ionic compound ADNQ is 197 °C. In addition, both compounds were tested according to the UN3c standard in a Systag, FlexyTSC Radex oven at 75 °C for 48 h with the result, that no weight loss or decomposition products were detected.

The detonation parameters calculated with the EXPLO5 program using the experimentally determined densities (X-ray) as well as some sensitivity values are summarized in Tab. 9.10. Especially the detonation parameters of APX are particularly promising – exceeding those of TNT and RDX and in part even those of HMX. The most important criteria of high explosives are the detonation velocity (D = APX: 9540, ADNQ: 9066, TNT: ~ 7178, RDX: 8906, HMX: 9324 m s^{-1}), the detonation pressure ($p_{C\text{-}J}$ = APX: 398, ADNQ: 327, TNT: 205, RDX: 346, HMX: 393 kbar) and the energy of explosion (Q_v = APX: –5943, ADNQ: –5193,TNT: 5,112, RDX: –6043, HMX: –6049 kJ kg^{-1}) [59a].

Tab. 9.10: Detonation parameters.

	APX	APX + 5% PVAA	ADNQ
ρ / g cm^{-3}	1.911	1.875	1.735
Ω / %	-8.33	-16.28	-9.63
Q_v / kJ kg^{-1}	-5,935	-5,878	-5,193
T_{ex} / K	4,489	4,377	3,828
$p_{C\text{-}J}$ / kbar	398	373	327
D / m s^{-1}	9,540	8,211	9,066
V_0 / l kg^{-1}	816	784	934
IS / J	3	5	10
FS / N	80	160	252
ESD / J	0.1	–	0.4

The sensitivity and performance values (Tab. 9.10) revealed that ADNQ may be a high-performing but still relatively insensitive high explosive, while APX needs to be classified as a primary explosive due to its high impact sensitivity.

9.5 Co-crystallization

Until recently, co-crystallization methods have only been used in pharmaceuticals. However, the hope in the energetics area is that due to strong intermolecular interactions (π stacking, H-bonding) one could reduce the sensitivity of a particular good performer which is too sensitive (e.g. CL-20, APX etc.) when co-crystallized with another compound. Unfortunately, when a high explosive is co-crystallized with a non-energetic compound, the energetic component inevitably sees its explosive power diluted. Quite recently, Matzger et al. reported the first energetic-energetic co-crystal composed of CL-20 and TNT in a 1:1 molar ratio [59c]. The co-crystal CL-20 · TNT also exemplifies the capabilities of aliphatic, highly nitrated compounds to co-crystallize despite their lack of strong, predictable interactions. With no available π-π stacking, this co-crystal forms based due to a series of CH hydrogen bonds between nitro group oxygens and aliphatic hydrogens as well as interactions between the electron-deficient ring of TNT and nitro groups of CL-20 (Fig. 9.29). The co-crystal CL-20 · TNT has a density of 1.91 g cm^{-3}, which is somewhat lower than those of CL-20 polymorphs (1.95–2.08 g cm^{-3}), but still substantially higher than that of TNT (1.70–1.71 g cm^{-3}). The impact sensitivity of CL-20 · TNT was measured using a drop test and determined to be approximately half as sensitive as CL-20. Therefore, incorporating insensitive TNT into a co-crystal with sensitive CL-20 greatly reduces its impact sensitivity, potentially improving the viability of CL-20 in explosive applications. This initial study holds promise that more energetic–energetic co-crystals may be discovered from among the nitro-rich, non-aromatic compounds that dominate the field.

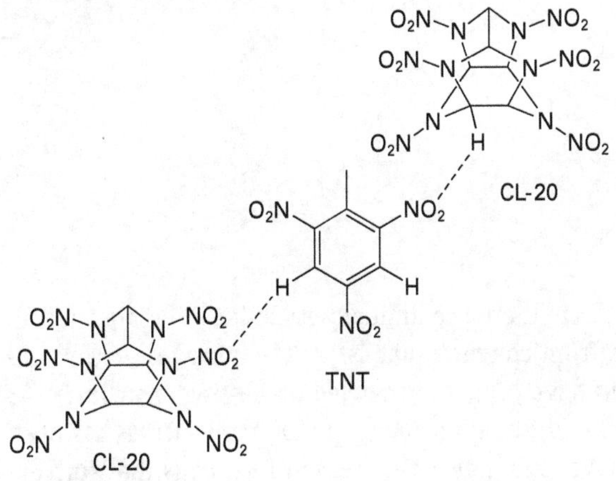

Fig. 9.29: Molecular assembly in the co-crystal CL-20·TNT.

9.6 Future energetics

Great strides have been made in increasing performance and decreasing sensitivity in energetic materials since the first commercialization of nitroglycerin (NG) in the form of dynamite in 1867 by Alfred Nobel. However, the DoD continues rely on traditional, half-century old energetics to meet their combat needs. New energetic materials must be developed to enable DoD to advance beyond traditional thought on energetic capabilities, to transition to the next stage in combat, and to maintain a lethality overmatch with all enemies.

Energetic materials are most commonly used in either high explosives or propellant formulations. Certain parameters are important in determining the effectiveness of new molecules in these formulations, including high densities (ρ), good oxygen balance (Ω) and high detonation/combustion temperatures and high specific impulses (I_{sp}) for rocket propellant formulations and lower combustion temperatures combined with a high force and pressure and a high N_2 / CO ratio of the reaction gases for gun propellants.

Using the heat of explosion (Q), the detonation velocity (D) and the detonation pressure (P) as a measure for the performance of a high explosive, one can clearly see from Fig. 9.30 that since the time NG was taken into service the performance of chemical explosives has improved substantially.

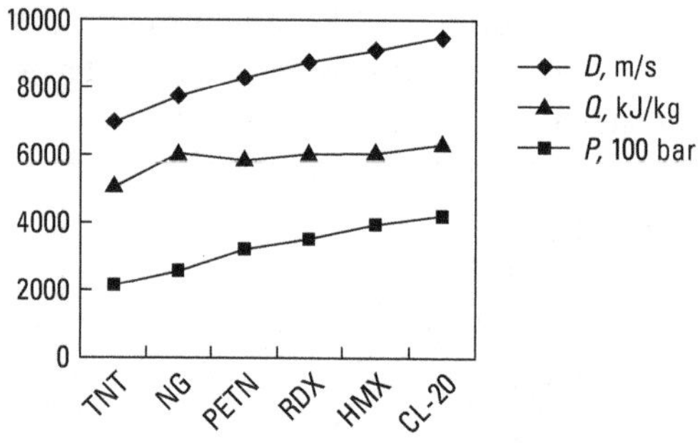

Fig. 9.30: Performance of chemical explosives.

In spite of the many years of research, there are limited possibilities to realize a substantial increase in performance from conventional C-H-N-O explosives. Recent advances in energetics energy output have come in improved processing or inclusion of energetic binders to increase overall formulation energy, but limited success has been realized in the development of novel energetics. One reason for this is that conventional nitramine and nitroaromatic explosives such as TNT, RDX, HMX and other similar molecules share the same three limitations (Tab. 9.11):

(i) they are not nitrogen-rich ($N \leq 50\%$)
(ii) the oxygen-balance is not close to zero
(iii) formulations (mixtures of various HEs) are required to achieve a good oxygen balance

Tab. 9.11: Nitrogen content (N) and oxygen balance (Ω) of conventional HEs.

	N / %	Ω / %
TNT	18.5	−73.9
PETN	17.7	−10.1
RDX, HMX	37.8	−21.6

Nitrogen rich molecules are desired as energetics because of the high energy content in N—N bonds. Oxygen balance is defined as the percentage of used oxygen that remains or is needed after an oxidation reaction and can therefore be positive or negative. Materials with an oxygen balance close to zero are typically, but not always, more effective energetics since all of the oxygen is used up in the reaction. An oxygen balance can be modified through formulation additions to bring the overall formulation oxygen balance close to zero.

Researchers have already realized the energy content limit for CHNO-based molecules. Research needs to expand beyond this way of thinking and increase efforts to explore different molecular structures and molecular make-up in order to realize the substantial increase in performance that will be required for future combat systems. Early research has shown that materials with a high-nitrogen content offer many advantages to those with carbon backbones, including the potential for vastly increased energy content. Research into molecules with high-nitrogen contents (> 50%) has shown potential for a substantial increase in available energy. The first generation of high-nitrogen compounds, such as hydrazinium azotetrazolate (HZT) and triaminoguanidinium azotetrazolate (TAGZT) (Fig. 9.31), did indeed meet the criteria for being nitrogen-rich and proved to be very desirable ingredients in erosion-reduced gun propellants (see Fig. 4.4a). However, due to the unfavorable oxygen balance such compounds are not suitable as energetic fillers in high explosive compositions (Tab. 9.12).

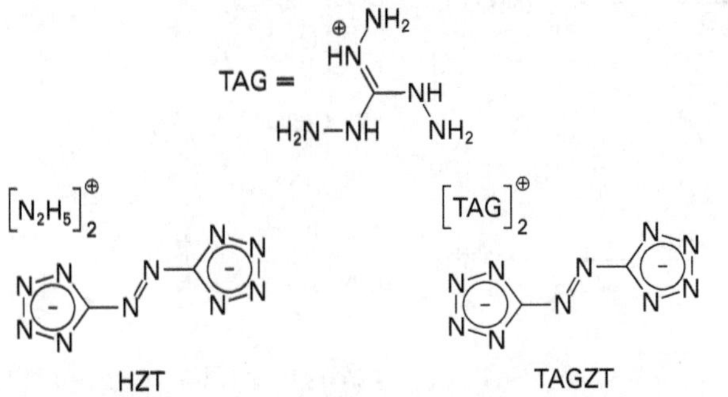

Fig. 9.31: Chemical structure of HZT and TAGZT.

Tab. 9.12: N content and Ω of high-N compounds.

		N / %	Ω / %
1st generation	HZT	85	-63
	TAGZT	82	-73
2nd generation	TKX-50	59	-27

The second generation of high-nitrogen compounds which have improved oxygen balances such as TKX-50 and ABTOX (Fig. 9.32), combine desirable high-nitrogen content with a good oxygen balance (Tab. 9.12). These compounds are therefore more suitable for use in high-explosive formulations. Moreover, materials with an oxygen balance close to zero are also suitable as powerful ingredients in solid rocket propellants. An increase of the I_{sp} of only 20 s would be expected to increase the payload or range by ca. 100%. Related to this, Fig. 4.4a shows the computed performance parameters for conventional and high-N gun propellants.

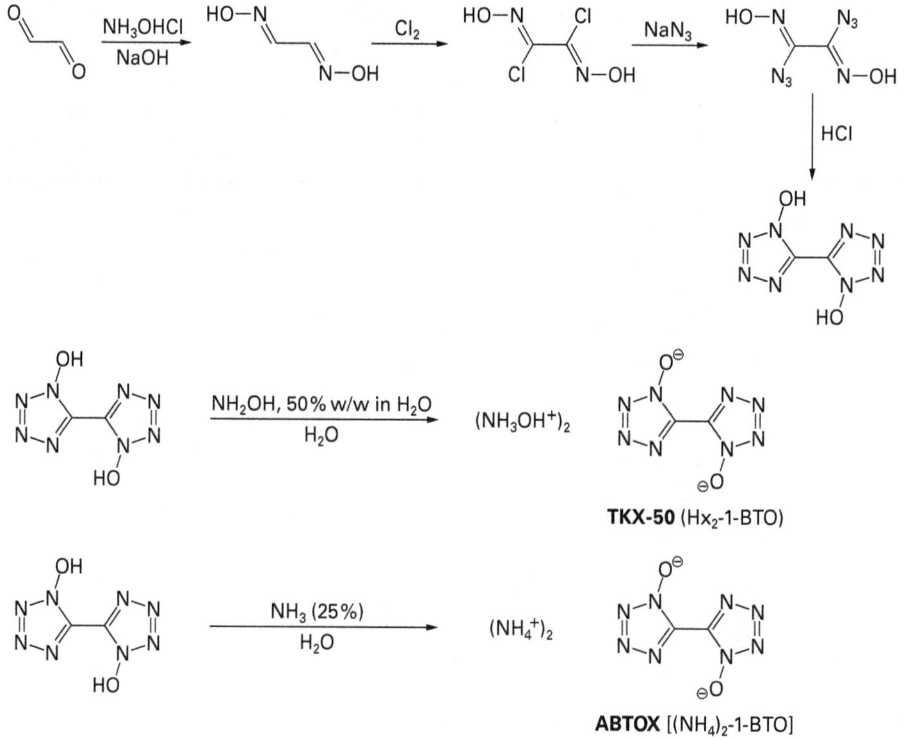

Fig. 9.32: Synthesis of 1-BTO from DAGL (top) and synthesis of TKX-50 and ABTOX (bottom).

The computed and predicted performance values not only exceed the first generation of high-nitrogen compounds (*e.g.* HZT) but, in the case of TKX-50, also the performance values of RDX and HMX (Fig. 9.33).

TKX-50 (bis(hydroxylammonium) 5,5′-bis(tetrazolate-1*N*-oxide)) is one of the most promising ionic salts being developed as a possible replacement for RDX [59d–f]. It can be prepared on a > 100 gram-scale scale by the reaction of 5,5′-(1-hydroxytetrazole) with dimethyl amine which forms the bis(dimethylammonium) 5,5′-(tetrazolate-1*N*-oxide) salt which is then isolated, purified and subsequently reacted in boiling water with two

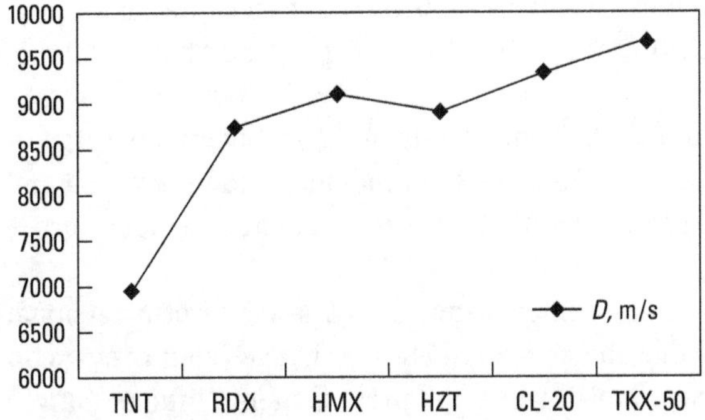

Fig. 9.33: Performance of conventional and high-nitrogen explosives (D = VoD).

equivalents of hydroxylammonium chloride to form TKX-50, dimethylammonium chloride and HCl (Fig. 9.32).

TKX-50 shows high performance with respect to its detonation velocity and C-J pressure, but a desirable low sensitivity towards impact and friction (Fig. 9.34).

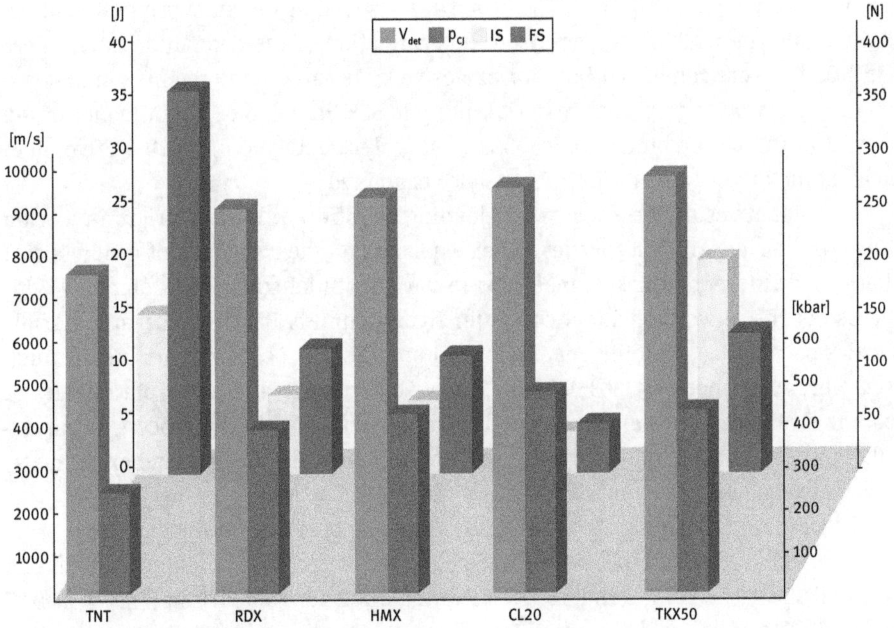

Fig. 9.34: Performance (detonation velocity and C-J pressure) and sensitivity data (impact and friction) of TKX-50 in comparison with other high explosives.

The thermal behavior of TKX-50 (bis(hydroxylammonium) 5,5′-(tetrazolate-1*N*-oxide)) and the kinetics of its thermal decomposition were studied using differential scanning calorimetry (DSC) and thermogravimetric analysis (TGA). The thermal decomposition of TKX-50 starts to occur within the range 210–250 °C depending on the heating rate used.

By applying multiple heating rate DSC measurements and Ozawa's isoconversional model free method, an activation energy of 34.2 kcal mol^{-1} and pre-exponential factor of $1.99 \cdot 1{,}012\ s^{-1}$ were calculated from the DSC peak maximum temperature – heating rate relationship.

By applying nonisothermal TGA experiments and the Flynn–Wall isoconversional model free method, it was found that the activation energy changes with conversion and lies between 34.7 and 43.3 kcal mol^{-1}, while the pre-exponential factor ranges from $9.81 \cdot 1{,}011$ to $1.79 \cdot 1{,}016\ s^{-1}$.

The enthalpy of formation of TKX-50 was calculated as (TKX-50) = +109 kcal mol^{-1}. The experimentally determined value based on bomb calorimetry measurements is ΔH_f° (TKX-50) = + 113 ± 2 kcal mol^{-1} (this value has subsequently been revised).

The calculated detonation parameters as well as the equations of state for the detonation products (EOS DP) of the explosive materials TKX-50 and MAD-X1 (and also for several of their derivatives) were obtained using the computer program EXPLO5 V.6.01. These values were also calculated for standard explosive materials which are commonly used such as TNT, PETN, RDX, HMX, as well as for the more powerful explosive material CL-20 for comparison. Determination of the detonation parameters and EOS DP were conducted both for explosive materials having the maximum crystalline density, and for porous materials of up to 50% in volume. The influence of the content of the plastic binder which was used (polyisobutylene up to 20% in volume) on all of the investigated properties was also examined.

Calculated results on shock wave loading of different inert barriers in a wide range of their dynamic properties under explosion on their surfaces of concrete size charges of different explosive materials in various initial states were obtained with the use of the one-dimensional computer hydrocode EP. Barriers due to materials such as polystyrene, textolite, magnesium, aluminum, zinc, copper, tantalum or tungsten were examined (Fig. 9.35). Initial values of pressure and other parameters of loading on the interface explosive-barrier were determined in the process of the calculations. Phenomena of propagation and attenuation of shock waves in barrier materials were considered too for all possible situations.

From these calculations, an essentially complete overview of the explosion properties and characteristics of shock wave action on barriers was obtained for several new and also for several standard explosive materials for comparison (Figs. 9.36, 9.37 and 9.38). The results which were obtained suggest that in a wide range of their initial states (porosity, inert binders), the new explosive materials TKX-50 and MAD-X1 possess better explosive properties and shock wave action on practically every compact barrier which was considered, in comparison with standard explosive materials including the military explosive RDX. The large number of the calculated results which were obtained can be used to study the influence of different factors on the explosion

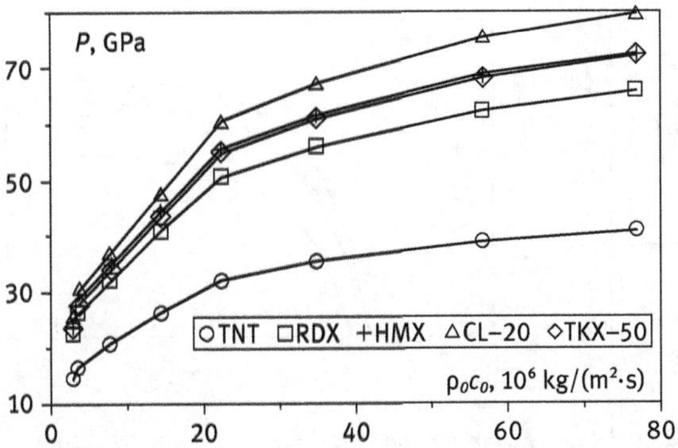

Fig. 9.35: The initial pressure on the interfaces of the explosive-barrier from different materials (from polystyrene to tungsten).

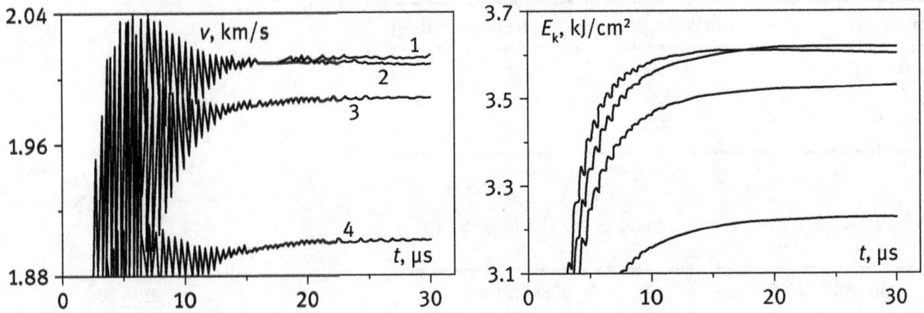

Fig. 9.36: Acceleration of copper plates by free charges of HMX (1), TKX-50 (2), MAD-X1 (3) and RDX (4).

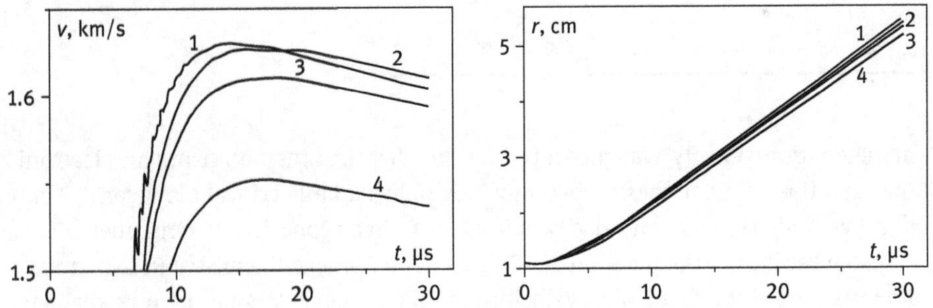

Fig. 9.37: Acceleration of Tantalum cylindrical layers by charges of HMX (1), TKX-50 (2), MAD-X1 (3) and RDX (4).

and shock wave action of new explosive materials. Furthermore, some of these results can then be used to plan experiments to confirm the predicted properties.

An experimental small-scale shock reactivity test (SSRT) (Tab. 9.13) was in good agreement with the calculated cylinder energies (Tab. 9.14).

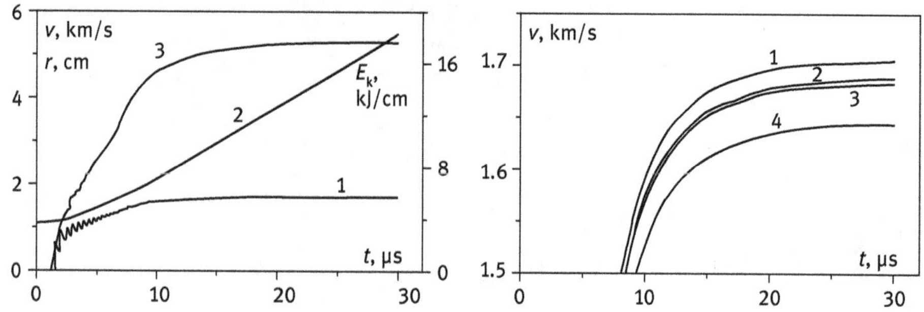

Fig. 9.38: Acceleration of a cylinder (1 = *v*; 2 = *r*; 3 = E_k) by a porous charge of TKX-50 (left) and by porous charges of HMX (1), TKX-50 (2), MAD-X1 (3) and RDX (4).

Tab. 9.13: Results of a small-scale shock reactivity test (SSRT).

explosive	mass [mg]	dent [mg SiO_2]
RDX	504	589
CL-20	550	947
TKX-50	509	857

Tab. 9.14: Calculated cylinder energies for TKX-50 and RDX.

compound V/V$_0$	E_C / kJ cm^{-3}	% of standard			
		TATB	PETN	HMX	CL-20
TKX-50					
2.2	-8.16	168	128	109	90
RDX					
2.2	-6.94	143	109	93	77

The detonation velocity was measured by the electrical method using an electronic timer and fiber optic probes. Apparently TKX-50 has a large critical diameter (> 3 cm), but shows a high detonation velocity of 9,460 m s^{-1} in a cylindrical 40 mm tube.

Quite recently, researchers from China's Xi'an Modern Chemistry Research Institute investigated the detonation velocities of TKX-50 depending on the tube diameter (Tab. 9.14a) (*Propellants Explos. Pyrotech.* **2019**, *44*, 408–412).

The detonation velocities were obtained for TKX-50-based explosive columns with different diameters, and an obvious scale effect was observed. Steady detonation can perhaps not be achieved when the column diameters are less than 30 mm.

The shock sensitivity was determined by the same authors using the gap test (Tab. 9.14b).

The critical initiation pressures of the TKX-50-based explosives that were examined are much higher than those for HMX and CL-20-based explosives.

Tab. 9.14a: Detonation velocities (VoD) of TKX-50 depending on the tube diameter.

sample	column diameter/mm	density/g cm^{-3}	VoD/m s^{-1}
TKX-50 / binder = 95.5 / 4.5	20 × 20	1.79	8,699
TKX-50 / binder = 95.5 / 4.5	30 × 30	1.80	8,774
TKX-50 / binder = 95.5 / 4.5	40 × 40	1.80	8,994
TKX-50 / binder = 95.5 / 4.5	50 × 50	1.81	8,996
TKX-50 / binder = 95.5 / 4.5	60 × 60	1.80	9,037

Tab. 9.14b: The thickness of the organic glass gap and the corresponding shock-wave pressure for TKX-50, HMX and CL-20 based explosives.

explosive	thickness / mm	pressure / GPa
TKX-50	18.45	6.24
RDX	40.50	3.02
CL-20	49.51	2.64

Study of the effect of fillers on the detonation characteristics of energetic (TKX-50) composites

The initial materials used for the arrangement of the investigated composite energetic materials with a filler content of up to 50 vol.% and for explosive compositions with a binder content of up to 10 wt.% are actually TKX-50 itself (dihydroxylammonium 5,5′-bistetrazole-1,1′-diolate) and several inert and energetic binders, such as paraffin, HTPB (hydroxyl-terminated polybutadiene), GAP (glycidyl azide polymer), poly-AMMO (poly-3-azidomethyl-3-methyloxetane) and poly-BAMO (poly-3,3-bis(azidomethyl) oxetane). The structural formulas of these substances are shown in Fig. 9.38a, and properties used in thermochemical calculations are given in Tab. 9.14c.

Software used and methods of computational study

Thermochemical and thermodynamic calculations for the composite energetic materials and explosive compositions were carried out using the EXPLO5 program. Calculations of the explosion action effect of the considered explosive compositions on metal barriers were carried out using the ANSYS Autodyn program. A description of these programs is given below.

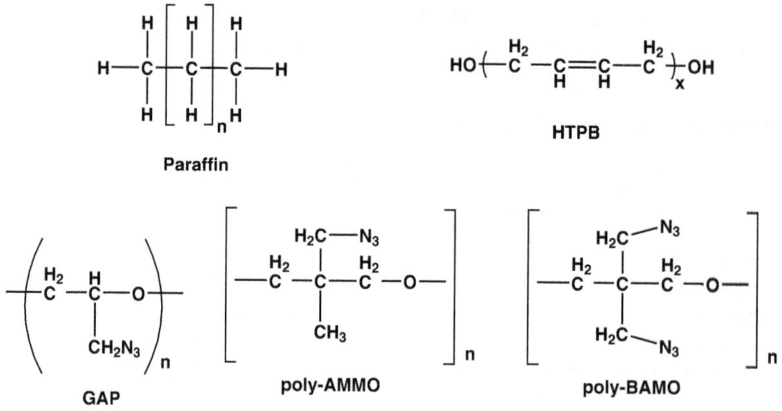

Fig. 9.38a: Structural formulae of paraffin, HTPB, GAP, poly-AMMO and poly-BAMO.

Tab.9.14c: Properties of substances used for performing thermochemical calculations.

Property	TKX-50	Paraffin	HTPB	GAP	AMMO	BAMO
Formula	$C_2H_8N_{10}O_4$	CH_2	$C_{10}H_{15.4}O_{0.07}$	$C_3H_5N_3O$	$C_5H_9N_3O$	$C_5H_8N_6O$
ρ_0, g/cm^3	1.877	0.90	0.92	1.29	1.06	1.30
ΔH_f, kJ/mol	194.1	30.6	-51.88	142.3	179	413

EXPLO5 is a thermochemical computer code that predicts the detonation (e.g. detonation velocity, pressure, energy, heat end temperature, etc.) and combustion (e.g. specific impulse, force, pressure, etc.) performance of energetic materials.

The calculation of detonation parameters is based on the chemical equilibrium steady-state model of detonation. The equilibrium composition of detonation and combustion products is calculated by applying the modified White, Johnson, and Dantzig free energy minimization technique.

The program uses the Becker-Kistiakowsky-Wilson (BKW) and Exp-6 EOS for gaseous detonation products, the ideal gas and virial equations of state of gaseous combustion products, and the Murnaghan equation of states for condensed products.

EXPLO5 is designed so that it enables calculation of the chemical equilibrium composition and thermodynamic parameters of the state along the shock adiabat of detonation products, the C-J state, and the detonation parameters at the C-J state, as well as the parameters of state along the expansion isentrope. The program has a non-linear curve fitting program built in to fit relative volume-pressure data along the expansion isentrope according to the Jones-Wilkins-Lee (JWL) model, enabling calculation of the detonation energy available for performing mechanical work.

ANSYS Autodyn is an explicit analysis tool for modeling nonlinear dynamics of solids, fluids, gases, and their interaction. Autodyn offers:

– Finite element solvers for computational structural dynamics (FE)

- Finite volume solvers for fast transient Computational Fluid Dynamics (CFD)
- Mesh-free particle solvers for high velocities, large deformation, and fragmentation (SPH)
- Multi-solver coupling for multi-physics solutions including coupling between FE, CFD, and SPH
- A wide suite of material models incorporating constitutive response and coupled thermodynamics
- Serial and parallel computation on shared and distributed memory systems

Study of the effect of fillers on the detonation characteristics of energetic composites

Composite energetic materials based on TKX-50 using fillers in the volume of up to 50% consisting of the inert and energetic binders shown in Fig. 9.38a were considered. The calculated detonation characteristics, composition of the formed detonation products, and expansion isentropes of the detonation products were obtained by performing thermochemical and thermodynamic calculations using the EXPLO5 program.

In filler content ranges of up to 50 vol.%, the dependences of the main detonation characteristics of the energetic material on the volume and mass content of the indicated binders were obtained, namely, the detonation velocity *D*, detonation pressure *P*, detonation temperature *T*, adiabatic index of the detonation products at the Jouguet point *k*, heat of explosion *Q* and the volume of the gaseous detonation products V_g. These thermochemical calculations also elucidated, the composition of the detonation products formed at the Jouguet point, and their evolution in the process of increasing the content of the filler in the composite material.

Tab.9.14d: Detonation characteristics of a composite energetic material depending on the volume and mass content of TKX-50 and AMMO.

TKX-50		ρ_0	*D*	*P*	*T*	*k*	*Q*	V_g
φ_t	ω_t	g/cm³	m/s	GPa	K		kJ/kg	dm³/kg
1	1	1.877	9,456	37.02	3,043	3.533	–4,711	924.2
0.95	0.9711	1.836	9,259	35.20	3,037	3.472	–4,668	922.3
0.90	0.9410	1.795	9,063	32.84	3,009	3.489	–4,621	920.2
0.85	0.9094	1.754	8,867	30.84	2,983	3.472	–4,572	917.6
0.80	0.8763	1.714	8,676	29.01	2,953	3.448	–4,520	914.6
0.75	0.8416	1.673	8,481	27.29	2,923	3.410	–4,467	911.5
0.70	0.8051	1.632	8,287	25.48	2,884	3.398	–4,411	908.4
0.65	0.7668	1.591	8,092	23.98	2,850	3.346	–4,354	904.9
0.60	0.7265	1.550	7,898	22.31	2,804	3.335	–4,296	901.4
0.55	0.6840	1.509	7,704	20.87	2,762	3.292	–4,236	897.4
0.50	0.6391	1.469	7,514	19.38	2,711	3.281	–4,175	893.2

Tabs. 9.14d and 9.14e show the detonation characteristics of composite energetic materials depending on the volume and mass content of TKX-50 with the fillers AMMO and BAMO respectively.

Tab. 9.14e: Detonation characteristics of a composite energetic material depending on the volume and mass content of TKX-50 and BAMO.

TKX-50		ρ_0 g/cm³	*D* m/s	*P* GPa	*T* K	*k*	*Q* kJ/kg	V_g dm³/kg
φ_t	ω_t							
1	1	1.877	9,456	37.02	3,043	3.533	−4,711	924.2
0.95	0.9648	1.848	9,310	35.74	3,043	3.481	−4,677	920.7
0.90	0.9285	1.819	9,163	34.04	3,028	3.487	−4,641	916.9
0.85	0.8911	1.790	9,016	32.47	3,011	3.482	−4,603	912.8
0.80	0.8524	1.762	8,874	31.01	2,992	3.475	−4,563	908.3
0.75	0.8124	1.733	8,727	29.51	2,970	3.473	−4,522	903.6
0.70	0.7711	1.704	8,580	28.13	2,948	3.460	−4,480	898.7
0.65	0.7284	1.675	8,433	26.89	2,928	3.430	−4,437	893.6
0.60	0.6841	1.646	8,285	25.65	2,904	3.405	−4,394	888.2
0.55	0.6383	1.617	8,137	24.38	2,877	3.391	−4,349	882.7
0.50	0.5908	1.588	7,988	23.01	2,843	3.404	−4,304	877.1

Comparative dependences for the velocity and pressure of detonation on the volumetric content of TKX-50 for these composite energetic materials are shown in Figs. 9.38b and 9.38c. In Figs. 9.38d and 9.38e, similar comparative relationships are shown for TKX-50 and composite energetic materials containing paraffin and GAP fillers.

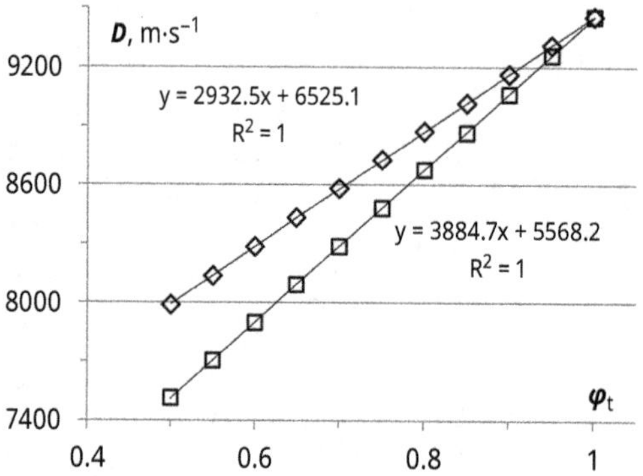

Fig. 9.38b: Influence of the volumetric content of TKX-50 on the detonation velocity of composite energetic materials with BAMO (rhombuses) and AMMO (squares).

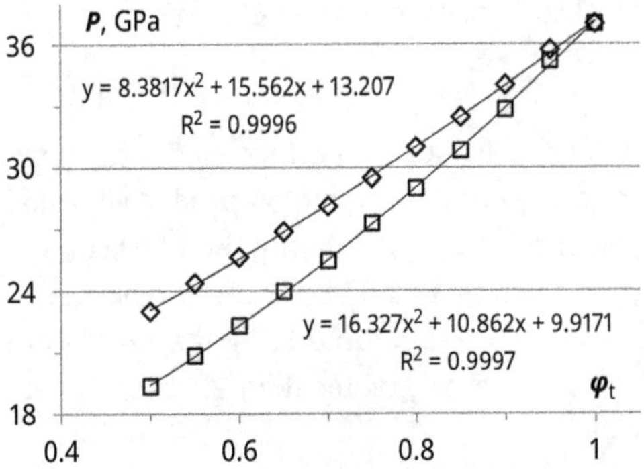

Fig.9.38c: Influence of the volumetric content of TKX-50 on the detonation pressure of composite energetic materials with BAMO (rhombuses) and AMMO (squares).

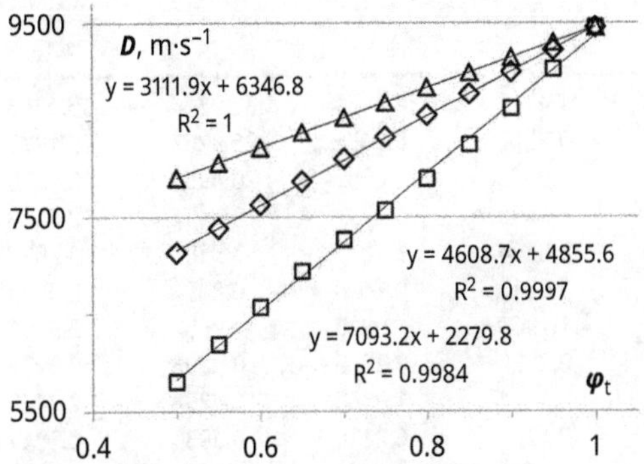

Fig.9.38d: Influence of the volumetric content of TKX-50 on the detonation velocity of the energetic material TKX-50 (squares) and composite energetic materials with paraffin (rhombuses) and GAP (triangles).

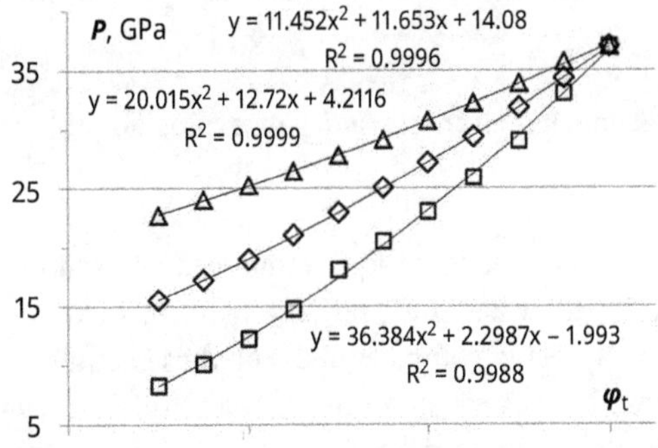

Fig. 9.38e: Influence of the volumetric content of TKX-50 on the detonation pressure of the energetic material TKX-50 (squares) and composite energetic materials with paraffin (rhombuses) and GAP (triangles).

Study of the effect of fillers on the chemical composition of the detonation products

As was previously mentioned, thermochemical calculations performed for the composite energetic materials showed the composition of the detonation products formed at the Jouguet point, as well as the changes observed on increasing the filler content in the composite material. These results are shown in Tab. 9.14f for the composite energetic material with the energetic filler GAP. In Fig. 9.38f and 9.38g the results are shown only for products whose molecular content was greater than 1% over the entire range of filler content under consideration.

Tab. 9.14f: Composition of detonation products at the Jouguet point for the composite energetic material TKX-50—GAP, depending on the volumetric content of the components.

TKX-50 Vol fraction Products	1.0	0.9	0.8	0.7	0.6	0.5
	Mol %					
N_2	46.4032	43.8070	41.4112	39.1831	37.0844	35.1203
H_2O	32.7348	31.6653	30.2529	28.6436	26.9263	25.1072
C (s,d)	15.8278	18.8724	21.6082	24.1232	26.4989	28.7260
CH_2O_2	1.7659	1.3229	1.0496	0.8734	0.7494	0.6562
NH_3	1.4825	2.0926	2.6864	3.1815	3.5500	3.7627
CO	0.5538	0.5877	0.6513	0.7423	0.8494	0.9903
H_2	0.5433	0.8396	1.1960	1.5818	1.9652	2.3361
CO_2	0.4531	0.3448	0.2979	0.2819	0.2806	0.2977
CH_4	0.1787	0.3581	0.6474	1.0547	1.5768	2.2425
C_2H_6	0.0244	0.0599	0.1267	0.2330	0.3831	0.5849
HCN	0.0240	0.0339	0.0451	0.0574	0.0697	0.0814
C_2H_4	0.0055	0.0118	0.0227	0.0389	0.0608	0.0889
N_2H_4	0.0024	0.0031	0.0033	0.0032	0.0028	0.0023
CH_3OH	0.0004	0.0007	0.0011	0.0016	0.0023	0.0032
H	0.0001	0.0002	0.0002	0.0003	0.0003	0.0003

Study of the effect of porosity on the detonation characteristics of explosive compositions

Calculations have also been performed for composite energetic materials based on TKX-50, but with 5 and 10 wt% binders. In this case, the energetic materials can already be considered as corresponding to explosive compositions. For these compositions, detonation conditions were considered in cases where they had an initial porosity of up to 10%. An example of such results is shown in Tab. 9.14g for an explosive composition with 5 wt% AMMO binder. In Figs. 9.38h and 9.38i, the results are shown for the dependence of the detonation velocity and pressure (for this composition) on the volumetric content of the explosive composition φ_c, or its porosity π_c ($\varphi_c + \pi_c = 1$).

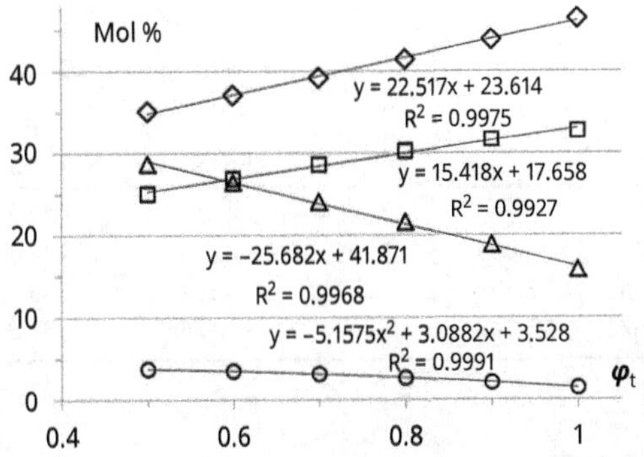

Fig. 9.38f: Influence of the volumetric content of TKX-50 in the composite energetic material TKX-50—GAP on the concentration of detonation products such as nitrogen (rhombuses), water (squares), carbon (triangles) and ammonia (circles).

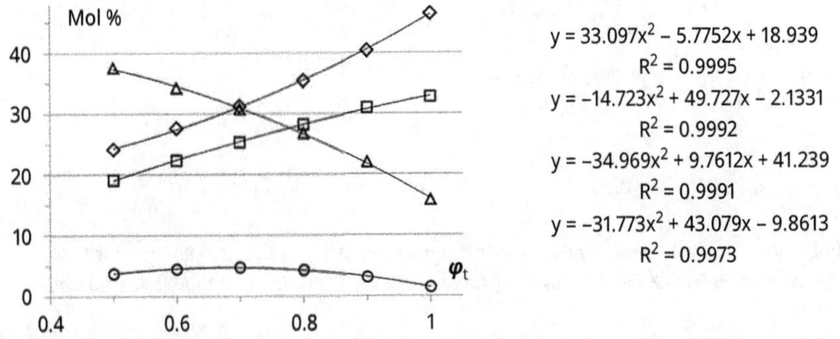

Fig. 9.38g: Influence of the volumetric content of TKX-50 in the composite energetic material TKX-50—HTPB on the concentration of detonation products such as nitrogen (rhombuses), water (squares), carbon (triangles) and ammonia (circles).

Tab. 9.14g: Detonation characteristics of the explosive composition 0.95 TKX-50–0.05 AMMO depending on its volumetric content in the sample (porosity).

0.95T–0.05A	ρ_0	*D*	*P*	*T*	*k*	*Q*	V_g
φ_c	**g/cm³**	**m/s**	**GPa**	**K**		**kJ/kg**	**dm³/kg**
1	1.807	9,121	33.54	3,018	3.482	–4,636	920.9
0.99	1.789	9,037	32.77	3,029	3.457	–4,632	921.5
0.98	1.771	8,954	32.02	3,040	3.434	–4,628	922.2
0.97	1.753	8,872	31.36	3,053	3.399	–4,624	923.0
0.96	1.735	8,791	30.53	3,060	3.391	–4,620	923.9
0.95	1.717	8,711	29.72	3,066	3.383	–4,615	925.0
0.94	1.699	8,633	29.08	3,078	3.354	–4,609	926.0

Tab. 9.14g (continued)

0.95T–0.05A	ρ_0	*D*	*P*	*T*	*k*	*Q*	V_g
φ_c	g/cm³	m/s	GPa	K		kJ/kg	dm³/kg
0.93	1.681	8,554	28.37	3,086	3.334	–4,604	927.2
0.92	1.662	8,477	27.82	3,099	3.295	–4,598	928.4
0.91	1.644	8,402	27.14	3,107	3.277	–4,591	929.7
0.90	1.626	8,327	26.42	3,122	3.268	–4,584	931.3

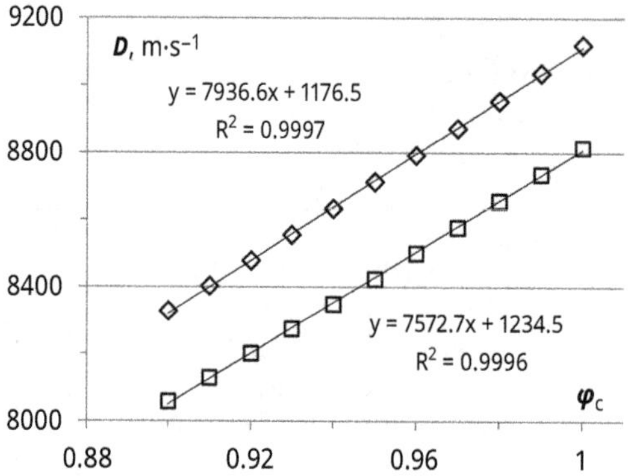

Fig. 9.38h: Influence of the volumetric content of the material (porosity) on the detonation velocity of explosive compositions 00.95 TKX–50-0.05 AMMO (rhombuses) and 0.90 TKX-50–0.10 AMMO (squares).

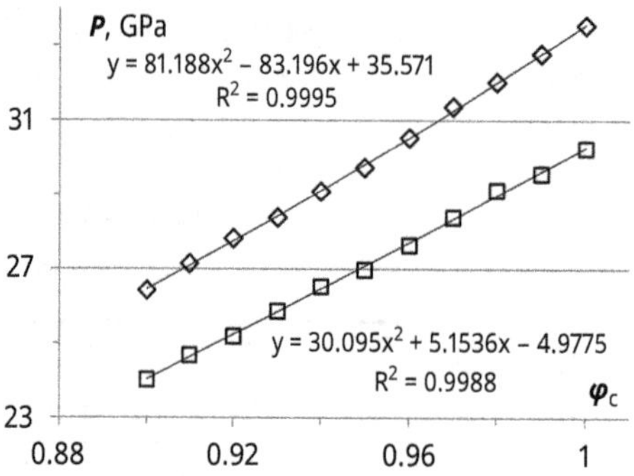

Fig.9.38i: Influence of the volumetric content of the material (porosity) on the detonation pressure of explosive compositions 0.95 TKX-50–0.05 AMMO (rhombuses) and 0.90 TKX-50–0.10 AMMO (squares).

Obtaining the coefficients of the equation of state for detonation products in the Jones-Wilkins-Lee (JWL) form

As noted earlier, in thermochemical calculations performed for all composite energetic materials, the coefficients of the equations of state in the Jones-Wilkins-Lee form were determined. This procedure is performed by approximating the calculated expansion isentrope of detonation products by the isentrope formula from this equation of state, which describes the relationship between the current specific volume of detonation products and their pressure

$$p = A\ exp(-R_1 v) + B\ exp(-R_2 v) + Cv^{-(1+\omega)}$$

The coefficients of the equation of state obtained in this way for two initial states of explosive compositions, compact (100%) and with two percent porosity (98%), are given in Tab. 9.14h. The results are for pure TKX-50, and explosive compositions based on TKX-50 with 5 wt.% of various binders. Detonation characteristics corresponding to the Jouguet conditions for these equations of state are also given.

Tab. 9.14h: Coefficients of the equation of state for detonation products in the JWL form for TKX-50 and explosive compositions based on it.

Coefficients EOS JWL	TKX-50		5% Paraffin		5% HTPB	
	100%	98%	100%	98%	100%	98%
ρ_0, g/cm^3	1.877	1.839	1.780	1.744	1.784	1.748
A, GPa	3,064	2,740	3,455	3,078	3,520	2,963
B, GPa	79.86	76.68	82.94	79.72	85.36	77.11
C, GPa	1.485	1.497	1.360	1.352	1.380	1.368
R_1	6.540	6.492	7.021	6.980	7.081	6.926
R_2	2.058	2.064	2.172	2.175	2.188	2.148
ω	0.508	0.497	0.526	0.518	0.522	0.517
D, m/s	9,456	9,282	9,002	8,830	8,973	8,805
Q, kJ/kg	–4,711	–4,705	–4,406	–4,401	–4,467	–4,460
P, GPa	37.02	35.47	31.89	30.37	31.83	30.25
ρ_0, g/cm^3	1.835	1.798	1.807	1.771	1.836	1.799
A, GPa	3,186	2,794	3,046	2,669	3,207	2,808
B, GPa	82.80	78.47	80.36	76.20	83.02	78.54
C, GPa	1.426	1.419	1.419	1.411	1.428	1.421
R_1	6.758	6.680	6.769	6.690	6.760	6.679
R_2	2.113	2.107	2.115	2.110	2.114	2.107
ω	0.541	0.533	0.533	0.526	0.541	0.534
D, m/s	9,236	9,066	9,121	8,954	9,249	9,079
Q, kJ/kg	–4,655	–4,648	–4,636	–4,628	–4,663	–4,655
P, GPa	35.01	33.17	33.54	32.03	35.12	33.28

In Figs. 9.38j and 9.38k, for example, the isentrope of expansion of the detonation products of pure, compact TKX-50 is shown in the form of pressure versus density and pressure versus specific volume dependences.

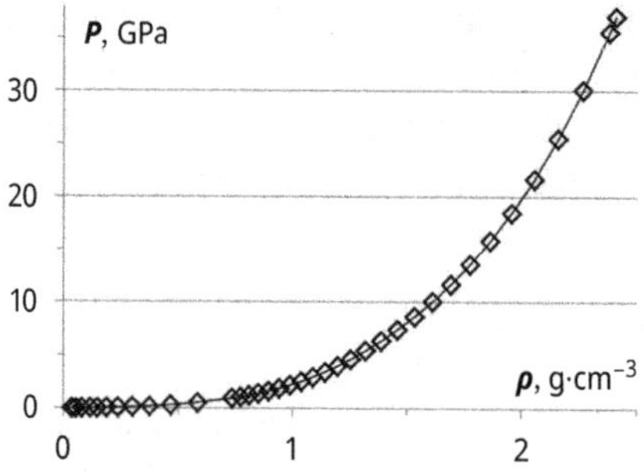

Fig. 9.38j: Expansion isentrope of TKX-50 detonation products as a function of pressure versus density.

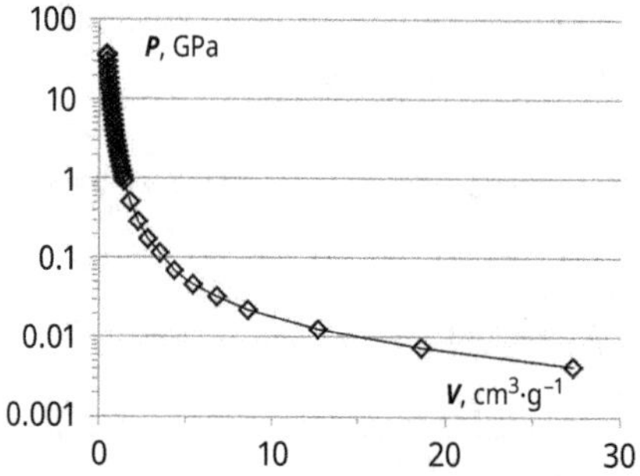

Fig. 9.38k: Expansion isentrope of TKX-50 detonation products as a function of pressure versus specific volume.

Analysis of experimental data on detonation velocity

Experimental data for the detonation velocity of several explosive compositions based on TKX-50 has been reported. An explosive composition of 97% TKX-50 and 3% paraffin was studied by W. P. Zhang et al. (*Chinese Journal of Explosives & Propellants.* **2015**, No. 06, 67–71). Good agreement with the results of calculations performed using the enthalpy of formation of TKX-50 (T. S. Konkova et al. 47th Annual Conference of ICT, Karlsruhe, Germany, ICT, **2016**, p. 90/1–90/8) was observed for the results obtained by Zhang. The results of this comparison are shown in Fig. 9.38l.

In the work of Zhang et al., the limiting conditions for compaction were characterized by the compression pressure $P_c = 300$ MPa. In order to reveal some general tendency, these results can be compared with the results of work by L. N. Yao et al. (*Phys. Conf. Ser.* **2020**. Vol. 1507, 022032), in which the compaction of two explosive compositions based on TKX-50 with two different binders was studied. Binders ETPE (energetic thermoplastic azide binder) and fluoroelastomer were present with a weight content of 5%. The results of this comparison are shown in Fig. 9.38m.

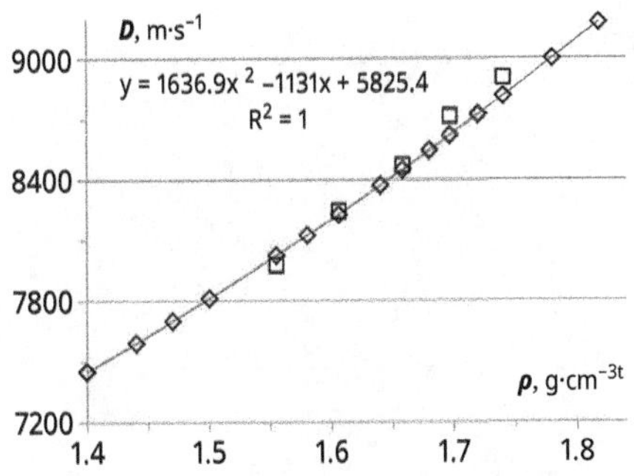

Fig.9.38l: Influence of the density of composition 1 on the detonation velocity; rhombuses – calculation, squares – experiment.

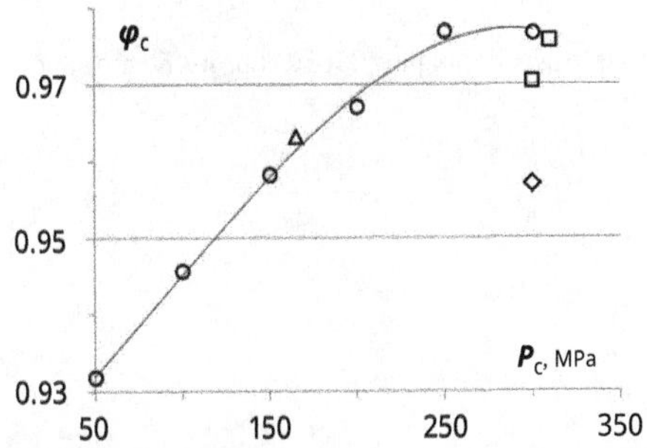

Fig.9.38m: Influence of compaction pressure on the relative density of samples with compositions 0.95 TKX-50 – 0.5 ETPE (circles, 1 (rhombus), 2 (squares) and 3 (triangle).

Based on these results, a tendency is observed for the influence of the relative mass content of TKX-50 in the samples on the arrangement of the experimental points in Fig. 9.38m. This trend indicates the influence of the relative content of the tough powder TKX-50 on the compaction of explosives based on it.

Study of the explosion action of the considered compositions on metal barriers

The explosion effect of 50 mm thick charges on various obstacles (primarily copper) was investigated. The ANSYS Autodyn program was used in the calculations for thick (50 mm) and thin (1 mm) copper specimens. Explosive compositions with 5 wt% binders were considered, having an ultimate relative density (100%) and an optimistic density close to real (98%). For copper, the shock EOS equation of state and the Steinberg-Guinan strength model from the program database were used.

The loading of thick specimens (50 mm) and thin specimens (1 mm) was considered. Every 1 mm of the substance was divided into 5 counting cells, that is, there were 250 of them in a thickness of 50 mm. Several results on loading thin samples with various explosive compositions are shown in Figs. 9.38n, 9.38o and 9.38p. The general relationship for pure TKX-50 as well as for charges containing binder is shown in Fig. 9.38q.

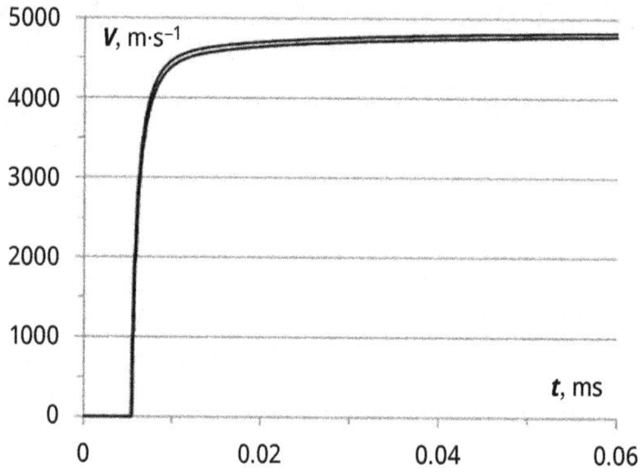

Fig. 9.38n: Velocities of copper plates 1 mm thick under explosive loading by TKX-50 charges 50 mm thick with a relative density of 100 and 98%.

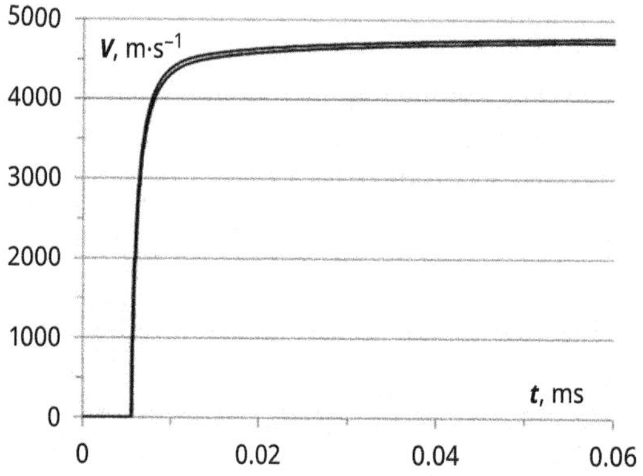

Fig. 9.38o: Velocities of copper plates 1 mm thick under explosive loading with charges containing BAMO binder 50 mm thick and a relative density of 100 and 98%.

The results on explosive loading of thick copper samples are given in Figs. 9.38r and 9.38s, which show the passage of shock waves at several coordinates in copper samples, and Figs. 9.38t and 9.38u show the passage of detonation waves at several coordinates in TKX-50 charges with a relative density of 100 and 98%. Similar results were obtained for all of the binders which were considered.

TKX-50 shows good compatibility with other energetic fillers, binders and plasticizers, as well as with most metals and metal oxides (Tab. 9.15). However, its purity is crucial in order for it to exhibit good compatibility with all of the above for use in formulations.

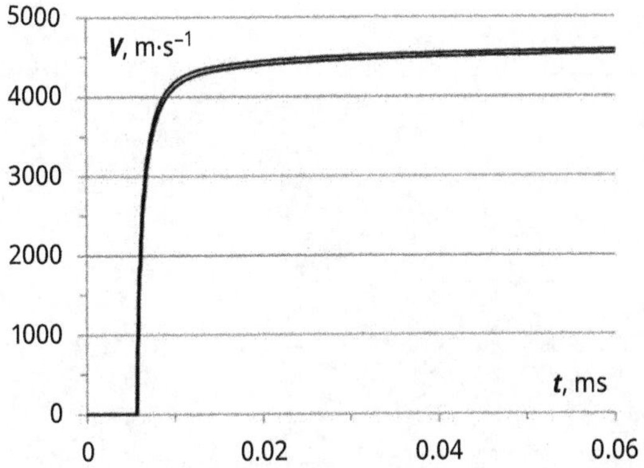

Fig. 9.98p: Velocities of copper plates 1 mm thick under explosive loading with charges containing paraffin binder 50 mm thick and a relative density of 100 and 98%.

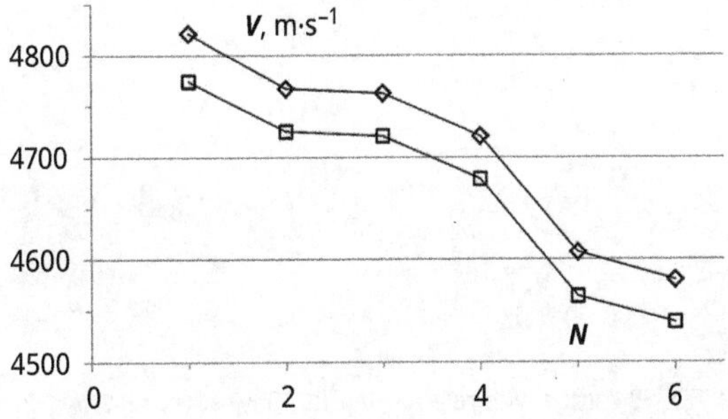

Fig. 9.98q: Velocities of copper plates 1 mm thick under explosive loading with charges containing different binders 50 mm thick and a relative density of 100 and 98%. 1 – pure TKX-50, 2 – BAMO, 3 – GAP, 4 – AMMO, 5 – HTPB, 6 – paraffin.

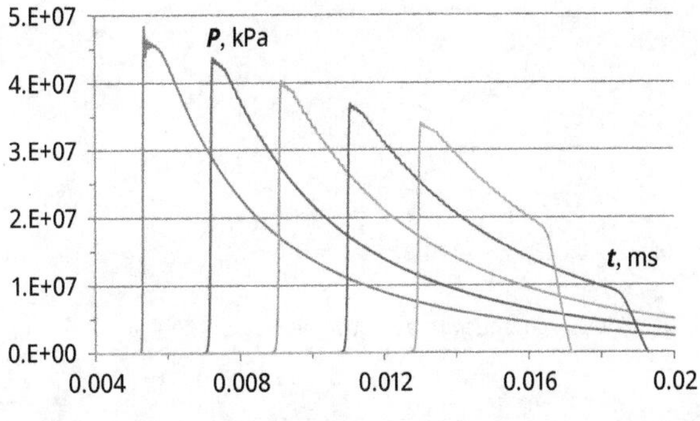

Fig. 9.38r: Diagrams of shock waves in a copper sample at coordinates 0, 10, 20, 30 and 40 mm under explosive loading with a TKX-50 charge with relative density 100%.

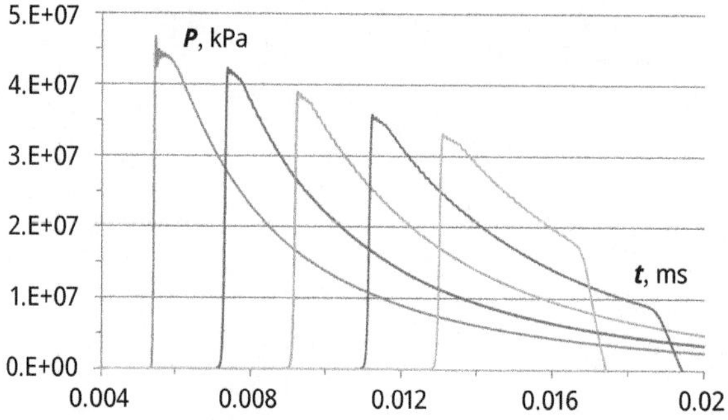

Fig. 9.38s: Diagrams of shock waves in a copper sample at coordinates 0, 10, 20, 30 and 40 mm under explosive loading with a TKX-50 charge with relative density 98%.

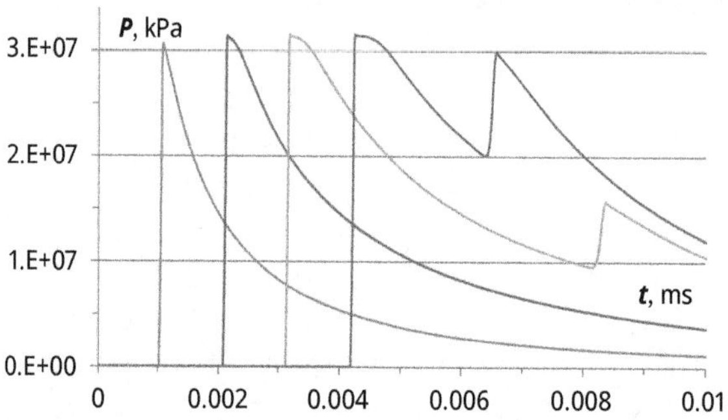

Fig. 9.38t: Diagrams of detonation waves in a TKX-50 charge with relative density 100% at coordinates 10, 20, 30 and 40 mm.

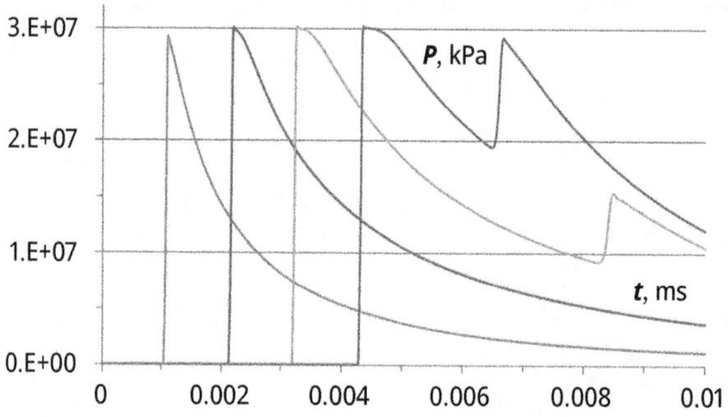

Fig. 9.38u: Diagrams of detonation waves in a TKX-50 charge with a relative density of 100% at coordinates 10, 20, 30 and 40 mm.

Tab. 9.15: Compatibility of TKX-50 performed according to a vacuum stability test (STANAG 4,556 ED.1), 40 h, 100 °C (compatible 0.0–3.0 mL).

Material	Reactivity of mix (mL)	Result
TKX-50	0.067	pass
TKX-50 & CuO	0.62	pass
TKX-50 & steel shavings	negligible	pass
TKX-50 & Al powder	0.13	pass
TKX-50 & R8002	2.4	pass
TKX-50 & BDNPA/F	0.94	pass
TKX-50 & chlorez wax	negligible	pass
TKX-50 & Viton	negligible	pass
TKX-50 & DNAN	negligible	pass
TKX-50 & TNT	negligible	pass
TKX-50 & NTO	negligible	pass
TKX-50 & HMX	negligible	pass
TKX-50 & PrNQ	1.2	pass

The vacuum stability test (VST, Tab. 9.15) is frequently used for determining the compatibility of energetic materials with contact materials as defined in STANAG 4,147, and for quality tests of energetic ingredients (other STANAGs). A dried and accurately weighed sample of an energetic material (or mixture of materials) is placed in the heating tube followed by its assembly and evacuation. The heating tube is immersed in a constant temperature bath (usually 100 °C or 120 °C) for a period of 40 h and the volume of evolved gas is recorded. Most explosives or explosive formulations yield less than 1 cm^3 of gas per gram of an explosive during 40 h at 120 °C. The vacuum stability test finds its wide application in qualification, surveillance, manufacture, quality control and R&D of a whole range of energetic materials.

Some of the relevant detonation performance parameters of TKX-50 were recently re-determined (T. M. Klapötke, S. Cudzilo, W. A. Trzcinski, *Prop. Explos. Pyrotech.*, **2022**, *47*, e202100358; https://doi.org/10.1002/prep.202100358) (Fig. 9.38v). The enthalpy of formation was obtained from the measured heat of combustion to be ΔH_f° (TKX-50(s)) = + 213.4 ± 1.2 kJ mol^{-1}. The heat of detonation was measured to be 4,650 ± 50 kJ kg^{-1}. The detonation velocity of a TKX-50/wax mixture (97:3) at a density of 1.74 g cm^{-3} was found experimentally to be 9,190 m s^{-1}. Using the experimentally obtained enthalpy of formation for TKX-50 of + 213.4 kJ mol^{-1} and the experimentally determined solid state density at room temperature of 1.887 g cm^{-3} (TMD), the computed performance parameters for TKX-50 at its theoretical maximum density at room temperature were calculated to be as follows: VoD = 9,642 m s^{-1}, $p_{C\text{-}J}$ = 37.0 GPa and Q_{det} = 4,770 kJ kg^{-1}.

In order to evaluate the long-term stability of TKX-50, isothermal predictions at different temperatures were calculated over a time period of 10 years [A. G. Harter, T. M. Klapötke, J. T. Lechner, J. Stierstorfer, *Prop Explos Pyrotech*, 2022, 47, e202200031]. Starting at 0 °C the mass loss in 20 °C steps, up to 100 °C was calculated, which gives a

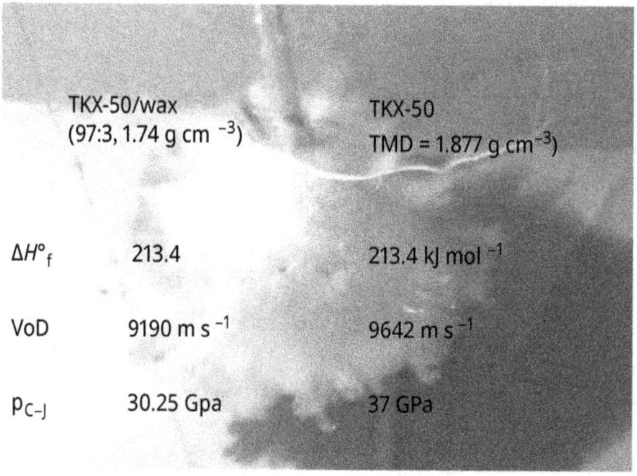

TKX-50/wax
(97:3, 1.74 g cm -3)
TKX-50
TMD = 1.877 g cm-3)
ΔH°f
213.4
213.4 kJ mol -1
VoD
9190 m s -1
9642 m s -1
pC-J
30.25 Gpa
37 GPa

Fig.9.38v: The photo shows an underwater explosion of a TKX-50 sample.

comprehensive insight into the thermal stabilities of the investigated compounds, ranging from frozen storage to high temperature stress. The isothermal prediction for TKX-50 is depicted in Fig. 9.38w.

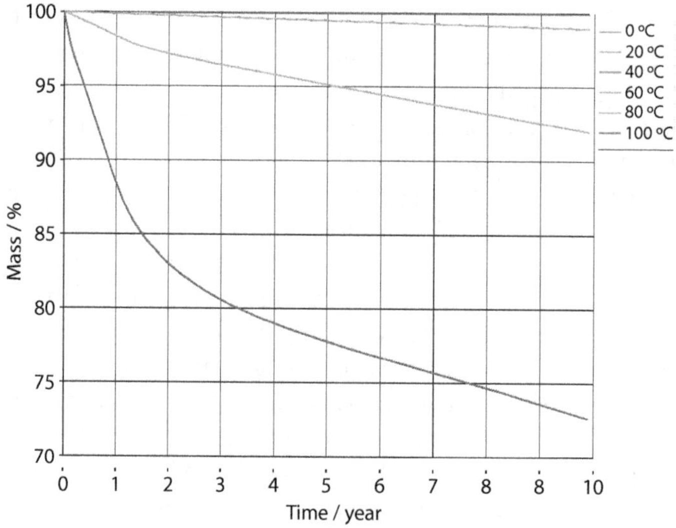

Fig. 9.38w: Isothermal predictions for TKX-50 at different temperatures over 10 years.

TKX-50 shows excellent thermal stabilities between 0 °C and 40 °C, with no notable mass loss over the predicted time period. Even when stored at 60 °C, which has been measured as a maximum temperature for sea shipping container, the predicted mass loss of TKX-50 is below 2% over 10 years. If TKX-50 is exposed to high temperature stress of 100 °C the prediction shows a steep mass loss of about 20% during the first 3 years, and a total mass loss of 27% over the entire 10 years.

RDX performs quite similar to TKX-50, with no notable mass loss between 0 °C and 40 °C, but it shows a slightly higher mass loss of about 35% when stored under 100 °C. When looking at HMX, a surprisingly low mass loss of only 5% can be observed, when stored at 100 °C over 10 years, with the majority taking place during the first year. In addition, HMX already has a notable mass loss of 1.5% at 40 °C, which makes it less stable than RDX and TKX-50 at lower temperatures. CL-20 possesses the best overall stability at all measured temperatures. No notable mass loss can be detected up to 60 °C, and even at 100 °C, the mass loss over ten years barely reaches the 4% mark. In contrast to this, PETN exhibits the lowest long-term stability of the tested compounds. In this case the prediction is likely to be less precise, since PETN has a low melting point and an overall higher vapor pressure, which leads to a slight, but constant mass loss up to the decomposition during the TGA measurement. This mass loss process is interpreted by the software as a decomposition, and therefore the prediction shows a lower stability than PETN exhibits in reality. Nonetheless the overall trend fits the expectations of PETN being the least stable of the investigated compound. According to the prediction, it is stable without mass loss between 0 °C and 20 °C, but already decomposes entirely after 3 years when stored at 40 °C. Above 60 °C, the decomposition takes place in under 3 months.

In addition to isothermal temperature predictions, it is also possible to calculate long term stabilities dependency of climatic data. Therefore, the climatic data of different cities were averaged over the last 30 years and used to predict long-term stabilities for the investigated energetic materials when stored under those conditions. The goal was to include a broad spectrum of different climatic conditions, while still using locations of high international importance. The final selection included six cities on four different continents, including Munich, Moscow, Washington D. C., Brasilia, Cairo and Beijing, representing Europe, North America, South America, Africa and Asia, respectively (Fig. 9.38x).

The climatic predictions for TKX-50 over 10 years are depicted in Fig. 9.38x. Similar to the isothermal predictions, TKX-50 shows excellent stabilities under all climatic conditions with a mass loss of below 0.01 mass % during the chosen time period. Even under harsh conditions like the hot summers in Cairo and the tropical climate in Brasilia, TKX-50 only loses around 0.003% of its mass. But it has to be mentioned that the humidity is not included in these predictions, as it is assumed that secondary explosives are stored under dry conditions. As expected, the milder regions like Europe and North America show even better stabilities, with Moscow having the best values, due to its cold winters. For the other compounds, the observed trend for the isothermal predictions is validated by the climatic predictions. RDX has a comparable performance to TKX-50, while PETN shows the highest mass loss, and CL-20 the overall best

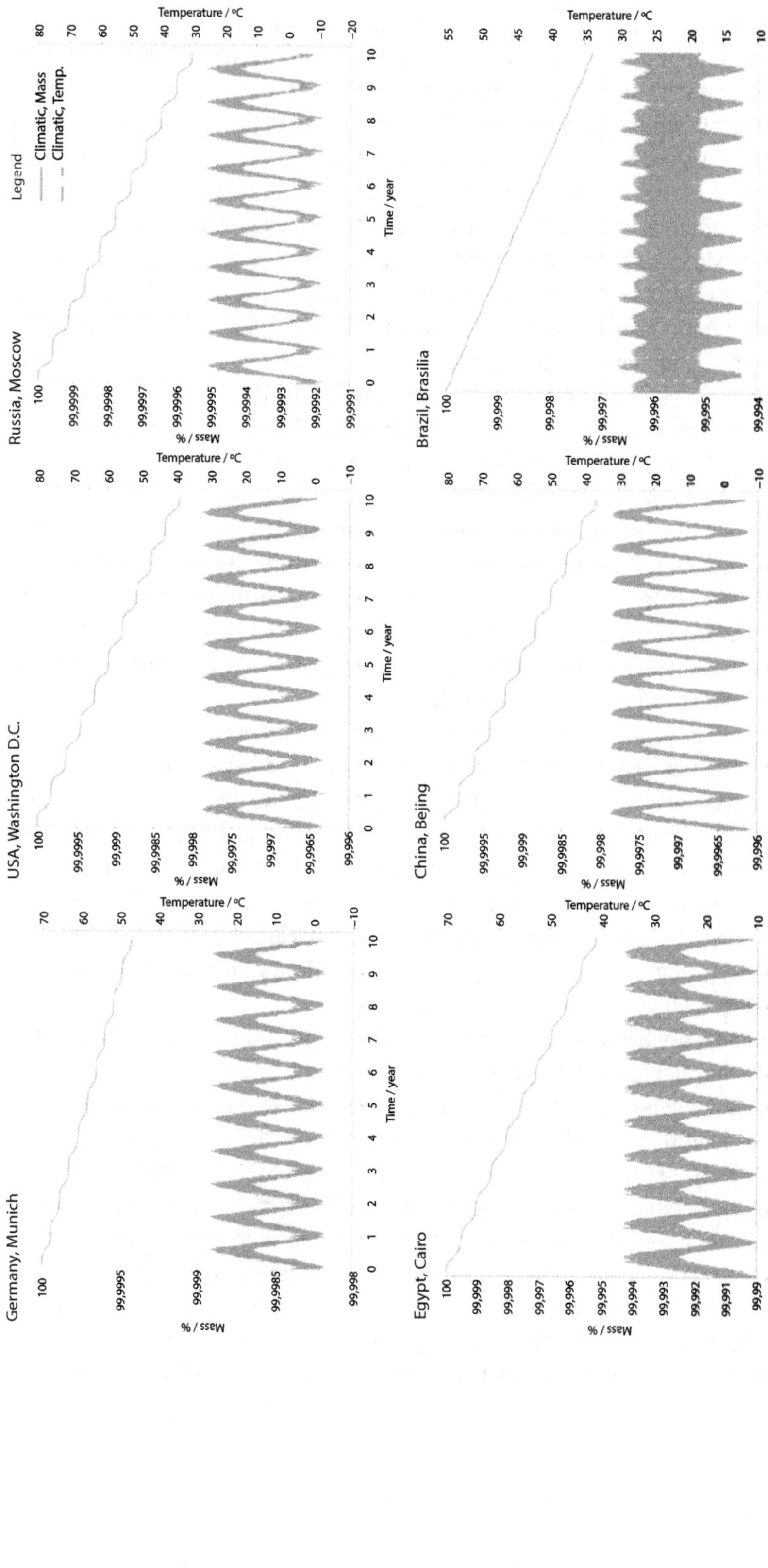

Fig.9.38x: Climatic prediction for TKX-50 over 10 years using the average temperature data of the last 30 years in various different climate zones.

stability. HMX also fits the isothermal data well, showing a high mass loss at low temperatures in comparison to the other compounds.
For many years, the synthesis of tetrazino-tetrazine 1,3,6,8-tetraoxide (TTTO) (see figure below) had been one of the holy grails of N–O-chemistry within the area of energetic materials. Recently, a ten-step synthesis of TTTO has been accomplished by Tartakokovsky et al. [*Angew. Chem. Int. Ed.* **2016**, *55*, 11,472–11,475]. The synthetic strategy was based on the sequential closure of two 1,2,3,4-tetrazine 1,3-dioxide rings by the generation of oxodiazonium ions and their intramolecular coupling with tertbutyl-NNO-azoxy groups. TTTO is considered as being an example of a new high-energy compound.

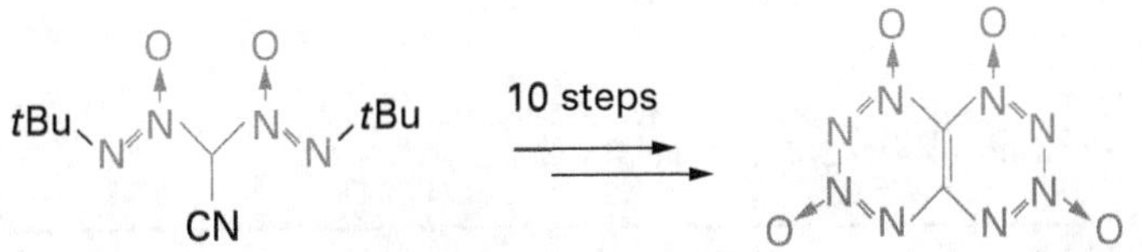

Another very promising ionic energetic material is carbonic dihydrazidinium *bis*[3-(5-nitroimino-1,2,4-triazolate)] (CBNT) which was first synthesized in 2010 by Shreeve et al. (*J. Am. Chem. Soc.* **2010**, *132*, 11,904) (Fig. 9.39). The salt is thermally stable up to 220 °C and has a high density of 1.95 g cm^{-3} and very good calculated detonation parameters of VoD = 9,399 m s^{-1} and $p_{C\text{-}J}$ = 360 kbar.

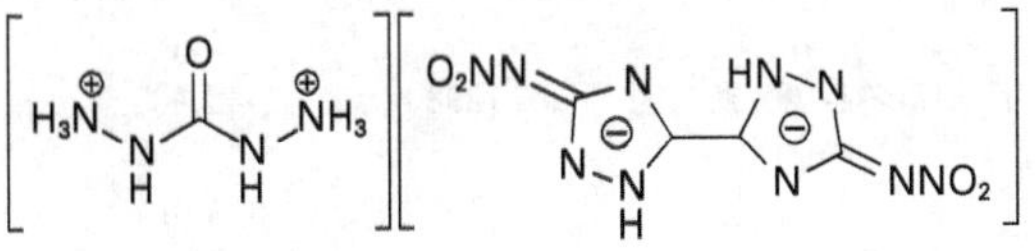

CBNT

Fig. 9.39: Structure of carbonic dihydrazidinium *bis*[3-(5-nitroimino-1,2,4-triazolate)] (CBNT).

Chemists at LMU have developed two new secondary explosives based on bispyrazolylmethane: one of which (**A, BDNAPM**) is stable up to 310 °C, and the other has an exceptional detonation velocity of over 9,300 m s^{-1} (**B, BTNPM**) (Tab. 9.16). In collaboration with the US Army Research Laboratory (ARL), the detonation velocities were experimentally determined using the LASEM method (Laboratory-Scale Method for Estimating Explosive Performance, see chapter 9.7).

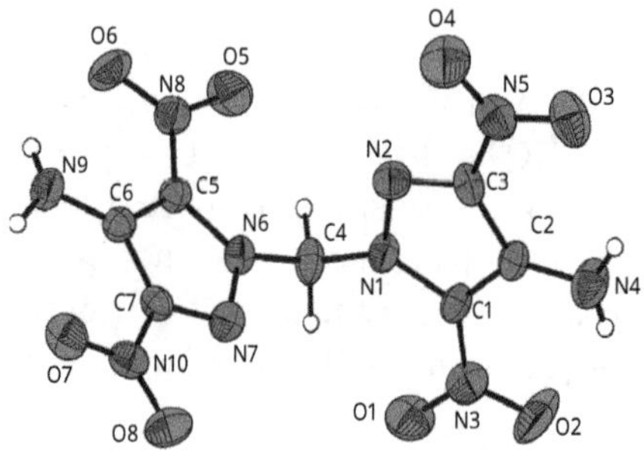

A

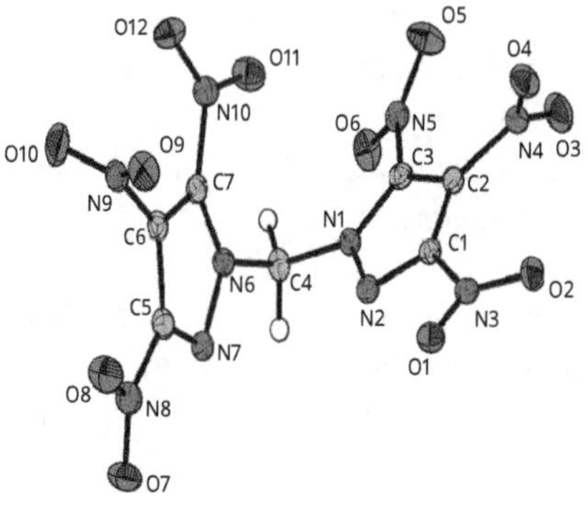

B

Tab. 9.16: Sensitivity and performance data for BDNAPM and BTNPM.

	BDNAPM	BTNPM	BTNPM + Al (85/15%)	HMX
Formula	$C_7H_6N_{10}O_8$	$C_7H_2N_{10}O_{12}$		$C_4H_8N_8O_8$
FW [g mol^{-1}]	358.19	418.15		296.16
IS [J]	11	4		7
FS [N]	> 360	144		112
ESD-test [J]	> 1	0.6		0.2
N [%]	39.10	33.50		37.84
Ω [%]	− 40.20	− 11.48	−23.1	− 21.61
$T_{dec.}$ [°C]	310	200		320
Density [g cm^{-3}]	1.836 (173 K) 1.802 (298 K)	1.934 (298 K)	2.02 (298 K)	1.905 (298 K)
$\Delta_f H°$ / kJ mol^{-1}	205.1	378.6		116.1
$\Delta_f U°$ / kJ kg^{-1}	655.8	976.8		492.5
EXPLO6.02 values:				
$-\Delta_E U°$ [kJ kg^{-1}]	5,052	6,254	8,003	5,794
T_E [K]	3,580	4,570	5,527	3,687
$p_{C\text{-}J}$ [kbar]	295	393	360	389
D [m s^{-1}]	8,372	9,300	8,583	9,235
Gas vol. [L kg^{-1}]	706	704	524	767

Prior to this work, methylene-bridged bispyrazolylmethane derivatives had not been described as high-energy materials. Compound **A** is thermally stable up to 310 °C and shows a detonation velocity of 8,372 m s^{-1}. HNS (VoD = 7,629 m s^{-1}) is currently the most commonly used explosive in oil exploration. However, compound **A** has a much better performance than HNS.

The detonation parameters of the hexanitro species **B** are even better, with a detonation velocity of 9,300 m s^{-1} and a detonation pressure of 39.3 GPa. Currently, the most commonly used explosive in military applications is RDX (hexogen), which has a significantly lower performance (8,803 m s^{-1} and 33.8 GPa) than compound **B**.

The synthesis of both compounds is relatively straightforward and is shown in the following equation:

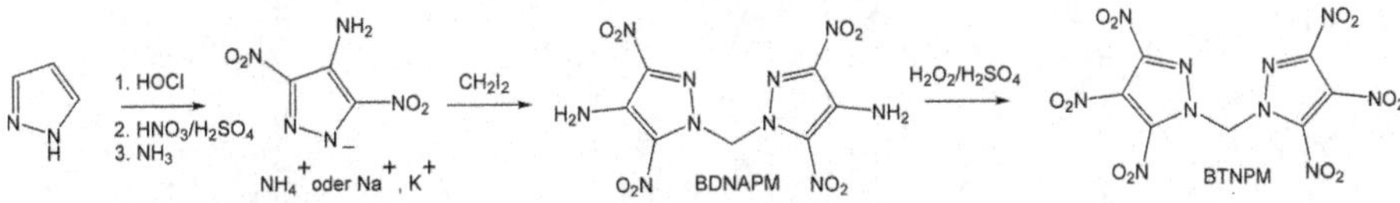

Some energetic pyrazoles (most notably dinitropyrazole and trinitro pyrazole, as well as the corresponding *N*-methylated derivatives) have proven to be sensitizing agents, with problematic levels of toxicity causing rashes, blisters, and very bad headaches over time. These symptoms have been found to be especially prevalent when larger scales of the energetic compounds are being handled.

Nitrate esters (e.g. NG) generally have a bad reputation because they show high sensitivities and often low thermal stabilities. However, they are essential ingredients in NC based propellant formulations.

A new and promising nitrate ester based on the bi-isooxazole unit has recently been reported by Sabatini et al. [*Eur. J. Org. Chem.* **2017**, 1765–1768]. The compound, bi-isoxazoletetrakis(methyl nitrate), can easily be synthesized in a metal-free [3 + 2] cycloaddition reaction as shown in Fig. 9.40.

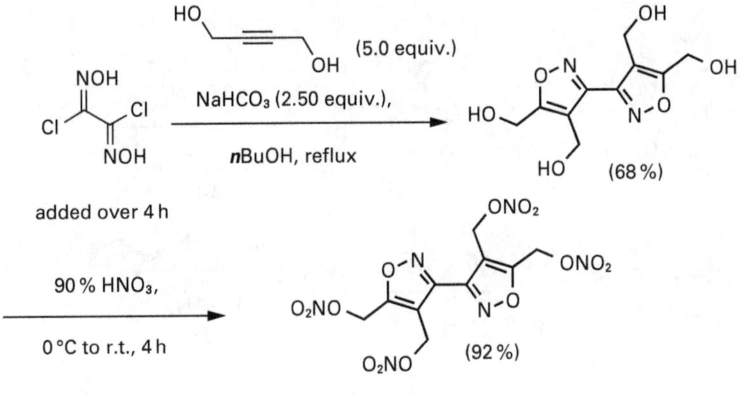

Fig. 9.40: Synthesis of bi-isoxazoletetrakis(methyl nitrate).

The sensitivities of the new tetranitrate ester are comparable to those of PETN (IS = 3 J, FS = 60 N, ESD = 0.0625 J), and the compound has a calculated detonation velocity and detonation pressure between those of NG and PETN (Tab. 9.16a).

Bi-isoxazoletetrakis(methyl nitrate) has the potential to be used in multiple functions. The characteristics and performance properties of the new nitrate ester suggest

Tab. 9.16a: Energetic properties of bi-isoxazoletetrakis(methyl nitrate).

	bi-isoxazoletetrakis(methyl nitrate)	PETN	NG
m.p. / °C	121.9	141.3	14
$T_{dec.}$ / °C	193.7	170.0	50.0
Ω (CO_2) / %	− 36.7	− 10.1	+ 3.5
Ω (CO) / %	0	+ 15.2	+ 24.7
ρ / g cm^{-3}	1.786	1.760	1.60
p_{C-J} / kbar	271	335	253
VoD / m s^{-1}	7,837	8,400	7,700
$\Delta H°_f$ / kJ mol^{-1}	−395	−539	−370

that this material has the potential to serve as: (a) a potential ingredient in melt-castable eutectic formulations; (b) a potential energetic plasticizing ingredient with nitrocellulose-based propellant formulations, in an effort to reduce the volatility/migration issues that arise during cook-off; (c) a potential ingredient in percussion primer compositions.

Another very promising melt-cast explosive is the compound bis(1,2,4-oxadiazole) bis(methylene) dinitrate (Tab. 9.16b). An improved synthesis for this compound has already been recently published by Sabatini, Chavez et al. (*OPR&D*, **2018**, *22*, 736–740) and is shown in Fig. 9.40a.

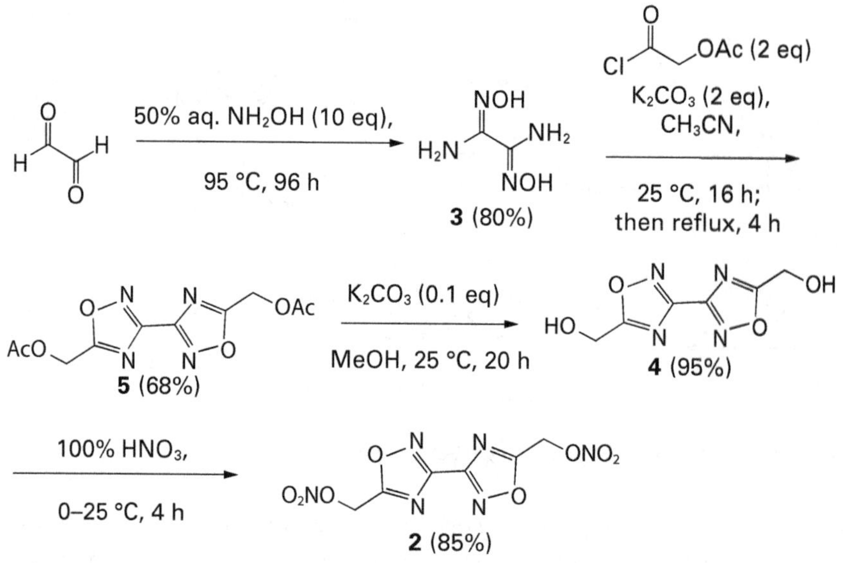

Fig. 9.40a: Synthesis of bis(1,2,4-oxadiazole)-bis(methylene) dinitrate.

Tab. 9.16b: Calculated physical properties of bis(1,2,4-oxadiazole)bis(methylene) dinitrate **(2)** in comparison with those of TNT and Composition B.

property	2	TNT	Comp. B
m.p./°C	84.5	80.4	78 – 80
$T_{dec.}$/°C	183	295	200
$\Omega(CO_2)$/%	– 33	– 74	–
ρ/g cm^{-3}	1.83	1.65	1.68 –1.74
$p_{C\text{-}J}$/GPa	29.4	20.5	26 – 28
VoD/m s^{-1}	8,180	6,950	7,800–8,000

The physical properties and energetic performance of bis(1,2,4-oxadiazole)bis(methylene) dinitrate (2) are summarized in Tab. 9.16b.

New melt-cast explosives

TNT (2,4,6-trinitrotoluene) is one of the most commonly used energetic materials ever synthesized, so much so, that it is often used colloquially as a general term for explosives. The compound has been produced on a large scale for more than 120 years and is still used today in many kinds of explosives, particularly in compositions such as Comp. B. By far, the main importance of TNT is its use as a so-called melt-castable explosive. The use of an energetic melt-castable component in explosive charges prevents cracks in the packing and results in an ideal filling of the cavity. To be considered for use as a melt-cast explosive, a secondary explosive must possess certain thermal properties. The thermal properties of TNT fit excellently with the requirements. For example, the melting point of TNT is around 80 °C. This low melting point is particularly advantageous for the industrial use of TNT as a melt-cast explosive, since the melting process can be carried out with the use of a steam bath. In addition, equally important is the large difference in the melting and decomposition temperatures of TNT, meaning that decomposition on melting does not occur. The decomposition temperature of TNT is more than 200 °C above (T_{dec} = 291 °C) the melting point of TNT and therefore in terms of safety, it is in a perfect range for explosives.

In addition to the suitable thermal properties of TNT, there are additional properties of TNT which have made it popular for use up until now. One crucial factor is the synthesis of TNT is simple and requires only readily available starting materials (toluene and a mixture of sulfuric acid and nitric acid). From an industrial point of view, this is highly important.

Furthermore, the sensitivity of TNT towards external stimuli such as impact (IS), friction (FS), and electrostatic discharge (ESD) are found to be within acceptable ranges, and therefore, this also aids the problem-free handling even of larger, industrial-scale quantities.

However, for modern-day use, TNT is not without its problems, and there are two key aspects which necessitate the urgent need for a TNT substitute to be identified.

The environmental impact of not only pure trinitrotoluene, but also its degradation products, partially (less) nitrated synthesis intermediates, and also of the by-products formed during its synthesis is high, and has resulted in an increase in the contamination levels of the groundwater and soil near production sites. With the increase in the demands on industrial chemistry with respect to green and environmentally friendly issues, which has increased in recent years, it is as a matter of urgency that compounds such as TNT which fail to meet these demands are replaced. In addition, the demand for military explosives, which is what TNT is mainly used for, is primarily driven by ever more powerful explosives.

Taking into account the positive properties of TNT with respect to its use as a melt-cast explosive, as well as its negative aspects in terms of its environmental sustainability, in addition to the ever-present demand for better explosive performance, a general list of properties and the corresponding desired specifications for future melt-cast explosives can be made:

Properties	Specification
Detonation velocity	>7,600 m s^{-1}
Density	>1.76 g cm^{-3}
m.p.	100–110 °C
T_{dec}	>180 °C
IS	>10 J
FS	>120 N
ESD	>700 mJ
Synthesis	Economic
Toxicity	Less than TNT, conforms with REACH
Mutagenicity	No
Cytotoxicity	No

Additionally, there are of course many other industrially relevant specifications which must also be considered. One such key issue for any new melt-cast candidate compound is to be able to develop a synthesis that is as simple as possible, is atom-economical, short – in terms of the number of reaction steps, and is as cheap as possible. Furthermore, for the compound to be of industrial-scale interest, no chemicals should be necessary for any aspect of the synthesis that are already banned by REACH or are on its list of candidates of high concern. In terms of toxicity/mutagenicity aspects, it is essential that not only the end-product (i.e. the new melt-cast explosive), but also the intermediates in the synthesis as well as all of the end products show no or low mutagenicity or toxicity.

Based on the above-mentioned selection parameters, *N*-heterocyclic azole systems have been extensively studied, since such systems have provided promising results for many of the benchmark property requirements listed above. In addition to establishing all of the sensitivity and energetic properties outlined above for a candidate compound, supplementary measurements such as compatibility tests with other common explosives and toxicity/mutagenicity determinations are equally important in order to determine, if a compound is to be considered further and proceed to the stage of possible synthesis optimization, as well as up-scaling to larger quantities.

So far, consideration of the detonation characteristics of possible future melt-cast candidate compounds, as well as the thermal behavior of these compounds, has shown that three compounds show particular promise as future melt-casts, with BNFF being the most promising, while MTNI and BODN also show potential (Fig. 9.40b). The relevant properties of these three compounds are summarized in Tab. 9.16c and compared with those of TNT.

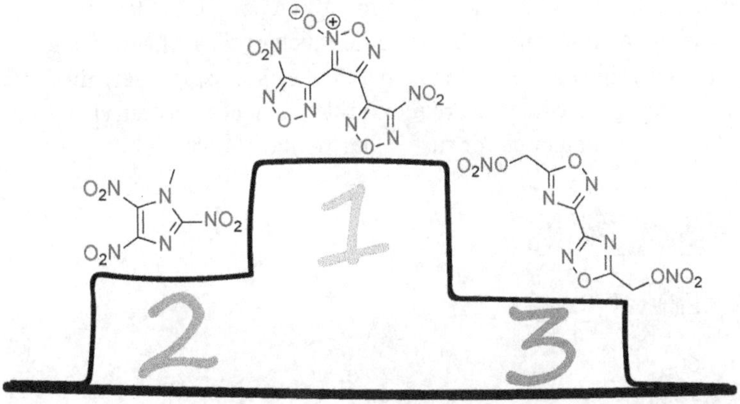

Fig. 9.40b: Three examples of compounds which have recently been reported in the literature which show promise as potential TNT substituents: (*bis*(nitrofurazano)furoxan (BNFF), 1-methyl-2,4,5-trinitroimidazole (MTNI) and 5,5′-*bis*-(1,2,4-oxadiazole)-3,3′-bismethylene dinitrate (BODN).

Tab. 9.16c: Energetically relevant physico-chemical parameters of BNFF, MTNI, BODN and TNT according to the data which has been previously reported in the scientific literature.

	BNFF	MTNI	BODN	TNT
IS / J	10	17	8.7	15
FS / N	120	360	282	>360
ρ / g cm^{-3}	1.875	1.82	1.832	1.65
m.p. /°C	109	95	84.5	80
$T_{dec.}$ /°C	279	310	183.4	295
VoD / m s^{-1}	8,970	8,541	8,180	6,823
number of steps in synthesis	4	4	4	1

The compound BNFF could be synthesized as described in the literature, however, unfortunately no melting point was observed in the thermal investigation and instead, sublimation was observed to occur. Attempts to investigate MTNI further caused drastic skin irritations and severe rashes for the synthetic chemists during the preparation and handling of 1,4-dinitroimidazole, making this compound unsuitable for further investigation using this synthetic route. Therefore, despite being the least favored candidate compound out of the three, BODN was investigated further. The synthesis reported in the literature is reproducible – which is a key first step, as are the physicochemical properties. Encouragingly, the synthesis of BODN can also be improved to give an overall yield of 73% (Fig. 9.40c). However, the difficulty in improving on TNT is evidenced by the disappointing result of the mutagenicity test, in which a positive Ames test was observed for BODN which effectively excludes BODN for further testing. It is the combination of properties as a whole which determines how suitable a candidate compound is. While a compound may show improved energetic performance, if the melting point is too high or the decomposition point is too low, the IS, FS or ESD sensitivity is too high or the synthetic route involves toxic chemicals, it will not be considered a suitable replacement candidate for TNT. It is precisely this difficult combination of properties which are required that has made TNT survive as the most widely used melt-cast compound for such a long period of time.

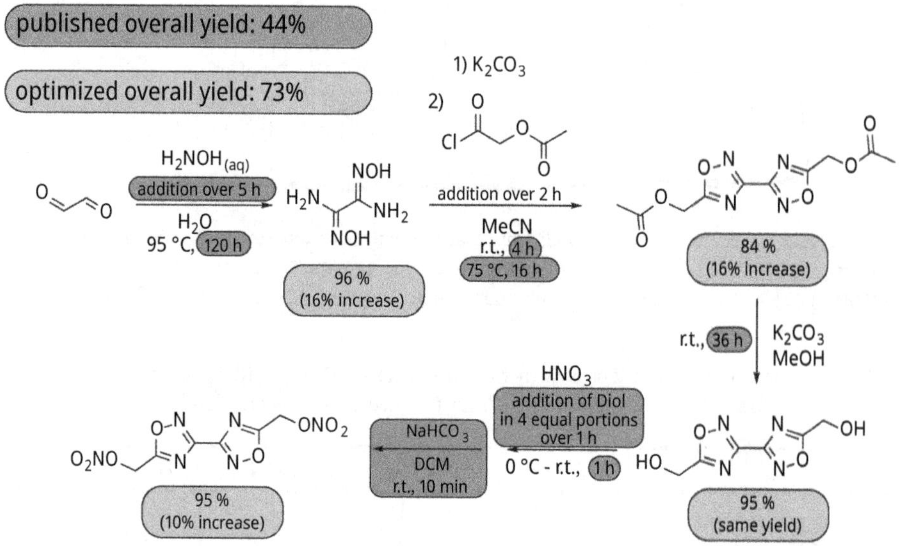

Fig.9.40c: Reaction scheme showing the optimized synthetic route for the preparation of BODN. The green boxes show the optimized yields and the percentage improvement in comparison with the original synthetic route. The blue boxes show added or modified reaction steps.

Hence, the hunt for new possible melt-cast replacements continues. One of the problems in finding such replacements is sifting through the large number of compounds which are published in the literature with the statement that the compound is a potential replacement for TNT, often based only on minimal analytic data. However, as this short section has shown, there is a lot more to finding a compound suitable for replacing TNT than just stating "potential TNT substitute" as a conclusion at the end of a publication. Figure 9.40d shows a schematic diagram showing the steps of the development (and failure) of a new melt-cast explosive.

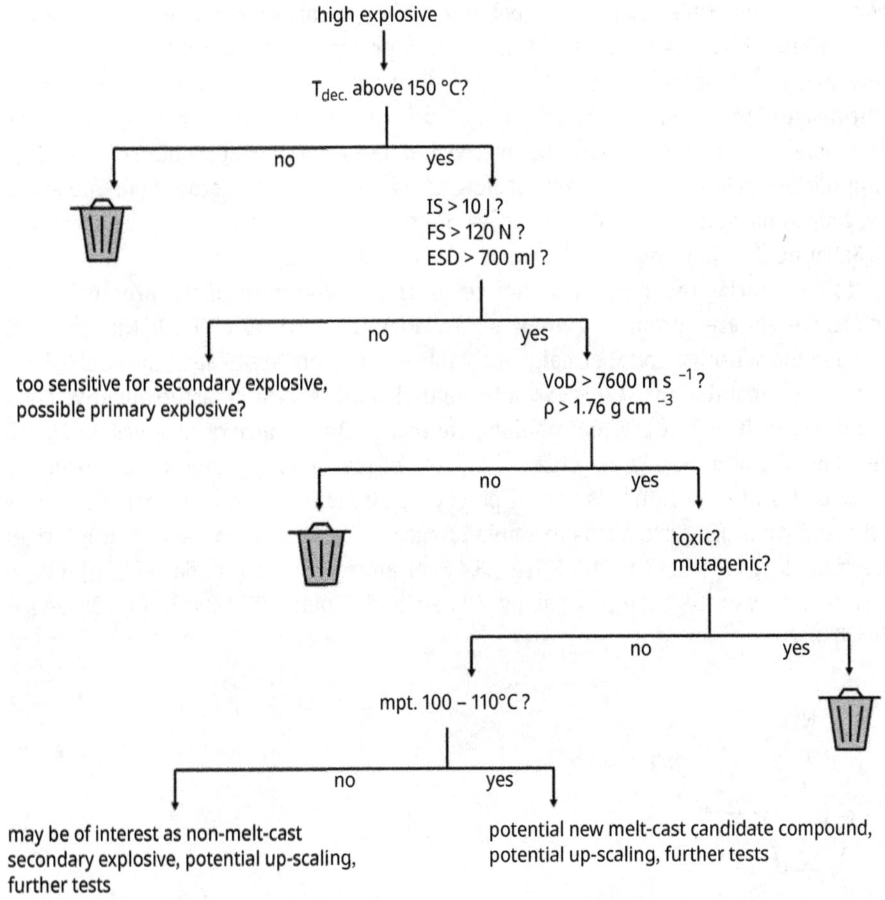

Fig.9.40d: Steps for the development of a new melt-cast explosive.

Literature

M. Benz, A. Delage, T. M. Klapötke, M. Kofen, J. Stierstorfer, A rocky road toward a suitable TNT replacement – A closer look at three promising azoles, *Propellants, Explos., Pyrotech.*, **2023**, *48*, e202300042; https://doi.org/10.1002/prep.202300042.

T. M. Klapötke, Casting TNT as an Explosive, *Chemistry Nature*, October **2023**, *15*, 1480; https://doi.org/10.1038/s41557-023-01337-4.

The nitric ester functionality not only contributes significantly to the overall energy of energetic molecules, its presence spans the many classes of energetic materials. Nitric ester materials may have the potential to serve as ingredients in primary explosives, pressable secondary explosives, melt-castable secondary explosives, double-base rocket propellants, and triple-base gun propellants. Although this functionality is believed to be linked to a high degree of sensitivity, with unpredictable chemical and thermal stability, many of the newer nitric ester materials discussed above have been found to exhibit a significant degree of insensitivity, and chemical/thermal stability. The nitric ester moiety – one of the earliest discovered and best known explosophores – still plays an important role in the discovery of new energetic materials, across many potential energetic applications. An excellent review on this topic can be found in the literature: J. J. Sabatini, E. C. Johnson, *ACS Omega* **2021**, *6, 18*, 11,813–11821.

To summarize this chapter, it can be concluded that most of the presently used high explosives are nitramine compounds, e.g. RDX and HMX (also CL-20). Nitrate esters are often too sensitive and thermally not stable enough, but are needed for double-base propellant formulations. Whereas nitromethyl and especially trinitromethyl compounds suffer from low thermal stability, the related fluoro-nitro derivatives $R_2CF(NO_2)$ are expensive and usually very toxic. One area of high-energy materials that needs to be focused on in the future is that of polycyclic aliphatic nitramines, including cages and fused rings. One promising example of such a compound, and a step in the right direction, is bicyclo-HMX (BCHMX [*cis*-1,3,4,6-tetranitrooctahydroimidazo-[4,5-d]imidazole]), which was first synthesized by Zeman and Elbeih (*PEP* **2013**, *38*, 238–243 & 379–385).

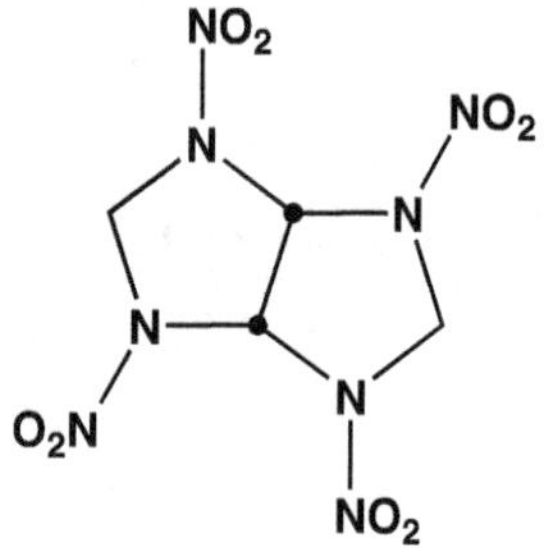

BCHMX (1,3,4,6-tetranitrooctahydroimidazo[4,5-d]imidazole) is usually prepared from KTACOS (potassium imidazo[4,5-d]imidazole-1,3,4,6-tetrasulfonate) which can easily be obtained from potassium sulfamate, formaldehyde, and glyoxal.

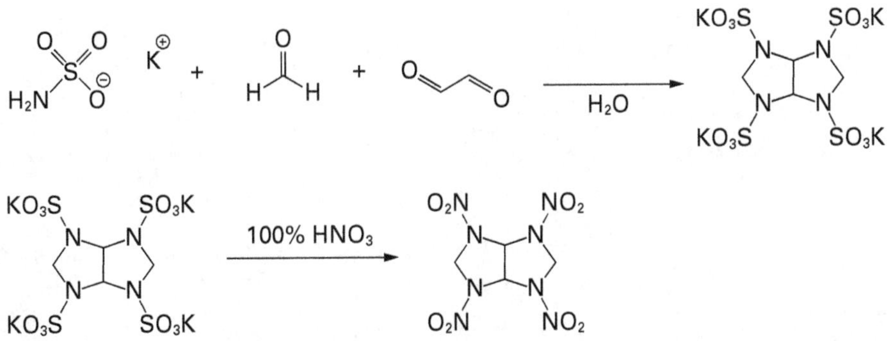

The design of novel high-energy density materials (HEDMs) is still one of the significant challenges in the field of applied chemistry. Newly discovered scaffolds are rare, but they open up new possibilities for the HEDM design. Recently, researchers from the Zelinsky Institute of Organic Chemistry, Russian Academy of Sciences reported a simple and effective two-step approach to previously unknown 1,4-dihydro-[1,2,3]triazolo[4,5-*d*][1,2,3]triazole (**1**) (Fig. 9.40e). This compound is an energetic material in its own right ($d = 1.84$ g·cm^{-3}, $\Delta H_f^\circ = +1{,}005$ kcal· kg^{-1}, $T_{onset} = 193$ °C (all experimental values), $D = 8.5$ km·s^{-1}, $P = 34$ GPa (calculated ones)), but also may serve as a scaffold for the synthesis of structurally diversified related compounds. The precursors of triazole-triazole **1** are the previously unknown 5-amino-4-diazo-4*H*-1,2,3-triazolium chloride (**4**) and 4-diazo-4*H*-1,2,3-triazol-5-amine (**5**), which can also serve as starting materials for the synthesis of a variety of novel heterocycles (Fig. 9.40e).

In a recent publication, Fershtat et al. from the Zelinsky Institute of Organic Chemistry, Russian Academy of Sciences reported a set of new heat-resistant high-energy materials incorporating the polynitrophenyl-1,2,5-oxadiazole scaffold enriched with azo/azoxy moieties (Fig. 9.40f). Due to a smart combination of explosophoric groups and 1,2,5-oxadiazole rings, the prepared high-energy substances have excellent thermal stability (up to 300 °C), good densities (up to 1.75 g cm^{-3}), high enthalpies of formation (340–538 kJ mol^{-1}) and high combined nitrogen-oxygen contents (63–68%) (Fig. 9.40g).

Future research will lead to even more powerful, high-nitrogen-high-oxygen explosives with enhanced and superior detonation parameters to fulfil the lethality requirements of all forces. New energetics will provide vastly increased energy content over RDX, up to several times the energetic performance. In addition, these materials will have a high energy density with a high activation energy.

With these capabilities, DoD will be able to develop new applications for energetic materials and vastly improve performance in current munitions. The increased performance of the next generation of energetics will enable DoD to meet the same strategic goals with fewer munitions and less energetic material. The increase in energy output can be harnessed in munitions that require a fraction the amount of energetics to deliver the same payload or thrust profile. This will allow engineers to put energetics in munitions that have never had them before.

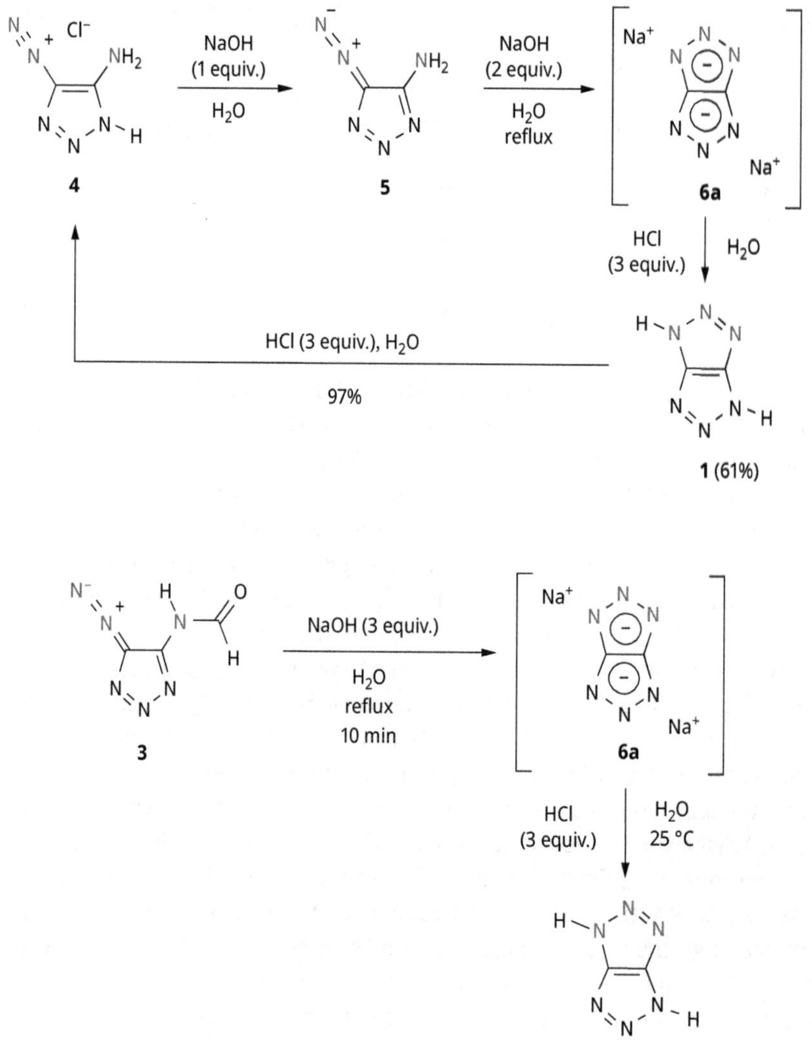

Fig. 9.40e: Synthesis of 1,4-dihydro-[1,2,3]triazolo[4,5-*d*][1,2,3]triazole (**1**) from the diazonium salt (**4**) (top) or via a one-pot reaction from diazotriazole (**3**) (bottom).

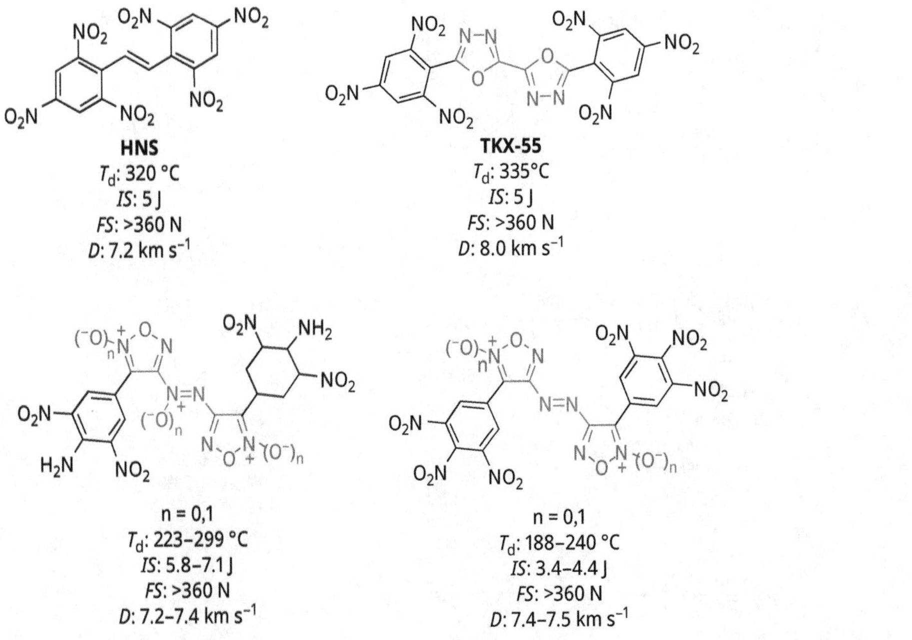

Fig. 9.40f: Previously known and newly prepared polynitrophenyl-derived heat-resistant explosives.

9.7 Detonation velocities from laser-induced air shock

Laser-induced air shock from energetic materials (LASEM) is a laboratory-scale technique for estimating the detonation velocity of energetic materials [see Gottfried et al.: *Angew Chem Int. Ed.*, **2016**, *55*, 16,132–16,135; *Prop. Expl. Pyrotech.*, **2017**, *42*, 353–359: DOI: 10.1002/prep.201600257 and *Phys. Chem. Chem. Phys.* **2014**, *16*, 21,452]. A focused nanosecond laser pulse is used to ablate and excite an energetic material residue, forming a laser-induced plasma (Fig. 9.41). Exothermic chemical reactions in the plasma accelerate the laser-induced shock wave generated by the plasma formation – thus, energetic materials produce faster shock waves than inert materials (as measured on the microsecond timescale by a high-speed camera). The measured laser-induced air shock velocity can be correlated to measured detonation velocities from large-scale testing, and used to estimate the detonation velocity of the material. However, the energetic material is not detonated by the laser, and therefore, LASEM only provides an estimate of the maximum attainable detonation velocity for an energetic material. Despite this, it has the distinct advantage of requiring only milligram quantities of the energetic material.

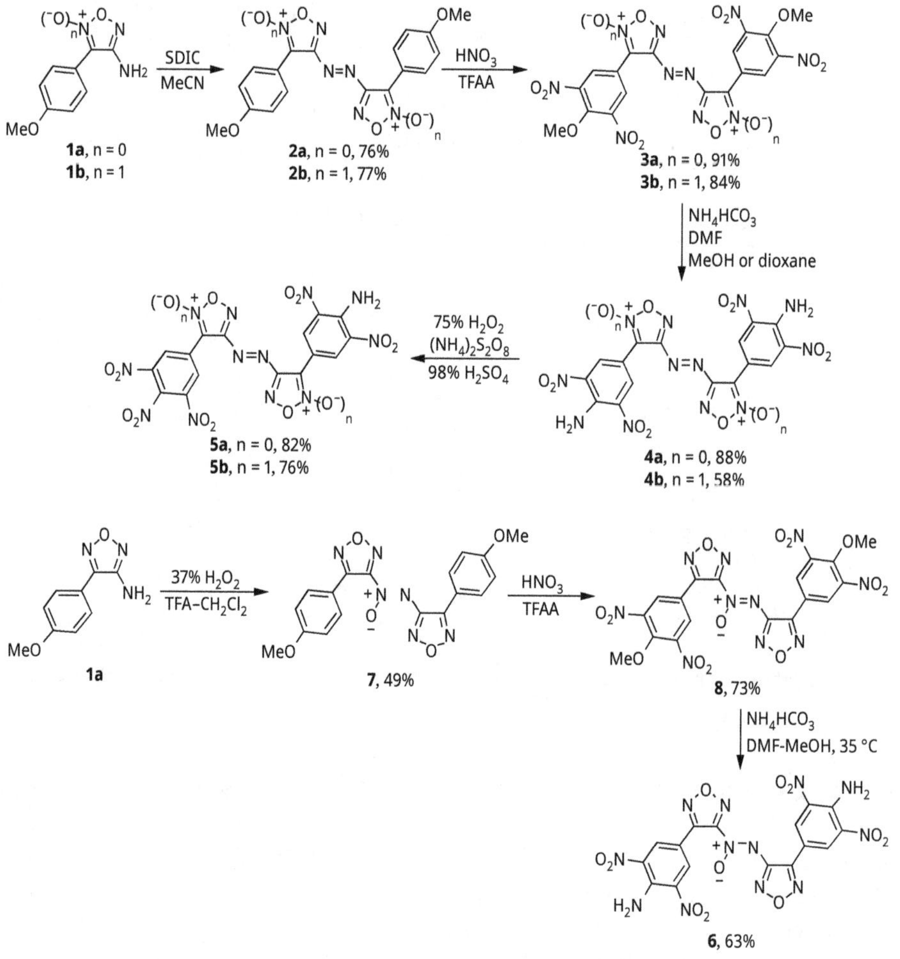

Fig. 9.40g: Synthesis of polynitrophenyl-1,2,5-oxadiazoles **4a,b**, **5a,b** and **6.**

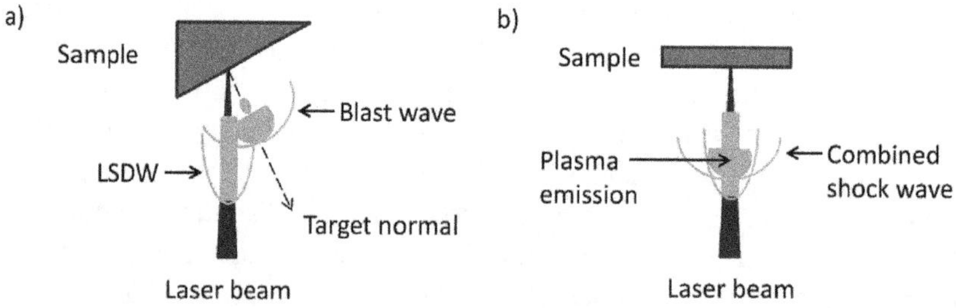

Fig. 9.41: Comparison between the LSDW (laser-supported detonation wave) and the blast wave, which are distinct when the laser interrogates the sample off-angle (a) and combined when the propagation of the laser is perpendicular to the sample (b).

New energetic material candidates are often initially synthesized in only small quantities for safety reasons, and the subsequent scale-up to multiple gram quantities can be costly. While the shot-to-shot variation in the measured laser-induced shock velocities can be significant due to the stochastic nature of the laser-material interaction, many laser shots can be acquired to obtain a reproducible average at minimal expense (especially compared to large-scale detonation testing, where significant safety hazards increase the cost of testing). The usefulness of the method was originally demonstrated by comparing the laser-induced air shock velocities of inert and energetic materials, followed by calibration of these values by correlating the measured laser-induced air shock velocities from conventional military explosives to the measured detonation velocities from large-scale detonation testing. The calibration fit was validated using conventional energetic materials (both monomolecular and composite) which were not used to develop the correlation. Estimated detonation velocities for materials where measured detonation velocities have not been reported were also calculated for nanoscale cyclotrimethylene trinitramine (nano-RDX) and three high-nitrogen explosives. This method has also recently been extended to mixtures of military explosives with metal additives (aluminum or boron).

In this chapter, the estimated detonation velocities of six new explosives, dihydroxylammonium 5,5′-bistetrazole-1,1′-diolate (TKX-50), dihydroxylammonium 5,5′-bis(3-nitro-1,2,4-triazolate-1*N*-oxide) (MAD-X1), bis(4-amino-3,5-dinitropyrazol-1-yl)methane (BDNAPM), bis(3,4,5-trinitropyrazol-1-yl)methane (BTNPM), 5,5′-bis(2,4,6-trinitrophenyl)-2,2′-bi-(1,3,4-oxadiazole) (TKX-55), and 3,3′-diamino-4,4′-azoxyfurazan (DAAF), are discussed and compared to the theoretically predicted detonation velocities obtained using two different thermochemical codes (EXPLO5 V6.01 and CHEETAH V8.0). TKX-50, MAD-X1, BDNAPM, BTNPM, TKX-55, and DAAF, are members of the group of recently synthesized explosives which are amongst the most promising prospective candidates for use in practical applications (Fig. 9.42).

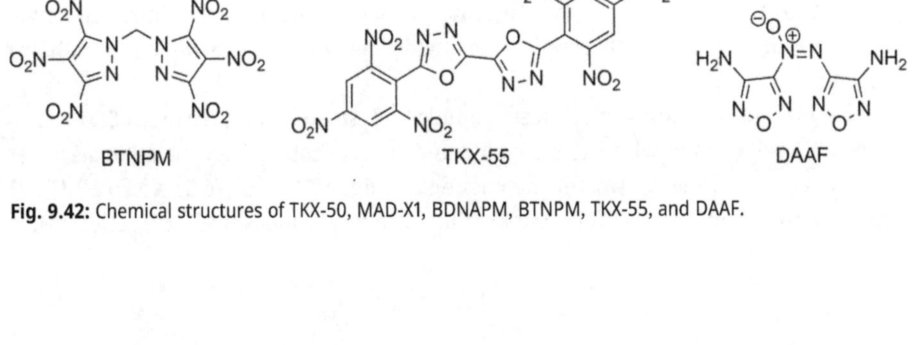

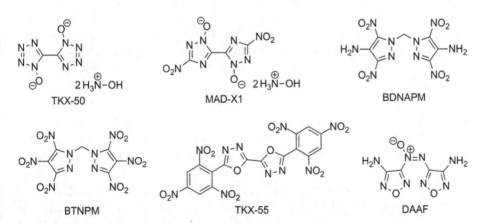

Fig. 9.42: Chemical structures of TKX-50, MAD-X1, BDNAPM, BTNPM, TKX-55, and DAAF.

Figures 9.43-I and 9.43-II show snapshots from the high-speed video of the laser excitation of the six explosives mentioned above. The selected photos represent the typical laser shot for each material. The brightness was increased (+40%) for all of the images in Fig. 9.43-I to enable enhanced visualization. Adjustment of the brightness and/or contrast for the first few frames was necessary during measurement of the shock wave position, in order to improve visualization of the shock wave in the presence of the defocused plasma emission. The few individual shots where the plasma emission completely obscured the shock wave position in the first frame were discarded (however, a total of 15 videos with measurable shock waves in the first frame were obtained for each sample). Since the camera was focused in front of the laser-induced plasma to enhance visualization of the shock wave, the plasma emission is slightly out of focus. The purple light in the snapshots is a result of the strong CN emission, while the white light is residual continuum emission from the laser-induced plasma. It was found that TKX-55 (Fig. 9.43-I-a) and DAAF (Fig. 9.43-I-d) produce the most intense plasma emission, followed by TKX-50 (Fig 9.43-I-b), BDNAPM (Fig. 9.43-I-e), and MAD-X1 (Fig. 9.43-I-f). BTNPM (Fig. 9.43-I-c) produced very little visible plasma emission following laser excitation. It has been previously observed that, in general, the most powerful energetic materials produce the smallest amount of visible plasma emission.

In Fig. 9.43-II-d–f, a thin, dark string of excited material coincident with the laser beam (center of images) is particularly visible in the later frames. It also appears, to a lesser extent, in Fig. 9.43-I-b (TKX-50) and in later frames of TKX-55, once the plasma emission has decreased (not shown). The structure of the plasma plume region is also distinct for these energetic materials (DAAF, BDNAPM, MAD-X1). The only conventional military explosive known to show similar features is TATB, which could indicate that these effects are related to the low thermal sensitivity of these energetic materials. BTNPM is the only explosive in this group that does not show a similar interaction with the laser beam, and it has the lowest decomposition temperature and a higher impact and friction sensitivity compared to most of the other materials.

Occasionally, unusual features appeared in the high-speed video for random laser shots of a few of the energetic materials. For example, in Fig. 9.43-II, distortion of the leading edge of the laser-induced shock wave in air is shown for (a) TKX-50, (b) BTNPM, and (c) MAD-X1 (in order of increasing degree of severity). For these, a second, much smaller plume of reacting material appeared to form near the top of main laser-induced plasma (coincident with the laser beam). This distortion is so severe in the MAD-X1 shot, that a bubble was formed in the main shock wave, which persisted for tens of microseconds. The cause of this behavior is currently unknown, but has so far never been observed in the high-speed videos of conventional military explosives.

The measured characteristic laser-induced air shock velocities (with 95% confidence intervals), along with the estimated and theoretically predicted detonation velocities calculated using two thermochemical codes (EXPLO5 V6.01 and CHEETAH V8.0), are shown in Tab. 9.17 for comparison. The average measured detonation veloc-

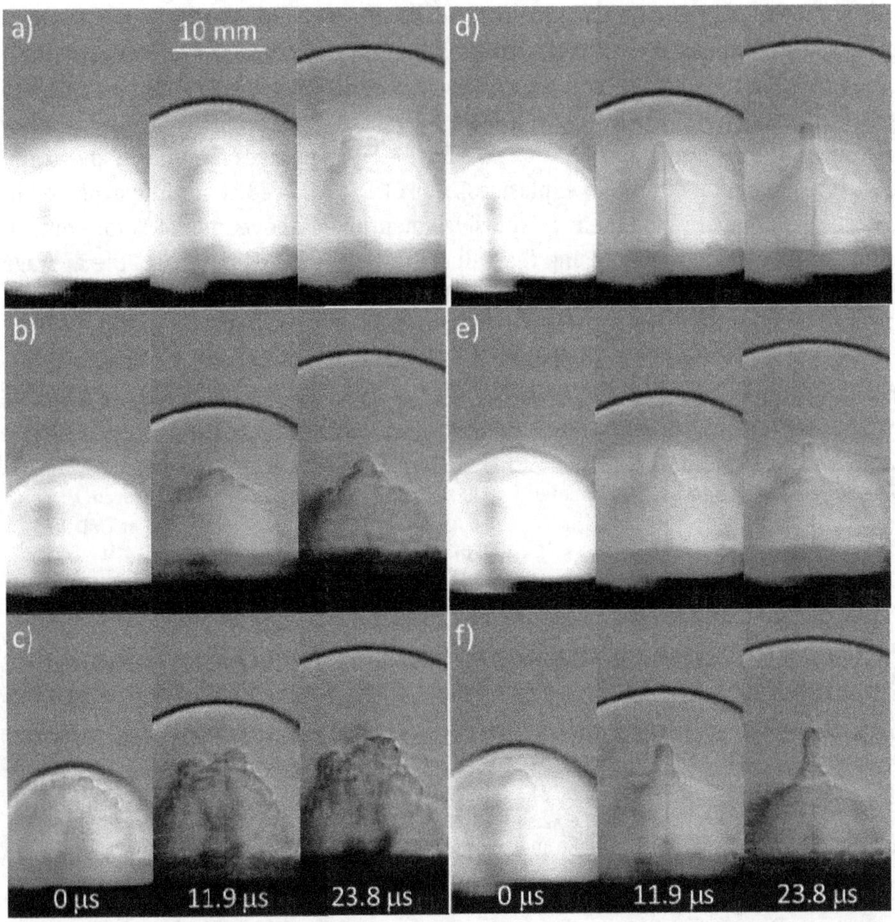

Fig. 9.43-I: Selected snapshots from the high-speed video of the laser excitation of (a) TKX-55, (b) TKX-50, (c) BTNPM, (d) DAAF, (e) BDNAPM, and (f) MAD-X1.

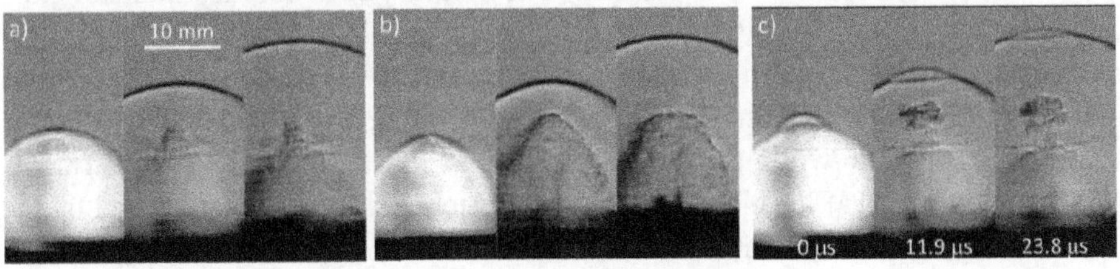

Fig. 9.43-II: Selected snapshots from the high-speed videos of the laser excitation of (a) TKX-50, (b) BTNPM, and (c) MAD-X1.

ities from the large-scale detonation testing of some conventional military explosives are also listed in Tab. 9.17. The percentage difference between the predicted or measured detonation velocities and the estimated detonation velocities from LASEM are given in parentheses. The average difference between the estimated detonation velocities and EXPLO5 predictions is 3.1% (3.6% for the six explosives mentioned above and 2.5% for conventional military explosives). For CHEETAH predictions, the average difference is 3.9% (4.8% for the six explosives mentioned above and 2.8% for conventional military explosives). Using the values in the CHEETAH database, the average difference between the estimated and predicted values for CHEETAH was 2.4% (1.0% for TXK-50 and DAAF; 3.0% for conventional military explosives).

Tab. 9.17: LASEM results in comparison with theoretical predictions and large-scale detonation test results (% difference from estimated detonation velocity indicated in parentheses).

Sample	**Laser-induced shock velocity [m · s^{-1}]**	**Estimated V_{LASEM} [km · s^{-1}]**	**EXPLO5 v6.01 $V_{C\text{-}J}$ [km · s^{-1}]**	**CHEETAH v8.0 $V_{C\text{-}J}$ [km · s^{-1}]**	**Measured $V_{det.}$ at TMD [km · s^{-1}]**
TKX-50	835 ± 11	9.56 ± 0.28	9.767 (2.1%)	9.735 (1.8%)	9.432 (−1.4%)
MAD-X1	807 ± 9	8.86 ± 0.22	9.195 (3.6%)	9.267 (4.4%)	*n/a*
BDNAPM	798 ± 9	8.63 ± 0.21	8.332 (−3.6%)	8.171 (−5.6%)	*n/a*
BTNPM	850 ± 13	9.91 ± 0.31	9.304 (−6.5%)	9.276 (−6.8%)	*n/a*
TKX-55	782 ± 11	8.23 ± 0.26	8.030 (−2.5%)	7.548 (−9.0%)	*n/a*
DAAF	774 ± 11	8.05 ± 0.26	8.316 (3.2%)	8.124 (0.9%)	8.11 ± 0.03 (1.0%)
TNT	731 ± 9	6.99 ± 0.23	7.286 (4.1%)	7.192 (2.8%)	7.026 ± 0.119 (0.5%)
HNS	740 ± 8	7.20 ± 0.21	7.629 (5.6%)	7.499 (4.0%)	7.200 ± 0.071 (0.0%)
NTO	784 ± 10	8.30 ± 0.25	8.420 (1.4%)	8.656 (4.1%)	8.335 ± 0.120 (0.4%)
RDX	807 ± 8	8.85 ± 0.19	8.834 (−0.2%)	8.803 (−0.5%)	8.833 ± 0.064 (−0.2%)
CL-20	835 ± 10	9.56 ± 0.24	9.673 (1.2%)	9.833 (2.8%)	9.57 (0.1%)

In contrast to the predicted detonation velocities, the difference between the estimated detonation velocities and measured large-scale detonation velocities for conventional military explosives is only 0.2%. The accuracy of LASEM for estimating the detonation velocities of high-nitrogen energetic materials is not known, since few large-scale results have been reported in the literature, other than for DAAF and TKX-50. The estimated detonation values agree with the reported detonation velocities for DAAF and TKX-50 to within less than 1.5%. The LASEM results suggest that TKX-55, BDNAPM, and BTNPM have higher detonation velocities than those predicted by the thermochemical codes, while the estimated detonation velocities for MAD-X1 and TKX-50 are slightly lower than those predicted by EXPLO5 or CHEETAH. A comparison of the two thermochemical codes shows that while the magnitude of the difference between the estimated detonation velocities and predicted detonation velocities may differ (especially for TKX-55), the predicted detonation velocities are typically either both higher or both lower than the estimated (or measured) detonation velocities.

High-speed video investigations confirm that the most powerful explosive among those investigated in this study is BTNPM (since it produced the least intense plasma emission); the weakest are TKX-55 and DAAF. Moreover, the structure of the plasma plume regions of DAAF, BDNAPM, MAD-X1, and TKX-55 appear to reflect the low thermal sensitivity of these energetic materials and the high thermal sensitivity of BTNPM.

The estimated detonation velocities of the six explosives which were mentioned above, which were based on the measured characteristic laser-induced air shock velocities, are in very good agreement with the detonation velocities calculated using the EXPLO5 V6.01 and CHEETAH V8.0 thermochemical codes. The average difference between the estimated detonation velocities and those calculated using EXPLO5 is 3.6%, while using CHEETAH it is 4.8%. The LASEM results show that TKX-55, BDNAPM, and BTNPM have higher detonation velocities than those calculated using EXPLO5 or CHEETAH, while the estimated detonation velocities for MAD-X1 and TKX-50 are somewhat lower than those calculated using the thermochemical codes. The detonation velocities calculated using EXPLO5 and CHEETAH differ slightly from each other (especially for TKX-55), nevertheless they show the same tendency (in general, the values are either both higher or both lower than the estimated – or measured – detonation velocities). In comparison to the predicted detonation velocities (EXPLO5: 2.5%; CHEETAH: 2.8%), the difference between the estimated detonation velocities and measured large-scale detonation velocities for conventional military explosives is only 0.2%. Moreover, for DAAF and TKX-50, the estimated detonation values are in agreement with the reported detonation velocities to within less than 1.5%.

9.8 Thermally stable explosives

A new field of secondary explosives was opened up and developed by the investigation of HNS by Shipp, and also with the investigation of TATB by Jackson and Wing, and is commonly referred to as "thermally stable explosives".

These explosives generally have lower energetic performances in comparison with RDX and would never be suitable replacements for RDX. However, they are exceptional since they show very high decomposition points and very good long-term stabilities. Therefore, such properties open-up the possibility for their use in special applications such as (i) developing new oil wells or renewable energy (geothermal energy) (Fig. 9.44), (ii) space applications and (iii) nuclear weapons.

Since it has a high decomposition temperature of 350 °C, TATB (Fig. 9.45) is a very interesting compound for these applications, but due to the high costs involved in its synthesis and its extreme insensitivity, it is mainly used in military (especially nuclear) applications. Although HNS has a lower decomposition temperature of 320 °C, it also has the advantage of lower production costs, which is why it is the commonly used explosive for civil applications. Two interesting, highly thermally stable explosives were synthesized by Coburn, namely *N*-(2,4,6-trinitrophenyl)-1*H*-1,2,4-triazol-3-

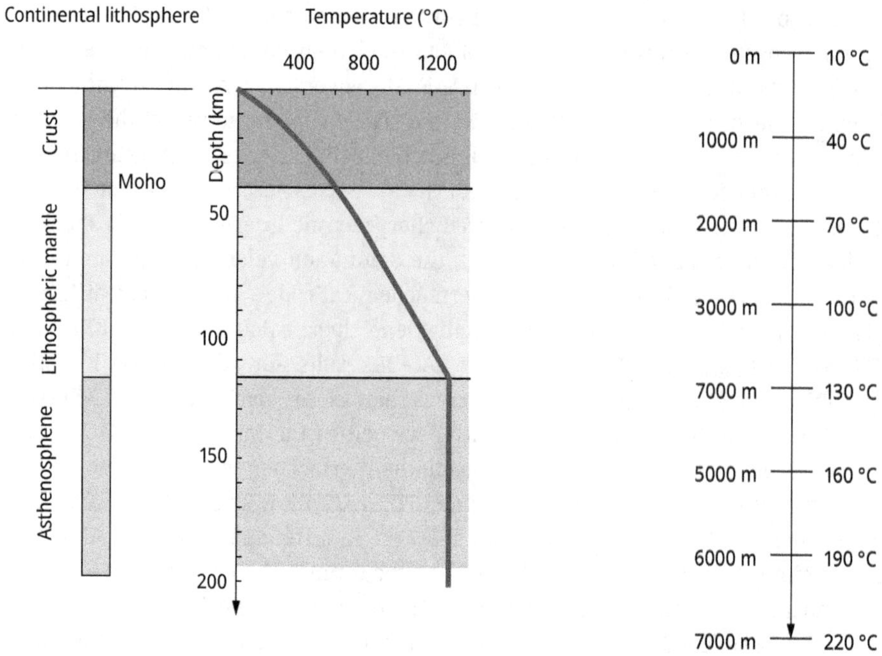

Fig. 9.44: Effect of the variation of temperature depending on the depth (in the ground).

amine (PATO) in 1969, and the 3,5-dinitro-*N*2,*N*6-bis(2,4,6-trinitrophenyl)pyridine-2,6-diamine (PYX) in 1972 (Fig. 9.45).

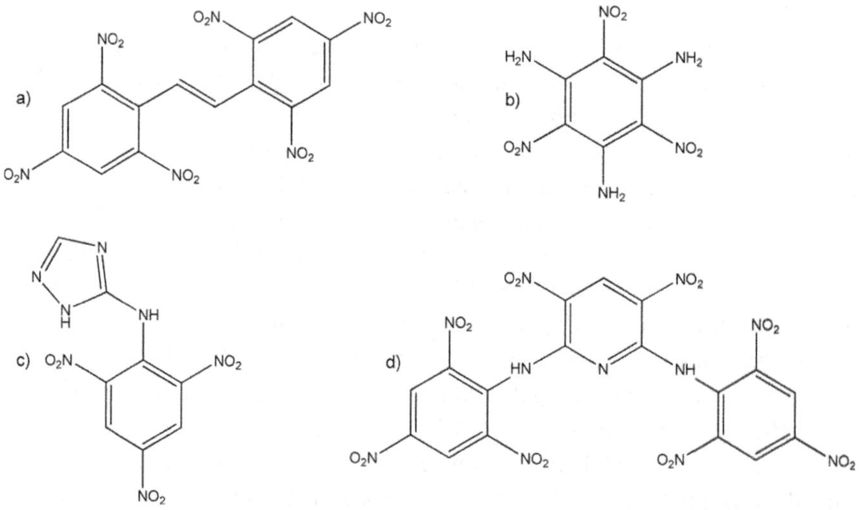

Fig. 9.45: Explosives which show high thermal stabilities a) HNS, b) TATB, c) PATO and d) PYX.

The structures in Fig. 9.45 show that thermally stable explosives commonly combine aromatic systems – normally benzene – with nitro groups. The combination of both systems in one molecule can result in an increase in the thermal stability. For example, HNS shows a decomposition temperature which is approximately 111 °C higher than that observed for RDX.

A short overview of the energetic performances of explosives with high thermal stabilities in comparison with that of RDX is given in Tab. 9.18.

Tab. 9.18: Energetic performance of RDX in comparison with those of HNS, TATB, PYX, and PATO.

	RDX	HNS	TATB	PYX	PATO
IS / J	7.5	5	50	25	8
FS / N	120	240	353	250	–
ESD / J	–	–	–	–	–
$T_{Dec.}$ / °C	210	320	350	350	310
ρ / g cm^{-3}	1.806	1.74	1.80	1.757	1.94
EXPLO5 6.01 values:					
T_{det} / K	3,800	3,676	3,526	3,613	3,185
$P_{C\text{-}J}$ / kbar	352	243	296	252	313
$V_{det.}$ / m s^{-1}	8,815	7,612	8,310	7,757	8,477
V_o / L kg^{-1}	792	602	700	633	624

The connection of two picryl moieties through a 1,3,4-oxadiazole bridge was reported by Dacons and Sitzmann, who described the synthesis of 2,5-bis(2,4,6-trinitrophenyl) -1,3,4-oxadiazole (DPO, Fig. 9.46), which is structurally related to TKX-55.

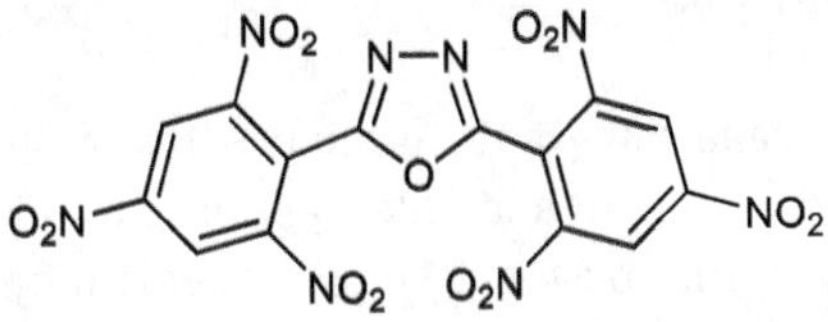

DPO

Fig. 9.46: Structure of DPO.

Due to the need for thermally highly stable secondary explosives for the oil industry, several possible HNS replacements have been suggested. However, HNS replacements should have the following properties:

1. Show an extremely high thermal stability – which means a decomposition point over 300 °C, as well as long-term stability with no decomposition at 260 °C over a period of 100 h
2. Detonation velocity greater than 7,500 m · s^{-1}
3. Specific energy greater than 975 kJ/kg. The specific energy is calculated according to the following equation:

$$F = p_e \cdot V = \boldsymbol{n} \cdot \boldsymbol{R} \cdot \boldsymbol{T}$$

(n is the number of products of decomposition, R is the gas constant, and T is the detonation temperature).

The specific energies for RDX, HNS, TATB, PYX, and PATO are given below.

	RDX	HNS	TATB	PYX	PATO
Specific energy / kJ/kg	1,026	752	839	777	675

4. Impact sensitivities > 7.4 J and friction sensitivities > 235 N. These values are slightly higher than those expected for RDX replacements (IS > 7 J; FS > 120 N).
5. Since these explosives are mostly for intended use in civil applications, the total costs should not be greater than 500 Euro/kg.
6. Finally, the critical diameter should be smaller than for HNS. The critical diameter indicates the minimum diameter of an explosive charge at which detonation can still take place. It also depends on the texture of a compound. In cast charges, it is larger than in pressed ones.

To obtain compounds with properties which correspond to the strict requirements outlined above for improved heat-resistant explosives, four general approaches have been outlined by J. P. Agrawal over the past few years which can be used to impart and increase the thermal stability:

- **Salt formation**
  The thermal stability can be enhanced via salt formation. The following example, in Fig. 9.47, is a good illustration of this approach.
- **Introduction of amino groups**
  Introducing amino groups into nitrated benzene rings in the *ortho* position influences the thermal stability. This is one of the simplest and oldest approaches for enhancing the thermal stability of explosives. The "push-pull" effect generated by the combination of electron withdrawing nitro groups next to electron donating amino groups, as well as the formation of hydrogen bonds, is responsible for the high stability towards heat or other external stimuli like friction or impact which is observed for such compounds. A good illustration is given by the introduction of amino groups to 1,3,5-trinitrobenzene in Fig. 9.48
- **Introduction of conjugation**
  A very good example for the introduction of conjugation in explosive molecules, and its ability to impart higher thermal stability, is given by hexanitrostilbene (HNS). Another interesting explosive with a similar structure to HNS is hexanitroazobenzene (HNAB). In HNAB, a conjugated system is also present, which increases the thermal stability in comparison with its monomer. In Fig. 9.49, these two examples are shown.

- **Condensation with triazole rings**
  Various studies on the synthesis of picryl- and picrylamino substituted were reported in detail by Coburn & Jackson. In these investigations, 1,2,4 -triazoles, 1,2,4-triazole or amino-1,2,4-triazole was reacted with picryl chloride (1-chloro-2,4,6-trinitrobenzene). One of the most interesting of these molecules is PATO (3-picrylamino-1,2,4-triazole), a well known, thermally stable explosive, which is obtained by the condensation of picryl chloride with 3-amino-1,2,4-triazole (Fig. 9.50). A similar compound was prepared by Agrawal and co-workers, namely 1,3-bis(1′,2′,4′-triazol-3′-ylamino)-2,4,6-trinitrobenzene (BTATNB), which was synthesized by the reaction of styphnyl chloride (1,5-dichloro-2,4,6-trinitrobenzene) with two equivalents of 3-amino-1,2,4-triazole. BTATNB exhibits a slightly higher thermal stability than PATO and is safer with respect to its impact and friction sensitivity.

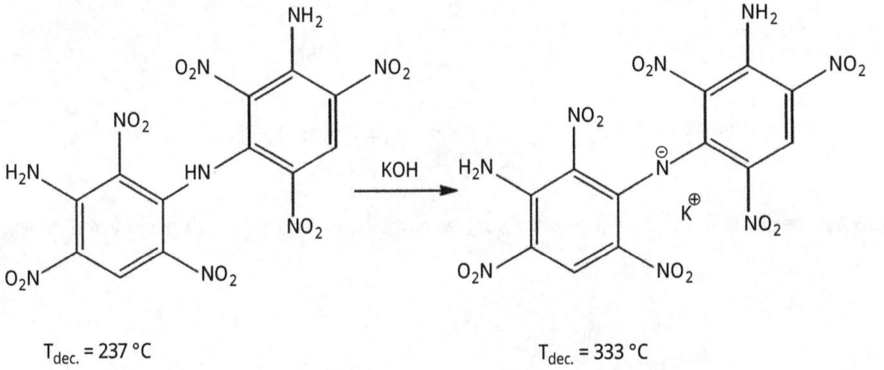

Fig. 9.47: Salt formation based on 3,3′-diamino-2,2′4,4′,6,6′-hexanitrophenylamine.

PATO can be obtained in almost quantitative yields from 3-amino-1,2,4-triazole – which is a common herbicide (Amitrole) – and picryl chloride which can be obtained from conversion of picric acid (*PEP* **2019**, *44*, 203–206):

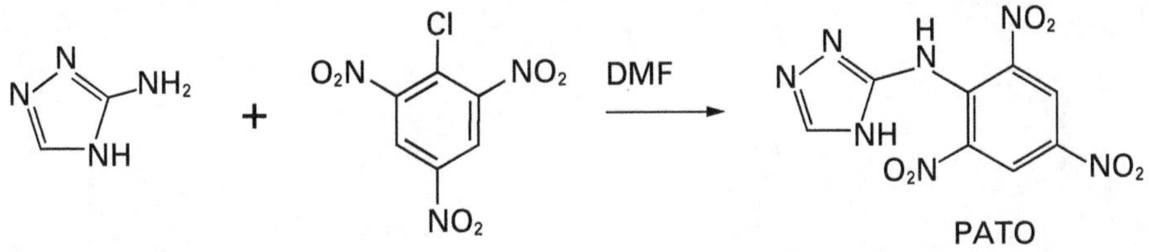

Agrawal et al. also reported the synthesis of BTDAONAB (Fig. 9.51), which does not melt below 550 °C and was proposed to be a better example of a thermally stable explosive than TATB. According to the authors, this material has a very low impact (21 J), and no friction sensitivity (> 360 N) and is thermally stable up to 550 °C. With

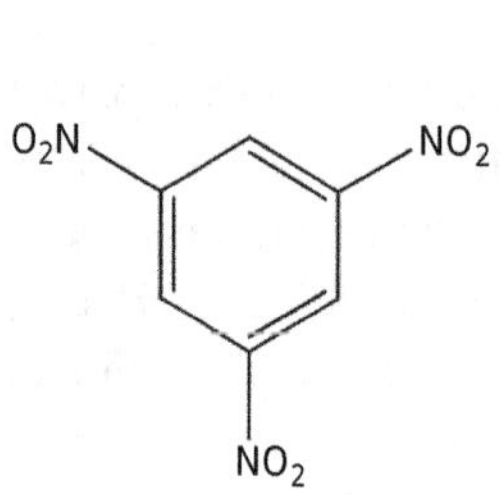

1, 3, 5-Trinitrobenzene
T_{dec} 122 °C

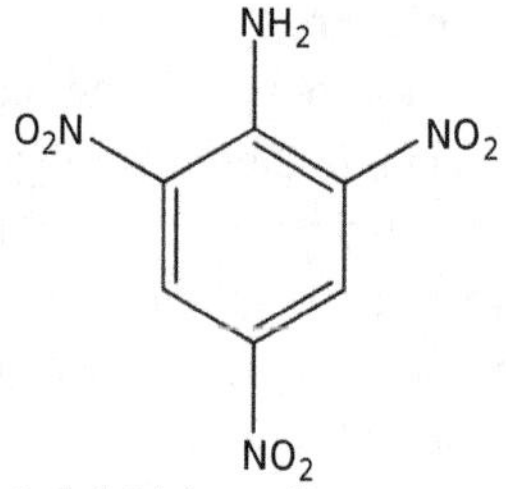

2, 4, 6-Trinitroaniline (TNA)
T_{dec} 192 °C

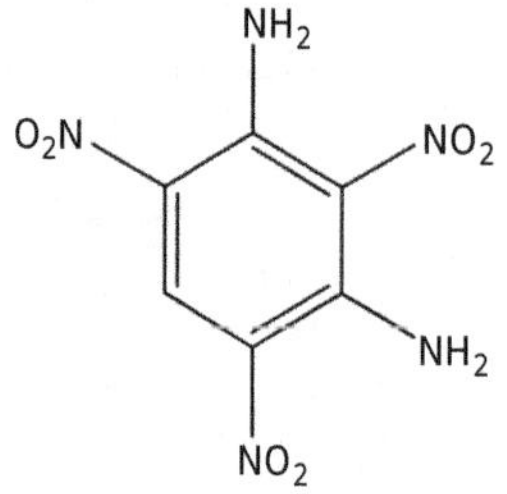

1, 3-Diamino-2, 4, 6-trinitrobenzene
T_{dec} 286 °C

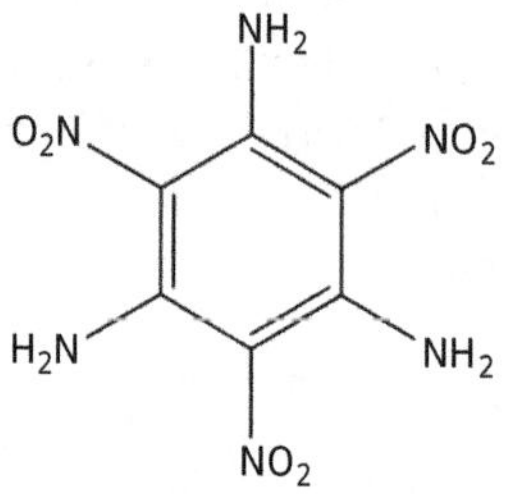

1, 3, 5-Triamino-2, 4, 6-trinitrobenzene (TATB)
T_{dec} >320 °C

Fig. 9.48: The introduction of amino groups into 1,3,5-trinitrobenzene results in an increase in the thermal stability.

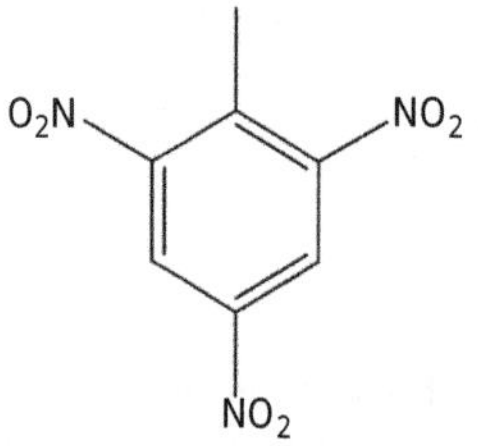

2, 4, 6-Trinitrotoluene (TNT)
T_{dec} 295 °C

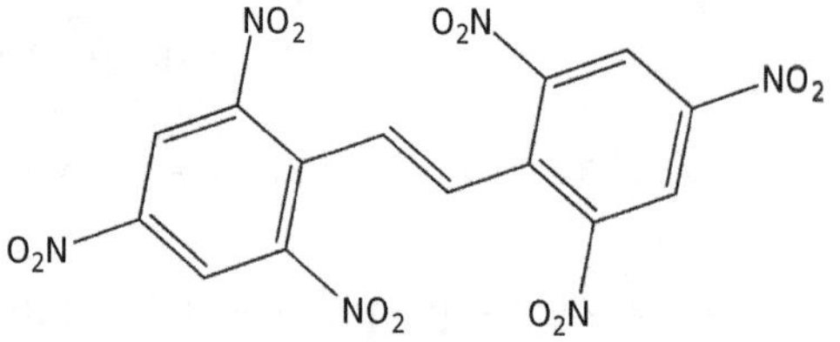

Hexanitrostilbene (HNS)
T_{dec} >320 °C

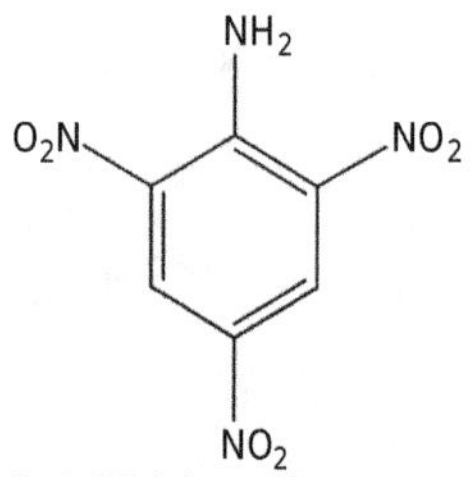

2, 4, 6-Trinitroaniline (TNA)
T_{dec} 192 °C

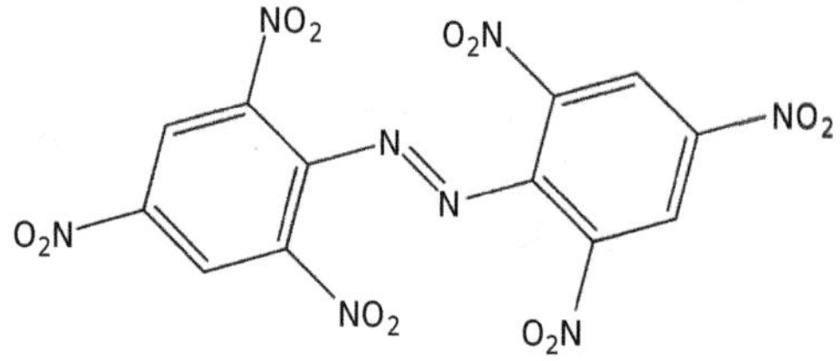

Hexanitroazobenzene (HNAB)
T_{dec} 215 °C

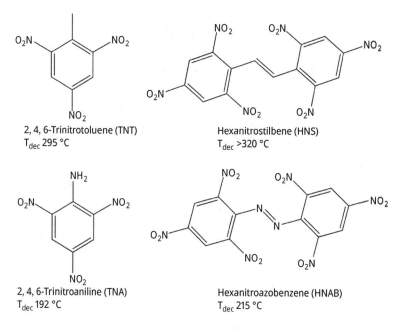

Fig. 9.49: Introduction of conjugation: monomer vs. dimer. On formation of the dimer, the thermal stability increases.

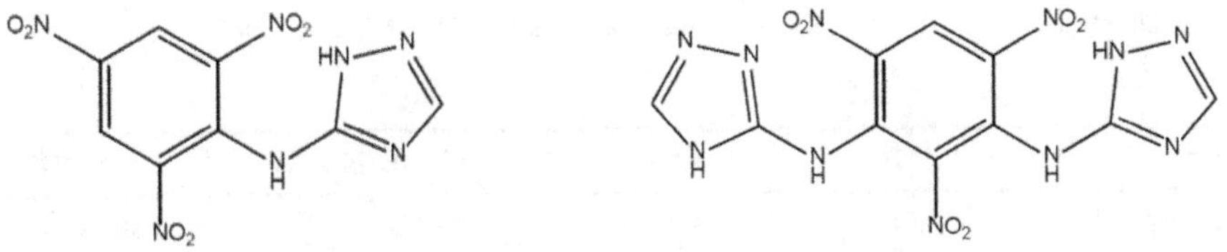

Fig. 9.50: Condensation with triazole rings PATO (left) and BTATNB (right).

these reported properties, BTDAONAB is superior to all of the nitro-aromatic compounds discussed above. BTDAONAB has a VoD of 8,300 m/s while TATB is about 8,000 m/s.

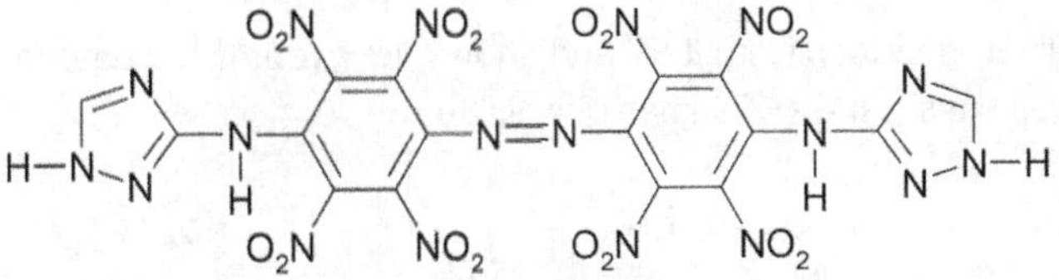

Fig. 9.51: Molecular structure of BTDAONAB.

Moreover, another nitro-aromatic compound (BeTDAONAB) – similar to Agrawal's BTDAONAB – has recently been published by Keshavarz et al., and is also reported to be very insensitive. In this compound, the terminal triazole moieties have been replaced by two energetic (more endothermic) tetrazole units (Fig. 9.52, Tab. 9.19).

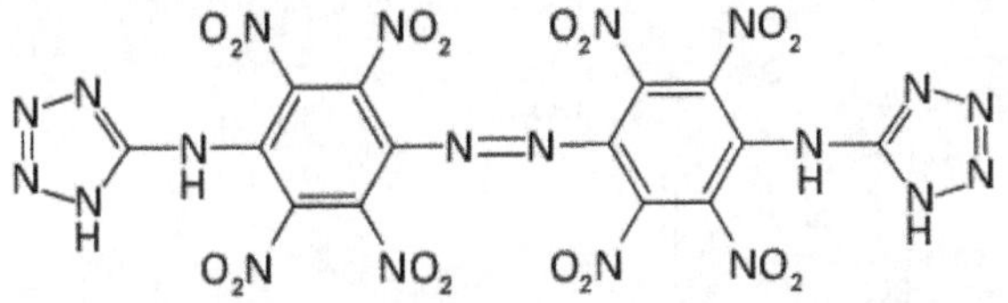

Fig. 9.52: Molecular structure of BeTDAONAB.

The drawback of these two thermally extremely stable compounds (BTDAONAB and BeTDAONAB, Tab. 9.19) is that they can't be initiated with presently used primary explosives such as lead azide (LA) or lead styphnate (LS).

New developments

Quite recently, the synthesis of the new heat-resistant explosive (T_{dec} = 335 °C) 5,5′-bis (2,4,6-trinitrophenyl)-2,2′-bi(1,3,4-oxadiazole) (TKX-55) has been reported. The com-

Tab. 9.19: Comparison of the thermal and explosive properties of TATB, HNS, BTDAONAB, and BeTDAONAB.

Property	TATB	HNS	BTDAONAB	BeTDAONAB
Density / g cm^{-3}	1.94	1.74	1.97	1.98
DTA (exo) / °C	360	353	550	275
DSC (exo) / °C	371	350		268
Ω_{CO} / %	− 18.6	− 17.8	− 6.8	− 5.9
IS / J	50	5	21	21
FS / N	> 353	240	353	362
VoD / m s^{-1}	7,900	7,600	8,600	8,700
$p_{C\text{-}J}$ / kbar	273	244	341	354

pound can be prepared from (demilitarized) TNT and shows very promising detonation parameters (VoD = 8,030 m s^{-1}, $p_{C\text{-}J}$ = 273 kbar (Fig. 9.53)).

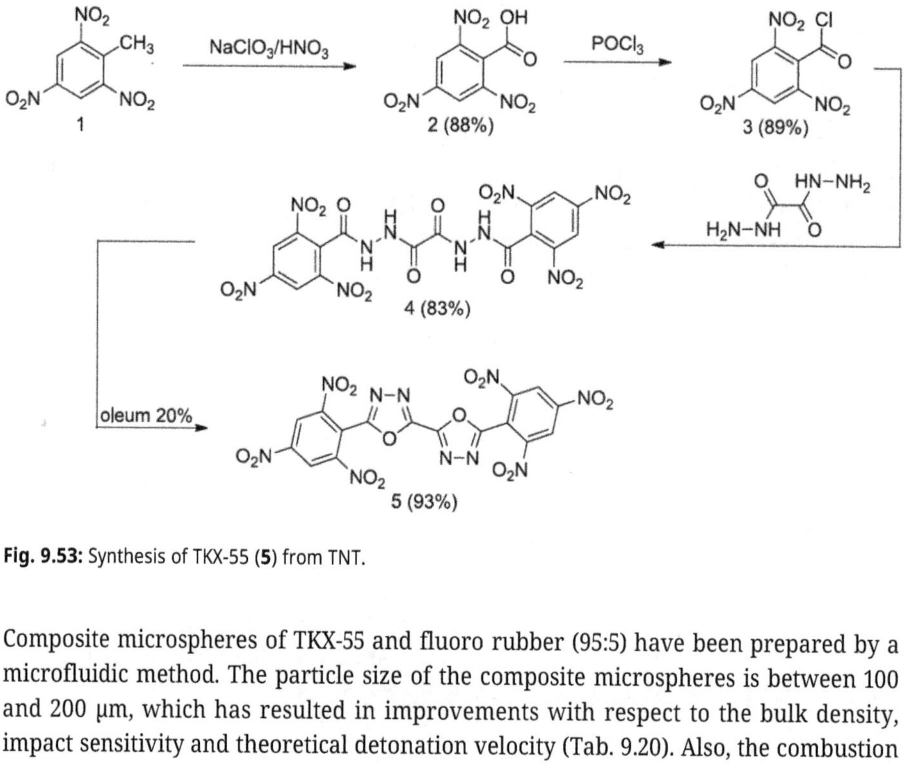

Fig. 9.53: Synthesis of TKX-55 (**5**) from TNT.

Composite microspheres of TKX-55 and fluoro rubber (95:5) have been prepared by a microfluidic method. The particle size of the composite microspheres is between 100 and 200 μm, which has resulted in improvements with respect to the bulk density, impact sensitivity and theoretical detonation velocity (Tab. 9.20). Also, the combustion performance of the composite TKX-55/fluoro rubber has also been investigated.

It was found that the addition of fluoro rubber could inhibit the combustion of TKX-55 and improve the safety of the composite. TKX-55 has the potential to be used in the future as the main explosive component of new heat-resistant PBX formulations.

Tab. 9.20: Sensitivities and performance of TKX-55 and TKX-55/ fluoro rubber (95:5).

	TKX-55	TKX-55/fluoro rubber
bulk density /g cm^{-3}	0.271	0.429
IS / J	5	25
FS / N	160	216
VoD / m s^{-1} (calcd.)	8,230	8,090

Lit.: Z. Lin, Z. Zhang, F. Gao, P. Zhou, Y. Liu, Y. Hou, C. An, J. Wang, B. Wu, New TKX-55-Based Heat resistant Microspheres, *Propellants, Explos., Pyrotech.*, **2023**, *48*, e202300057. https://doi.org/10.1002/prep.202300057

Even more thermally stable ($T_{dec.}$ = 360 °C) than TKX-55 is a new calixarene explosive (Fig. 9.54) which can be prepared by a facile synthetic procedure. However, the performance (VoD = 7,865 m s^{-1}, $p_{C\text{-}J}$ = 289 kbar) is somewhat lower than that of TKX-55.

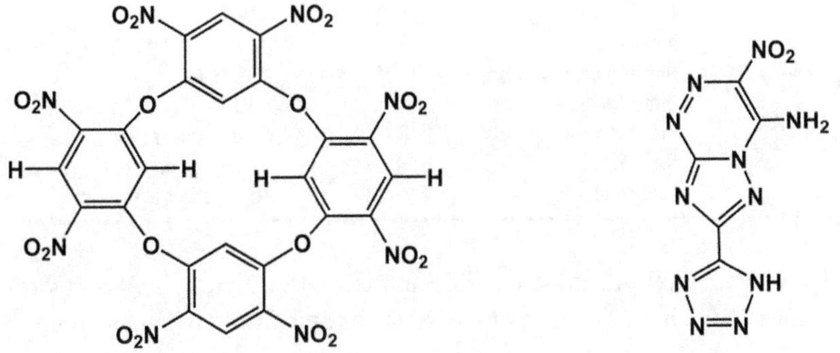

Fig. 9.54: Molecular structures of a new calixarene explosive (left) and of tetrazolyl triazolotriazine (right).

Tetrazolyl triazolotriazine (Fig. 9.54) was also recently suggested as a possible candidate for use as a TATB replacement. However, its decomposition temperature of 305 °C is somewhat lower than that of TATB (360 °C).

There is clearly demand for new thermally highly stable explosives, and the race is still on to find better and cheaper examples for use in oil and gas exploration, space applications, and nuclear weapons.

Check your knowledge: See Chapter 14 for Study Questions.

10 Synthesis of energetic materials

10.1 Molecular building blocks

Most inorganic or organic energetic materials contain at least one of the building blocks shown in Tab. 10.1 and can therefore be classified by their functional groups.

Tab. 10.1: Molecular building blocks for energetic materials.

group	energetic substance	example
—O—O—	organic peroxo compounds	TATP
ClO_3^-	chlorate	$KClO_3$
ClO_4^-	perchlorate	AP
$—NF_2$	organic difluoramine derivatives	HNFX-1
$—NO_2$	organic nitro compounds	TNT
$—O—NO_2$	organic nitrato- or nitroesters	NG, PETN, NG-A
NO_3^-	nitrate	AN
—N=N—	organic or inorganic diazo compounds	TAGzT
$—N_3$	covalent azides	GAP
N_3^-	ionic azides	$[N_2H_5][N_3]$, hydrazinium azide
$—NH—NO_2$, $—NR—NO_2$	nitramino compounds	RDX, HMX
$N(NO_2)_2^-$	dinitramide	ADN
$=N—NO_2$	nitrimino compounds	nitriminotetrazole
—CNO	fulminates	$Hg(CNO)_2$, mercury fulminate
—C≡C—	acetylides	Ag_2C_2, silver acetylide

The molecular structures of nitriminotetrazole and tetrakis(difluoramino)octahydro-dinitro-diazocine (HNFX) have not been discussed so far and are shown in Fig. 10.1.

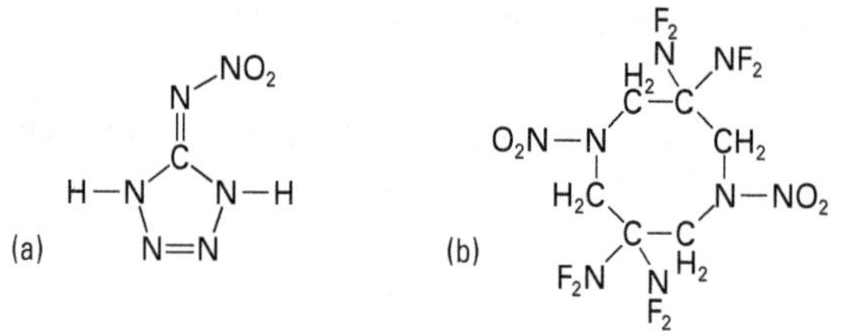

Fig. 10.1: Molecular structures of nitriminotetrazole (a) and tetrakis(difluoramino)octahydro-dinitro-diazocine (HNFX) (b).

It is interesting that nitriminotetrazole contains an oxidizing nitrimino group in addition to the endothermic ring, which helps to create a more positive oxygen balance (Ω = –12.3%). HNFX works similarly as it corresponds to a HMX molecule in which two of the nitramine groups are exchanged for $C(NF_2)_2$ groups. In this example, the two

https://doi.org/10.1515/9783111447087-010

remaining nitramine groups and the introduced difluoroamine groups act as strong oxidizers.

10.2 Nitration reactions

As we have seen in Tab. 1.2, practically all formulations used by the military contain TNT, as well as RDX and/or HMX. Additionally, PETN and nitroglycerin (in dynamite) play an important role. All these substances are obtained by nitration reactions. An overview of different types of nitration is given in Tab. 10.2.

Tab. 10.2: Possible nitration reactions.

reaction	C nitration	O nitration	N nitration
product	nitro compounds, $R{-}NO_2$	nitrate ester, $R{-}O{-}NO_2$	nitramine, $R{-}NH{-}NO_2$
examples	TNT PA Tetryl TATB HNS	NC NG PETN NG-A	RDX HMX NQ
nitrating agent	mixed acid: HNO_3/H_2SO_4	mixed acid: HNO_3/H_2SO_4	mixed acid: HNO_3/H_2SO_4, HNO_3 (100%)

The nitration of organic compounds still occurs almost exclusively using mixed acid or 65% to 100% HNO_3. Mixed acid is a mixture of concentrated nitric acid and concentrated sulfuric acid, which acts as a very strong nitrating agent due to the presence of the NO_2^+ ion.

$$HNO_3 + H_2SO_4 \rightarrow H_2ONO_2^+ + HSO_5^-$$

$$H_2ONO_2^+ + HSO_4^- + H_2SO_4 \rightarrow NO_2^+ + H_3O^+ + 2HSO_4^-$$

Aromatic hydrocarbons can be converted into nitro compounds via an electrophilic substitution using mixed acid. For example, trinitrotoluene is obtained by the direct nitration of toluene using mixed acid. Other good nitrating agents are $NO_2^+BF_4^-$ and $NO_2^+OSO_2CF_3^-$ (nitronium triflate) in CH_2Cl_2 with ultrasonic activation.

A good example of an *N*-nitration which can be undertaken using mixed acid or 100% HNO_3, is the nitration of aminotetrazole to nitriminotetrazole (Fig. 10.2):

Nitriminotetrazole can also be obtained in good yield by the diazotization of nitroaminoguanidine to nitroguanylazide and subsequent cyclization under alkaline conditions (Fig. 10.3).

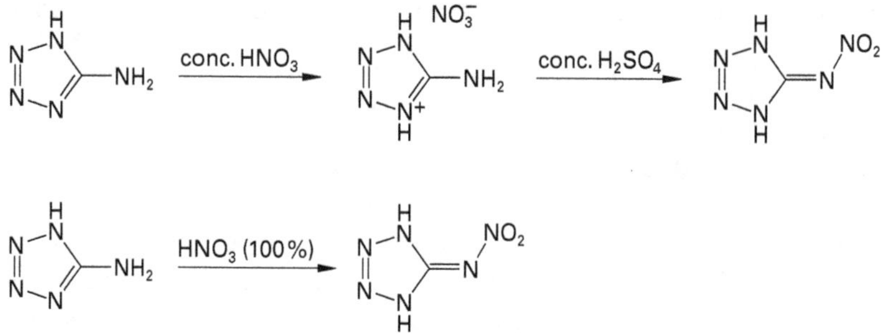

Fig. 10.2: Nitration of aminotetrazole to nitriminotetrazole.

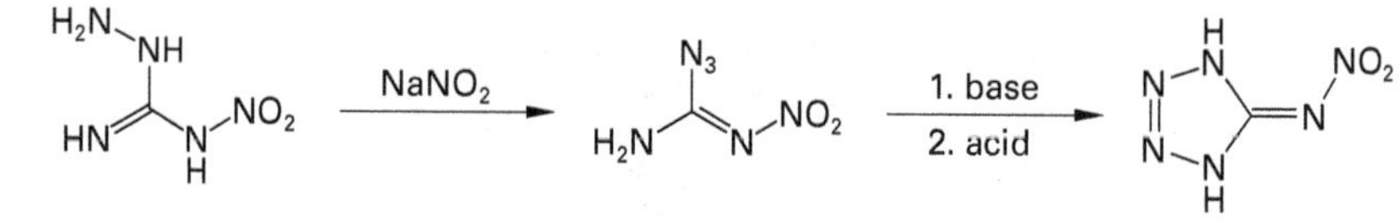

nitroaminoguanidine nitroguanylazide

Fig. 10.3: Synthesis of nitriminotetrazole from nitroaminoguanidine.

A further example of a nitration using nitric acid is the preparation of RDX from hexamethylentetramine (HMTA) via triacetyltriazine (TART) in a two-step reaction (Fig. 10.4):

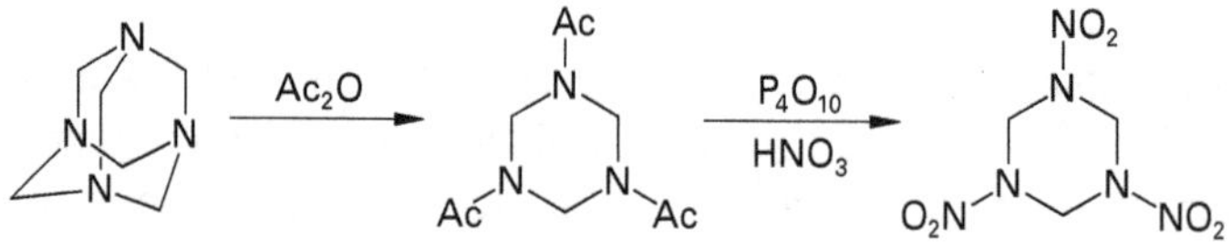

HMTA TART RDX

Fig. 10.4: Synthesis of hexogen (RDX) from hexamethylentetramine (HMTA) via triacetyltriazine (TART).

In modern synthetic chemistry N_2O_5 can also be used as a nitrating agent; it is advantageous because anhydrous conditions can be used and, if pure N_2O_5 is used, there are no acidic impurities present. If acidic impurities do not matter, a mixture of N_2O_5 and HNO_3 is also a good and strong nitrating agent.

Previously, technical N_2O_5 was mainly obtained by the dehydration of nitric acid at –10 °C, as N_2O_5 is the anhydride of nitric acid. It is an easily sublimed solid (subl. 32 °C, 1 bar).

$$12\,HNO_3 + P_4O_{10} \xrightarrow{-10\ ^\circ C} 6\,N_2O_5 + 4\,H_3PO_4$$

Since 1983, the technical synthesis usually used follows that developed by Lawrence Livermore National Laboratory, in which the electrolysis of nitric acid in the presence of N_2O_4 results in the formation of an approx. 15–20% solution of N_2O_5 in anhydrous nitric acid.

$$2\,HNO_3 \xrightarrow{N_2O_4,\ 2e^-} N_2O_5 + H_2O$$

In the following, we want to look at a further way to produce pure and almost acid-free N_2O_5. This was developed into a semi-industrial process (pilot plant) as late as 1992 in the DRA Laboratories (Defence Research Agency). This method involves the gas-phase ozonization of N_2O_4 using an ozone-oxygen mixture with an approx. 5–10% ozone content.

$$N_2O_4 + O_3 \longrightarrow N_2O_5 + O_2$$

Solutions of pure N_2O_5 in chlorinated organic solvents (CH_2Cl_2, $CFCl_3$) are mild nitrating agents, which have recently found a wide range of applications (Tab. 10.3).

Tab. 10.3: Synthetic application of CH_2Cl_2 / N_2O_5 solutions.

reaction type	product
aromatic nitration	$C{-}NO_2$
nitrolysis	$N{-}NO_2$
ring-opening reactions	$N{-}NO_2$ or $O{-}NO_2$
selective nitration	$O{-}NO_2$ (rarer $N{-}NO_2$)

An example of a ring-opening nitration reaction is shown below.

$$\text{cyclo}-(CH_2)_nX \xrightarrow{N_2O_5,\ CH_2Cl_2,\ 0-10\ °C} O_2NO-(CH_2)_n-X-NO_2$$
$$(n = 2 \text{ or } 3;\ X = O \text{ or NR } (R = \text{Alkyl}))$$

If the substituent on nitrogen is an H atom and not an alkyl group, the nitration occurs with preservation of the cyclic, four-membered ring structure.

$$\text{cyclo}-(CH_2)_3NH \xrightarrow{N_2O_5,\ CH_2Cl_2,\ 0-10\ °C} \text{cyclo}-(CH_2)_3N(NO_2)$$

In a similar synthetic route, the energetic compounds TNAZ (trinitroazetidine) and CL-20 were prepared (Fig. 1.7).

ADN is a unique highly energetic material, because it does not contain carbon, chlorine or fluorine. ADN is an environmentally benign high energy inorganic oxidizer salt, first synthesized in 1971 at the Zelinsky Institute of Organic Chemistry in Moscow. It is claimed that ADN-based solid propellants are in operational use in Russian TOPOL intercontinental ballistic missiles and that ADN was previously produced in ton-scale quantities in the former USSR. ADN forms colorless crystals and is a very oxygen-rich nitrogen oxide, which possesses excellent properties for use as an explosive (in combination with strong reducing agents such as aluminum, aluminum hydride or organic compounds) or solid rocket propellant. Additionally, the absence of halogens makes ADN an environmentally friendly solid rocket propellant and makes RADAR detection of the rocket emission more difficult. ADN is currently being tested as one of the most promising replacements for ammonium perchlorate.

Ab initio calculations predict a structure with C_2 symmetry for the free dinitramide ion, $N(NO_2)_2^-$ (Fig. 10.5), while in solution and in the solid-state the local symmetry is essentially C_1. This can be explained on the basis of weak cation-anion interactions or interactions with the solvent, since the dinitramide ion is very easy to deform because of the very small barrier to rotation of the NNO_2 moiety (< 13 kJ mol^{-1}).

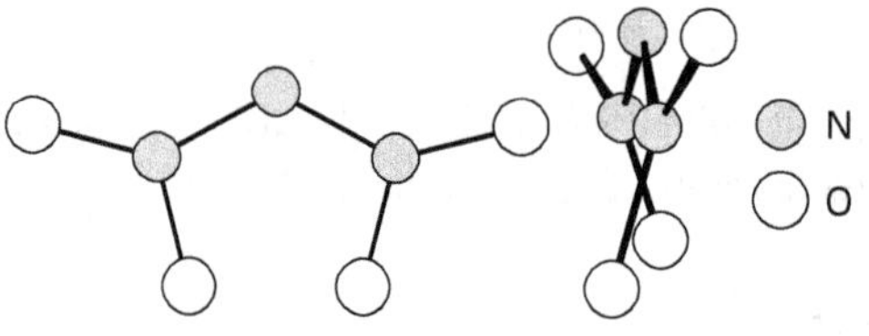

Fig. 10.5: Two different views of the dinitramide ion with C_2 symmetry.

Although ADN is already industrially prepared, the lab synthesis is not trivial and depends on the precise control of the reaction conditions. Normally, ADN is synthesized by the nitration of ammonia using N_2O_5 (prepared by the ozonization of NO_2) in a chlorinated solvent (Fig. 10.6):

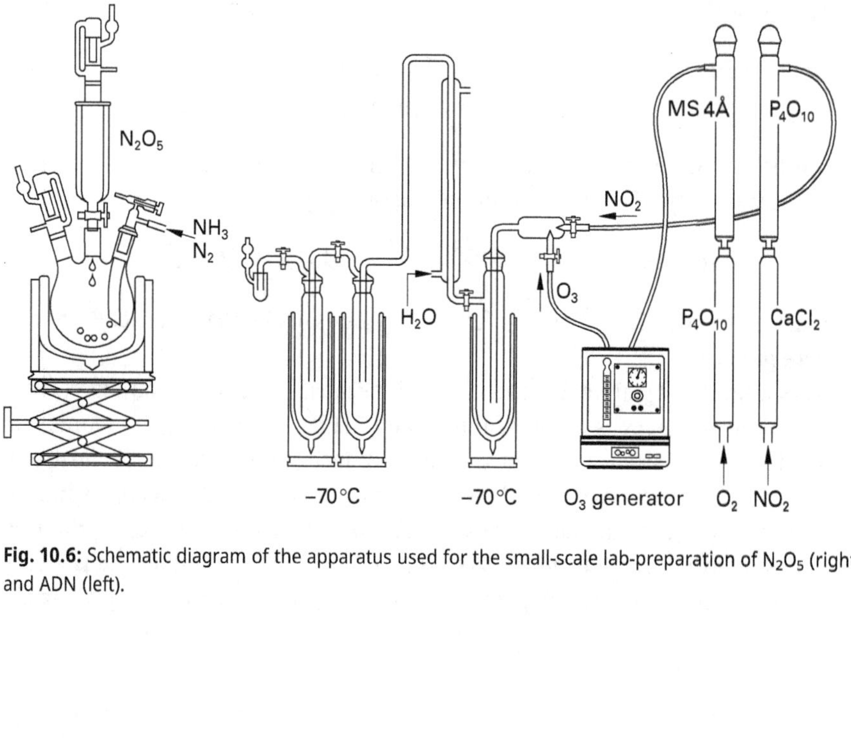

Fig. 10.6: Schematic diagram of the apparatus used for the small-scale lab-preparation of N_2O_5 (right) and ADN (left).

Industrially, ADN is commonly prepared by the nitration of urea. In the first stage, urea is converted into urea nitrate using dilute nitric acid, which is then dehydrated to form nitrourea by reaction with concentrated sulfuric acid. The 1,1-dinitrourea isomer can subsequently be obtained by nitrating nitrourea with a strong nitrating agent ($NO_2^+BF_4^-$, N_2O_5). The dinitrourea then reacts with gaseous ammonia to form ADN and urea:

synthesis of N_2O_5:		
$2\,NO_2 + O_3$	→	$N_2O_5 + O_2$
synthesis of ADN:		
$NH_3 + N_2O_5$	→	$(O_2N)NH_2 + HNO_3$
$(O_2N)NH_2 + N_2O_5$	→	$(O_2N)_2NH + HNO_3$
$(O_2N)_2NH + NH_3$	→	$[NH_4]^+[N(NO_2)_2]^-$
$2\,HNO_3 + 2\,NH_3$	→	$2\,[NH_4]^+[NO_3]^-$
$4\,NH_3 + 2\,N_2O_5$	→	$[NH_4]^+[N(NO_2)_2]^- + [NH_4]^+[NO_3]^-$

Note: Free dinitraminic acid is a very strong acid with a pK_a value of − 5.6. It is unstable, however, and decomposes quickly to N_2O and HNO_3.

$$H_2N{-}CO{-}NH_2 \xrightarrow{\text{dil. } HNO_3} [H_2N{-}CO{-}NH_3]^+[NO_3]^-$$

$$[H_2N{-}CO{-}NH_3]^+[NO_3]^- \xrightarrow{\text{conc. } H_2SO_4} H_2N{-}CO{-}NH(NO_2)$$

$$(O_2N)HN{-}CO{-}NH_2 \xrightarrow{N_2O_5} (O_2N)_2N{-}CO{-}NH_2$$

$$(O_2N)_2N{-}CO{-}NH_2 + 2\,NH_3 \longrightarrow [NH_4^+][N(NO_2)_2]^- + H_2N{-}CO{-}NH_2$$

Using a new process, ADN can be prepared in a very environmentally friendly manner (without the use of chlorinated organic solvents) by the direct nitration of salts (potassium or ammonium) of sulfamic acid (amidosulfonic acid, $H_2N{-}SO_3H$) using mixed acid (HNO_3 / H_2SO_4):

$$H_2N{-}SO_3 \xrightarrow{HNO_3/H_2SO_4,\, T < -25\,°C} N(NO_2)_2^-$$

Sulfamic acid (amidosulfonic acid) itself can be obtained simply from urea and pyrosulfuric acid (disulfuric acid, $H_2S_2O_7$):

$$OC(NH_2)_2 + H_2S_2O_7 \rightarrow 2\,H_2N{-}SO_3H + CO_2$$

Conversion of ADN into different metal salts of ADN is facile by reaction with the corresponding metal hydroxide:

$$MOH + [NH_4][N(NO_2)_2] \rightarrow [M][N(NO_2)_2] + NH_3 + H_2O$$

Table 10.4 gives an overview of different starting materials and nitrating agents which can be used for the synthesis of dinitramide salts.

Tab. 10.4: Different substrates and nitrating agents for the synthesis of dinitramide salts.

Starting material	Nitrating agent	Reaction temperature /°C	Yield /%
$Me_3Si(CH_2)_2NCO$	NO_2BF_4/HNO_3	< 0	25
H_2NHNO_2	NO_2BF_4	−20	60
H_2NNO_2	N_2O_5	−20	<1
H_2NNO_2	$NO_2HS_2O_7$	−40	53
NH_4COONH_2	NO_2BF_4	0	15
NH_3	NO_2BF_4	−78	25
NH_3	N_2O_5	−78	15
NH_3	$NO_2HS_2O_7$	−78	15
$H_2NCONHNO_2$	NO_2BF_4	−40	20
$H_2NCOOC_2H_5$	N_2O_5	−48	70
$H_2NCOOC_2H_5$	NO_2BF_4	−60	60
$HN(CH_2CH_2COOEt)_2$	NO_2BF_4	<0	31
$HN(CH_2CH_2CN)_2$	NO_2BF_4	−10	60
$NH_4H_2NSO_3$	HNO_3/H_2SO_4	−45	60

10.3 Processing

Most explosives which are used by the military are solids under normal conditions and are usually obtained as granules on a technical scale. These granules are then mixed with other explosives and energetic (e.g. binder, plasticizer) or non-energetic additives (anti-oxidants, wetters, wax) and made into their final form using one of the following technical processes:

1. melting, mostly under vacuum
2. pressing, often under vacuum
3. extruding at high pressures

Melting (melt-casting) is one of the oldest technical processes and is normally used for formulations which contain TNT as the melt-castable component and RDX or HMX as the energetic filler. This is essentially the reason why TNT is still often used today. TNT melts at approx. 80 °C and has a very high ignition temperature of 240 °C (Fig. 10.7). The melting and decomposition points of RDX and HMX lie close together, while other energetic compounds such as NQ or TATB decompose at much lower temperatures than their estimated melting points (Fig. 10.7). The main disadvantage of the melt process is, that during cooling cracks can form in the explosive, which has a negative influence on its performance and sensitivity. To overcome the shortcomings of melt-casting, liquid polymers (e.g. HTPB, GAP etc.) are used as binders while explosives such as RDX or

HMX are used as fillers for PBXs. In this so-called **normal-casting** process, the explosive (RDX, HMX) is mixed with a binder (HTPB) which may be crosslinked with a diisocyanate to provide a cured and dimensionally stable matrix. This process is often performed under vacuum, and filling can also be agitated by vibration in order to remove any gas bubbles trapped in the matrix. This type of casting is generally adopted for large caliber warheads.

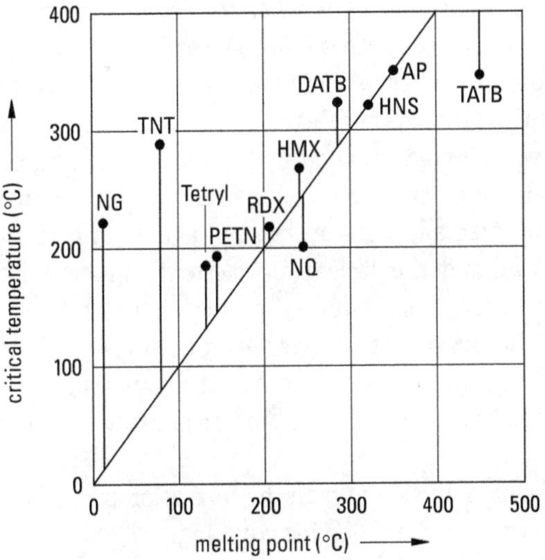

Fig. 10.7: Melting points and critical temperatures of some energetic materials. (This diagram is reproduced in slightly modified form from the original with permission from Prof. Dr. Manfred Held.).

A powder is generally used in **pressing**, which is mainly undertaken under vacuum. The advantage in comparison with the melt process is the low temperature. However, this process is also more hazardous which is why it is always automated or remote controlled. The subsequent mechanical processing of the finished explosive block is also relatively dangerous and should always be carried out using remote-control.

Extruding is usually carried out using a screw-type extruding machine as it is known from the plastic industry. In this process, the explosive and polymer binder are pre-mixed and then mixed, compressed and extruded through a small hole under high pressure using the extruder. This process is particularly suitable for the preparation of polymer bound explosives (PBXs).

10.4 Safe handling of spent acid

A number of severe accidents in the explosives industry have occurred in connection with the chemical instability of spent acids which are generated in industrial-scale nitration units. In order to counter this hazard, safety charts have been constructed by several authors, including Biazzi.

Nitration reactions are not only exothermic (energy releasing), they also produce under controlled reaction conditions nitrated organic compounds which can be highly explosive. This makes the nitration of organic compounds a potentially hazardous process, and it is therefore perhaps not surprising that the accidental explosion of many industrial nitration plants has been reported in the literature over the years. In addition to the above-mentioned aspects, it also has to be taken into account that unlike common explosives such as TNT, compounds such as nitroglycerin and nitroglycerol (both nitric esters) are known to become very unstable if they are subjected to prolonged contact with acids. During the production of each of these compounds, a considerable number of accidents have occurred predominantly during handling or storing of the spent acid. Spent acid is used to denote the residual acid which remains after the nitration process. The main problematic steps in the production of nitric esters on an industrial scale are the separation of the desired nitric ester from the spent acid and the undesired accumulation of the nitric esters in points such as transfer lines and collecting areas, as well as the build-up of dangerous quantities of nitric ester over periods of time. It is paramount that the unintentional accumulation of nitric acid esters is avoided at all costs, since acidic nitric esters are chemically unstable and sensitive mixtures.

Acidic solutions hydrolyze nitric esters which results in the formation of the corresponding alcohol and nitric acid according to the following equation:

$$RCH_2NO_3 + H_2O \rightarrow RCH_2OH + HNO_3 \qquad (10.1)$$

Alcohols show a higher sensitivity to wet oxidation than esters. The alcohol group is the most sensitive part and importantly, the alcohol functional group doesn't exist in a completely esterified molecule. Oxidation of alcohols occurs as follows:

$$RCH_2OH + O \rightarrow RCHO + H_2O \qquad (10.2)$$

$$RCH_2NO_3 + O \rightarrow RCHO + HNO_3 \qquad (10.3)$$

Based on reaction (10.2), it can be concluded that an increase in the number of OH groups (dinitroglycerin or mononitroglycol content) results in a decrease in the stability. At this stage it must be pointed out that eq. (10.3) is probably of importance at elevated temperatures, where the decomposition is a combination of not only thermal but also of oxidative aspects. The latter involves the breaking-up of radicals in the dinitroglycerin or mononitroglycol molecule.

Decomposition may also be affected by the following:

$$RCH_2ONO_2 + NO^+ + 3\,H_2O \rightarrow RCH_2ONO + NO_3^- + 2\,H_3O^+ \qquad (10.4)$$

$$RCH_2ONO_2 + HNO_2 \rightarrow RCH_2NO_2 + HNO_3 \qquad (10.5)$$

Organic nitrites may also be formed by the reaction of nitrogen dioxide with the organic alcohol:

$$RCH_2OH + 2NO_2 \rightarrow RCH_2NO_2 + HNO_3 \quad (10.6)$$

Organic nitrites are more unstable than nitric acid esters and undergo oxidation as follows:

$$RCH_2NO_2 + O \rightarrow RCHO + HNO_2 \quad (10.7)$$

Considering reactions (10.4)–(10.7), it can be summarized that it is the formation of nitrites that explains the strong decomposition effect of HNO_2 and lower nitrogen oxides, in addition to the autocatalytic character of the decomposition reactions.

Since the separation of nitric esters from spent acid as well as the unwanted accumulation of nitric esters have been shown to be the main causes of accidents at nitroglycerin and nitroglycol units, clearly the safety of industrial nitration plants depends hugely on the composition of spent acid – assuming that the raw materials are in accordance with the specifications. The content of dissolved ester in the acid phase, the content of HNO_3 dissolved in the oil phase, as well as the content of hydroxyl groups arising from mononitroglycol or dinitroglycol in the acid phase are all determined by the composition and temperature of the spent acid.

As a result, it can be concluded that in the manufacture of nitroglycerin and nitroglycol it is imperative that the spent acid composition regions which are safe (or dangerous) in contact with nitric esters are known. The so-called “high safety area” in Fig. 10.8 corresponds to the area in which a maximum of 0.2% HNO_2 is formed after

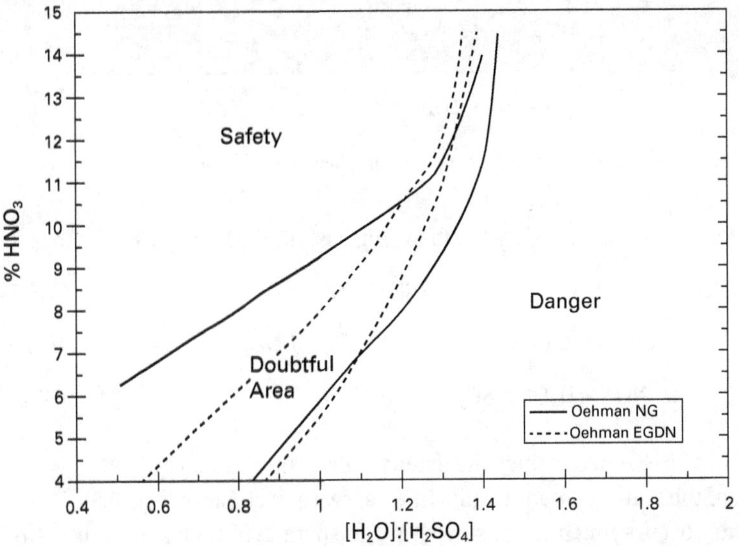

Fig. 10.8: Safety chart for spent acid handling from a NG/EGDN plant at 70 °C (from Biazzi).

heating for 2 h at 70 °C. The region designated the “normal safety area” is where 0.2–1% HNO_2 is formed under the same conditions. These statements are only valid when pure materials are used. Systems which contain a high HNO_2 content widen the danger area. Figure 10.9 has been specifically constructed for a system of 35 °C and a HNO_2 content of 0.2 wt %. Fig. 10.10 depicts a safety chart for spent acid handling from a NG plant at 35 °C.

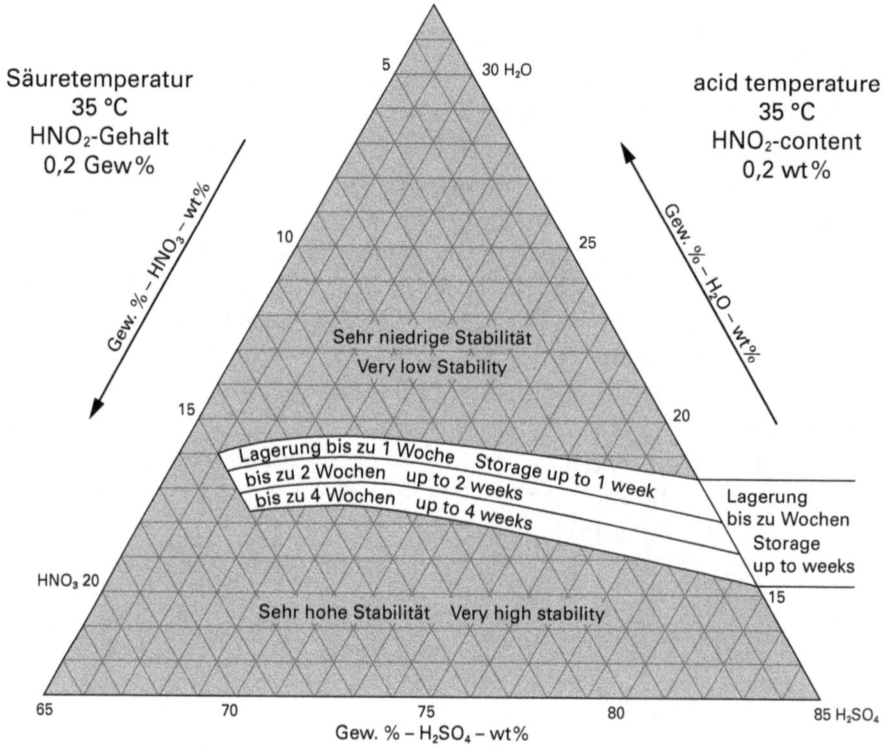

Fig. 10.9: Safety chart for spent acid handling from a NG/EGDN plant at 35 °C (from Biazzi).

In short, Fig. 10.9 can be summarized by the following eqs. (10.8–10.10) and statements:

$$RCH_2ONO_2 + NO^+ + 3\,H_2O \rightleftharpoons RCH_2ONO + NO_3^- + 2\,H_3O^+ \tag{10.8}$$

$$RCH_2ONO_2 + HNO_2 \rightleftharpoons RCH_2NO_2 + HNO_3 \tag{10.9}$$

$$NO^+ + H_3O^+ + SO_4^{2-} \rightleftharpoons NOHSO_4 + H_2O \tag{10.10}$$

- the formation of HNO_2 determines the extent of decomposition (eq. 10.9)
- an increase of the water content results in a decrease in stability (eq. 10.8)
- decomposition occurs much more slowly in the spent acid phase than in the oil phase (HNO_2 is stabilized through H_2SO_4) (eq. 10.10)

– Nitric acid acts not only as a stabilizer causing a lower degree of hydrolysis, but also influences the equilibrium (eq. 10.8)

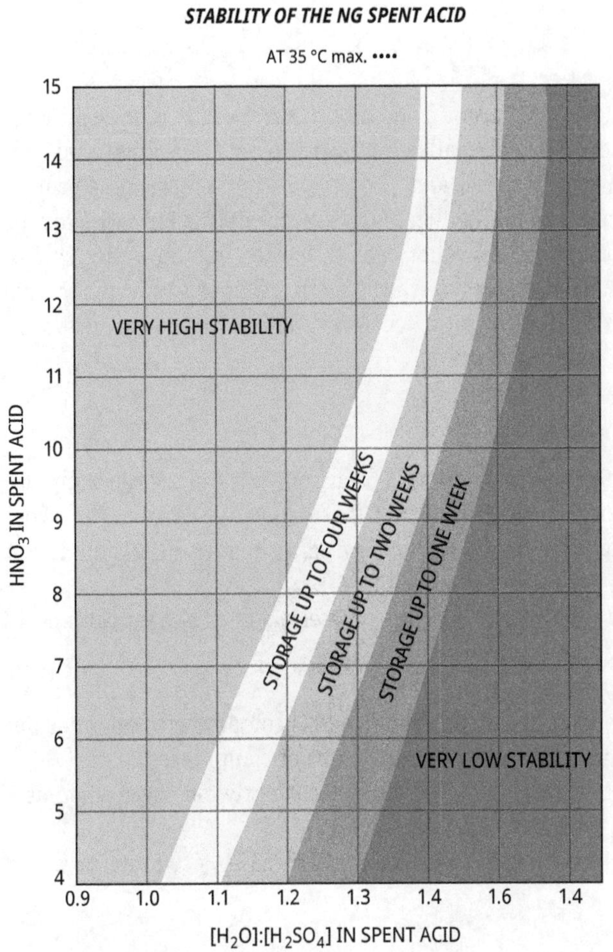

Fig. 10.10: Safety chart for spent acid handling from a NG plant at 35 °C (from BIAZZI).

Check your knowledge: See Chapter 14 for Study Questions.

Literature

F. Rigas, I. Sebos, D. Doulia, Journal of Loss Prevention in the Process Industries 11 (1998) 161–168.

11 Safe handling of energetic materials in the laboratory

11.1 General

A chemical research laboratory is different from an industrial production facility for many reasons. First of all, only much smaller quantities of energetic materials are handled in a chemical research laboratory and secondly, during research, the properties of new, energetic materials are not known and therefore particular care must be taken. One of the most important safety rules can be worked out from the "rule of thumb", which relates the distance D (in m), which offers a chance of survival, to the mass w (in kg) of the explosive. For a typical secondary explosive at large distances, the proportionality constant is approximately 2:

$$D = cw^{0.33} \approx 2\,w^{0.33}$$

It is important to note that this approximation is based only on the pressure/impulse of the shock wave and does not take into account the fragment formation (e.g. from confined charges, fume-cupboard front shields in the laboratory, reaction vessel glass).

In order to work safely with highly energetic materials in the chemical laboratory, the following rules must be obeyed:

- use smallest possible quantities
- keep maximum distance from the experiment (vessels containing compounds not to be transported by hand, but with well fitting tongs or clamp instead)
- mechanical manipulation is to be used if possible, particularly for larger amounts
- vessels never to be encased by hand (confinement)
- protective clothing is to be used (gloves, leather or Kevlar vest, ear protectors, face shield)

Additionally, one must bear in mind that although primary explosives are much more dangerous to handle – based on the probability of unexpected initiation due to their higher sensitivity – secondary explosives exhibit a much higher performance and if initiated, result in much larger damage than a primary explosive.

With regards to research into new energetic materials, a rule of thumb is that all synthetic attempts should first of all be carried out in 250 mg quantities while various sensitivity data is obtained (impact, friction and electrostatic sensitivity, thermal stability). Only when all of these values are known, can increasing the synthesis to 1 g be considered (only if the properties are suitable). Later, increasing it to 5 g and perhaps eventually to 10 g quantities could be possible depending on the sensitivity data ob-

https://doi.org/10.1515/9783111447087-011

tained. Further tests such as the long-term stability (DSC, long-term heat-flux measurements etc.) must then be carried out.

In general, the maximum overpressure resulting from a detonation decreases with increasing distance from the center of the detonation as shown in Fig. 11.1. The schematic representation of the overpressure over the time is shown in Fig. 11.2.

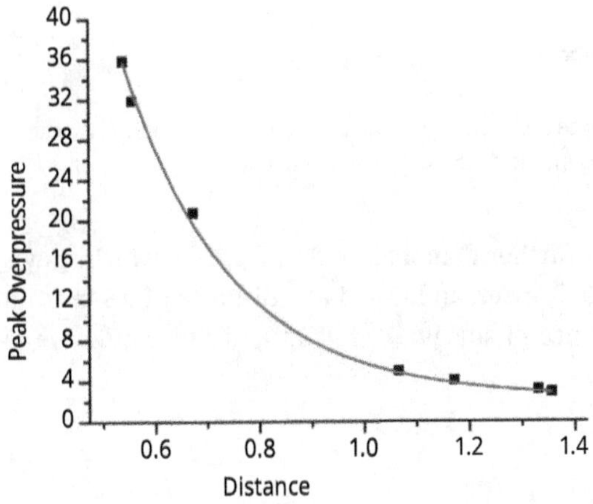

Fig. 11.1: Schematic plot of the overpressure vs. distance from a detonation (arbitrary units).

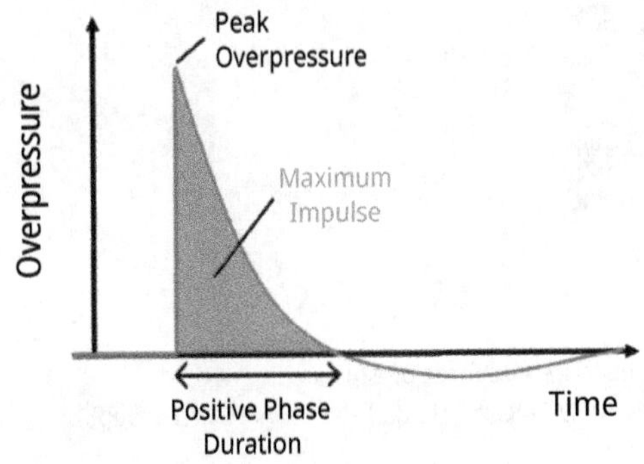

Fig. 11.2: Schematic plot of the overpressure vs. time for a detonation.
(Reproduced with kind permission of Dr. Aris Makris, MED-EMG, a brand of the SAFARILAND group).

This explains why at very short distances (Fig. 11.3) from a detonation ($d = 0$) the effects of fire and heat, overpressure, impact and fragments are all highly relevant, while with increasing distance from a detonation ($d > 0$), the danger to people decreases most for fire/heat, followed by overpressure, impact and fragmentation (Fig. 11.3). In other words, at large distances, fragment impact is the highest risk to people, and decreases with $1/d^2$. On the other hand, the impact is somewhat proportional to the acceleration by the blast wave.

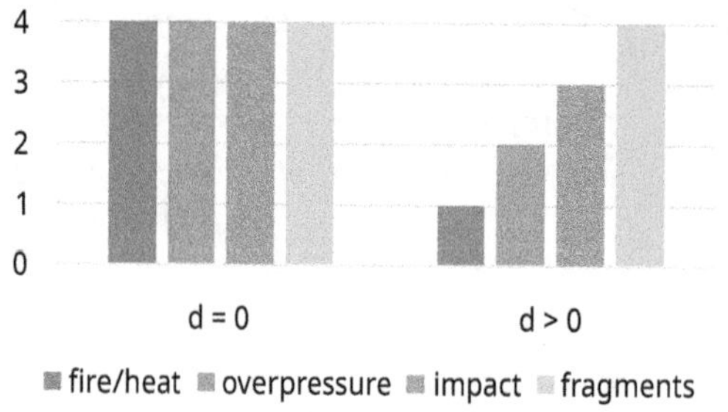

Fig. 11.3: Schematic presentation of the relevance (risk, in arbitrary units) to people caused by fire/heat, overpressure, impact and fragments depending on the distance from a detonation.

The importance of the distance rule is further demonstrated in Fig. 11.4, which shows that on detonation of a 2 kg mass of TNT, a person located at a distance of 0.8 m from the detonation has a less than 1% chance of survival, while at a distance of 2 m, the survivability increases to over 99%.

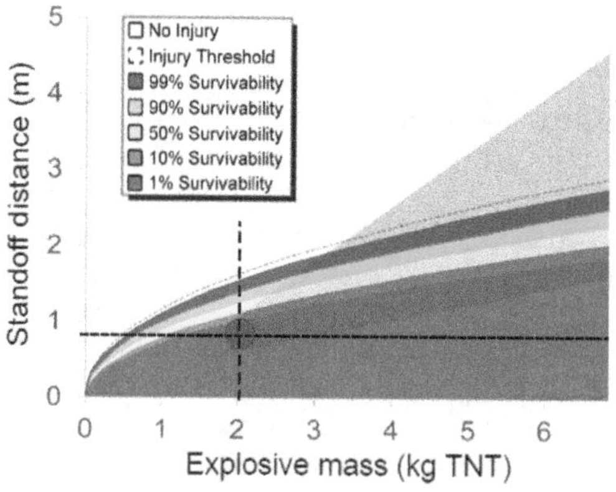

Fig. 11.4: Survivability as a function of the standoff distance and mass of explosive.
(Reproduced with kind permission of Dr. Aris Makris, MED-EMG, a brand of the SAFARILAND group).

The effect the mass has on the overpressure at a certain standoff distance is illustrated in Fig. 11.5. If a 50 g charge (e.g. an A-P mine) is detonated, at a standoff distance of 1 m the pressure has essentially already dropped to ambient. However, if a 10 kg charge (e.g. a 10 kg IED) is used, this is NOT the case.

It is also important to note that not only the distance from a detonation (with the same mass), but also the position of a person is relevant. Figure 11.6 shows that if a person's head is at the same distance from a detonation, at a lower position the head is further from a mine along the ground, the experienced acceleration is lower, and the survivability therefore higher.

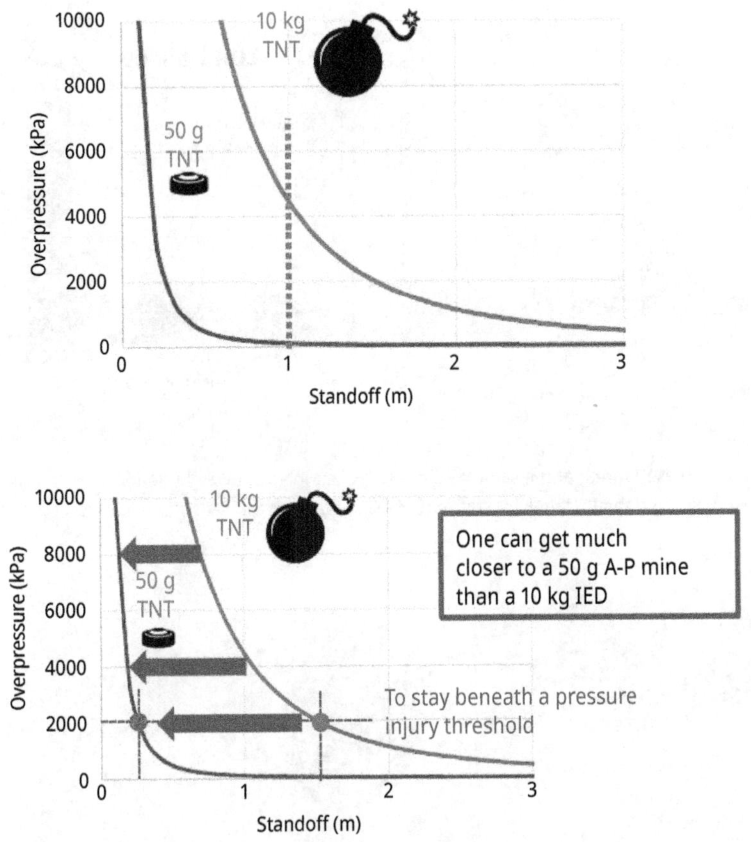

Fig. 11.5: Threat caused by different masses of explosives.
(Reproduced with kind permission of Dr. Aris Makris, MED-EMG, a brand of the SAFARILAND group).

Usually, an LDH helmet (lightweight demining helmet) gives much better protection than a helmet with goggles, but not as good protection as a combat helmet + visor (Fig. 11.7).

In terms of safety, the most important in the lab is:

- Prevention of explosive accidents
- Protection of people, property and environment
- Exchange of information by reporting incidents, near events and accidents

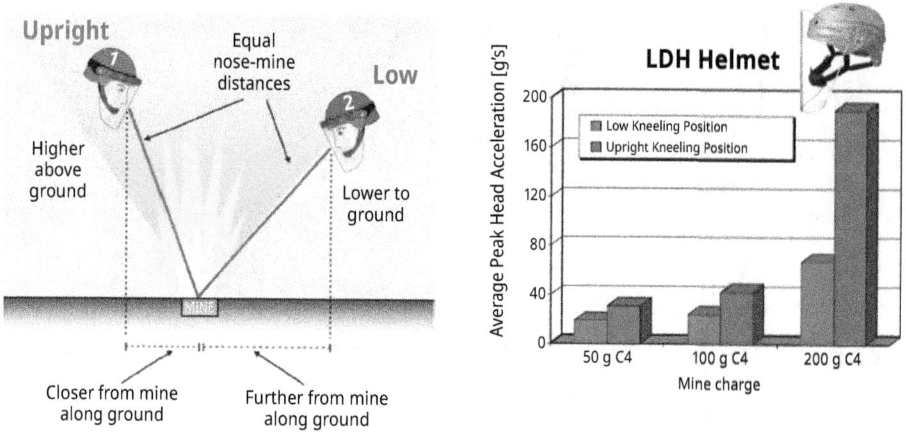

Fig. 11.6: Positions of increased threat at the same distance (LDH helmet = lightweight demining helmet). (Reproduced with kind permission of Dr. Aris Makris, MED-EMG, a brand of the SAFARILAND group).

Fig. 11.7: Combat helmet with goggles (left), LDH helmet, combat helmet + visor (right).

As a rule of thumb, in terms of the Basis of Safety (BOS) the (accidental) initiation of an explosive can be rationalized by the FISH rules which are illustrated the Fig. 11.8.

Furthermore, it is important to remember that for the prevention of accidents, a hierarchy of controls applies in which eliminating the hazard is the most effective measure, protection of the worker (or student in the lab) with personal protective equipment is least effective (Fig. 11.9).

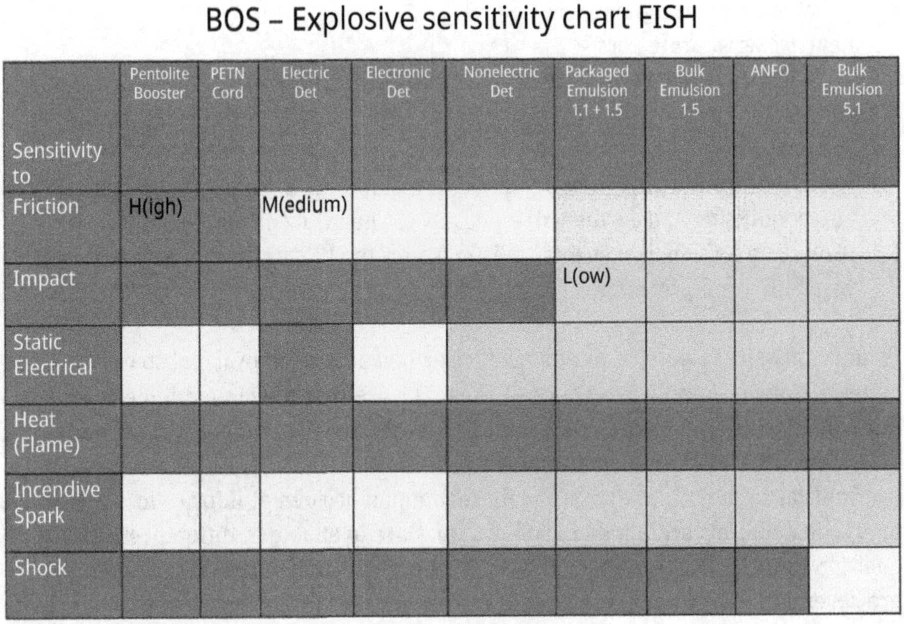
BOS – Explosive sensitivity chart FISH
Pentolite Booster
PETN Cord
Electric Det
Electronic Det
Nonelectric Det
Packaged Emulsion 1.1 + 1.5
Bulk Emulsion 1.5
ANFO
Bulk Emulsion 5.1
Sensitivity to
Friction
H(igh)
M(edium)
Impact
L(ow)
Static Electrical
Heat (Flame)
Incendive Spark
Shock

Fig. 11.8: Explosive sensitivity chart.

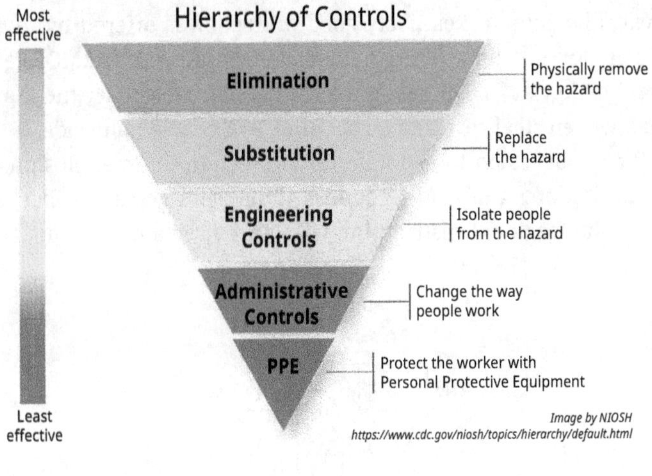

Fig. 11.9: Hierarchy of controls.

11.2 Protective equipment

In addition to the clothing that every chemist should wear in the lab (lab coat, safety glasses, enclosed shoes, non-flammable clothing), one can be protected further during work with small quantities (see above) of explosives. Possibilities include:

- Suitable protective gloves (see below)
- Kevlar® wrist protectors
- Full-visor face shield with polycarbonate shield (to be worn over normal safety glasses)
- Ear protectors
- Shoes with conducting soles against electrostatics (it is important that the floor is also conducting and earthened, especially for handling primary explosives)
- Protective leather jacket (made from 2 mm thick leather) or Kevlar protective vest with groin protector.

Finally, only spatulas made out of conducting plastic instead of metal spatulas should be used. Furthermore, it is important to make sure that the lab equipment, and certainly the vessels for storing the energetic materials are also made out of conducting plastic and stored on a conducting surface.

Injuries to hands are one of the most common accidental injuries in a lab where energetic materials are handled. Safety gloves are available in different materials. In addition to the protective effect, the fine-feel also has to be taken into account. There are several opinions with respect to the most suitable protective gloves. Some chemists prefer thick leather gloves (welding gloves), while others prefer Kevlar gloves. According to the DIN EN 388 (mechanical risks), protective gloves have been classified into three groups based on their cut resistance (I = low, II = medium, III = high). Gloves from class III should always be chosen. Relatively good protection is offered by the combination of two pairs of gloves: a tight-fitting pair made from Kevlar® ARMOR (a fiber developed by DuPont) or Kevlar® ES gloves and a second pair worn over the top made from Kevlar® with a woven steel core and a rubber or PVC coated palm side.

Different tests have been suggested to estimate the protection in an explosion. [66] For example, the explosion of 1 g of lead azide in a 10 ml glass vessel has been investigated. The results of the fragment distribution (size and direction dependent) are shown in Fig. 11.10.

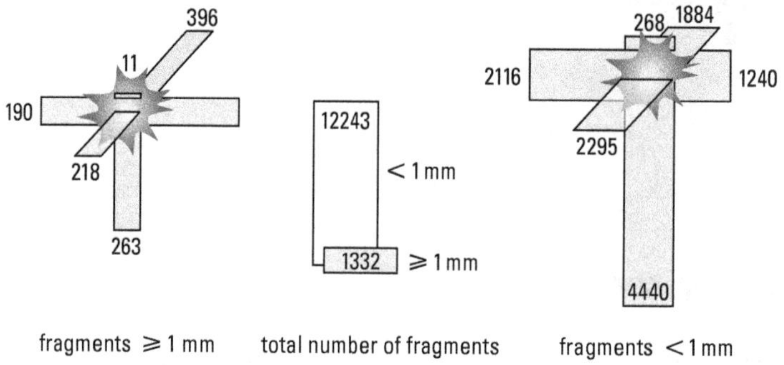

Fig. 11.10: Size and fragment distribution after an explosion with a quantity of 1 g $Pb(N_3)_2$ in a 10 ml glass vessel.

Figure 11.11 shows a typical experimental set-up to investigate the protective effect of different gloves. Pictures a and b show the experimental set-up with a hand (gelatine in one-way gloves) at a distance of 10 cm from a 10 ml glass vessel containing 1 g $Pb(N_3)_2$. Figure 11.11c shows the perforations in an unprotected glove and Fig. 11.11d shows the

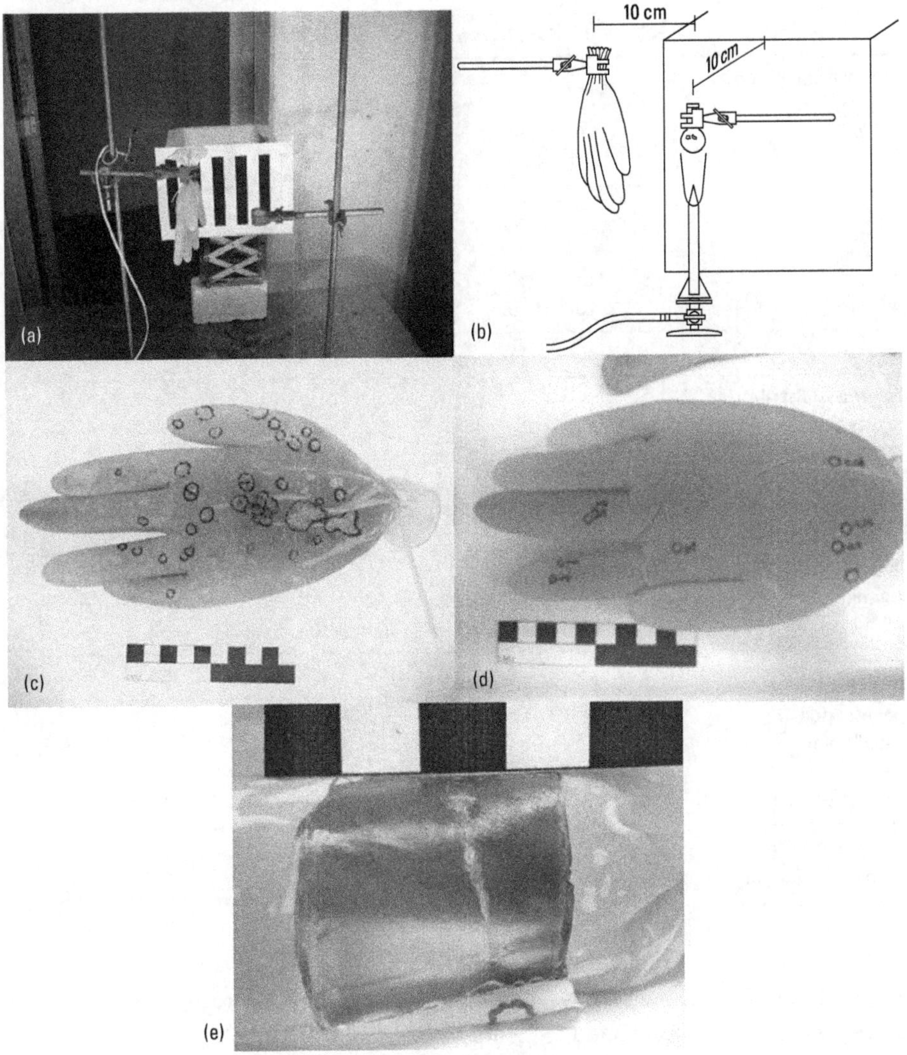

Fig. 11.11: Experimental set-up for investigating different protective gloves, scale unit 1 cm.

fragment impact on a hand protected with a Kevlar® glove. In Fig. 11.11e, the penetration depth of a glass splinter is enlarged, despite the protection offered by the Kevlar glove (from Fig. 11.11d, below right).

Tab. 11.1: Number of splinters and penetration depth using different gloves (exptl. set-up see Figs. 11.11a and 11.11b).

	unprotected hand	Kevlar® glove		Kevlar® glove with steel core		Kevlar® glove with steel reinforcement		leather welding gloves	
	Exp. I	Exp. I	Exp. II	Exp. I	Exp. II	Exp. I	Exp. II	Exp. I	Exp. II
number of splinters									
in glove									
≥ 1 mm	0	10	13	17	18	10	14	5	4
< 1 mm	4	45	12	38	16	8	21	18	13
total	4	55	25	55	32	18	35	23	17
in hand									
≥ 1 mm	30[a]	14[b]	13	3	2	0	1	1[c]	0
< 1 mm	20	0	5	3	0	2	1	1	0
total	50	14	18	6	2	2	2	2	0
fragment distribution									
in glove									
≥ 1 mm	0%	5%	9%	7%	9%	5%	6%	3%	2%
< 1 mm	0%	4%	1%	3%	1%	1%	2%	1%	1%
total	0%	2%	2%	4%	2%	1%	2%	1%	2%
in hand									
≥ 1 mm	13%	9%	9%	1%	1%	0%	0%	1%	0%
< 1 mm	1%	0%	0%	0%	0%	0%	0%	0%	0%
total	3%	1%	1%	0%	0%	0%	0%	0%	0%
penetration									
depth (mm)									
deepest	15.0	7.0	8.6	5.0	6.0	1.0	6.0	5.0	–
average value	4.9	2.5	2.8	1.8	3.8	0.8	4.5	4.0	–

[a]5 splinters > 5 mm
[b]1 splinter penetrated the whole palm
[c]1 splinter > 5 mm

Table 11.1 shows the penetration depths of glass splinters with different types of gloves and an unprotected hand for comparison.

11.3 Laboratory equipment

In addition to the personal protective equipment discussed above, one can/must equip the laboratory according to the current safety standards. Apart from the standards that every chemical laboratory must correspond to anyway, particular attention must be paid to the following:

- conducting floor covering which is earthened
- humidity over 60%, better 70% (using a humidifier, in particular during winter in rooms with central heating)
- short-term "intermediate storage" of small quantities of energetic materials in appropriate protective containers
- storing of larger quantities of energetic materials in a specially designated storeroom and not in the laboratory
- separate fridges/cold rooms for oxidizers (HNO_3, NTO, H_2O_2, MON) and fuels (MMH, UDMH)
- mechanical manipulators for the handling of larger quantities of energetic materials
- autoclave rooms or bunkers for performing different tests (Koenen test, detonation velocity, . . .).

Check your knowledge: See Chapter 14 for Study Questions.

12 Energetic materials of the future

Since the introduction of nitroglycerin as an explosive by Alfred Nobel in 1867 in the form of Dynamite, large advances have been made in the performance and reduction of sensitivities of highly energetic materials (Fig. 12.1). However, most of the formulations currently used are over 50 years old and do not fulfill all of today's requirements, in particular with regard to their performance, collateral damage, insensitivity, toxicity, compatibility with the environment and use in special operations.

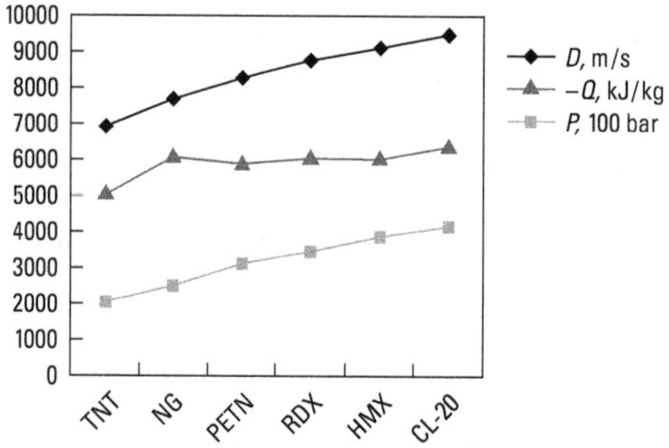

Fig. 12.1: Performance of explosives (*Q*, heat of explosion; *D*, detonation velocity; $P_{C\text{-}J}$ detonation pressure).

Energetic materials are used mainly in explosives or propellant formulations. As we have already discussed, there are different parameters that characterize the performance of explosives and propellants and which are important in the development of new energetic materials. These include high densities (ρ, more important for explosives than for propellant charges), good oxygen balance (Ω) and high detonation and combustion temperatures (not for propellant charges). However, substances which are suitable for use as secondary explosives are not always suitable for use as propellants and vice versa. While the performance of explosives is particularly influenced by the heat of explosion ($Q_{C\text{-}J}$), the detonation pressure ($p_{C\text{-}J}$) and the detonation velocity (D), good rocket propellants possess a large specific impulse (I_{sp}) and thrust (F). For propellant charges, in addition to a high specific impulse, a high force (f), a low combustion temperature and the highest possible N_2 / CO ratio in the products of the reaction are desirable.

When we consider the heat of explosion ($Q_{C\text{-}J}$), the detonation pressure ($p_{C\text{-}J}$) and the detonation velocity (D) as a measure that determines the performance of explosives, then Fig. 12.1 shows that large advances have been made since the development of nitroglycerin.

https://doi.org/10.1515/9783111447087-012

Despite these advances, it appears to be difficult to make significant improvements to the performance with conventional C/H/N/O explosives based on nitro and nitrato compounds. This is partly due to the fact that classic explosives suffer from the following three limitations (Tab. 12.1):
1. they are not nitrogen-rich enough (N-rich > 60% N),
2. the oxygen balance is often too negative,
3. to reach a good oxygen balance, different explosives have to be mixed.

Tab. 12.1: Nitrogen content and oxygen balance of some conventional explosives.

	nitrogen content / %	oxygen balance / %
TNT	18.5	−73.9
PETN	17.7	−10.1
RDX	37.8	−21.6
HMX	37.8	−21.6

More recent research on nitrogen-rich molecules (N > 60%) has shown, that these compounds often have a higher energy content because of a much more positive enthalpy of formation in comparison with their C analogues. The first generation of nitrogen-rich compounds such as hydrazinium azotetrazolate (HZT) and triaminoguanidinium azotetrazolate (TAGzT) (Fig. 12.2) have already generated interest as components of propellant charges because of their low erosion properties. However, due to their poor oxygen balance, they are unsuitable for use as high-performance explosives (Tab. 12.2).

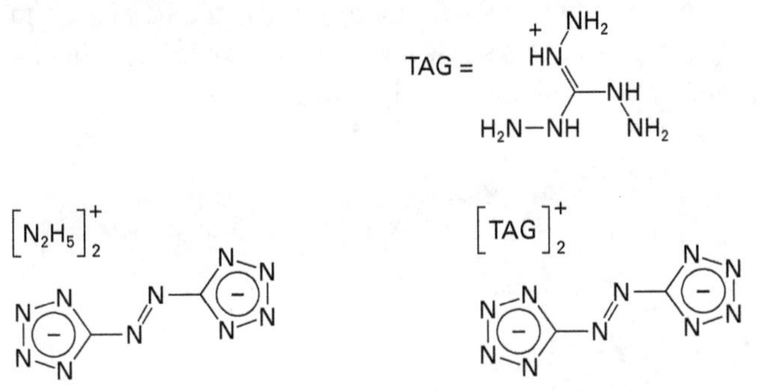

HZT TAGZT

Fig. 12.2: Structures of hydrazinium azotetrazolate (HZT) and triaminoguanidinium azotetrazolate (TAGzT).

This could change with the introduction of the second generation of nitrogen-rich explosives which possess strongly oxidizing groups, and a (nearly) neutral oxygen bal-

Tab. 12.2: Nitrogen content and oxygen balance of first and second generation N-rich compounds.

	nitrogen content / %	oxygen balance / %
1st generation		
HZT	85	-63
TAGZT	82	-73
2nd generation		
TAG-DN	57	-18
HAT-DN	58	0

ance. Examples of these are the compounds triaminoguanidinium dinitramide (TAG-DN) and aminotetrazolium dinitramide (HAT-DN) (Fig. 12.3, Tab. 10.3) [64, 65].

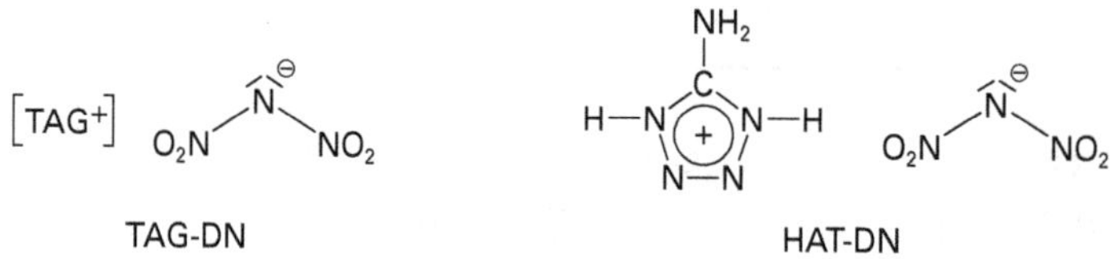

Fig. 12.3: Structures of triaminoguanidinium dinitramide (TAG-DN) and aminotetrazolium dinitramide (HAT-DN).

TAG-DN and HAT-DN are therefore potentially good candidates for use as explosives. But compounds with good oxygen balances are also suitable as components in rocket propellants. An increase in the specific impulse I^*_{sp} of only 20 s should be enough to allow almost a doubling of the payload. Figure 12.4 shows the calculated specific impulse (isobaric, $p = 70$ bar) for AP /Al (AP /Al) (70 : 30), $HAT\text{-}NO_3$, HAT-DN and TAG-DN.

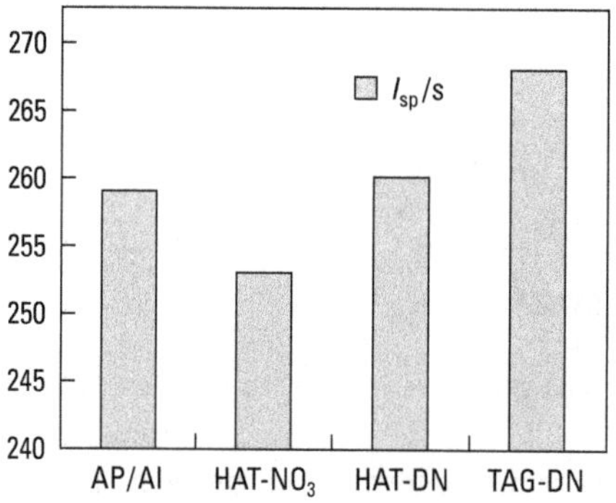

Fig. 12.4: Calculated specific impulses (I^*_{sp} / s) for AP /Al (70 : 30), $HAT\text{-}NO_3$, HAT-DN and TAG-DN (isobaric combustion at p = 70 bar).

The calculated detonation velocities for the second generation of nitrogen-rich compounds (which is one of the criteria for the performance of explosives) are not only better than those of the first generation, but in the case of HAT-DN it is even better than the values for RDX and HMX (Fig. 12.5).

Possible directions for future research are (see also Tab. 12.3):

1. metallized nitrogen-rich compounds with positive oxygen balances
2. metastable poly-nitrogen element modifications
3. organometallic compounds with nitrogen-rich or poly-N-ligands
4. boron-nitrogen compounds with good oxygen balances
5. substitution of oxygen (in $—NO_2$ groups) by fluorine ($—NF_2$ groups)
6. new, better and perchlorate-free oxidizers
7. new synthetic, energy-rich polymers as nitrocellulose replacements
8. more effective, F, Cl and peroxide containing agent-defeat substances
9. nano-materials (carbon nanotubes)
10. N_4, N_8 fixation in SWCNTs
11. new high-energy oxidizers (HEDO) (Fig. 12.6)

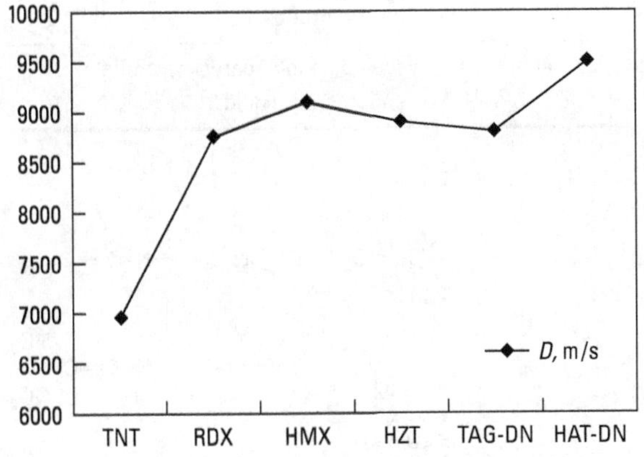

Fig. 12.5: Detonation velocities (*D*) of classic CHNO and nitrogen-rich explosives.

Tab. 12.3: Future areas of research.

need	possible solution	research
downsizing of warheads: 4 → 1	increase of ΔU_{ex}: today: 5–6 kJ g^{-1} future: 10–20 kJ g^{-1}; up to 75% of the mass of a warhead gets “wasted” (non-reactive structure material)	super high explosives, novel HEDMs: poly-N, poly-N_2O; reactive fragments: thermites, Al-PTFE composites, Al-Zn-Zr, Al-W and Al-U alloys

Tab. 12.3 (continued)

need	possible solution	research
UxV's	increase of ΔU_{ex}: today: 5–6 kJ g^{-1} future: 10–20 kJ g^{-1}; amalgamation of warhead and weapon into one no-return system	super high explosives, novel HEDMs: poly-N, poly-N_2O; reactive structures: novel materials
countermeasures against chemical WMDs	ADWs	peroxides, ozone, Cl-containing compounds, hypochlorites
countermeasures against biological WMDs	ADWs	peroxides, ozone, Cl-containing compounds, hypochlorites, F-containing compounds (HF formation)
more efficient fuels	nano	research concerning aging problems, coating with strongly oxidizing groups (e.g. pernitro alcohols)
perchlorate-free oxidizers	new HEDOs	high-nitro compounds, compounds with peroxo groups
specialized GWT missions	thermobaric systems; Fuel Air Explosives (FAE)	coated nano-Al particles; strongly endothermic liquids

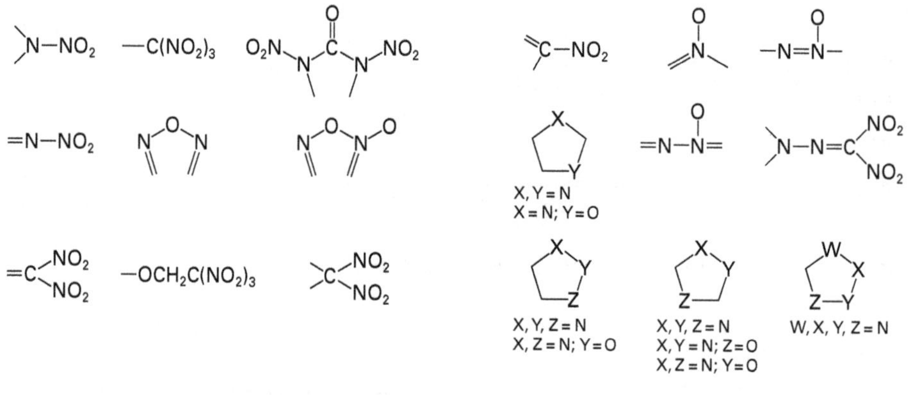

Fig. 12.6: Energetic synthons for "advanced oxidizers".

In this context it is worthwhile to stress that for certain applications various detonation parameters are of particular interest. For example, if the blast effect is the most important property, the detonation pressure ($p_{C\text{-}J}$) is one of the most relevant parameters. However, for shaped charge applications, high detonation velocities are required.

Since lethal weapons predominantly rely on overpressure and blast effect, new high-performance secondary explosives can help to down-size the weapon system. On the other hand, if weapon systems rely on the formation of fragments (fragmenting warheads), new, more energetic high explosives can only marginally help to improve the performance. In this case, chemically reactive fragments could be an answer, since in conventional warheads up to 75% of the mass corresponds to the non-reactive casing (steel casing). Research on chemically stable, but highly reactive materials could help to solve this problem. Such reactive structural materials (RSM) might be thermites, Al-PTFE composite materials or Al-Zn-Zr, Al-W or Al-U based alloys. Such RSMs should possess the following properties:

- High density
- High hardness
- Fast, highly exothermic reaction on hitting the target

A new system which might be extremely interesting in this context is the intermetallic phase ZrW_2. ZrW_2 is a high-density, very hard intermetallic compound that reacts/burns in a highly exothermic reaction with air at high temperatures. This intermetallic phase should provide a very suitable reactive material for warhead applications.

ZrW_2 adopts the $MgCu_2$ structure in the solid state, is very hard and also densely packed. One can also expect that this intermetallic compound should burn at very high temperature in a strongly exothermic reaction due to the zirconium content of 33%. The heat of formation of ZrO_2 (–1,100 kJ/mol) is as high as that of CeO_2 (–1,089 kJ/mol). For Cermisch metals it is very well-known that burning occurs with a strong white flame at high temperature and is very exothermic.

Metallized formulations are also suitable for special missions (*e.g.* caves, tunnels e.g. in GWT), since in enclosed spaces, the oxygen in the air needed for breathing for survival is removed.

DFOX, which is a new secondary explosive (Fig. 12.7) and possesses a positive oxygen balance (Ω_{CO} = +52%, Ω_{CO_2} = +41%), is unique in so far that, when aluminized, not only do the heat of detonation (Q_{ex}) and the detonation temperature (T_{ex}) increase significantly, but also the detonation velocity (VoD) and detonation pressure ($p_{C\text{-}J}$) increase in the DFOX/Al 80/20 formulation, whereas RDX shows "normal" behavior with decreasing VoD and $p_{C\text{-}J}$ (Tab. 12.4). The DFOX/Al 70/30 mixture performs essentially equal to the RDX/Al 70/30 mixture. However, it possesses a detonation temperature 1,000 K higher and heat of detonation 500 kJ kg^{-1} higher than the corresponding aluminized RDX formulation.

A further approach for the synthesis of high performance/low sensitivity secondary explosives which has recently been extensively investigated is the use of fused nitrogen-rich heterocycles. Fused heterocycles consist of two (or more) heterocycles which share two adjacent ring atoms and the bond between these two atoms. A well-known, non-energetic example of a fused-ring compound is naphthalene (Fig. 12.8). Well-known examples of energetic materials based on fused heterocyclic systems do

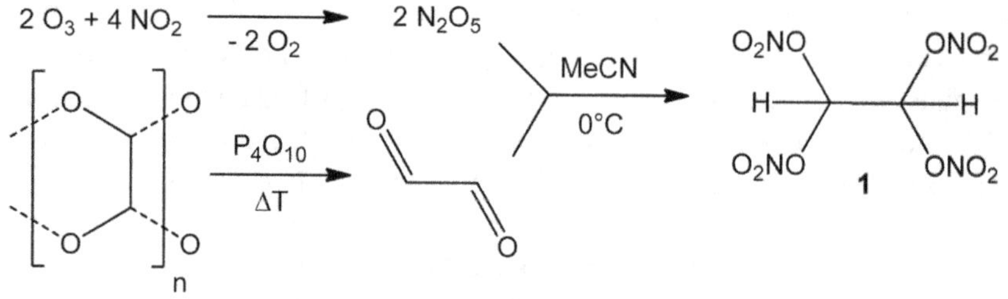

Fig. 12.7: Synthesis of DFOX (**1**) from glyoxal and N_2O_5. DFOX has a higher oxygen content than NO_2, Li_2O_2, $C(NO_2)_4$, ADN, $KClO_4$, NH_4ClO_4 and tetranitroacetimidic acid.

Tab. 12.4: Detonation parameters for DFOX and RDX in aluminized formulations.

	DFOX	DFOX/Al (80/20)	DFOX/Al (70/30)	RDX	RDX/Al (80/20)	RDX/Al (70/30)
VoD / m s^{-1}	7,678	7,798	7,156	8,919	8,201	7,723
p_{C-J} / GPa	23.6	28.3	24.2	34.3	31.7	26.6
T_{ex} / K	2,375	5,358	6,345	3,749	4,978	5,391
Q_{ex} / kJ kg^{-1}	-2,408	-7,784	−10,142	-5,742	-8,259	-9,559

already exist such as DINGU, SORGUYL and TACOT (Fig. 12.8). However, until recently, relatively few were thoroughly investigated for use as potential secondary explosives.

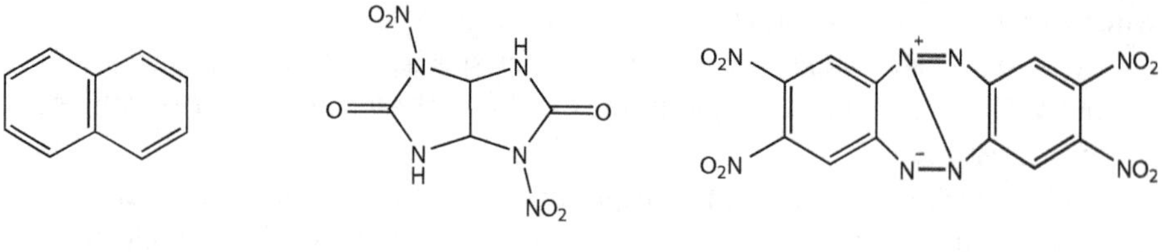

Fig. 12.8: Naphthalene is a simple example of a fused-ring compound. In fused-ring compounds, a pair of rings share two neighboring ring atoms and the corresponding bond (left). Two well-known examples of energetic compounds which contain fused rings: 1,4-dinitroglycourile (DINGU) (middle) and TACOT (right).

By using nitrogen-rich rings, fused-ring systems can be prepared which are planar conjugated structures with high nitrogen contents. The fused heterocycle system then acts as a backbone to which energetic functional groups (explosophores) can be attached, to (hopefully) tailor the properties of the resulting energetic material. In addition to designing a material with a high heat of formation and high energetic performance, careful construction of the hydrogen bond network and noncovalent interactions between layers in the crystal can result in higher thermal stabilities, lower sensitivity to mechanical stimuli and higher densities. The stability of these polycyclic scaffolds arises predominantly from hydrogen bonding, planar geometries and π-π stacking interactions. This is a considerable change from the traditional approaches used for synthesizing energetic materials which were generally based on a cyclic or acyclic carbon backbone or carbon cage, to which energetic groups were attached. Prominent examples of such compounds are TNT, PETN and ONC, respectively [2].

Recently, a large number of energetic compounds based on fused heterocycles have been prepared as either neutral compounds, or as energetic salts and have been summarized in a recent review [1]. The progress in this area has been swift and a number of different combinations of heterocyclic rings have been used to form the fused heterocyclic ring scaffolds. Important to note in these compounds are the ring sizes involved (*i.e.* five membered or six membered rings etc.), as well as the total number of rings that make up the fused-ring scaffold. In particular, energetic compounds based on [5,5]-, [5,6]- or [6,6]-bicycles, [5,6,5]- or [5,7,5]-tricycles, or – less commonly – tetracycles have been investigated.

One of the drawbacks of such compounds is that they often require a synthetic route involving a large number of steps. For example, 3,6-dinitropyrazolo[4,3-*c*]pyrazole shown above in Fig. 12.9 requires a five-step synthetic procedure, and TTTO (Fig. 12.9) involves an eight-step synthetic process. Another consideration that is important in the synthesis of energetic fused heterocycles is whether explosive compounds must be isolated at different stages in the preparation. Despite these issues, energetic compounds based on fused heterocyclic rings have been prepared showing promising properties for use as secondary explosives. Some examples are given in Tab. 12.5 [1]. However, as with many classes of secondary explosives, it has also proven very difficult with these secondary explosives to obtain the ideal combination of high performance, high thermal stability and low sensitivity towards impact and friction. Table 12.5 shows that many high-performing compounds (in terms of energetic performance) are sadly not viable for use as secondary explosives in the field, due to a thermal stability that is too low, or sensitivity to external stimuli that is too high.

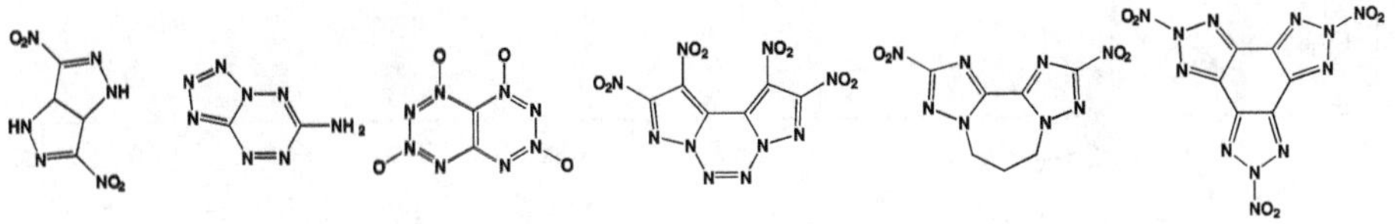

Fig. 12.9: Examples of fused-ring energetic materials. The fused-ring systems which act as scaffolds can be based on (from left to right) [5,5]-, [5,6]- or [6,6]-bicycles, [5,6,5]- or [5,7,5]-tricycles, or tetracycles: 3,6-dinitropyrazolo[4,3-c]pyrazole, tetrazolo[1,5-b][1,2,4,5]-tetrazine-6-amine, TTTO, 1,2,9,10-tetranitro-dipyrazolo[1,5-d:5′,1′-f][1,2,3,4]tetrazine, 2,10-dinitro-6,7-dihydro-5*H*-bis([1,2,4]triazolo)[1,5a:50,10-c][1,4] diazepine and 2,5,8-trinitro-benzo[1,2-d:3,4-d′:5,6-d″]tris[1,2,3]triazole[1].

Tab. 12.5: Examples of fused ring energetic materials which have been recently prepared and show good energetic performance. Unfortunately, the sensitivity data or thermal stability is often unsatisfactory [1].

Compound	ρ/g cm^{-3}	VoD/ms^{-1}	P_{CJ}/kbar	$T_{dec.}$ (°C)	IS (J)	FS (N)
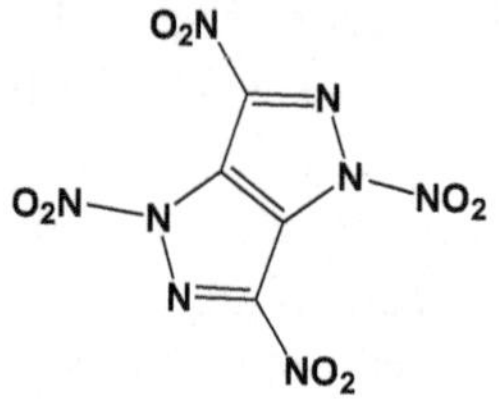	1.95	9,460	409	145	3	20
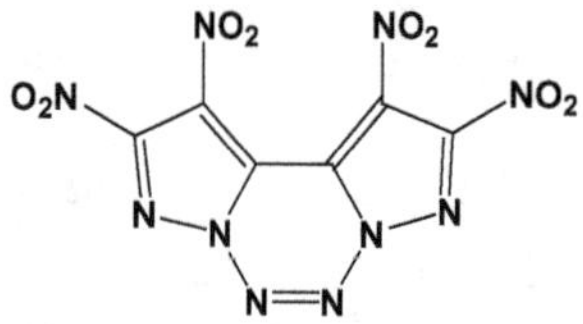	1.96	9,631	440	233	10	240
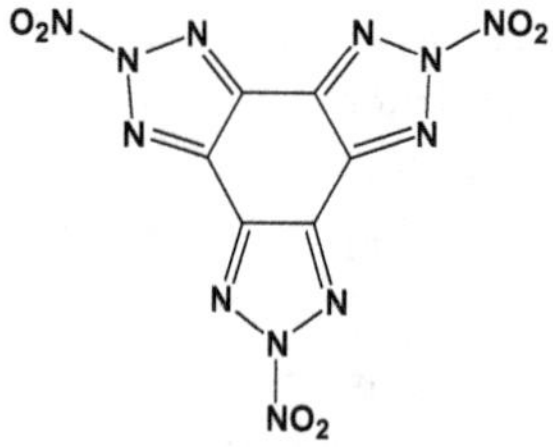	1.82	8,376	312	38	3	
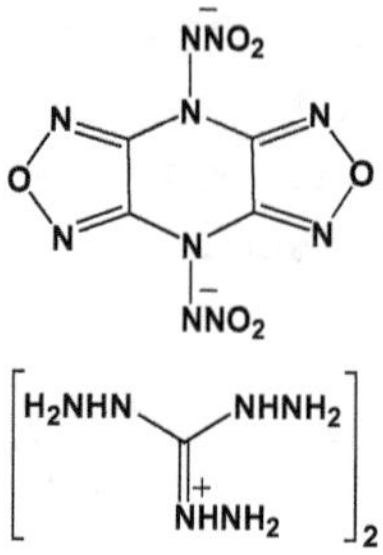	1.71	9,413	330	207	3	54
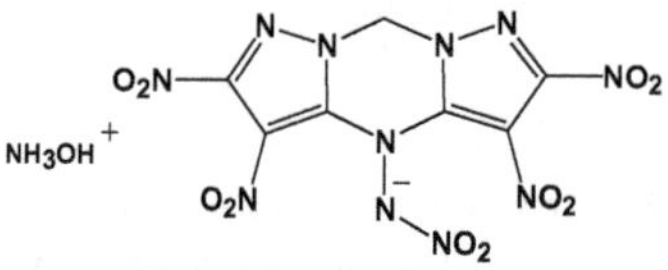	1.87	9,034	371	138	10	80

Tab. 12.5 (continued)

Compound	ρ/g cm^{-3}	VoD/ms^{-1}	P_{CJ}/kbar	$T_{dec.}$ (°C)	IS (J)	FS (N)
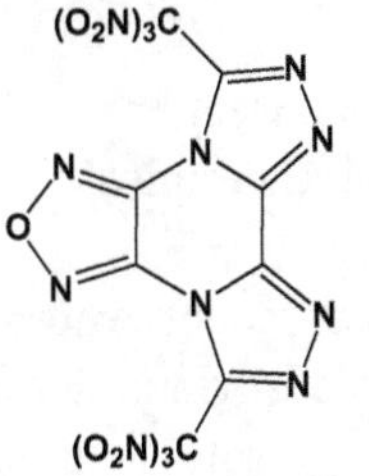	1.96	8,838	360	128	3	100
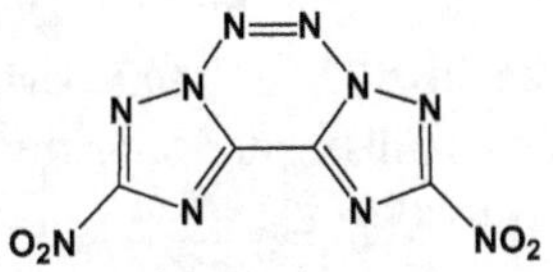	1.91	9,400	380	138	5.3	92
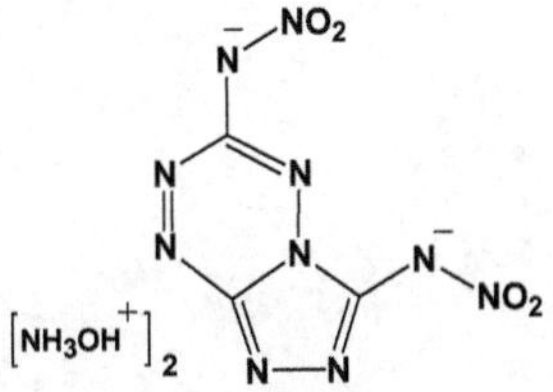	1.92	9,712	429	154	25	> 360
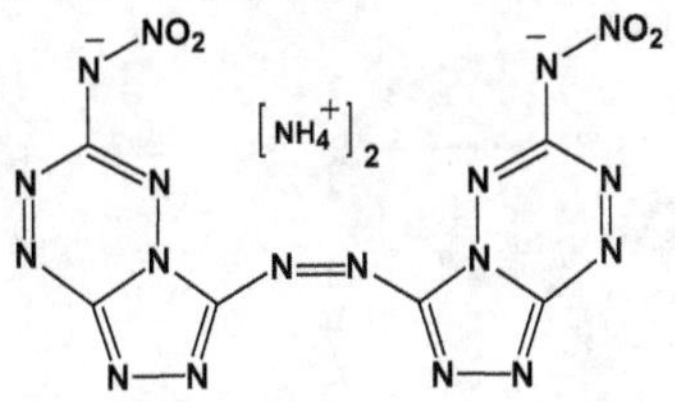	1.82	9,566	388	240	10	> 40

Literature

[1] H. Gao, Q. Zhang, J. M. Shreeve, *J. Mater. Chem. A*, **2020**, *8*, 4193–4216.

[2] T. M. Klapötke, *Energetic Materials Encyclopedia, Volumes 1–3*, 2nd edn., de Gruyter, Berlin/Boston, **2021**.

Check your knowledge: See Chapter 14 for Study Questions.

13 Related topics

13.1 Thermobaric weapons

There are two related weapon systems in this category: thermobaric weapons and Fuel Air Systems, also known in German as aerosol bombs (**FAE, Fuel Air Explosive**). Both function according to the same principle.

An **aerosol bomb**, Fuel Air Explosive (FAE) or Fuel Air Bomb, is a weapon whose effect on the detonation of an aerosol or substance distributed as a dust cloud does not depend on an oxidizer being present in the molecule. A FAE bomb consists of a container with a flammable substance (e.g. ethylene oxide). Two explosive charges are used as detonators: the first explosion causes the fuel to be distributed as fine particles into the air as an aerosol. Following this, microseconds – milliseconds later the aerosol is detonated, which results in the release of high pressure (Fig. 13.1).

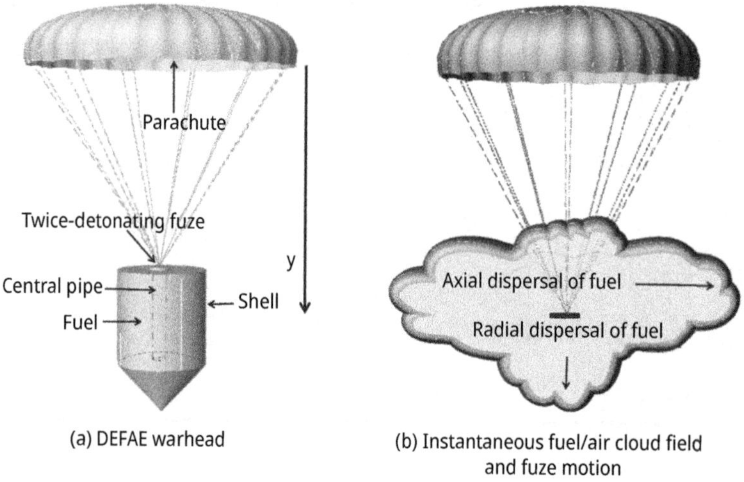

Fig. 13.1: Schematic diagram of a DEFAE (double-event FAE) weapon system and the instantaneous fuel/air cloud formation (fuel: Al/propylene oxide = 50:50; central pipe contains high explosive) [reproduced from C. Ye, Q. Zhang, *Prop. Expl. Pyrotech.*, **2021**, *47*, e202100202: doi.org/10.1002/prep.202100202, with friendly permission of Wiley].

The pressure wave which results from the explosive deflagration is considerably weaker than that of a comparable explosive such as RDX. However, the explosive deflagration occurs almost simultaneously in a large volume of 10–50 m diameter. The fuel can penetrate caves, tunnels or bunkers, which makes this weapon effective when used on targets in the global war against terror (GWT), unlike conventional explosive charges which only have a limited effect due to their poor pressure effect.

https://doi.org/10.1515/9783111447087-013

Furthermore, aerosol bombs have a considerably larger heat effect than conventional explosive charges. Therefore, these bombs are more effective at killing people and in the destruction of other soft targets such as unarmed vehicles and other targets.

The explosion removes the oxygen from the air, because the explosive composition does not contain its own oxidizer and uses the oxygen present in the air instead. This is however, not the deadly mechanism of the aerosol bombs. Much more significant is death by suffocation which is observed as the result of an aerosol bomb. The reason for this is not the lack of oxygen, but the damage which is done to the lungs through a so-called barotrauma. The negative pressure phase which occurs after the positive pressure phase (see Ch. 2.2) causes an expansion of the air in the lungs; resulting in damage. The properties of an aerosol bomb – long, relatively flat pressure wave with a corresponding distinct "partial vacuum", as well as the use of atmospheric oxygen are beneficial effects.

For **thermobaric weapons** (also known as **EBX** = enhanced blast explosives, see below), in addition to a conventional explosion, a flammable substance (usually Al), with little or no oxidizer (e.g. oxygen) distributed in the air detonates immediately as a result of the explosion. This post-detonative reaction ("fireball" of Al with air) usually occurs within μs after the detonation of the high explosive. This causes the effect of the original explosion to be magnified, which results in an even larger heat and pressure effect. A nice diagram explaining how thermobaric systems work has been published by Sandia (Fig. 13.1a). It is interesting to note that when an EBX is detonated under anaerobic (experimental) conditions, the Al does not react and therefore does not contribute to the reaction enthalpy ($\Delta H_{comb.}$ or $\Delta H_{expl.}$).

The schematic construction of a thermobaric system is shown in Fig. 13.1b.

The initial pressure wave and the subsequent underpressure, follow a phase in which the underpressure resulting from the explosion causes a flux of surrounding air to the centre of the explosion. The released, non-exploded, burning substance is as a result of the underpressure sucked into where the explosion occurred, and therefore enters all non air-tight objects (caves, bunker, tunnels) and burns these. Suffocation and internal damage can result to both humans and animals that are located out of the immediate vicinity: e.g. in deeper tunnels as a result of the pressure wave, the burning of oxygen, or the subsequent underpressure.

As discussed above, thermobaric weapons contain monopropellants or secondary explosives and energetic particles. Boron, aluminum, silicon, titanium, magnesium, zirconium and carbon can be considered to be energetic particles. The main advantage of thermobaric systems is that they release large quantities of heat and pressure, often in amounts larger than for only secondary explosives.

In thermobaric weapons, highly aluminized secondary explosives can be used instead of monopropellants. For example, RDX in combination with a binder and a large quantity of aluminum (fuel-rich) can be used. Research is currently being undertaken to investigate energetic polymers which could possibly be metallized for potential applications.

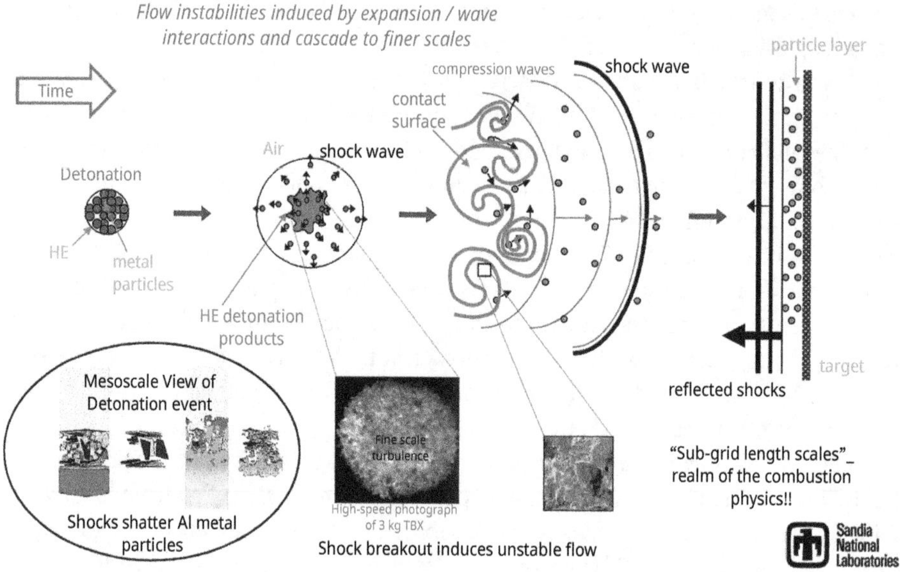

Fig. 13.1a: How thermobaric systems work. (With kind permission of Cooper, Kaneshige, Pahl, Baer, Schmitt, Sandia Natl. Labs, SAND2007–3791C).

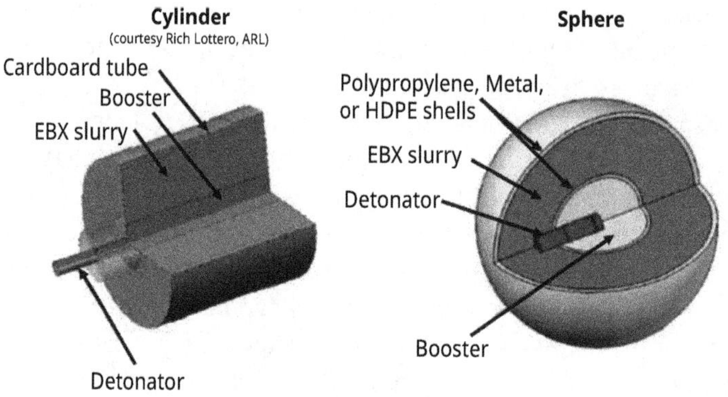

Fig. 13.1b: General construction of a thermobaric system (With kind permission of Cooper, Kaneshige, Pahl, Baer, Schmitt, Sandia Natl. Labs, SAND2007–3791C).

In thermobaric weapons, first anaerobic detonation occurs within the microsecond time-frame, followed by a post-detonative combustion which is also anaerobic and occurs in the hundredth of a microsecond time-frame (Fig. 13.1c). Only then, the post detonative burning occurs, which lasts several milliseconds and generates strong heat radiation even at small shock-wave pressures (approx. 10 bar).

The following relationship gives a conservative approximation for the distance from the detonation of a thermobaric weapon, which can still experience large damage as a

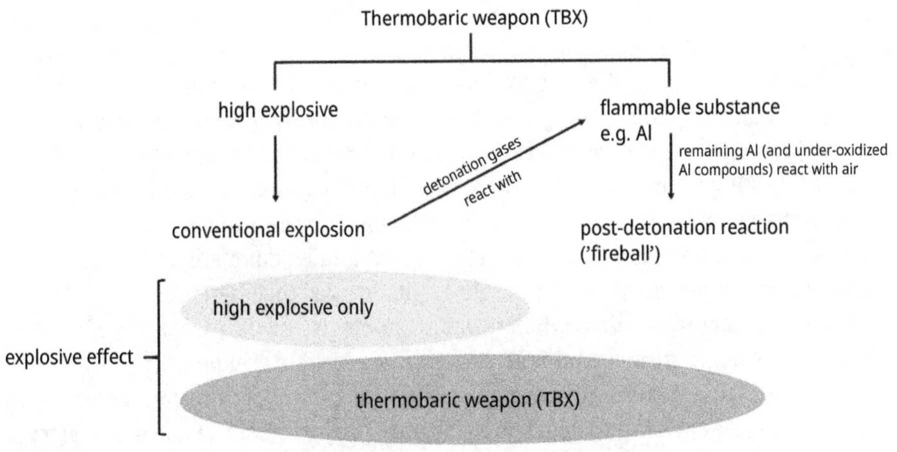

Fig. 13.1c: Scheme illustrating important aspects of a TBX.

result of the heat and pressure-wave generated. D is the distance (in meters), C is a constant (0.15 for an assumed overpressure of 70 mbar), n is the so-called yield factor from the burning (normally 0.1), and E is the energy (in J) released on the explosion and combustion.

$$D = C(\mathrm{n}E)^{0.33}$$

If the energy E is replaced by "$m\,Q$" in which m is the mass (in kg) of energetic material used and Q is the heat of combustion (in J kg^{-1}), the critical distance can be represented by:

$$D = 0.15\mathrm{kg}^{-0.33}\mathrm{s}^{0.66}\mathrm{m}^{0.33}(0.1m\,Q)^{0.33}$$

Thermobaric weapons can be used as bombs or as shoulder-launch mobile systems. There are two fields of operation for which the thermobaric weapons are superior to other systems:

(i) Areas of conflict involving caves, tunnels or other difficult to access areas and deep, confined targets (HDBTs, hard and/or deeply buried targets).
(ii) Combat and destruction of chemical and biological (CB) weapons through the strong heat effect (see also Ch. 13.2).

Shock waves generated by the detonation of thermobaric weapons have a much longer duration than those which are generated by conventional high explosives. In addition, they also show a larger lethal radius. Under confinement, thermobaric detonations cause a series of reflected shock waves which maintain the fireball, and can extend its duration to between 10 and 50 ms, as exothermic recombination reactions occur. Further damage can result on cooling of the gases since this causes the pressure to drop sharply, which results in a partial under-pressure which is powerful enough to cause physical damage to people and structures. Consequently, thermobaric weapons are highly effective for use in the destruction of tunnels, bunkers, underground structures and buildings. A thermobaric explosion occurs in stages: in the initial stage, a cloud of flammable vapor or flammable metal is released (e.g. a finely powdered Al/AP mixture), which is then subsequently ignited. This effectively turns the air in the region of this cloud into a huge bomb. The resulting explosion has fragmentation, concussion, and incendiary elements. Actually, detonation of thermobaric explosives can be viewed as three discrete events merged together:

1. An initial anaerobic detonation reaction (microseconds duration) – defines the high-pressure performance characteristics of the system: armor penetrating ability
2. Post-detonation anaerobic combustion reactions (hundreds of microseconds in duration) – defines the system's intermediate pressure performance characteristics: wall/bunker breaching capability
3. Post-detonation aerobic combustion reaction (milliseconds in duration) – defines the personnel/material defeat capability of the system – impulse and thermal delivery.

Thermobaric effects are obtained by a long duration overpressure and heat due to the afterburning of the detonation products in air. Immediately after detonation occurs the ambient pressure increases, forming an overpressure pulse. The peak overpressure and positive phase duration determine the specific impulse of the blast wave. A negative phase succeeds the positive phase with its negative pressure (suction).

Today, cast composite thermobaric explosives most often consist of a high explosive (RDX, HMX), polymeric binder (e.g. HTPB), fuel component (metal powder for enhancement of the blast effect, (e.g. Al, Mg) and an oxidizer (AP or AN).

Table 13.1 compares the detonation temperatures and combustion temperatures with afterburn of a conventional explosive formulation (HMX-HTPB) and a thermobaric composition.

An excellent paper on thermobaric detonations by Danica Simić et al. can be found in the literature (*Central European Journal of Energetic Materials*, **2016**, *13*(1), 161–182).

The difference between TBX and EBX

Ideal molecular high explosives (HE) such as TNT, RDX, PETN and HMX generate fast decaying blast waves with high peak pressure and of very short duration during detonation, and are designed to throw shrapnel, shatter structures or penetrate armors. However, they are lethal only within their immediate vicinity, and show visible shortcomings for defeating hardened targets such as tunnels and caves. To overcome these shortcomings, great effort has been focused on the development of new weapons

Tab. 13.1: Comparison of the heat heats of reaction (Q_{ex}) of a conventional high-explosive HMX-based formulation with two thermobaric compositions.

	anaerob	aerob[a,b]	anaerob	aerob[a,b]	anaerob	aerob[a,b]
HMX	85	85	45	45	45	45
AP			10	10		
TNEF					10	10
Al			15	15	15	15
Mg			15	15	15	15
HTPB	15	15	15	15	15	15
$\Omega(CO_2)$/%	−66.2	0	−77.4	0	−79.8	0
ρ/g cm^{-3}	1.63	1.63	1.68	1.68	1.67	1.67
Q_{ex}/kJ kg^{-1}	4,878	13,769	8,301	18,113	8,539	18,916
T/K	3,158[d]	6,118[c]	4,583[d]	7,408[c]	4,705[d]	7,486[c]
$p_{C\text{-}J}$/GPa	22.4		15.8		15.4	
D/m s^{-1}	7,645		6,443		6,372	

[a]Aerobic calculation performed using isochoric combustion run, loading density 0.001 g/cm^3.
[b]Al_2O_3 and MgO presumed to be solids.
[c]Adiabatic combustion (with oxygen) temperature.
[d]Detonation temperature.

which are able to generate a higher impulse, higher blast and are able to use their energy not to destroy corners or walls, but to travel around it efficiently and defeat hardened targets. Scientists had obviously been activated by the dramatic and devastating explosions occurring in the chemical, petrochemical, and food industries that do not directly involve explosives. For instance, wheat or soy flour may form an explosive mixture, if a critical concentration is reached, and for which a spark is enough to initiate explosion. In such a manner, many accidents occur around the world each year. Another example is coal dust in mines, which can also generate the same type of explosion. In the early 1960s, scientists began experimenting with this concept to produce a weapon that uses the same principle, but employs volatile gases and finely powered explosives. Thermobaric weapons have been part of that development. As the name suggests, they are optimized for the effects of heat and pressure, while the typical advancement of weapons focuses on achieving better results in fragmentation, penetration, and brisance. Thermobaric weapons are a subgroup of a larger family of weapon systems commonly known as volumetric weapons. The characteristics of this weapon category are the creation of a large fireball and good blast performance. Russia was the first to develop such kinds of weapons. The RPO-A Schmel rocket infantry flame-thrower fielded in 1984 was the first successful thermobaric weapon, in which a self-deflagrating mixture consisting of magnesium (Mg) and isopropyl nitrate (IPN) was applied. This simple thermobaric explosive sent a devastating pressure wave through the caves and tunnels of Afghanistan, resulting in deep damage within the subterranean mazes. Recent conflicts have seen increased use of thermobaric weapons. Russia has employed this type of weaponry extensively in Afghanistan and Chechnya, however, this new class of energetic material did not receive extensive attention until March 5th, 2003, when U. S. Forces bombarded the Gardez mountains, in eastern Afghanistan. This event was widely reported by the news media, and the new term “thermobaric bomb” was used. Recently, great efforts have focused on the development of these weapons. The Russian formulation is effective, but not without problems. Liquid IPN is both volatile and detonable, and has been known to leak and spill, causing storage and toxicity problems. The US National Academy of Sciences has suggested that the fundamental physical phenomenon of thermobaric explosives should be recognized to rapidly enforce the development of thermobaric weapons. Thus, a lot of research has focused on the comprehension of thermobaric effects, in order to enhance – or prevent them. It is general knowledge, that chemically active metals such as aluminum and magnesium are added to condensed explosives in order to increase the total energy liberated during the explosion. Although these metals don’t react quickly enough to take part in the reactions in the detonation zone, they may react with the gaseous products of the explosion or with oxygen from the surrounding air and significantly increase the strength of the blast wave. Comparing the combustion energy of metals in the air (20–30 kJ per gram of metal) with the detonation energy of classic explosives (6 kJ per gram of explosive) suggests that it is possible to significantly increase the effective energy delivered from the explosive charge, if the burning of the metal particles is so programmed to proceed with adequate rate, then the energy released could strengthen the blast wave.

The ability of an explosive to produce a blast wave is called the blast ability. The impact of a blast wave can be enhanced further by debris (fragments) driven by the expanding detonation products. Thermobaric Explosives (TBX) or explosives with high-blast (Enhanced Blast Explosives, EBX) produce long-lasting pressure waves with moderate intensity. The effects of an explosion of such materials are more similar to the effects of a detonation of fuel-air mixtures (Fuel-Air Explosives, FAE) than those of ideal, condensed explosives. A characteristic feature of this type of explosives in the initial phase of the explosion is the phenomenon of dispersion of the detonation products and unreacted fuel to the surrounding air. In the next stages of the explosion, a mixture of fuel and oxygen from the air is spontaneously initiated to an exothermic reaction which enhances the pressure wave. Even without debris, such a wave is a serious threat to humans. The composition materials used in TBX and EBX explosives may consist of an explosive matrix, metal powders or other fuel, a binder, and an oxidant. Charges in the form of cylindrical layers containing these components are also used. The detonation of such explosives is accompanied by three stages of combustion:

1) In the first stage, anaerobic (without the participation of oxygen from the air) reactions are present in the detonation wave and last a few microseconds. They are mainly redox reactions of explosive molecules. Phenomena in the detonation wave affect the crushing ability of an explosive composition, such as the ability to destroy the casing.

2) Anaerobic combustion reactions, lasting hundreds of microseconds, are mainly reactions involving the detonation products with fuel particles, too large to be burned in the detonation wave. Combustion reactions after the detonation affect the blast wave characteristics that determine the blasting capability of an explosive, i.e. the ability to break down the walls of buildings, bunkers, etc.
3) Lastly, the aerobic (with oxygen from the air) combustion stage takes tens of milliseconds and is carried out in the air and fuel-rich products after mixing them by the shock-wave generated by the explosion and the accompanying turbulence. The main effect of the afterburning process is the heat, which raises the temperature and pressure of gases, and strengthens the blast wave.

Generally, there are two types of "afterburning":

1 – reactions during the 2nd stage of detonation-driven propulsion, such as when "inert" aluminum – which has been shock-driven to a denser phase – releases its "excess-energy" from the 100 kbar phase transition to help "heat" the expanding gases (a small contribution to propulsion due to how the gases are already expanding to lower density) and chemical "combustion" with detonation-products as redox reactions (this is also true for ultra-fine-diamond formed from the carbon rings during the 1st stage and which is later burned) and

2 – reactions with external, turbulent-mixed gases which can occur in two forms:
  a – rapid-combustion-explosion (such as occurs with a gun-muzzle blast) or
  b – stagnation-compression-ignition upon impact with a surface.

A review of the available literature shows that very often the terms TBX and EBX are used interchangeably, as the phenomena accompanying their explosions are similar. However, an EBX is used primarily to strengthen the blast wave, while a TBX is used to achieve an increase in temperature and pressure of the explosion. The classification of charges to a specific type depends on how the fuel is burned after the detonation ends.

If the main goal is to increase the strength of the blast wave, then usually an EBX is used. If, however, an increase in the temperature and pressure of an explosion is the most wanted effect, then a TBX is used. Anaerobic and aerobic reactions take place in both EBXs and TBXs. However, there are some important differences. In an EBX formulation, the metallic fuel reacts mainly in an anaerobic manner (without oxygen from the air). As a consequence, the energy which is liberated contributes to sustaining the initial blast wave and impulse. In contrast, in a TBX formulation, the most important reactions involving the metallic fuel are aerobic (requiring oxygen from the air). The combustion energy which is released from the aerobic processes results in the generation of moderate pressure and high temperature with long durations in the final stage of the explosion, after detachment of the shock wave.

In materials such as EBXs, there is talk about anaerobic combustion reactions, or combustion without oxygen from the air. This means that after the passing of the detonation wave, most of the fuel burns in the products of detonation. In materials such as TBXs, reactions of the fuel and oxygen from the air dominate. Explosives such as a TBX or EBX can therefore provide a much higher total energy than conventional explosives. It should be pointed out, that the main part of the additional energy is used to heat the gaseous medium and to strengthen the total pressure pulse of the blast wave, but this wave is still characterized by a relatively low amplitude (Fig. 13.1d). In the case of an internal detonation, multiple shock reflections enhance the afterburning reactions, generating a high level of quasi-static pressure and substantial damage to the surrounding structures. In recent years, scientists from various countries have conducted parallel research to find solid explosives, which would cause effects similar to the model liquid explosive (IPN/Mg), as well as in the development of tests for better characterization of the explosion phenomenon of such materials. In principle, first small-scale tests are carried out, which allow a rough assessment of the explosive composition in terms of its sensitivity and characteristics of the blast wave. Larger-scale tests allow for subsequent selection of the components and selection of the best composition in terms of blast capacity. Studies on the composition focus on several types of binders, various explosives, and metallic fuels. This includes, for example, the influence of the size and shape of the particles on the characteristics of the detonation, the process of particle dispersion, and blast wave parameters. The final composition must be a compromise between reaction rate, optimal dispersion of fuel and non-oxidized detonation products, and the ability to initiate combustion of the mixture of fuel and air. The explosion process must be sufficiently slow to disperse the fuel, but fast enough to guarantee that the combustion process is not interrupted. If the process is too fast then the fuel is too widely dispersed and the heat density generated is too small to initiate the subsequent oxidation of the fuel particles.

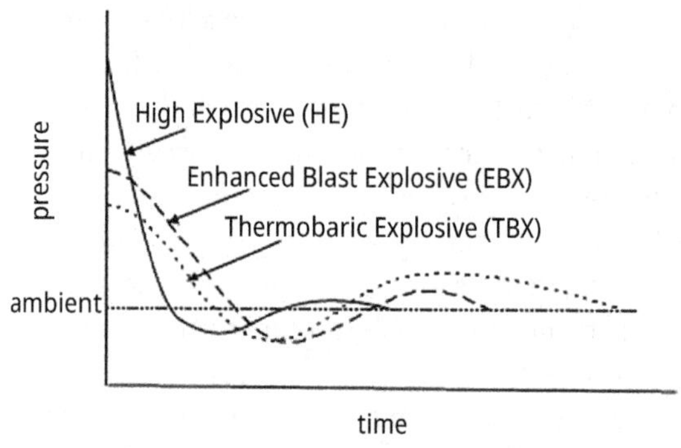

Fig. 13.1d: Pressure history inside the blast wave; high explosive vs. TBX and EBX detonations.

In a recent study, the blast performance of isopropyl nitrate-based thermobaric explosives vs. cast-cured PBX was investigated. The results can be summarized as follows:

- Open-field tests performed on large-scale thermobaric charges of gelled IPN-based compositions demonstrated that there is a significant difference in the performance of the prepared mixtures with Al and Mg regarding the blast and thermal output.
- Compositions containing Al and Mg have significantly higher blast parameters and thermal output than the non-metalized IPN.
- Comparison between cast-cured PBX TBE and IPN-based formulations revealed that PBX TBE had the highest blast parameters from 4 m to 7 m from the sample, as well as the highest maximum temperature of the fireball (~3,500 °C compared to ~2,800 °C).
- These measurements provided crucial insights into the spatial distribution and intensity of the blast effects, which are essential for understanding the potential destructive capability of thermobaric explosives that may have practical purposes in military and industrial applications.

Thermobaric explosives (TBX) and enhanced blast explosives (EBX) constitute a subfamily of volumetric weapons. They are fuel-enriched heterogeneous explosives. Unlike ideal high explosives, they are designed to produce a long lasting pressure wave able to travel along long corridors, propagate around corners and pass through obstacles. They are extremely effective and destructive in enclosed spaces, due to their ability to produce a high level of quasi-static pressure (QSP). A much higher total energy is produced by TBX and EBX explosions compared to conventional explosives. This energy may significantly strengthen the generated blast wave and/or magnify the explosion fireball and resulting thermal effect. A good recipe for a TBX or EBX composition could consist of an explosive matrix and a metal powder or other fuel, a binder, and an oxidant. Composites in the form of macroscopic energetic granules containing these components are also used. TBXs and EBXs are divided into solid explosive mixtures, liquid explosive mixtures, and the very promising composite and layered charges. The current state of research and features of each type are thoroughly described in the literature. If the fundamental physical and chemical phenomena of TBXs and EBXs can be understood and consistently controlled, a new weapons system of significant efficiency – or series of weapons systems – may become available to the warfighter in the future.

The detonation of a thermobaric system with three discrete events merged together may be summarized as follows (Fig. 13.1e):

(1) Initial anaerobic detonation (microseconds)
  - defines systems' high-pressure performance characteristics, armor penetrating ability

(2) Post-detonation anaerobic combustion recations (100s of μs)
  - defines systems' intermediate pressure performance characteristics, wall/bunker breaching capability

(3) Post-detonation aerobic combustion reaction (milliseconds)

– defines systems' personnel/material defeat capability, impulse and thermal delivery

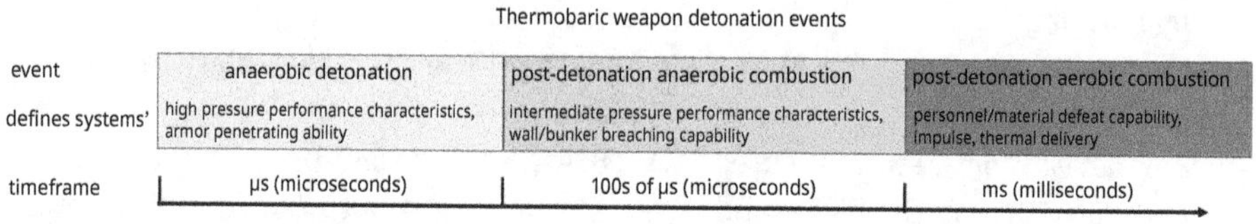

Fig. 13.1e: Events during a thermobaric detonation.

Typically, a metallized thermobaric composition is made up of the following ingredients:

12–20% polymeric binder
15–20% oxidizer
25–35% metallic fuel
35–50% high explosive (RDX or HMX)

While a classic binder could be HTPB, it is advantageous to change this in TBX formulations to a Si-based polymer (e.g. poly(dimethylsiloxame (PDMS)), in order to increase the heat of the reaction as a result of the very exothermic formation of SiO_2. The oxidizer could be ammonium perchlorate (AP) or potassium perchlorate or, alternatively, a new energetic oxidizer, e.g. TNEF (see below). Usually Al, Mg or magnalium (Ag/Mg alloy) is used as the metal fuel. However, the exothermicity of the reaction could be dramatically increased by changing to other fuels such as boron or beryllium. Finally, as the high explosive, white crystalline hexogen (HMX) or octogen (HMX) powder can be used, or a high explosive with even higher performance (e.g. CL-20 or TKX-50).

(see also: S. K. Kolev, T. T. Tsonev, *Propellants Explos. Pyrotech.*, **2022**, *47*, e202100195; doi.org/10.1002/prep.202100195).

The post-detonation burning time of the Al particles can be calculated using the D^n formula. The exponent in the D^n law is shown to be lower than two, with nominal values of ≈1.5 to 1.8 being typical. The effect of the ambient medium on the burning time is considered – oxygen as an oxidizer being twice as effective as water and about five times more effective than carbon dioxide. The effect of pressure and the initial temperature is only of minor importance. The burning times of aluminum particles can be estimated by the equation:

$$t_b = \frac{aD^n}{X_{eff}\, p^{0.1} T_0^{0.2}}$$

where $X_{eff} = C_{O2} + 0.6\ C_{H2O} + 0.22\ C_{CO2}$; $a = 0.0244$ for $n = 1.5$ and $a = 0.00735$ for $n = 1.8$; the pressure is measured in atmospheres, the temperature in kelvins, the diameter in micrometers, and the time in milliseconds.

Lit.: M. W. Beckstead, Correlating Aluminum Burning Times, *Combustion, Explosion and Shock Waves*, **2005**, *41*, 533–546.

While usually aluminum or magnesium (or Al/Mg alloys) are used as metal fuels in thermobaric formulations, other metals and non-metals are possible (Fig. 13.1f).

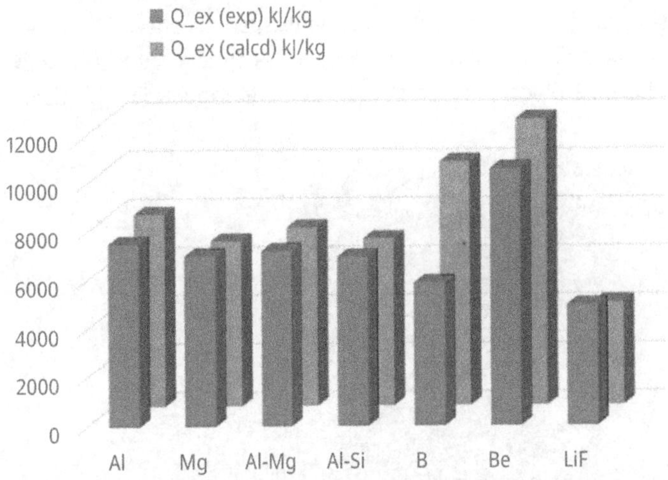

Fig. 13.1f: Calculated and measured heats of detonation of TBX formulations containing 60% RDX, 20% AP and 20% fuel (metal). LiF was included as an inert reference material.

From the diagram above (Fig. 13.1f), it is clear that theoretically (calculated), boron and beryllium should deliver the highest heats of detonation (Q_{ex}). This could be confirmed experimentally in the case of Be, but not for B. The likely reason that the experimentally determined value of Q_{ex} is lower than the calculated Q_{ex} value for boron is that boron does not react completely to form B_2O_3, but reacts only to form the intermediate HOBO, which delivers much less energy than fully oxidized B_2O_3. In the case of beryllium, the calculated and measured values for Q_{ex} agree reasonably well.

In the first published study ever, the authors made a direct comparison between the effectiveness of Al, Mg, and Be powders as fuels in model thermobaric compositions containing 20% fuel, 20% ammonium perchlorate, and 60% RDX (1,3,5-trinitro-1,3,5-triazacyclohexane) passivated with wax (Fig. 13.1g). Experimentally determined calorimetric measurements of the heat of detonation along with the overpressure histories in an explosion chamber filled with nitrogen were used to determine the quasi-static pressure (QSP) under anaerobic conditions. Overpressure measurements were also performed in a semi-closed bunker, and all blast wave parameters generated after the detonation of 500 g charges of the tested explosives were determined. Detonation calorimetry results, QSP values, and blast wave parameters (pressure amplitude, specific and total impulses) clearly showed that Be is much more effective as a fuel than either

Al or Mg in both anaerobic post-detonation reactions, as well as the subsequent aerobic combustion. The heat of detonation of the RDXwax/AP/Be explosive mixture is over 40% higher than that of the corresponding mixture containing aluminum instead of beryllium, and 50% higher than that of the mixture containing magnesium instead of beryllium. Furthermore, the TNT equivalent of the Be-containing composition due to the overpressure in the nitrogen-filled explosion chamber was determined to be 1.66, while the equivalent calculated using an air shock wave-specific impulse at a distance of 2.5 m was found to be equal to 1.69. The high values of these parameters confirm the high reactivity of beryllium in both the anaerobic and aerobic stages of the thermobaric explosion.

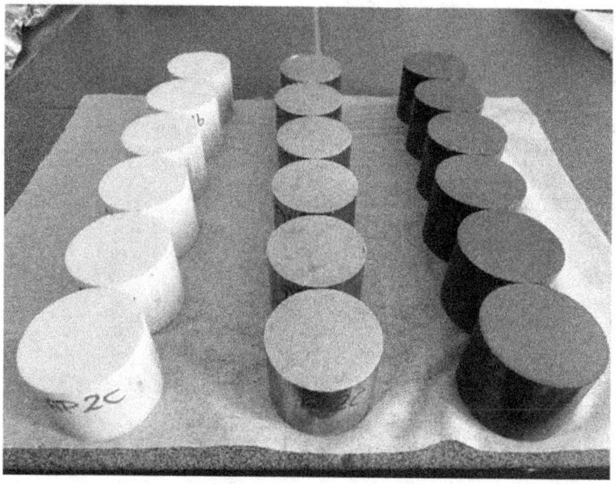

Fig. 13.1g: Pressed TBX charges: RDX/AP (left), RDX/AP/Al (center) and RDX/AP/Be (right).

In this context, it should also be mentioned that if various metallic fuels are used in one TBX formulation (e.g. 80% Al and 20% Mg), it is better to use the corresponding alloy of the metal fuels rather than just a mixture of the elements. This is for the following reasons: (i) higher homogenicity, (ii) no de-mixing possible, and (iii) usually reduced sensitivity towards the atmosphere (especially if the lower percentage metal is highly reactive, e.g. Ce, Li etc.).

Overall, the following conclusions were made based on the observations discussed above:

- The heat of detonation of RDXwax/AP/Be is over 40% higher than either RDXwax/AP/Al or RDXwax/AP/Mg.
- The highest quasi-static overpressure in the N_2-filled chamber was observed for the RDXwax/AP/Be mixture.
- All of the blast wave parameters after detonation in a semi-closed bunker were found to be the highest for the RDXwax/AP/Be mixture.

- The RDXwax/AP/Be mixture was found to show a high heat of combustion, and is highly reactive towards detonation products of RDX and AP in detonation wave.
- The highest values for TNT equivalents were observed for the Be-containing mixtures.

Lit.: T. M. Klapötke, S. Cudziło, W. A. Trzciński, J. Paszula, L. Bauer, C. Riedelsheimer, J. T. Lechner, *Prop. Expl. Pyrotech.*, **2023**, *48*, e202300010, https://doi.org/10.1002/prep.202300010.

T. M. Klapötke, S. Cudziło, W. A. Trzciński, J. Paszula, *Defence Technology*, **2024**, *36*, 13 – 19. https://doi.org/10.1016/j.dt.2024.02.011.

N. G. Yen, L. Y. Wang, *Prop. Expl. Pyrotech.*, **2012**, *37*, 143 – 155, DOI: 10.1002/prep.200900050.

It could possibly be argued that beryllium is rather expensive and toxic. However, the performance is outstanding and the toxicity according to NIOSH comparable to that of platinum (Tab. 13.1a).

Tab. 13.1a: IDLH values for different metals, source: NIOSH via: https://www.cdc.gov/niosh/idlh/intridl4.html.

Immediately dangerous to life or health (IDLH) values				
	Substance	**CAS no.**	**IDLH Value (1994)**	
Be	Beryllium compounds (as Be)	7440-41-7 (metal)	4	mg Be/m^3
Pt	Platinum (soluble salts, as Pt)	n/a	4	mg Pt/m^3
Ag	Silver (metal dust and soluble compounds, as Ag)	7440-22-4 (metal)	10	mg Ag/m^3
Ag	Silver (metal dust and fume, as Ag)	7440-22-4 (metal)	10	mg Ag/m^3
Zr	Zirconium compounds (as Zr)	7440-67-7 (metal)	25	mg Zr/m^3
Ba	Barium (soluble compounds, as Ba)	7440-39-3	50	mg Ba/m^3
Zn	Zinc chloride fume	7646-85-7	50	mg Zn/m^3
Cu	Copper (dust and mists, as Cu)	7440-50-8	100	mg Cu/m^3
Cu	Copper (dust and mists, as Cu and CuO)	1317-38-0 (CuO)	100	mg Cu/m^3
Cu	Copper fume (as Cu)	1317-38-0 (CuO)	100	mg Cu/m^3
Pb	Lead compounds (as Pb)	7439-92-1	100	mg Pb/m^3
Zn	Zinc oxide	1314-13-2	500	mg Zn/m^3
Mg	Magnesium oxide fume	1309-48-4	750	mg Mg/m^3

In another recent study, the possibility of replacing AP in TBX formulations was investigated. Out of all of the oxidizers which have been published by the energetic materials group at LMU, the best three examples are probably *bis*(2,2,2-trinitroethyl)oxalate [BTNEO] (**1**), 2,2,2-trinitroethyl-nitrocarbamate [TNENC] (**2**) and 2,2,2-trinitroethyl formate [TNEF] (**3**) (Fig. 13.1h).

Common to all three of these compounds is the absence of any halogen, presence of oxygen-rich $-C(NO_2)_3$ groups, as well as the incorporation of O atoms into the carbon skeleton. All three compounds possess a positive oxygen balance with respect to

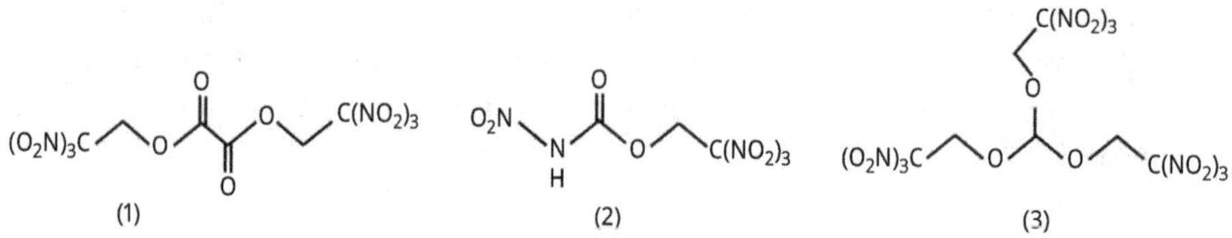

Fig. 13.1h: New CHNO oxidizers.

CO and CO_2 and have acceptable sensitivity parameters towards impact, friction and ESD (Tab. 13.1b). Unfortunately, the decomposition temperature of the nitrocarbamate **(2)** is only 153 °C, which is too low for application. Therefore, the oxalate **(1)** and the formate **(3)** are the more promising candidates for replacing ammonium perchlorate and potassium perchlorate in thermobaric formulations. TNEF **(3)** has a highly positive oxygen balance of 30.4% (with respect to CO), a high density of 1.81 g cm^{-3} and a decomposition temperature of 192 °C. Therefore, TNEF has been investigated the most intensively out of these compounds, and has already been scaled up in the lab to 100 g scale.

Tab. 13.1b: Relevant data for the three oxidizers BTNEO **(1)**, TNENC **(2)** and TNEF **(3)**.

	BTNEO (1)	TNENC (2)	TNEF (3)
Formula	$C_6H_4N_6O_{16}$	$C_3H_3N_5O_{10}$	$C_7H_7N_9O_{21}$
M / g mol^{-1}	416	269	553
IS / J	10	10	5
FS / N	>360	96	9
ESD / J	0.7	0.1	0.2
Ω(CO) / %	15.4	32.7	30.4
Ω(CO_2) / %	7.7	14.9	10.1
m.p. / °C	115	109	128
T_{dec} / °C	186	153	192
ρ / g cm^{-3}	1.84	1.73	1.81
ΔH°_f / kJ mol^{-1}	−688	−366	−519

The suitability of TNEF **(3)** as a replacement for ammonium perchlorate (AP) in thermobaric (TBX) formulations has already been shown computationally and experimentally. A thermobaric composition consisting of HMX, Al, binder and TNEF (tris(2,2,2-trinitroethyl) orthoformate) has been prepared, and the energetic and thermal stability properties established and compared with those of the same thermobaric composition in which TNEF was replaced by the traditional oxidizer ammonium perchlorate (AP). The detonation velocities of the two thermobaric mixtures were measured and showed the mixture containing TNEF to be slightly higher than that for the mixture containing AP. If the classical TBX formulation (TBX-1: 40% HMX, 20% Al, 20% AP, 20% binder) is compared with a new formulation containing TNEF instead of AP (TBX-2: 40% HMX, 20% Al, 20% TNEF, 20% binder), it is clear that both TBX-formulations perform essentially equally, with TBX-2 even possessing a slightly higher heat of detonation (Q_{ex}) and detonation temperature (T_{ex}) (Tab. 13.1c, Fig. 13.1i).

Tab. 13.1c: Performance parameters of two TBX formulations (calculated with EXPLO5_V6.05.02).

	TBX-1	TBX-2
VoD / m s^{-1}	6,699	6,593
p_{C-J} / GPa	18.2	17.6
Q_{ex} / kJ kg	8,202	8,355
T_{ex} / K	4,371	4,451

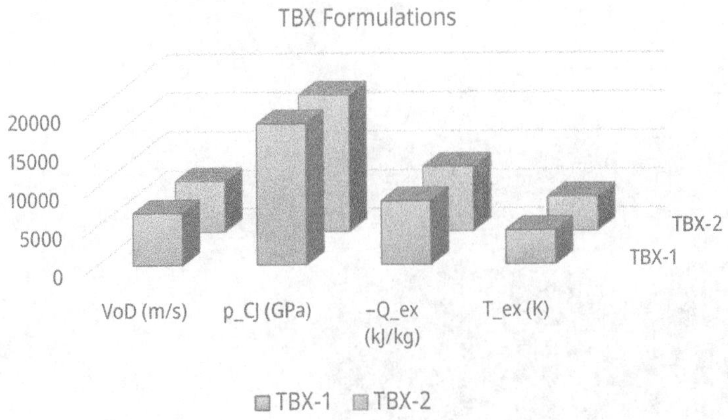

Fig. 13.1i: Graphical representation of the performance parameters of the TBX formulations TBX-1 (containing AP) and TBX-2 (containing TNEF instead of AP).

The maximum overpressures generated from 600 g each of the formulations TBX-1 and TBX-2 were also calculated and experimentally recorded as a function of the measuring spot distance from the center of the detonation (Fig. 13.1j).

In conclusion, so far, results indicate that TNEF in the TBX formulation described above is essentially as good as AP, but has the advantage of being chlorine-free (Fig. 13.1k).

Two excellent papers on this topic can be found in the literature (W. A. Trzciński, L. Maiz, *Prop. Expl. Pyrotech.*, **2015**, *40*, 632 – 644; L. Türker, *Defence Technology*, **2016**, *12*, 423 – 445).

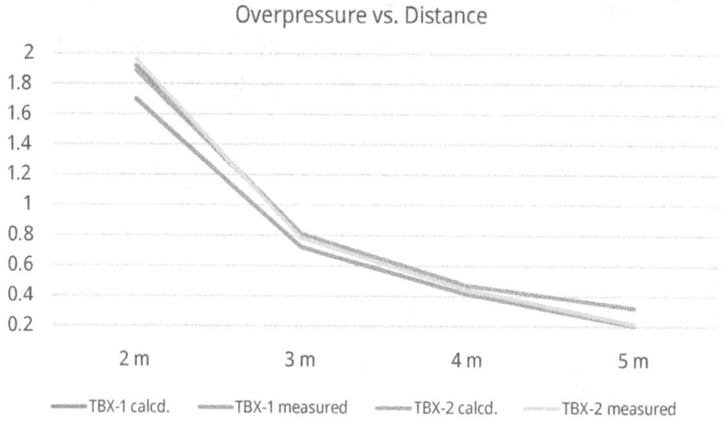

Fig. 13.1j: The maximum overpressures (in bar) generated from 600 g each of the formulations TBX-1 and TBX-2 as a function of the distance.

Fig. 13.1k: Detonation of a TBX-2 charge.

13.2 Agent defeat weapons

Agent defeat weapons (also known as agent defeat warheads, ADW) are airborne warheads, which can be used in the fight against chemical and biological (CB) weapons of mass destruction (WMD) to minimize the amount of collateral damage that occurs. Collateral damage can largely be minimized when the active agents in the chemical or biological weapons are destroyed or neutralized upon being attacked, before they can be spread into the environment.

An ADW is constructed similarly to a multiprojectile warhead (see Ch. 7.1), but is additionally combined with a thermobaric weapon (see Ch. 13.1). After penetration of the chemical or biological weapon system, the active substances are destroyed as a result of the large heat effect of the thermobaric explosion. In addition, some ADWs also contain Cl-releasing systems, which also provide further destruction through oxidation in particular of the biological warfare agents. Other chemical compounds which are suitable for fighting biological agents contain peroxide and hypochlorite groups as well as chlorine.

It has been shown that although Cl_2 is a better biocide than HCl, the biocidal activity of HF exceeds both Cl_2 and HCl. Only 200 ppm HF destroy most bacteria, including Anthrax spores. The most recent research in particular by Bob Chapman (NAWC, China Lake) has shown that ADW containing fluorinated payloads ($—NF_2$ groups) are effective in destroying biological threats. For example, tetrakis-(difluoramino)octahydro-dinitro-diazocine, HNFX (see Fig. 10.1) [67] can very quickly remove the dangers of Anthrax spores. The biocidal activity of HNFX is not due to heat or pressure, but due to the harsh conditions of exposure to biocidal detonation products (i.e. HF). A control experiment using HMX as the explosive showed an order of magnitude lower killing of spores (86% spore survival). Also octa-F-PETA, i.e. $C(—CH_2—NF_2)_4$ has successfully been tested against Anthrax spores. A possible synthesis for octa-F-PETA is shown in Fig. 13.2.

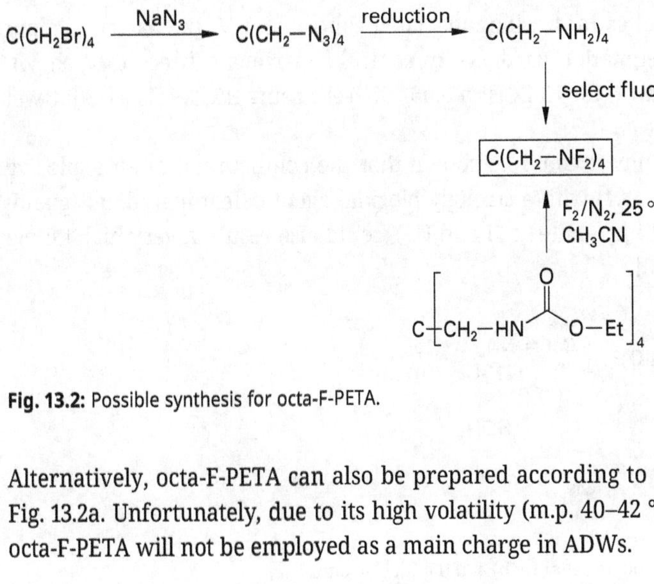

Fig. 13.2: Possible synthesis for octa-F-PETA.

Alternatively, octa-F-PETA can also be prepared according to the procedure shown in Fig. 13.2a. Unfortunately, due to its high volatility (m.p. 40–42 °C; sublimes at ca. 30 °C), octa-F-PETA will not be employed as a main charge in ADWs.

Fig. 13.2a: Alternative synthesis for octa-F-PETA.

A chlorine containing energetic material which also shows use as a potential agent defeat explosive is hexachloromelamine, [—N=C(NCl$_2$)—]$_3$ (Fig. 13.3).

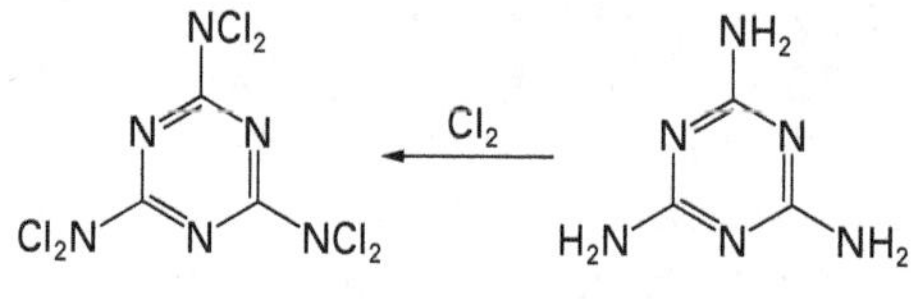

hexachloromelamine

Fig. 13.3: Synthesis of hexachloromelamine.

This compound is known to be biocidal and can be prepared by direct chlorination of melamine (Fig. 13.2).

The killing efficiency *R* is usually defined as

$$R = \log(N_0 + 1) - \log(N_E + 1)$$

Where N_0 is the bacterial count for the inoculated control and N_E the bacterial count after exposure to the detonation products. In control experiments it could be shown that the killing efficiency (*R*) for HMX is ca 0.06, whereas using HNFX it is ca. 3 for wet and up to 9 for dry spores.

In a more recent study it could be shown that the combination of an explosive (diazidoglyoxime, **2**, Fig. 13.4) with a strongly biocidal agent defeat ingredient (guanidinium difluoroiodate, **1b** generates I_2, I and HF), could also result in very high killing efficiencies of $R = 6–8$ [68].

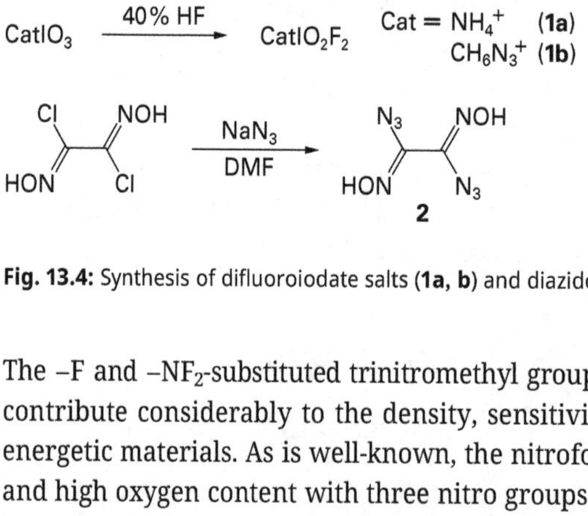

Fig. 13.4: Synthesis of difluoroiodate salts (**1a, b**) and diazidoglyoxime, **2.**

The –F and –NF$_2$-substituted trinitromethyl groups such as –C(NO$_2$)$_2$F and –C(NO$_2$)$_2$NF$_2$ contribute considerably to the density, sensitivity and energy of fluorine-containing energetic materials. As is well-known, the nitroform group has a large dipole moment and high oxygen content with three nitro groups. However, it has poor thermal stability and is sensitive toward impact and friction because of not only the large steric hindrance due to the crowded nitro groups, but also due to the low C–N bond energy. The introduction of fluorine generally enhances the thermostability and improves the sensitivity. Although $-C(NF_2)_2NO_2$, $-CF_2NO_2$, $-C(F)NO_2NF_2$, and other similar groups exist theoretically, only the $-C(NO_2)_2F$ and $-C(NO_2)_2NF_2$ groups have been reported in energetic material compounds.

The combination of the fluorodinitromethyl group (which is an excellent explosophoric group) with an amine oxime group as a benign precursor has been shown to be an excellent synthetic strategy to create novel fluorine-containing energetic materials. In 2018, Sheremetev's group successfully synthesized 2-fluoro-2,2-dinitroacetamidoxime, as well as several derivatives based on the amine oxime compound, as depicted in Fig. 13.4a. The potassium salt, compound **1** reacted with 4-chloromethyl-1-fluoro-1,4-diazoniabicyclo[2.2.2]octane ditetra-fluoroborate (Selectfluor) in acetonitrile to yield **2**. Compounds **3**, **5**, and **6** were synthesized via the reactions of **2** and acetic anhydride, sodium nitrite/HCl, and sodium nitrite/HBr, respectively. Furthermore, the reaction of **3** with sodium azide (NaN_3) was shown to generate the corresponding covalent azide compound **4** (Fig. 13.4a). The synthetic strategy of using fluoronitromethyl derivatives as demonstrated by this example provides new ideas for the production of other energetic materials with fluorine-containing functional groups, in which the amine oxime group functions as a precursor.

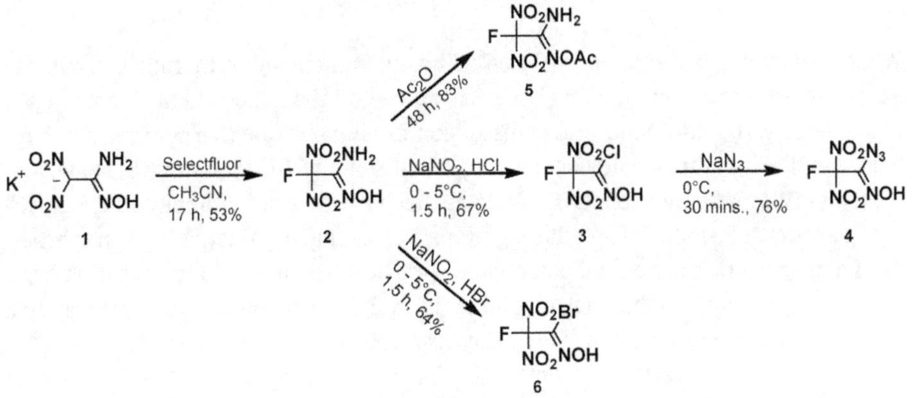

Fig. 13.4a: Synthetic route for the preparation of 2-fluoro-2,2-dinitroacetamidoxime and corresponding derivatives.

Lit.: Z. Guo, Q. Yu, Y. Chen, J. Liu, T. Li, Y. Peng, W. Yi, *Chem. Rec.*, **2023**, *23*, e202300108, doi.org/10.1002/tcr.202300108.

13.3 Nanothermites

Research into nanothermites is currently a hot topic in the field of energetic materials. For thermite-type reactions, a metal oxide is the oxidizer, and aluminum is the fuel. Thermite-type reactions on the nanoscale have been called many names including nanothermites, metastable intermolecular composites (MICs) or superthermites. The phrase "nanothermites" comes from the particle sizes used in these energetic mixtures, in contrast to the more familiar thermite type reaction in which the particle sizes are in the micron region. The term "metastable intermolecular composites" comes from the fact that the mixtures of metal oxide and aluminum are stable up to their ignition temperature, at which point self-propagating high-temperature synthesis (SHS) occurs and the thermodynamic products of a metal and aluminum oxide are produced. Finally, the term "superthermite" comes from the fact that thermites composed of nano-sized materials exhibit very different combustion characteristics when compared to those mixed with micron-sized precursors. Superior combustion velocites and explosive behavior compared to the usually observed deflagration often characterize thermites made with nanoscale precursors.

$$M_xO_y + Al \longrightarrow M + Al_2O_3. \qquad (1)$$

While most metal oxides can be reduced with aluminum in an exothermic thermite reaction, and the reactivity of many has been studied on the nanoscale (including Fe_2O_3, MnO_2, CuO, WO_3, MoO_3, and Bi_2O_3), discussion of all nanoscale thermite reactions reported in the literature would be too broad. Discussion will focus on iron(III) oxide, molybdenum(VI) oxide, and copper(II) oxide as they provide good examples of either differing synthetic methodologies tailoring properties, exemplary final product properties, or of a material that has a high potential for practical use. We will limit the discussion of thermite examples to those where aluminum is the fuel because of the practical uses of such mixtures. However, other fuels such as Zr, Hf, Mg etc. have been used.

An energetic material's particle size traditionally ranges from 1–100 μm. As is known, generally smaller particles give higher reaction rates and nanothermites are not an exception. The reaction rate of a nanoscale thermite is generally several orders of magnitude larger than those on micron scales, and the much larger surface area can significantly change combustion behavior, as well as ignition behavior by increasing sensitivity. The changes in properties are all associated with reduced diffusion distances since the associated surface area increases.

The initial reactions in a binary fuel-oxidizer system such as a thermite reaction are assumed to be diffusion-limited solid-solid reactions. As a result, the rate of reaction and therefore the combustion velocity can be increased dramatically as the particle size decreases and particle contacts increase. A mathematical study undertaken by Brown et al. compared observed combustion rates for a Si/Pb_3O_4 pyrotechnic mixture with the calculated fuel-oxidizer contact points for silicon particles of given sizes mixed with 5 µm Pb_3O_4. Contact points were calculated assuming the approximation that the fuel and oxidizer are hard spheres. Results from this study are presented in Tab. 13.2, and show that a small change in particle size has a very strong effect on fuel-oxidizer contact points, and a correspondingly large effect on the combustion velocity.

Tab. 13.2: Combustion rate as a function of contact points in an Si/Pb_3O_4 composite.

Si particle diameter (µm)	contact points (× 10^9)	combustion rate (mm / s)
2	30.2	257.4
4	8.7	100.6
5	6.1	71.5

Despite this study not being on the nanoscale, it does show the dramatic effect particle size has on combustion rate, since doubling the particle size causes the combustion rate to be more than halved. The Si/Pb_3O_4 mixture has found use as a pyrotechnic delay mixture (slow burning mixture used for timing in pyrotechnics), as its combustion rate is easily tailored.

As particle sizes decrease, the sensitivity of the thermite mixture to impact and friction increases. The micron scale thermites are usually quite insensitive to impact and shock, but thermites on the nanoscale can be quite sensitive to both or one of the two depending on the metal oxide. This is exemplified in the work of Spitzer, where a tungsten(VI) oxide and aluminum thermite was prepared by mixing nano and micron aluminum with nano and micron WO_3. The results of this are presented in Tab. 13.3.

Tab. 13.3: Sensitivities for a nano vs. micron Al / WO_3 thermite.

Al diameter (nm)	WO_3 diameter (nm)	impact (J)	friction (N)	combustion rate (m / s)
1,912	724	> 49 (insensitive)	> 353 (insensitive)	< 0.08
51	50	42 (insensitive)	< 4.9 (very sensitive)	7.3

While increased sensitivity can make nanothermites more dangerous to handle, the increased friction or impact sensitivity is also beneficial in some practical applications such as in percussion primers. However, some nanothermites have increased ESD sensitivity, which has no current practical application and is only a safety hazard. For example, a Bi_2O_3/Al nanothermite has sufficient impact and friction sensitivity to be considered for use in ammunition primers, however, it has an ESD sensitivity of 0.125 µJ (40 nm Bi_2O_3, 41 nm Al). This is a static potential, which is easily achieved by the human body. This makes handling hazardous, especially since the Bi_2O_3/Al thermite has a combustion velocity of over 750 m/s which means that this thermite explodes rather than burns. The increased ESD sensitivity of nano vs. micron composites is believed to be the result of the increased ability of high surface areas to develop charges.

A final property affected by the reduction in particle size from micron to nano scale is the ignition temperature of the thermites. For example, a thermite composed of 100 nm MoO_3 and 40 nm aluminum exhibits an ignition temperature of 458 °C, whereas the same MoO_3 with 10–14 µm Al exhibits an ignition temperature of 955 °C. This indicates a change in mechanism between the two, as the micron composite exhibits melting and volatilization of Al and MoO_3 (from DSC) before the thermite reaction occurs, while the nano-thermite reaction occurs before the Al melting could take place. This indicates that the reaction for the nano-composite is based on solid-state diffusion, while for the micron composite, the reaction is a gas (MoO_3)-liquid (Al) reaction. This change in mechanism may apply to other thermite systems as well, however detailed DSC data for many other systems has not been published so far.

In thermites using aluminum metal as the fuel, the passivation of the metal surface with oxide must be taken into account. For micrometer sized particles of aluminum, the oxide passivation layer is negligible, but on the nano-scale, this passivated layer of alumina begins to account for a significant mass portion of the nanoparticles. In addition, the precise nature of the oxide layer is not the same for all manufacturers of aluminum nanoparticles, so the researcher must use TEM to measure oxide thickness to allow calculation of the active aluminum content before stoichiometric calculations are carried out for the mixing of thermites. Table 13.4 shows details of some of the percentages of aluminum in aluminum nanoparticles, and shows just how significant and inconsistent the oxide layer can be.

Tab. 13.4: Active aluminum content of aluminum nanoparticles.

average Al nanoparticle size (nm)	active aluminum content (%)
30	30
45	64
50	43
50	68
79	81
80	80
80	88

The aluminum oxide layer is also known to reduce the propagation of thermite reactions, since alumina is an effective absorber of thermal energy. A study by Weismiller et al. of a 49% active aluminum nanopowder in a thermite with copper oxide supports the idea that too much oxide can actually reduce thermite performance. Table 13.5

shows Weismiller's data, which indicates the negative effect of a thick oxide layer on aluminum nanoparticles.

Tab. 13.5: CuO /Al thermites with highly oxidized (49% Al) aluminum nanoparticles.

CuO	Al	combustion velocity (m / s)	mass burning rate (g / s)
micron	micron	220	2,700
micron	nano (49% Al)	200	1100
nano	micron	630	4,850
nano	nano (49% Al)	900	4,000

The combustion velocities given in Tab. 13.5 show that the use of nano CuO is more of a factor towards high velocities than the Al when the Al is highly oxidized, and even slightly decreases performance when the velocity of μm-CuO / μm-Al is compared to μm-CuO / nm-Al. The mass burning rate shows this even more starkly, as with a given particle size of CuO, the switch from micron to heavily oxidized nano aluminum decreases the mass burning rate.

In general, there are three common ways of preparing nanothermites; these include arrested reactive milling, physical mixing, and sol-gel methodology. Changing from the microscale to the nano, these methods provide accessibility and use of different particle sizes, as well as materials with different degrees of contact between oxidizer and fuel.

Arrested reactive milling (ARM) is a technique for the preparation of nanoenergetic materials by milling the metal oxide and aluminum in a ball or shaker mill. While it may or may not involve particle sizes on the nanoscale, the energetic composites prepared through this method have properties more similar to those of nanothermites as mixing is on the nanoscale; fuel and oxidizer may be contained in the same particle after milling. The particles produced by arrested reactive milling are in the range of 1–50 μm, and are composed of layers of aluminum and oxidizer on the scale of 10–100 nm. The size of the particles obtained is a function of the milling time. However, due to the reactivity of the mixture, after a certain milling time (a function of initial particle sizes, milling media used, and type of metal oxide) when the particle size is reduced below a certain value, the milling causes ignition of the thermite mix. A liquid such as hexane is usually added to the mill to reduce static build up. The term arrested reactive milling comes from the fact that milling is stopped before the mix ignites, resulting in a useable thermite mixture. Advantages of this method are that the particles produced approach their maximum density. Aluminum is hidden within the particle matrix which reduces the presence of nonreactive alumina, the ability to start with non-nanoscale precursors, and milling time offer convenient control of the degree of intermixing and therefore the reactivity. Disadvantages include the fact that only few thermite mixtures can be prepared by this method, as many mixtures are too sensitive and ignite before sufficient intermixing can occur.

Physical mixing is the simplest and most common method for the preparation of nanothermites. Nano-powders of metal oxide and aluminum are mixed in a volatile, inert liquid (to reduce static charge) and they are then sonicated to ensure good fuel-oxidizer mixing and to break up macroscale agglomerates. The liquid then evaporates and the thermite is ready to be used. Advantages of the physical mixing method is the simplicity and its wide range of applicability to many thermite systems. The only major disadvantage is the necessity of starting with nanoscale powders which may or may not be commercially available.

Sol-gel nanothermite preparations take advantage of the unique structural and mixing properties available from sol-gel chemistry. In these nanothermite mixtures, the aluminum nanoparticles reside in the pores of the metal oxide matrix, which is widely assumed to increase thermite power in comparison to physical mixing by huge reductions in diffusion distances between the fuel and oxidizer, and by increasing the contact area. The synthesis of energetic sol-gel thermites involves the addition of a suspension of aluminum nanopowder in a solvent to a metal oxide sol just prior to gellation. After gellation, the gel can be processed to an energetic xerogel or aerogel. Through the sol-gel methodology, interfacial area, pore size, and matrix geometry can be controlled, which results in different, tunable energetic properties. In addition, the sol-gel methodology allows the molecular incorporation of organic agents to the matrix which further tune the thermite's properties by acting as a gas generating agent. Other than the ability to incorporate organics, advantages of the sol-gel method include its ability to form low-density aerogels or xerogels, which brings with it the ability to form energetic surface coatings. Disadvantages include potential oxidation of the aluminum nanoparticles by water in the metal oxide gel before the solvent can be removed. This disadvantage can be overcome by using sol-gel chemistry to only prepare aerogel or xerogel oxide precursors, which can then be followed by physical mixing with aluminum. However, this option comes at the cost of reducing the high-level of interfacial contact achievable by sol-gel chemistry.

Due to the range of properties that nanothermites can be tailored to have, nanothermites find applications in a correspondingly large field. As research on nanothermites is a relatively new topic, many applications have only been proposed so far, and testing in such areas has not yet been undertaken. Many of the potential applications of nano-thermites are a direct result of their high energy densities, which are comparable to those of lithium batteries, leading to applications of the power-generating variety such as microscale propulsion, energetic surface coatings and nano-scale welding. Other applications are a direct result of the pyrotechnic behavior of nanothermites, and such materials may find use as gas generating agents for automobile airbags, exploding-on-contact missiles, environmentally friendly ammunition primers and electric igniters. The fastest thermites even have potential application as primary explosives. Their application in microscale propulsion, ammunition primers, and electric igniters has been more extensively investigated experimentally than in other potential applications for nanothermites, and show very promising results.

Microscale propulsion involves the production of thrust on the microscale (< 1 mm). This is referred to as micropyrotechnics or microenergetics, and applications include rapid switching and propulsion for small spacecraft. Performance energetic materials used for propulsion on the macroscale including RDX (hexahydro-1,3,5-trinitro -1,3,5-triazine, hexogen) or HMX (1,3,5,7-tetranitro-1,3,5,7-tetrazocane, octogen) are unable to function on the microscale, as they cannot sustain combustion in such small diameters, since too much energy is lost to the combustion chamber which inhibits propagation through the energetic material. Nanothermites have no such problem as they have a much higher energy density, so energy lost to the chamber becomes insignificant.

An ammunition primer is the part of a round of ammunition that is impacted by the firing pin of the firearm. Traditional primers contain lead containing compounds including lead azide and lead styphnate, as these sensitive explosives detonate from the impact of the firing pin, and the resultant flame ignites the propellant in the cartridge, firing the bullet. Figure 13.5 shows a shotgun shell primer. The toxicity of lead means the use of such primers is both an environmental and personal hazard to the user. Nanothermites have been shown to be an effective replacement for the lead salts, as they have been tailored to have properties similar to the currently used lead azide/styphnate mixture.

Fig. 13.5: Shotgun shell primer.

Electric igniters find use throughout the energetic material industry in all areas of propellants, explosives and pyrotechnics. They are used wherever an electric current is required to initiate an energetic material. As a result of the precise timing they afford, they can be used for igniting everything from rockets to blasting caps, to fireworks displays. Electric igniters are also known as electric matches, and consist of a flammable head of material around a resistive bridgewire, igniting when a certain electrical current is passed through the match. Figure 13.6 shows a general diagram of an electric match. Like primers, commonly used electric matches contain toxic lead compounds including the dioxide, thiocyanate and nitroresorcinate. Therefore, finding non-toxic replacements is desirable. Again, the use of nano-thermites has shown significant promise as non-toxic, green replacements for the lead compounds.

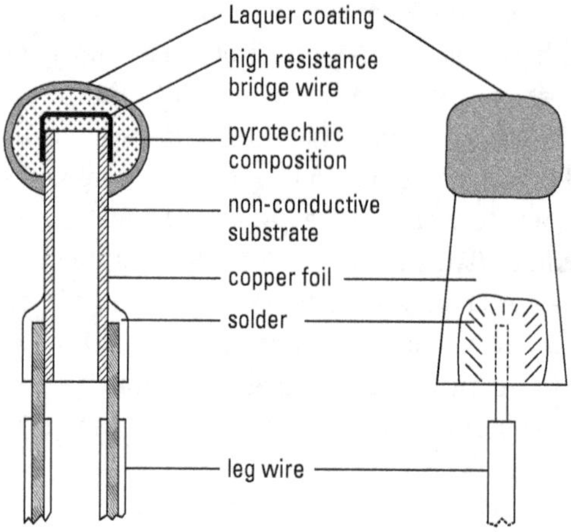

Fig. 13.6: Electric igniter/match diagram.

13.3.1 Example: Iron oxide / aluminum thermite

On the macroscale, the iron oxide / aluminum thermite is used for the high temperature welding of iron railway rails. This is made possible through its slow rate of energy release and high combustion temperature. On the nanoscale, however, the iron oxide / aluminum thermite behaves very differently. In the literature, this thermite system is often synthesized by sol-gel methods as iron oxide sol-gel chemistry is well known, however, it has also been made by arrested reactive milling and physical mixing as well.

The iron oxide thermite is a very good example of the versatility of the sol-gel thermite methodology. Table 13.6 shows the huge effect the synthetic strategy used can have on the properties of the thermite.

From these numbers, the wide range of thermite properties that become available is apparent. The ease of tuning energetic properties is also apparent from the high velocity (320 and 895 m/s) xerogels and aerogels show as the xerogel and aerogel have different sensitivities to ESD and to impact. These differences are shown in Tab. 13.7.

Tab. 13.6: Effect of the synthetic methodology used to prepare the Fe_2O_3/Al thermite on the combustion velocity.

synthetic methodology	velocity (m/s)
arrested reactive milling (max mill time before ignition)	0.5
physical mixing (80 nm Al, oxidizer size not given)	9
sol gel (aerogel oxide physically mixed with 80 nm Al)	80
sol gel (xerogel, 70 nm Al added before gellation)	320
sol gel (aerogel, 70 nm Al added before gellation)	895

Tab. 13.7: Impact and ESD sensitivities of aerogel and xerogel Fe_2O_3/Al thermites.

type	impact (J)	ESD (mJ)
xerogel	36.6	> 1,000
aerogel	21.9	30

The impact sensitivity of the aerogel thermite may make it suitable for use in primers, because all other preparatory methods for the iron oxide thermite produce thermites which are unsuitable for this application. The especially sensitive nature of aerogel thermites stems from the inability of the aerogel matrix to conduct heat, a characteristic which is only applicable to aerogels, due to their insulating property.

A further development on the synthesis-dependent application of the iron oxide thermite is that of Clapsaddle et al. Tailoring of energetic properties was obtained by the addition of organo-silicon precursors to the sol leading to aerogel oxidizers containing any desired percentage of organic-functionalized silica. The silica functioned to reduce the combustion velocity, and the organic addition to increase the amount of gas released upon combustion. Research in this field is still in the initial stages, but the thermites may have application in the propulsive and gas-generating fields.

13.3.2 Example: Copper oxide / aluminum thermite

The copper oxide and aluminum thermite is well known for its high combustion velocity – even on the micron scale. This high combustion velocity is even more pronounced on the nano-scale, where the thermites reach the highest combustion velocities known for a mixture of a metal oxide and aluminum.

Although in the literature CuO/Al thermites have not been prepared by reactive milling or sol-gel methods, unique to the CuO/Al thermite is a new synthetic methodology by Gangopadhyay et al. called self-assembly. In this method, the authors coated copper nanorods (20 × 100 nm) with a coordinating polymer (P4VP), and then followed this by coating the coated rods with 80 nm aluminum powder. The result of this self-assembly process is the formation of the thermite with the highest combustion velocity known of 2,400 m / s. Tab. 13.8 gives the combustion velocity of the self-assembled thermite compared with physically mixed thermites.

Tab. 13.8: Combustion velocities of the CuO / Al thermite (80 nm Al, 2.2 nm oxide layer).

synthetic method	CuO morphology	combustion velocity (m/s)
physical mixing	micron	675
physical mixing	nanorods	1,650
self assembly	nanorods	2,400

Gangopadhyay et al. state that the reason for the increased combustion velocity is the highly improved interfacial contact between the fuel and oxidizer. However, the presence of an organic material binding the fuel and oxidizer will increase the combustion pressure, which has been shown by Weismiller and Bulian to result in an increase in the combustion rates. The dependence of the combustion velocity on pressure has been suggested to result from the decomposition of CuO to gaseous oxygen at 1,000 °C, which may be the active oxidizer.

Despite the reason for the high combustion velocity being under debate, the high combustion velocity is in the range of explosive velocities, therefore, the CuO /Al thermite may have application as a primary explosive. The more easily prepared, physically mixed CuO nanothermite has been patented for use in ammunition primers. Sensitivity data for the CuO /Al thermite is unavailable, but it is expected to be relatively high. High impact sensitivity may be the reason why there are no literature reports of its preparation by arrested reactive milling.

13.3.3 Example: Molybdenum trioxide / aluminum thermite

MoO_3 /Al has been prepared only by physical mixing and arrested reactive milling. Due to the availability of both MoO_3 and aluminum nanopowders, the preparation used exclusively for practical applications of these thermites is physical mixing. With combustion velocities ranging from 150–450 m/s, MoO_3 /Al thermites are perhaps some of the most widely studied nano-thermites, and extensive work has been reported on their practical applications in electric igniters, primers (patented), and microscale propulsion.

A study reported by Son et al. on the feasibility of the use of a MoO_3 /Al nanothermite (79 nm spherical Al, 30 × 200 nm sheet MoO_3) for microscale propulsion and microscale pyrotechnics shows great promise for the application of these materials. It was found that when confined in tubes with a diameter of 0.5 mm, the MoO_3/Al thermite would combust at rates as high as 790 m/s. This is an advantage over conventional energetic materials such as HMX, which cannot propagate in tubes under 1 mm.

Naud et al. have preformed extensive work on the use of the MoO_3 /Al nanothermite for application in electric matches. Electric matches prepared from this thermite mixture have been shown to have lower friction, impact, thermal and ESD sensitivities than the currently used matches, which contain toxic lead compounds making them both safer and more environmentally friendly to use.

13.4 Homemade explosives

Usually explosives can be categorized as military, commercial and homemade (HME) explosives. HMEs are energetic formulations that can be created "at home". The term HME has been used to cover a wide range of materials from pure explosive compounds, such as triacetone triperoxide (TATP), that can be synthesized from readily available chemicals, or pentaerythritol tetranitrate (PETN) since pentaerythritol can be purchased in bulk for paint use, to home-made variants of explosive formulations, such as ammonium nitrate fuel oil (ANFO), that are used in very large commercial blasting operations.

A major difference between military and commercial explosives on the one hand, and HMEs on the other is how they are produced: "made in a factory with rigorous reproducibility" (military/commercial) vs. "made anywhere else" (HME).

HMEs such as black powder, triacetonetriperoxide (TATP), hexamethylene-triperoxidediamine (HMTD), chlorates, and perchlorates mixed with sugar or other fuels have come to be used by terrorists instead of commercial explosives.

13.5 Explosive welding

Explosion welding or explosive welding (EXW) is a solid-state joining process in which welding is accomplished by acceleration of one of the components at extremely high velocity through the use of chemical explosives [see also: Lancaster, J. F. (1999). *Metallurgy of welding* (6th ed.). Abington, Cambridge: Abington Pub., **1999**, ISBN 1-85573-428-1]. Explosive welding uses a controlled explosive detonation to force two metals together at high pressure. When an explosive is detonated on the surface of a metal, a high-pressure pulse is generated which propels the metal at very high speed. If this piece of metal collides at an angle with another piece of metal, welding may occur. The composite system which results is joined by a durable, metallurgical bond. For welding to occur, a jetting action is required at the collision interface. This jet is the product of the surfaces of the two pieces of metals colliding. This cleans the metals and allows two pure metallic surfaces to join under extremely high pressure. The metals do not commingle; they are atomically bonded. Due to this fact, any metal can be welded to any metal (i.e. copper to steel; titanium to stainless steel). Typical impact pressures are millions of psi. Fig. 13.7 shows the explosive welding process.

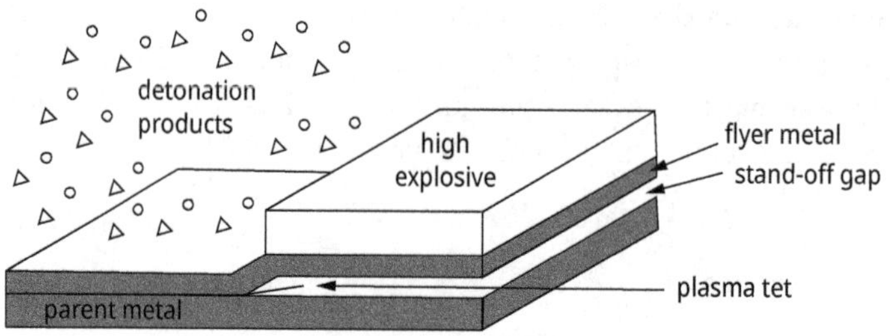

Fig. 13.7: Schematic diagram of the explosive welding process (planar geometry).

This process is most commonly utilized to clad carbon steel plates with a thin layer of a corrosion resistant material (e.g. stainless steel, nickel alloy, titanium, or zirconium). Due to the nature of this process, the geometries which can be produced are very limited and must be simple. Typical geometries produced include plates, tubing, and tubesheets.

Explosion welding can produce a bond between two metals that cannot necessarily be welded by conventional means. It is important to point out that the process does not melt either metal, instead the surfaces of both metals are plasticized, causing them to come into contact, which is sufficiently intimate to create a weld.

13.6 Gas generators (Airbags)

An airbag system consists of/functions as follows:
- an electrical impulse heats the bridgewire in the igniter
- the primary charge of the igniter (pyrotechnic mixture *e.g.* zirconium-potassium-perchlorate, ZPP) ignites and produces hot gases
- the booster charge (*e.g.* titanium-hydride-potassium-perchlorate, THPP, or boron and potassium nitrate) is subsequently ignited and further hot gases are produced
- the resulting hot gas ignites the propellant (tablets)
- the propellant (main gas-generator, mostly N_2) burns, and the resulting gaseous combustion products which are produced are directed into the airbag and fill it (see Fig. 13.8)
- this all occurs within approximately 60 ms

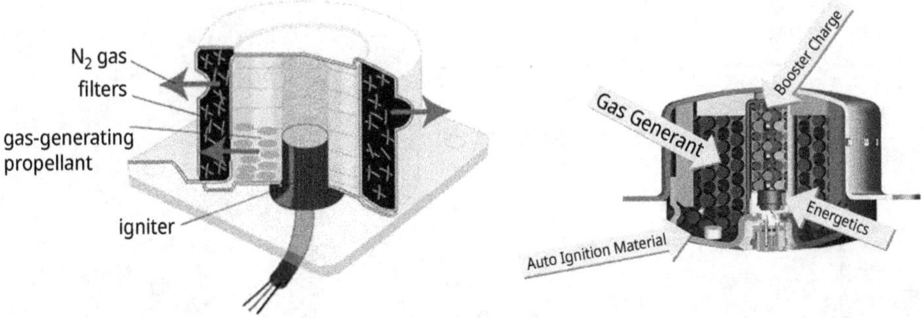

Fig. 13.8: Airbag inflation device (gas generator) (left) and gas generator with autoignition tablet (right): booster charge e.g. THPP, energetics = igniter, e.g. ZPP, autoignition material.

Sodium azide was used as the propellant in the past, however it is not used nowadays due to its toxicity, and also because of security concerns (terrorism: NaN_3 from scrap + $Pb(NO_3)_2 \rightarrow \ldots$).

$$42\,NaN_3 + 2\,KNO_3 + 8\,Fe_2O_3 + 5\,SiO_2$$
$$\rightarrow 21\,Na_2O + K_2O + 8\,Fe + 8\,FeO + 5\,SiO_2 + 64\,N_2$$

The iron oxide lowers the burning temperature by about 600 °C to 1,330 °C, the silicon dioxide acts as a slag up-taker. Within 30 milliseconds the propellant is completely combusted and produces approximately 40% nitrogen gas and 60% slag (weight percent). Nowadays, modern propellants have a gas yield of 60–99%.

Ammonium nitrate has also been used, however, this resulted in several accidents (explosion, not combustion) since in warm and moist areas (Florida) many phase transitions occurred with the corresponding density change → cracks, cavities . . . (Fig. 13.9, Tab. 13.9). Figure 13.9 shows the LT-DSC scan of commercial AN powder. The first endothermic peak at 52.8 °C and an exothermic peak at 54 °C indicate the transformation of phase IV to phase III. The second, third and fourth endothermic peaks at 85.8 °C, 125.7 °C and 169.8 °C correspond to the phase III to phase II transformation, phase II to phase I transformation and the melting of phase I at 1 atm respectively.

Currently, usually guanidinium nitrate (GN) or nitroguanidine (NiGu) are used as the main propellant together with a suitable oxidizer (e.g. basic copper nitrate). A typical gas-generating propellant composition could be:

guanidinium nitrate (GN)	52%
SiO_2	3%
basic copper nitrate	45%

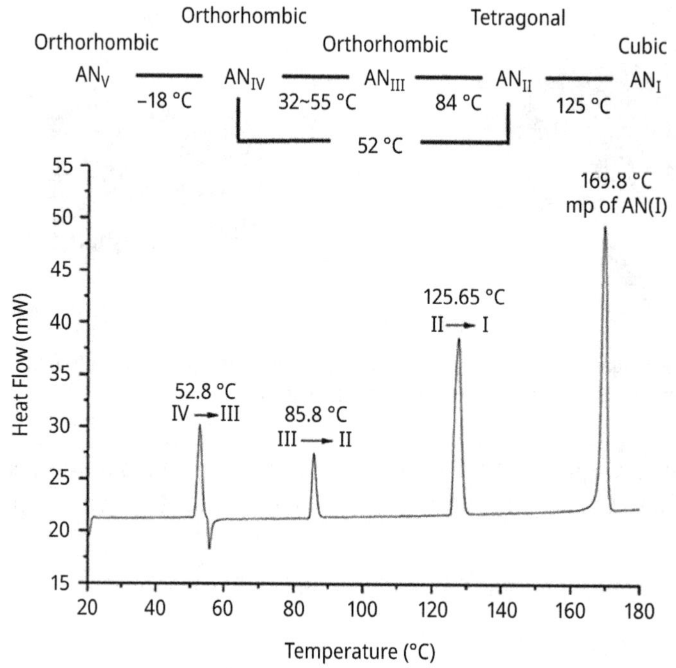

Fig. 13.9: Phase transitions in ammonium nitrate (AN) determined by DSC.

Tab. 13.9: Summary of crystal structures of AN.

Phase	**V**	**IV**	**III**	**II**	**I**
Crystal structure	orthorhombic	orthorhombic	orthorhombic	tetragonal	cubic
Phase transition temperature / K		255	305	357	398 441(melt)
Density / g cm^{-3}	1.75	1.70	1.57	1.60	1.55
Volume change / %		−2.8	+ 3.7	− 1.7	+ 2.3

Instead of basic copper nitrate, one can also use copper diammine dinitrate, e.g.:

guanyl urea nitrate	38%
SiO_2	3%
Copper diammine dinitrate	59%

Gas-producing compounds develop a large volume of gas when they are ignited. One category of gas-producing compounds uses a compound based on guanidine (HN=C $(NH_2)_2$) mixed with a sensitizing agent and/or oxidizer. For example, US Patent Nr.

2,165,236 from Holm reports on a gas-producing compound which contains Nitroguanidine in a binder. Part of the nitroguanidine can be replaced by guandinium nitrate ($H_2NC(NH)NH_2{\cdot}HNO_3$). Examples of typical binders are nitrocellulose and cellulose acetate.

Some advantages and disadvantages of different propellants in gas generators are summarized in Tab. 13.9a.

Tab. 13.9a: Advantages and disadvantages of some common fuels and oxidizers for gas generators.

fuel	formula	+	–
sodium azide	NaN_3	cheap	toxic
		reliable	may form HN_3
			potential of terrorist abuse
			does NOT produce CO, CO_2
			risk of formation of copper azide, $Cu(N_3)_2$
guanidinium nitrate	$[C(NH_2)_3]^+[NO_3]^-$	good performance	rel. expensive
			less energetic than NaN_3 and ATz
5-aminotetrazole	5-ATz	high energy per mass	expensive
			hygroscopic
di(ammonium)-5,5′-bis-1*H*-tetrazolate	BHT·2 NH_3	high energy per mass	expensive
			hygroscopic (?)
oxidizer	**formula**	**+**	**–**
phase-stabilized ammonium nitrate	$[NH_4]^+[NO_3]^-$, PSAN	cheap	hygroscopic
		high gas yield	multiple polymorphism near room temperature
		produces non-toxic gases: H_2O, N_2, O_2	endothermic decomposition: poor ignitability and burning performance
		safe to handle	
strontium nitrate	$Sr(NO_3)_2$	no phase issues	more expensive than PSAN
		less hygroscopic than PSAN	lower gas yield than PSAN
basic copper nitrate	$Cu(OH)(NO_3)$, BCN	no phase issues	more expensive than PSAN
		less hygroscopic than PSAN	lower gas yield than PSAN
		more energetic	potentially hazardous

Usually the phase stabilization of ammonium nitrate (AN) is achieved by adding potassium nitrate (KN) in an amount of 9–10 w-% to obtain phase-stabilized ammonium nitrate (PSAN). KN has a slightly smaller cationic radius ($r(K^+) = 1.38$ Å) in comparison with ammonium $\left(r\left(NH_4^+\right) = 1.48\,\text{Å}\right)$. Therefore, by replacing NH_4^+ ions with K^+ ions, the crystal structure of AN (originally phase IV) becomes close to phase III. KN doped PSAN is stable as phase III at room temperature, and phase transition and volume expansions are suppressed over a wide temperature range (193 ~ 373 K).

Guanidinium nitrate, however, also suffers from polymorphism. There are three structures known for guanidinium nitrate which are abbreviated as follows:

GuNi_GN1 (space group *Cm*),
GuNi_GN2 (space group *C*2),
GuNi_GN3 (space group *C*2/*m*).

GuNi_GN1 is the low-temperature phase. Figure 13.9a shows the cell volumes per GuNi unit as a function of the temperature.

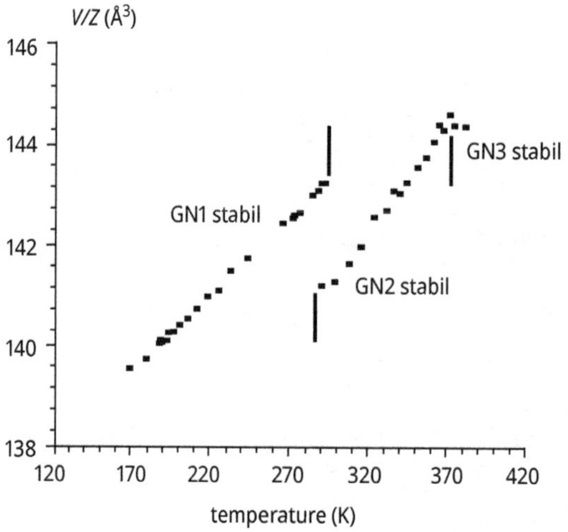

Fig. 13.9a: Cell volume per GuNi unit as a function of the temperature.

Figure 13.9a shows for GN1 a steady increase in volume with increasing temperature up to 292 K. Above 292 K there is a phase transition to GN2, which is associated with an abrupt decrease in cell volume (V) from 143.72 to 140.38 Å^3. This corresponds to a decrease of 2.4%. It is a first-order phase transition associated with a major structural change. In the case of a first-order phase transition, not only the molar volume changes, but also the entropy. The change in entropy becomes clear in a DTA experiment, whereby an endothermic peak appears upon heating. The transition from GN2 to GN3 (Fig. 13.9a) is a phase transition of second-order. The volume graph shows no discontinuity, just the thermal expansion, compressibility and specific heat (C_p) divided by the absolute temperature T (C_p / T) change. In a DTA experiment, this phase change would only be evident by kinking in the baseline. A second-order phase transition is mostly a disorder-order conversion. Other examples are the conversion from magnetic to ferromagnetic, or normal electrical conductivity to superconducting. In these examples, only the electronic order changes. Structurally, for example, the body-centered cubic structure of alpha iron is retained. As already mentioned, the volume does not change, but the thermal expansion does. An even more recent composition for a car airbag would be the following:

GuNi	25 – 35%
$KClO_4$	30 – 40%
NQ	25 – 35%
Basic copper nitrate	$Cu(NO_3)_2 \cdot 3\ Cu(OH)_2$

Literature

M. J. Hermann, W. Engel, Phase Transitions and Lattice Dynamics of Ammonium Nitrate, *Propellants, Expl. Pyrotech.*, **1997**, *22*, 143–147.

C.-O. Lieber, Aspects of the Mesoscale of Crystalline Explosives, *Propellants, Expl. Pyrotech.*, **2000**, *25*, 288–301.

T. Lee, J. W. Chen, H. L. Lee, T. Y. Lin, Y. C. Tsai, S.-L. Cheng, S.-W. Lee, J.-C. Hu, L.-T. Chen, Stabilization and Spheroidization of Ammonium Nitrate: Co-Crystallization with Crown Ethers and Spherical Crystallization by Solvent Screening, *Chem. Engineer. J.*, **2013**, *225*, 809–817.

13.7 Toxicity measurements

The **luminescent bacteria inhibition test** using **vibrio fischeri** is a well-established method to determine the aquatic toxicity. The determined parameter is the effective concentration (EC_{50}) after 15 and 30 minutes. The inhibition of the natural bioluminescence of these bacteria gives a first idea of the toxicity of compounds towards different ecosystems. It is a cost and time efficient experimental method, which does not involve animals.

The EC_{50} (effective concentration) values are usually determined using a spectrometer. The measurement is based on the bioluminescence of *Vibrio fischeri* NRRL-B-11177 bacteria strains, whereby the EC_{50} is the concentration level at which the bioluminescence is decreased by 50%. Prior to the measurement, a ten-point dilution series is prepared according to DIN/EN/ISO 11348 (without G1 level) with a known weight of the compounds and a 2% NaCl stock solution. The measurements, which must be strictly carried out at 15 °C, are then started by determination of the bioluminescence of untreated reactivated bacteria. After 15 and 30 minutes exposure time with a specific amount of component, the bioluminescence is determined again. The toxicity of the compounds can then be classified according to their EC_{50} values (non-toxic > 1.00 g L^{-1}; toxic 0.10–1.00 g L^{-1}; very toxic < 0.10 g L^{-1}).

To investigate trends in the aquatic toxicity towards *Vibrio fischeri*, different types of energetic materials were tested (Fig. 13.10). A small selection of possible primary and secondary explosives, oxidizers and pyrotechnical materials with different chemical constitution have been tested. Coordination (e.g. $[Cu(dtp)_3](ClO_4)_2$), neutral (e.g. azidoethanol and propyl-linked ditetrazoles) and ionic (e.g. polynitropyrazoles, -triazoles and tetrazoles) compounds, as well as compounds with different substitution patterns and energetic functionalities such as azido-, nitro-, flourodinitro- and nitramino-groups have been investigated. Finally, the results for these compounds were compared to those of commonly used materials, like RDX, ammonium perchlorate (AP) and azide salts.

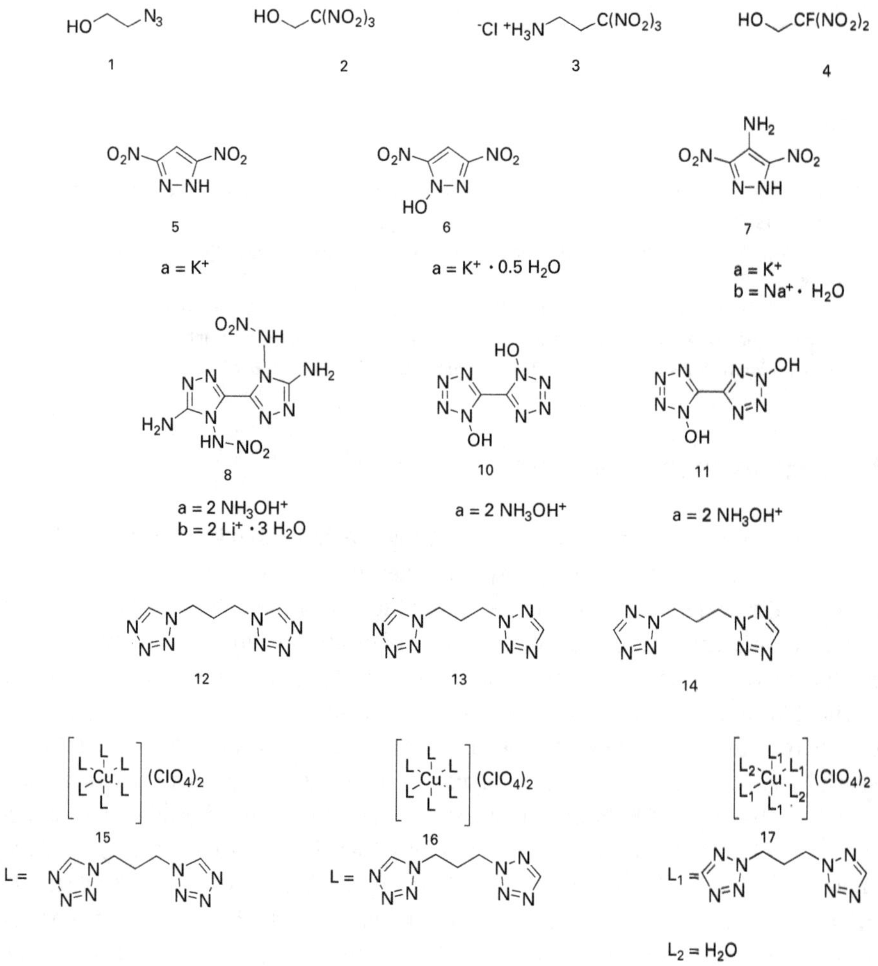

Fig. 13.10: Overview of measured compounds 1–17.

Comparing 2-azidoethanol (**1**), 2,2,2-trinitroethanol (**2**), 3,3,3-trinitropropylammonium chloride (**3**) and 2-fluoro-2,2-dinitroethanol (**4**), **1** is the least toxic compound towards *Vibrio fischeri* (Tab. 13.10). It is also relatively low in toxicity compared to the ionic azide compounds NaN_3 and NH_4N_3. Once adding a trinitroalkyl moiety the toxicity increases dramatically, and is even worse with a flourodinitroethyl moiety. This is consistent with previous toxicity measurements using gram positive bacteria such as *Staphylococcus aureus* and *Intestinal bacillus*. 2,4-Dintropyrozoles show toxicity towards the aquatic bacteria, though it is considered as moderately toxic. The findings were comparable, regardless if they were measured on a neutral compound (1-oxide or 3-amine), or measured on salts. The neutral nitraminotriazole **8** is very toxic towards *Vibrio fischeri*, whereas the hydroxylammonium salt **8a** is moderatly toxic, and the lithium salt **8b** is considered not toxic in this test. The substitution pattern can have a varying effect on the toxicity of tetrazoles. Tetrazoles **10a** and **11a** are both moderately toxic, whereas TKX-50 (salt **10a**) still is less toxic. Furthermore, it is even less toxic than the commonly used secondary explosive RDX. For tetrazoles **12–14**, which also have varying substitution patterns, the toxicity ranges from EC_{50} values of 0.36 g L^{-1} (**14**) to 10.30 g L^{-1} (**12**). On adding the copper(II) metal (which is known for being toxic to microorganisms), the toxicities of the tetrazole complexes **12–14** increases dramatically. Nonetheless, complexes **15–17** (which could be used as potential primary explosives) are less toxic compared to the measured azide salts.

Tab. 13.10: EC_{50} values of measured compounds after 15 and 30 min in g L^{-1} and their considered toxicity level after 30 min (+ / o / −).

Compound	EC_{50} (15 min) [g L^{-1}]	EC_{50} (30 min) [g L^{-1}]	Toxicity level	Compound	EC_{50} (15 min) [g L^{-1}]	EC_{50} (30 min) [g L^{-1}]	Toxicity level
NaN_3	0.25	0.18	o	**7a**	0.75	0.74	o
NH_4N_3	0.26	0.15	o	**7b**	0.60	0.58	o
NH_4NO_3	10.49	6.39	+	**8**	0.13	0.07	−
$NH_4N(NO_2)_2$	7.25	4.50	+	**8a**	0.75	0.35	o
NH_4ClO_4	14.58	11.13	+	**8b**	> 1.58	> 1.58	+
RDX	0.33	0.24	o	**10a**	1.17	0.58	o
1	8.70	8.55	+	**11a**	0.32	0.24	o
2	0.29	0.22	o	**12**	13.90	10.30	+
3	< 0.10	< 0.10	−	**13**	0.81	0.79	o
4	0.002	0.001	−	**14**	0.36	0.36	o
5	0.27	0.19	o	**15**	0.44	0.35	o
5a	1.21	0.95	+	**16**	0.64	0.44	o
6a	0.70	0.43	o	**17**	0.34	0.28	o

Another well-established method for evaluating the harmful potential of a certain substance is the **AMES test**. This OECD accepted method uses bacteria to test if mutations can be caused in the DNA of the organism and is the minimum test carried out in the course of REACH (Registration, Evaluation and Authorization of Chemicals).

The Ames test was first reported in the early 1970s by Bruce Ames at UC Berkeley, and is a reverse mutation test using bacteria to detect mutagenic properties of a test substance, and is conducted in vitro. It uses bacteria to test whether a given chemical can cause mutations in the DNA of the test organism and provides an animal-free, inexpensive, quick and easy method to estimate the carcinogenic potential of the test compound. A positive result in the Ames test suggests that the chemical may be mutagenic and could potentially act as a carcinogen, since cancer is often associated with genetic mutations. However, with the Ames test, false positives and false negative results can potentially occur.

In the Ames test, bacterial strains such as *Salmonella typhimurium* or *Escherichia coli* are introduced into a histidine-free nutrient solution. The bacteria used cannot synthesize histidine due to a defective gene, despite histidine being essential for its growth. If a mutagenic substance is added, this substance can cause a reverse mutation of the defective gene in the bacteria, which enables the bacteria to produce histidine and grow – even though no external histidine has been added (Fig. 13.10a). Since different strains of bacteria show varying specificity in their response to mutagens, multiple test strains are required to detect the various classes of mutagens.

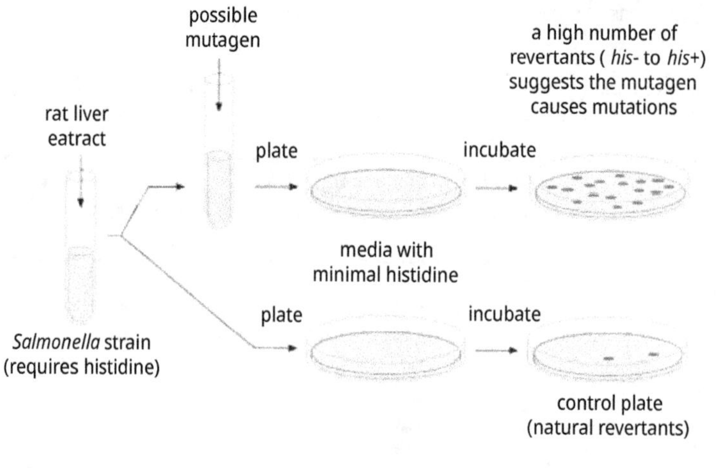

Fig. 13.10a: Schematic representation of the AMES test.

The Ames test is used to detect point mutations, which involves substitution, addition, or deletion of one or a few DNA base pairs. The OECD Guidelines for Testing of Chemicals (No. 471) form the basis for the test, which employs auxotrophic strains of *Salmonella typhimurium* (TA98, TA100, TA1535, TA1537) and *Escheria coli* (wp2[pKM101] + wp2 uvrA mixed 1 : 2). The biosynthesis of the corresponding amino acids is blocked by point mutations made in the histidine (*Salmonella typhimurium*), or the tryptophan (*Escheria coli*) operon. The mutagenic potential is evaluated by detecting the appearance of the reverse mutants of the auxotrophic strains, making them prototrophs, able to grow in corresponding deficient media.

The mutagenic potential of salts **8b** (Li_2ANAT) and **10** (TKX-50) were evaluated using the AMES test with S9 mixture. The toxicological potential of salt **7a** was evaluated using *Vibrio fischeri*, due to its high water-solubility and because it is the water-soluble monomer of BDNAPM, a highly potential secondary explosive. BDNAPM (Fig. 13.11) was also experimentally tested with the AMES test, but cannot be evaluated using *Vibrio fischeri* itself, because of its low water-solubility. All experimental findings were calculated using a smart algorithm and are listed in Tab. 13.11. The toxicity and mutagenic potential of the compounds in Tabs. 13.10 and 13.11 are rated as non-toxic/mutagenic (+), moderately toxic/mutagenic (o) and toxic/mutagenic (−).

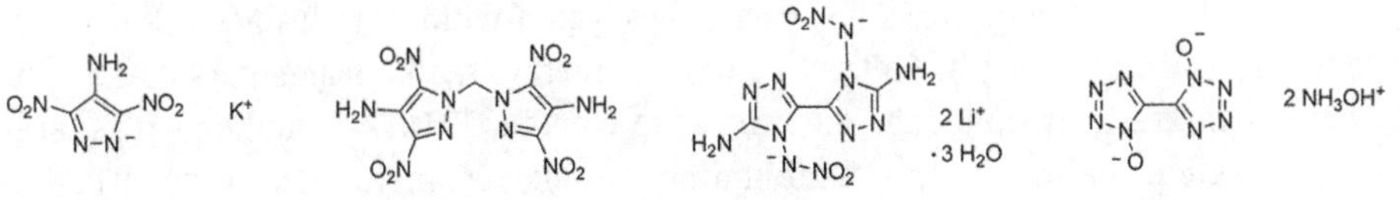

Fig. 13.11: Compounds evaluated using AMES test and *Vibrio fischeri*.

Tab. 13.11: Results of experimental and in silicio AMES test and the considered mutagenicity/toxicity level (+ / o / −).

Compound	AMES test	Vibrio fischeri
7a	n.a.	o
BDNAPM	–	n.a.
Li_2ANAT	o	+
TKX-50	+	o

Check your knowledge: See Chapter 14 for Study Questions.

Literature

C. Unger et al., *ICT Annual Conference* **2019**, P 85.

13.8 Energetic MOFs (EMOFs)

The term MOF is an abbreviation that is used to describe a metal organic framework. Metal organic frameworks are coordination networks that are constructed from metal coordination sites (inorganic joints) and organic ligands (struts) through strong bonds. This results in extended, open, infinite, robust network architectures, which contain potential permanent voids. [1–3] Typically, MOFs are formed by self-assembly of the components in solution, using organic ligands which contain oxygen or nitrogen donor atoms. MOFs with pore sizes in the 2–47 Å range with 1,000–6,000 m^2/g surface areas have been described in the literature [3]. As a result of their properties, MOFs have been intensively investigated over the last decade for use in many fields – from selective heterogeneous catalysis [2] to selective CO_2 capture [2]. A typical method for preparing a MOF on a < 1 g scale involves heating a mixture of the metal salt and organic ligand in a solvent for extended periods of time (12–48 h). A mechanochemical method can also be used, in which no solvent is involved and the two components are simply ground together using a ball mill [3].

Recently, the idea of MOFs has been investigated within the framework of energetic materials, resulting in EMOFs – energetic metal organic frameworks. And considerable interest surrounds the question as to whether EMOFs could be next generation energetic materials. The combination of selected metal ions with bridging energetic linkers can result in the formation of 1-D, 2-D or 3-D networks that are highly energetic (Fig. 13.12) [4]. Whether a 1-D, 2-D or 3-D network is formed, depends on the metal ion geometry (joint) and also on the denticity of the energetic organic ligand (strut).

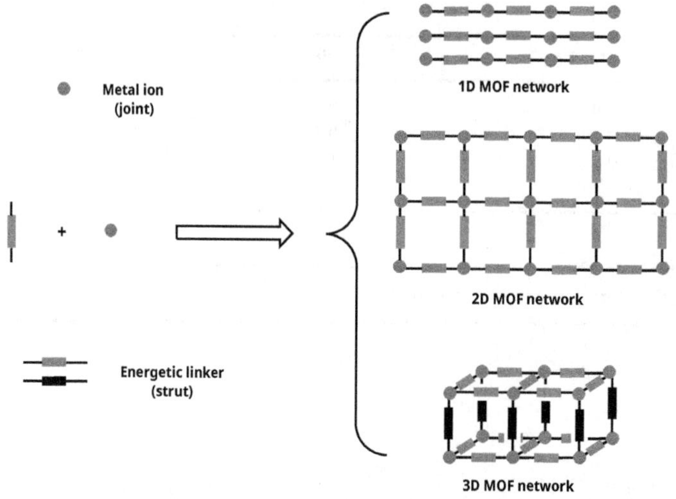

Fig. 13.12: Metal ions (joints) and energetic molecules (struts) can undergo self-assembly to form energetic MOFs with 1D, 2D or 3D network structures [4].

Generally, it has so far been observed that 2-D MOFs show lower sensitivity to friction, impact and spark, but crucially, retain the energy content of similar 1-D EMOFs and show high thermal stabilities [4]. Several examples of 3-D energetic MOFs have been shown to be air stable and insoluble in most common organic solvents, while demonstrating good thermal stabilities. Significantly, they also show much lower sensitivities to friction, impact and electrostatic ignition sources, and have higher heats of detonation than 1-D or 2-D MOFs. One drawback however, is that 3-D MOFs can form structures in which pores of various shapes and sizes are present, which can result in a lower density [4]. Such properties are of great significance when attempting to prepare future primary explosives with improved stability and safety. Although hydrazine is not strictly an organic compound, 1-D materials with hydrazine (N_2H_4) as the struts and either Co^{2+} or Ni^{2+} metal ions as the joints in the compounds $[Ni(N_2H_4)(ClO_4)_2]_n$, $[Ni(N_2H_4)_3(NO_3)_2]_n$ (Fig. 13.13) and $[Co(N_2H_4)_5(ClO_4)_2]_n$ have been described in the literature as being early examples of energetic MOFs [5]. Meanwhile, a considerable number of EMOFs have also been prepared using energetic ligands such as triazole and tetrazole-based struts instead of hydrazine.

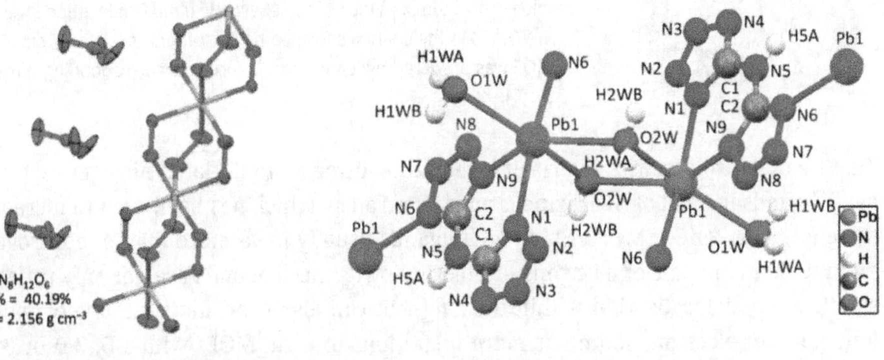

Fig. 13.13: Nickel hydrazine nitrate (NHN) an early example of a 1-D EMOF (left), $[Pb(bta)\cdot 2H_2O]$ excerpt of the polymeric $[Pb(bta)\cdot 2H_2O]_n$ 3-D energetic framework (right) [5].

In addition to categorizing EMOFs based on whether they are 1-D, 2-D or 3-D network structures, they can also be classified based on whether they are neutral, cationic or anionic. This categorization depends on whether the main skeleton is neutral or has a positive/negative charge. Currently, EMOFs most commonly exhibit a neutral or cationic skeleton. EMOFs with a cationic skeleton are formed by the coordination of neutral energetic ligands (often through lone pairs of electrons on N atoms in the ligand) with metal ions. Neutral skeletons are formed by the coordination of negatively charged energetic ligands with positively charged metal ions. However, anionic skeletons are rarer, and are formed by coordination of metal ions with negatively charged energetic ligands, but which results in an overall negatively charged main body skeleton. Cations are incorporated in pores of the skeleton to balance out the charge. Suitable cations include (amongst others) NH_4^+, $N_2H_5^+$, NH_3OH^+, $C(NH_2)_3^+$ (GU), $C(NH_2)_2(NHNH_2)^+$ (AG), $C(NHNH_2)_3^+$ (TAG), diamino-1,2,4-triazolium or triamino-1,2,4-triazolium.

Exchanging nitrate or perchlorate anions for more energetic and halogen-free anions such as the 5-nitrotetrazolate anion is another strategy that has been used to increase the energy content of MOFs. Using ion exchange, it has been shown that ions such as NO_3^-, ClO_4^- or BF_4^-, which are located in the pores within the cationic metal-ligand framework, can be exchanged by higher energy anions such as the dinitramide ($N(NO_2)_2^-$) or nitroformate ($C(NO_2)_3^-$) anions. This is possible since the ions located in the skeleton pores only exhibit weak electrostatic interactions with the cationic skeleton framework (Fig. 13.14).

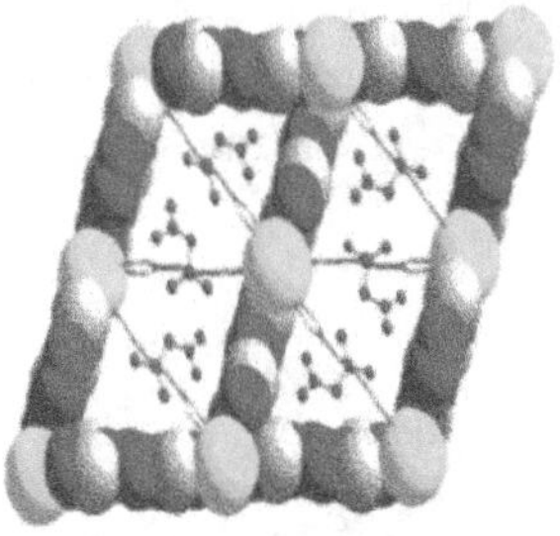

Fig. 13.14: The structure of the 3-D EMOF $\{Cu(atrz)_3[N(NO_2)_2]_2\}_n$ which shows the presence of the energetic dinitramide anion ($N(NO_2)_2^-$) in the channels formed by the cationic skeleton. This EMOF was prepared by ion exchange form the corresponding nitrate salt [6].

The use of triazole, tetrazole, triazine, tetrazine, furazan and related nitrogen-rich organic ligands instead of hydrazine is another strategy which has been used to increase the energy content of MOFs. Since such ligands usually have more than one nitrogen atom with a lone pair of electrons located in a ring, these ligands are not only usually stabilized by delocalization within the ring, but can also coordinate via one or more lone pairs of electrons to one or more metal ions to form MOFs with 1-D, 2-D or 3-D structures. A summary of the main features of these EMOFs is given in Tab. 13.12 [6].

As was mentioned above, obviously, the energetic properties of EMOFs can, in principle, be tuned by changing the central metal ion and the energetic ligands used. Some of the nitrogen-rich heterocycles which are used as struts in energetic MOFs are shown in Fig. 13.15. The use of such ligand results in the formation of nitrogen-rich EMOFs. Ultimately, by carefully selecting the metal salt and organic ligand, it is hoped to be able to tune the properties of future MOF-based primary explosives [6].

The main areas of application for which EMOFs are currently being considered are: green primary explosives, highly thermally stable secondary explosives, as well as energetic material catalysts. Furthermore, the development of new, green, heavy-metal free primary explosives has been a goal for many years. Cyanuric azide and DDNP are both metal-free primary explosives, however, there are stability issues. Examples of EMOFs with the properties of primary explosives are given in Tab. 13.13, and their properties compared with those of lead azide, however, many still contain undesirable heavy metals.

Tab. 13.12: Summary of the main features of EMOFs categorized based on the charge of the main body skeleton [6].

property	charge of main body skeleton		
	cationic	**neutral**	**anionic**
components	metal ion (cation) and neutral energetic ligand	metal ion (cation) and negatively charged ligands	metal cation and negatively charged ligands
coordination	energetic ligands via N lone pairs of electrons to the M^{x+} cation	energetic ligands via N lone pairs of electrons to the M^{x+} cation	energetic ligands via N lone pairs of electrons to the M^{x+} cation
pores	anion e.g. NO_3^-, ClO_4^- which balances positive charge of main body skeleton	EMOFs often stabilized by small solvent molecules in channels	contain cations to balance the negative charge of the main body skeleton
EMOF example	energetic anions used to replace NO_3^- or ClO_4^- e.g. 5-nitrotetrazolate anion ($N_4C-NO_2^-$) in planar, 2-D, $[Cu(atrz)_2(CN_5O_2)_2 2H_2O]_n$ 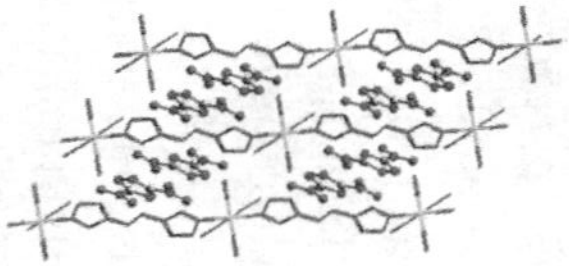	$[Cu(bta)(NH_3)_2]_n$ (1-D), $[Pb(Htztr)_2(H_2O)]_n$ (2-D), $[Cu(Htztr)]_n$ (3-D). In the latter two, tetrazoliumtriazole loses two H^+ to form the 2− charged anion	e.g. $[(AG)_3(Co(BTM)_3)]$ in which BTM = Bitetrazomethane, AG = Aminoguanidinium 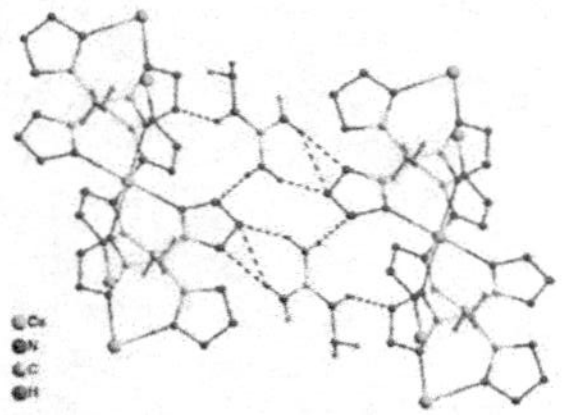

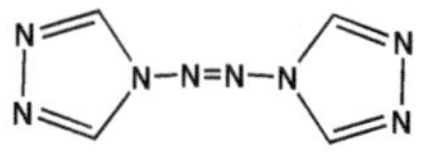

Diazo-bis(triazole)
% N = 60.1

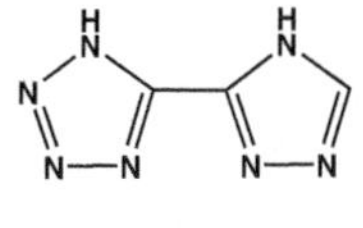

3-(Tetrazol-5-yl)triazole (H_2tztr)
% N = 71.54

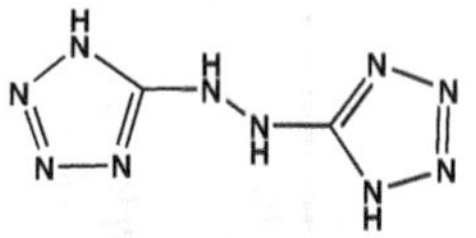

5,5'-Hydrazinebistetrazole
% N = 83.31

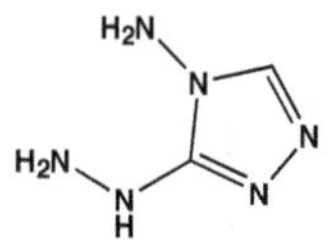

3-Hydrazino-4-amino-1,2,4-triazole (HATr)
% N = 73.65

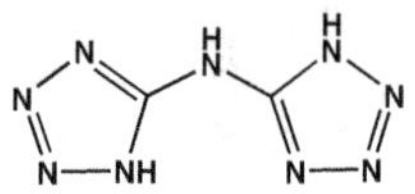

Bis(1*H*-tetrazol-5-yl)amine (H2BTA)
% N = 82.34

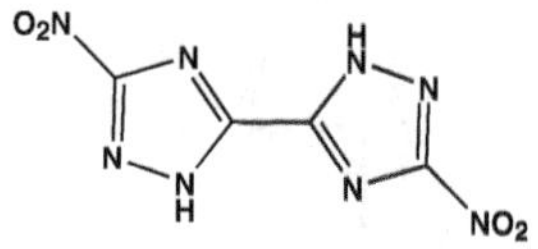

Bi-5,5'-dinitro-2*H*,2*H*'-3,3'-bi-1,2,4-triazole (DNBT)
% N = 68.3

Fig. 13.15: Examples of nitrogen-rich heterocycles which are used as struts to form nitrogen-rich EMOFs [6].

Tab. 13.13: Examples of EMOFs investigated as possible new primary explosives [6] (PA = picrate).

compound	ρ (g cm^{-3})	% N	T_{dec} (°C)	IS (J)	FS (N)
$Pb(N_3)_2$	4.800	28.9	315	2.5–4.0	0.1 – 1
$[Pb(bta)(H_2O)]_n$	3.412	33.5	314		
$\{[Cu(ATZ)](ClO_4)_2\}_n$	1.400	32.7	> 250	1	10
$[Cu(atrz)_3(NO_3)_2]_n$	1.680	53.4	243	22.5	112
$[Cu(tztr)]_n$	2.216		360	> 360	> 24.75
$\{[Zn(ATZ)_3](PA)_2 \cdot 2.5H_2O\}_n$	1.757	30.9	276.3	27.8	

Thermally highly stable secondary explosives are used for applications in which extreme resistance to heat is required, such as in deep oil wells. Several examples of EMOFs with such thermal stability have been reported in the past decade. For example, the EMOF which results from the one-step regioselective coupling of 3,4,5-trinitropyrazole with ANTA in aqueous KOH, namely, potassium 4-(5-amino-3-nitro-1*H*-1,2,4-triazol-1-yl)-3,5-dinitropyrazole was shown to possess a thermal stability ($T_{dec.}$ = 323 °C) similar to that of even TATB ($T_{dec.}$ = 330 °C), whilst exhibiting a relatively high density of 1.980 g cm^{-3} and better explosive performance parameters (heat of detonation = 5,965 kJ/kg, VoD = 8,457 m/s, $P_{C\text{-}J}$ = 32.5 GPa) than TATB [6].

Another potential area of application for EMOFs is that of metastable intermolecular (or intermixed) composites (MICs). MICs are used as additives in solid propellant formulations and liquid fuels and generally contain aluminum and an oxidant. However, the oxidation of nano-aluminum is a problem, and attempts to improve the performance focus on decreasing the particle size. Finally, it has been proposed that EMOFs with energetic bistetrazole ligands may be useful as catalysts in propellant and pyrotechnic formulations. However, clearly, the chemistry of EMOFs remains in its infancy.

Literature

[1] S. R. Batten, N. R. Champness, X. - M. Chen, J. Garcia-Martinez, S. Kitagawa, L. Öhrström, M. O'Keeffe, M. P. Suh, J. Reedijk, *Pure Appl. Chem.*, **2013**, *85*, 1715–1724.
[2] J. R. Long, O. M. Yaghi, *Chem. Soc. Revs.*, **2009**, *38*, 1213–1214.
[3] K. Naka, *Encyclopedia of Polymeric Nanomaterials*, **2014**, Springer-Verlag, Berlin, Heidelberg, pp. 1–6.
[4] Q. Zhang, J. M. Shreeve, *Angew. Chem. Int. Ed.*, **2014**, *53*, 2540–2542.
[5] a) Q. Liu, B. Jin, Q. Zhang, Y. Shang, Z. Guo, B. Tan, R. Peng, *Materials*, **2016**, *9*, 681. b) O. S. Bushuyev, P. Brown, A. Maiti, R. H. Gee, G. R. Peterson, B. L. Weeks, L. J. Hope-Weeks, *J. Am. Chem. Soc.*, **2012**, *134*, 1422–1425.
[6] L. Xu, J. Qiao, S. Xu, X. Zhao, W. Gong, T. Huang, *Catalysts*, **2020**, *10*, 690.

13.9 Civil explosives and boosters

Civil explosives

Civil explosives are defined as all those, which have application outwith military purposes. Examples of such areas of application are stone quarries, mining or tunneling, as well as the oil and gas industry, building and demolition industries. Civil explosives can be classified as being either molecular explosives or composite explosives. The former are usually single secondary explosives such as PETN (Pentaerythritol tetranitrate) or RDX (Hexogen), whereas the latter are mixtures of two or more compounds such as ANFOs or emulsion explosives.

An example of a group of composite explosives is ANFOs. Ammonium nitrate fuel oils are mixtures that mainly consist of ammonium nitrate (AN) and approximately 5% of an oil, which is absorbed by special very porous AN prills. Capillary forces hold the oil in the pores. The detonation velocity of ANFOs depends on the diameter of the borehole, but is usually between 2,500 – 4,000 m s^{-1}. The detonation velocity increases with increasing porosity of the prills, whereas the density of the mixture decreases with increasing porosity. ANFOs are very safe with respect to handling and transport. They are transported either in trucks or packed in bags. Mixing of the prills and oil can be carried out directly at the location, and also immediately before blasting. Overall, ANFOs are very cheap and easy-to-prepare explosives. However, due to their good solubility in water, they are not suitable for use in moist boreholes. In addition, ANFOs have low densities which limit the blasting energy.

Another group of composite explosives is slurries, so called watergels. They consist of an oxidizing nitrate salt such as ammonium nitrate, sodium nitrate or calcium nitrate and a carbon-rich fuel such as oil. By adding cross-linking agents, the mixture becomes gel-like and resistant to water. The mixture can be sensitized by adding glass micro-balloons, or by gassing the mixture. By adding aluminum, the energy content of the mixture can be increased further. Slurries are very safe in terms of handling and transport and can be transported either using tank trucks or packed in cartridges.

The sensitization process, for example by chemical gassing, can also take place at the blasting location, which ensures even safer transport. The cartridges are packaged in waxed paper or plastic foil. Depending on the borehole diameter, such slurries have an average detonation velocity of 4,000 – 5,000 m s^{-1}.

The so-called emulsion explosives are very similar to slurry explosives. By definition, an emulsion is a mixture of two liquids that are usually not miscible. Emulsion explosives consist of a continuous and a discontinuous phase. The continuous phase is comprised of a carbon-rich fuel and an emulsifier, while the discontinuous phase consists of a saturated aqueous solution of oxidizing nitrate salts. To blend these two phases, an emulsifier and mechanical energy supply are necessary. Emulsifiers belong to the class of surfactants, and have both hydrophilic and lipophilic properties, meaning they can interact with both phases, thus resulting in a lowering of the interfacial tension and enabling miscibility. In emulsion explosives, mainly the non-ionic sorbitan monooleate (SMO), or ionic polyisobutylene succinic anhydride (PIBSA) emulsifiers are used, which are shown in Fig. 13.16.

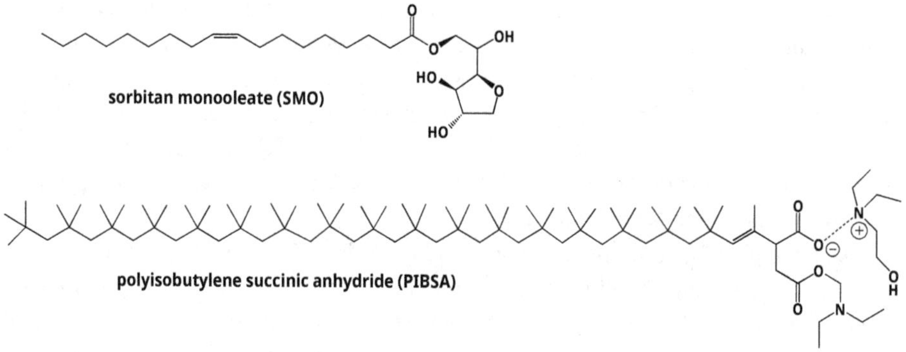

Fig. 13.16: Chemical structures of SMO and PIBSA.

The emulsions are prepared by heating the discontinuous phase until all of the salts are completely dissolved. While stirring vigorously, the still warm salt solution is added to the discontinuous phase, and the water-in-oil-emulsion matrix forms, as is shown in Fig. 13.17.

By applying mechanical energy, the aqueous solution becomes finely dispersed into small drops, and the emulsifier arranges itself to stabilize the emulsion, as shown in Fig. 13.18.

This process is called emulsification. By further refinements, the drops can be made smaller and the viscosity of the emulsion can be adjusted to the desired value. By extending the duration of stirring, the water drops become smaller, and therefore the viscosity increases. The water drops in the emulsion matrix are coated by an oil film, which makes the mixture water resistant. Emulsion explosives can also be sensitized by creating voids, as was described previously for slurries. During initiation, the

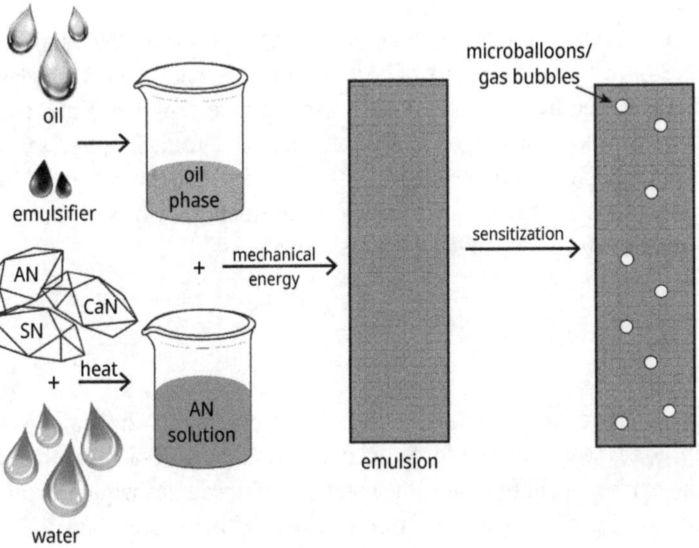

Fig. 13.17: Preparation and composition of an emulsion explosive.

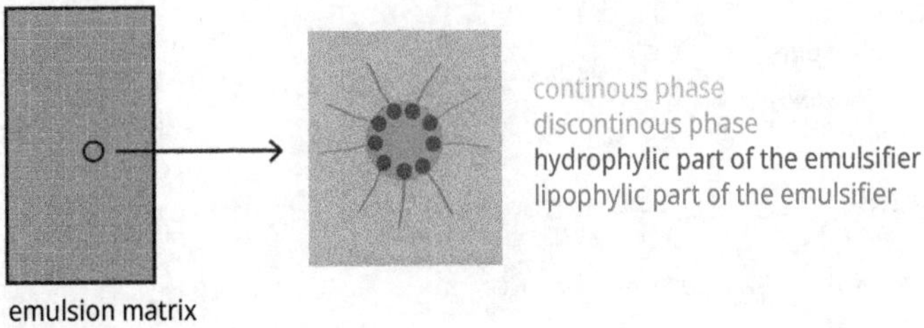

Fig. 13.18: Arrangement of the emulsifier in a water-in-oil emulsion.

air holes in the hollow glass micro-balloons (usual size about 60–70 microns) or the inserted gas bubbles are compressed and a local high pressure is created. This compression is almost adiabatic, which creates hotspots at the interface between the bubbles and the mixture. The energy released at these hotspots is used as activation energy for the explosives, and therefore results in the detonation of the whole explosive system. If an emulsion explosive is pumpable, it can be transported in tank-trucks. In addition, emulsion explosives can also be packaged in cartridges. Emulsion explosives are very safe to transport, use and handle, due to their low sensitivity. The energy of the emulsion increases with decreasing water content of the mixture. The average detonation velocity for emulsion explosives is between 5,000 – 5,200 m s^{-1}.

Mixtures of porous ammonium nitrate prills and a bulk emulsion (pumpable emulsion matrix) are called heavy ANFOs (HANFOs). These mixtures contain less than or equal to 50% emulsion. The advantage of HANFOs compared to ANFOs is their higher resistance towards water which means that they can be used in wet boreholes. They also have a higher density and higher energy in comparison with ANFOs.

A further distinction within the category of civil explosives is made between high explosives, which can be ignited by a detonator alone (cap-sensitive explosives), and blasting agents, which require a booster charge to be ignited (booster-sensitive explosives).

Some composite explosives, are blasting agents. They are not cap sensitive, and for reliable ignition, a booster charge must be placed between the cap and the main charge. Such explosive mixtures are called boosters or primers.

Boosters

A booster is an energetic composite placed between the detonator and the main explosive charge. It helps to transmit and reinforce the detonation wave created by the detonator, towards the explosive charge (blasting agents), which can consequently more easily ignite. Boosters are mainly prepared from mixtures of high explosives such as TNT, PETN, RDX, Tetryl or NG, which are shown in Fig. 13.19.

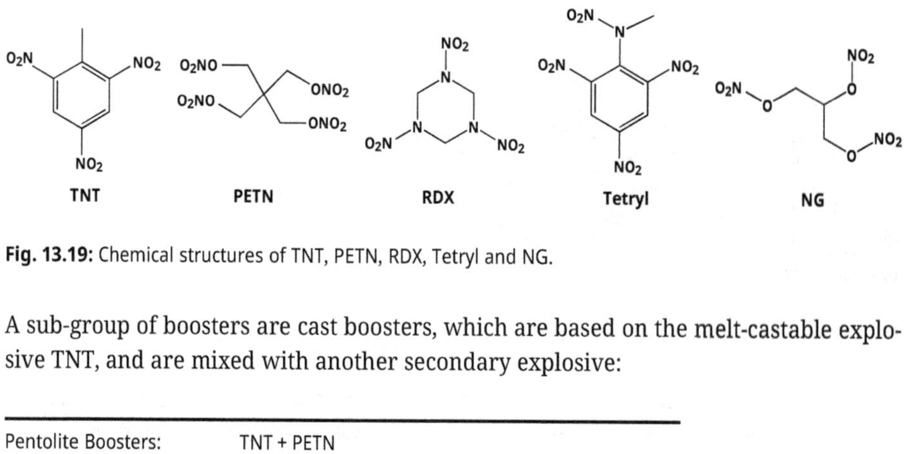

Fig. 13.19: Chemical structures of TNT, PETN, RDX, Tetryl and NG.

A sub-group of boosters are cast boosters, which are based on the melt-castable explosive TNT, and are mixed with another secondary explosive:

Pentolite Boosters:	TNT + PETN
Composition B Booster:	TNT + RDX
Torpex Booster:	TNT + RDX and Al
Tetrytol Booster:	TNT + Tetryl
Amatol/Sodatol Booster:	Pentolite or Composition B which contains AN or SN

In addition, there is also the group of plasticized boosters. Plasticized boosters mainly contain PETN and an elastomeric binder. Due to the binder, they have the appearance and texture of rubber. They have high resistance to moisture and are therefore often used for work on underground boreholes. In addition, they show long-term stability and are very safe to handle.

Typical characteristics and desired properties for a booster are as follows:
High densities ≥ 1.5 g cm^{-3}
High detonation velocity ≥ 7,000 m s^{-1}
High detonation pressure ≥ 15 GPa
High energy of formation ≥ 3,500 kJ kg^{-1}
Long shelf-life even under difficult storage conditions (> 2 years)
Good and safe handling characteristics (CEN 13631: IS > 2 J, FS > 80 N)
Environmentally friendly decomposition products
Cheap, simple and safe production

As was previously mentioned, boosters are used to increase the energy of the detonator during a blast, in order to ignite the main charge – such as an emulsion explosive or ANFO.

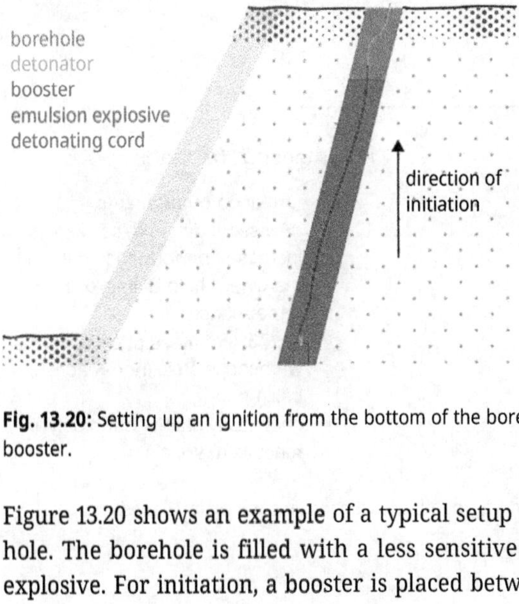

Fig. 13.20: Setting up an ignition from the bottom of the borehole using a detonator, detonating cord and booster.

Figure 13.20 shows an example of a typical setup of a blast from the bottom of a borehole. The borehole is filled with a less sensitive explosive, in this case an emulsion explosive. For initiation, a booster is placed between the detonator and the explosive in order to propagate the initiation.

Literature

E. G. Mahadevan, Ammonium Nitrate Explosives for Civil Applications, Wiley-VCH, Weinheim, 2013.
M. A. Cook, Science of High Explosives, Reinhold, USA, 1958.

13.10 Structured reactive materials (SRM)

The effectiveness of lethal weapons is predominantly dependent on the overpressure generated and blast effect. Therefore, if such weapons employed new secondary explosives which show significantly higher performance, this would help to down-size the weapon system. If however, the weapon system is one which relies predominantly on the formation of fragments (fragmenting warheads, Fig. 13.21) for its effectiveness, incorporating new high explosives – even those which show a significant improvement in performance – can only marginally help to improve the effectiveness of the weapon. In cases such as these, the key to improving the effectiveness could lie in enabling these weapons to generate chemically reactive fragments derived from structured reactive materials. The reason for this, is that in conventional warheads, up to 75% of the mass is due to the non-reactive casing (steel casing). The use of chemically stable, but highly reactive materials for the casing instead of steel could help to solve this problem.

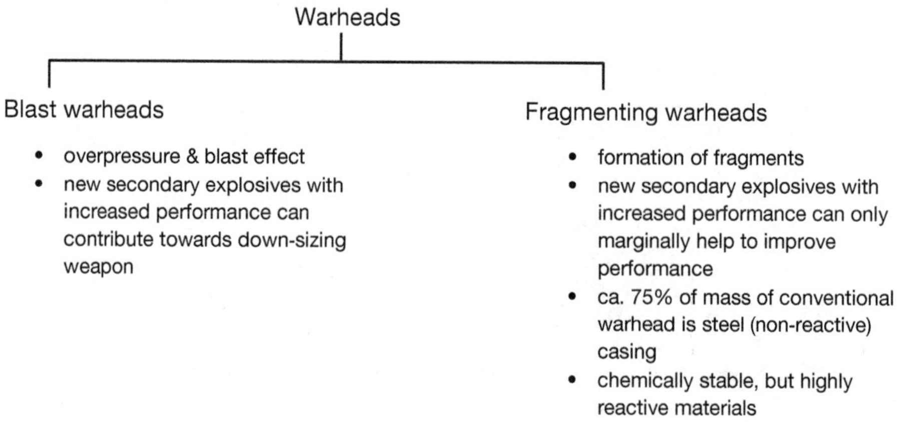

Fig. 13.21: Comparison of aspects of blast and fragmenting warheads (in addition to shaped charge warheads).

Reactive materials (or structured reactive materials) are a new class of materials that are being investigated for use in weapons systems, in order to increase the lethality of direct-hit or fragmentation warheads (see also chapters 7.1 and 12). Reactive materials (usually thermites or thermite-like compositions) consist of two or more nonexplosive solid materials, which – crucially – do not react with each other until they are exposed to a strong mechanical stimulus. After the appropriate stimulus has been applied, the solid materials undergo fast burning or detonation which releases a large amount of energy (combustion, detonation) in addition to the kinetic energy. This is nicely illustrated by Fig. 13.22, which shows that using TNT as the explosive (30% mass percent), the use of a chemically non-inert casing can add an additional 40–70% of the total chemical energy produced by the explosive munition.

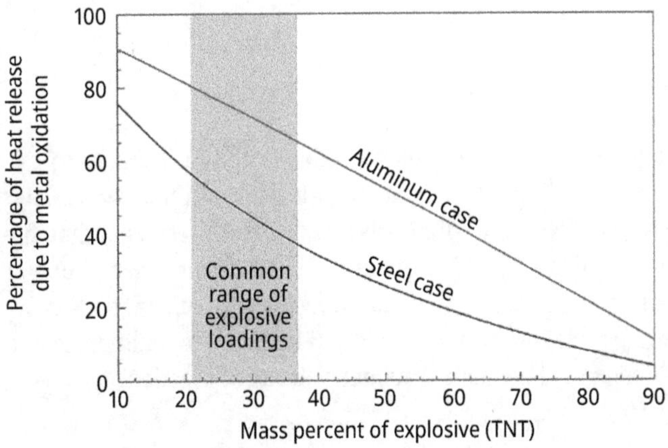

Fig. 13.22: Estimation of the contribution to the heat released if complete oxidation of a metal case occurs for a munition containing 30% TNT as explosive. Diagram copied from [Hastings and Dreizin, see below] used with kind permission of the publisher.

Consequently, fragments or projectiles made of reactive materials show a significantly larger damaging effect than non-reactive counterparts. Structured reactive materials (SRM) are therefore expected to enhance the lethality of weapons, while maintaining or reducing the mass of the payload.

Possible candidate classes of materials under investigation as SRMs are:

- thermites
- intermetallic phases (alloys)
- metal-polymer mixtures (e.g. Al-PTFE)
- composites (MIC, metastable intermolecular composites)

Such SRMs should possess the following properties:

- High density
- High hardness
- Fast, highly exothermic reaction on hitting the target
- High mechanical strength (strong enough to act as structural component)
- Stable to survive extreme processes of launch, penetration
- Easily prepared, easily transported, safe to handle
- Unstable enough to allow reliable ignition when stimulus is applied
- Undergo rapid conversion from being a structural component to fine powder

Possible applications for SRMs are expected to include the following:
- Kinetic penetrators (high density and high reactivity are important)
- Reactive fragments
- Reactive bullets
- Reactive armor
- Munitions casings

At the first stage of designing new SRMs, one initial selection criteria is the heat of oxidation of the metals since the greater the energy released the better. In this context, it is interesting to compare the heats of reactions of metals forming their corresponding fluorides, oxides, borides and carbides (Fig. 13.23). Several trends become clear, such as that the fluorination and oxidation reactions of metals are generally substantially more exothermic than those forming metal borides or carbides (Fig. 13.23). In addition, generally, the metals with lower densities show high gravimetric heats of reaction.

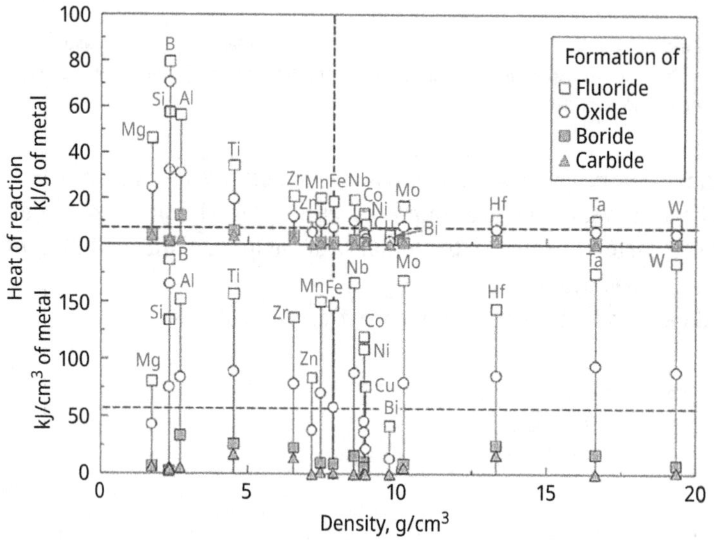

Fig. 13.23: Comparison of the heats of reaction for the formation of metal fluorides, oxides, borides and carbides.

Obviously, obtaining the maximum energy release from the formation of metal fluorides or oxides is hugely important.

Metal-polymer mixtures

The topic of metal-polymer mixtures is usually associated with shaped charge (SC) liners or explosively formed projectiles (EFP). The reactive material liner is a kind of integrated solid energetic liner (and a special type of structured reactive material) prepared by filling a metal powder in a polymeric binder, such as PTFE/Al and PTFE/Cu. Such liners are usually produced by cold-pressing followed by a sintering process under an inert N_2 atmosphere. Generally, the reactive material liner under the shaped charge effects can form a reactive jet that has a high enough velocity to enable it to penetrate the target, and subsequently, its violent chemical energy is released inside the penetration crater. A recent example of such a mixture is the system PTFE/W/Cu/Pb which shows higher penetration depth compared to the original PTFE/Al liner. Figure 13.24 shows an example of a pressed and sintered PTFE/W/Cu/Pb SC liner. Some of the disadvantages of metal-fluoropolymer SRMs such as composites of Al/PTFE are that they often exhibit low densities and low strength. One example of such an Al/PTFE composite is 26.5 wt.% Al/73.5 wt.% PTFE, which can be prepared by blending the powders followed by consolidation by compression at ca. 375 °C to ensure structural integrity. Al/PTFE composites have also been prepared using nano Al.

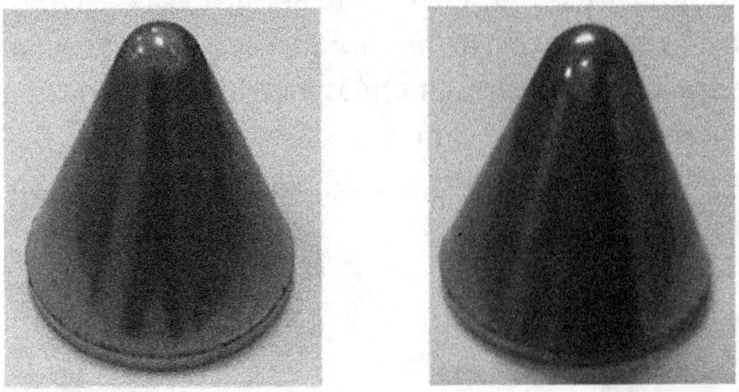

Fig. 13.24: A pressed (left) and sintered (right) PTFE/W/Cu/Pb SC liner.

Intermetallic phases (Alloys)

In the selection of metal components for the preparation of an SRMs alloy, there are many considerations that have to be undertaken, such as the melting point and density of the metal, toxicity, cost and of course the $\Delta H_{f,298}$ values for the corresponding metal oxides, to name but a few. A comparison of selected relevant properties for the metals Ti, Zr and Hf, as well as of Ta, W, Re and Os are given in Tab. 13.14.

Two physical properties which are important determinants of the performance of a projectile are the hardness and density of the penetrator. This is because the ballistic limit velocity (*i.e.* the minimum velocity of a penetrator that is required to pene-

Tab. 13.14: Comparison of properties of one group of transition metals (Ti, Zr, Hf) and selected 5d transition metals (Hf, Ta, W, Re, Os) which are relevant when selecting candidate metals to act as components in new SRMs.

metal/property	mpt. (°C)	density (g/cm^3)	metal toxicity	$\Delta H_{f,298}$ metal oxide (kJ/mol)
Ti	1,667	4.51	non-toxic	− 940 (TiO_2)
Zr	1,857	6.50	non-toxic	−1,100 (ZrO_2)
Hf	2,227	13.28	non-toxic	−1,118 (HfO_2)
Ta	3,000	16.65	non-toxic	−2,046 (Ta_2O_5)
W	3,410	19.26	non-toxic	− 835 (WO_3)
Re	3,180	21.03	non-toxic	−1,128 (Re_2O_7)
Os	3,045	22.61	non-toxic (OsO_4 is highly toxic)	− 386 (OsO_4)

trate an object) drops significantly when the hardness of the projectile is greater than that of the target. Examples of recently investigated alloys are the systems Al-Zn-Zr and Al-U.

Two examples of intermetallic phases which have been recently reported and which may be two extremely interesting new examples of SRMs are ZrW_2 and HfW_2 (Fig. 13.25). These alloys were both prepared from the constituent metals (Zr and W; Hf and W), which were molten in an arc-furnace. The furnace was evacuated down to 10^{-4} mbar, before re-filling the furnace with an argon atmosphere of 100 mbar and melting the metals. The melting process lasted 1–2 min, and was repeated about five times in order to obtain a homogeneous sample. In Fig. 13.22, two reguli with an approximate mass of 5 g are shown.

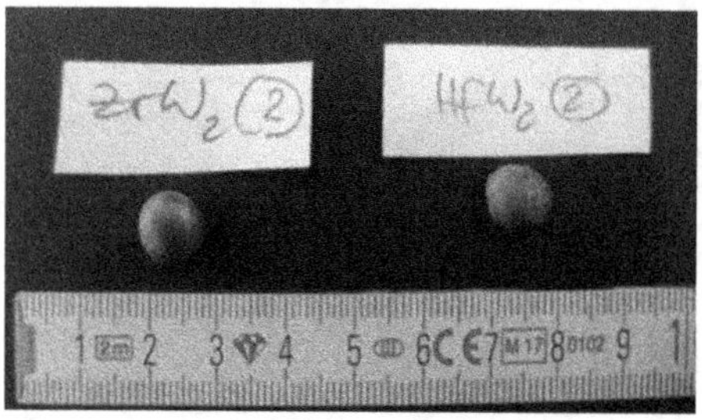

Fig. 13.25: Two samples of approximately 5 g of ZrW_2 and HfW_2. ZrW_2 has a more "golden" color, whereas HfW_2 is silvery-metallic.

Both ZrW_2 and HfW_2 possess a high-density (ρ = 14.31 g/cm^3 ZrW_2) and are very hard intermetallic compounds (Cu_2Mg-type crystal structure), that react/burn in highly exothermic reactions with air at high temperatures. Since both compounds have been shown to possess remarkable combinations of properties such as high melting temperatures, high densities, high hardness and high ignition energies on burning in air, these intermetallic phases could prove to be effective SRMs for warhead applications. A further important aspect of these compounds is that both alloys consist of non-toxic metals. Future research in SRMs will focus on materials with high mechanical strength, high density, and high energy density, that can rapidly convert from a consolidated structural material to fine powder with a large surface area, and be dispersed and then ignited to produce a large thermobaric blast.

The main advantages of ZrW_2 and HfW_2 as SRMs would be (Fig. 13.26):

- ZrW_2 and HfW_2 have very high melting points (m.p. ZrW_2: 2,210 °C, HfW_2: 2,512 °C; both compounds melt peritectically)
- Both compounds react with air (O_2 and N_2!) at high temperatures in extremely exothermic reactions (Tabs. 13.15 and 13.17)
- The combustion products formed from the reaction with oxygen and nitrogen are the corresponding metal oxides and nitrides respectively (largely: ZrO_2 + W and HfO_2 + W), which are non-toxic
- ZrW_2 and HfW_2 are intermetallic phases, which are significantly harder than the component metals. The values for the Vickers hardness HV 10 (in N mm^{-2}) are: Zr 140, Hf 228, W 385, ZrW_2 704, HfW_2 970 (Tab. 13.16). The higher Vickers hardness value for HfW_2 in comparison with ZrW_2 may be the result of stronger bonding in the former. The hardness of HfW_2 lies between that of quartz and orthoclase!
- Both materials have an extremely high density. In the (110) plane of the Laves phases, all atoms are in contact with each other, therefore no higher density is possible

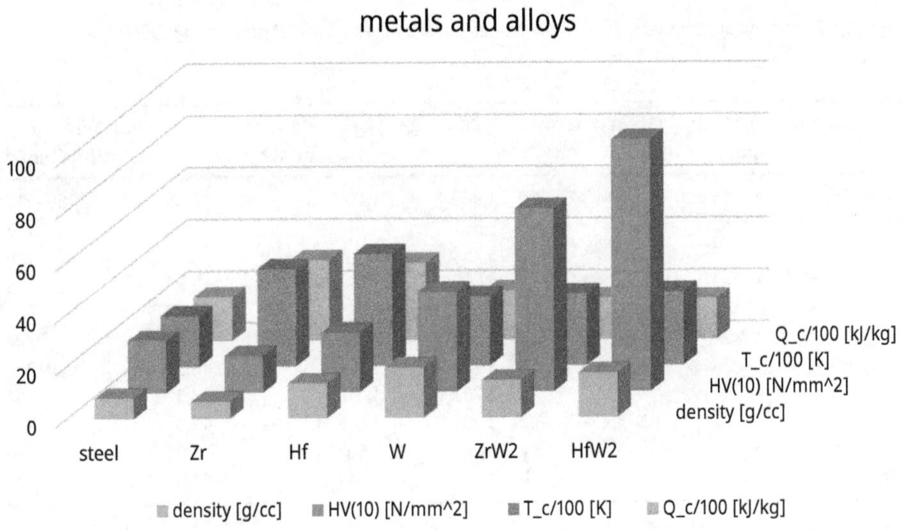

Fig. 13.26: Selected properties of metals and alloys.

Tab. 13.15: Heats of formation of the oxides and nitrides of the reactive tetravalent 4d and 5d transition metals Zr and Hf.

Metal M	Dioxide MO_2 (kJ/mol)	Nitride MN (kJ/mol)
Zr	−1,100	−291
Hf	−1,118	−374

Tab. 13.16: Hardness according to the Moh's and Vickers scale.

substance	Moh's scale	Vickers hardness ($N \cdot mm^{-2}$)
Diamond	10	10,000
Corundum	9	2,035
Topaz	8	1,567
Quartz	7	1,161
Orthoclase	6	817
HfW_2		970
ZrW_2		704
W	7.5	385
Hf	5.5	228
Zr	5.0	140

Tab. 13.17: Combustion of metals and intermetallic phases (alloys) calculated with EXPLO5, isobaric combustion, 0.1 MPa.

Fuel, mass-%	Oxidizer, mass-%	Oxygen balance	T_c/K	$-Q_c$/kJ kg^{-1}	Density of metal or alloy/g cm^{-3}	Hardness (HV10)/N mm^{-2}
Steel, 35	Air, 65	0	1,960	1,703	7.85	160–210
Aluminum, 20	Air, 80	0	3,581	4,411	2.70	167
Zirconium, 39	Air, 69	0	3,754	3,143	6.49	140
Hafnium, 55	Air, 45	0	4,322	2,983	13.28	228
Tungsten, 46	Air, 54	6.6	2,681	1,918	19.28	385
ZrW_2, 76	Air, 24	−7.6	3,541	1,545	14.31	704
ZrW_2, 66	Air, 34	−3.6	3,150	1,572	14.31	704
ZrW_2, 56	Air, 44	0	2,767	1,618	14.31	704
HfW_2, 79	Air, 21	−6.7	3,609	1,458	17.03	970
HfW_2, 70	Air, 30	−3.2	3,061	1,403	17.03	970
HfW_2, 61	Air, 39	0	2,716	1,447	17.03	970

Milling

Reactive and energetic materials are generally metastable, with the potential to undergo transformation into more stable products, accompanied by significant energy release. Mechanochemical methods have been used to produce a wide variety of reactive materials, including intermetallics, thermites, metal-metalloid and metal-polymer composites. However, mechanical milling presents challenges due to the sensitivity of these materials, which can be triggered by impact or friction. Despite this, milling remains an efficient, scalable technique for achieving nanoscale mixing of reactive components. High-energy ball mills, such as shaker, planetary and attritor mills (as outlined in Tab. 13.18), are commonly employed to prepare energetic materials, including metastable alloys and reactive composites. These mills provide enough energy to induce plastic deformation, causing particle fracture, cold welding and composite formation.

Tab. 13.18: Milling equipment used for the preparation of energetic powders.

Mill type	Quantities	Advantages	Disadvantages
Shaker Mills 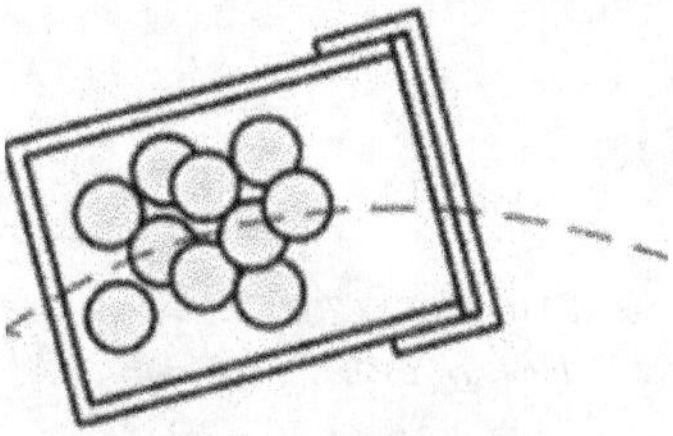	1–10 g	+ High collision energy + Short milling times + Pressure-proof	– Batch process – Small capacity – Milling speed not variable – No temperature control
Planetary Mills 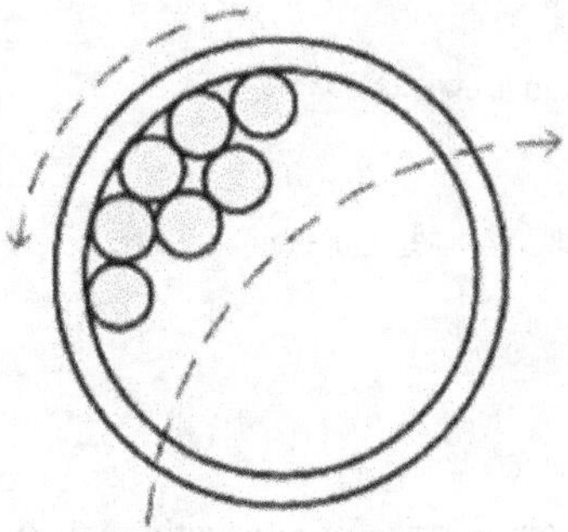	Up to 100 g	+ Larger batch sizes than shaker mills + Variable milling speed + Pressure-proof + Can be cooled	– Batch process

Tab. 13.18 (continued)

Mill type	Quantities	Advantages	Disadvantages
Attritor Mills 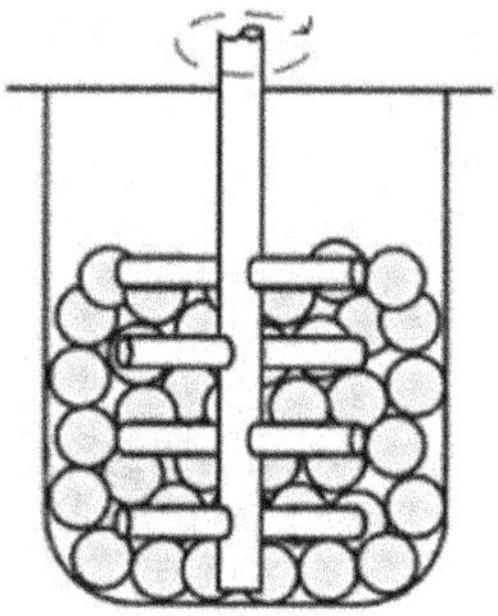	Several kg	+ Can be cooled + Batch or continuous processes possible	– Usually not pressure-proof – Difficult to set under inert gas

Literature

J. Chen, Y. J. Chen, X. Li, Zh. F. Liang, T. Zhou, Ch. Xiao, *J. Phys.: Conf. Ser.*, **2020**, *1507*, 062004.
H. - G. Guo, Y. - F. Zheng, L. Tang, Q.-B. Yu, C. Ge, H.-F. Wang, *Defence Technology*, **2019**, *15*, 495–505.
H. Guo, J. Xie, H. Wang, Q. Yu, Y. Zheng, *Materials*, **2019**, *12*, 3486.
D. E. Eakins, N. N. Thadhani, *Intl. Mat. Rev*, **2009**, *54*, 181–213.
N. Du, W. Xiong, T. Wang, X. - F. Zhang, H. – H. Chen, M. - T. Tan, *Defence Technology*, **2021**, 17, 1791–1803. https://doi.org/10.1016/j.dt.2020.11.008.
J. Evers, T. M. Klapötke, *Chin J. Explos. Propellant* **2015**, *38*, 8–10.
J. Evers, S.Huber, G. Oehlinger, T. M. Klapötke, *Z. Anorg. Allg. Chem.*, **2020**, *646*, 1805–1811.
D. L. Hastings, E. L. Dreizin, *Adv. Eng. Mater.*, **2018**, *20*, 1700631.
E. L. Dreizin, M. Schoenitz, *J. Mater. Sci.*, **2017**, *52*, 11789–11809.
See also: https://www.army.mil/article/77589/new_technology_holds_promise_of_greater_lethality.

13.11 3D Printing of explosives

Arguably, one of the most exciting advances in the broader area of energetic materials research is the application of 3D printing technology to print explosives and propellants with – ideally – tailor-made properties such as specific shapes or porosity. Additive manufacturing of energetic materials is a rapidly expanding area of energetic materials research and has been thoroughly reviewed recently by *Muravyev* and *co-authors* [1]. Additive manufacturing is a term used for processes that build parts from a 3D model layer-by-layer. This is opposed to subtractive manufacturing, in which unwanted material is cut away to produce the desired 3D structure [2]. Specifically, 3D printing is defined as consisting of the fabrication of objects through the deposition of a material using a print head, nozzle or another printing technology [2]. There are seven categories that describe the 3D printing methods available: (i) material extrusion, (ii) material jetting, (iii) vat photopolymerization, (iv) powder bed fusion (PBF), (v) directed energy deposition (DED), (vi) binder jetting and (vii) sheet lamination. A number of 3D printing methods have been applied to energetic materials, such as inkjet printing, direct ink writing (DIW), light/laser-assisted curing technology and fusion deposition [3].

An example of a printable explosive ink that was recently reported is that based on CL-20/HTPB [3]. The CL-20/HTPB composite propellant has been shown to possess low viscosity and good mechanical properties and has been studied in-depth. The explosive ink was obtained as a white, milky liquid containing 85% CL-20 by adding submicron CL-20 powder to a mixture of HTPB and N100 in xylene/chloroform under stirring [3]. After printing the explosive ink onto a glass surface in the desired structure, the HTPB and N100 reacted to form a polyurethane network on curing (Fig. 13.27) [3].

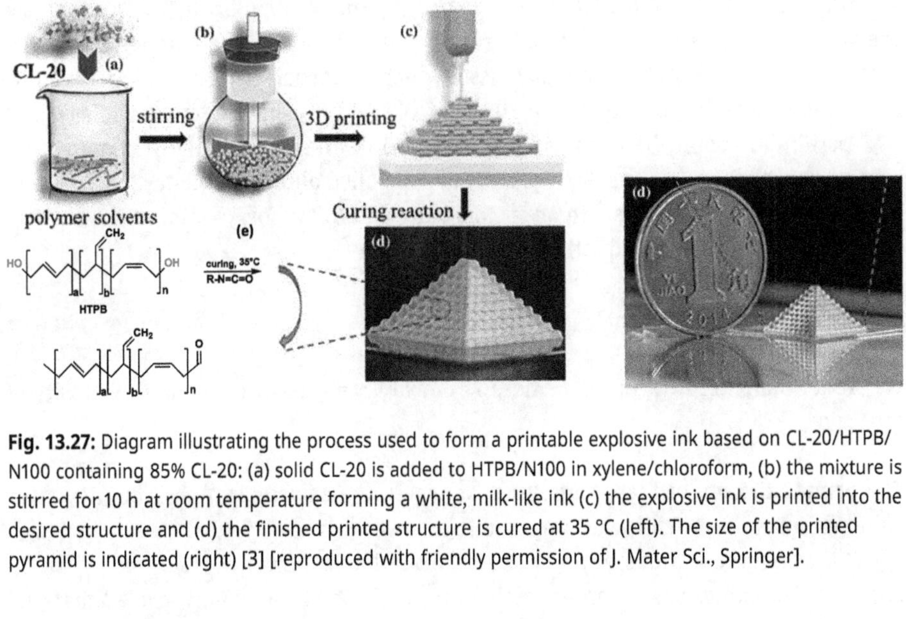

Fig. 13.27: Diagram illustrating the process used to form a printable explosive ink based on CL-20/HTPB/N100 containing 85% CL-20: (a) solid CL-20 is added to HTPB/N100 in xylene/chloroform, (b) the mixture is stitrred for 10 h at room temperature forming a white, milk-like ink (c) the explosive ink is printed into the desired structure and (d) the finished printed structure is cured at 35 °C (left). The size of the printed pyramid is indicated (right) [3] [reproduced with friendly permission of J. Mater Sci., Springer].

Crucially, during and after printing, the 3D structures were found to maintain exactly the shape that was printed, with shrinkage of approximately 5% after curing for several days, and also showed good homogeneity, with SEM showing the absence of unwanted cracks or voids (Fig. 13.28) [3].

Sub-micro CL-20 mixed with an energetic binder comprising GAP/N100 has been printed using direct writing deposition and shown to produce steady detonation in grooves at a size of less than 0.4 × 0.4 mm [4]. Excitingly, EDF-11 which is a CL-20-based explosive ink formulation (PVA, ethylcellulose, H_2O, EtOH, CL-20) has been patented by the US Army as a booster explosive [5]. Another example of the application

Fig. 13.28: Images showing the 3D printed microstructure woodpile scaffold (left), SEM image of a woodpile scaffold (middle) and high-magnification SEM image of a filament (right) [3] [reproduced with friendly permission of J. Mater Sci., Springer].

of DIW and inkjet printing involving energetic materials has been illustrated in the loading of energetic materials in Micro Electronic Mechanical Systems (MEMS) safe-arm devices, which pose traditional press and cast methods considerable problems. Recently, Xu and co-authors stated that inkjet printing is simpler than DIW since the preparation of the explosive ink only requires dissolution of the components in organic solvents and avoids problematic issues such as particle agglomeration, clogging, etc. [6]. It was demonstrated that mixtures of DBTF/RDX using EC (ethylcellulose) and GAP as binders could be used as an explosive ink after dissolving in acetone and printed using inkjet printing devices on heat controlled aluminum plates [6]. The fluid properties of the inkjet inks – in order to ensure suitability for printing – can be evaluated using the following equation:

$$Z = \frac{(\alpha \rho \gamma)^{1/2}}{n}$$

Where α = diameter of printing orifice, ρ = density, γ = surface tension, η = viscosity of the fluid

The printed density of these samples depends on the specific composition, however, values of up to 96.9% TMD were reached [6].

One very recent report reports on the formation of RDX-based aluminized gradient structure explosives, as an example of an energetic functionally graded material (FGM). Functionally graded materials show gradual differences in composition and structure over a specific volume [7]. This means that a composite material can be prepared in which the components can be varied within the structure, for example, by constructing ABAB layers of the two components. Using such an approach, the aim is to engineer the properties of the explosive mixture to meet the demands of its application. Direct ink writing (DIW) and fused deposition modeling (FDM) are two techniques that are being investigated for such purposes. The FDM additive manufacturing approach has the disadvantage of requiring the substance to be heated until it melts, and is therefore not applicable for many explosives. This problem is overcome by using the DIW technique. The DIW technique has the advantage of performing the whole process under ambient conditions. In the example discussed here, the distribution of the aluminum powder in an RDX-based PBX can be controlled using the DIW technique. To achieve this, explosive inks containing 0–30% aluminum as well as RDX were prepared by dispersing the correct amount of Al and RDX in *n*-hexane solution followed by vacuum drying before subsequent addition to the polymer-solvent mixture (HTPB and N100 (polyisocyanate) in xylene/chloroform). The ink containing aluminum was deposited on a glass substrate using DIW, followed by DIW of the corresponding aluminum-free ink layer as shown in Fig. 13.29 resulting in the printing of an ABAB layer structure in which layer A contains Al whereas layer B doesn't, and thereby, the distribution of aluminum in the material is controlled [7]. The samples were cured by leaving at 30 °C for several days, to ensure curing of the ink occurred as well as evaporation of the solvent (Fig. 13.29) [7]. The traditional methods for producing explosives with gradient structures are press loading and cast curing processes, which are less exact, the explosives are subjected to non-ambient conditions and cavities and cracks in the explosives are often formed using these techniques resulting in decreased explosive stability. The use of DIW techniques aims to eliminate these issues.

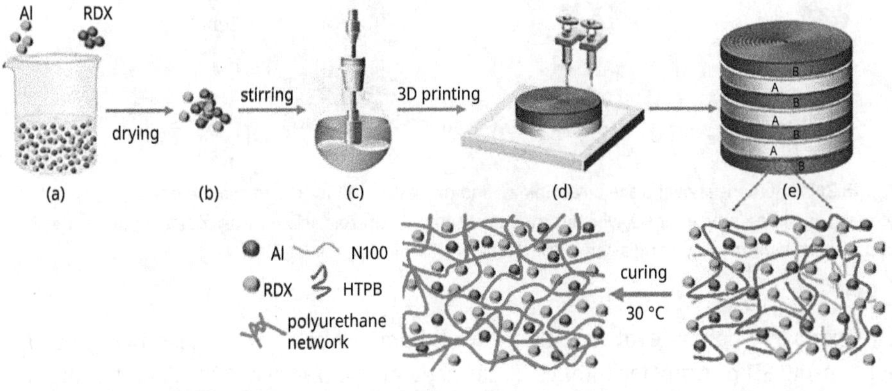

Fig. 13.29: Schematic diagram showing the direct ink writing (DIW) technique for preparation of an Al-containing (layer A), aluminum-free (layer B) composite [7] [reproduced with friendly permission of J. Mater Sci., Springer].

It was reported that the critical detonation diameter and thickness for the DIW printed composite was found to be higher, than for RDX containing the same percentage content of aluminum in the whole sample [7].

The direct ink writing technique has not only been applied to secondary explosives, but also to highly sensitive primary explosives. One example is the preparation of energetics-on-a-chip using the primary explosive copper azide [8]. This was achieved *in situ* on the chip by printing nanoporous copper metal on a chip, followed by exposing it to gaseous hydrazoic acid, which results in a solid-gas reaction forming copper azide on the chip [8]. By adding a foaming agent to the ink, a 3D porous structure for copper azide could be obtained, which is necessary in order for the reaction with hydrazoic acid to proceed throughout the material [8]. The precursor ink consisted of water/polyvinyl alcohol and isopropanol/ethyl cellulose solutions with a small quantity of CATB added to act as the foaming agent, as well as a very small quantity of dodecanol which acts as the foam stabilizer and also 40% nano-copper powder was added (Fig. 13.30) [8]. This ink can then be printed onto, for example, a silicon wafer and left to dry at a slightly elevated temperature (Fig. 13.30) [8]. Exposure of the printed material to gaseous HN_3 (generated from stearic acid and sodium azide) resulted in the formation of predominantly $Cu(N_3)_2$ containing Cu(II), but also a smaller quantity of Cu(I) azide CuN_3 [8]. This technique allows the formation of primary explosive materials with specific structure and porosity and thereby highly sensitive explosive compounds such as primary explosives can be produced with complex architectures not only on the macroscale but also on the microscale.

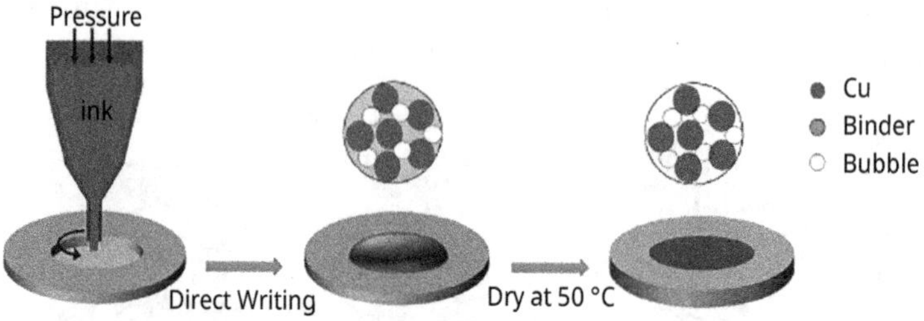

Fig. 13.30: Schematic showing the direct ink writing process (DIW) for the formation of the copper-containing precursor ink which is then exposed to hydrazoic acid to form *in situ* copper azide [8] [reproduced with friendly permission of Mater. Letts., Elsevier].

The explosive behavior of traditional, well-known explosives is also being investigated using 3D printing techniques. It has been known for over 100 years that the sensitivity of TNT is altered by the presence of sand, which makes TNT more sensitive to certain external stimuli and consequently easier to detonate. Using 3D printing, *Mueller, Dattelbaum* and *co-workers* at Los Alamos National Laboratory are investigating the possibility of designing the hollow spaces which are present in a TNT explosive sample. The rationale for this is that the shock sensitivity of a secondary explosive depends on the manner of interaction of hot-spots within the material. The hot-spots result from exposing an explosive material to a shockwave, which causes voids inside the material to collapse, and the subsequent non-uniform heating around these voids results in the formation of hotspots. By introducing voids with a specific distribution in the explosive material, the researchers hope that they will be able to specifically determine the chemical reaction zone during detonation for a particular explosive, and ultimately be able to design blast properties of an explosive material. To achieve their goals, the researchers use 3D printing to define the voids in each mesoscale layer, and then print the area around the voids with the explosive material, resulting in a highly defined structure. Two additive manufacturing methods are being used by the group to achieve this, namely fused-deposition modeling, and optically cured additive manufacturing. In the former method, the explosive material is melted and a thin layer is printed and allowed to solidify before a second layer is added. In the latter process, thin layers are also deposited, however, solidification of each layer is accelerated to a manner of seconds by using UV light [9].

Literature

[1] N. V. Muravyev, K. A. Monogarov, U. Schaller, I. V. Fomenkov, A. N. Pivkina, *Propellants, Expl. Pyrotech.*, **2019**, *44*, 941–969.

[2] Homeland Security Advisory Council, *Final Report of the Emerging Technologies Subcommittee 3D-Printing*, February 24 **2020**, USA.

[3] W. Dunju, G. Changping, W. Ruihao, Z. Baohui, G. Bing, N. Fude, *J. Mater. Sci.*, **2020**, *55*, 2836–2845.

[4] D. Wang, B. Zheng, C. Guo, B. Gao, J. Wang, G. Yang, H. Huang, F. Nie, *RSC Adv.*, **2016**, *6*, 112325–112331.

[5] D. Stec III, A. Wilson, B. E. Fuchs, N. Mehta, P. Cook, *High Explosive Fills for MEMS Devices*, US Patent US 8636861B1, 28th January **2014**.

[6] C. Xu, C. An, Y. Long, Q. Li, H. Guo, S. Wang, J. Wang, *RSC Adv.*, **2018**, *8*, 35863–35869.

[7] X. Zhou, Y. Mao, D. Zheng, L. Zhong, R. Wang, B. Gao, D. Wang, *J. Mater. Sci.*, **2021**, *56*, 9171–9182.

[8] L. Zhang, F. Zhang, Y. Wang, R. Han, J. Chen, R. Zhang, *Mater. Letts.*, **2019**, *238*, 130–133.

[9] E. Hutterer, Explosiv3Design – 3D Printing Could Revolutionize the High-Explosives Industry, March **2016**, 1663, pp. 1–5.

13.12 Continuous flow chemistry

Flow chemistry has more and more become recognized as a disruptive innovation, since it is considered to broaden the scope of chemistry and also create new market opportunities. Perhaps the most well-known current use of flow chemistry is in pharmaceutical applications, however, the use of flow chemistry strategies is becoming increasingly popular in a wide range of areas such as organometallic chemistry, fine chemicals, polymers, peptides, nanomaterials, and also recently in the synthesis of energetic materials.

Compared to the traditional batch synthesis technique, there are several advantages to adopting a flow chemistry approach. Specifically, for energetic materials synthesis, one example of these advantages is the fact that flow reactors provide efficient heat dissipation due to their high surface-to-volume ratio. In energetic materials synthesis where highly exothermic reactions such as the mixing of sulfuric and nitric acid are of core importance, and nitration reactions, or potential side reactions such as the oxidation of nitro-aromatic compounds are also central themes, this is a clear advantage. The rate of heat transfer in flow reactors can be up to several orders of magnitude faster than in batch reactors, and this can be of key importance in the prevention of hot spots, which are always to be avoided since they can lead to side reactions or runaway reactions.

Flow chemistry is also well-suited for the optimization and design of experiments, and uses a statistical approach that enhances the efficiency of chemical processes by thoroughly exploring the parameter space of a reaction, including interactions between factors. A further advantage of flow chemistry is that in principle, variables such as temperature, pressure, flow rate, and reagent quantities can be precisely and automatically controlled meaning that a sequence of experiments can be planned, programmed and performed.

Traditionally, the industrial production of explosives has relied almost exclusively on the batch processing strategy, however, this may possibly change in the near future due to the improved safety, reproducibility, and efficiency of flow chemistry techniques. It can likely be expected that flow chemistry will advance more and more from the laboratory to being adopted on a larger scale for the industrial production of certain energetic materials, with the primary goal being to improve the safety of industrial energetic material production.

If we compare the pros and cons of classical batch synthesis with those of continuous flow synthesis, the following conclusions can be drawn:

The **benefits of batch synthesis**:
- robust, well-established process
- very versatile and relatively affordable
- simple design and relative ease of production
- compatible with all branches of chemistry
- compatible with mass production of chemicals
- established procedures for product traceability based on batch numbers
- widely used in research laboratories and for industrial production

The **limitations of batch synthesis**:
- low surface-to-volume ratio that affects the heat exchange efficiency
- presence of potential hot spots due to poor mixing efficiency and/or poor heat exchange (can trigger side reactions and lower the product purity)
- safety concerns when handling toxic and hazardous materials
- scale-up is resource intensive and time consuming
- large footprint, especially for mass production
- production variability
- only conventional reaction conditions can be applied

The **benefits of continuous flow synthesis**:
- high surface-to-volume ratio and efficient heat exchange
- excellent control over reaction conditions

- efficient mixing which prevents formation of hot spots and thermal runaway
- unconventional reaction conditions possible, e.g. high temperature and pressure (superheating)
- reduced reaction time
- reduced footprint
- reduced material consumption
- potentially higher product quality and reaction yield
- fast transfer from R&D to mass production
- scaling-up is generally easy (by increasing reactor volume or using more flow devices)

The **limitations of continuous flow synthesis**:
- relatively new technology with need for highly specialized personnel and know-how
- high cost of flow chemistry systems
- processing high viscosity liquids is challenging
- processing of heterogeneous reactions can be challenging
- clogging or rupture of the system due to the formation of solid precipitates
- requires selection of appropriate dosing/pumping devices and connectors
- lack of established procedures for product traceability
- transposing batch processes to flow requires research effort and revalidation of established methods

Microflow reactors have been shown to provide several advantages, e.g. the excellent cooling ability, which allows exothermic reactions such as nitrations to be performed safely. An example of a reaction that could easily be carried out in a flow reactor is shown below (Fig. 13.31) and is the synthesis of nitroguanidine. Recently, the synthesis of nitroguanidine (NQ) using a microflow reactor has been performed on a lab scale and was found to be a significant improvement over the previously reported route – which involved batch synthesis – in terms of safety and production rate. A scale-up to an industrial scale has therefore become feasible.

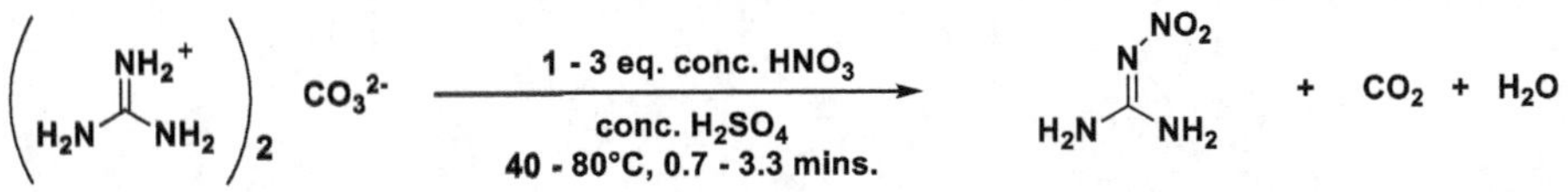

Fig. 13.31: Synthesis of NQ in a flow system (not balanced).

Another example which would nicely illustrate the advantages of flow chemistry for energetic materials would be the preparation of NENAs and DINAs. Those compounds can be prepared in a two-stage process as shown in Fig. 13.32 or also in a one-stage one-pot process. However, the problem with these processes is that both the final compounds and the corresponding intermediates are sensitive to external stimuli and need to be handled with care. Therefore, using a flow chemistry set-up for the synthesis of alkyl-NENAs and DINA would have considerable advantages, since it offers a potentially safer and more controllable method (Fig. 13.33). The flow system which has been developed for these classes of compounds comprises two coiled tube reactors in succession, in which the first reactor is maintained at room temperature, while the second reactor is kept at 35–40 °C. A flow of amino alcohol, either neat or as an amine acetate solution in glacial acetic acid, is reacted with a flow of nitric acid in the first reactor. Immediately downstream, the exit flow is combined with a flow of acetic anhydride-acetic chloride mixture and led into the second reactor. After a few

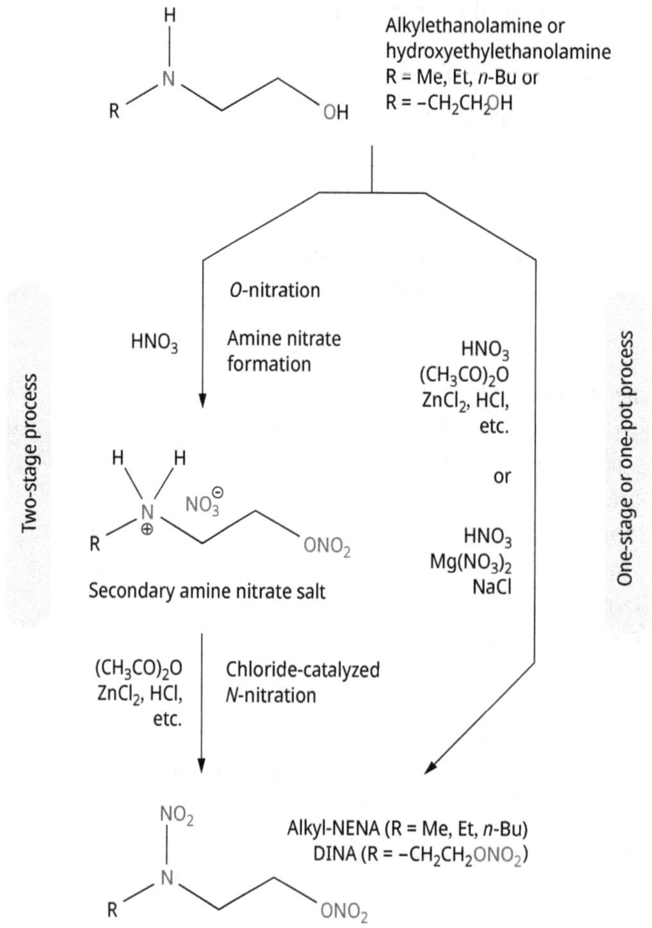

Fig. 13.32: General synthetic routes for NENAs and DINAs.

minutes, the flow effluent is led into a beaker containing water and simmered for 30 minutes at 60 °C. By adjusting the temperature of the second reactor and the reagent ratios, alkyl-NENAs or DINA having excellent purity could be obtained in high yields – in a reproducible manner and without the possibility of thermal runaways.

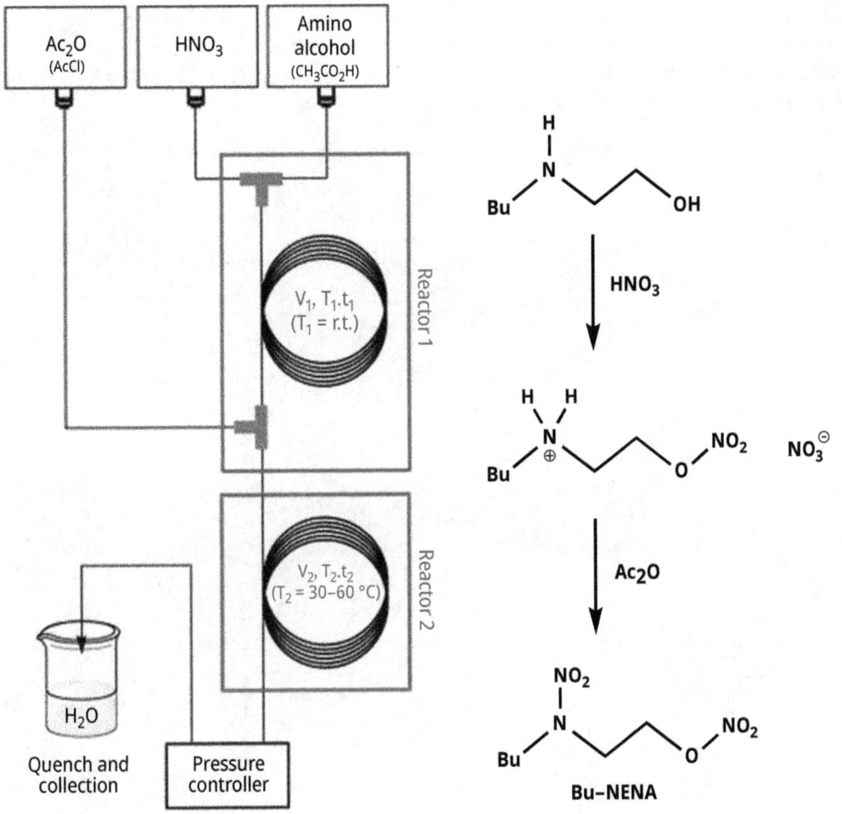

Fig. 13.33: General set-up for the flow-syntheses for NENAS and DINAs (left) and synthesis for Bu-NENA (right).

The recent rise in popularity of "flow chemistry" in the broader sense has resulted in microflow reactors and flow reactors generally being considered as one and the same thing. However, this is not the case. Flow reactors have been described in the literature for a long time by chemical engineers, and for reactions on the scale of producing 200 kg/h of RDX, a flow reactor is used. Simply scaling up a microflow reactor process is generally always unsuccessful. Despite this, microflow reactors are highly useful in a laboratory setting, especially in research.

Literature

D. Kyprianou, G. Rarata, G. Emma, G. Diaconu, M. Vahčič, D. Anderson D., Flow Chemistry and the Synthesis of Energetic Materials, JRC Technical Report, European Commission, **2022**; https://publications.jrc.ec.europa.eu/repository/handle/JRC128574

K. Karaghiosoff, T. M. Klapötke, M. Völkl, *Org. Process Res. Dev.*, **2024**, *28*, 1458–1463; https://doi.org/10.1021/acs.oprd.2c00398.

K. A. Fredriksen, T. E. Kristensen, Synthesis of nitroxyethylnitramine (NENA) plasticizers and dinitroxydiethylnitramine (DINA) in continuous flow, *Propellants, Explos., Pyrotech.*, **2023**, *48*, e202200321; https://doi.org/10.1002/prep.202200321.

14 Study questions

Answers can be found online under: https://www.degruyter.com/document/isbn/9783111446981/html

1. Please write suitable Lewis-type structures for the following molecules,

   FOX-7:

   PETN:

   RDX:

2. If 10 kg of a common high explosive is detonated, the minimum distance from the explosion which gives a chance of survival is

   O 2.2 m O 4.3 m O 8.6 m

3. Give examples for the following:
   - single-base gun propellant:
   - double-base gun propellant:
   - triple-base gun propellant:

4. Why does Mg nowadays get replaced with Zr in smoke munition?

5. Name one commercially available computer code to calculate
   - enthalpies of formation:
   - detonation parameters:

6. Calculate the oxygen balance with respect to CO_2 of ethylene glycol dinitrate, $O_2N—O—CH_2—CH_2—O—NO_2$.

7. If a high explosive has a VoD of 9,000 m s^{-1}. What is its approximate Gurney velocity?

8. Which of the following materials is more suitable for a kinetic energy penetrator, and why?

   O SS-316 O DU

   reason:

9. In order to reduce the erosive effects of a gun propellant
   - the N_2 / CO ratio should be O high O low
   - T_c should be O high O low

10. In a drophammer experiment, a mass of 5 kg hits the sample from a height of 50 cm. Calculate the impact energy.

11. Give five different nitrating agents.

https://doi.org/10.1515/9783111447087-014

12. For which application would hypergolic behavior be desirable (give two applications)?

13. Name a possible replacement for TNT.

14. Name a possible replacement for Composition B.

15. How does the specific impulse I^*_{sp} depend on the combustion temperature (T_c) and the average molecular weight (M) of the combustion gases formed?

16. What is the relationship between the average thrust ($\bar{F}$) of a rocket and its specific impulse (I_{sp})?

17. Give one example each of a solid rocket propellant which belongs to the following classes and state whether the material is homogeneous or heterogeneous:
    - double-base:
    - composite:

18. As a "rule of thumb" by how much would one need to increase the specific impulse I^*_{sp} of a rocket in order to double its payload?

19. Which metals are clandestine signalling and illuminating formulations in the range 700–1,000 nm usually based on?

20. How does the detonation pressure ($p_{C\text{-}J}$) scale with the loading density?

21. How can the enthalpies of sublimation ($\Delta H_{sub.}$) and vaporization ($\Delta H_{vap.}$) be estimated from the melting and boiling points of a substance?

22. How does the penetration depth of kinetic energy munition depend on the material of the penetrator?

23. How does the Gurney velocity ($\sqrt{2E}$) of an explosive depend on the detonation velocity (D = VoD)?

24. Describe three different methods suitable for the synthesis of N_2O_5.

25. Describe the laboratory scale synthesis of ADN starting from ammonia.

26. Which detonation product seems to be most effective for biological agent defeat explosives (e.g. against anthrax spores)?

27. Which general reaction describes the combustion of Al based nanothermites?

28. Name one lead-free tetrazolate based primary explosive.

29. Name one high explosive which has the same chemical composition ($C_aH_bN_cO_d$) as RDX.

30. What are the environmental/health concerns of AP?

31. For which type of rocket/missile propulsion could chemical thermal propulsion (CTP) be very valuable?

32. In which pressure range do the combustion chambers of a solid rocket motor and a large calibre gun operate?

33. What are the most important performance parameters for
    a) high explosives (secondary explosives)
    b) gun propellants
    c) rocket propellants?

34. Which two parameters influence the erosion problems in gun barrels the most?

35. Compare the perforation of a shaped charge device (SC) with that of an explosively formed projectile (EFP) for ideal standoff distances.

36. Which experimental test is suitable to assess the shockwave sensitivity of an explosive?

37. Which experimental test (IM test) would be suitable to assess the behaviour of an explosive when subjected to a fuel fire?

38. Compare the high explosives RDX and HNS. What are the most important advantages and disadvantages of the two compounds?

39. What prevents TATP from having any application in the defense world (two reasons)?

40. Which condition is a good/reasonable approximation (e.g. isochoric, isobaric) for the following:
    a) gun propellants,
    b) rocket propellants,
    c) secondary explosives?

41. What are the main light emitting species in visible signal flare formulations which emit in the following regions:
    a) red,
    b) green?

42. How does the maximum wavelength of a blackbody radiator (λ_{max}) depend on/change with the temperature? Which law describes this phenomenon?

43. How does the linear burn rate generally depend on the pressure?

44. How does the electrostatic discharge sensitivity (ESD) usually depend on the particle (grain) size?

45. Are shaped charges (SC) or explosively formed projectiles (EFP) more sensitive to the standoff distance?

46. Why do detonations or combustions often produce CO and not exclusively CO_2 despite a positive oxygen balance?

47. What are the three main "strategies" of designing CHNO based high explosives?

48. For each of the following, give one example of an explosive that only contains the following energetic groups,
    - nitro:
    - nitrate (nitrate ester):
    - nitramino:
    - nitrimino:
    - azide:
    - peroxo:

49. What are the disadvantages of using nitroglycerine for military applications?

50. According to Kamlet and Jacobs the detonation velocity (*D*) and the detonation pressure ($p_{C\text{-}J}$) can be estimated. Give the relevant equations.

51. Calculate the spark energy (ESD test) for a spark generated under the following conditions: C = 0.005 μF, V = 10,000 V

52. Give two reagents which can be used to convert an amine into an *N*-oxide.

53. Give the structures of:

    MMH: UDMH: DMAZ:

    To which class(es) (in terms of application) do these compounds belong?
    What might be a practical use of DMAZ and why?

54. A Hawker Siddeley Harrier is a single engine fighter plain with the capability of performing vertical take-off. How much thrust would the engine need to deliver to perform a vertical take-off if we assume a fighter mass of 17 t.

55. Name two types of electric detonators.

56. Give five advantages of an EBW detonator compared to a conventional hot-wire detonator.

57. Name a suitable replacement for tetrazene.

58. Name a suitable replacement for MMH in gelled hypergolic propellants.

59. The shock-wave energy and bubble-energy correspond to which important property?

60. Give two empirical relationships that correlate the VoD and $p_{C\text{-}J}$ with the specific impulse.

61. Name a possible replacement formulation for that of Composition B that does not contain TNT.

62. Give one possible chemical synthesis of nitroform (NF) starting from commercially available ingredients.

63. Give one possible synthesis of trinitroethanol (TNE) starting from NF.

64. In August 2020, in a storage facility located at Beirut harbor, 2,750 t of ammonium nitrate were stored in large bags on the ground. Suddenly, due to an external stimulus (fire), a detonation occurred in which the entire amount of AN reacted. An estimation of the mass which exploded based on the surface explosion crater is possible based on the dependence of the crater volume V_c (in m^3) on the explosive mass (M_{ex} in kg as TNT equivalents):

$$M_{ex} = 10.65V_c^{0.504}$$

The Trauzl test (lead block test) results for TNT and AN are as follows (per 10 g):
TNT 300 cm^3
AN 225 cm^3
Assuming the crater which was formed corresponds to a half-sphere, which diameter and which depth of crater would you expect?
Please discuss your results and compare them with the observed crater dimension: depth 40 m and ca. 75 m diameter.
Name three important applications for AN.
Name two other major accidents with AN that occurred in history.

65. Name (semi)empirical and thermodynamic equilibrium codes for the prediction of detonation parameters (name two each).

66. Order the following compounds according to their critical diameters (from small to large): RDX, LA, TATB.

67. Give a typical composition (in %) for a flash composition.

68. How can the enthalpy of sublimation be estimated according to Trouton's rule?

69. Calculate the lattice energy (U_L) for guanidinium nitrate ($CH_6N_4O_3$) with a crystal density of 1.44 g cm^{-3} according to Jenkins and also according to Gutowski.

70. What is the relationship between the cavity volume (ΔV_T) in the Trauzl test and the heat of detonation (Q_{ex}) and volume of detonation gases (V_0)?

71. What is the relationship between the C–J pressure (p_{C-J}) and the loading density (ρ_0) and detonation velocity (D)?

72. In the case of a large distance from the center of an explosion, arrange in ascending order the risk for a person to be injured due to: overpressure, fire/heat, fragments, impact.

73. Order according to the heat output the following fuels in TBX weapons (from lowest to highest values in kJ kg^{-1}): B, Al, Be.

74. What is the difference between AN and PSAN?

75. Give a typical composition of a TBX formulation.

15 Literature

Additional reading

General

Agrawal, J. P., *High Energy Materials*, Wiley-VCH, Weinheim, **2010**.

Agrawal, J. P.; Hodgson, R. D., *Organic Chemistry of Explosives*, Wiley-VCH, Weinheim, **2007**.

Akhavan, J., *The Chemistry of Explosives*, 4th edn., The Royal Society of Chemistry, Cambridge, **2022**.

Bailey, A.; Murray, S. G., *Explosives, Propellants & Pyrotechnics*, Brassey's, London, **1989**.

Brinck, T. (ed.), *Green Energetic Materials*, Wiley, Weinheim, **2014**.

Cook, J. R., *The Chemistry and Characteristics of Explosive Materials*, 1st edn., Vantage Press, New York, **2001**.

Dolan, J. E.; Langer, S. S. (ed.), *Explosives in the Service of Man*, The Royal Society of Chemistry, Cambridge, **1997**.

Fair, H. D.; Walker, R. F., *Energetic Materials*, Vol. 1 & 2, Plenum Press, New York, **1977**.

Fedoroff, B. T.; Aaronson, H. A.; Reese, E. F.; Sheffield, O. E.; Clift, G. D., Encyclopedia of Explosives, in: *U.S. Army Research and Development Command (TACOM, ARDEC) Warheads*, Energetics and Combat Support Center, Picatinny Arsenal, New Jersey, **1960**.

Fordham, S., *High Explosives and Propellants*, 2nd edn., Pergamon Press, Oxford, **1980**.

Jonas, J. A.; Walters, W. P., *Explosive Effects and Applications*, Springer, New York, **1998**.

Keshavarz, M. H.; Klapötke, T. M., *Energetic Compounds*, De Gruyter, Berlin/Boston, **2017**.

Keshavarz, M. H.; Klapötke, T. M., *The Properties of Energetic Materials*, De Gruyter, Berlin/Boston, **2018**.

Keshavarz, M. H., *Liquid Fuels as Jet Fuels and Propellants*, Nova, New York, **2018**.

Keshavarz, M. H., *Combustible Organic Materials*, De Gruyter, Berlin/Boston, **2018**.

Klapötke, T. M. (ed.), High Energy Density Materials, in: *Structure and Bonding*, Vol. 125, Springer, Berlin, **2007**.

Klapötke, T. M., in: E. Riedel (ed.), *Moderne Anorganische Chemie*, 3rd edn., ch. 1, Nichtmetallchemie, Walter de Gruyter, Berlin, **2007**.

Klapötke, T. M., *Energetic Materials Encyclopedia*, 2. Aufl., De Gruyter, Berlin/Boston, **2021**.

Koch, E.-C., *Sprengstoffe, Treibmittel, Pyrotechnika*, 2. Aufl., De Gruyter, Berlin/Boston, **2019**.

Köhler, J.; Meyer, R.; Homburg, A., *Explosivstoffe*, 10. Auflage, Wiley-VCH, Weinheim, **2008**.

Kubota, N., *Propellants and Explosives*, Wiley-VCH, Weinheim, **2002**.

Olah, G. A.; Squire, D. R., *Chemistry of Energetic Materials*, Academic Press, San Diego, **1991**.

Urbanski, T., *Chemistry and Technology of Explosives*, Vol. 1–4, Pergamon Press, Oxford, **1964/1984**.

Venugopalan, S., *Demystifying Explosives: Concepts in High Energy Materials*, Elsevier, Amsterdam, **2015**.

Synthetic chemistry

Agrawal, J. P.; Hodgson, R. D., *Organic Chemistry of Explosives*, Wiley, Hoboken, NJ, **2007**.

Ang, H. G.; Santhosh, G., *Advances in Energetic Dinitramides*, World Scientific, New Jersey, London, Singapore, **2007**.

Bailey, P. D.; Morgan, K. M., *Organonitrogen Chemistry*, Oxford Science Publications, Oxford, **1996**.

Boyer, J. H., *Nitroazoles*, VCH, Weinheim, **1986**.

Feuer, H.; Nielsen, A. T., *Nitro Compounds*, VCH, Weinheim, **1990**.

Feuer, H., *Nitrile Oxides, Nitrones, and Nitronates in Organic Synthesis*, Wiley, Hoboken, NJ, **2008**.

https://doi.org/10.1515/9783111447087-015

Hammerl, A.; Klapötke, T. M., Pentazoles. in: Katritzky, A. R.; Ramsden C. A.; Scriven, E. F. V.; Taylor, R. J. K. (ed.); *Comprehensive Heterocyclic Chemistry III*, Elsevier, Oxford, **2008**; Vol. 6, pp. 739–758.
Ledgard, J., *The Preparatory Manual of Explosives*, 3rd edn., J. B. Ledgard, **2007**.
Nielsen, A. T., *Nitrocarbons*, VCH, Weinheim, **1995**.
Ono, N., *The Nitro Group in Organic Synthesis*, Wiley-VCH, New York, **2001**.
Torssell, K. B. G., *Nitrile Oxides, Nitrones, and Nitronates in Organic Synthesis*, VCH, Weinheim, **1988**.
Gao, H.; Shreeve, J. M., *Chem. Rev.* **2011**, 111, 7377–7436.

Detonation physics

Dremin, A. N., *Toward Detonation Theory*, Springer, Berlin, New York, **1999**.
Fickett, W.; Davis, W. C., *Detonation Theory and Experiment*, Dover Publications, Mineola, New York, **1979**.
Hidding, B., *Untersuchung der Eignung von Silanen als Treibstoffe in der Luft- und Raumfahrt* (Diplomarbeit), UniBW, München, **2004**.
Holian, B. L.; Baer, M. R.; Brenner, D.; Dlott, D. D.; Redondo, A.; Rice, B. M.; Sewell, T.; Wight, C. A., *Molecular Dynamics Simulations of Detonation Phenomena*, ITRI Press, Laurel, MD, **2004**.
Liberman, M. A., *Introduction to Physics and Chemistry of Combustion*, Springer, Berlin, **2008**.
Penney, W. G.; Taylor, G.; Jones, H.; Evans, W. M.; Davies, R. M.; Owen, J. D.; Edwards, D. H.; Thomas, D. E.; Adams, C. A.; Bowden, F. P.; Ubbelohde, A. R.; Courtnev-Pratt, J. S.; Paterson, S.; Taylor, J.; Gurton, O. A., *Proc. Royal Soc.* **1950**, *204 A*, 3.
Zel'dovich, Y. B., Raizer, Y. P., *Physics of Shock Waves and High-Temperature Hydrodynamic Phenomena*, Dover, Mineola, New York, **2002**.

Pyrotechnics

Conkling, J. A., *Chemistry of Fireworks*, Marcel Dekker, New York, **1985**.
Hardt, A. P., *Pyrotechnics*, Pyrotechnica Publications, Post Falls, **2001**.
Koch, E.-C., *J. Pyrotech.* **2002**, *15*, 9–23.
Koch, E.-C., *J. Pyrotech.* **2005**, *21*, 39–50.
Koch, E.-C., *Propellants Explo. Pyrotech.* **2001**, *26(1)*, 3–11.
Koch, E.-C., *Propellants Explos. Pyrotech.* **2002**, *27(5)*, 262–266.
Koch, E.-C., *Propellants Explos. Pyrotech.* **2002**, *27*, 340.
Koch, E.-C., *Propellants Explos. Pyrotech.* **2004**, *29(2)*, 67–80.
Koch, E.-C., *Propellants Explos. Pyrotech.* **2006**, *31(1)*, 3–19.
Koch, E.-C., *Propellants Explos. Pyrotech.* **2008**, *33(3)*, 165–176.
Koch, E.-C., *Propellants Explos. Pyrotech.* **2008**, *33(5)*, 333–335.
Koch, E.-C., *Propellants Explos. Pyrotech.* **2009**, *34*, 6–12.
Russell, M. S., *The Chemistry of Fireworks*, The Royal Society of Chemistry, Cambridge, **2000**.
Steinhauser, G.; Klapötke, T. M., *Angew. Chem.* **2008**, *120*, 2–21; *Intl. Ed.* **2008**, *47*, 2–20.
Steinhauser, G.; Klapötke, T. M., *J. Chem. Educ.* **2010**, *87(2)*, 150–156.
Koch, E.-C., *Metal-Fluorocarbon Based Energ. Materials*, Wiley-VCH, Weinheim, **2012**.
Crow, J. M., *Chem. World* **2012**, 46–49.

Material sciences

Teipel, U. (ed.), *Energetic Materials*, Wiley-VCH, Weinheim, **2005**.

Engineering and technology

Cooper, P. W.; Kurowski, S. R., *Technology of Explosives*, Wiley-VCH, New York, **1996**.
Cooper, P. W., *Explosives Engineering*, Wiley-VCH, New York, **1996**.

Rocket propulsion

Ciezki, H. K., *Rocket Propellants – An Introduction and Handbook for Students, Propulsion Engineers and Scientists*, Wiley-VCH, Weinheim, **2025**.
Kubota, N., *Propellants and Explosives*, Wiley, VCH, Weinheim, **2002**.
Jacobs, P. W. M.; Whitehead, H. M., *Chem. Rev.* **1969**, *69(4)*, 551–590.
Schmidt, E. W., *Hydrazine and its Derivatives*, 2. Aufl., Band 1 und 2, Wiley, New York, **2001**.
Yang, V.; Brill, T. B.; Ren, W.-Z. (ed.), *Solid Propellant Chemistry, Combustion, and Motor Interior Ballistics*, Vol. 185, AAIA, Reston, VA, **2000**.
Ulas, A.; Kuo, K. K., *Comb. Flame* **2001**, *127*, 1935–1957.
Beckstead, M. W.; Puduppakkama, K.; Thakre, P.; Yang, V., *Prog. Energy Comb. Sci.* **2007**, *33*, 497.
DeLuca, L. (ed.), *Chemical Rocket Propulsion – A Comprehensive Survey of Energetic Materials*, Springer, ISBN: 978-3-319-27746-2, **2017**.
UY, H. E. A., Double-base Propellants, in: *Solid Rocket Propulsion Technology*, **2012**, 369.
Naminosuke, K., Propellant and Explosives, in: *Thermochemical Aspects of Combustion in Germany*, Wiley, **2007**.
Epishina, M. A.; Ovchinnikov, I. V.; Kulikov, A. S.; Makhova, N. N.; Tartakovsky, V. A., *Mendeleev Commun.* **2011**, *21*, 21.
Ermakov, A.; Bulatov, P.; Vinogradov, D.; Tartakovskii, V., *Russ. J. Org. Chem.*, **2004**, *40*, 1062.

Gun propulsion

Jahnuk, H., *Prop. Expl. Pyrotech.* **1976**, *1*, 47–50.
Dürschner, M. G., *Prop. Expl. Pyrotech.* **1976**, *1*, 81–85.
Zentner, B. A.; Reed, R., *Calculated Performance of Gun Propellant Compositions Containing High Nitrogen Ingredients*, 5th Intl. Gun Propellant Symp., Picatinny Arsenal, NJ, November 19–21, **1991**, IV–278.

Weapons

Carlucci, D. E.; Jacobson, S. S., *Ballistics – Theory and Design of Guns and Ammunition*, CRC Press, Boca Raton, Fl., **2008**.
Dathan, H., *Waffenlehre für die Bundeswehr*, 4th edn, Mittler, Herford, Bonn, **1980**.
Heard, B. J., *Handbook of Firearms and Ballistics*, 2nd edn, Wiley-Blackwell, Chichester, **2008**.

Testing methods

Kneisl, P., *13th NTREM*, **2010**, 184–193.
Sućeska, M., *Test Methods for Explosives*, Springer, New York, **1995**.

Gurney velocity

Backofen, J. E., *New Trends Res. Energetic Mater.* **2006**, 9th Seminar, 76–89.
Backofen, J. E.; Weickert, C. A., in: Furnish, M. D.; Thadhani, N. N.; Horie, Y. (eds.), *Shock Compression of Condensed Matter*, **2001** Amer. Inst. Phys. **2002**, 954–957.
Cooper, P. W., *Explosives Engineering*, Wiley-VCH, New York, Chichester, Weinheim, Kap. 27, **1996**, 385–403.
Gurney, R. W., *Internal Report BRL-405*, **1943**, Ballistic Research Laboratory, Aberdeen, MD.
Hardesty, D. R.; Kennedy, J. E., *Combust. Flame* **1977**, *28*, 45.
Jaansalu, K. M.; Dunning, M. R.; Andrews, W. S., *Prop. Expl. Pyrotech.* **2007**, *32*, 80.
Kamlet, M. J.; Finger, M., *Combust. Flame* **1979**, *34*, 213.
Keshavarz, M. H., *Prop. Expl. Pyrotech.* **2008**, *33*, 316.
Koch, A.; Arnold, N.; Estermann, M., *Prop. Expl. Pyrotech.* **2002**, *27*, 365.
Reaugh, J. E.; Souers, P. C., *Prop. Expl. Pyrotech.* **2004**, *29*, 124.
Stronge, W. J.; Xiaoqing, M.; Lanting, Z., *Int J. Mech. Sci.* **1989**, *31*, 811.

Computational chemistry (quantum mechanics)

Politzer, P.; Murray, J. S. N., Energetic Materials, Part 1: Decomposition, Crystal and Molecular Properties, in: *Theoretical and Computational Chemistry.*, Vol. 12, Elsevier, Amsterdam, **2003**.
Politzer, P.; Murray, J. S. N., Energetic Materials, Part 2, Energetic Materials, Part 2: Detonation, Combustion, in: *Theoretical and Computational Chemistry*, Vol. 13, Elsevier, Amsterdam, **2003**.
Rice, B. M.; Hare, J. J.; Byrd, E. F. C., *J. Phys. Chem.* **2007**, *III*, 10874–10879.
Seminario, J. M.; Politzer, P., Modern Density Functional Theory: A Tool for Chemistry, in: *Theoretical and Computational Chemistry.*, Vol. 2, Elsevier, Amsterdam, **1995**.
Politzer, P. A.; Murray, J. S., *Energetic Materials*, Part 1 & 2, Elsevier, Amsterdam, **2003**.

Nanothermites

Brown, M. E.; Taylor, S. J.; Tribelhorn, M. J., *Propellants Explos. Pyrotech.* **1998**, *23*, 320–327.
Bulian, C. J.; Kerr, T. T.; Puszynski, J. A., *Proc. Int. Pyrotech. Semin.* **2004**, 327–338.
Clapsaddle, B. J.; Zhao, L.; Gash, A. E.; Satcher, J. H.; Shea, K. J.; Pantoya, M. L.; Simpson, R. L., *Res. Soc. Symp. Proc. 800.* **2004**, AA2.7.1–AA2.7.6.
Clapsaddle, B. J.; Zhao, L.; Prentice, D.; Pantoya, M. L.; Gash, A. E.; Satcher, J. H.; Shea, K. J.; Simpson, R. L., *Int. Annu. Conf. ICT* **2005**, 36th (Energetic Materials), 39/1–39/10.
Comet, M.; Pichot, V.; Spitzer, D.; Siegert, B.; Ciszek, F.; Piazzon, N.; Gibot, P., *Int. Annu. Conf. ICT*, **2008**, 39th, V38/1–V38/8.
Gash, A. E.; Satcher, J. H.; Simpson, R. L.; Clapsaddle, B. J., *Mat. Res. Soc. Symp. Proc. 800.* **2004**, AA2.2.1–AA2.2.12.
Naud, D. L.; Hiskey, M.; Son, S. F.; Busse, J. R.; Kosanke, K., *J. Pyrotech.* **2003**, *17*, 65–75.

Pantoya, M. L.; Granier, J. J., *Propellants Explosi. Pyrotech.* **2005**, *30(1)*, 53–62.
Perry, W. L.; Smith, B. L.; Bulian, C. J.; Busse, J. R.; Macomber, C. S.; Dye, R. C.; Son, S. F., *Propellants Explos. Pyrotech.* **2004**, *29*, 99–105.
Piercey, D. G.; Klapötke, T. M., *Central Europ. J. Energ. Mat.* **2010**, *7*, 115–129.
Prakash, A.; McCormick, A. V.; Zachariah, M. R., *Chem. Mater.* **2004**, *16(8)*, 1466–1471.
Puszynski, J. A.; Bichay, M. M.; Swiatkiewicz, J. J., United States Patent Application Number *20060113014*. June 1, 2006.
Puszynski, J. A.; Bulian, C. J.; Swiatkiewicz, J. J., *J. Propul. Power* **2007**, *23(4)*, 698–706.
Puszynski, J. A., *Mat. Res. Soc. Symp. Proc.*, **2004**, *800*, AA6.4.1–AA6.4.10.
Sanders, V. E.; Asay, B. W.; Foley, T. J.; Tappan, B. C.; Pacheco, A. N.; Son, S. F., *Propul. Power.* **2007**, *23(4)*, 707–714.
Schoenitz, M.; Ward, T.; Dreizin, E. L., *Res. Soc. Symp. Proc.* **2004**, *800*, AA2.6.1–AA2.6.6.
Shende, R.; Subramanian, S.; Hasan, S.; Apperson, S.; Thiruvengadathan, R.; Gangopadhyay, K.; Gangopadhyay, S.; Redner, S.; Kapoor, D.; Nicolich, S.; Balas, W., *Propellants Explos. Pyrotech.* **2008**, *33(2)*, 122–130.
Son, S. F., *Mat. Res. Soc. Symp. Proc.* **2004**, *800*, AA5.2.1–AA5.2.12.
Son, S. F.; Yetter, R. A.; Yang, Y., *J. Propul. Power*, **2007**, *23(4)*, 643–644.
Son, S. F.; Asay, B. W.; Foley, T. J.; Yetter, R. A.; Wu, M. H.; Risha, G. A., *Propul. Power.* **2007**, *23(4)*, 714–721.
Spitzer, D.; Comet, M.; Moeglin, J.-P.; Stechele, E.; Werner, U.; Suma, A., *Yves Int. Annu. Conf. ICT* **2006**, (Energetic Materials), 117/1–117/10.
Tillotson, T. M.; Gash, A. E.; Simpson, R. L.; Hrubesh, L. W.; Satcher, J. H.; Poco, J. F., *J. Non-Cryst. Solids*, **2001**, *285*, 338–345.
Umbrajkar, S. M.; Schoenitz, M.; Dreizin, E. L., *Propellants Explos. Pyrotech.* **2006**, *31(5)*, 382–389.
Walker, J.; Tannenbaum, R., *Res. Soc. Symp. Proc.* **2004**, *800*, AA7.8.1–AA7.8.10.
Walter, K. C.; Pesiri, D. R.; Wilson, D. E., *Propul. Power*, **2007**, *23(4)*, 645–650.
Weismiller, M. R.; Malchi, J. Y.; Yetter, R. A.; Foley, T. J., *Chem. Phys. Process. Comb.* **2007**, 595, 600.

Primary explosives

Cartwright, M.; Wilkinson, J., *Propellants Explos. Pyrotech.* **2011**, *36*, 530–540.
Cartwright, M.; Wilkinson, J., *Propellants Explos. Pyrotech.* **2010**, *35*, 326–332.
Fronabarger, J. W.; Williams, M. D.; Sanborn, W. B.; Bragg, J. G.; Parrish, D. A.; Bichay, M., *Propellants Explos. Pyrotech* **2011**, *36*, 541–550.
Ugryumov, I A.; Ilyushin, M. A.; Tselinskii, I. V.; Kozlov, A. S., *Russ. J. Appl. Chem.* **2003**, *76*, 439–441.
Fronabarger, J. W.; Williams, M. D.; Sanborn, W. B.; Parrish, D. A.; Bichay, M., *Propellants Explos. Pyrotech.* **2011**, *36*, 459–470.
Fischer, D.; Klapötke, T. M.; Piercey, D. G.; Stierstorfer, J., *J. Energ. Mat.* **2012**, *30*, 40–54.
Mehta, N.; Damavarapu, R.; Cheng, G.; Dolch, T.; Rivera, J.; Duddu, R.; Yang, K., *Proc. Int. Pyrotechn. Semin.* **2009**, *36*, 175–180.
Matyáš, R.; Pachman, J., *Primary Explosives*, Springer, Berlin, **2012**.
Ahmad, S. R.; Cartwright M., *Laser Ignition of Energetic Materials*, Wiley, Weinheim, **2015**.
Ilyshin, M. A.; Tselinsky, I. V.; Shugalei, I. V., *Centr. Europ. J. Energ. Mat.* **2012**, *9(4)*, 293–327.

Thermally stable explosives

Baird, T.; Drummond, R.; Langseth, B.; Silipigno, L., *Oilfield Review*, **1998**, 50.
Sikder, A. K.; Sikder, N., *J. Hazard. Mater.* **2004**. *112*, 1.
Wang, X., Current situation of study on insensitive composite explosives in USA. *Huozhayao Xuebao*, **2007**, *30*, 78–80.
Albright, D., South Africa and the affordable bomb, in: *Bulletin of the Atomic Scientists*, Jul **1994**, Educational Foundation for Nuclear Science, Inc., 37–47.
Nuclear notebook, in *Bulletin of the Atomic Scientists*, May **1988**, Educational Foundation for Nuclear Science, Inc., 55.
Rieckmann, T.; Völker, S.; Lichtblau, L.; Schirra, R., *Chem. Eng. Sci.*, **2001**, *56*, 1327.
Coburn, M. D.; Jackson, T. E., *J. Heterocycl. Chem.* **1968**, *5*, 199.
Meyer, R.; Köhler, J.; Homburg, A., *Explosives*, 7th edn ., **2008**, Wiley, Weinheim.
Sitzmann, M. E., *J. Energ. Mater.* **1988**, *6*, 129.
Agrawal, J. P., *Progress Energy Comb. Sci.* **1998**, *24*, 1.
Agrawal, J. P., *Propellants, Explos. Pyrotech.*, **2005**, *30*, 316.
Agrawal, J. P., *Centr. Eur. J. of Energ Mater.* **2012**, *9*, 273.
Agrawal, J. P., Prasad, U. S.; Surve, R. N., *New. J. Chem.* **2000**, *24*, 583.
Sikder, N.; Sikder, A. K.; Agrawal, J. P., *Ind. J. Eng. Mater Sci.* **2004**, *11*, 516.
Mousavi, S.; Esmaelipour, K.; Keshavaraz, M. H., *Centr. Europ. J. of Energ. Mat.* **2013**, *10*, 455.
Klapötke, T. M.; Witkowski, T., *Chem. Plus Chem.* **2016**, *81*, 357.
Gottfried, J. L.; Klapötke, T. M.; Witkowski, T. G., *Propellants Explos. Pyrotech.*, **2017**, *42*, 353–359.
Zhang, X.; Zou, F.; Yang, P.; Gao, C.; Zhang, C.; Zou, M.; Cao, X.; Hu, W.; Zhou, Q., *Propellants, Explos. Pyrotech.*, **2020**, *45*, 1853–1858.
Snyder, C. J.; Myers, T. W.; Imler, G. H.; Chavez, D. E.; Parrish, D. A.; Veauthier, J. M.; Scharff, R. J., *Propellants Explos. Pyrotech.* **2017**, *42*, 238–242.

Cited Literature

[1] (a) Beck, W.; Evers, J.; Göbel, M.; Oehlinger, G.; Klapötke, T. M., *Z. Anorg. Allg. Chem.* **2007**, *633*, 1417–1422. (b) Beck, W.; Klapötke, T. M., *J. Mol. Struct.* (THEOCHEM) **2008**, *848*, 94–97. (c) Karaghiosoff, K.; Klapötke, T. M.; Michailovski, A.; Holl, G., *Acta Cryst. C* **2002**, *C 58*, o580–o581. (d) Gökçınar, E.; Klapötke, T. M.; Bellamy, A. J., *J. Mol. Struct.* (THEOCHEM), **2010**, *953*, 18–23. (e) Majano, G.; Mintova, S.; Bein, T.; Klapötke, T. M., *Advanced Materials* **2006**, *18*, 2240–2243. (f) Majano, G.; Mintova, S.; Bein, T.; Klapötke, T. M., *J. Phys. Chem. C*, **2007**, *111*, 6694–6699. (g) Matyushin, Y. N.; KonkovaT. S.; Korchatova, L. I., *To the Question of the Enthalpy of Formation for the Nitrocellulose*, ICT, **1998**, Karlsruhe, Germany, 197-1–179-8.

[2] (a) Evers, J.; Klapötke, T. M., Mayer, P.; Oehlinger, G., Welch, J., *Inorg. Chem.* **2006**, *45*, 4996–5007. (b) Crawford, M.-J.; Evers, J.; Göbel, M.; Klapötke, T. M.; Mayer, P.; Oehlinger, G.; Welch, J. M., *Propellants Explos. Pyrotechn.* **2007**, *32*, 478–495.

[3] Huynh, M. H. V.; Hiskey, M. A.; Meyer, T. J.; Wetzler, M., *PNAS*, **2006**, *103*, 5409–5412. See also: Klapötke, T. M.; Miro Sabate, C.; Welch, J. M., *J. Chem. Soc., Dalton Trans.* **2008**, 6372–6380.

[4] Geisberger, G.; Klapötke, T. M.; Stierstorfer, J., *Eur. J. Inorg. Chem.* **2007**, 4743–4750.

[5] SERDP information: Cleanup CU-1164: http://www.p2pays.org/ref/19/18164.pdf (2/21/03).

[6] The official SERDP source for perchlorate information: http://www.serdp.org/research/er-perchlorate.cfm

[7] SERDP & ESTCP Annual Symposium **2007**: http://www.serdp.org/Symposium/.

[8] Göbel, M.; Klapötke, T. M., *Z. Anorg. Allg. Chem.* **2007**, *633*, 1006–1017.
[9] Klapötke, T. M.; Penger, A.; Scheutzow, S.; Vejs, L., *Z. Anorg. Allg. Chem.* **2008**, *634*, 2994–3000.
[10] Karaghiosoff, K.; Klapötke, T. M.; Sabate, C. Miro, *Chemistry Europ. J.* **2009**, *15*, 1164–1176.
[11] Doherty, R. M., Novel Energetic Materials for Emerging Needs, 9th-IWCP on Novel Energetic Materials and Applications, Lerici (Pisa), Italy, September 14–18, **2003**.
[12] Fischer, N.; Klapötke, T. M.; Scheutzow, S.; Stierstorfer, J., *Centr. Europ. J. Energ. Mat.* **2008**, *5*, 3–18.
[13] Steinhauser, G.; Klapötke, T. M., *Angew. Chem.* **2008**, *120*, 2–21; *Intl. Ed.* **2008**, *47*, 2–20.
[14] Klapötke, T. M., Tetrazoles for the safe detonation, *Nachr. Chem.* **2008**, *56(6)*, 645–648.
[15] Klapötke, T. M.; Stierstorfer, J.; Tarantik, K. R., Neue, umweltverträglichere Formulierungen für Pyrotechnika, *Die aktuelle Wochenschau, GDCh*, 15.09. **2008**, Woche 37: http://www.aktuelle-wochenschau.de/woche37/woche37.html.
[16] (a) Halford, B., *Chem. Eng. News*, **2008**, June 30, 14–18. (b) Sabatini, J. J.; Poret, J. C.; Broad, R. N., *Angew. Chem. Int. Ed.* **2011**, *50*, 4624–4626.
[17] Kamlet, M. J.; Jacobs, S. J., *J. Chem. Phys.* **1968**, *48*, 23.
[18] Kamlet, M. J.; Ablard, J. E., *J. Chem. Phys.* **1968**, *48*, 36.
[19] Kamlet, M. J.; Dickinson, C., *J. Chem. Phys.* **1968**, *48*, 43.
[20] Sućeska, M., Proceedings of 32nd Int. in: Annual Conference of ICT, July 3–6, Karlsruhe, **2001**, S. 110/1–110/11.
[21] Curtiss, L. A.; Raghavachari, K.; Redfern, P. C.; Pople, J. A., *J. Chem. Phys.* **1997**, *106*, 1063.
[22] Byrd, E. F. C.; Rice, B. M., *J. Phys. Chem. A.* **2006**, *110*, 1005.
[23] Rice, B. M.; Pai, S. V.; Hare, J., *Combust. Flame* **1999**, *118*, 445.
[24] Linstrom, P. J.; Mallard, W. G. (eds.), NIST Chemistry WebBook, NIST Standard Reference Database Number 69, June **2005**, National Institute of Standards and Technology, Gaithersburg MD, 20899 (http://webbook.nist.gov).
[25] Westwell, M. S.; Searle, M. S.; Wales, D. J.; Williams, D. H., *J. Am. Chem. Soc.* **1995**, *117*, 5013.
[26] Jenkins, H. D. B.; Roobottom, H. K.; Passmore, J.; Glasser, L., *Inorg. Chem.* **1999**, *38(16)*, 3609–3620.
[27] Jenkins, H. D. B.; Tudela, D.; Glasser, L., *Inorg. Chem.* **2002**, *41(9)*, 2364–2367.
[28] Jenkins, H. D. B.; Glasser, L., *Inorg. Chem.* **2002**, *41(17)*, 4378–88.
[29] Jenkins, H. D. B., *Chemical Thermodynamics at a Glance*, Blackwell, Oxford, **2008**.
[30] Ritchie, J. P.; Zhurova, E. A.; Martin, A.; Pinkerton, A. A., *J. Phys. Chem. B*, **2003**, *107*, 14576–14589.
[31] Sućeska, M., Calculation of Detonation Parameters by EXPLO5 Computer Program, *Mater. Sci. Forum*, **2004**, 465–466, 325–330.
[32] Sućeska, M., *Propellants Explos. Pyrotech.* **1991**, *16*, 197.
[33] Sućeska, M., *Propellants Explos. Pyrotech.* **1999**, *24*, 280.
[34] Hobbs, M. L.; Baer, M. R., Calibration of the BKW-EOS with a Large Product Species Data Base and Measured C-J Properties, in: Proc. of the 10th Symp. (International) on Detonation, ONR 33395-12, Boston, MA, July 12–16, **1993**, p. 409.
[35] (a) Fischer, N.; Klapötke, T. M.; Scheutzow, S.; Stierstorfer, J., *Central Europ. J. of Energ. Mat.*, **2008**, *5(3–4)*, 3–18; (b) Gökçınar, E.; Klapötke, T. M.; Kramer, M. P., *J. Phys. Chem.* **2010**, *114*, 8680–8686.
[36] Trimborn, F., *Propellants Explos. Pyrotech.* **1980**, *5*, 40.
[37] Held, M., *NOBEL-Hefte*, **1991**, *57*, 14.
[38] Sućeska, M., *Intl. J. Blasting Fragm.*, **1997**, *1*, 261.
[39] (a) Politzer, P.; Murray, J. S., Computational Characterization of Energetic Materials, in: Maksic, Z. B.; Orville-Thomas, W. J. (eds.), *Pauling's Legacy: Modern Modelling of the Chemical Bond, Theoretical and Computational Chemistry*, Vol. 6, **1999**, 347–363; (b) Politzer, P.; Murray, J. S.; Seminario, J. M.; Lane, P.; Grice, M. E.; Concha, M.C., *J. Mol. Struct.* **2001**, *573*, 1. (c) Murray, J. S.; Lane, P.; Politzer, P., *Molec. Phys.* **1998**, *93*, 187; (d) Murray, J. S.; Lane, P.; Politzer, P., *Mol. Phys.* **1995**, *85*, 1. (e) Murray, J. S.; Concha, M. C.; Politzer, P., *Molec. Phys.* **2009**, *107(1)*, 89; (f) Politzer, P.; Martinez, J.; Murray, J. S.;

Concha, M. C.; Toro-Labbe, A., *Mol. Phys.* **2009**, *107(19)*, 2095–2101; (g) Pospisil, M.; Vavra, P.; Concha, M. C.; Murray, J. S.; Politzer, P., *J. Mol. Model.* **2010**, *16*, 985–901.
[40] (a) Rice, B. M.; Hare, J. J.; Byrd, E. F., *J. Phys. Chem. A*, **2007**, *111(42)*, 10874. (b) Rice, B. M.; Byrd, E. F. C., *J. Mater. Res.* **2006**, *21(10)*, 2444.
[41] Byrd, E. F.; Rice, B. M., *J. Phys. Chem. A*, **2006**, *110(3)*, 1005.
[42] Rice, B. M.; Hare, J., *Thermochimica Acta*, **2002**, *384(1–2)*, 377.
[43] Rice, B. M., Pai, S. V.; Hare, J., *Combust. Flame*, **1999**, *118(3)*, 445.
[44] Rice, B. M.; Hare, J. J., *J. Phys. Chem. A*, **2002**, *106(9)*, 1770.
[45] Engelke, R., *J. Phys. Chem.*, **1992**, *96*, 10789.
[46] Glukhovtsev, M. N.; Schleyer, P. v. R., *Chem. Phys. Lett.*, **1992**, *198*, 547.
[47] Eremets, M. I.; Gavriliuk, A. G.; Serebryanaya, N. R.; Trojan, I. A.; Dzivenko, D. A.; Boehler, R.; Mao, H. K.; Hemley, R. J., *J. Chem. Phys.* **2004**, *121*, 11296.
[48] Eremets, M. I.; Gavriliuk, A. G.; Trojan, I. A.; Dzivenko, D. A.; Boehler, R., *Nat. Mater.* **2004**, *3*, 558.
[49] Eremets, M. I.; Popov, M. Y.; Trojan, I. A.; Denisov, V. N.; Boehler, R.; Hemley, R. J., *J. Chem. Phys.* **2004**, *120*, 10618.
[50] Klapötke, T. M.; Holl, G., *Green Chem.* **2001**, G75.
[51] Klapötke, T. M.; Holl, G., *Chem. Aust.* **2002**, 11.
[52] Klapötke, T. M.; Stierstorfer, J., *Central Europ. J. Energ. Mat.*, **2008**, *5(1)*, 13.
[53] Klapötke, T. M.; Stierstorfer, J., *J. Chem. Soc. Dalton Trans.* **2008**, 643–653.
[54] Göbel, M.; Klapötke, T. M., *Adv. Funct. Mat.* **2009**, *19*, 347–365.
[55] (a) Singh, R. P.; Verma, R. D.; Meshri, D. T.; Shreeve, J. M., *Angew. Chem. Int. Ed.*, **2006**, *45*, 3584. (b) Sikder, A. K.; Sikder, N., *J. Haz. Mater.* **2004**, *A112*, 1.
[56] (a) Joo, Y.-H.; Twamley, B.; Grag, S.; Shreeve, J. M., *Angew. Chem.* **2008**, *120*, 6332; *Angew. Chem. Int. Ed.* **2008**, *47*, 6236; (b) Yoo, Y.-H.; Shreeve, J. M., *Angew. Chem.* **2009**, *48(3)*, 564–567. (c) Klapötke, T. M.; Mayer, P.; Stierstorfer, J.; Weigand, J. J., *J. Mat. Chem.* **2008**, *18*, 5248–5258. (d) Saly, M. J.; Heeg, M. J.; Winter, C. H., *Inorg. Chem.* **2009**, *48(12)*, 5303–5312.
[57] (a) Chambreau, S. D.; Schneider, S.; Rosander, M.; Hawkins, T.; Gallegos, C. J.; Pastewait, M. F.; Vaghjiani, G. L., *J. Phys. Chem. A*, **2008**, *112*, 7816. (b) Schneider, S.; Hawkins, T.; Ahmed, Y.; Rosander, M.; Hudgens, L.; Mills, J., *Angew. Chem. Int. Ed.* **2011**, *50*, 5886–5888.
[58] (a) Amariei, D.; Courthéoux, L.; Rossignol, S.; Kappenstein, C., *Chem. Eng. Process.* **2007**, *46*, 165. (b) Singh, R. R.; Verma, R. D.; Meshri, D. T.; Shreeve, J. M., *Angew Chem.* **2006**, *118(22)*, 3664–3682. (c) Gao, H.; Joo, Y.-H.; Twamley, B.; Zhou, Z.; Shreeve, J. M., *Angew. Chem.* **2009**, *121(15)*, 2830–2833. (d) Zhang, Y.; Gao, H.; Joo, Y.-H.; Shreeve, J. M., *Angew. Chem. Int. Ed.* **2011**, *50*, 9554–9562.
[59] (a) Altenburg, T.; Penger, A.; Klapötke, T. M.; Stierstorfer, J., *Z. Anorg. Allg. Chem.* **2010**, *636*, 463–471. (b) Klapötke, T. M.; Stierstorfer, J., *Helv. Chim. Acta*, **2007**, *90*, 2132–2150. (c) Bolton, O.; Matzger, A. J., *Angew. Chem. Int. Ed.* **2011**, *50*, 8960–8963. (d) Fischer, N.; Fischer, D.; Klapötke, T. M.; Piercey, D. G.; Stierstorfer, J., How high can you go? Pushing the limits of explosive performance, *J. Mater. Chem.* **2012**, *22(38)*, 20418–20422. (e) Fischer, N.; Klapötke, T. M.; Matecic Musanic, S.; Stierstorfer, J.; Sućeska, M., *TKX-50, New Trends in Research of Energetic Materials*, Part II, Czech Republic, **2013**, 574–585. (f) Golubev, V.; Klapötke, T. M.; Stierstorfer, J., TKX-50 and MAD-X1 – A Progress Report, *40th Intl. Pyrotechnics Seminar*, Colorado Springs, CO, July 13–18, **2014**.
[60] Millar, R. W.; Colclough, M. E.; Golding, P.; Honey, P. J.; Paul, N. C.; Sanderson, A. J.; Stewart, M. J., *Phil. Trans. R. Soc. Lond. A*, **1992**, *339*, 305.
[61] Harris, A. D.; Trebellas, J. C.; Jonassen, H. B., *Inorg. Synth.* **1967**, *9*, 83.
[62] Borman, S., *Chem. Eng. News*, **1994**, *Jan. 17*, 18.
[63] Christe, K. O.; Wilson, W. W.; Petrie, M. A.; Michels, H. H.; Bottaro, J. C.; Gilardi, R., *Inorg. Chem.* **1996**, *35*, 5068.
[64] Klapötke, T. M.; Stierstorfer, J., *Phys. Chem. Chem. Phys.* **2008**, *10*, 4340.
[65] Klapötke, T. M.; Stierstorfer, J.; Wallek, A. U., *Chem. Mater.* **2008**, *20*, 4519.

[66] Klapötke, T. M.; Krumm, B.; Mayr, N.; Steemann, F. X.; Steinhauser, G., *Safety Sci.* **2010**, 48, 28–34.
[67] Chapman, R. D.; Gilardi, R. D.; Welker, M. F.; Kreutzberger, C. B., *J. Org. Chem.* **1999**, *64*, 960–965.
[68] Fischer, D.; Klapötke, T. M.; Stierstorfer, J., *Z. Anorg. Allg. Chem.* **2011**, *637*, 660–665.
[69] Langlet, A.; Latypov, N.; Wellmar, U.; Goede, P., *Pyrotechics*, **2004**, *29*, No. 6.
[70] Schöyer, H. F. R.; Welland-Veltmans, W. H. M.; Louwers, J.; Korting, P. A. O. G.; van der Heijden, A. E. D. M.; Keizers, H. L. J.; van der Berg, R. P., *J. Propuls. Power*, **2002**, *18*, 131.
[71] Schöyer, H. F. R.; Welland-Veltmans, W. H. M.; Louwers, J.; Korting, P. A. O. G.; van der Heijden, A. E. D. M.; Keizers, H. L. J.; van der Berg, R. P., *J. Propuls. Power*, **2002**, *18*, 138.
[72] Shechter, H.; Cates Jr., H., *J. Org. Chem.* **1961**, *26*, 51.
[73] Feuer, H.; Swarts, W. A., *J. Org. Chem.* **1962**, *27*, 1455.
[74] Ovchinnikov, I. V.; Kulikov, A. S.; Epishina, M. A.; Makhova, N. N.; Tartakovsky, V. A., *Russ. Chem. Bull. Int. Ed.* **2005**, *54*, 1346.
[75] Metelkina, E. L., Novikova, T. A., *Russ. J. Org. Chem.* **2002**, *38*, 1378.
[76] Klapötke, T. M.; Krumm, B.; Scherr, M.; Spieß, G.; Steemann, F. X., *Z. Anorg. Allg. Chem.* **2008**, *8*, l244.
[77] Göbel, M.; Klapötke, T. M.; Mayer, P. M., *Z. Anorg. Allg. Chem.* **2006**, *632*, 1043.
[78] Welch, D. E., *Process for Producing Nitroform*, US 3491160, **1970**.
[79] Frankel, M. B.; Gunderloy, F. C.; Woolery, D. O., US 4122124, **1978**.
[80] Langlet, A.; Latypov, N.; Wellmar, U., *Method of preparing Nitroform*, WO 03/018514 A1, 2003.
[81] Behrend, R.; Meyer, E.; Rusche, F., *Liebigs Ann. Chem.* **1905**, *339*, 1.
[82] M. H. Keshavarz, H. Motamedoshariati, R. Moghayadnia, H. R. Nazari, J. Azarniamehraban, A new computer code to evaluate detonation performance of high explosives and their thermochemical properties, part I., *J. Haz. Mat.* **2009**, *172*, 1218–1228.
[83] M. H. Keshavarz, M. Jafari, H. Motamedoshariati, R. Moghayadnia, Energetic Materials Designing Bench (EMDB), Version 1.0, Malek-Aschtar University, Schahinschahr, **2017**.
[84] J. L. Gottfried, T. M. Klapötke, T. G. Witkowski, Estimated detonation velocities for TKX-50, MAD-X1, BDNAPM, BTNPM, TKX-55 and DAAF using the laser-induced air shock from energetic materials technique, *Prop. Expl. Pyrotech.* **2017**, *42*, 353–359.
[85] M. H. Keshavarz, T. M. Klapötke, M. Suceska, *Prop. Expl. Pyrotech.*, **2017**, *42*, 854–856.

Computer codes

Sućeska, M., *EXPLO5 Program, Version 7.01*, Zagreb, Croatia, **2024**:https://www.ozm.cz/explosives-performance-tests/thermochemical-computer-code-explo5/.
Fraunhofer Institut Chemische Technologie, ICT, ICT Thermodynamic Code, Version 1.00, **1998–2000**.
Frisch, M. J. et al., *Gaussian 16*, Gaussian Inc., Wallingford, CT, **2016**.
Fried, L.E.; Glaesemann, K.R.; Howard, W.M.; Souers, P.C., *CHEETAH 4.0 User's Manual*, Lawrence Livermore National Laboratory, **2004**.
Lu, J. P., Evaluation of the Thermochemical Code–CHEETAH 2.0 for Modelling Explosives Performance, DSTO-DR-1199, Defence Science and Technology Organization (Australian Government), Edinburgh, **2001**.
HyperChem, *Release 7.0, Molecular Modeling System for Windows*, Hypercube Inc., **2002**.
Kee, R. J.; Rupley, F. M.; Miller, J. A.; Coltrin, M. E.; Grcar, J. F.; Meeks, E.; Moffat, H. K.; Lutz, A. E.; Dixon-Lewis, G.; Smooke, M. D.; Warnatz, J.; Evans, G. H.; Larson, R. S.; Mitchell, R. E.; Petzold, L. R.; Reynolds, W. C.; Caracotsios, M.; Stewart, W. E.; Glarborg, P., *CHEMKIN Collection, Release 3.5*, Reaction Design, Inc., San Diego, CA, **1999**.

McBride, B.; Gordon, S., NASA-Glenn Chemical Equilibrium Program CEA, **1999**.

N. F. Gavrilov, G. G. Ivanova, V. N. Selin, V. N. Sofronov, UP-OK Program for Solving Continuum Mechanics Problems in One-Dimensional Complex, VANT. In: *Procedures and Programs for Numerical Solution of Mathematical Physics Problems*, Vol. 3, **1982**, 11–21.

ANSYS Autodyn® *Academic Research, Release 15.0, Help System, Equations of State*, ANSYS, Inc., **2015**.

S. Wahler, RoseBoom 2.3, **2023**, https://www.roseexplosive.com/.

WEB pages

ARL https://arl.devcom.army.mil/
SERDP https://serdp-estcp.mil/
ESTCP https://serdp-estcp.mil/
WIWEB https://www.bundeswehr.de/de/organisation/ausruestung-baainbw/organisation/wiweb
DEVCOM https://www.t2.army.mil/T2-Laboratories/Designated-Laboratories/DEVCOM-Armaments-Center/
NSWC https://www.navsea.navy.mil/Home/Warfare-Centers/NSWC-Indian-Head/ https://www.navsea.navy.mil/Home/Warfare-Centers/NSWC-Crane/
ICT https://www.ict.fraunhofer.de/en.html
DTRA https://www.dtra.mil/
EPA https://www.epa.gov/
HEMSI https://www.hemsindia.co.in/
NIOSH https://www.cdc.gov/niosh/index.html
NIST https://webbook.nist.gov/chemistry/
IMEMG https://imemg.org/
MSIAC https://www.msiac.nato.int/
FHG https://www.ict.fraunhofer.de/
OZM https://www.ozm.cz/
BIAZZI https://www.biazzi.com/nitration/
MTI http://www.vti.mod.gov.rs/index.php?view=aboutus
MTA https://mta.ro/
MTC https://www.mtc.edu.eg/mtcwebsite/
EMP http://www.emp.mdn.dz/
BAAINBw https://www.bundeswehr.de/de/organisation/ausruestung-baainbw/organisation/baainbw
WTD91 https://www.bundeswehr.de/de/organisation/ausruestung-baainbw/organisation/wtd-91
(All updated on September 3rd 2024).

Journals

Propellants, Explosives, Pyrotechnics: https://onlinelibrary.wiley.com/journal/15214087.

Central European Journal of Energetic Materials: https://web.archive.org/web/20200713224131/http://www.wydawnictwa.ipo.waw.pl/CEJEM.html.

Journal of Energetic Materials: https://www.tandfonline.com/journals/uegm20.

Journal of Pyrotechnics: https://www.jpyro.co.uk/.

The American Institute of Aeronautics and Astronautics: https://arc.aiaa.org/.

Journal of Hazardous Materials: https://www.sciencedirect.com/journal/journal-of-hazardous-materials.

International Journal of Energetic Materials and Chemical Propulsion (IJEMCP):https://www.begellhouse.com/en/journals/energetic-materials-and-chemical-propulsion.html.
Engineering Science and Military Technologies: https://ejmtc.journals.ekb.eg/.
Defense Technology (Elsevier): https://www.sciencedirect.com/journal/defence-technology.
FirePhysChem: https://www.sciencedirect.com/journal/firephyschem.
Chinese Journal of Explosives & Propellants: http://hzyxb.paperopen.com/en/#/.
(All updated on September 3rd 2024).

Conferences

Gordon Research Conference: Energetic Materials, every other year (2022, 2024, . . .), usually June, in New Hampshire (USA): https://www.grc.org/energetic-materials-conference/
ICT International Annual Conference, annually, June/July, in Karlsruhe (Germany):https://www.ict.fraunhofer.de/en/comp/landingpages/annualconference.html
New Trends in Research of Energetic Materials (NTREM), annually, April, in Pardubice (Czech Republic): https://www.ntrem.com/
PARARI – Australian Ordnance Symposium, every other year (2022, 2024, . . .), usually November, (Australia): https://www.parari.org/
International Pyrotechnics Seminar, every other year (2022, 2024, . . .), July, CO (USA): https://www.ipsusa-seminars.org/
High Energy Materials Society of India (HEMSI), bi-annually (2021, 2023, . . .), usually November (India): https://hemce2024.in/
Insensitive Munitions (IM) and Energetic Materials Technology Symposium (usually every 2 years): https://imemg.org/
Biennial APS Conference on Shock Compression of Condensed Matter – SCCM (every other year, odd years): https://www.aps.org/events/2025/24th-topical-group-compression-condensed-matter
International Conference IPOEX: Explosives Research – Application – Safety (yearly), Institute of Industrial Organic Chemistry, Ustroń Zawodzie, Poland:https://ipo.lukasiewicz.gov.pl/IPOEX2024/en/miedzynarodowa-konferencja-naukowa-ipoex/general-information.html
HEMs (High Energy and Special Materials), every year, even years (2024, 2026, . . .) in Tomsk, Russia, odd years (2023, 2025) in other countries, usually in India (https://infobrics.org/post/41649/): https://forms.yandex.ru/cloud/6621e96943f74f317da6f634/
Explosive Ordnance Seminar, every year in Europe: https://intelligence-sec.com/about-us/https://intelligence-sec.com/events/explosive-ordnance-seminar-europe-2025/
EUROPYRO, usually every other year (odd years) in Europe: https://europyro2023.org/
OTEH (Serbia), every second (even) year: http://www.vti.mod.gov.rs/oteh/
(All updated on September 3rd 2024).

16 Appendix

I am one of those who think like Nobel,
that humanity will draw more good
than evil from new discoveries.
Marie Curie

Important reaction types in organic nitrogen chemistry

$$\text{R–NH}_2 \xrightarrow{\text{HNO}_2\text{, aq.}} [\text{R–N}_2]^+ \xrightarrow{\text{H}_2\text{O}} \text{R–OH}$$

$$\text{R–NH}_2 \xrightarrow{\text{HNO}_2\text{, aq.}} [\text{R–N}_2]^+ \xrightarrow{\text{NaN}_3} \text{R–N}_3$$

$$\text{R}_2\text{NH} \xrightarrow{\text{HNO}_2\text{, aq.}} \text{R}_2\text{N–N=O}$$

$$\text{R–NH}_2 \xrightarrow{KO_2/THF} \text{R–NO}_2$$

$$\text{Ph–NH}_2 \xrightarrow{\text{HNO}_2\text{, aq.}} [\text{Ph–N}_2]^+ \xrightarrow{\text{PhH, –H}^+} \text{Ph–N=N–Ph}$$

$$\text{Ph–NH}_2 \xrightarrow{\text{HNO}_2\text{, aq.}} [\text{Ph–N}_2]^+ \xrightarrow{\text{NaN}_3} \text{Ph–N}_3$$

$$\text{Ph–NH}_2 \xrightarrow{\text{HNO}_2\text{, aq.}} [\text{Ph–N}_2]^+ \xrightarrow{\text{NaBF}_4} [\text{Ph–N}_2]^+[\text{BF}_4]^- \xrightarrow{\text{NaN}_3} \text{Ph–N}_5 \xrightarrow{-\text{N}_2} \text{PhN}_3$$

$$\text{Ph–NH–NH–Ph} \xrightarrow{\text{HgO, –H}_2\text{O, –Hg}} \text{Ph–N=N–Ph}$$

$$\text{Ph–NH}_2 \xrightarrow[\text{diazotation}]{\text{HNO}_2\text{, aq.}} [\text{Ph–N}_2]^+ \xrightarrow[\text{Sandmeyer}]{\text{Cu}^\text{I}\text{X}} \text{Ph–X}$$

$$\text{NaN}_3 + \text{R–Br} \longrightarrow \text{R–N}_3 \xrightarrow{\text{H}_2\text{/Pd–C}} \text{R–NH}_2$$

$$\text{Ph–CH}_2\text{–Br} + \text{NaN}_3 \longrightarrow \text{Ph–CH}_2\text{–N}_3$$

$$\text{Ph–CH}_2\text{–N}_3 \xrightarrow{\text{H}_2\text{/Pd–C}} \text{Ph–CH}_2\text{–NH}_2 + \text{N}_2$$

$$\text{Ph–C(O)–Cl} + \text{NaN}_3 \longrightarrow \text{Ph–C(O)–N}_3$$

$$\text{Ph–CH}_2\text{–CN} \xrightarrow{\text{H}_2\text{/cat. or LiAlH}_4} \text{Ph–CH}_2\text{–CH}_2\text{–NH}_2$$

$$\text{R–CN} \xrightarrow{\text{H}_2\text{/cat. or LiAlH}_4} \text{R–CH}_2\text{–NH}_2$$

$$\text{R}_1\text{R}_2\text{HC–NO}_2 \xrightarrow{\text{H}_2\text{/cat. or LiAlH}_4} \text{R}_1\text{R}_2\text{HC–NH}_2$$

$$\text{R}_1\text{R}_2\text{HC–CN} \xrightarrow{\text{H}_2\text{/cat. or LiAlH}_4} \text{R}_1\text{R}_2\text{HC–CH}_2\text{–NH}_2$$

$$\text{R–NO}_2 \xrightarrow{\text{H}_2\text{/cat. or LiAlH}_4} \text{R–NH}_2$$

https://doi.org/10.1515/9783111447087-016

$$R{-}C(O){-}NH_2 \xrightarrow{PCl_5} R{-}CN$$

$$R{-}C(O){-}NH_2 \xrightarrow{LiAlH_4} R{-}CH_2{-}NH_2$$

$$HMTA \xrightarrow{HNO_3} RDX \quad (+\ HMX)$$

$$C({-}CH_2{-}OH)_4 \xrightarrow{HNO_3} PETN$$

$$R{-}C(O){-}R \xrightarrow{N_2H_4} R_2C{=}N{-}NH_2$$

$$R{-}X \xrightarrow{N_2H_4} R{-}NH{-}NH_2 \quad (X = Hal,\ OMe)$$

$$RC(O)R + H_2NR' \longrightarrow RC(=NR')R$$

$$R_2NH + CH_2O \xrightarrow[\text{Mannich}]{H^+} \underset{\text{iminium ion (reacts with C nucleophiles)}}{[R_2N{=}CH_2]^+ \longleftrightarrow [R_2N{-}CH_2]^+}$$

$$R_2N{-}C(O)H + POCl_3 \longrightarrow \underset{\text{Vilsmeyer iminium ion}}{[R_2N{=}CHCl]^+ \longleftrightarrow [R_2N{-}CHCl]^+}$$

$$R{-}COOH \xrightarrow{SOCl_2} R{-}C(O){-}Cl$$

$$R{-}C(O){-}NH{-}R' \xrightarrow{N_2H_4} R{-}C(O){-}NH{-}NH_2 + R'{-}NH_2$$

hydrazinolysis

$$R{-}C(O){-}OR \xrightarrow{N_2H_4} R{-}C(O){-}NH{-}NH_2 \xrightarrow{HNO_2} R{-}C(O){-}N_3$$

$$R{-}X + KCN \longrightarrow R{-}CN \quad (X = Cl,\ Br,\ I,\ OTs,\ \ldots)$$

$$Ph{-}C(O){-}NH_2 \xrightarrow{PCl_5} Ph{-}CN$$

$$Ph{-}Cl \xrightarrow{N_2H_4} Ph{-}NH{-}NH_2$$

$$R{-}C(O){-}Cl \xrightarrow{N_2H_4} R{-}C(O){-}NH{-}NH_2$$

$$\begin{aligned} R{-}C(O){-}NH{-}NH_2 + NO^+ &\xrightarrow{-H^+} R{-}C(O){-}NH{-}NH{-}N{=}O \\ &\longrightarrow R{-}C(O){-}NH{-}N{=}N{-}OH \\ &\longrightarrow R{-}C(O){-}N_3 + H_2O \end{aligned}$$

$$R{-}C(O){-}N_3 \xrightarrow{T,\ -N_2} R{-}N{=}C{=}O$$

$$R{-}C(O){-}R \xrightarrow{N_2H_4} R_2C{=}N{-}NH_2$$

$$R_3N \xrightarrow{PhCO_3H \text{ or } H_2O_2/AcOH} R_3N^+{-}O^- \quad (\text{also with pyridines instead of } R_3N)$$

$$R_2N{-}H \xrightarrow{\text{MSH or DPA, Base}} R_2N{-}NH_2$$

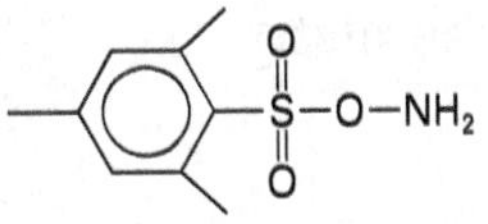

MSH = mesitylsulfonylhydroxylamine

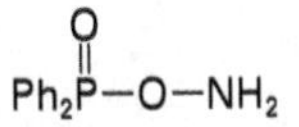

DPA = diphenylphosphenylhydroxylamine

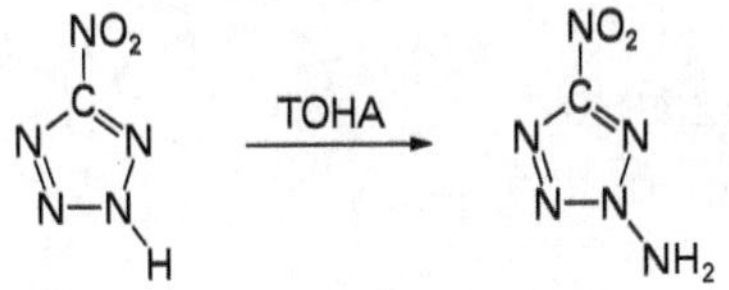

TOHA = 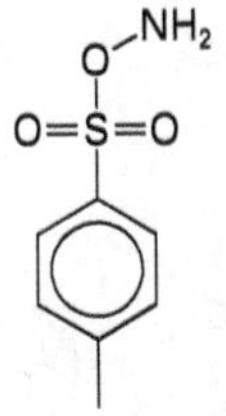

HOSA = $H_3N^+ - O - SO_3^-$

N amination: TOHA = O-tosylhydroxylamine

C amination: TMHI = trimethylhydrazinium iodide

Curing of a diol (HTPB) with isocyanate binder

$O=C=N-R^1-N=C=O$ + $H-O-R^2-OH$ ⟶ $O=C=N-R^1-N-\overset{O}{\underset{H}{C}}-O-R^2-OH$

⟶ ⟶ $-\overset{O}{C}-N-R^1-N-\overset{O}{C}-O-R^2-O-\overset{O}{C}-N-R^1-N-\overset{O}{C}-O-R^2-O-$

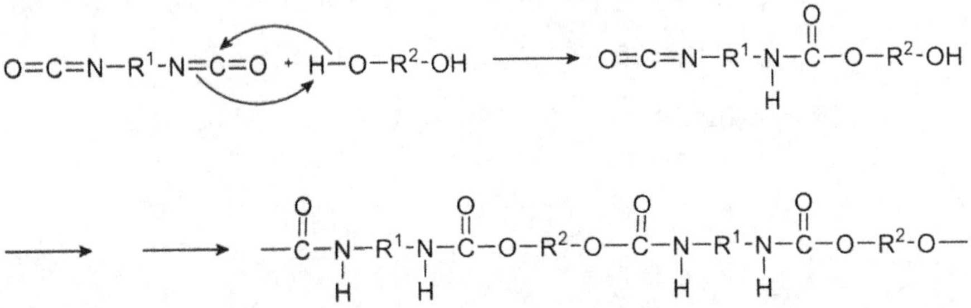

Important reaction types in inorganic nitrogen chemistry

azide transfer reagents: NaN_3
AgN_3
Me_3SiN_3

HN_3 synthesis:

$$\underset{\text{stearic acid}}{RCO_2H} + NaN_3 \xrightarrow[\text{melt}]{\sim 140\,°C} RCO_2Na + HN_3 \uparrow$$

$$HBF_4 + NaN_3 \xrightarrow{Et_2O} NaBF_4 \downarrow + HN_3 \uparrow$$

$$H_2SO_4 + NaN_3 \longrightarrow HN_3 + NaHSO_4$$

NaN_3 synthesis:

$$NaNO_3 + 3\,NaNH_2 \xrightarrow{100\,°C,\ NH_3(liq.)} NaN_3 + 3\,NaOH + NH_3$$

$$N_2O + 2\,NaNH_2 \xrightarrow{190\,°C} NaN_3 + NaOH + NH_3$$

$$R'{-}NH_2 \xrightarrow{HC(OR)_3,\ NaN_3,\ H^+} R'{-}N_4CH$$

hydrazine synthesis:

Raschig process (SNPE):

$$NaOCl_{aq.} + NH_{3\,aq.} \xrightarrow{0\,°C} NH_2Cl_{aq.} + NaOH$$

$$NH_2Cl_{aq.} + NaOH + NH_3 \xrightarrow{130\,°C} N_2H_4 + H_2O + NaCl$$

$$NH_2Cl + H_2N(CH_3) \xrightarrow[-NaCl/-H_2O]{NaOH} \underset{(MMH)}{H_2N{-}NH(CH_3)}$$

laboratory-scale preparation of anhydrous hydrazine:

$$N_2H_4 \cdot H_2O \xrightarrow{BaO} N_2H_4 \xrightarrow{Na} N_2H_4\ \text{anh.}$$

$$(HOCN)_3 + N_2H_4/H_2O \longrightarrow (HOCN)_3 \cdot N_2H_4 \xrightarrow{200\,°C} (HOCN)_3 + N_2H_4$$

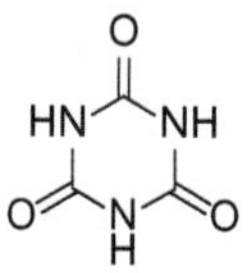

ammonium perchlorate (AP):

$$NH_3 + HClO_4 \longrightarrow AP$$

$$R{-}NH{-}NH_2 + HNO_2(NaNO_2/HCl) \longrightarrow R{-}N_3$$

$$R{-}C(O){-}Cl + N_2H_4 \longrightarrow R{-}C(O){-}NH{-}NH_2 + HCl$$

or:

$$R{-}C(O){-}OMe + N_2H_4 \longrightarrow R{-}C(O){-}NH{-}NH_2 + MeOH$$

$$R{-}C(O){-}NH{-}NH_2 + HNO_2 \longrightarrow R{-}C(O){-}N_3 \longrightarrow R{-}NCO + N_2$$

$$R{-}NH_2 + H_2O_2/H_2SO_4 \longrightarrow R{-}NO_2$$

ammonium nitrate (AN):

$$NH_3 + HNO_3 \longrightarrow AN$$

N-oxide formation:

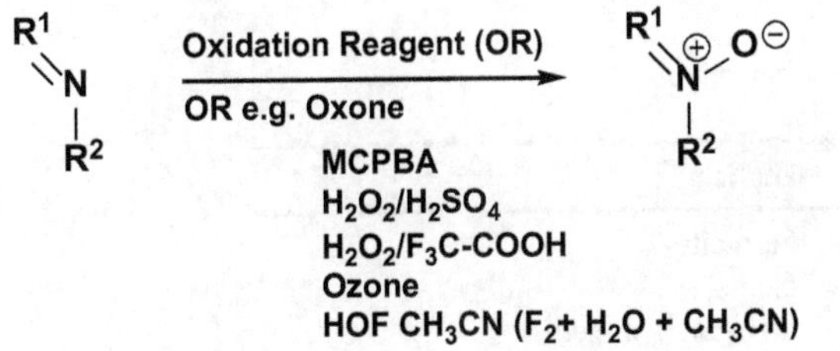

R – CN – oxone → R – NO_2 (R = azole ring)

For organic azides the following rule of thumb applies: organic, covalently bound azide compounds can be handled and are not expected to be explosive if the number (#) of N atoms is smaller than the number of C atoms and if the ratio of (C atoms + O atoms) divided by the number of N atoms is greater than three:

$$\#(N) < \#(C)$$

$$[\#(C) + \#(O)] / \#(N) > 3$$

(a) P. A. S. Smith, Open-Chain Nitrogen Compounds, Bd. 2, Benjamin, New York, **1966**, 211–256;

(b) J. H. Boyer, R. Moriarty, B. de Darwent, P. A. S. Smith, *Chem. Eng. News* **1964**, *42*, 6.

Laboratory-scale synthesis of ADN starting from potassium sulfamate and cyanoguanidine:

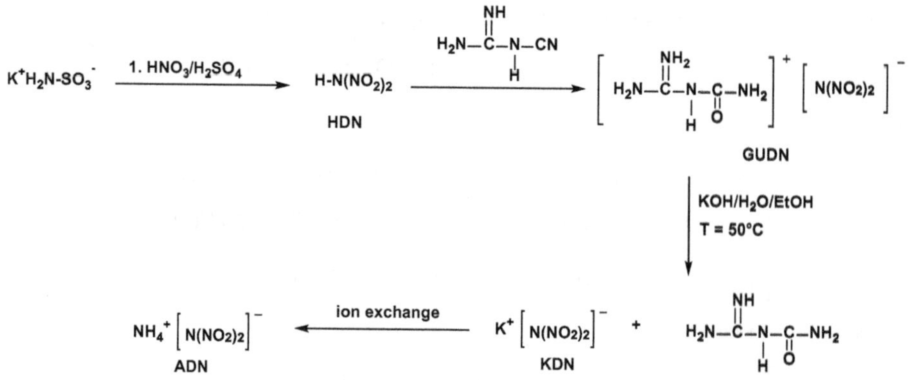

(according to: K. Nosratzadegan, M. Mahdavi, K. Ghani, K. Barati, Prop. Explos. Pyrotech. 2019, 44, 830–836.

Tab. 16.1: Technical terms in German and English language.

German	English
Abbrandgeschwindigkeit	burn rate
Abgaswolke (Flugzeug, Rakete)	plume
Aufschlagzünder oder Perkussionszünder	percussion primer
Binder	binder
Brandwaffe	incendiary device
Brennkammer	combustion chamber
Brennstoff, Treibstoff	fuel
Detonationsgeschwindigkeit	velocity of detonation, detonation velocity
Dreibasig	triple-base
Einbasig	single-base
Elektrostatische Empfindlichkeit	electrostatic sensitivity
Elektrostatisches Potential	electrostatic potential
Energetischer Stoff	energetic material
Expansionsdüse	(expansion) nozzle
Fallhammer	drophammer
Flachladung	flat cone charge
Gasgenerator	gas generator
Hohlladung	shaped charge
Ionische Flüssigkeit	ionic liquid
Köder-Munition oder Täuschkörper	decoy flares / countermeasure munition
Leuchtsatz	illuminating device
Mündungsfeuer	muzzle flash
Oxidator	oxidizer
Plastifizierungsmittel	plasticizer

Tab. 16.1 (continued)

German	English
Primärer Explosivstoff	primary explosive
Projektilbildende Ladung	explosively formed projectile
Pyrotechnische Mischung	pyrotechnic composition
Raketentreibstoff	rocket propellant
Raucherzeuger	smoke generator
Reibeapparat	friction tester
Reibeempfindlichkeit	friction sensitivity
Schlagempfindlichkeit	impact sensitivity
Schmelzgießen	melt cast
Schwarzpulver	blackpowder
Sekundärer Explosivstoff	secondary explosive, high explosive
Signalfackel	signal flare
Spezifische Energie	force
Sprenggelatine	blasting gelatine
Sprengkapsel (Schockwelle)	blasting cap / detonator / primer
Stahlhülsentest	steel shell test
Täuschkörper	decoy flare
Treibladungspulver	gun propellant
Universalbombe	general purpose bomb
Verdämmung	confinement
Verzögerungssatz	delay composition
Zünder (durch Hitze)	initiator
Zweibasig	double-base

Tab. 16.2: Abbreviations.

AA	ammonium azide
ADN	ammonium dinitramide
ADW	agent defeat warhead
AF	azidoformamidinium
AF-N	azidoformamidinium nitrate
AG	aminoguanidinium
AG-N	aminoguanidinium nitrate
ALEX	electrically exploded aluminum, nano material (20–60 nm)
AN	ammonium nitrate
ANFO	ammonium nitrate fuel oil
AT, 5-AT	aminotetrazole, 5-aminotetrazole
BI	bullet impact
BP	black powder (gun powder)
BTA	bis(tetrazolyl)amine
BTAT	bis(trinitroethyl)-1,2,4,5-tetrazine-3,6-diamine
BTH	bistetrazolylhydrazine
BTTD	bis(trinitroethyl)-tetrazole-1,5-diamine
C-J point	Chapman-Jouguet point

Tab. 16.2 (continued)

AA	ammonium azide
CL-20	HNIW, hexanitrohexaazaisowurtzitane
CTP	chemical thermal propulsion
DADNE	1,1-diamino-2,2-dinitro-ethene
DADP	diacetone diperoxide
DAG	diaminoguanidinium
DAG-Cl	diaminoguanidinium chloride
DAT	diaminotetrazole
DDNP	diazodinitrophenole
DDT	deflagration to detonation transition
DINGU	dinitroglycoluril
DNAN	2,4-dinitroanisole
EDD	explosive detection dog
EFP	explosively formed projectile
EPA	environmental protection agency
ESD	electrostatic discharge sensitivity
ESP	electrostatic potential
FAE	fuel-air explosives
FCO	fast cook-off
FI	fragment impact
FLOX	fluorine – liquid oxygen
FOX-12	guanylurea dinitramide
FOX-7	DADNE, diaminodinitroethene
FOX-7	1,1-diamino-2,2-dinitro-ethene
GUDN	guanylurea dinitramide
GWT	global war on terror
H_2BTA	bis(tetrazolyl)amine
HA	hydrazinium azide
HAA	hydroxylammonium azide
HAN	hydroxylammonium nitrate
HAT-DN	aminotetrazolium dinitramide
HME	homemade explosive
HMTA	hexamethylenetetramine
HMTD	hexamethylene triperoxide diamine
HMX	octogen, high melting explosive, her/his Majesty's explosive
HN	hydrazinium nitrate
HNFX-1	tetrakis(difluoramino)octahydro dinitro diazocine
HNS	hexanitrostilbene
HTPB	hydroxy-terminated polybutadiene
Hy-At	hydrazinium aminotetrazolate
HZT	hydrazinium azotetrazolate
IED	improvised explosive device
IM	insensitive munition
IRFNA	inhibited red fuming nitric acid
LOVA	low-vulnerability ammunition
LOX	liquid oxygen
MEKP	methylethylketone-peroxide

Tab. 16.2 (continued)

AA	ammonium azide
MF	mercury fulminate
MMH	monomethylhydrazine
MOAB	massive ordnance air blast bomb
MON-XX	mixture of dinitrogen tetroxide (NTO) with XX % NO
MTV	magnesium-Teflon-Viton
Na_2ZT	sodium azotetrazolate
NC	nitrocellulose
NG	nitroglycerine
NIOSH	National Institute for Occupational Safety and Health
NQ	nitroguanidine
NT	nitrotetrazole
NTO	5-nitro-1,2,4-triazol-3-one
NTP	nuclear thermal propulsion
ONC	octanitrocubane
PA	picric acid
PBX	polymer bonded explosive
PETN	nitropenta, pentaerythritoltetranitrate
RDX	hexogen, research department explosive, Royal demolition explosive
RFNA	red fuming nitric acid
SC	shaped charge
SCI	shaped charge impact
SCJ	shaped charge jet
SCO	slow cook-off
SCRAM-Jet	supersonic combustion RAM jet
SR	sympathetic reaction
STP	solar thermal propulsion
TAG	triaminoguanidine
TAG-Cl	triaminoguanidinium chloride
TAG-DN	triamonoguanidinium dinitramide
TAGzT	triaminoguanidinium azotetrazolate
TART	triacetyltriazine
TATB	triaminotrinitrobenzene
TATP	triacetone triperoxide
TMD	theoretical maximum density (in the crystal)
TNAZ	1,3,3-trinitroazetidine
TNT	trinitrotoluene
TO	triazol-5-one
TTD	trinitroethyl-tetrazol-1,5-diamine
TTD	trinitroethyl-1*H*-tetrazole-1,5-diamine
UDMD	unsymmetrically substituted dimethylhydrazine
UXO	unexploded ordnance
VOD	velocity of detonation
WFNA	white fuming nitric acid
WMD	weapon of mass distruction
ZT	azotetrazole

Tab. 16.3: CBS-4M calculated gas-phase enthalpies of formation.

M	M	$\Delta_f H°(g, M)$ / kcal mol^{-1}	$\Delta_f H°(g, M)$ / kcal mol^{-1}, lit. value
1*H*-nitrotetrazole, 1-H-NT	CHN_5O_2	+87.1	
1-Me-aminotetrazolium, 1-MeHAT$^+$	$C_2H_6N_5^-$	+224.0	
1-Me-nitriminotetrazole, 1-MeHAtNO$_2$	$C_2H_4N_6O_2$	+86.5	
1-Me-nitrotetrazole, 1-Me-NT	$C_2H_3N_5O_2$	+80.7	
2*H*-nitrotetrazole, 2-H-NT	CHN_5O_2	+84.0	
2-Me-nitrotetrazole, 2-Me-NT	$C_2H_3N_5O_2$	+74.5	
aminoguanidinium, AG$^+$	$CH_7N_4^+$	+161.0	
aminotetrazolium, HAT$^+$	$CH_4N_5^+$	+235.0	
ammonium, NH_4^+	NH_4^+	+151.9	+147.9[a]
azidoformidinium, AF$^+$	$(H_2N)_2CN_3^+$, $CH_4N_5^+$	+235.4	
azidotetrazolate, CN_7^-	CN_7^-	+114.1	
bistetrazolylamine, BTA	$C_2H_3N_9$	+178.2	
diaminotetrazolium, HDAT$^+$	$CH_5N_6^+$	+252.2	
dinitramide, DN$^-$	$N(NO_2)_2^-$, $N_3O_4^-$	−29.6	
guanizinium, Gz$^+$	$C_2H_7N_6$	+207.7	
hydrazinium, $N_2H_5^+$	$N_2H_5^+$	+185.1	+184.6[b]
NG-A	$C_3H_7N_5O_7$	−13.1	
nitrate, NO_3^-	NO_3^-	−74.9	−71.7[c,d]
nitriminotetrazole, H_2AtNO_2	$CH_2N_6O_2$	+95.0	
nitroglycerine, NG	$C_3H_5N_3O_9$	−67.2	
perchlorate, ClO_4^-	ClO_4^-	−66.1	
triaminoguanidinium, TAG$^+$	$C(NHNH_2)_3^+$, $CH_9N_6^+$	+208.8	

[a] D. A. Johnson, *Some thermodynamic aspects of inorganic chemistry*, Cambridge Univ. Press, 2nd edn., Cambridge, **1982**, appendix 5.
[b] P. J. Linstrom and W. G. Mallard, Eds., *NIST Chemistry WebBook*, NIST Standard Reference Database Number 69, June **2005**, National Institute of Standards and Technology, Gaithersburg MD, 20899 (http://webbook.nist.gov).
[c] J. A. Davidson, F. C. Fehsenfeld, C. J. Howard, *Int. J. Chem. Kinet.*, **1977**, *9*, 17.
[d] D. A. Dixon, D. Feller, C.-G. Zhan, J. S. Francisco, *International Journal of Mass Spectrometry*, **2003**, *227(3)*, 421.

Tab. 16.4: Molecular volumes[e,f,g].

	V_M / Å^3	V_M / nm^3
aminotetrazolium cation, [HAT]$^+$	69	0.069
ammonium, NH_4^+	21	0.021
azide, N_3^-	58	0.058
Ba^{2+}	12.3	0.0123
chloride, Cl^-	47	0.047
diaminotetrazolium cation, [HDAT]$^+$	93	0.093
dinitramide, DN$^-$	89	0.089

Tab. 16.4 (continued)

	V_M / Å^3	V_M / nm^3
guanizinium, $[Gz]^+$	117.6	0.118
H_2O, hydrate water	25	0.025
H_2O, structural water	14	0.014
hydrazinium, $N_2H_5^+$	28	0.028
Mg^{2+}	2.0	0.00199
nitrate, NO_3^-	64	0.064
nitriminotetrazolate anion, $[HAtNO_2]^-$	136	0.136
nitriminotetrazolate dianion, $[AtNO_2]^{2-}$	136	0.136
perchlorate, $[ClO_4]^-$	89.0	0.089
Sr^{2+}	8.6	0.00858
triaminoguanidinium, $[TAG]^+$	108	0.108
[1-MeHAT] $[NO_3]$	208	0.208
[2-MeHAt] [DN]	206.4	0.206
$[2\text{-MeHAt}]^+$	117.4	0.117
[AF] [DN]	174	0.174
[AF] [DN] · H_2O	199	0.199
[AG] $[CN_7]$	201.8	0.202
$[CH_3N_4]$ [DN]	161.3	0.161
[Gz] [DN]	206.6	0.206
[HAT] [DN]	172	0.172
[HAT] $[NO_3]$	133	0.133
[HDAT] [DN]	182	0.182
$[N_2H_5]$ $[CN_7]$	151.6	0.152
$[N_2H_5]$ $[CH_2N_5]$	125.7	0.126
$[N_2H_5]_2$ [OD]	224.4	0.224
$[NH_4]$ [1,5-BT]	164	0.164
$[NH_4]$ $[CN_7]$	132.3	0.132
$[NH_4]$ [DN], ADN	110	0.110
$[NH_4]$ $[NO_3]$, AN	77	0.077
$[NH_4]_2$ [OD]	195.0	0.195
[Sr] $[AtNO_2]$ · 2 H_2O	172.5	0.173
[TAG] $[1\text{-Me-AtNO}_2]$	262	0.262
[TAG] [Cl]	153.5	0.154
[TAG] [DN]	215	0.215
[TAG] $[HAtNO_2]$	244	0.244
[TAG] $[NO_3]$	174.1	0.174
$[TAG]_2$ $[AtNO_2]$	352	0.352

[e] H. D. B. Jenkins, H. K. Roobottom, J. Passmore, L. Glasser, *Inorg. Chem.*, **1999**, *38(16)*, 3609.
[f] H. D. B. Jenkins, D. Tudela, L. Glasser, *Inorg. Chem.*, **2002**, *41(9)*, 2364.
[g] H. D. B. Jenkins, L. Glasser, *Inorg. Chem.*, **2002**, *41(17)*, 4378.

Tab. 16.5: Classification of some weapon systems and war-heads.

weapon	description	example
bomb	war-head respectively HE in a shell, solely ballistic	MK 80 series
rocket	bomb + propulsion system, solid rocket motor	Hydra 70 (70 mm)
missile	guided rocket, guidance: sensor, IR seeker, radar, GPS	AIM (air intercept missile), e.g. AIM-9-Sidewinder
strategic missile	missile for larger targets, target: military base or city; large often intercontinental; war-head often nuclear; propulsion: AP / Al = high I_{sp}	Peacekeeper
tactical missile	missile for tactical use, target: localized in battlefield; often conventional war-head with HE; only occasionally nuclear, relatively small system; 3 types: 1. high I_{sp}: AP / Al 2. smoke-reduced: AP only 3. minimum signature (smoke): NG / NC or NG / NC / NQ	Cruise Missile
ballistic missile	no propulsion system for most of the flight (except launch), free fall (ballistic) to target, leaves atmosphere during flight, liquid or solid boosters	SCUD (Russian)
cruise missile	remains in atmosphere, own propulsion system during entire flight, smaller than ballistic missile, for tactical purpose	Tomahawk

Tab. 16.6: Classification of low- and high-explosives.

	low-explosive	high-explosive
example	black powder, gun powder	RDX, HMX
initiation	flame, spark	detonation, shock-wave
confinment	necessary	not necessary
propagation	particle-to-particle burning (thermal)	shock-wave

Tab. 16.6 (continued)

	low-explosive	high-explosive
velocity	subsonic	supersonic
effect	bursting, deformation, cracking	shattering
chemical analysis after reaction	uncombusted particles	traces (25–100 ppb), MS, LC/MSMS (electrospray)

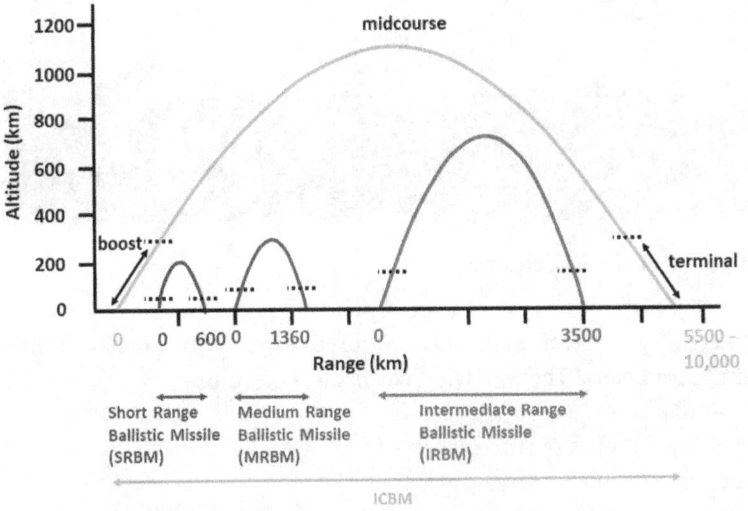

Fig. 16.1: Ranges and altitudes of different types of ballistic missiles.

Kingery-Bulmash Equation

Among the various blast load equations, the Kingery-Bulmash equation enables calculation of the parameters of a pressure-time-history curve, and can be applied to both a free-air burst as well as a surface burst (Fig. 16.2).

The Kingery-Bulmash (K-B) equations are a widely accepted means of predicting air-blast parameters from surface explosive detonations. The parameters predicted from these calculations include incident and reflected pressure, time of arrival, shock front velocity, incident impulse, reflected impulse, and positive phase duration. K-B calculations are based on curve fitting of experimental data, and can be used to predict air-blast parameters for explosive weights of less than 1 kg (2.2 pounds) up to 400,000 kg (440.9 US Tons), and ranges from 0.05 to 40 meters (1.9 inches to 131.2 feet). The K-B equations can also determine the parameters for multiple explosive types using scaled distances and relative equivalency to TNT.

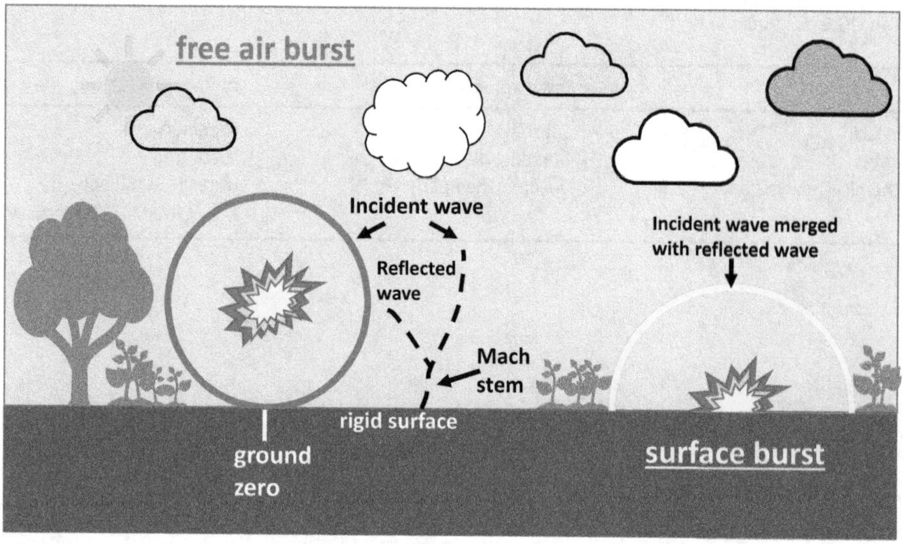

Fig. 16.2: Free air burst and surface burst explosions.

As shown in Fig. 16.3, a pressure-time history curve of a blast load may be represented by several parameters. The following parameters were used to estimate the blast load:

Ps: Maximum incident blast overpressure;

Pr: Maximum reflected blast overpressure;

Is: Impulse due to the incident blast overpressure;

Ir: Impulse due to the reflected blast overpressure;

Ta: Arrival time of the shock wave;

To: Duration of the positive pressure phase;

U: Velocity of the shock wave at arrival.

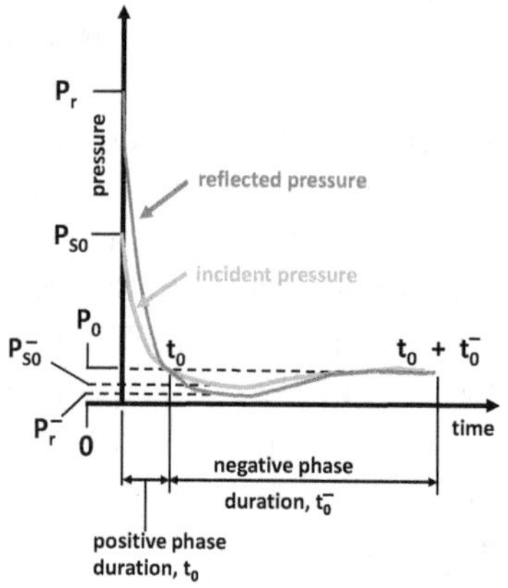

Fig. 16.3: Pressure-time curve of a blast load.

The Kingery-Bulmash equations have been developed for TNT-explosion events. The concept of TNT-equivalency is widely-used to reflect the differences in the explosive strength of different explosive formulations. The amount of an explosive is transferred to the equivalent amount of TNT by multiplication with a certain factor, derived theoretically or from experimental values. The implemented values in this tool are taken from AASTP-1 (ALLIED AMMUNITION STORAGE AND TRANSPORT PUBLICATION).

A calculator based on the Kingery-Bulmash equations to model a hemispheric, surface explosion is available on the internet: https://unsaferguard.org/un-saferguard/kingery-bulmash. However, this calculator should not be used for explosions corresponding to a spherical burst in the air.

Literature

[1] D. Jeon, K. Kim, S. Han, *Computation* **2017**, *5*, 41; doi:10.3390/computation5030041.

[2] United States Department of Defense (U.S DoD). Structures to Resist the Effects of Accidental Explosions; Report to Army Armament Research and Development Command; United States Department of Defense: Arlington County, VA, USA, **2008**; Volume 4.

[3] M. Langenderfer, K. Williams, A. Douglas, B. Rutter, C. E. Johnson, *Shock Waves,* **2021**, *31*, 175–192.

[4] https://doi.org/10.1007/s00193-021-00993-0

JWL model (Jones-Wilkins-Lee)

The detonation energy can be evaluated by applying the so-called JWL model (from Jones-Wilkins-Lee). The model assumes that the detonation of an explosive may be completely described in pressure-volume space. It also assumes that the detonating explosive compresses instantly from the room temperature and pressure value up the Rayleigh line to the C-J point. Then, it expands down the isentrope given by the JWL equation of state:

$$p = A\exp(-R_1 v) + B\exp(-R_2 v) + Cv^{-(1+\omega)}$$

where v is volume relative to the undetonated state (V/V_0), the parameters A, B, R_1, and R_2 are constant fitting parameters associated with the model, and the parameter ω is an assumed constant material parameter, which can characterize the Grüneisen function #. Finally, the parameter C is dependent only on the entropy S, such that individual isentropes may be obtained by holding the value of C constant at particularly selected values. In short: A, B, C, R_1, R_2 and ω are JWL coefficients that can be derived experimentally using the cylinder test, or theoretically from thermochemical calculations (e.g., EXPLO5).

Integrating the JWL equation of state, the equation for calculation of the energy of detonation products on the isentrope at any relative volume is obtained ($E_s(v)$):

$$E_s(v) = -\int_{\infty}^{v} p\,dv = \frac{A}{R_1}\exp(-R_1 v) + \frac{B}{R_2}\exp(-R_2 v) + \frac{C}{\omega}v^{-\omega}$$

The energy of detonation at any relative volume ($E_d(v)$) is given by the equation:

$$E_d(v) = -\{[E_s(\mathrm{CJ}) - E_s(v)] - E_c\} \qquad (E_c = \text{shock energy})$$

Since at infinite volume the energy on isentrope equals zero, the detonation energy is

$$E_d(v \to \infty) = -[E_s(\mathrm{CJ}) - E_c] = E_0$$

To calculate the detonation energy according to the JWL model, the coefficients in the JWL equation of state should be known, that is, previously determined. The usual way to derive the JWL coefficients is from the cylinder test data: by matching the wall velocity-time history using large computer finite-element hydrocodes. The coefficients are finally obtained by nonlinear fitting of the *p*-*v* values of the gas expansion progress.

Another possibility is theoretical evaluation of the detonation energy from a thermochemical calculation (EXPLO5). This method consists in calculating pressure-volume data along the expansion isentrope (assuming equilibrium or frozen expansion), and then fitting the data that is obtained to the JWL equation to obtain the JWL coefficients. The JWL coefficients are then used to calculate detonation energy.

It should be noted that the energy of detonation products at 1 bar and 298 K is usually called "the total detonation energy". The total detonation energy corresponds to the heat of detonation determined using calorimetry.

Based on experience, the JWL parameters for ideal explosives calculated by EXPLO5 deviate less than 5% compared to the JWL parameters obtained from the cylinder test. However, for non-ideal explosives, the deviation can be much larger. For some examples of the JWL parameters, see Tab. 16.7. Figure 16.4 shows the EXPLO5 (V8.01.01) calculated JWL coefficients for HMX at a density of 1.85 g/cm^3. The results agree well with those shown in Tab. 16.7 (note: 1 Mbar = 100 GPa).

Tab. 16.7: JWL parameters according to LLNL (taken from PEP **1989**, *14*, 199–211).

Explosive	*A* (Mbar)	*B* (Mbar)	*C* (Mbar)	R_1	R_2	ω
ANFO	0.87611	0.00798	0.00711	4.30566	0.89071	0.35
Aquanal	0.92469	0.00630	0.00787	4.37413	0.77124	0.17
Comp. B (64:36)	4.96376	0.03944	0.01288	4.06244	0.94846	0.35
Cyclotol	5.60038	0.05131	0.01361	4.12004	0.99514	0.35
HMX	7.78311	0.16808	0.01075	4.27291	1.33644	0.35
HNS (ρ = 1.2 g/cm^3)	2.47935	0.03573	0.01292	4.85865	1.166	0.40
HNS (ρ = 1.6 g/cm^3)	4.09184	0.01831	0.01454	4.30417	0.84383	0.35
LX-04	7.33674	0.0407	0.01506	4.29717	0.94499	0.35
LX-07	6.72651	0.0755	0.01179	4.15542	1.09109	0.35

Tab. 16.7 (continued)

Explosive	*A* (Mbar)	*B* (Mbar)	*C* (Mbar)	R_1	R_2	ω
LX-09	7.50812	0.07757	0.01492	4.27855	1.08086	0.35
LX-10	7.59532	0.05691	0.01545	4.25178	1.00075	0.35
LX-14	7.21469	0.05026	0.01513	4.20084	0.96953	0.35
Nitromethane	2.03532	0.03605	0.00984	4.28602	1.08386	0.35
PBX-9011	6.04767	0.05049	0.0117	4.10059	1.99915	0.35
PBX-9404	7.4685	0.06909	0.01447	4.26914	1.05592	0.35
PETN	6.7036	0.104	0.01564	4.44191	1.19107	0.35
Pourvex	2.72361	0.01287	0.00786	4.29924	0.85678	0.31
TNT	3.62033	0.02492	0.00887	4.02757	0.88784	0.25
Unigel	1.21831	0.01857	0.00954	3.6015	0.86185	0.20
Comp. B (60:40)	4.55278	0.04535	0.01253	4.07523	0.9841	0.35
TNT-AN	3.31047	0.0653	0.00669	4.33197	1.25752	0.35
TNT-Nigu (65:35)	3.89889	0.0834	0.00946	4.25879	1.2302	0.35
TNT-Nigu (50:50)	4.69918	0.08327	0.00839	4.36999	1.25346	0.35
TNT-Nigu-Al (42:31:27)	6.11576	0.04676	0.00571	4.58795	1.18041	0.35
TNT-Nigu-Al (50:35:15)	4.98035	0.06424	0.00726	4.4154	1.20885	0.35

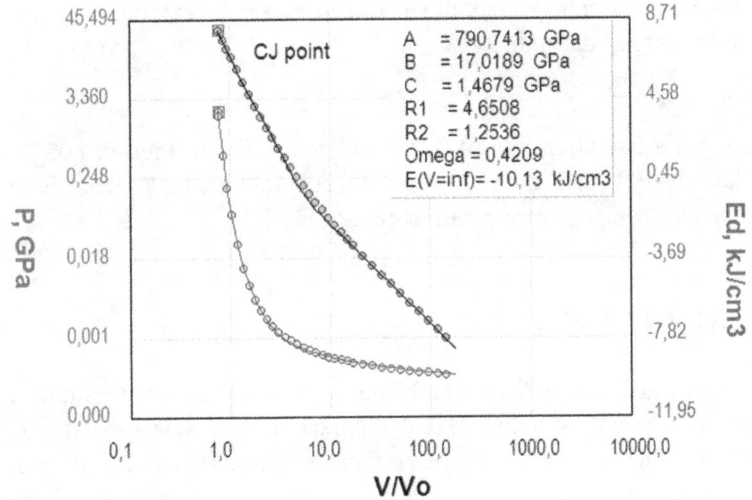

Fig. 16.4: Expansion isentrope for HMX with a density of 1.85 g/cm^3.

Literature

M. Suceska, Test Methods for Explosives, Springer, New York, **1995**, chapter 5.5.
H. Hornberg, Prop. Explos. Pyrotech. **1986**, *11*, 23 –31.
H. Hornberg, F. Volk, Prop. Explos. Pyrotech. **1989**, *14*, 199 –211.
S. B. Segletes, An Examination of the JWL Equation of State, ARL-TR-8403, July **2018**.

Grüneisen function

An important thermodynamic parameter associated with the EOS – essential for use in high-pressure simulations – is the Grüneisen function, Γ. In its most general manifestation, it is a state function, so that Γ = Γ(*S*, *V*). However, as employed in simulations for explosive-metal interactions, it is almost universally treated in a constrained manner, as Γ = Γ(*V*). This restricted case conforms to what is known as the "Grüneisen assumption", and materials obeying it are called "Grüneisen materials". In the general case, Γ is thermodynamically defined as

$$\Gamma(S, V) \equiv V \frac{(\partial p/\partial S)_v}{(\partial E/\partial S)_v}$$

However, when the Grüneisen assumption applies, it is often more convenient to eliminate entropy altogether and express the above equation as

$$\Gamma(V) \equiv V(\partial p/\partial E)_V$$

If we assume an additional thermodynamic constraint, namely, that the Grüneisen function is not a function of entropy, but, in fact everywhere constant, then the Grüneisen function reduces to a constant,

$$\Gamma = \omega$$

Therefore, under this assumption, the JWL EOS will satisfy the Grüneisen EOS assumption (i.e., that $(\partial p/\partial E)V$, and therefore Γ, is a function of volume *V* alone), with the Grüneisen function reducing to a constant parameter ω.

Disposal of Energetic Materials

For the disposal of energetic materials, the MuniRem reagents (https://www.munirem.com/) can be used, for example MuniRem-OPT-D. MuniRem is a scalable and versatile solution for the chemical neutralization and destruction (or disposal) of a broad range of explosives and chemical warfare agents. The MuniRem Reagents utilize reduction chemistry to instantly neutralize and degrade explosives and chemical warfare agents (CWA), while metals present are stabilized as insoluble metal sulfides. MuniRem is reported to be a safe, scalable, and versatile solution which can be used for the destruction of compounds such as Sarin gas, Mustard (CWA), HMX, RDX, TNT, TNR, DNTs, ADNTs, NBs, NDMA, black powder, ammonium nitrate, lead styphnate, lead azide, nitrocellulose, nitroguanidine, nitroglycerine, PBX, PETN, primer mix, and reactive aluminum. A further important aspect of these disposal reagents is the short clean-up time required.

Collection of formulae

Berthelot-Rot value:

$$B_R\ [\mathrm{kJ\,m^{-3}}] = \rho_0^2\ V_0\ Q_v$$

Burn rate:

$$r = \beta(T)\ p^{a} \quad a < 1 \text{ deflagrating explosive}$$
$$a > 1 \text{ detonating explosive}$$

Brisance:

$$B = \rho \times F \times D \qquad D = \mathrm{VoD}$$

Specific energy:

$$F = p_{ex} \qquad V = n\ RT$$

Distance rule:

$$D = cw^{0.33} \approx 2\ w^{0.33} \quad w\text{: mass of high expl.}$$
$$D\text{: distance}$$

Projectile velocity:

$$v = \sqrt{\frac{2\ m\ Q\ \eta}{M}}$$

m: mass of propellant [g]
Q: heat of combustion $[\mathrm{J\,g^{-1}}]$
M: mass of projectile [g]
η: constant

Specific impulse:

$$I_{sp} = \frac{\bar{F} \cdot t_b}{m} = \frac{1}{m} \cdot \int_0^{t_b} F(t)\mathrm{d}t$$

$$I_{sp}^{*} = \frac{I_{sp}}{g}$$

$$I_{sp} = \sqrt{\frac{2\ \gamma\ R\ T_c}{(\gamma - 1)M}}$$

$$\gamma = \frac{C_p}{C_V}$$

$$I_{sp}^{*} = \frac{1}{g}\sqrt{\frac{2\ \gamma\ R\ T_c}{(\gamma - 1)M}}$$

$$F = I_{sp} \frac{\Delta m}{\Delta t}$$

$$I_{sp} \propto \sqrt{\frac{T_c}{M}}$$

$$F_{impulse} = \frac{dm}{dt} v_e$$

$$F = F_{impulse} + F_{pressure} = \frac{dm}{dt} v_e + (p_e - p_a) A_e$$

Planck's rule:

$$W_\lambda = \frac{2 \pi h c^2}{\lambda^5} \frac{1}{e^{\frac{hc}{\lambda k T}} - 1}$$

Rule of Wien:

$$\lambda_{max} = 2897.756\, \mu m\, K\, T^{-1}$$

VoD for infinitely large diameter:

$$D = D_\infty \left(1 - A_L \frac{1}{d}\right)$$

Density at room temp.:

$$.d_{298\,K} = \frac{d_\tau}{1 + \alpha_V (298 - T_O)} \quad \text{with} \quad \alpha_V = 1.5 \cdot 10^{-4}\ K^{-1}$$

Kamlet-Jacobs equations:

$$p_{C-J}\ [\text{kbar}] = K \rho_0^2\ \Phi$$

$$D\left[\text{mm}\ \mu s^{-1}\right] = A \Phi^{0.5} (1 + B \rho_0)$$

$$\Phi = N (M)^{0.5} (Q)^{0.5}$$

$$K = 15.88$$

$$A = 1.01$$

$$B = 1.30$$

N: number of moles gas per gram explosive
M: mass of gas in gram per mol of gas
Q: heat of explosion in cal per gram

Oxygen balance for $C_aH_bN_cO_d$:

$$\Omega_{CO_2} = \frac{\left[d-(2a)-\left(\frac{b}{2}\right)\right] \times 1600}{M}$$

$$\Omega_{CO} = \frac{\left[d-a-\left(\frac{b}{2}\right)\right] \times 1600}{M}$$

Power index:

$$PI = \frac{Q \times V_0}{Q_{PA} \times V_{PA}} \times 100 \qquad \text{PA: picric acid}$$

Explosion temperature:

$$T_{ex} = \frac{Q_{ex}}{\sum C_v} + T_i$$

Atomization method:

$$\Delta_f H^o(g, m) = H^o_{(molecule)} - \sum H^o_{(atoms)} + \sum \Delta_f H^o_{(atoms)}$$

$H^o_{(molecule)}$: from CBS–4M

$\sum H^o_{(atoms)}$: from CBS–4M

$\sum \Delta_f H^o_{(atoms)}$: from NIST

enthalpy of sublimation:

$$\Delta H_{sub.}\ [\text{J mol}^{-1}] = 188\ T_m\ [\text{K}]$$

enthalpy of vaporization:

$$\Delta H_{sub.}\ [\text{J mol}^{-1}] = 90\ T_b\ [\text{K}]$$

lattice energies & enthalpies:

$$\left(1\text{Å}^3 = 10^{-3}\ \text{nm}^3\right)$$

$$U_L = |z_+|\ |z_-|\ v\left(\alpha / (V_M)^{0.33} + \beta\right)$$

		α [kJ/mol]	β [kJ/mol]
(V_M in nm^3)	MX	117.3	51.9
	MX_2	133.5	60.9
	M_2X	165.3	–29.8

$$\Delta H_L\ (M_pX_q) = U_L + [p\ (n_M/2 - 2) + q\ (n_X/2 - 2)]\ RT$$

n_M, n_X: 3 monoatomic ions

5 linear polyatomic ions
6 non-linear polyatomic

Nobel-Abel equation:

$$p\ (v - b_E) = n\,\mathrm{RT}$$

b_E: covolume

Spark energy:

$$E = \frac{1}{2} C U^2$$

Gurney velocity:

$$\sqrt{2\,E} = \frac{3\sqrt{3}}{16}\ D \approx \frac{D}{3.08} \quad [\text{in km s}^{-1}]$$

Cylindrical charge: $\frac{V}{\sqrt{2\,E}} = \left(\frac{M}{C} + \frac{1}{2}\right)^{-0.5}$

Spherical charge: $\frac{V}{\sqrt{2\,E}} = \left(2\frac{M}{C} + \frac{3}{5}\right)^{-0.5}$

Sandwich: $\frac{V}{\sqrt{2\,E}} = \left(2\frac{M}{C} + \frac{1}{3}\right)^{-0.5}$

M: mass of metal confinement

C: mass of explosive

Enthalpy of formation vs. energy of formation:

$$H_m = U_m + \Delta n\ RT$$

$$e.g. \quad 7\,C + 5/2\ H_2 + 3/2\ N_2 + 3\ O_2 \longrightarrow C_7H_5N_3O_6\ (s)$$

$$\Delta n = -7$$

$$U_m = H_m - \Delta n\ RT$$

Conversion of electronic energies (Gaussian) into enthalpies:

$$p\Delta V = \sum v_i\, RT \qquad \Delta U^{tr} = \sum v_i\, 3/2RT$$

$$\Delta U^{rot} = \sum v_i\ (F_i^{rot}/2)RT \qquad \Delta U^{vib} = \sum v_i\ \mathrm{zpe}$$

Author

Thomas M. Klapötke, born in 1961, studied chemistry at TU Berlin and received his PhD in 1986 under the supervision of Hartmut Köpf. After his postdoctoral studies in Fredericton, New Brunswick with Jack Passmore he finished his habilitation at TU Berlin in 1990. From 1995 until 1997 Klapötke was Ramsay Professor of Chemistry at the University of Glasgow in Scotland. Since 1997 he is Professor and holds the Chair of Inorganic Chemistry at LMU Munich. In 2009 Klapötke was appointed a permanent visiting Professor for Mechanical Engineering and Chemistry at the Center of Energetic Concepts Development (CECD) in the University of Maryland (UM), College Park. In 2014 Klapötke was appointed an Adjunct Professor at the University of Rhode Island. Klapötke is a Fellow of the Royal Society of Chemistry (C.Sci., C.Chem. F.R.S.C., U.K.), a member of the American Chemical Society (ACS) and the Fluorine Division of the ACS, a member of the Gesellschaft Deutscher Chemiker (GDCh), a Life Member of the International Pyrotechnics Society (IPS), a Life Member of the National Defense Industrial Association (NDIA, USA) and was appointed a Honorary Fellow of the Indian High Energy Materials Society (HEMSI) in 2011. Most of Klapötke's scientific collaborations are between LMU and the US Army Research Laboratory (ARL) in Aberdeen, MD and the US Army Armament Research, Development and Engineering Center (ARDEC) in Picatinny, NJ. Klapötke also collaborates with the US Army Engineer Research and Development Center (ERDC) in Champaign, IL, the German Bundeswehr (WTD-91) and several industrial partners in Germany and the USA. From 2011–2012 Klapötke was a Technical Team Member of NATO's Munitions Related Contamination team (AVT-197). He is Editor-in-Chief of the *Journal of Engineering Science and Military Technologies*, the subject Editor in the area of explosives synthesis of the *Central European Journal of Energetic Materials* (CEJEM) and an editorial board member of *Propellants, Explosives and Pyrotechnics* (PEP), *Journal of Energetic Materials*, *Chinese Journal of Explosives and Propellants* and Life Safety/Security Technologies (TSU). In 2023 Klapötke received an honorary doctorate (Dr. h.c.) from the Military Technical Academy Ferdinand I in Bucharest. Klapötke has published over 950 scientific papers in international peer reviewed journals, 37 book chapters, 18 books and holds 16 patents.

https://doi.org/10.1515/9783111447087-017

Index

https://doi.org/10.1515/9783111447087-018

www.ingramcontent.com/pod-product-compliance
Lightning Source LLC
Jackson TN
JSHW060734080525
84031JS00007B/30